DATA CLASSIFICATION
Algorithms and Applications

Chapman & Hall/CRC
Data Mining and Knowledge Discovery Series

SERIES EDITOR
Vipin Kumar
University of Minnesota
Department of Computer Science and Engineering
Minneapolis, Minnesota, U.S.A.

AIMS AND SCOPE

This series aims to capture new developments and applications in data mining and knowledge discovery, while summarizing the computational tools and techniques useful in data analysis. This series encourages the integration of mathematical, statistical, and computational methods and techniques through the publication of a broad range of textbooks, reference works, and handbooks. The inclusion of concrete examples and applications is highly encouraged. The scope of the series includes, but is not limited to, titles in the areas of data mining and knowledge discovery methods and applications, modeling, algorithms, theory and foundations, data and knowledge visualization, data mining systems and tools, and privacy and security issues.

PUBLISHED TITLES

ADVANCES IN MACHINE LEARNING AND DATA MINING FOR ASTRONOMY
Michael J. Way, Jeffrey D. Scargle, Kamal M. Ali, and Ashok N. Srivastava

BIOLOGICAL DATA MINING
Jake Y. Chen and Stefano Lonardi

COMPUTATIONAL BUSINESS ANALYTICS
Subrata Das

COMPUTATIONAL INTELLIGENT DATA ANALYSIS FOR SUSTAINABLE DEVELOPMENT
Ting Yu, Nitesh V. Chawla, and Simeon Simoff

COMPUTATIONAL METHODS OF FEATURE SELECTION
Huan Liu and Hiroshi Motoda

CONSTRAINED CLUSTERING: ADVANCES IN ALGORITHMS, THEORY, AND APPLICATIONS
Sugato Basu, Ian Davidson, and Kiri L. Wagstaff

CONTRAST DATA MINING: CONCEPTS, ALGORITHMS, AND APPLICATIONS
Guozhu Dong and James Bailey

DATA CLASSIFICATION: ALGORITHMS AND APPLICATIONS
Charu C. Aggarawal

DATA CLUSTERING: ALGORITHMS AND APPLICATIONS
Charu C. Aggarawal and Chandan K. Reddy

DATA CLUSTERING IN C++: AN OBJECT-ORIENTED APPROACH
Guojun Gan

DATA MINING FOR DESIGN AND MARKETING
Yukio Ohsawa and Katsutoshi Yada

DATA MINING WITH R: LEARNING WITH CASE STUDIES
Luís Torgo

FOUNDATIONS OF PREDICTIVE ANALYTICS
James Wu and Stephen Coggeshall

GEOGRAPHIC DATA MINING AND KNOWLEDGE DISCOVERY,
SECOND EDITION
Harvey J. Miller and Jiawei Han

HANDBOOK OF EDUCATIONAL DATA MINING
Cristóbal Romero, Sebastian Ventura, Mykola Pechenizkiy, and Ryan S.J.d. Baker

INFORMATION DISCOVERY ON ELECTRONIC HEALTH RECORDS
Vagelis Hristidis

INTELLIGENT TECHNOLOGIES FOR WEB APPLICATIONS
Priti Srinivas Sajja and Rajendra Akerkar

INTRODUCTION TO PRIVACY-PRESERVING DATA PUBLISHING: CONCEPTS
AND TECHNIQUES
Benjamin C. M. Fung, Ke Wang, Ada Wai-Chee Fu, and Philip S. Yu

KNOWLEDGE DISCOVERY FOR COUNTERTERRORISM AND
LAW ENFORCEMENT
David Skillicorn

KNOWLEDGE DISCOVERY FROM DATA STREAMS
João Gama

MACHINE LEARNING AND KNOWLEDGE DISCOVERY FOR
ENGINEERING SYSTEMS HEALTH MANAGEMENT
Ashok N. Srivastava and Jiawei Han

MINING SOFTWARE SPECIFICATIONS: METHODOLOGIES AND APPLICATIONS
David Lo, Siau-Cheng Khoo, Jiawei Han, and Chao Liu

MULTIMEDIA DATA MINING: A SYSTEMATIC INTRODUCTION TO
CONCEPTS AND THEORY
Zhongfei Zhang and Ruofei Zhang

MUSIC DATA MINING
Tao Li, Mitsunori Ogihara, and George Tzanetakis

NEXT GENERATION OF DATA MINING
Hillol Kargupta, Jiawei Han, Philip S. Yu, Rajeev Motwani, and Vipin Kumar

RAPIDMINER: DATA MINING USE CASES AND BUSINESS ANALYTICS
APPLICATIONS
Markus Hofmann and Ralf Klinkenberg

RELATIONAL DATA CLUSTERING: MODELS, ALGORITHMS,
AND APPLICATIONS
Bo Long, Zhongfei Zhang, and Philip S. Yu

SERVICE-ORIENTED DISTRIBUTED KNOWLEDGE DISCOVERY
Domenico Talia and Paolo Trunfio

SPECTRAL FEATURE SELECTION FOR DATA MINING
Zheng Alan Zhao and Huan Liu

STATISTICAL DATA MINING USING SAS APPLICATIONS, SECOND EDITION
George Fernandez

SUPPORT VECTOR MACHINES: OPTIMIZATION BASED THEORY,
ALGORITHMS, AND EXTENSIONS
Naiyang Deng, Yingjie Tian, and Chunhua Zhang

TEMPORAL DATA MINING
Theophano Mitsa

TEXT MINING: CLASSIFICATION, CLUSTERING, AND APPLICATIONS
Ashok N. Srivastava and Mehran Sahami

THE TOP TEN ALGORITHMS IN DATA MINING
Xindong Wu and Vipin Kumar

UNDERSTANDING COMPLEX DATASETS: DATA MINING WITH MATRIX
DECOMPOSITIONS
David Skillicorn

DATA CLASSIFICATION
Algorithms and Applications

Edited by

Charu C. Aggarwal

IBM T. J. Watson Research Center
Yorktown Heights, New York, USA

CRC Press
Taylor & Francis Group
Boca Raton   London   New York

CRC Press is an imprint of the
Taylor & Francis Group, an **informa** business

A CHAPMAN & HALL BOOK

CRC Press
Taylor & Francis Group
6000 Broken Sound Parkway NW, Suite 300
Boca Raton, FL 33487-2742

First issued in paperback 2020

© 2015 by Taylor & Francis Group, LLC
CRC Press is an imprint of Taylor & Francis Group, an Informa business

No claim to original U.S. Government works

ISBN-13: 978-1-4665-8674-1 (hbk)
ISBN-13: 978-0-367-65914-1 (pbk)

Library of Congress Cataloging-in-Publication Data

Data classification : algorithms and applications / edited by Charu C. Aggarwal.
    pages cm -- (Chapman & Hall/CRC data mining and knowledge discovery series ; 35)
    Summary: "This book homes in on three primary aspects of data classification: the core methods for data classification including probabilistic classification, decision trees, rule-based methods, and SVM methods; different problem domains and scenarios such as multimedia data, text data, biological data, categorical data, network data, data streams and uncertain data: and different variations of the classification problem such as ensemble methods, visual methods, transfer learning, semi-supervised methods and active learning. These advanced methods can be used to enhance the quality of the underlying classification results"-- Provided by publisher.
    Includes bibliographical references and index.
    ISBN 978-1-4665-8674-1 (hardback)
    1. File organization (Computer science) 2. Categories (Mathematics) 3. Algorithms. I. Aggarwal, Charu C.

QA76.9.F5.D38 2014
005.74'1--dc23   2013050912

Visit the Taylor & Francis Web site at
http://www.taylorandfrancis.com

and the CRC Press Web site at
http://www.crcpress.com

To my wife Lata, and my daughter Sayani

Contents

16 Uncertain Data Classification   417

Reynold Cheng, Yixiang Fang, and Matthias Renz

17 Rare Class Learning   445

Charu C. Aggarwal

23 Visual Classification   607
Giorgio Maria Di Nunzio

24 Evaluation of Classification Methods   633
Nele Verbiest, Karel Vermeulen, and Ankur Teredesai

Editor Biography

Charu C. Aggarwal is a Research Scientist at the IBM T. J. Watson Research Center in Yorktown Heights, New York. He completed his B.S. from IIT Kanpur in 1993 and his Ph.D. from Massachusetts Institute of Technology in 1996. His research interest during his Ph.D. years was in combinatorial optimization (network flow algorithms), and his thesis advisor was Professor James B. Orlin. He has since worked in the field of performance analysis, databases, and data mining. He has published over 200 papers in refereed conferences and journals, and has applied for or been granted over 80 patents. He is author or editor of ten books. Because of the commercial value of the aforementioned patents, he has received several invention achievement awards and has thrice been designated a *Master Inventor* at IBM. He is a recipient of an *IBM Corporate Award* (2003) for his work on bio-terrorist threat detection in data streams, a recipient of the *IBM Outstanding Innovation Award* (2008) for his scientific contributions to privacy technology, a recipient of the *IBM Outstanding Technical Achievement Award* (2009) for his work on data streams, and a recipient of an *IBM Research Division Award* (2008) for his contributions to System S. He also received the *EDBT 2014 Test of Time Award* for his work on condensation-based privacy-preserving data mining.

He served as an associate editor of the *IEEE Transactions on Knowledge and Data Engineering* from 2004 to 2008. He is an associate editor of the *ACM Transactions on Knowledge Discovery and Data Mining*, an action editor of the *Data Mining and Knowledge Discovery Journal*, editor-in-chief of the *ACM SIGKDD Explorations*, and an associate editor of the *Knowledge and Information Systems Journal*. He serves on the advisory board of the *Lecture Notes on Social Networks*, a publication by Springer. He serves as the vice-president of the *SIAM Activity Group on Data Mining*, which is responsible for all data mining activities organized by SIAM, including their main data mining conference. He is a fellow of the IEEE and the ACM, for "*contributions to knowledge discovery and data mining algorithms.*"

Contributors

Charu C. Aggarwal
IBM T. J. Watson Research Center
Yorktown Heights, New York

Salem Alelyani
Arizona State University
Tempe, Arizona

Mohammad Al Hasan
Indiana University - Purdue University
Indianapolis, Indiana

Alain Biem
IBM T. J. Watson Research Center
Yorktown Heights, New York

Shiyu Chang
University of Illinois at Urbana-Champaign
Urbana, Illinois

Yi Chang
Yahoo! Labs
Sunnyvale, California

Reynold Cheng
The University of Hong Kong
Hong Kong

Hongbo Deng
Yahoo! Research
Sunnyvale, California

Giorgio Maria Di Nunzio
University of Padua
Padova, Italy

Wei Fan
Huawei Noah's Ark Lab
Hong Kong

Yixiang Fang
The University of Hong Kong
Hong Kong

Jing Gao
State University of New York at Buffalo
Buffalo, New York

Quanquan Gu
University of Illinois at Urbana-Champaign
Urbana, Illinois

Dimitrios Gunopulos
University of Athens
Athens, Greece

Jiawei Han
University of Illinois at Urbana-Champaign
Urbana, Illinois

Wei Han
University of Illinois at Urbana-Champaign
Urbana, Illinois

Thomas S. Huang
University of Illinois at Urbana-Champaign
Urbana, Illinois

Ruoming Jin
Kent State University
Kent, Ohio

Xiangnan Kong
University of Illinois at Chicago
Chicago, Illinois

Dimitrios Kotsakos
University of Athens
Athens, Greece

Victor E. Lee
John Carroll University
University Heights, Ohio

Qi Li
State University of New York at Buffalo
Buffalo, New York

Xiao-Li Li
Institute for Infocomm Research
Singapore

Yaliang Li
State University of New York at Buffalo
Buffalo, New York

Bing Liu
University of Illinois at Chicago
Chicago, Illinois

Huan Liu
Arizona State University
Tempe, Arizona

Lin Liu
Kent State University
Kent, Ohio

Xianming Liu
University of Illinois at Urbana-Champaign
Urbana, Illinois

Ben London
University of Maryland
College Park, Maryland

Sinno Jialin Pan
Institute for Infocomm Research
Sinpapore

Pooya Khorrami
University of Illinois at Urbana-Champaign
Urbana, Illinois

Chih-Jen Lin
National Taiwan University
Taipei, Taiwan

Matthias Renz
University of Munich
Munich, Germany

Kaushik Sinha
Wichita State University
Wichita, Kansas

Yizhou Sun
Northeastern University
Boston, Massachusetts

Jiliang Tang
Arizona State University
Tempe, Arizona

Ankur Teredesai
University of Washington
Tacoma, Washington

Hanghang Tong
City University of New York
New York, New York

Nele Verbiest
Ghent University
Belgium

Karel Vermeulen
Ghent University
Belgium

Fei Wang
IBM T. J. Watson Research Center
Yorktown Heights, New York

Po-Wei Wang
National Taiwan University
Taipei, Taiwan

Ning Xu
University of Illinois at Urbana-Champaign
Urbana, Illinois

Philip S. Yu
University of Illinois at Chicago
Chicago, Illinois

ChengXiang Zhai
University of Illinois at Urbana-Champaign
Urbana, Illinois

Preface

The problem of classification is perhaps one of the most widely studied in the data mining and machine learning communities. This problem has been studied by researchers from several disciplines over several decades. Applications of classification include a wide variety of problem domains such as text, multimedia, social networks, and biological data. Furthermore, the problem may be encountered in a number of different scenarios such as streaming or uncertain data. Classification is a rather diverse topic, and the underlying algorithms depend greatly on the data domain and problem scenario.

Therefore, this book will focus on three primary aspects of data classification. The first set of chapters will focus on the core methods for data classification. These include methods such as probabilistic classification, decision trees, rule-based methods, instance-based techniques, SVM methods, and neural networks. The second set of chapters will focus on different problem domains and scenarios such as multimedia data, text data, time-series data, network data, data streams, and uncertain data. The third set of chapters will focus on different variations of the classification problem such as ensemble methods, visual methods, transfer learning, semi-supervised methods, and active learning. These are advanced methods, which can be used to enhance the quality of the underlying classification results.

The classification problem has been addressed by a number of different communities such as pattern recognition, databases, data mining, and machine learning. In some cases, the work by the different communities tends to be fragmented, and has not been addressed in a unified way. This book will make a conscious effort to address the work of the different communities in a unified way. The book will start off with an overview of the basic methods in data classification, and then discuss progressively more refined and complex methods for data classification. Special attention will also be paid to more recent problem domains such as graphs and social networks.

The chapters in the book will be divided into three types:

- **Method Chapters:** These chapters discuss the *key techniques* that are commonly used for classification, such as probabilistic methods, decision trees, rule-based methods, instance-based methods, SVM techniques, and neural networks.

- **Domain Chapters:** These chapters discuss the specific methods used for different *domains* of data such as text data, multimedia data, time-series data, discrete sequence data, network data, and uncertain data. Many of these chapters can also be considered application chapters, because they explore the specific characteristics of the problem in a particular domain. Dedicated chapters are also devoted to large data sets and data streams, because of the recent importance of the big data paradigm.

- **Variations and Insights:** These chapters discuss the *key variations* on the classification process such as classification ensembles, rare-class learning, distance function learning, active learning, and visual learning. Many variations such as transfer learning and semi-supervised learning use side-information in order to enhance the classification results. A separate chapter is also devoted to evaluation aspects of classifiers.

This book is designed to be comprehensive in its coverage of the entire area of classification, and it is hoped that it will serve as a knowledgeable compendium to students and researchers.

Chapter 1

An Introduction to Data Classification

Charu C. Aggarwal

IBM T. J. Watson Research Center
Yorktown Heights, NY
charu@us.ibm.com

1.1 Introduction

The problem of data classification has numerous applications in a wide variety of mining applications. This is because the problem attempts to learn the relationship between a set of *feature variables* and a *target variable* of interest. Since many practical problems can be expressed as associations between feature and target variables, this provides a broad range of applicability of this model. The problem of classification may be stated as follows:

Given a set of training data points along with associated training labels, determine the class label for an unlabeled test instance.

Numerous variations of this problem can be defined over different settings. Excellent overviews on data classification may be found in [39, 50, 63, 85]. Classification algorithms typically contain two phases:

- *Training Phase:* In this phase, a model is constructed from the training instances.

- *Testing Phase:* In this phase, the model is used to assign a label to an unlabeled test instance.

In some cases, such as lazy learning, the training phase is omitted entirely, and the classification is performed directly from the relationship of the training instances to the test instance. Instance-based methods such as the nearest neighbor classifiers are examples of such a scenario. Even in such cases, a pre-processing phase such as a nearest neighbor index construction may be performed in order to ensure efficiency during the testing phase.

The output of a classification algorithm may be presented for a test instance in one of two ways:

1. *Discrete Label:* In this case, a label is returned for the test instance.

2. *Numerical Score:* In this case, a numerical score is returned for each class label and test instance combination. Note that the numerical score can be converted to a discrete label for a test instance, by picking the class with the highest score for that test instance. The advantage of a numerical score is that it now becomes possible to compare the relative propensity of different test instances to belong to a particular class of importance, and rank them if needed. Such methods are used often in rare class detection problems, where the original class distribution is highly imbalanced, and the discovery of some classes is more valuable than others.

The classification problem thus segments the unseen test instances into groups, as defined by the class label. While the segmentation of examples into groups is also done by clustering, there is a key difference between the two problems. In the case of clustering, the segmentation is done using similarities between the feature variables, with no prior understanding of the structure of the groups. In the case of classification, the segmentation is done on the basis of a *training data* set, which encodes knowledge about the structure of the groups in the form of a *target variable*. Thus, while the segmentations of the data are usually related to notions of similarity, as in clustering, significant deviations from the similarity-based segmentation may be achieved in practical settings. As a result, the classification problem is referred to as *supervised learning*, just as clustering is referred to as *unsupervised learning*. The supervision process often provides significant application-specific utility, because the class labels may represent important properties of interest.

Some common application domains in which the classification problem arises, are as follows:

- **Customer Target Marketing:** Since the classification problem relates feature variables to target classes, this method is extremely popular for the problem of customer target marketing.

In such cases, feature variables describing the customer may be used to predict their buying interests on the basis of previous training examples. The target variable may encode the buying interest of the customer.

- **Medical Disease Diagnosis:** In recent years, the use of data mining methods in medical technology has gained increasing traction. The features may be extracted from the medical records, and the class labels correspond to whether or not a patient may pick up a disease in the future. In these cases, it is desirable to make disease predictions with the use of such information.

- **Supervised Event Detection:** In many temporal scenarios, class labels may be associated with time stamps corresponding to unusual events. For example, an intrusion activity may be represented as a class label. In such cases, time-series classification methods can be very useful.

- **Multimedia Data Analysis:** It is often desirable to perform classification of large volumes of multimedia data such as photos, videos, audio or other more complex multimedia data. Multimedia data analysis can often be challenging, because of the complexity of the underlying feature space and the semantic gap between the feature values and corresponding inferences.

- **Biological Data Analysis:** Biological data is often represented as discrete sequences, in which it is desirable to predict the properties of particular sequences. In some cases, the biological data is also expressed in the form of networks. Therefore, classification methods can be applied in a variety of different ways in this scenario.

- **Document Categorization and Filtering:** Many applications, such as newswire services, require the classification of large numbers of documents in real time. This application is referred to as document categorization, and is an important area of research in its own right.

- **Social Network Analysis:** Many forms of social network analysis, such as collective classification, associate labels with the underlying nodes. These are then used in order to predict the labels of other nodes. Such applications are very useful for predicting useful properties of actors in a social network.

The diversity of problems that can be addressed by classification algorithms is significant, and covers many domains. It is impossible to exhaustively discuss all such applications in either a single chapter or book. Therefore, this book will organize the area of classification into key topics of interest. The work in the data classification area typically falls into a number of broad categories;

- **Technique-centered:** The problem of data classification can be solved using numerous classes of techniques such as decision trees, rule-based methods, neural networks, SVM methods, nearest neighbor methods, and probabilistic methods. This book will cover the most popular classification methods in the literature comprehensively.

- **Data-Type Centered:** Many different data types are created by different applications. Some examples of different data types include text, multimedia, uncertain data, time series, discrete sequence, and network data. Each of these different data types requires the design of different techniques, each of which can be quite different.

- **Variations on Classification Analysis:** Numerous variations on the standard classification problem exist, which deal with more challenging scenarios such as rare class learning, transfer learning, semi-supervised learning, or active learning. Alternatively, different variations of classification, such as ensemble analysis, can be used in order to improve the effectiveness of classification algorithms. These issues are of course closely related to issues of model evaluation. All these issues will be discussed extensively in this book.

This chapter will discuss each of these issues in detail, and will also discuss how the organization of the book relates to these different areas of data classification. The chapter is organized as follows. The next section discusses the common techniques that are used for data classification. Section 1.3 explores the use of different data types in the classification process. Section 1.4 discusses the different variations of data classification. Section 1.5 discusses the conclusions and summary.

1.2   Common Techniques in Data Classification

In this section, the different methods that are commonly used for data classification will be discussed. These methods will also be associated with the different chapters in this book. It should be pointed out that these methods represent the most *common* techniques used for data classification, and it is difficult to comprehensively discuss all the methods in a single book. The most common methods used in data classification are decision trees, rule-based methods, probabilistic methods, SVM methods, instance-based methods, and neural networks. Each of these methods will be discussed briefly in this chapter, and all of them will be covered comprehensively in the different chapters of this book.

1.2.1   Feature Selection Methods

The first phase of virtually all classification algorithms is that of feature selection. In most data mining scenarios, a wide variety of features are collected by individuals who are often not domain experts. Clearly, the irrelevant features may often result in poor modeling, since they are not well related to the class label. In fact, such features will typically worsen the classification accuracy because of overfitting, when the training data set is small and such features are allowed to be a part of the training model. For example, consider a medical example where the features from the blood work of different patients are used to predict a particular disease. Clearly, a feature such as the *Cholesterol* level is predictive of heart disease, whereas a feature[1] such as *PSA* level is not predictive of heart disease. However, if a small training data set is used, the *PSA* level may have freak correlations with heart disease because of random variations. While the impact of a single variable may be small, the cumulative effect of many irrelevant features can be significant. This will result in a training model, that *generalizes* poorly to unseen test instances. Therefore, it is critical to use the correct features during the training process.

There are two broad kinds of feature selection methods:

1. *Filter Models:* In these cases, a crisp criterion on a single feature, or a subset of features, is used to evaluate their suitability for classification. This method is independent of the specific algorithm being used.

2. *Wrapper Models:* In these cases, the feature selection process is embedded into a classification algorithm, in order to make the feature selection process sensitive to the classification algorithm. This approach recognizes the fact that different algorithms may work better with different features.

In order to perform feature selection with filter models, a number of different measures are used in order to quantify the relevance of a feature to the classification process. Typically, these measures compute the imbalance of the feature values over different ranges of the attribute, which may either be discrete or numerical. Some examples are as follows:

---

[1]This feature is used to measure prostate cancer in men.

- *Gini Index:* Let $p_1 \ldots p_k$ be the fraction of classes that correspond to a particular *value* of the discrete attribute. Then, the gini-index of that value of the discrete attribute is given by:

$$G = 1 - \sum_{i=1}^{k} p_i^2 \tag{1.1}$$

The value of G ranges between 0 and $1 - 1/k$. Smaller values are more indicative of class imbalance. This indicates that the feature value is more discriminative for classification. The overall gini-index for the attribute can be measured by weighted averaging over different values of the discrete attribute, or by using the maximum gini-index over any of the different discrete values. Different strategies may be more desirable for different scenarios, though the weighted average is more commonly used.

- *Entropy:* The entropy of a particular value of the discrete attribute is measured as follows:

$$E = - \sum_{i=1}^{k} p_i \cdot \log(p_i) \tag{1.2}$$

The same notations are used above, as for the case of the gini-index. The value of the entropy lies between 0 and $\log(k)$, with smaller values being more indicative of class skew.

- *Fisher's Index:* The Fisher's index measures the ratio of the between class scatter to the within class scatter. Therefore, if p_j is the fraction of training examples belonging to class j, μ_j is the mean of a particular feature for class j, μ is the global mean for that feature, and σ_j is the standard deviation of that feature for class j, then the Fisher score F can be computed as follows:

$$F = \frac{\sum_{j=1}^{k} p_j \cdot (\mu_j - \mu)^2}{\sum_{j=1}^{k} p_j \cdot \sigma_j^2} \tag{1.3}$$

A wide variety of other measures such as the χ^2-statistic and mutual information are also available in order to quantify the discriminative power of attributes. An approach known as the Fisher's discriminant [61] is also used in order to *combine* the different features into directions in the data that are highly relevant to classification. Such methods are of course feature *transformation* methods, which are also closely related to feature selection methods, just as unsupervised dimensionality reduction methods are related to unsupervised feature selection methods.

The Fisher's discriminant will be explained below for the two-class problem. Let $\overline{\mu_0}$ and $\overline{\mu_1}$ be the d-dimensional row vectors representing the means of the records in the two classes, and let Σ_0 and Σ_1 be the corresponding $d \times d$ covariance matrices, in which the (i, j)th entry represents the covariance between dimensions i and j for that class. Then, the equivalent Fisher score $FS(\overline{V})$ for a d-dimensional row vector $\overline{V}$ may be written as follows:

$$FS(\overline{V}) = \frac{(\overline{V} \cdot (\overline{\mu_0} - \overline{\mu_1}))^2}{\overline{V}(p_0 \cdot \Sigma_0 + p_1 \cdot \Sigma_1)\overline{V}^T} \tag{1.4}$$

This is a generalization of the *axis-parallel* score in Equation 1.3, to an arbitrary direction $\overline{V}$. The goal is to determine a direction $\overline{V}$, which maximizes the Fisher score. It can be shown that the optimal direction $\overline{V^*}$ may be determined by solving a generalized eigenvalue problem, and is given by the following expression:

$$\overline{V^*} = (p_0 \cdot \Sigma_0 + p_1 \cdot \Sigma_1)^{-1} (\overline{\mu_0} - \overline{\mu_1})^T \tag{1.5}$$

If desired, successively orthogonal directions may be determined by iteratively projecting the data onto the residual subspace, after determining the optimal directions one by one.

More generally, it should be pointed out that many features are often closely correlated with one another, and the *additional* utility of an attribute, once a certain set of features have already been selected, is different from its standalone utility. In order to address this issue, the *Minimum Redundancy Maximum Relevance* approach was proposed in [69], in which features are incrementally selected on the basis of their incremental gain on adding them to the feature set. Note that this method is also a filter model, since the evaluation is on a subset of features, and a crisp criterion is used to evaluate the subset.

In wrapper models, the feature selection phase is embedded into an iterative approach with a classification algorithm. In each iteration, the classification algorithm evaluates a particular set of features. This set of features is then augmented using a particular (e.g., greedy) strategy, and tested to see of the quality of the classification improves. Since the classification algorithm is used for evaluation, this approach will generally create a feature set, which is sensitive to the classification algorithm. This approach has been found to be useful in practice, because of the wide diversity of models on data classification. For example, an SVM would tend to prefer features in which the two classes separate out using a linear model, whereas a nearest neighbor classifier would prefer features in which the different classes are clustered into spherical regions. A good survey on feature selection methods may be found in [59]. Feature selection methods are discussed in detail in Chapter 2.

1.2.2 Probabilistic Methods

Probabilistic methods are the most fundamental among all data classification methods. Probabilistic classification algorithms use statistical inference to find the best class for a given example. In addition to simply assigning the best class like other classification algorithms, probabilistic classification algorithms will output a corresponding *posterior* probability of the test instance being a member of each of the possible classes. The posterior probability is defined as the probability after observing the specific characteristics of the test instance. On the other hand, the *prior* probability is simply the fraction of training records belonging to each particular class, with no knowledge of the test instance. After obtaining the posterior probabilities, we use decision theory to determine class membership for each new instance. Basically, there are two ways in which we can estimate the posterior probabilities.

In the first case, the posterior probability of a particular class is estimated by determining the class-conditional probability and the prior class separately and then applying Bayes' theorem to find the parameters. The most well known among these is the Bayes classifier, which is known as a generative model. For ease in discussion, we will assume discrete feature values, though the approach can easily be applied to numerical attributes with the use of discretization methods. Consider a test instance with d different features, which have values $X = \langle x_1 \ldots x_d \rangle$ respectively. Its is desirable to determine the posterior probability that the class $Y(T)$ of the test instance T is i. In other words, we wish to determine the posterior probability $P(Y(T) = i | x_1 \ldots x_d)$. Then, the Bayes rule can be used in order to derive the following:

$$P(Y(T) = i | x_1 \ldots x_d) = P(Y(T) = i) \cdot \frac{P(x_1 \ldots x_d | Y(T) = i)}{P(x_1 \ldots x_d)} \tag{1.6}$$

Since the denominator is constant across all classes, and one only needs to determine the class with the maximum posterior probability, one can approximate the aforementioned expression as follows:

$$P(Y(T) = i | x_1 \ldots x_d) \propto P(Y(T) = i) \cdot P(x_1 \ldots x_d | Y(T) = i) \tag{1.7}$$

The key here is that the expression on the right can be evaluated more easily in a data-driven way, as long as the *naive Bayes assumption* is used for simplification. Specifically, in Equation 1.7, the expression $P(Y(T) = i | x_1 \ldots x_d)$ can be expressed as the product of the feature-wise conditional probabilities.

$$P(x_1 \ldots x_d | Y(T) = i) = \prod_{j=1}^{d} P(x_j | Y(T) = i) \qquad (1.8)$$

This is referred to as *conditional independence*, and therefore the Bayes method is referred to as "naive." This simplification is crucial, because these individual probabilities can be estimated from the training data in a more robust way. The naive Bayes theorem is crucial in providing the ability to perform the product-wise simplification. The term $P(x_j | Y(T) = i)$ is computed as the fraction of the records in the portion of the training data corresponding to the ith class, which contains feature value x_j for the jth attribute. If desired, Laplacian smoothing can be used in cases when enough data is not available to estimate these values robustly. This is quite often the case, when a small amount of training data may contain few or no training records containing a particular feature value. The Bayes rule has been used quite successfully in the context of a wide variety of applications, and is particularly popular in the context of text classification. In spite of the naive independence assumption, the Bayes model seems to be quite effective in practice. A detailed discussion of the naive assumption in the context of the effectiveness of the Bayes classifier may be found in [38].

Another probabilistic approach is to directly model the posterior probability, by learning a discriminative function that maps an input feature vector directly onto a class label. This approach is often referred to as a discriminative model. Logistic regression is a popular discriminative classifier, and its goal is to directly estimate the posterior probability $P(Y(T) = i | X)$ from the training data. Formally, the logistic regression model is defined as

$$P(Y(T) = i | X) = \frac{1}{1 + e^{-\theta^T X}}, \qquad (1.9)$$

where θ is the vector of parameters to be estimated. In general, maximum likelihood is used to determine the parameters of the logistic regression. To handle overfitting problems in logistic regression, regularization is introduced to penalize the log likelihood function for large values of θ. The logistic regression model has been extensively used in numerous disciplines, including the Web, and the medical and social science fields.

A variety of other probabilistic models are known in the literature, such as probabilistic graphical models, and conditional random fields. An overview of probabilistic methods for data classification are found in [20, 64]. Probabilistic methods for data classification are discussed in Chapter 3.

1.2.3 Decision Trees

Decision trees create a hierarchical partitioning of the data, which relates the different partitions at the leaf level to the different classes. The hierarchical partitioning at each level is created with the use of a *split criterion*. The split criterion may either use a condition (or predicate) on a single attribute, or it may contain a condition on multiple attributes. The former is referred to as a univariate split, whereas the latter is referred to as a multivariate split. The overall approach is to try to recursively split the training data so as to maximize the discrimination among the different classes over different nodes. The discrimination among the different classes is maximized, when the level of skew among the different classes in a given node is maximized. A measure such as the gini-index or entropy is used in order to quantify this skew. For example, if $p_1 \ldots p_k$ is the fraction of the records belonging to the k different classes in a node N, then the gini-index $G(N)$ of the node N is defined as follows:

$$G(N) = 1 - \sum_{i=1}^{k} p_i^2 \qquad (1.10)$$

The value of $G(N)$ lies between 0 and $1 - 1/k$. The smaller the value of $G(N)$, the greater the skew. In the cases where the classes are evenly balanced, the value is $1 - 1/k$. An alternative measure is

TABLE 1.1: Training Data Snapshot Relating Cardiovascular Risk Based on Previous Events to Different Blood Parameters

Patient Name	CRP Level	Cholestrol	High Risk? (Class Label)
Mary	3.2	170	Y
Joe	0.9	273	N
Jack	2.5	213	Y
Jane	1.7	229	N
Tom	1.1	160	N
Peter	1.9	205	N
Elizabeth	8.1	160	Y
Lata	1.3	171	N
Daniela	4.5	133	Y
Eric	11.4	122	N
Michael	1.8	280	Y

the entropy $E(N)$:

$$E(N) = -\sum_{i=1}^{k} p_i \cdot \log(p_i) \tag{1.11}$$

The value of the entropy lies[2] between 0 and $\log(k)$. The value is $\log(k)$, when the records are perfectly balanced among the different classes. This corresponds to the scenario with maximum entropy. The smaller the entropy, the greater the skew in the data. Thus, the gini-index and entropy provide an effective way to evaluate the quality of a node in terms of its level of discrimination between the different classes.

While constructing the training model, the split is performed, so as to minimize the weighted sum of the gini-index or entropy of the two nodes. This step is performed recursively, until a termination criterion is satisfied. The most obvious termination criterion is one where all data records in the node belong to the same class. More generally, the termination criterion requires either a minimum level of skew or purity, or a minimum number of records in the node in order to avoid overfitting. One problem in decision tree construction is that there is no way to *predict* the best time to stop decision tree growth, in order to prevent overfitting. Therefore, in many variations, the decision tree is pruned in order to remove nodes that may correspond to overfitting. There are different ways of pruning the decision tree. One way of pruning is to use a minimum description length principle in deciding when to prune a node from the tree. Another approach is to hold out a small portion of the training data during the decision tree growth phase. It is then tested to see whether replacing a subtree with a single node improves the classification accuracy on the hold out set. If this is the case, then the pruning is performed. In the testing phase, a test instance is assigned to an appropriate path in the decision tree, based on the evaluation of the split criteria in a hierarchical decision process. The class label of the corresponding leaf node is reported as the relevant one.

Figure 1.1 provides an example of how the decision tree is constructed. Here, we have illustrated a case where the two measures (features) of the blood parameters of patients are used in order to assess the level of cardiovascular risk in the patient. The two measures are the *C-Reactive Protein (CRP)* level and *Cholesterol* level, which are well known parameters related to cardiovascular risk. It is assumed that a training data set is available, which is already labeled into high risk and low risk patients, based on previous cardiovascular events such as myocardial infarctions or strokes. At the same time, it is assumed that the feature values of the blood parameters for these patients are available. A snapshot of this data is illustrated in Table 1.1. It is evident from the training data that

---

[2]The value of the expression at $p_i = 0$ needs to be evaluated at the limit.

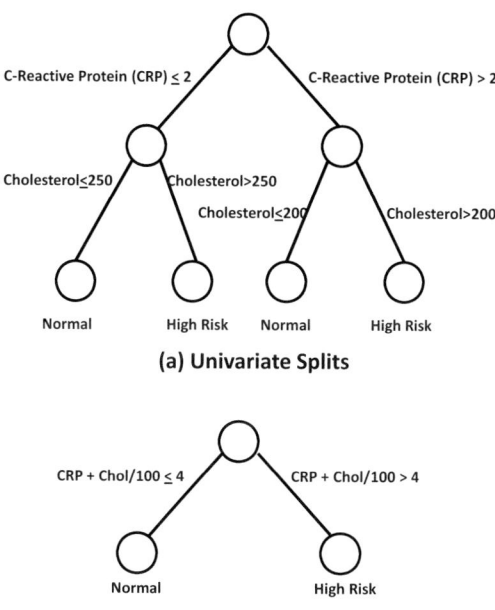

FIGURE 1.1: Illustration of univariate and multivariate splits for decision tree construction.

higher CRP and Cholesterol levels correspond to greater risk, though it is possible to reach more definitive conclusions by combining the two.

An example of a decision tree that constructs the classification model on the basis of the two features is illustrated in Figure 1.1(a). This decision tree uses univariate splits, by first partitioning on the CRP level, and then using a split criterion on the Cholesterol level. Note that the Cholesterol split criteria in the two CRP branches of the tree are different. In principle, different features can be used to split different nodes at the same level of the tree. It is also sometimes possible to use conditions on multiple attributes in order to create more powerful splits at a particular level of the tree. An example is illustrated in Figure 1.1(b), where a linear combination of the two attributes provides a much more powerful split than a single attribute. The split condition is as follows:

```
CRP + Cholestrol/100 ≤ 4
```

Note that a single condition such as this is able to partition the training data very well into the two classes (with a few exceptions). Therefore, the split is more powerful in discriminating between the two classes in a smaller number of levels of the decision tree. Where possible, it is desirable to construct more compact decision trees in order to obtain the most accurate results. Such splits are referred to as multivariate splits. Some of the earliest methods for decision tree construction include C4.5 [72], ID3 [73], and CART [22]. A detailed discussion of decision trees may be found in [22, 65, 72, 73]. Decision trees are discussed in Chapter 4.

1.2.4 Rule-Based Methods

Rule-based methods are closely related to decision trees, except that they do not create a strict hierarchical partitioning of the training data. Rather, overlaps are allowed in order to create greater robustness for the training model. Any path in a decision tree may be interpreted as a rule, which assigns a test instance to a particular label. For example, for the case of the decision tree illustrated in Figure 1.1(a), the rightmost path corresponds to the following rule:

`CRP> 2 & Cholestrol > 200 ⇒ HighRisk`

It is possible to create a set of disjoint rules from the different paths in the decision tree. In fact, a number of methods such as *C4.5*, create related models for both decision tree construction and rule construction. The corresponding rule-based classifier is referred to as *C4.5Rules*.

Rule-based classifiers can be viewed as more general models than decision tree models. While decision trees require the induced rule sets to be *non-overlapping*, this is not the case for rule-based classifiers. For example, consider the following rule:

`CRP> 3 ⇒ HighRisk`

Clearly, this rule overlaps with the previous rule, and is also quite relevant to the prediction of a given test instance. In rule-based methods, a set of rules is mined from the training data in the first phase (or training phase). During the testing phase, it is determined which rules are relevant to the test instance and the final result is based on a combination of the class values predicted by the different rules.

In many cases, it may be possible to create rules that possibly conflict with one another on the right hand side for a particular test instance. Therefore, it is important to design methods that can effectively determine a resolution to these conflicts. The method of resolution depends upon whether the rule sets are ordered or unordered. If the rule sets are ordered, then the top matching rules can be used to make the prediction. If the rule sets are unordered, then the rules can be used to vote on the test instance. Numerous methods such as *Classification based on Associations (CBA)* [58], *CN2* [31], and *RIPPER* [26] have been proposed in the literature, which use a variety of rule induction methods, based on different ways of mining and prioritizing the rules.

Methods such as *CN2* and *RIPPER* use the *sequential covering paradigm*, where rules with high accuracy and coverage are sequentially mined from the training data. The idea is that a rule is grown corresponding to specific target class, and then all training instances matching (or *covering*) the antecedent of that rule are removed. This approach is applied repeatedly, until only training instances of a particular class remain in the data. This constitutes the *default class*, which is selected for a test instance, when no rule is fired. The process of mining a rule for the training data is referred to as *rule growth*. The growth of a rule involves the successive addition of conjuncts to the left-hand side of the rule, after the selection of a particular consequent class. This can be viewed as growing a single "best" path in a decision tree, by adding conditions (split criteria) to the left-hand side of the rule. After the rule growth phase, a rule-pruning phase is used, which is analogous to decision tree construction. In this sense, the rule-growth of rule-based classifiers share a number of conceptual similarities with decision tree classifiers. These rules are ranked in the same order as they are mined from the training data. For a given test instance, the class variable in the consequent of the first matching rule is reported. If no matching rule is found, then the default class is reported as the relevant one.

Methods such as *CBA* [58] use the traditional association rule framework, in which rules are determined with the use of specific support and confidence measures. Therefore, these methods are referred to as associative classifiers. It is also relatively easy to prioritize these rules with the use of these parameters. The final classification can be performed by either using the majority vote from the matching rules, or by picking the top ranked rule(s) for classification. Typically, the confidence of the rule is used to prioritize them, and the support is used to prune for statistical significance. A single catch-all rule is also created for test instances that are not covered by any rule. Typically, this catch-all rule might correspond to the majority class among training instances not covered by any rule. Rule-based methods tend to be more robust than decision trees, because they are not restricted to a strict hierarchical partitioning of the data. This is most evident from the relative performance of these methods in some sparse high dimensional domains such as text. For example, while many rule-based methods such as *RIPPER* are frequently used for the text domain, decision trees are used rarely for text. Another advantage of these methods is that they are relatively easy to generalize to different data types such as sequences, XML or graph data [14, 93]. In such cases, the left-hand side of the rule needs to be defined in a way that is specific for that data domain. For example, for a sequence classification problem [14], the left-hand side of the rule corresponds to a sequence of symbols. For a graph-classification problem, the left-hand side of the rule corresponds to a frequent structure [93]. Therefore, while rule-based methods are related to decision trees, they have significantly greater expressive power. Rule-based methods are discussed in detail in Chapter 5.

1.2.5 Instance-Based Learning

In instance-based learning, the first phase of constructing the training model is often dispensed with. The test instance is directly related to the training instances in order to create a classification model. Such methods are referred to as *lazy learning methods*, because they wait for knowledge of the test instance in order to create a locally optimized model, which is specific to the test instance. The advantage of such methods is that they can be directly tailored to the particular test instance, and can avoid the information loss associated with the incompleteness of any training model. An overview of instance-based methods may be found in [15, 16, 89].

An example of a very simple instance-based method is the nearest neighbor classifier. In the nearest neighbor classifier, the top k nearest neighbors in the training data are found to the given test instance. The class label with the largest presence among the k nearest neighbors is reported as the relevant class label. If desired, the approach can be made faster with the use of nearest neighbor index construction. Many variations of the basic instance-based learning algorithm are possible, wherein aggregates of the training instances may be used for classification. For example, small clusters can be created from the instances of each class, and the centroid of each cluster may be used as a new instance. Such an approach is much more efficient and also more robust because of the reduction of noise associated with the clustering phase which aggregates the noisy records into more robust aggregates. Other variations of instance-based learning use different variations on the distance function used for classification. For example, methods that are based on the Mahalanobis distance or Fisher's discriminant may be used for more accurate results. The problem of distance function design is intimately related to the problem of instance-based learning. Therefore, separate chapters have been devoted in this book to these topics.

A particular form of instance-based learning, is one where the nearest neighbor classifier is not explicitly used. This is because the distribution of the class labels may not match with the notion of proximity defined by a particular distance function. Rather, a locally optimized classifier is constructed using the examples in the neighborhood of a test instance. Thus, the neighborhood is used only to define the neighborhood in which the classification model is constructed in a lazy way. Local classifiers are generally more accurate, because of the simplification of the class distribution within the locality of the test instance. This approach is more generally referred to as *lazy learning*. This is a more general notion of instance-based learning than traditional nearest neighbor classifiers. Methods for instance-based classification are discussed in Chapter 6. Methods for distance-function learning are discussed in Chapter 18.

1.2.6 SVM Classifiers

SVM methods use linear conditions in order to separate out the classes from one another. The idea is to use a linear condition that separates the two classes from each other as well as possible. Consider the medical example discussed earlier, where the risk of cardiovascular disease is related to diagnostic features from patients.

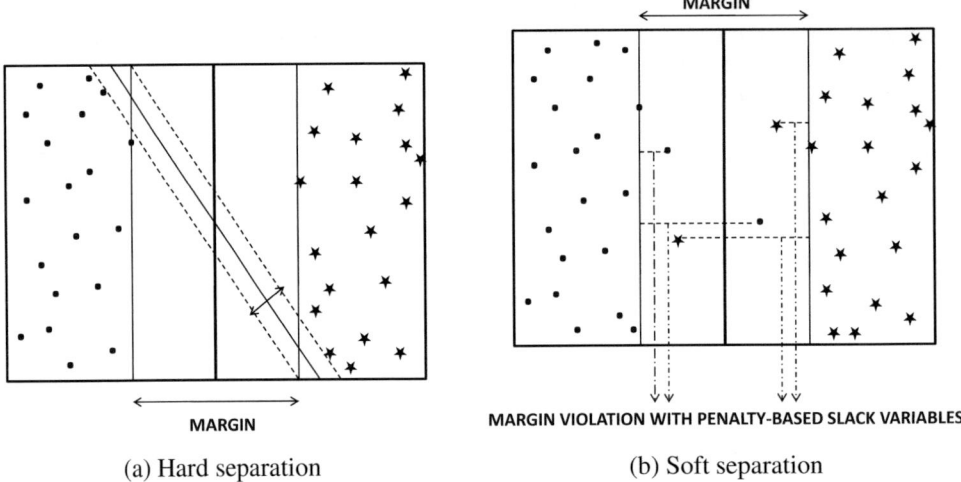

(a) Hard separation   (b) Soft separation

FIGURE 1.2: Hard and soft support vector machines.

```
CRP + Cholestrol/100 ≤ 4
```

In such a case, the split condition in the multivariate case may also be used as stand-alone condition for classification. This, a SVM classifier, may be considered a single level decision tree with a very carefully chosen multivariate split condition. Clearly, since the effectiveness of the approach depends only on a single separating hyperplane, it is critical to define this separation carefully.

Support vector machines are generally defined for binary classification problems. Therefore, the class variable y_i for the ith training instance $\overline{X_i}$ is assumed to be drawn from $\{-1,+1\}$. The most important criterion, which is commonly used for SVM classification, is that of the *maximum margin hyperplane*. In order to understand this point, consider the case of linearly separable data illustrated in Figure 1.2(a). Two possible separating hyperplanes, with their corresponding *support vectors* and *margins* have been illustrated in the figure. It is evident that one of the separating hyperplanes has a much larger margin than the other, and is therefore more desirable because of its greater generality for unseen test examples. Therefore, one of the important criteria for support vector machines is to achieve maximum margin separation of the hyperplanes.

In general, it is assumed for d dimensional data that the separating hyperplane is of the form $\overline{W} \cdot \overline{X} + b = 0$. Here $\overline{W}$ is a d-dimensional vector representing the coefficients of the hyperplane of separation, and b is a constant. Without loss of generality, it may be assumed (because of appropriate coefficient scaling) that the two symmetric support vectors have the form $\overline{W} \cdot \overline{X} + b = 1$ and $\overline{W} \cdot \overline{X} + b = -1$. The coefficients $\overline{W}$ and b need to be learned from the training data $\mathcal{D}$ in order to maximize the margin of separation between these two parallel hyperplanes. It can shown from elementary linear algebra that the distance between these two hyperplanes is $2/||\overline{W}||$. Maximizing this objective function is equivalent to minimizing $||\overline{W}||^2/2$. The problem constraints are defined by the fact that the training data points for each class are on one side of the support vector. Therefore, these constraints are as follows:

$$\overline{W} \cdot \overline{X_i} + b \geq +1 \quad \forall i : y_i = +1 \tag{1.12}$$

$$\overline{W} \cdot \overline{X_i} + b \leq -1 \quad \forall i : y_i = -1 \tag{1.13}$$

This is a constrained convex quadratic optimization problem, which can be solved using Lagrangian methods. In practice, an off-the-shelf optimization solver may be used to achieve the same goal.

In practice, the data may not be linearly separable. In such cases, soft-margin methods may be used. A slack $\xi_i \geq 0$ is introduced for training instance, and a training instance is allowed to violate the support vector constraint, for a penalty, which is dependent on the slack. This situation is illustrated in Figure 1.2(b). Therefore, the new set of constraints are now as follows:

$$\overline{W} \cdot \overline{X_i} + b \geq +1 - \xi_i \quad \forall i : y_i = +1 \tag{1.14}$$

$$\overline{W} \cdot \overline{X_i} + b \leq -1 + \xi_i \quad \forall i : y_i = -1 \tag{1.15}$$

$$\xi_i \geq 0 \tag{1.16}$$

Note that additional non-negativity constraints also need to be imposed in the slack variables. The objective function is now $||\overline{W}||^2/2 + C \cdot \sum_{i=1}^{n} \xi_i$. The constant C regulates the importance of the margin and the slack requirements. In other words, small values of C make the approach closer to soft-margin SVM, whereas large values of C make the approach more of the hard-margin SVM. It is also possible to solve this problem using off-the-shelf optimization solvers.

It is also possible to use transformations on the feature variables in order to design non-linear SVM methods. In practice, non-linear SVM methods are learned using kernel methods. The key idea here is that SVM formulations can be solved using only pairwise dot products (similarity values) between objects. In other words, the optimal decision about the class label of a test instance, from the solution to the quadratic optimization problem in this section, can be expressed in terms of the following:

1. Pairwise dot products of different training instances.

2. Pairwise dot product of the test instance and different training instances.

The reader is advised to refer to [84] for the specific details of the solution to the optimization formulation. The dot product between a pair of instances can be viewed as notion of similarity among them. Therefore, the aforementioned observations imply that it is possible to perform SVM classification, with pairwise similarity information between training data pairs and training-test data pairs. The actual feature values are not required.

This opens the door for using transformations, which are represented by their similarity values. These similarities can be viewed as kernel functions $K(\overline{X}, \overline{Y})$, which measure similarities between the points $\overline{X}$ and $\overline{Y}$. Conceptually, the kernel function may be viewed as dot product between the pair of points in a newly transformed space (denoted by mapping function $\Phi(\cdot)$). However, this transformation does not need to be explicitly computed, as long as the kernel function (dot product) $K(\overline{X}, \overline{Y})$ is already available:

$$K(\overline{X}, \overline{Y}) = \Phi(\overline{X}) \cdot \Phi(\overline{Y}) \tag{1.17}$$

Therefore, all computations can be performed in the original space using the dot products implied by the kernel function. Some interesting examples of kernel functions include the Gaussian radial basis function, polynomial kernel, and hyperbolic tangent, which are listed below in the same order.

$$K(\overline{X_i}, \overline{X_j}) = e^{-||\overline{X_i} - \overline{X_j}||^2/2\sigma^2} \tag{1.18}$$

$$K(\overline{X_i}, \overline{X_j}) = (\overline{X_i} \cdot \overline{X_j} + 1)^h \tag{1.19}$$

$$K(\overline{X_i}, \overline{X_j}) = \tanh(\kappa \overline{X_i} \cdot \overline{X_j} - \delta) \tag{1.20}$$

These different functions result in different kinds of nonlinear decision boundaries in the original space, but they correspond to a linear separator in the transformed space. The performance of a classifier can be sensitive to the choice of the kernel used for the transformation. One advantage of kernel methods is that they can also be extended to arbitrary data types, as long as appropriate pairwise similarities can be defined.

The major downside of SVM methods is that they are slow. However, they are very popular and tend to have high accuracy in many practical domains such as text. An introduction to SVM methods may be found in [30, 46, 75, 76, 85]. Kernel methods for support vector machines are discussed in [75]. SVM methods are discussed in detail in Chapter 7.

1.2.7 Neural Networks

Neural networks attempt to simulate biological systems, corresponding to the human brain. In the human brain, neurons are connected to one another via points, which are referred to as *synapses*. In biological systems, learning is performed by changing the strength of the synaptic connections, in response to impulses.

This biological analogy is retained in an artificial neural network. The basic computation unit in an artificial neural network is a *neuron* or *unit*. These units can be arranged in different kinds of architectures by connections between them. The most basic architecture of the neural network is a perceptron, which contains a set of input nodes and an output node. The output unit receives a set of inputs from the input units. There are d different input units, which is exactly equal to the dimensionality of the underlying data. The data is assumed to be numerical. Categorical data may need to be transformed to binary representations, and therefore the number of inputs may be larger. The output node is associated with a set of weights $\overline{W}$, which are used in order to compute a function $f(\cdot)$ of its inputs. Each component of the weight vector is associated with a connection from the input unit to the output unit. The weights can be viewed as the analogue of the synaptic strengths in biological systems. In the case of a perceptron architecture, the input nodes do not perform any computations. They simply transmit the input attribute forward. Computations are performed only at the output nodes in the basic perceptron architecture. The output node uses its weight vector along with the input attribute values in order to compute a function of the inputs. A typical function, which is computed at the output nodes, is the signed linear function:

$$z_i = \text{sign}\{\overline{W} \cdot \overline{X_i} + b\} \tag{1.21}$$

The output is a predicted value of the binary class variable, which is assumed to be drawn from $\{-1, +1\}$. The notation b denotes the bias. Thus, for a vector $\overline{X_i}$ drawn from a dimensionality of d, the weight vector $\overline{W}$ should also contain d elements. Now consider a binary classification problem, in which all labels are drawn from $\{+1, -1\}$. We assume that the class label of $\overline{X_i}$ is denoted by y_i. In that case, the sign of the predicted function z_i yields the class label. An example of the perceptron architecture is illustrated in Figure 1.3(a). Thus, the goal of the approach is to *learn* the set of weights $\overline{W}$ with the use of the training data, so as to minimize the least squares error $(y_i - z_i)^2$. The idea is that we start off with random weights and gradually update them, when a mistake is made by applying the current function on the training example. The magnitude of the update is regulated by a learning rate λ. This update is similar to the updates in gradient descent, which are made for least-squares optimization. In the case of neural networks, the update function is as follows.

$$\overline{W}^{t+1} = \overline{W}^t + \lambda(y_i - z_i)\overline{X_i} \tag{1.22}$$

Here, $\overline{W}^t$ is the value of the weight vector in the tth iteration. It is not difficult to show that the incremental update vector is related to the negative gradient of $(y_i - z_i)^2$ with respect to $\overline{W}$. It is also easy to see that updates are made to the weights, only when mistakes are made in classification. When the outputs are correct, the incremental change to the weights is zero.

The similarity to support vector machines is quite striking, in the sense that a linear function is also learned in this case, and the sign of the linear function predicts the class label. In fact, the perceptron model and support vector machines are closely related, in that both are linear function approximators. In the case of support vector machines, this is achieved with the use of maximum margin optimization. In the case of neural networks, this is achieved with the use of an incremental

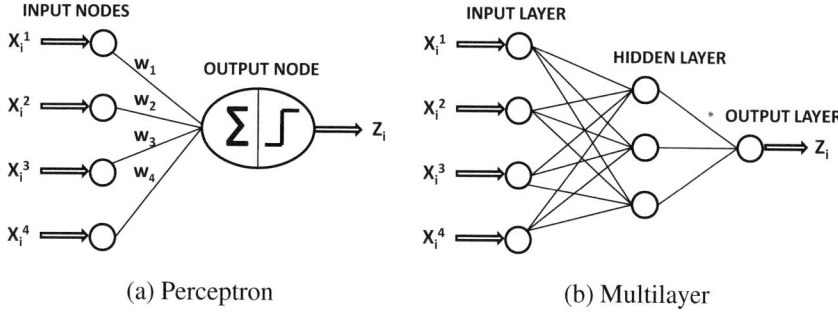

(a) Perceptron   (b) Multilayer

FIGURE 1.3: Single and multilayer neural networks.

learning algorithm, which is approximately equivalent to least squares error optimization of the prediction.

The constant λ regulates the learning rate. The choice of learning rate is sometimes important, because learning rates that are too small will result in very slow training. On the other hand, if the learning rates are too fast, this will result in oscillation between suboptimal solutions. In practice, the learning rates are fast initially, and then allowed to gradually slow down over time. The idea here is that initially large steps are likely to be helpful, but are then reduced in size to prevent oscillation between suboptimal solutions. For example, after t iterations, the learning rate may be chosen to be proportional to $1/t$.

The aforementioned discussion was based on the simple perceptron architecture, which can model only linear relationships. In practice, the neural network is arranged in three layers, referred to as the *input layer*, *hidden layer*, and the *output layer*. The input layer only transmits the inputs forward, and therefore, there are really only two layers to the neural network, which can perform computations. Within the hidden layer, there can be any number of layers of neurons. In such cases, there can be an arbitrary number of layers in the neural network. In practice, there is only one hidden layer, which leads to a 2-layer network. An example of a multilayer network is illustrated in Figure 1.3(b). The perceptron can be viewed as a very special kind of neural network, which contains only a single layer of neurons (corresponding to the output node). Multilayer neural networks allow the approximation of nonlinear functions, and complex decision boundaries, by an appropriate choice of the network topology, and non-linear functions at the nodes. In these cases, a logistic or sigmoid function known as a *squashing function* is also applied to the inputs of neurons in order to model non-linear characteristics. It is possible to use different non-linear functions at different nodes. Such general architectures are very powerful in approximating arbitrary functions in a neural network, given enough training data and training time. This is the reason that neural networks are sometimes referred to as *universal function approximators*.

In the case of single layer perceptron algorthms, the training process is easy to perform by using a gradient descent approach. The major challenge in training multilayer networks is that it is no longer known for intermediate (hidden layer) nodes, what their "expected" output should be. This is only known for the final output node. Therefore, some kind of "error feedback" is required, in order to determine the changes in the weights at the intermediate nodes. The training process proceeds in two phases, one of which is in the forward direction, and the other is in the backward direction.

1. *Forward Phase:* In the forward phase, the activation function is repeatedly applied to propagate the inputs from the neural network in the forward direction. Since the final output is supposed to match the class label, the final output at the output layer provides an error value, depending on the training label value. This error is then used to update the weights of the output layer, and propagate the weight updates backwards in the next phase.

2. *Backpropagation Phase:* In the backward phase, the errors are propagated backwards through the neural network layers. This leads to the updating of the weights in the neurons of the different layers. The gradients at the previous layers are learned as a function of the errors and weights in the layer ahead of it. The learning rate λ plays an important role in regulating the rate of learning.

In practice, any arbitrary function can be approximated well by a neural network. The price of this generality is that neural networks are often quite slow in practice. They are also sensitive to noise, and can sometimes overfit the training data.

The previous discussion assumed only binary labels. It is possible to create a k-label neural network, by either using a multiclass "one-versus-all" meta-algorithm, or by creating a neural network architecture in which the number of output nodes is equal to the number of class labels. Each output represents prediction to a particular label value. A number of implementations of neural network methods have been studied in [35,57,66,77,88], and many of these implementations are designed in the context of text data. It should be pointed out that both neural networks and SVM classifiers use a linear model that is quite similar. The main difference between the two is in how the optimal linear hyperplane is determined. Rather than using a direct optimization methodology, neural networks use a *mistake-driven* approach to data classification [35]. Neural networks are described in detail in [19,51]. This topic is addressed in detail in Chapter 8.

1.3 Handing Different Data Types

Different data types require the use of different techniques for data classification. This is because the choice of data type often qualifies the kind of problem that is solved by the classification approach. In this section, we will discuss the different data types commonly studied in classification problems, which may require a certain level of special handling.

1.3.1 Large Scale Data: Big Data and Data Streams

With the increasing ability to collect different types of large scale data, the problems of scale have become a challenge to the classification process. Clearly, larger data sets allow the creation of more accurate and sophisticated models. However, this is not necessarily helpful, if one is computationally constrained by problems of scale. Data streams and big data analysis have different challenges. In the former case, real time processing creates challenges, whereas in the latter case, the problem is created by the fact that computation and data access over extremely large amounts of data is inefficient. It is often difficult to compute summary statistics from large volumes, because the access needs to be done in a distributed way, and it is too expensive to shuffle large amounts of data around. Each of these challenges will be discussed in this subsection.

1.3.1.1 Data Streams

The ability to continuously collect and process large volumes of data has lead to the popularity of data streams [4]. In the streaming scenario, two primary problems arise in the construction of training models.

- *One-pass Constraint:* Since data streams have very large volume, all processing algorithms need to perform their computations in a single pass over the data. This is a significant challenge, because it excludes the use of many iterative algorithms that work robustly over static data sets. Therefore, it is crucial to design the training models in an efficient way.

- *Concept Drift:* The data streams are typically created by a generating process, which may change over time. This results in *concept drift*, which corresponds to changes in the underlying stream patterns over time. The presence of concept drift can be detrimental to classification algorithms, because models become stale over time. Therefore, it is crucial to adjust the model in an incremental way, so that it achieves high accuracy over current test instances.

- *Massive Domain Constraint:* The streaming scenario often contains discrete attributes that take on millions of possible values. This is because streaming items are often associated with discrete identifiers. Examples could be email addresses in an email addresses, IP addresses in a network packet stream, and URLs in a click stream extracted from proxy Web logs. The massive domain problem is ubiquitous in streaming applications. In fact, many synopsis data structures, such as the count-min sketch [33], and the Flajolet-Martin data structure [41], have been designed with this issue in mind. While this issue has not been addressed very extensively in the stream *mining* literature (beyond basic synopsis methods for counting), recent work has made a number of advances in this direction [9].

Conventional classification algorithms need to be appropriately modified in order to address the aforementioned challenges. The special scenarios, such as those in which the domain of the stream data is large, or the classes are rare, pose special challenges. Most of the well known techniques for streaming classification use space-efficient data structures for easily updatable models [13, 86]. Furthermore, these methods are explicitly designed to handle concept drift by making the models temporally adaptive, or by using different models over different regions of the data stream. Special scenarios or data types need dedicated methods in the streaming scenario. For example, the massive-domain scenario can be addressed [9] by incorporating the count-min data structure [33] as a synopsis structure within the training model. A specially difficult case is that of rare class learning, in which rare class instances may be mixed with occurrences of completely new classes. This problem can be considered a hybrid between classification and outlier detection. Nevertheless it is the most common case in the streaming domain, in applications such as intrusion detection. In these cases, some kinds of rare classes (intrusions) may already be known, whereas other rare classes may correspond to previously unseen threats. A book on data streams, containing extensive discussions on key topics in the area, may be found in [4]. The different variations of the streaming classification problem are addressed in detail in Chapter 9.

1.3.1.2 The Big Data Framework

While streaming algorithms work under the assumption that the data is too large to be stored explicitly, the big data framework leverages advances in storage technology in order to actually store the data and process it. However, as the subsequent discussion will show, even if the data can be explicitly stored, it is often not easy to process and extract insights from it.

In the simplest case, the data is stored on disk on a single machine, and it is desirable to scale up the approach with disk-efficient algorithms. While many methods such as the nearest neighbor classifier and associative classifiers can be scaled up with more efficient subroutines, other methods such as decision trees and SVMs require dedicated methods for scaling up. Some examples of scalable decision tree methods include *SLIQ* [48], *BOAT* [42], and *RainForest* [43]. Some early parallel implementations of decision trees include the *SPRINT* method [82]. Typically, scalable decision tree methods can be performed in one of two ways. Methods such as *RainForest* increase scalability by storing attribute-wise summaries of the training data. These summaries are sufficient for performing single-attribute splits efficiently. Methods such as *BOAT* use a combination of bootstrapped samples, in order to yield a decision tree, which is very close to the accuracy that one would have obtained by using the complete data.

An example of a scalable SVM method is *SVMLight* [53]. This approach focusses on the fact that the quadratic optimization problem in SVM is computationally intensive. The idea is to always optimize only a small working set of variables while keeping the others fixed. This working set is selected by using a steepest descent criterion. This optimizes the advantage gained from using a particular subset of attributes. Another strategy used is to discard training examples, which do not have any impact on the margin of the classifiers. Training examples that are away from the decision boundary, and on its "correct" side, have no impact on the margin of the classifier, even if they are removed. Other methods such as *SVMPerf* [54] reformulate the SVM optimization to reduce the number of slack variables, and increase the number of constraints. A cutting plane approach, which works with a small subset of constraints at a time, is used in order to solve the resulting optimization problem effectively.

Further challenges arise for extremely large data sets. This is because an increasing size of the data implies that a distributed file system must be used in order to store it, and distributed processing techniques are required in order to ensure sufficient scalability. The challenge here is that if large segments of the data are available on different machines, it is often too expensive to shuffle the data across different machines in order to extract integrated insights from it. Thus, as in all distributed infrastructures, it is desirable to exchange intermediate insights, so as to minimize communication costs. For an application programmer, this can sometimes create challenges in terms of keeping track of where different parts of the data are stored, and the precise ordering of communications in order to minimize the costs.

In this context, Google's *MapReduce* framework [37] provides an effective method for analysis of large amounts of data, especially when the nature of the computations involve linearly computable statistical functions over the elements of the data streams. One desirable aspect of this framework is that it abstracts out the precise details of where different parts of the data are stored to the application programmer. As stated in [37]: "*The run-time system takes care of the details of partitioning the input data, scheduling the program's execution across a set of machines, handling machine failures, and managing the required inter-machine communication. This allows programmers without any experience with parallel and distributed systems to easily utilize the resources of a large distributed system.*" Many classification algorithms such as *k*-means are naturally linear in terms of their scalability with the size of the data. A primer on the *MapReduce* framework implementation on *Apache Hadoop* may be found in [87]. The key idea here is to use a *Map* function in order to distribute the work across the different machines, and then provide an automated way to shuffle out much smaller data in (key,value) pairs containing intermediate results. The *Reduce* function is then applied to the aggregated results from the *Map* step in order to obtain the final results.

Google's original *MapReduce* framework was designed for analyzing large amounts of Web logs, and more specifically deriving linearly computable statistics from the logs. It has been shown [44] that a declarative framework is particularly useful in many *MapReduce* applications, and that many existing classification algorithms can be generalized to the *MapReduce* framework. A proper choice of the algorithm to adapt to the *MapReduce* framework is crucial, since the framework is particularly effective for linear computations. It should be pointed out that the major attraction of the *MapReduce* framework is its ability to provide application programmers with a cleaner abstraction, which is independent of very specific run-time details of the distributed system. It should not, however, be assumed that such a system is somehow inherently superior to existing methods for distributed parallelization from an *effectiveness* or *flexibility* perspective, especially if an application programmer is willing to design such details from scratch. A detailed discussion of classification algorithms for big data is provided in Chapter 10.

1.3.2 Text Classification

One of the most common data types used in the context of classification is that of text data. Text data is ubiquitous, especially because of its popularity, both on the Web and in social networks. While a text document can be treated as a string of words, it is more commonly used as a bag-of-words, in which the ordering information between words is not used. This representation of text is much closer to multidimensional data. However, the standard methods for multidimensional classification often need to be modified for text.

The main challenge with text classification is that the data is extremely high dimensional and sparse. A typical text lexicon may be of a size of a hundred thousand words, but a document may typically contain far fewer words. Thus, most of the attribute values are zero, and the frequencies are relatively small. Many common words may be very noisy and not very discriminative for the classification process. Therefore, the problems of feature selection and representation are particularly important in text classification.

Not all classification methods are equally popular for text data. For example, rule-based methods, the Bayes method, and SVM classifiers tend to be more popular than other classifiers. Some rule-based classifiers such as *RIPPER* [26] were originally designed for text classification. Neural methods and instance-based methods are also sometimes used. A popular instance-based method used for text classification is Rocchio's method [56, 74]. Instance-based methods are also sometimes used with centroid-based classification, where frequency-truncated centroids of class-specific clusters are used, instead of the original documents for the k-nearest neighbor approach. This generally provides better accuracy, because the centroid of a small closely related set of documents is often a more stable representation of that data locality than any single document. This is especially true because of the sparse nature of text data, in which two related documents may often have only a small number of words in common.

Many classifiers such as decision trees, which are popularly used in other data domains, are not quite as popular for text data. The reason for this is that decision trees use a strict hierarchical partitioning of the data. Therefore, the features at the higher levels of the tree are implicitly given greater importance than other features. In a text collection containing hundreds of thousands of features (words), a single word usually tells us very little about the class label. Furthermore, a decision tree will typically partition the data space with a very small number of splits. This is a problem, when this value is orders of magnitude less than the underlying data dimensionality. Of course, decision trees in text are not very balanced either, because of the fact that a given word is contained only in a small subset of the documents. Consider the case where a split corresponds to presence or absence of a word. Because of the imbalanced nature of the tree, most paths from the root to leaves will correspond to word-absence decisions, and a very small number (less than 5 to 10) word-presence decisions. Clearly, this will lead to poor classification, especially in cases where word-absence does not convey much information, and a modest number of word presence decisions are required. Univariate decision trees do not work very well for very high dimensional data sets, because of disproportionate importance to some features, and a corresponding inability to effectively leverage all the available features. It is possible to improve the effectiveness of decision trees for text classification by using multivariate splits, though this can be rather expensive.

The standard classification methods, which are used for the text domain, also need to be suitably modified. This is because of the high dimensional and sparse nature of the text domain. For example, text has a dedicated model, known as the multinomial Bayes model, which is different from the standard Bernoulli model [12]. The Bernoulli model treats the presence and absence of a word in a text document in a symmetric way. However, in a given text document, only a small fraction of the lexicon size is present in it. The absence of a word is usually far less informative than the presence of a word. The symmetric treatment of word presence and word absence can sometimes be detrimental to the effectiveness of a Bayes classifier in the text domain. In order to achieve this goal, the multinomial Bayes model is used, which uses the frequency of word presence in a document, but ignores non-occurrence.

In the context of SVM classifiers, scalability is important, because such classifiers scale poorly both with number of training documents and data dimensionality (lexicon size). Furthermore, the sparsity of text (i.e., few non-zero feature values) should be used to improve the training efficiency. This is because the training model in an SVM classifier is constructed using a constrained quadratic optimization problem, which has as many constraints as the number of data points. This is rather large, and it directly results in an increased size of the corresponding Lagrangian relaxation. In the case of kernel SVM, the space-requirements for the kernel matrix could also scale quadratically with the number of data points. A few methods such as *SVMLight* [53] address this issue by carefully breaking down the problem into smaller subproblems, and optimizing only a few variables at a time. Other methods such as *SVMPerf* [54] also leverage the sparsity of the text domain. The *SVMPerf* method scales as $O(n \cdot s)$, where s is proportional to the average number of non-zero feature values per training document.

Text classification often needs to be performed in scenarios, where it is accompanied by linked data. The links between documents are typically inherited from domains such as the Web and social networks. In such cases, the links contain useful information, which should be leveraged in the classification process. A number of techniques have recently been designed to utilize such side information in the classification process. Detailed surveys on text classification may be found in [12, 78]. The problem of text classification is discussed in detail in Chapter 11 of this book.

1.3.3 Multimedia Classification

With the increasing popularity of social media sites, multimedia data has also become increasingly popular. In particular sites such as *Flickr* or *Youtube* allow users to upload their photos or videos at these sites. In such cases, it is desirable to perform classification of either portions or all of either a photograph or a video. In these cases, rich meta-data may also be available, which can facilitate more effective data classification. The issue of data representation is a particularly important one for multimedia data, because poor representations have a large semantic gap, which creates challenges for the classification process. The combination of text with multimedia data in order to create more effective classification models has been discussed in [8]. Many methods such as semi-supervised learning and transfer learning can also be used in order to improve the effectiveness of the data classification process. Multimedia data poses unique challenges, both in terms of data representation, and information fusion. Methods for multimedia data classification are discussed in [60]. A detailed discussion of methods for multimedia data classification is provided in Chapter 12.

1.3.4 Time Series and Sequence Data Classification

Both of these data types are temporal data types in which the attributes are of two types. The first type is the contextual attribute (time), and the second attribute, which corresponds to the time series value, is the behavioral attribute. The main difference between time series and sequence data is that time series data is continuous, whereas sequence data is discrete. Nevertheless, this difference is quite significant, because it changes the nature of the commonly used models in the two scenarios.

Time series data is popular in many applications such as sensor networks, and medical informatics, in which it is desirable to use large volumes of streaming time series data in order to perform the classification. Two kinds of classification are possible with time-series data:

- *Classifying specific time-instants:* These correspond to specific events that can be inferred at particular instants of the data stream. In these cases, the labels are associated with instants in time, and the behavior of one or more time series are used in order to classify these instants. For example, the detection of significant events in real-time applications can be an important application in this scenario.

- *Classifying part or whole series:* In these cases, the class labels are associated with portions or all of the series, and these are used for classification. For example, an ECG time-series will show characteristic shapes for specific diagnostic criteria for diseases.

Both of these scenarios are equally important from the perspective of analytical inferences in a wide variety of scenarios. Furthermore, these scenarios are also relevant to the case of sequence data. Sequence data arises frequently in biological, Web log mining, and system analysis applications. The discrete nature of the underlying data necessitates the use of methods that are quite different from the case of continuous time series data. For example, in the case of discrete sequences, the nature of the distance functions and modeling methodologies are quite different than those in time-series data.

A brief survey of time-series and sequence classification methods may be found in [91]. A detailed discussion on time-series data classification is provided in Chapter 13, and that of sequence data classification methods is provided in Chapter 14. While the two areas are clearly connected, there are significant differences between these two topics, so as to merit separate topical treatment.

1.3.5 Network Data Classification

Network data is quite popular in Web and social networks applications in which a variety of different scenarios for node classification arise. In most of these scenarios, the class labels are associated with nodes in the underlying network. In many cases, the labels are known only for a subset of the nodes. It is desired to use the known subset of labels in order to make predictions about nodes for which the labels are unknown. This problem is also referred to as *collective classification*. In this problem, the key assumption is that of *homophily*. This implies that edges imply similarity relationships between nodes. It is assumed that the labels vary smoothly over neighboring nodes. A variety of methods such as Bayes methods and spectral methods have been generalized to the problem of collective classification. In cases where content information is available at the nodes, the effectiveness of classification can be improved even further. A detailed survey on collective classification methods may be found in [6].

A different form of graph classification is one in which many small graphs exist, and labels are associated with individual graphs. Such cases arise commonly in the case of chemical and biological data, and are discussed in detail in [7]. The focus of the chapter in this book is on very large graphs and social networks because of their recent popularity. A detailed discussion of network classification methods is provided in Chapter 15 of this book.

1.3.6 Uncertain Data Classification

Many forms of data collection are uncertain in nature. For example, data collected with the use of sensors is often uncertain. Furthermore, in cases when data perturbation techniques are used, the data becomes uncertain. In some cases, statistical methods are used in order to infer parts of the data. An example is the case of link inference in network data. Uncertainty can play an important role in the classification of uncertain data. For example, if an attribute is known to be uncertain, its contribution to the training model can be de-emphasized, with respect to an attribute that has deterministic attributes.

The problem of uncertain data classification was first studied in [5]. In these methods, the uncertainty in the attributes is used as a first-class variable in order to improve the effectiveness of the classification process. This is because the relative importance of different features depends not only on their correlation with the class variable, but also the uncertainty inherent in them. Clearly, when the values of an attribute are more uncertain, it is less desirable to use them for the classification process. This is achieved in [5] with the use of a density-based transform that accounts for the varying level of uncertainty of attributes. Subsequently, many other methods have been proposed to account for the uncertainty in the attributes during the classification process. A detailed description of uncertain data classification methods is provided in Chapter 16.

1.4 Variations on Data Classification

Many natural variations of the data classification problem correspond to either small variations of the standard classification problem or are enhancements of classification with the use of additional data. The key variations of the classification problem are those of rare-class learning and distance function learning. Enhancements of the data classification problem make use of meta-algorithms, more data in methods such as transfer learning and co-training, active learning, and human intervention in visual learning. In addition, the topic of model evaluation is an important one in the context of data classification. This is because the issue of model evaluation is important for the design of effective classification meta-algorithms. In the following section, we will discuss the different variations of the classification problem.

1.4.1 Rare Class Learning

Rare class learning is an important variation of the classification problem, and is closely related to outlier analysis [1]. In fact, it can be considered a supervised variation of the outlier detection problem. In rare class learning, the distribution of the classes is highly imbalanced in the data, and it is typically more important to correctly determine the positive class. For example, consider the case where it is desirable to classify patients into malignant and normal categories. In such cases, the majority of patients may be normal, though it is typically much more costly to misclassify a truly malignant patient (false negative). Thus, false negatives are more *costly* than false positives. The problem is closely related to cost-sensitive learning, since the misclassification of different classes has different classes. The major difference with the standard classification problem is that the objective function of the problem needs to be modified with costs. This provides several avenues that can be used in order to effectively solve this problem:

- *Example Weighting:* In this case, the examples are weighted differently, depending upon their cost of misclassification. This leads to minor changes in most classification algorithms, which are relatively simple to implement. For example, in an SVM classifier, the objective function needs to be appropriately weighted with costs, whereas in a decision tree, the quantification of the split criterion needs to weight the examples with costs. In a nearest neighbor classifier, the k nearest neighbors are appropriately weighted while determining the class with the largest presence.

- *Example Re-sampling:* In this case, the examples are appropriately re-sampled, so that rare classes are over-sampled, whereas the normal classes are under-sampled. A standard classifier is applied to the re-sampled data without any modification. From a technical perspective, this approach is equivalent to example weighting. However, from a computational perspective, such an approach has the advantage that the newly re-sampled data has much smaller size. This is because most of the examples in the data correspond to the normal class, which is drastically under-sampled, whereas the rare class is typically only mildly over-sampled.

Many variations of the rare class detection problem are possible, in which either examples of a single class are available, or the normal class is contaminated with rare class examples. A survey of algorithms for rare class learning may be found in [25]. This topic is discussed in detail in Chapter 17.

1.4.2 Distance Function Learning

Distance function learning is an important problem that is closely related to data classification. In this problem it is desirable to relate pairs of data instances to a distance value with the use of either supervised or unsupervised methods [3]. For example, consider the case of an image collection, in which the similarity is defined on the basis of a user-centered semantic criterion. In such a case, the use of standard distance functions such as the Euclidian metric may not reflect the semantic similarities between two images well, because they are based on human perception, and may even vary from collection to collection. Thus, the best way to address this issue is to explicitly incorporate human feedback into the learning process. Typically, this feedback is incorporated either in terms of pairs of images with explicit distance values, or in terms of rankings of different images to a given target image. Such an approach can be used for a variety of different data domains. This is the training data that is used for learning purposes. A detailed survey of distance function learning methods is provided in [92]. The topic of distance function learning is discussed in detail in Chapter 18.

1.4.3 Ensemble Learning for Data Classification

A meta-algorithm is a classification method that re-uses one or more currently existing classification algorithm by applying either multiple models for robustness, or combining the results of the same algorithm with different parts of the data. The general goal of the algorithm is to obtain more robust results by combining the results from multiple training models either sequentially or independently. The overall error of a classification model depends upon the bias and variance, in addition to the intrinsic noise present in the data. The bias of a classifier depends upon the fact that the decision boundary of a particular model may not correspond to the true decision boundary. For example, the training data may not have a linear decision boundary, but an SVM classifier will assume a linear decision boundary. The variance is based on the random variations in the particular training data set. Smaller training data sets will have larger variance. Different forms of ensemble analysis attempt to reduce this bias and variance. The reader is referred to [84] for an excellent discussion on bias and variance.

Meta-algorithms are used commonly in many data mining problems such as clustering and outlier analysis [1, 2] in order to obtain more accurate results from different data mining problems. The area of classification is the richest one from the perspective of meta-algorithms, because of its crisp evaluation criteria and relative ease in combining the results of different algorithms. Some examples of popular meta-algorithms are as follows:

- *Boosting:* Boosting [40] is a common technique used in classification. The idea is to focus on successively difficult portions of the data set in order to create models that can classify the data points in these portions more accurately, and then use the ensemble scores over all the components. A hold-out approach is used in order to determine the incorrectly classified instances for each portion of the data set. Thus, the idea is to sequentially determine better classifiers for more difficult portions of the data, and then combine the results in order to obtain a meta-classifier, which works well on all parts of the data.

- *Bagging:* Bagging [24] is an approach that works with random data samples, and combines the results from the models constructed using different samples. The training examples for each classifier are selected by sampling with replacement. These are referred to as *bootstrap* samples. This approach has often been shown to provide superior results in certain scenarios, though this is not always the case. This approach is not effective for reducing the bias, but can reduce the variance, because of the specific random aspects of the training data.

- *Random Forests:* Random forests [23] are a method that use sets of decision trees on either splits with randomly generated vectors, or random subsets of the training data, and compute the score as a function of these different components. Typically, the random vectors are generated from a fixed probability distribution. Therefore, random forests can be created by either random *split* selection, or random *input* selection. Random forests are closely related to bagging, and in fact bagging with decision trees can be considered a special case of random forests, in terms of how the sample is selected (bootstrapping). In the case of random forests, it is also possible to create the trees in a lazy way, which is tailored to the particular test instance at hand.

- *Model Averaging and Combination:* This is one of the most common models used in ensemble analysis. In fact, the random forest method discussed above is a special case of this idea. In the context of the classification problem, many Bayesian methods [34] exist for the model combination process. The use of different models ensures that the error caused by the bias of a particular classifier does not dominate the classification results.

- *Stacking:* Methods such as stacking [90] also combine different models in a variety of ways, such as using a second-level classifier in order to perform the combination. The output of different first-level classifiers is used to create a new feature representation for the second level classifier. These first level classifiers may be chosen in a variety of ways, such as using different bagged classifiers, or by using different training models. In order to avoid overfitting, the training data needs to be divided into two subsets for the first and second level classifiers.

- *Bucket of Models:* In this approach [94] a "hold-out" portion of the data set is used in order to decide the most appropriate model. The most appropriate model is one in which the highest accuracy is achieved in the held out data set. In essence, this approach can be viewed as a competition or bake-off contest between the different models.

The area of meta-algorithms in classification is very rich, and different variations may work better in different scenarios. An overview of different meta-algorithms in classification is provided in Chapter 19.

1.4.4 Enhancing Classification Methods with Additional Data

In this class of methods, *additional labeled or unlabeled data* is used to enhance classification. Both these methods are used when there is a direct paucity of the underlying training data. In the case of transfer learning, additional training (labeled) data from a different domain or problem is used to supervise the classification process. On the other hand, in the case of semi-supervised learning, unlabeled data is used to enhance the classification process. These methods are briefly described in this section.

1.4.4.1 Semi-Supervised Learning

Semi-supervised learning methods improve the effectiveness of learning methods with the use of *unlabeled* data, when only a small amount of labeled data is available. The main difference between semi-supervised learning and transfer learning methods is that *unlabeled* data with the same features is used in the former, whereas external labeled data (possibly from a different source) is used in the latter. A key question arises as to why unlabeled data should improve the effectiveness of classification in any way, when it does not provide any additional labeling knowledge. The reason for this is that unlabeled data provides a good idea of the manifolds in which the data is embedded, as well as the density structure of the data in terms of the clusters and sparse regions. The key assumption is that the classification labels exhibit a smooth variation over different parts of the manifold structure of the underlying data. This manifold structure can be used to determine feature correlations, and joint feature distributions, which are very helpful for classification. The semi-supervised setting is also sometimes referred to as the *transductive* setting, when the test instances must be specified together with the training instances. Some problem settings such as collective classification of network data are naturally transductive.

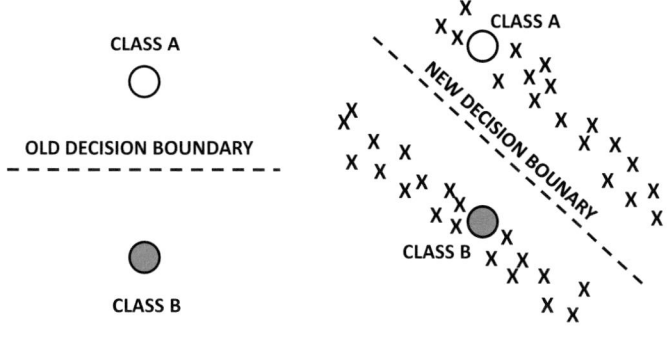

(a) only labeled examples   (b) labeled and unlabeled examples

FIGURE 1.4: Impact of unsupervised examples on classification process.

The motivation of semi-supervised learning is that knowledge of the dense regions in the space and correlated regions of the space are helpful for classification. Consider the two-class example illustrated in Figure 1.4(a), in which only a single training example is available for each class. In such a case, the decision boundary between the two classes is the straight line perpendicular to the one joining the two classes. However, suppose that some additional unsupervised examples are available, as illustrated in Figure 1.4(b). These unsupervised examples are denoted by 'x'. In such a case, the decision boundary changes from Figure 1.4(a). The major assumption here is that the classes vary *less* in dense regions of the training data, because of the *smoothness* assumption. As a result, even though the added examples do not have labels, they contribute significantly to improvements in classification accuracy.

In this example, the *correlations* between feature values were estimated with unlabeled training data. This has an intuitive interpretation in the context of text data, where *joint* feature distributions can be estimated with unlabeled data. For example, consider a scenario, where training data is available about predicting whether a document is the "*politics*" category. It may be possible that the word "*Obama*" (or some of the less common words) may not occur in any of the (small number of) training documents. However, the word "*Obama*" may often co-occur with many features of the "*politics*" category in the unlabeled instances. Thus, the unlabeled instances can be used to learn the relevance of these less common features to the classification process, especially when the amount of available training data is small.

Similarly, when the data is clustered, each cluster in the data is likely to predominantly contain data records of one class or the other. The identification of these clusters only requires unsupervised data rather than labeled data. Once the clusters have been identified from unlabeled data, only a small number of labeled examples are required in order to determine confidently which label corresponds to which cluster. Therefore, when a test example is classified, its clustering structure provides critical information for its classification process, even when a smaller number of labeled examples are available. It has been argued in [67] that the accuracy of the approach may increase exponentially with the number of labeled examples, as long as the assumption of smoothness in label structure variation holds true. Of course, in real life, this may not be true. Nevertheless, it has been shown repeatedly in many domains that the addition of unlabeled data provides significant advantages for the classification process. An argument for the effectiveness of semi-supervised learning that uses the spectral clustering structure of the data may be found in [18]. In some domains such as graph data, semi-supervised learning is the only way in which classification may be performed. This is because a given node may have very few neighbors of a specific class.

Semi-supervised methods are implemented in a wide variety of ways. Some of these methods directly try to label the unlabeled data in order to increase the size of the training set. The idea is to incrementally add the most confidently predicted label to the training data. This is referred to as *self training*. Such methods have the downside that they run the risk of overfitting. For example, when an unlabeled example is added to the training data with a specific label, the label might be incorrect because of the specific characteristics of the feature space, or the classifier. This might result in further propagation of the errors. The results can be quite severe in many scenarios.

Therefore, semi-supervised methods need to be carefully designed in order to avoid overfitting. An example of such a method is *co-training* [21], which partitions the attribute set into two subsets, on which classifier models are independently constructed. The top label predictions of one classifier are used to augment the training data of the other, and vice-versa. Specifically, the steps of co-training are as follows:

1. Divide the feature space into two disjoint subsets f_1 and f_2.

2. Train two independent classifier models $\mathcal{M}_1$ and $\mathcal{M}_2$, which use the disjoint feature sets f_1 and f_2, respectively.

3. Add the unlabeled instance with the most confidently predicted label from $\mathcal{M}_1$ to the training data for $\mathcal{M}_2$ and vice-versa.

4. Repeat all the above steps.

Since the two classifiers are independently constructed on different feature sets, such an approach avoids overfitting. The partitioning of the feature set into f_1 and f_2 can be performed in a variety of ways. While it is possible to perform random partitioning of features, it is generally advisable to leverage redundancy in the feature set to construct f_1 and f_2. Specifically, each feature set f_i should be picked so that the features in f_j (for $j \neq i$) are redundant with respect to it. Therefore, each feature set represents a different view of the data, which is sufficient for classification. This ensures that the "confident" labels assigned to the other classifier are of high quality. At the same time, overfitting is avoided to at least some degree, because of the disjoint nature of the feature set used by the two classifiers. Typically, an erroneously assigned class label will be more easily detected by the disjoint feature set of the other classifier, which was not used to assign the erroneous label. For a test instance, each of the classifiers is used to make a prediction, and the combination score from the two classifiers may be used. For example, if the naive Bayes method is used as the base classifier, then the product of the two classifier scores may be used.

The aforementioned methods are generic meta-algorithms for semi-supervised leaning. It is also possible to design variations of existing classification algorithms such as the EM-method, or transductive SVM classifiers. EM-based methods [67] are very popular for text data. These methods attempt to model the joint probability distributions of the features and the labels with the use of partially supervised clustering methods. This allows the estimation of the conditional probabilities in the Bayes classifier to be treated as missing data, for which the EM-algorithm is very effective. This approach shows a connection between the partially supervised clustering and partially supervised classification problems. The results show that partially supervised classification is most effective, when the clusters in the data correspond to the different classes. In transductive SVMs, the labels of the unlabeled examples are also treated as integer decision variables. The SVM formulation is modified in order to determine the maximum margin SVM, with the best possible label assignment of unlabeled examples. Surveys on semi-supervised methods may be found in [29, 96]. Semi-supervised methods are discussed in Chapter 20.

1.4.4.2 Transfer Learning

As in the case of semi-supervised learning, transfer learning methods are used when there is a direct paucity of the underlying training data. However, the difference from semi-supervised learning is that, instead of using unlabeled data, labeled data from a different domain is used to enhance the learning process. For example, consider the case of learning the class label of Chinese documents, where enough training data is not available about the documents. However, similar English documents may be available that contain training labels. In such cases, the knowledge in training data for the English documents can be *transferred* to the Chinese document scenario for more effective classification. Typically, this process requires some kind of "bridge" in order to relate the Chinese documents to the English documents. An example of such a "bridge" could be pairs of similar Chinese and English documents though many other models are possible. In many cases, a small amount of auxiliary training data in the form of labeled Chinese training documents may also be available in order to further enhance the effectiveness of the transfer process. This general principle can also be applied to cross-category or cross-domain scenarios where knowledge from one classification category is used to enhance the learning of another category [71], or the knowledge from one data domain (e.g., text) is used to enhance the learning of another data domain (e.g., images) [36, 70, 71, 95]. Broadly speaking, transfer learning methods fall into one of the following four categories:

1. *Instance-Based Transfer:* In this case, the feature spaces of the two domains are highly overlapping; even the class labels may be the same. Therefore, it is possible to transfer knowledge from one domain to the other by simply re-weighting the features.

2. *Feature-Based Transfer:* In this case, there may be some overlaps among the features, but a significant portion of the feature space may be different. Often, the goal is to perform a transformation of each feature set into a new low dimensional space, which can be shared across related tasks.

3. *Parameter-Based Transfer:* In this case, the motivation is that a good training model has typically learned a lot of structure. Therefore, if two tasks are related, then the structure can be transferred to learn the target task.

4. *Relational-Transfer Learning:* The idea here is that if two domains are related, they may share some similarity relations among objects. These similarity relations can be used for transfer learning across domains.

The major challenge in such transfer learning methods is that *negative* transfer can be caused in some cases when the side information used is very noisy or irrelevant to the learning process. Therefore, it is critical to use the transfer learning process in a careful and judicious way in order to truly improve the quality of the results. A survey on transfer learning methods may be found in [68], and a detailed discussion on this topic may be found in Chapter 21.

1.4.5 Incorporating Human Feedback

A different way of enhancing the classification process is to use some form of human supervision in order to improve the effectiveness of the classification process. Two forms of human feedback are quite popular, and they correspond to active learning and visual learning, respectively. These forms of feedback are different in that the former is typically focussed on *label acquisition* with human feedback, so as to enhance the training data. The latter is focussed on either visually creating a training model, or by visually performing the classification in a diagnostic way. Nevertheless, both forms of incorporating human feedback work with the assumption that the active input of a user can provide better knowledge for the classification process. It should be pointed out that the feedback in active learning may not always come from a user. Rather a generic concept of an *oracle* (such as *Amazon Mechanical Turk*) may be available for the feedback.

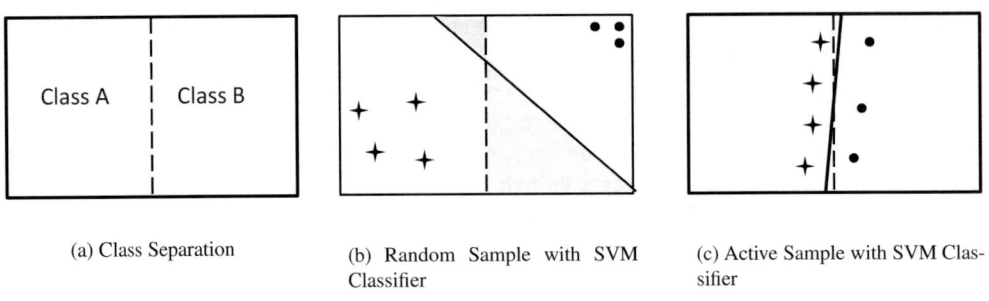

(a) Class Separation   (b) Random Sample with SVM Classifier   (c) Active Sample with SVM Classifier

FIGURE 1.5: Motivation of active learning.

1.4.5.1   Active Learning

Most classification algorithms assume that the learner is a passive recipient of the data set, which is then used to create the training model. Thus, the data collection phase is cleanly separated out from modeling, and is generally not addressed in the context of model construction. However, data collection is costly, and is often the (cost) bottleneck for many classification algorithms. In active learning, the goal is to collect more labels *during the learning process* in order to improve the effectiveness of the classification process at a low cost. Therefore, the learning process and data collection process are tightly integrated with one another and enhance each other. Typically, the classification is performed in an interactive way with the learner providing well-chosen examples to the user, for which the user may then provide labels.

For example, consider the two-class example of Figure 1.5. Here, we have a very simple division of the data into two classes, which is shown by a vertical dotted line, as illustrated in Figure 1.5(a). The two classes here are labeled by A and B. Consider the case where it is possible to query only seven examples for the two different classes. In this case, it is quite possible that the small number of allowed samples may result in a training data which is unrepresentative of the true separation between the two classes. Consider the case when an SVM classifier is used in order to construct a model. In Figure 1.5(b), we have shown a total of seven samples randomly chosen from the underlying data. Because of the inherent noisiness in the process of picking a small number of samples, an SVM classifier will be unable to accurately divide the data space. This is shown in Figure 1.5(b), where a portion of the data space is incorrectly classified, because of the error of modeling the SVM classifier. In Figure 1.5(c), we have shown an example of a well chosen set of seven instances along the decision boundary of the two classes. In this case, the SVM classifier is able to accurately model the decision regions between the two classes. This is because of the careful choice of the instances chosen by the active learning process. An important point to note is that it is particularly useful to sample instances that can clearly demarcate the decision boundary between the two classes.

In general, the examples are typically chosen for which the learner has the greatest level of uncertainty based on the current training knowledge and labels. This choice evidently provides the greatest additional information to the learner in cases where the greatest uncertainty exists about the current label. As in the case of semi-supervised learning, the assumption is that unlabeled data is copious, but acquiring labels for it is expensive. Therefore, by using the help of the learner in choosing the appropriate examples to label, it is possible to greatly reduce the effort involved in the classification process. Active learning algorithms often use support vector machines, because the latter are particularly good at determining the boundaries between the different classes. Examples that lie on these boundaries are good candidates to query the user, because the greatest level of uncertainty exists for these examples. Numerous criteria exist for training example choice in active learning algorithms, most of which try to either reduce the uncertainty in classification or reduce the error associated with the classification process. Some examples of criteria that are commonly used in order to query the learner are as follows:

- *Uncertainty Sampling:* In this case, the learner queries the user for labels of examples, for which the greatest level of uncertainty exists about its correct output [45].

- *Query by Committee (QBC):* In this case, the learner queries the user for labels of examples in which a committee of classifiers have the greatest disagreement. Clearly, this is another indirect way to ensure that examples with the greatest uncertainty are queries [81].

- *Greatest Model Change:* In this case, the learner queries the user for labels of examples, which cause the greatest level of change from the current model. The goal here is to learn new knowledge that is not currently incorporated in the model [27].

- *Greatest Error Reduction:* In this case, the learner queries the user for labels of examples, which causes the greatest reduction of error in the current example [28].

- *Greatest Variance Reduction:* In this case, the learner queries the user for examples, which result in greatest reduction in output variance [28]. This is actually similar to the previous case, since the variance is a component of the total error.

- *Representativeness:* In this case, the learner queries the user for labels that are most *representative* of the underlying data. Typically, this approach combines one of the aforementioned criteria (such as uncertainty sampling or QBC) with a representativeness model such as a density-based method in order to perform the classification [80].

These different kinds of models may work well in different kinds of scenarios. Another form of active learning queries the data *vertically*. In other words, instead of examples, it is learned which *attributes* to collect, so as to minimize the error at a given cost level [62]. A survey on active learning methods may be found in [79]. The topic of active learning is discussed in detail in Chapter 22.

1.4.5.2 Visual Learning

The goal of visual learning is typically related to, but different from, active learning. While active learning collects examples from the user, visual learning takes the help of the user in the classification process in either creating the training model or using the model for classification of a particular test instance. This help can be received by learner in two ways:

- *Visual feedback in construction of training models:* In this case, the feedback of the user may be utilized in constructing the best training model. Since the user may often have important domain knowledge, this visual feedback may often result in more effective models. For example, while constructing a decision tree classifier, a user may provide important feedback about the split points at various levels of the tree. At the same time, a visual representation of the current decision tree may be provided to the user in order to facilitate more intuitive choices. An example of a decision tree that is constructed with the use of visual methods is discussed in [17].

- *Diagnostic classification of individual test instances:* In this case, the feedback is provided by the user during classification of test instances, rather than during the process of construction of the model. The goal of this method is different, in that it enables a better understanding of the *causality* of a test instance belonging to a particular class. An example of a visual method for diagnostic classification, which uses exploratory and visual analysis of test instances, is provided in [11]. Such a method is not suitable for classifying large numbers of test instances in batch. It is typically suitable for understanding the classification behavior of a small number of carefully selected test instances.

A general discussion on visual data mining methods is found in [10, 47, 49, 55, 83]. A detailed discussion of methods for visual classification is provided in Chapter 23.

1.4.6 Evaluating Classification Algorithms

An important issue in data classification is that of *evaluation* of classification algorithms. How do we know how well a classification algorithm is performing? There are two primary issues that arise in the evaluation process:

- *Methodology used for evaluation:* Classification algorithms require a training phase and a testing phase, in which the test examples are cleanly separated from the training data. However, in order to *evaluate* an algorithm, some of the labeled examples must be removed from the training data, and the model is constructed on these examples. The problem here is that the removal of labeled examples implicitly underestimates the power of the classifier, as it relates to the set of labels already available. Therefore, how should this removal from the labeled examples be performed so as to not impact the learner accuracy too much?

  Various strategies are possible, such as *hold out*, *bootstrapping*, and *cross-validation*, of which the first is the simplest to implement, and the last provides the greatest accuracy of implementation. In the hold-out approach, a fixed percentage of the training examples are "held out," and not used in the training. These examples are then used for evaluation. Since only a subset of the training data is used, the evaluation tends to be pessimistic with the approach. Some variations use stratified sampling, in which each class is sampled independently in proportion. This ensures that random variations of class frequency between training and test examples are removed.

  In bootstrapping, sampling with replacement is used for creating the training examples. The most typical scenario is that n examples are sampled with replacement, as a result of which the fraction of examples not sampled is equal to $(1 - 1/n)^n \approx 1/e$, where e is the basis of the natural logarithm. The class accuracy is then evaluated as a weighted combination of the accuracy a_1 on the unsampled (test) examples, and the accuracy a_2 on the full labeled data. The full accuracy A is given by:

$$A = (1 - 1/e) \cdot a_1 + (1/e) \cdot a_2 \qquad (1.23)$$

  This procedure is repeated over multiple bootstrap samples and the final accuracy is reported. Note that the component a_2 tends to be highly optimistic, as a result of which the bootstrapping approach produces highly optimistic estimates. It is most appropriate for smaller data sets.

  In cross-validation, the training data is divided into a set of k disjoint subsets. One of the k subsets is used for testing, whereas the other $(k-1)$ subsets are used for training. This process is repeated by using each of the k subsets as the test set, and the error is averaged over all possibilities. This has the advantage that all examples in the labeled data have an opportunity to be treated as test examples. Furthermore, when k is large, the training data size approaches the full labeled data. Therefore, such an approach approximates the accuracy of the model using the entire labeled data well. A special case is "leave-one-out" cross-validation, where k is chosen to be equal to the number of training examples, and therefore each test segment contains exactly one example. This is, however, expensive to implement.

- *Quantification of accuracy:* This issue deals with the problem of quantifying the error of a classification algorithm. At first sight, it would seem that it is most beneficial to use a measure such as the absolute classification accuracy, which directly computes the fraction of examples that are correctly classified. However, this may not always be appropriate in all cases. For example, some algorithms may have much lower variance across different data sets, and may therefore be more desirable. In this context, an important issue that arises is that of the statistical significance of the results, when a particular classifier performs better than another on a data set. Another issue is that the output of a classification algorithm may either be presented as a discrete label for the test instance, or a numerical score, which represents the propensity of the test instance to belong to a specific class. For the case where it is presented as a discrete label, the accuracy is the most appropriate score.

In some cases, the output is presented as a numerical score, especially when the class is rare. In such cases, the Precision-Recall or ROC curves may need to be used for the purposes of classification evaluation. This is particularly important in imbalanced and rare-class scenarios. Even when the output is presented as a binary label, the evaluation methodology is different for the rare class scenario. In the rare class scenario, the misclassification of the rare class is typically much more costly than that of the normal class. In such cases, cost sensitive variations of evaluation models may need to be used for greater robustness. For example, the cost sensitive accuracy weights the rare class and normal class examples differently in the evaluation.

An excellent review of evaluation of classification algorithms may be found in [52]. A discussion of evaluation of classification algorithms is provided in Chapter 24.

1.5   Discussion and Conclusions

The problem of data classification has been widely studied in the data mining and machine learning literature. A wide variety of methods are available for data classification, such as decision trees, nearest neighbor methods, rule-based methods, neural networks, or SVM classifiers. Different classifiers may work more effectively with different kinds of data sets and application scenarios.

The data classification problem is relevant in the context of a variety of data types, such as text, multimedia, network data, time-series and sequence data. A new form of data is probabilistic data, in which the underlying data is uncertain and may require a different type of processing in order to use the uncertainty as a first-class variable. Different kinds of data may have different kinds of representations and contextual dependencies. This requires the design of methods that are well tailored to the different data types.

The classification problem has numerous variations that allow the use of either additional training data, or human intervention in order to improve the underlying results. In many cases, meta-algorithms may be used to significantly improve the quality of the underlying results.

The issue of scalability is an important one in the context of data classification. This is because data sets continue to increase in size, as data collection technologies have improved over time. Many data sets are collected continuously, and this has lead to large volumes of data streams. Even in cases where very large volumes of data are collected, big data technologies need to be designed for the classification process. This area of research is still in its infancy, and is rapidly evolving over time.

Bibliography

[1] C. Aggarwal. *Outlier Analysis*, Springer, 2013.

[2] C. Aggarwal and C. Reddy. *Data Clustering: Algorithms and Applications*, CRC Press, 2013.

[3] C. Aggarwal. Towards Systematic Design of Distance Functions in Data Mining Applications, *ACM KDD Conference*, 2003.

[4] C. Aggarwal. *Data Streams: Models and Algorithms*, Springer, 2007.

[5] C. Aggarwal. On Density-based Transforms for Uncertain Data Mining, *ICDE Conference*, 2007.

[6] C. Aggarwal. *Social Network Data Analytics*, Springer, Chapter 5, 2011.

[7] C. Aggarwal and H. Wang. *Managing and Mining Graph Data*, Springer, 2010.

[8] C. Aggarwal and C. Zhai. *Mining Text Data*, Chapter 11, Springer, 2012.

[9] C. Aggarwal and P. Yu. On Classification of High Cardinality Data Streams. *SDM Conference*, 2010.

[10] C. Aggarwal. Towards Effective and Interpretable Data Mining by Visual Interaction, *ACM SIGKDD Explorations*, 2002.

[11] C. Aggarwal. Toward exploratory test-instance-centered diagnosis in high-dimensional classification, *IEEE Transactions on Knowledge and Data Engineering*, 19(8):1001–1015, 2007.

[12] C. Aggarwal and C. Zhai. A survey of text classification algorithms, *Mining Text Data*, Springer, 2012.

[13] C. Aggarwal, J. Han, J. Wang, and P. Yu. A framework for classification of evolving data streams. In *IEEE TKDE Journal*, 2006.

[14] C. Aggarwal. On Effective Classification of Strings with Wavelets, *ACM KDD Conference*, 2002.

[15] D. Aha, D. Kibler, and M. Albert. Instance-based learning algorithms, *Machine Learning*, 6(1):37–66, 1991.

[16] D. Aha. Lazy learning: Special issue editorial. *Artificial Intelligence Review*, 11:7–10, 1997.

[17] M. Ankerst, M. Ester, and H.-P. Kriegel. Towards an Effective Cooperation of the User and the Computer for Classification, *ACM KDD Conference*, 2000.

[18] M. Belkin and P. Niyogi. Semi-supervised learning on Riemannian manifolds, *Machine Learning*, 56:209–239, 2004.

[19] C. Bishop. *Neural Networks for Pattern Recognition*, Oxford University Press, 1996.

[20] C. Bishop. *Pattern Recognition and Machine Learning*, Springer, 2007.

[21] A. Blum and T. Mitchell. Combining labeled and unlabeled data with co-training. *Proceedings of the Eleventh Annual Conference on Computational Learning Theory*, pages 92–100, 1998.

[22] L. Breiman. *Classification and regression trees*. CRC Press, 1993.

[23] L. Breiman. Random forests. *Journal Machine Learning Archive*, 45(1):5–32, 2001.

[24] L. Breiman. Bagging predictors. *Machine Learning*, 24(2):123–140, 1996.

[25] N. V. Chawla, N. Japkowicz, and A. Kotcz. Editorial: Special Issue on Learning from Imbalanced Data Sets, *ACM SIGKDD Explorations Newsletter*, 6(1):1–6, 2004.

[26] W. Cohen and Y. Singer. Context-sensitive learning methods for text categorization, *ACM Transactions on Information Systems*, 17(2):141–173, 1999.

[27] D. Cohn, L. Atlas, and R. Ladner. Improving generalization with active learning, *Machine Learning*, 5(2):201–221, 1994.

[28] D. Cohn, Z. Ghahramani and M. Jordan. Active learning with statistical models, *Journal of Artificial Intelligence Research*, 4:129–145, 1996.

[29] O. Chapelle, B. Scholkopf, and A. Zien. *Semi-supervised learning. Vol. 2, Cambridge*: MIT Press, 2006.

[30] N. Cristianini and J. Shawe-Taylor. *An Introduction to Support Vector Machines and Other Kernel-based Learning Methods*, Cambridge University Press, 2000.

[31] P. Clark and T. Niblett. The CN2 Induction algorithm, *Machine Learning*, 3(4):261–283, 1989.

[32] B. Clarke. Bayes model averaging and stacking when model approximation error cannot be ignored, *Journal of Machine Learning Research*, pages 683–712, 2003.

[33] G. Cormode and S. Muthukrishnan, An improved data-stream summary: The count-min sketch and its applications, *Journal of Algorithms*, 55(1), (2005), pp. 58–75.

[34] P. Domingos. Bayesian Averaging of Classifiers and the Overfitting Problem. *ICML Conference*, 2000.

[35] I. Dagan, Y. Karov, and D. Roth. Mistake-driven Learning in Text Categorization, *Proceedings of EMNLP*, 1997.

[36] W. Dai, Y. Chen, G.-R. Xue, Q. Yang, and Y. Yu. Translated learning: Transfer learning across different feature spaces. *Proceedings of Advances in Neural Information Processing Systems*, 2008.

[37] J. Dean and S. Ghemawat. MapReduce: A flexible data processing tool, *Communication of the ACM*, 53:72–77, 2010.

[38] P. Domingos and M. J. Pazzani. On the optimality of the simple Bayesian classifier under zero-one loss. *Machine Learning*, 29(2–3):103–130, 1997.

[39] R. Duda, P. Hart, and D. Stork, *Pattern Classification*, Wiley, 2001.

[40] Y. Freund, R. Schapire. A decision-theoretic generalization of online learning and application to boosting, *Lecture Notes in Computer Science*, 904:23–37, 1995.

[41] P. Flajolet and G. N. Martin. Probabilistic counting algorithms for data base applications. *Journal of Computer and System Sciences*, 31(2):182–209, 1985.

[42] J. Gehrke, V. Ganti, R. Ramakrishnan, and W.-Y. Loh. BOAT: Optimistic Decision Tree Construction, *ACM SIGMOD Conference*, 1999.

[43] J. Gehrke, R. Ramakrishnan, and V. Ganti. Rainforest—a framework for fast decision tree construction of large datasets, *VLDB Conference*, pages 416–427, 1998.

[44] A. Ghoting, R. Krishnamurthy, E. Pednault, B. Reinwald, V. Sindhwani, S. Tatikonda, T. Yuanyuan, and S. Vaithyanathan. SystemML: Declarative Machine Learning with MapReduce, *ICDE Conference*, 2011.

[45] D. Lewis and J. Catlett. Heterogeneous Uncertainty Sampling for Supervised Learning, *ICML Conference*, 1994.

[46] L. Hamel. *Knowledge Discovery with Support Vector Machines*, Wiley, 2009.

[47] C. Hansen and C. Johnson. *Visualization Handbook*, Academic Press, 2004.

[48] M. Mehta, R. Agrawal, and J. Rissanen. SLIQ: A Fast Scalable Classifier for Data Mining, *EDBT Conference*, 1996.

[49] M. C. F. de Oliveira and H. Levkowitz. Visual Data Mining: A Survey, *IEEE Transactions on Visualization and Computer Graphics*, 9(3):378–394. 2003.

[50] T. Hastie, R. Tibshirani, and J. Friedman. *The Elements of Statistical Learning: Data Mining, Inference, and Prediction*, Springer, 2013.

[51] S. Haykin. *Neural Networks and Learning Machines*, Prentice Hall, 2008.

[52] N. Japkowicz and M. Shah. *Evaluating Learning Algorithms: A Classification Perspective*, Cambridge University Press, 2011.

[53] T. Joachims. Making Large scale SVMs practical, *Advances in Kernel Methods, Support Vector Learning*, pages 169–184, Cambridge: *MIT Press*, 1998.

[54] T. Joachims. Training Linear SVMs in Linear Time, *KDD*, pages 217–226, 2006.

[55] D. Keim. Information and visual data mining, *IEEE Transactions on Visualization and Computer Graphics*, 8(1):1–8, 2002.

[56] W. Lam and C. Y. Ho. Using a Generalized Instance Set for Automatic Text Categorization. *ACM SIGIR Conference*, 1998.

[57] N. Littlestone. Learning quickly when irrelevant attributes abound: A new linear-threshold algorithm. *Machine Learning*, 2:285–318, 1988.

[58] B. Liu, W. Hsu, and Y. Ma. Integrating Classification and Association Rule Mining, *ACM KDD Conference*, 1998.

[59] H. Liu and H. Motoda. *Feature Selection for Knowledge Discovery and Data Mining*, Springer, 1998.

[60] R. Mayer. *Multimedia Learning*, Cambridge University Press, 2009.

[61] G. J. McLachlan. *Discriminant analysis and statistical pattern recognition*, Wiley-Interscience, Vol. 544, 2004.

[62] P. Melville, M. Saar-Tsechansky, F. Provost, and R. Mooney. An Expected Utility Approach to Active Feature-Value Acquisition. *IEEE ICDM Conference*, 2005.

[63] T. Mitchell. *Machine Learning*, McGraw Hill, 1997.

[64] K. Murphy. *Machine Learning: A Probabilistic Perspective*, MIT Press, 2012.

[65] S. K. Murthy. Automatic construction of decision trees from data: A multi-disciplinary survey, *Data Mining and Knowledge Discovery*, 2(4):345–389, 1998.

[66] H. T. Ng, W. Goh and K. Low. Feature Selection, Perceptron Learning, and a Usability Case Study for Text Categorization. *ACM SIGIR Conference*, 1997.

[67] K. Nigam, A. McCallum, S. Thrun, and T. Mitchell. Text classification from labeled and unlabeled documents using EM, *Machine Learning*, 39(2–3):103–134, 2000.

[68] S. J. Pan and Q. Yang. A survey on transfer learning, *IEEE Transactons on Knowledge and Data Engineering*, 22(10):1345–1359, 2010.

[69] H. Peng, F. Long, and C. Ding. Feature selection based on mutual information: Criteria of max-dependency, max-relevance, and min-redundancy, *IEEE Transactions on Pattern Analysis and Machine Intelligence*, 27(8):1226–1238, 2005.

[70] G. Qi, C. Aggarwal, and T. Huang. Towards Semantic Knowledge Propagation from Text Corpus to Web Images, *WWW Conference*, 2011.

[71] G. Qi, C. Aggarwal, Y. Rui, Q. Tian, S. Chang, and T. Huang. Towards Cross-Category Knowledge Propagation for Learning Visual Concepts, *CVPR Conference*, 2011.

[72] J. Quinlan. *C4.5: Programs for Machine Learning*, Morgan-Kaufmann Publishers, 1993.

[73] J. R. Quinlan. Induction of decision trees, *Machine Learning*, 1(1):81–106, 1986.

[74] J. Rocchio. Relevance feedback information retrieval. *The Smart Retrieval System - Experiments in Automatic Document Processing*, G. Salton, Ed. Englewood Cliffs, Prentice Hall, NJ: pages 313–323, 1971.

[75] B. Scholkopf and A. J. Smola. Learning with Kernels: *Support Vector Machines, Regularization, Optimization, and Beyond*, Cambridge University Press, 2001.

[76] I. Steinwart and A. Christmann. *Support Vector Machines*, Springer, 2008.

[77] H. Schutze, D. Hull, and J. Pedersen. A Comparison of Classifiers and Document Representations for the Routing Problem. *ACM SIGIR Conference*, 1995.

[78] F. Sebastiani. Machine learning in automated text categorization, *ACM Computing Surveys*, 34(1):1–47, 2002.

[79] B. Settles. *Active Learning*, Morgan and Claypool, 2012.

[80] B. Settles and M. Craven. An analysis of active learning strategies for sequence labeling tasks, *Proceedings of the Conference on Empirical Methods in Natural Language Processing (EMNLP)*, pages 1069–1078, 2008.

[81] H. Seung, M. Opper, and H. Sompolinsky. Query by Committee. Fifth Annual Workshop on Computational Learning Theory, 1992.

[82] J. Shafer, R. Agrawal, and M. Mehta. SPRINT: A scalable parallel classfier for data mining, *VLDB Conference*, pages 544–555, 1996.

[83] T. Soukop and I. Davidson. *Visual Data Mining: Techniques and Tools for Data Visualization*, Wiley, 2002.

[84] P.-N. Tan, M. Steinbach, and V. Kumar. *Introduction to Data Mining*. Pearson, 2005.

[85] V. Vapnik. *The Nature of Statistical Learning Theory*, Springer, New York, 1995.

[86] H. Wang, W. Fan, P. Yu, and J. Han. Mining Concept-Drifting Data Streams with Ensemble Classifiers, *KDD Conference*, 2003.

[87] T. White. *Hadoop: The Definitive Guide*. Yahoo! Press, 2011.

[88] E. Wiener, J. O. Pedersen, and A. S. Weigend. A neural network approach to topic spotting. *SDAIR*, pages 317–332, 1995.

[89] D. Wettschereck, D. Aha, and T. Mohri. A review and empirical evaluation of feature weighting methods for a class of lazy learning algorithms, *Artificial Intelligence Review*, 11(1–5):273–314, 1997.

[90] D. Wolpert. Stacked generalization, *Neural Networks*, 5(2):241–259, 1992.

[91] Z. Xing and J. Pei, and E. Keogh. A brief survey on sequence classification. *SIGKDD Explorations*, 12(1):40–48, 2010.

[92] L. Yang. Distance Metric Learning: A Comprehensive Survey, 2006. `http://www.cs.cmu.edu/~liuy/frame_survey_v2.pdf`

[93] M. J. Zaki and C. Aggarwal. XRules: A Structural Classifier for XML Data, *ACM KDD Conference*, 2003.

[94] B. Zenko. Is combining classifiers better than selecting the best one? *Machine Learning*, 54(3):255–273, 2004.

[95] Y. Zhu, S. J. Pan, Y. Chen, G.-R. Xue, Q. Yang, and Y. Yu. Heterogeneous Transfer Learning for Image Classification. *Special Track on AI and the Web, associated with The Twenty-Fourth AAAI Conference on Artificial Intelligence*, 2010.

[96] X. Zhu and A. Goldberg. *Introduction to Semi-Supervised Learning*, Morgan and Claypool, 2009.

Chapter 2

Feature Selection for Classification: A Review

Jiliang Tang

Arizona State University
Tempe, AZ
`Jiliang.Tang@asu.edu`

Salem Alelyani

Arizona State University
Tempe, AZ
`salelyan@asu.edu`

Huan Liu

Arizona State University
Tempe, AZ
`huan.liu@asu.edu`

2.1 Introduction

Nowadays, the growth of the high-throughput technologies has resulted in exponential growth in the harvested data with respect to both dimensionality and sample size. The trend of this growth of the UCI machine learning repository is shown in Figure 2.1. Efficient and effective management of these data becomes increasing challenging. Traditionally, manual management of these datasets has been impractical. Therefore, data mining and machine learning techniques were developed to automatically discover knowledge and recognize patterns from these data.

However, these collected data are usually associated with a high level of noise. There are many reasons causing noise in these data, among which imperfection in the technologies that collected the data and the source of the data itself are two major reasons. For example, in the medical images domain, any deficiency in the imaging device will be reflected as noise for the later process. This kind of noise is caused by the device itself. The development of social media changes the role of online users from traditional content consumers to both content creators and consumers. The quality of social media data varies from excellent data to spam or abuse content by nature. Meanwhile, social media data are usually informally written and suffers from grammatical mistakes, misspelling, and improper punctuation. Undoubtedly, extracting useful knowledge and patterns from such huge and noisy data is a challenging task.

Dimensionality reduction is one of the most popular techniques to remove noisy (i.e., irrelevant) and redundant features. Dimensionality reduction techniques can be categorized mainly into feature extraction and feature selection. Feature extraction methods project features into a new feature space with lower dimensionality and the new constructed features are usually combinations of original features. Examples of feature extraction techniques include Principle Component Analysis (PCA), Linear Discriminant Analysis (LDA), and Canonical Correlation Analysis (CCA). On the other hand, the feature selection approaches aim to select a small subset of features that minimize redundancy and maximize relevance to the target such as the class labels in classification. Representative feature selection techniques include Information Gain, Relief, Fisher Score, and Lasso.

Both feature extraction and feature selection are capable of improving learning performance, lowering computational complexity, building better generalizable models, and decreasing required storage. Feature extraction maps the original feature space to a new feature space with lower dimensions by combining the original feature space. It is difficult to link the features from original feature space to new features. Therefore further analysis of new features is problematic since there is no physical meaning for the transformed features obtained from feature extraction techniques. Meanwhile, feature selection selects a subset of features from the original feature set without any transformation, and maintains the physical meanings of the original features. In this sense, feature selection is superior in terms of better readability and interpretability. This property has its significance in many practical applications such as finding relevant genes to a specific disease and building a sentiment lexicon for sentiment analysis. Typically feature selection and feature extraction are presented separately. Via sparse learning such as ℓ_1 regularization, feature extraction (transformation) methods can be converted into feature selection methods [48].

For the classification problem, feature selection aims to select subset of highly discriminant features. In other words, it selects features that are capable of discriminating samples that belong to different classes. For the problem of feature selection for classification, due to the availability of label information, the relevance of features is assessed as the capability of distinguishing different classes. For example, a feature f_i is said to be relevant to a class c_j if f_i and c_j are highly correlated.

In the following subsections, we will review the literature of data classification in Section (2.1.1), followed by general discussions about feature selection models in Section (2.1.2) and feature selection for classification in Section (2.1.3).

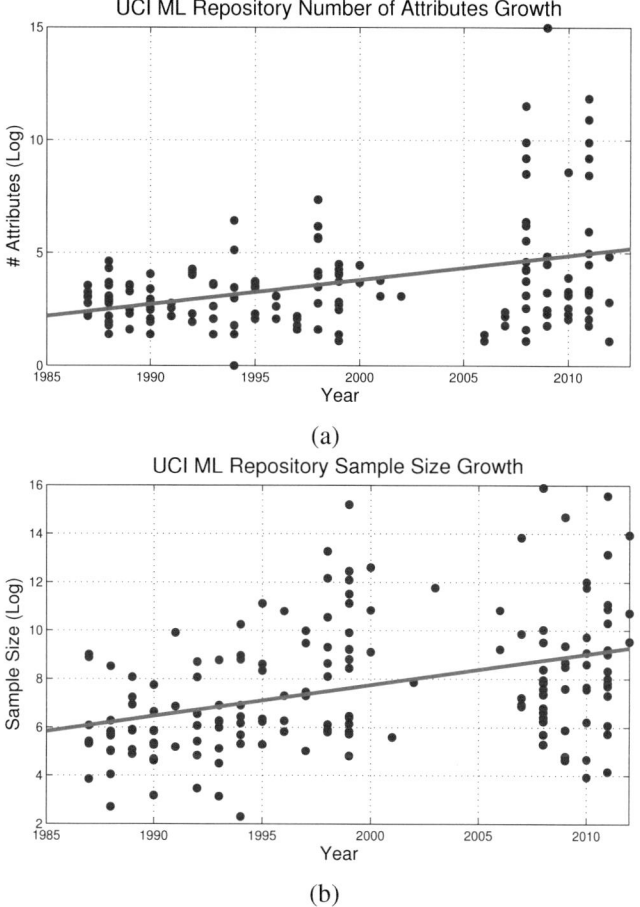

FIGURE 2.1: Plot (a) shows the dimensionality growth trend in the UCI Machine Learning Repository from the mid 1980s to 2012 while (b) shows the growth in the sample size for the same period.

2.1.1 Data Classification

Classification is the problem of identifying to which of a set of categories (sub-populations) a new observation belongs, on the basis of a training set of data containing observations (or instances) whose category membership is known. Many real-world problems can be modeled as classification problems, such as assigning a given email into "spam" or "non-spam" classes, automatically assigning the categories (e.g., "Sports" and "Entertainment") of coming news, and assigning a diagnosis to a given patient as described by observed characteristics of the patient (gender, blood pressure, presence or absence of certain symptoms, etc.). A general process of data classification is demonstrated in Figure 2.2, which usually consists of two phases — the training phase and the prediction phase.

In the training phase, data is analyzed into a set of features based on the feature generation models such as the vector space model for text data. These features may either be categorical (e.g., "A", "B", "AB" or "O", for blood type), ordinal (e.g., "large", "medium", or "small"), integer-valued (e.g., the number of occurrences of a part word in an email) or real-valued (e.g., a measurement of blood pressure). Some algorithms work only in terms of discrete data such as ID3 and require that real-valued or integer-valued data be discretized into groups (e.g., less than 5, between 5 and 10, or greater than 10). After representing data through these extracted features, the learning algorithm

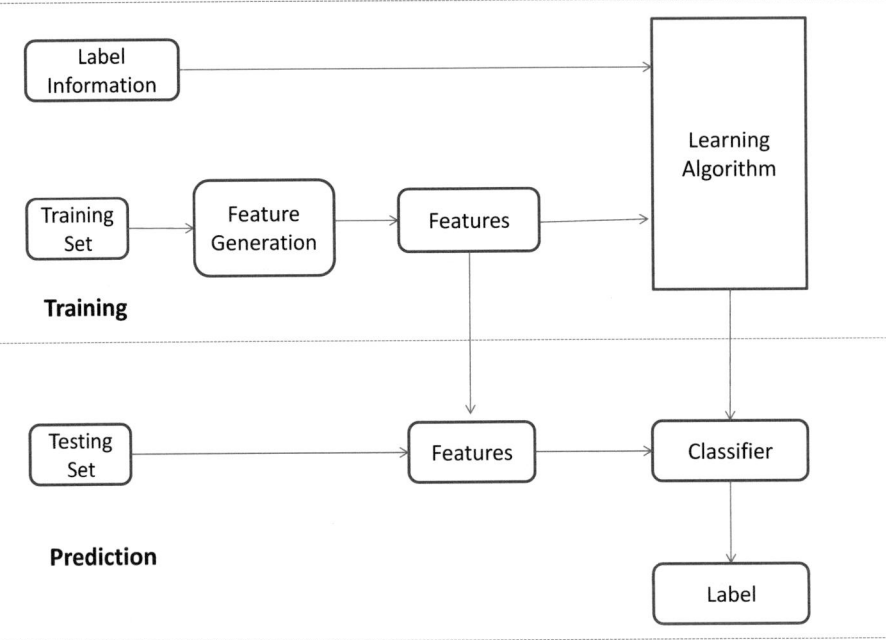

FIGURE 2.2: A general process of data classification.

will utilize the label information as well as the data itself to learn a map function f (or a classifier) from features to labels, such as,

$$f(features) \rightarrow labels. \tag{2.1}$$

In the prediction phase, data is represented by the feature set extracted in the training process, and then the map function (or the classifier) learned from the training phase will perform on the feature represented data to predict the labels. Note that the feature set used in the training phase should be the same as that in the prediction phase.

There are many classification methods in the literature. These methods can be categorized broadly into linear classifiers, support vector machines, decision trees, and Neural networks. A linear classifier makes a classification decision based on the value of a linear combination of the features. Examples of linear classifiers include Fisher's linear discriminant, logistic regression, the naive bayes classifier, and so on. Intuitively, a good separation is achieved by the hyperplane that has the largest distance to the nearest training data point of any class (so-called functional margin), since in general the larger the margin the lower the generalization error of the classifier. Therefore, the support vector machine constructs a hyperplane or set of hyperplanes by maximizing the margin. In decision trees, a tree can be learned by splitting the source set into subsets based on a feature value test. This process is repeated on each derived subset in a recursive manner called recursive partitioning. The recursion is completed when the subset at a node has all the same values of the target feature, or when splitting no longer adds value to the predictions.

2.1.2 Feature Selection

In the past thirty years, the dimensionality of the data involved in machine learning and data mining tasks has increased explosively. Data with extremely high dimensionality has presented serious challenges to existing learning methods [39], i.e., the curse of dimensionality [21]. With the presence of a large number of features, a learning model tends to overfit, resulting in the degeneration of performance. To address the problem of the curse of dimensionality, dimensionality reduction techniques have been studied, which is an important branch in the machine learning and data mining research area. Feature selection is a widely employed technique for reducing dimensionality among practitioners. It aims to choose a small subset of the relevant features from the original ones according to certain relevance evaluation criterion, which usually leads to better learning performance (e.g., higher learning accuracy for classification), lower computational cost, and better model interpretability.

According to whether the training set is labeled or not, feature selection algorithms can be categorized into supervised [61, 68], unsupervised [13, 51], and semi-supervised feature selection [71, 77]. Supervised feature selection methods can further be broadly categorized into filter models, wrapper models, and embedded models. The filter model separates feature selection from classifier learning so that the bias of a learning algorithm does not interact with the bias of a feature selection algorithm. It relies on measures of the general characteristics of the training data such as distance, consistency, dependency, information, and correlation. Relief [60], Fisher score [11], and Information Gain based methods [52] are among the most representative algorithms of the filter model. The wrapper model uses the predictive accuracy of a predetermined learning algorithm to determine the quality of selected features. These methods are prohibitively expensive to run for data with a large number of features. Due to these shortcomings in each model, the embedded model was proposed to bridge the gap between the filter and wrapper models. First, it incorporates the statistical criteria, as the filter model does, to select several candidate features subsets with a given cardinality. Second, it chooses the subset with the highest classification accuracy [40]. Thus, the embedded model usually achieves both comparable accuracy to the wrapper and comparable efficiency to the filter model. The embedded model performs feature selection in the learning time. In other words, it achieves model fitting and feature selection simultaneously [15,54]. Many researchers also paid attention to developing unsupervised feature selection. Unsupervised feature selection is a less constrained search problem without class labels, depending on clustering quality measures [12], and can eventuate many equally valid feature subsets. With high-dimensional data, it is unlikely to recover the relevant features without considering additional constraints. Another key difficulty is how to objectively measure the results of feature selection [12]. A comprehensive review about unsupervised feature selection can be found in [1]. Supervised feature selection assesses the relevance of features guided by the label information but a good selector needs enough labeled data, which is time consuming. While unsupervised feature selection works with unlabeled data, it is difficult to evaluate the relevance of features. It is common to have a data set with huge dimensionality but small labeled-sample size. High-dimensional data with small labeled samples permits too large a hypothesis space yet with too few constraints (labeled instances). The combination of the two data characteristics manifests a new research challenge. Under the assumption that labeled and unlabeled data are sampled from the same population generated by target concept, semi-supervised feature selection makes use of both labeled and unlabeled data to estimate feature relevance [77].

Feature weighting is thought of as a generalization of feature selection [69]. In feature selection, a feature is assigned a binary weight, where 1 means the feature is selected and 0 otherwise. However, feature weighting assigns a value, usually in the interval [0,1] or [-1,1], to each feature. The greater this value is, the more salient the feature will be. Most of the feature weight algorithms assign a unified (global) weight to each feature over all instances. However, the relative importance, relevance, and noise in the different dimensions may vary significantly with data locality. There are local feature selection algorithms where the local selection of features is done specific to a test instance, which is is common in lazy leaning algorithms such as kNN [9, 22]. The idea is that feature selection or weighting is done at classification time (rather than at training time), because knowledge of the test instance sharpens the ability to select features.

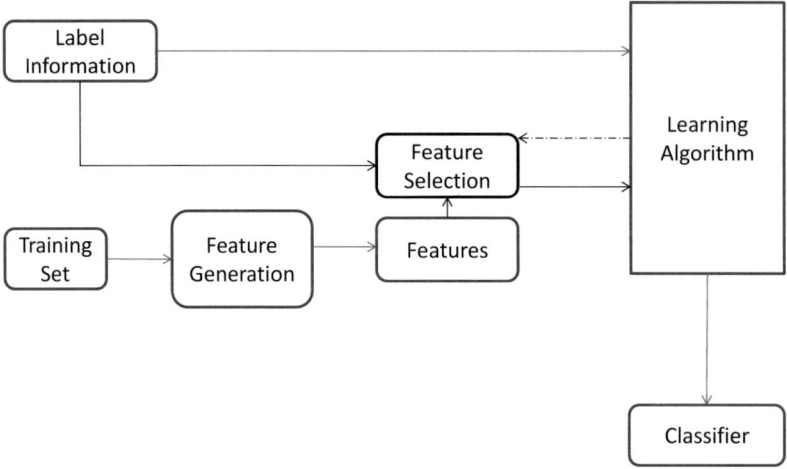

FIGURE 2.3: A general framework of feature selection for classification.

Typically, a feature selection method consists of four basic steps [40], namely, subset generation, subset evaluation, stopping criterion, and result validation. In the first step, a candidate feature subset will be chosen based on a given search strategy, which is sent, in the second step, to be evaluated according to a certain evaluation criterion. The subset that best fits the evaluation criterion will be chosen from all the candidates that have been evaluated after the stopping criterion are met. In the final step, the chosen subset will be validated using domain knowledge or a validation set.

2.1.3   Feature Selection for Classification

The majority of real-world classification problems require supervised learning where the underlying class probabilities and class-conditional probabilities are unknown, and each instance is associated with a class label [8]. In real-world situations, we often have little knowledge about relevant features. Therefore, to better represent the domain, many candidate features are introduced, resulting in the existence of irrelevant/redundant features to the target concept. A relevant feature is neither irrelevant nor redundant to the target concept; an irrelevant feature is not directly associated with the target concept but affects the learning process, and a redundant feature does not add anything new to the target concept [8]. In many classification problems, it is difficult to learn good classifiers before removing these unwanted features due to the huge size of the data. Reducing the number of irrelevant/redundant features can drastically reduce the running time of the learning algorithms and yields a more general classifier. This helps in getting a better insight into the underlying concept of a real-world classification problem.

A general feature selection for classification framework is demonstrated in Figure 2.3. Feature selection mainly affects the training phase of classification. After generating features, instead of processing data with the whole features to the learning algorithm directly, feature selection for classification will first perform feature selection to select a subset of features and then process the data with the selected features to the learning algorithm. The feature selection phase might be independent of the learning algorithm, as in filter models, or it may iteratively utilize the performance of the learning algorithms to evaluate the quality of the selected features, as in wrapper models. With the finally selected features, a classifier is induced for the prediction phase.

Usually feature selection for classification attempts to select the minimally sized subset of features according to the following criteria:

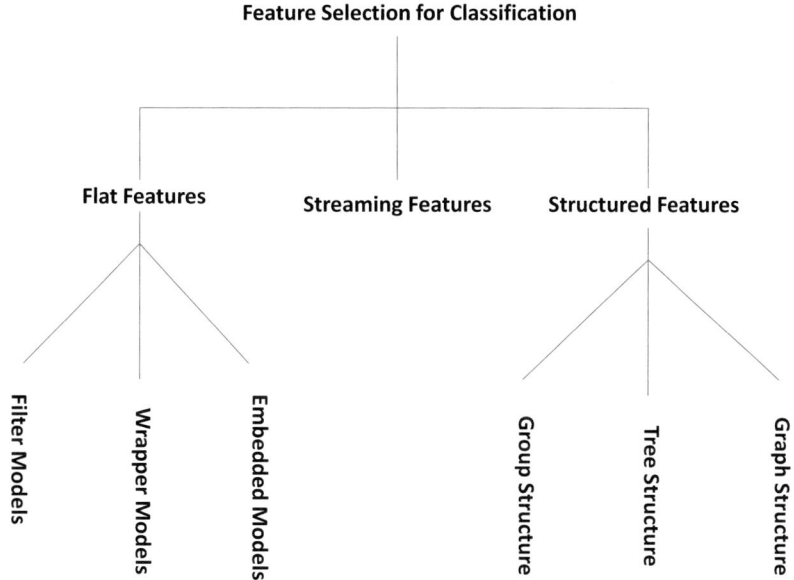

FIGURE 2.4: A classification of algorithms of feature selection for classification.

- the classification accuracy does not significantly decrease; and

- the resulting class distribution, given only the values for the selected features, is as close as possible to the original class distribution, given all features.

Ideally, feature selection methods search through the subsets of features and try to find the best one among the competing 2^m candidate subsets according to some evaluation functions [8]. However, this procedure is exhaustive as it tries to find only the best one. It may be too costly and practically prohibitive, even for a medium-sized feature set size (m). Other methods based on heuristic or random search methods attempt to reduce computational complexity by compromising performance. These methods need a stopping criterion to prevent an exhaustive search of subsets.

In this chapter, we divide feature selection for classification into three families according to the feature structure — methods for flat features, methods for structured features, and methods for streaming features as demonstrated in Figure 2.4. In the following sections, we will review these three groups with representative algorithms in detail.

Before going to the next sections, we introduce notations we adopt in this book chapter. Assume that $\mathcal{F} = \{f_1, f_2, \ldots, f_m\}$ and $C = \{c_1, c_2, \ldots, c_K\}$ denote the feature set and the class label set where m and K are the numbers of features and labels, respectively. $\mathbf{X} = \{\mathbf{x}_1, \mathbf{x}_2, \ldots, \mathbf{x}_3\} \in \mathbb{R}^{m \times n}$ is the data where n is the number of instances and the label information of the i-th instance $\mathbf{x}_i$ is denoted as y_i.

2.2   Algorithms for Flat Features

In this section, we will review algorithms for flat features, where features are assumed to be independent. Algorithms in this category are usually further divided into three groups — filter models, wrapper models, and embedded models.

2.2.1 Filter Models

Relying on the characteristics of data, filter models evaluate features without utilizing any classification algorithms [39]. A typical filter algorithm consists of two steps. In the first step, it ranks features based on certain criteria. Feature evaluation could be either *univariate* or *multivariate*. In the univariate scheme, each feature is ranked independently of the feature space, while the multivariate scheme evaluates features in an batch way. Therefore, the multivariate scheme is naturally capable of handling redundant features. In the second step, the features with highest rankings are chosen to induce classification models. In the past decade, a number of performance criteria have been proposed for filter-based feature selection, such as Fisher score [10], methods based on mutual information [37, 52, 74], and ReliefF and its variants [34, 56].

Fisher Score [10]: Features with high quality should assign similar values to instances in the same class and different values to instances from different classes. With this intuition, the score for the i-th feature S_i will be calculated by Fisher Score as,

$$S_i = \frac{\sum_{k=1}^{K} n_j (\mu_{ij} - \mu_i)^2}{\sum_{k=1}^{K} n_j \rho_{ij}^2}, \tag{2.2}$$

where μ_{ij} and ρ_{ij} are the mean and the variance of the i-th feature in the j-th class, respectively, n_j is the number of instances in the j-th class, and μ_i is the mean of the i-th feature.

Fisher Score evaluates features individually; therefore, it cannot handle feature redundancy. Recently, Gu et al. [17] proposed a generalized Fisher score to jointly select features, which aims to find a subset of features that maximize the lower bound of traditional Fisher score and solve the following problem:

$$\|\mathbf{W}^\top diag(\mathbf{p})\mathbf{X} - \mathbf{G}\|_F^2 + \gamma \|\mathbf{W}\|_F^2,$$
$$s.t., \quad \mathbf{p} \in \{0,1\}^m, \quad \mathbf{p}^\top \mathbf{1} = d, \tag{2.3}$$

where $\mathbf{p}$ is the feature selection vector, d is the number of features to select, and $\mathbf{G}$ is a special label indicator matrix, as follows:

$$\mathbf{G}(i,j) = \begin{cases} \sqrt{\frac{n}{n_j}} - \sqrt{\frac{n_j}{n}} & \text{if } \mathbf{x}_i \in c_j, \\ -\sqrt{\frac{n_j}{n}} & \text{otherwise.} \end{cases} \tag{2.4}$$

Mutual Information based on Methods [37, 52, 74]: Due to its computational efficiency and simple interpretation, information gain is one of the most popular feature selection methods. It is used to measure the dependence between features and labels and calculates the information gain between the i-th feature f_i and the class labels C as

$$IG(f_i, C) = H(f_i) - H(f_i|C), \tag{2.5}$$

where $H(f_i)$ is the entropy of f_i and $H(f_i|C)$ is the entropy of f_i after observing C:

$$H(f_i) = -\sum_j p(x_j) log_2(p(x_j)),$$
$$H(f_i|C) = -\sum_k p(c_k) \sum_j p(x_j|c_k) log_2(p(x_j|c_k)). \tag{2.6}$$

In information gain, a feature is relevant if it has a high information gain. Features are selected in a univariate way, therefore, information gain cannot handle redundant features. In [74], a fast filter method FCBF based on mutual information was proposed to identify relevant features as well as redundancy among relevant features and measure feature-class and feature-feature correlation.

Given a threshold ρ, FCBC first selects a set of features S that is highly correlated to the class with $SU \geq \rho$, where SU is symmetrical uncertainty defined as

$$SU(f_i, C) = 2 \frac{IG(f_i, C)}{H(f_i) + H(C)}. \qquad (2.7)$$

A feature f_i is called predominant *iff* $SU(f_i, c_k) \geq \rho$ and there is no $f_j(f_j \in S, j \neq i)$ such as $SU(j, i) \geq SU(i, c_k)$. f_j is a redundant feature to f_i if $SU(j, i) \geq SU(i, c_k)$. Then the set of redundant features is denoted as $S_{(P_i)}$, which is further split into $S_{p_i}^+$ and $S_{p_i}^-$. $S_{p_i}^+$ and $S_{p_i}^-$ contain redundant features to f_i with $SU(j, c_k) > SU(i, c_k)$ and $SU(j, c_k) \leq SU(i, c_k)$, respectively. Finally FCBC applied three heuristics on $S(P_i)$, $S_{p_i}^+$ and $S_{p_i}^-$ to remove the redundant features and keep the features that are most relevant to the class. FCBC provides an effective way to handle feature redundancy in feature selection.

Minimum-Redundancy-Maximum-Relevance (mRmR) is also a mutual information based method and it selects features according to the maximal statistical dependency criterion [52]. Due to the difficulty in directly implementing the maximal dependency condition, mRmR is an approximation to maximizing the dependency between the joint distribution of the selected features and the classification variable. *Minimize Redundancy* for discrete features and continuous features is defined as:

$$\text{for Discrete Features: } \min W_I, \quad W_I = \frac{1}{|S|^2} \sum_{i,j \in S} I(i, j),$$

$$\text{for Continuous Features: } \min W_c, \quad W_c = \frac{1}{|S|^2} \sum_{i,j \in S} |C(i, j)| \qquad (2.8)$$

where $I(i, j)$ and $C(i, j)$ are mutual information and the correlation between f_i and f_j, respectively. Meanwhile, *Maximize Relevance* for discrete features and continuous features is defined as:

$$\text{for Discrete Features: } \max V_I, \quad V_I = \frac{1}{|S|^2} \sum_{i \in S} I(h, i),$$

$$\text{for Continuous Features: } \max V_c, \quad V_c = \frac{1}{|S|^2} \sum_i F(i, h) \qquad (2.9)$$

where h is the target class and $F(i, h)$ is the F-statistic.

ReliefF [34,56]: Relief and its multi-class extension ReliefF select features to separate instances from different classes. Assume that ℓ instances are randomly sampled from the data and then the score of the i-th feature S_i is defined by Relief as,

$$S_i = \frac{1}{2} \sum_{k=1}^{\ell} d(\mathbf{X}_{ik} - \mathbf{X}_{iM_k}) - d(\mathbf{X}_{ik} - \mathbf{X}_{iH_k}), \qquad (2.10)$$

where M_k denotes the values on the i-th feature of the nearest instances to $\mathbf{x}_k$ with the same class label, while H_k denotes the values on the i-th feature of the nearest instances to $\mathbf{x}_k$ with different class labels. $d(\cdot)$ is a distance measure. To handle multi-class problem, Equation (2.10) is extended as,

$$S_i = \frac{1}{K} \sum_{k=1}^{\ell} \left(-\frac{1}{m_k} \sum_{\mathbf{x}_j \in M_k} d(\mathbf{X}_{ik} - \mathbf{X}_{ij}) + \sum_{y \neq y_k} \frac{1}{h_{ky}} \frac{p(y)}{1 - p(y)} \sum_{\mathbf{x}_j \in H_k} d(\mathbf{X}_{ik} - \mathbf{X}_{ij}) \right) \qquad (2.11)$$

where M_k and H_{ky} denotes the sets of nearest points to $\mathbf{x}_k$ with the same class and the class y with sizes of m_k and h_{ky} respectively, and $p(y)$ is the probability of an instance from the class y. In [56], the authors related the relevance evaluation criterion of ReliefF to the hypothesis of margin maximization, which explains why the algorithm provides superior performance in many applications.

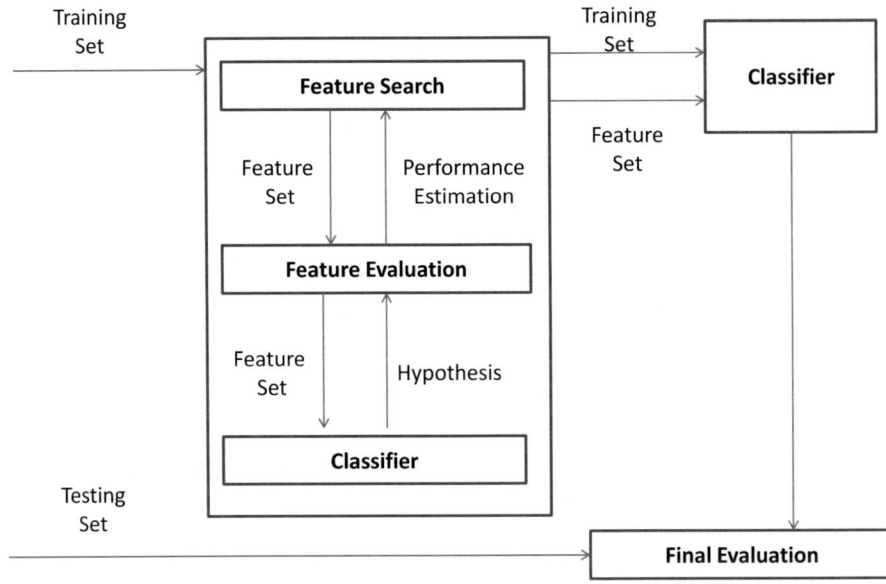

FIGURE 2.5: A general framework for wrapper methods of feature selection for classification.

2.2.2 Wrapper Models

Filter models select features independent of any specific classifiers. However the major disadvantage of the filter approach is that it totally ignores the effects of the selected feature subset on the performance of the induction algorithm [20, 36]. The optimal feature subset should depend on the specific biases and heuristics of the induction algorithm. Based on this assumption, wrapper models utilize a specific classifier to evaluate the quality of selected features, and offer a simple and powerful way to address the problem of feature selection, regardless of the chosen learning machine [26, 36]. Given a predefined classifier, a typical wrapper model will perform the following steps:

- *Step 1:* searching a subset of features,

- *Step 2:* evaluating the selected subset of features by the performance of the classifier,

- *Step 3:* repeating *Step 1* and *Step 2* until the desired quality is reached.

A general framework for wrapper methods of feature selection for classification [36] is shown in Figure 2.5, and it contains three major components:

- feature selection search — how to search the subset of features from all possible feature subsets,

- feature evaluation — how to evaluate the performance of the chosen classifier, and

- induction algorithm.

In wrapper models, the predefined classifier works as a black box. The feature search component will produce a set of features and the feature evaluation component will use the classifier to estimate the performance, which will be returned back to the feature search component for the next iteration of feature subset selection. The feature set with the highest estimated value will be chosen as the final set to learn the classifier. The resulting classifier is then evaluated on an independent testing set that is not used during the training process [36].

The size of search space for m features is $O(2^m)$, thus an exhaustive search is impractical unless m is small. Actually the problem is known to be NP-hard [18]. A wide range of search strategies can be used, including hill-climbing, best-first, branch-and-bound, and genetic algorithms [18]. The hill-climbing strategy expands the current set and moves to the subset with the highest accuracy, terminating when no subset improves over the current set. The best-first strategy is to select the most promising set that has not already been expanded and is a more robust method than hill-climbing [36]. Greedy search strategies seem to be particularly computationally advantageous and robust against overfitting. They come in two flavors — forward selection and backward elimination. Forward selection refers to a search that begins at the empty set of features and features are progressively incorporated into larger and larger subsets, whereas backward elimination begins with the full set of features and progressively eliminates the least promising ones. The search component aims to find a subset of features with the highest evaluation, using a heuristic function to guide it. Since we do not know the actual accuracy of the classifier, we use accuracy estimation as both the heuristic function and the valuation function in the feature evaluation phase. Performance assessments are usually done using a validation set or by cross-validation.

Wrapper models obtain better predictive accuracy estimates than filter models [36,38]. However, wrapper models are very computationally expensive compared to filter models. It produces better performance for the predefined classifier since we aim to select features that maximize the quality; therefore the selected subset of features is inevitably biased to the predefined classifier.

2.2.3 Embedded Models

Filter models select features that are independent of the classifier and avoid the cross-validation step in a typical wrapper model; therefore they are computationally efficient. However, they do not take into account the biases of the classifiers. For example, the relevance measure of Relief would not be appropriate as feature subset selectors for Naive-Bayes because in many cases the performance of Naive-Bayes improves with the removal of relevant features [18]. Wrapper models utilize a predefined classifier to evaluate the quality of features and representational biases of the classifier are avoided by the feature selection process. However, they have to run the classifier many times to assess the quality of selected subsets of features, which is very computationally expensive. Embedded Models embedding feature selection with classifier construction, have the advantages of (1) wrapper models — they include the interaction with the classification model and (2) filter models — and they are far less computationally intensive than wrapper methods [40, 46, 57].

There are three types of embedded methods. The first are pruning methods that first utilize all features to train a model and then attempt to eliminate some features by setting the corresponding coefficients to 0, while maintaining model performance such as recursive feature elimination using a support vector machine (SVM) [5, 19]. The second are models with a built-in mechanism for feature selection such as ID3 [55] and C4.5 [54]. The third are regularization models with objective functions that minimize fitting errors and in the meantime force the coefficients to be small or to be exactly zero. Features with coefficients that are close to 0 are then eliminated [46]. Due to good performance, regularization models attract increasing attention. We will review some representative methods below based on a survey paper of embedded models based on regularization [46].

Without loss of generality, in this section, we only consider linear classifiers $\mathbf{w}$ in which classification of $\mathbf{Y}$ can be based on a linear combination of $\mathbf{X}$ such as SVM and logistic regression. In regularization methods, classifier induction and feature selection are achieved simultaneously by estimating $\mathbf{w}$ with properly tuned penalties. The learned classifier $\mathbf{w}$ can have coefficients exactly equal to zero. Since each coefficient of $\mathbf{w}$ corresponds to one feature such as $\mathbf{w}_i$ for f_i, feature selection is achieved and only features with nonzero coefficients in $\mathbf{w}$ will be used in the classifier.

Specifically, we define $\hat{\mathbf{w}}$ as,

$$\hat{\mathbf{w}} = \min_{\mathbf{w}} c(\mathbf{w}, \mathbf{X}) + \alpha \, penalty(\mathbf{w}) \tag{2.12}$$

where $c(\cdot)$ is the classification objective function, $penalty(\mathbf{w})$ is a regularization term, and α is the regularization parameter controlling the trade-off between the $c(\cdot)$ and the penalty. Popular choices of $c(\cdot)$ include quadratic loss such as least squares, hinge loss such as ℓ_1 SVM [5], and logistic loss such as BlogReg [15].

- *Quadratic loss:*

$$c(\mathbf{w}, \mathbf{X}) = \sum_{i=1}^{n} (y_i - \mathbf{w}^\top \mathbf{x}_i)^2, \tag{2.13}$$

- *Hinge loss:*

$$c(\mathbf{w}, \mathbf{X}) = \sum_{i=1}^{n} \max(0, 1 - y_i \mathbf{w}^\top \mathbf{x}_i), \tag{2.14}$$

- *Logistic loss:*

$$c(\mathbf{w}, \mathbf{X}) = \sum_{i=1}^{n} log(1 + exp(-y_i(\mathbf{w}^\top \mathbf{x}_i + b))). \tag{2.15}$$

Lasso Regularization [64]: Lasso regularization is based on ℓ_1-norm of the coefficient of $\mathbf{w}$ and defined as

$$penalty(\mathbf{w}) = \sum_{i=1}^{m} |\mathbf{w}_i|. \tag{2.16}$$

An important property of the ℓ_1 regularization is that it can generate an estimation of $\mathbf{w}$ [64] with exact zero coefficients. In other words, there are zero entities in $\mathbf{w}$, which denotes that the corresponding features are eliminated during the classifier learning process. Therefore, it can be used for feature selection.

Adaptive Lasso [80]: The Lasso feature selection is consistent if the underlying model satisfies a non-trivial condition, which may not be satisfied in practice [76]. Meanwhile the Lasso shrinkage produces biased estimates for the large coefficients; thus, it could be suboptimal in terms of estimation risk [14].

The adaptive Lasso is proposed to improve the performance of as [80]

$$penalty(\mathbf{w}) = \sum_{i=1}^{m} \frac{1}{\mathbf{b}_i} |\mathbf{w}_i|, \tag{2.17}$$

where the only difference between Lasso and adaptive Lasso is that the latter employs a weighted adjustment $\mathbf{b}_i$ for each coefficient $\mathbf{w}_i$. The article shows that the adaptive Lasso enjoys the oracle properties and can be solved by the same efficient algorithm for solving the Lasso.

The article also proves that for linear models $\mathbf{w}$ with $n \gg m$, the adaptive Lasso estimate is selection consistent under very general conditions if $\mathbf{b}_i$ is a $\sqrt{n}$ consistent estimate of $\mathbf{w}_i$. Complimentary to this proof, [24] shows that when $m \gg n$ for linear models, the adaptive Lasso estimate is also selection consistent under a partial orthogonality condition in which the covariates with zero coefficients are weakly correlated with the covariates with nonzero coefficients.

Bridge regularization [23, 35]: Bridge regularization is formally defined as

$$penalty(\mathbf{w}) = \sum_{i=1}^{n} |\mathbf{w}_i|^{\gamma}, \quad 0 \le \gamma \le 1. \tag{2.18}$$

Lasso regularization is a special case of bridge regularization when $\gamma = 1$.

For linear models, the bridge regularization is feature selection consistent, even when the Lasso is not [23] when $n \gg m$ and $\gamma < 1$; the regularization is still feature selection consistent if the features associated with the phenotype and those not associated with the phenotype are only weakly correlated when $n \gg m$ and $\gamma < 1$.

Elastic net regularization [81]: In practice, it is common that a few features are highly correlated. In this situation, the Lasso tends to select only one of the correlated features [81]. To handle features with high correlations, elastic net regularization is proposed as

$$penalty(\mathbf{w}) = \sum_{i=1}^{n} |\mathbf{w}_i|^{\gamma} + (\sum_{i=1}^{n} \mathbf{w}_i^2)^{\lambda}, \tag{2.19}$$

with $0 < \gamma \le 1$ and $\lambda \ge 1$. The elastic net is a mixture of bridge regularization with different values of γ. [81] proposes $\gamma = 1$ and $\lambda = 1$, which is extended to $\gamma < 1$ and $\lambda = 1$ by [43].

Through the loss function $c(\mathbf{w}, \mathbf{X})$, the above-mentioned methods control the size of residuals. An alternative way to obtain a sparse estimation of $\mathbf{w}$ is Dantzig selector, which is based on the normal score equations and controls the correlation of residuals with $\mathbf{X}$ as [6],

$$\min \quad \|\mathbf{w}\|_1, \quad s.t. \quad \|\mathbf{X}^{\top}(\mathbf{y} - \mathbf{w}^{\top}\mathbf{X})\|_{\infty} \le \lambda, \tag{2.20}$$

$\|\cdot\|_{\infty}$ is the ℓ_{∞}-norm of a vector and Dantzig selector was designed for linear regression models. Candes and Tao have provided strong theoretical justification for this performance by establishing sharp non-asymptotic bounds on the ℓ_2-error in the estimated coefficients, and showed that the error is within a factor of $\log(p)$ of the error that would be achieved if the locations of the non-zero coefficients were known [6, 28]. Strong theoretical results show that LASSO and Dantzig selector are closely related [28].

2.3 Algorithms for Structured Features

The models introduced in the last section assume that features are independent and totally overlook the feature structures [73]. However, for many real-world applications, the features exhibit certain intrinsic structures, e.g., spatial or temporal smoothness [65, 79], disjoint/overlapping groups [29], trees [33], and graphs [25]. Incorporating knowledge about the structures of features may significantly improve the classification performance and help identify the important features. For example, in the study of arrayCGH [65, 66], the features (the DNA copy numbers along the genome) have the natural spatial order, and incorporating the structure information using an extension of the ℓ_1-norm outperforms the Lasso in both classification and feature selection. In this section, we review feature selection algorithms for structured features and these structures include group, tree, and graph.

Since most existing algorithms in this category are based on linear classifiers, we focus on linear classifiers such as SVM and logistic classifiers in this section. A very popular and successful approach to learn linear classifiers with structured features is to minimize a empirical error penalized by a regularization term as

$$\min_{\mathbf{w}} \quad c(\mathbf{w}^{\top}\mathbf{X}, \mathbf{Y}) + \alpha \, penalty(\mathbf{w}, \mathcal{G}), \tag{2.21}$$

where $\mathcal{G}$ denotes the structure of features, and α controls the trade-off between data fitting and regularization. Equation (2.21) will lead to sparse classifiers, which lend themselves particularly well to interpretation, which is often of primary importance in many applications such as biology or social sciences [75].

2.3.1 Features with Group Structure

In many real-world applications, features form group structures. For example, in the multifactor analysis-of-variance (ANOVA) problem, each factor may have several levels and can be denoted as a group of dummy features [75]; in speed and signal processing, different frequency bands can be represented by groups [49]. When performing feature selection, we tend to select or not select features in the same group simultaneously. Group Lasso, driving all coefficients in one group to zero together and thus resulting in group selection, attracts more and more attention [3, 27, 41, 50, 75].

Assume that features form k disjoint groups $\mathcal{G} = \{G_1, G_2, \ldots, G_k\}$ and there is no overlap between any two groups. With the group structure, we can rewrite $\mathbf{w}$ into the block form as $\mathbf{w} = \{\mathbf{w}_1, \mathbf{w}_2, \ldots, \mathbf{w}_k\}$ where $\mathbf{w}_i$ corresponds to the vector of all coefficients of features in the i-th group G_i. Then, the group Lasso performs the $\ell_{q,1}$-norm regularization on the model parameters as

$$penalty(\mathbf{w}, \mathcal{G}) = \sum_{i=1}^{k} h_i \|\mathbf{w}_{G_i}\|_q, \tag{2.22}$$

where $\| \cdot \|_q$ is the ℓ_q-norm with $q > 1$, and h_i is the weight for the i-th group. Lasso does not take group structure information into account and does not support group selection, while group Lasso can select or not select a group of features as a whole.

Once a group is selected by the group Lasso, all features in the group will be selected. For certain applications, it is also desirable to select features from the selected groups, i.e., performing simultaneous group selection and feature selection. The sparse group Lasso takes advantages of both Lasso and group Lasso, and it produces a solution with simultaneous between- and within-group sparsity. The sparse group Lasso regularization is based on a composition of the $\ell_{q,1}$-norm and the ℓ_1-norm,

$$penalty(\mathbf{w}, \mathcal{G}) = \alpha \|\mathbf{w}\|_1 + (1 - \alpha) \sum_{i=1}^{k} h_i \|\mathbf{w}_{G_i}\|_q, \tag{2.23}$$

where $\alpha \in [0, 1]$, the first term controls the sparsity in the feature level, and the second term controls the group selection.

Figure 2.6 demonstrates the different solutions among Lasso, group Lasso, and sparse group Lasso. In the figure, features form four groups $\{G_1, G_2, G_3, G_4\}$. Light color denotes the corresponding feature of the cell with zero coefficients and dark color indicates non-zero coefficients. From the figure, we observe that

- Lasso does not consider the group structure and selects a subset of features among all groups;

- group Lasso can perform group selection and select a subset of groups. Once the group is selected, all features in this group are selected; and

- sparse group Lasso can select groups and features in the selected groups at the same time.

In some applications, the groups overlap. One motivation example is the use of biologically meaningful gene/protein sets (groups) given by [73]. If the proteins/genes either appear in the same pathway, or are semantically related in terms of gene ontology (GO) hierarchy, or are related from

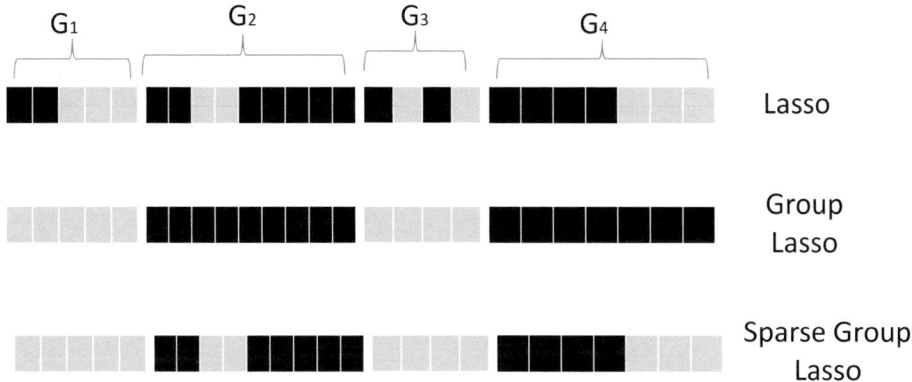

FIGURE 2.6: Illustration of Lasso, group Lasso and sparse group Lasso. Features can be grouped into four disjoint groups $\{G_1, G_2, G_3, G_4\}$. Each cell denotes a feature and light color represents the corresponding cell with coefficient zero.

gene set enrichment analysis (GSEA), they are related and assigned to the same groups. For example, the canonical pathway in MSigDB has provided 639 groups of genes. It has been shown that the group (of proteins/genes) markers are more reproducible than individual protein/gene markers and modeling such group information as prior knowledge can improve classification performance [7]. Groups may overlap — one protein/gene may belong to multiple groups. In these situations, group Lasso does not correctly handle overlapping groups and a given coefficient only belongs to one group. Algorithms investigating overlapping groups are proposed as [27, 30, 33, 42]. A general overlapping group Lasso regularization is similar to that for group Lasso regularization in Equation (2.23)

$$penalty(\mathbf{w}, \mathcal{G}) = \alpha \|\mathbf{w}\|_1 + (1 - \alpha) \sum_{i=1}^{k} h_i \|\mathbf{w}_{G_i}\|_q, \qquad (2.24)$$

however, groups for overlapping group Lasso regularization may overlap, while groups in group Lasso are disjoint.

2.3.2 Features with Tree Structure

In many applications, features can naturally be represented using certain tree structures. For example, the image pixels of the face image can be represented as a tree, where each parent node contains a series of child nodes that enjoy spatial locality; genes/proteins may form certain hierarchical tree structures [42]. Tree-guided group Lasso regularization is proposed for features represented as an index tree [30, 33, 42].

In the index tree, each leaf node represents a feature and each internal node denotes the group of the features that correspond to the leaf nodes of the subtree rooted at the given internal node. Each internal node in the tree is associated with a weight that represents the height of the subtree, or how tightly the features in the group for that internal node are correlated, which can be formally defined as follows [42].

For an index tree $\mathcal{G}$ of depth d, let $\mathcal{G}_i = \{G_1^i, G_2^i, \ldots, G_{n_i}^i\}$ contain all the nodes corresponding to depth i where n_i is the number of nodes of the depth i. The nodes satisfy the following conditions:

- the nodes from the same depth level have non-overlapping indices, i.e., $G_j^i \cap G_k^i = \emptyset$, $\forall i \in \{1, 2, \ldots, d\}$, $j \neq k$, $1 \leq j, k \leq n_i$;

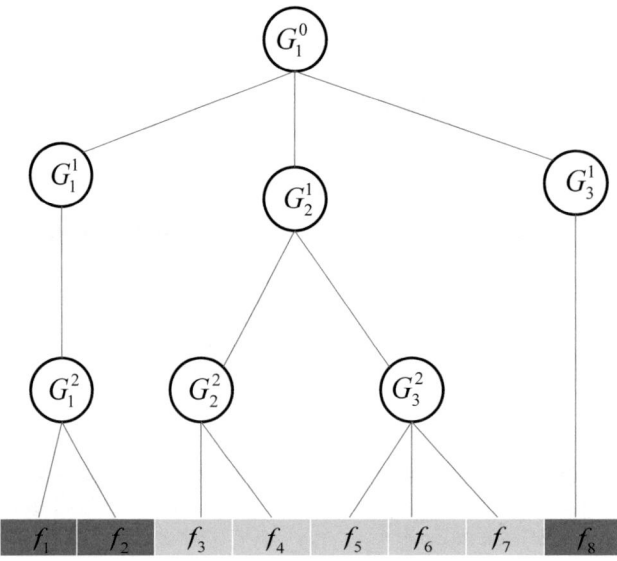

FIGURE 2.7: An illustration of a simple index tree of depth 3 with 8 features.

- let $G_{j_0}^{i-1}$ be the parent node of a non-root node G_j^i, then $G_j^i \subseteq G_{j_0}^{i-1}$.

Figure 2.7 shows a sample index tree of depth 3 with 8 features, where G_j^i are defined as

$$G_1^0 = \{f_1, f_2, f_3, f_4, f_5, f_6, f_7, f_8\},$$
$$G_1^1 = \{f_1, f_2\}, \ G_1^2 = \{f_3, f_4, f_5, f_6, f_7\}, \ G_1^3 = \{f_8\},$$
$$G_1^2 = \{f_1, f_2\}, \ G_2^2 = \{f_3, f_4\} \ G_2^3 = \{f_5, f_6, f_7\}.$$

We can observe that

- G_1^0 contains all features;

- the index sets from different nodes may overlap, e.g., any parent node overlaps with its child nodes;

- the nodes from the same depth level do not overlap; and

- the index set of a child node is a subset of that of its parent node.

With the definition of the index tree, the tree-guided group Lasso regularization is,

$$penalty(\mathbf{w}, \mathcal{G}) = \sum_{i=0}^{d} \sum_{j=1}^{n_i} h_j^i \|\mathbf{w}_{G_j^i}\|_q, \tag{2.25}$$

Since any parent node overlaps with its child nodes, if a specific node is not selected (i.e., its corresponding model coefficient is zero), then all its child nodes will not be selected. For example, in Figure 2.7, if G_2^1 is not selected, both G_2^2 and G_3^2 will not be selected, indicating that features $\{f_3, f_4, f_5, f_6, f_7\}$ will be not selected. Note that the tree structured group Lasso is a special case of the overlapping group Lasso with a specific tree structure.

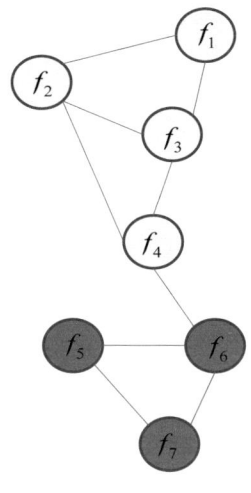

	1	1				
1		1	1			
1	1		1			
	1	1			1	
					1	1
			1	1		1
				1	1	

Graph Feature Representation: **A**

Graph Features

FIGURE 2.8: An illustration of the graph of 7 features $\{f_1, f_2, \ldots, f_7\}$ and its corresponding representation **A**.

2.3.3 Features with Graph Structure

We often have knowledge about pair-wise dependencies between features in many real-world applications [58]. For example, in natural language processing, digital lexicons such as WordNet can indicate which words are synonyms or antonyms; many biological studies have suggested that genes tend to work in groups according to their biological functions, and there are some regulatory relationships between genes. In these cases, features form an undirected graph, where the nodes represent the features, and the edges imply the relationships between features. Several recent studies have shown that the estimation accuracy can be improved using dependency information encoded as a graph.

Let $\mathcal{G}(N, E)$ be a given graph where $N = \{1, 2, \ldots, m\}$ is a set of nodes, and E is a set of edges. Node i corresponds to the i-th feature and we use $\mathbf{A} \in \mathbb{R}^{m \times m}$ to denote the adjacency matrix of $\mathcal{G}$. Figure 2.8 shows an example of the graph of 7 features $\{f_1, f_2, \ldots, f_7\}$ and its representation **A**.

If nodes i and j are connected by an edge in E, then the i-th feature and the j-th feature are more likely to be selected together, and they should have similar weights. One intuitive way to formulate graph Lasso is to force weights of two connected features close by a squre loss as

$$penalty(\mathbf{w}, \mathcal{G}) = \lambda \|\mathbf{w}\|_1 + (1 - \lambda) \sum_{i,j} \mathbf{A}_{ij} (\mathbf{w}_i - \mathbf{w}_j)^2, \qquad (2.26)$$

which is equivalent to

$$penalty(\mathbf{w}, \mathcal{G}) = \lambda \|\mathbf{w}\|_1 + (1 - \lambda) \mathbf{w}^\top \mathcal{L} \mathbf{w}, \qquad (2.27)$$

where $\mathcal{L} = \mathbf{D} - \mathbf{A}$ is the Laplacian matrix and $\mathbf{D}$ is a diagonal matrix with $\mathbf{D}_{ii} = \sum_{j=1}^{m} \mathbf{A}_{ij}$. The Laplacian matrix is positive semi-definite and captures the underlying local geometric structure of the data. When $\mathcal{L}$ is an identity matrix, $\mathbf{w}^\top \mathcal{L} \mathbf{w} = \|\mathbf{w}\|_2^2$ and then Equation (2.27) reduces to the elastic net penalty [81]. Because $\mathbf{w}^\top \mathcal{L} \mathbf{w}$ is both convex and differentiable, existing efficient algorithms for solving the Lasso can be applied to solve Equation (2.27).

Equation (2.27) assumes that the feature graph is unsigned, and encourages positive correlation between the values of coefficients for the features connected by an edge in the unsigned graph. However, two features might be negatively correlated. In this situation, the feature graph is signed, with both positive and negative edges. To perform feature selection with a signed feature graph, GFlasso employs a different ℓ_1 regularization over a graph [32],

$$penalty(\mathbf{w}, \mathcal{G}) = \lambda\|\mathbf{w}\|_1 + (1-\lambda)\sum_{i,j}\mathbf{A}_{ij}\|\mathbf{w}_i - sign(r_{ij})\mathbf{x}_j\|, \qquad (2.28)$$

where r_{ij} is the correlation between two features. When f_i and f_j are positively connected, $r_{ij} > 0$, i.e., with a positive edge, $penalty(\mathbf{w}, \mathcal{G})$ forces the coefficients $\mathbf{w}_i$ and $\mathbf{w}_j$ to be similar; while f_i and f_j are negatively connected, $r_{ij} < 0$, i.e., with a negative edge, the penalty forces $\mathbf{w}_i$ and $\mathbf{w}_j$ to be dissimilar. Due to possible graph misspecification, GFlasso may introduce additional estimation bias in the learning process. For example, additional bias may occur when the sign of the edge between f_i and f_j is inaccurately estimated.

In [72], the authors introduced several alternative formulations for graph Lasso. One of the formulations is defined as

$$penalty(\mathbf{w}, \mathcal{G}) = \lambda\|\mathbf{w}\|_1 + (1-\lambda)\sum_{i,j}\mathbf{A}_{ij}\max(|\mathbf{w}_i|, |\mathbf{w}_j|), \qquad (2.29)$$

where a pairwise ℓ_∞ regularization is used to force the coefficients to be equal and the grouping constraints are only put on connected nodes with $\mathbf{A}_{ij} = 1$. The ℓ_1-norm of $\mathbf{w}$ encourages sparseness, and $\max(|\mathbf{w}_i|, |\mathbf{w}_j|)$ will penalize the larger coefficients, which can be decomposed as

$$\max(|\mathbf{w}_i|, |\mathbf{w}_j|) = \frac{1}{2}(|\mathbf{w}_i + \mathbf{w}_j| + |\mathbf{w}_i - \mathbf{w}_j|), \qquad (2.30)$$

which can be further represented as

$$\max(|\mathbf{w}_i|, |\mathbf{w}_j|) = |\mathbf{u}^\top\mathbf{w}| + |\mathbf{v}^\top\mathbf{w}|, \qquad (2.31)$$

where $\mathbf{u}$ and $\mathbf{v}$ are two vectors with only two non-zero entities, i.e., $\mathbf{u}_i = \mathbf{u}_j = \frac{1}{2}$ and $\mathbf{v}_i = -\mathbf{v}_j = \frac{1}{2}$.

The GOSCAR formulation is closely related to OSCAR in [4]. However, OSCAR assumes that all features form a complete graph, which means that the feature graph is complete. OSCAR works for $\mathbf{A}$ whose entities are all 1, while GOSCAR can work with an arbitrary undirected graph where $\mathbf{A}$ is any symmetric matrix. In this sense, GOSCAR is more general. Meanwhile, the formulation for GOSCAR is much more challenging to solve than that of OSCAR.

The limitation of the Laplacian Lasso — that the different signs of coefficients can introduce additional penalty — can be overcome by the grouping penalty of GOSCAR. However, GOSCAR can easily overpenalize the coefficient $\mathbf{w}_i$ or $\mathbf{w}_j$ due to the property of the max operator. The additional penalty would result in biased estimation, especially for large coefficients. As mentioned above, GFlasso will introduce estimation bias when the sign between $\mathbf{w}_i$ and $\mathbf{w}_j$ is wrongly estimated. This motivates the following non-convex formulation for graph features,

$$penalty(\mathbf{w}, \mathcal{G}) = \lambda\|\mathbf{w}\|_1 + (1-\lambda)\sum_{i,j}\mathbf{A}_{ij}\big|\,|\mathbf{w}_i| - |\mathbf{w}_j|\,\big|, \qquad (2.32)$$

where the grouping penalty $\sum_{i,j}\mathbf{A}_{ij}\big|\,|\mathbf{w}_i| - |\mathbf{w}_j|\,\big|$ controls only magnitudes of differences of coefficients while ignoring their signs over the graph. Via the ℓ_1 regularization and grouping penalty, feature grouping and selection are performed simultaneously where only large coefficients as well as pairwise difference are shrunk [72].

For features with graph structure, a subset of highly connected features in the graph is likely to be selected or not selected as a whole. For example, in Figure 2.8, $\{f_5, f_6, f_7\}$ are selected, while $\{f_1, f_2, f_3, f_4\}$ are not selected.

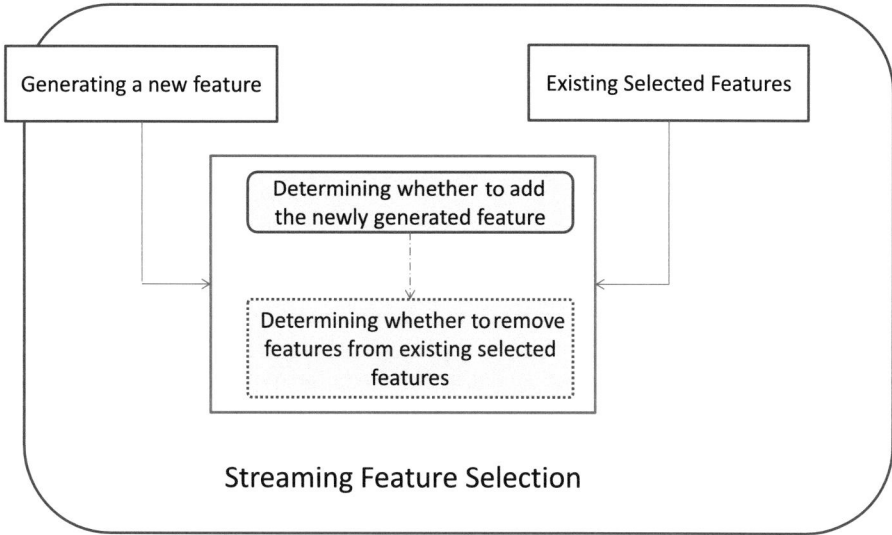

FIGURE 2.9: A general framework for streaming feature selection.

2.4   Algorithms for Streaming Features

Methods introduced above assume that all features are known in advance, and another interesting scenario is taken into account where candidate features are sequentially presented to the classifier for potential inclusion in the model [53,67,70,78]. In this scenario, the candidate features are generated dynamically and the size of features is unknown. We call this kind of features streaming features and feature selection for streaming features is called streaming feature selection. Streaming feature selection has practical significance in many applications. For example, the famous microblogging Web site Twitter produces more than 250 millions tweets per day and many new words (features) are generated, such as abbreviations. When performing feature selection for tweets, it is not practical to wait until all features have been generated, thus it could be more preferable to streaming feature selection.

A general framework for streaming feature selection is presented in Figure 2.9. A typical streaming feature selection will perform the following steps:

- *Step 1:* Generating a new feature.

- *Step 2:* Determining whether to add the newly generated feature to the set of currently selected features.

- *Step 3:* Determining whether to remove features from the set of currently selected features.

- *Step 4:* Repeat *Step 1* to *Step 3*.

Different algorithms may have different implementations for *Step 2* and *Step 3*, and next we will review some representative methods in this category. Note that *Step 3* is optional and some streaming feature selection algorithms only implement *Step 2*.

2.4.1 The Grafting Algorithm

Perkins and Theiler proposed a streaming feature selection framework based on grafting, which is a general technique that can be applied to a variety of models that are parameterized by a weight vector $\mathbf{w}$, subject to ℓ_1 regularization, such as the Lasso regularized feature selection framework in Section 2.2.3, as

$$\hat{\mathbf{w}} = \min_{\mathbf{w}} c(\mathbf{w}, \mathbf{X}) + \alpha \sum_{j=1}^{m} |\mathbf{w}_j|. \tag{2.33}$$

When all features are available, *penalty*$(\mathbf{w})$ penalizes all weights in $\mathbf{w}$ uniformly to achieve feature selection, which can be applied to streaming features with the grafting technique. In Equation (2.33), every one-zero weight $\mathbf{w}_j$ added to the model incurs a regularize penalty of $\alpha|\mathbf{w}_j|$. Therefore the feature adding to the model only happens when the loss of $c(\cdot)$ is larger than the regularizer penalty. The grafting technique will only take $\mathbf{w}_j$ away from zero if:

$$\frac{\partial c}{\partial \mathbf{w}_j} > \alpha, \tag{2.34}$$

otherwise the grafting technique will set the weight to zero (or exclude the feature).

2.4.2 The Alpha-Investing Algorithm

α-investing controls the false discovery rate by dynamically adjusting a threshold on the p-statistic for a new feature to enter the model [78], which is described as

- Initialize w_0, $i = 0$, selected features in the model $SF = \emptyset$

- *Step 1:* Get a new feature f_i

- *Step 2:* Set $\alpha_i = w_i/(2i)$

- *Step 3:*

$$w_{i+1} = w_i - \alpha_i \qquad \text{if } pvalue(x_i, SF) \geq \alpha_i.$$
$$w_{i+1} = w_i + \alpha_\Delta - \alpha_i, \;\; SF = SF \cup f_i \quad \text{otherwise}$$

- *Step 4:* $i = i + 1$

- *Step 5:* Repeat *Step 1* to *Step 3*

where the threshold α_i is the probability of selecting a spurious feature in the i-th step and it is adjusted using the wealth w_i, which denotes the current acceptable number of future false positives. Wealth is increased when a feature is selected to the model, while wealth is decreased when a feature is not selected to the model. The p-value is the probability that a feature coefficient could be judged to be non-zero when it is actually zero, which is calculated by using the fact that Δ-Loglikelihood is equivalent to t-statistics. The idea of α-investing is to adaptively control the threshold for selecting features so that when new features are selected to the model, one "invests" α increasing the wealth raising the threshold, and allowing a slightly higher future chance of incorrect inclusion of features. Each time a feature is tested and found not to be significant, wealth is "spent," reducing the threshold so as to keep the guarantee of not selecting more than a target fraction of spurious features.

2.4.3 The Online Streaming Feature Selection Algorithm

To determine the value of α, the grafting algorithm requires all candidate features in advance, while the α-investing algorithm needs some prior knowledge about the structure of the feature space to heuristically control the choice of candidate feature selection, and it is difficult to obtain sufficient prior information about the structure of the candidate features with a feature stream. Therefore the online streaming feature selection algorithm (OSFS) is proposed to solve these challenging issues in streaming feature selection [70].

An entire feature set can be divided into four basic disjoint subsets — (1) irrelevant features, (2) redundant feature, (3) weakly relevant but non-redundant features, and (4) strongly relevant features. An optimal feature selection algorithm should have non-redundant and strongly relevant features. For streaming features, it is difficult to find all strongly relevant and non-redundant features. OSFS finds an optimal subset using a two-phase scheme — online relevance analysis and redundancy analysis. A general framework of OSFS is presented as follows:

- Initialize $BCF = \emptyset$;

- *Step 1:* Generate a new feature f_k;

- *Step 2:* Online relevance analysis

$$
\begin{aligned}
&\text{Disregard } f_k, \text{ if } f_k \text{ is irrelevant to the class labels.} \\
&BCF = BCF \cup f_k \qquad \text{otherwise;}
\end{aligned}
$$

- *Step 3:* Online Redundancy Analysis;

- *Step 4:* Alternate *Step 1* to *Step 3* until the stopping criteria are satisfied.

In the relevance analysis phase, OSFS discovers strongly and weakly relevant features, and adds them into best candidate features (BCF). If a new coming feature is irrelevant to the class label, it is discarded, otherwise it is added to BCF.

In the redundancy analysis, OSFS dynamically eliminates redundant features in the selected subset. For each feature f_k in BCF, if there exists a subset within BCF making f_k and the class label conditionally independent, f_k is removed from BCF. An alternative way to improve the efficiency is to further divide this phase into two analyses — inner-redundancy analysis and outer-redundancy analysis. In the inner-redundancy analysis, OSFS only re-examines the feature newly added into BCF, while the outer-redundancy analysis re-examines each feature of BCF only when the process of generating a feature is stopped.

2.5 Discussions and Challenges

Here are several challenges and concerns that we need to mention and discuss briefly in this chapter about feature selection for classification.

2.5.1 Scalability

With the tremendous growth of dataset sizes, the scalability of current algorithms may be in jeopardy, especially with these domains that require an online classifier. For example, data that cannot be loaded into the memory require a single data scan where the second pass is either un-available or very expensive. Using feature selection methods for classification may reduce the issue of scalability for clustering. However, some of the current methods that involve feature selection in the classification process require keeping full dimensionality in the memory. Furthermore, other methods require an iterative process where each sample is visited more than once until convergence.

On the other hand, the scalability of feature selection algorithms is a big problem. Usually, they require a sufficient number of samples to obtain, statically, adequate results. It is very hard to observe feature relevance score without considering the density around each sample. Some methods try to overcome this issue by memorizing only samples that are important or a summary. In conclusion, we believe that the scalability of classification and feature selection methods should be given more attention to keep pace with the growth and fast streaming of the data.

2.5.2 Stability

Algorithms of feature selection for classification are often evaluated through classification accuracy. However, the stability of algorithms is also an important consideration when developing feature selection methods. A motivated example is from bioinformatics: The domain experts would like to see the same or at least a similar set of genes, i.e., features to be selected, each time they obtain new samples in the presence of a small amount of perturbation. Otherwise they will not trust the algorithm when they get different sets of features while the datasets are drawn for the same problem. Due to its importance, stability of feature selection has drawn the attention of the feature selection community. It is defined as the sensitivity of the selection process to data perturbation in the training set. It is found that well-known feature selection methods can select features with very low stability after perturbation is introduced to the training samples. In [2] the authors found that even the underlying characteristics of data can greatly affect the stability of an algorithm. These characteristics include dimensionality m, sample size n, and different data distribution across different folds, and the stability issue tends to be data dependent. Developing algorithms of feature selection for classification with high classification accuracy and stability is still challenging.

2.5.3 Linked Data

Most existing algorithms of feature selection for classification work with generic datasets and always assume that data is independent and identically distributed. With the development of social media, linked data is available where instances contradict the i.i.d. assumption.

Linked data has become ubiquitous in real-world applications such as tweets in Twitter[1] (tweets linked through hyperlinks), social networks in Facebook[2] (people connected by friendships), and biological networks (protein interaction networks). Linked data is patently not independent and identically distributed (i.i.d.), which is among the most enduring and deeply buried assumptions of traditional machine learning methods [31,63]. Figure 2.10 shows a typical example of linked data in social media. Social media data is intrinsically linked via various types of relations such as user-post relations and user-user relations.

Many linked-data-related learning tasks are proposed, such as collective classification [47,59], and relational clustering [44,45], but the task of feature selection for linked data is rarely touched. Feature selection methods for linked data need to solve the following immediate challenges:

- How to exploit relations among data instances; and

- How to take advantage of these relations for feature selection.

Two attempts to handle linked data w.r.t. feature selection for classification are LinkedFS [62] and FSNet [16]. FSNet works with networked data and is supervised, while LinkedFS works with

---

[1] http://www.twitter.com/
[2] https://www.facebook.com/

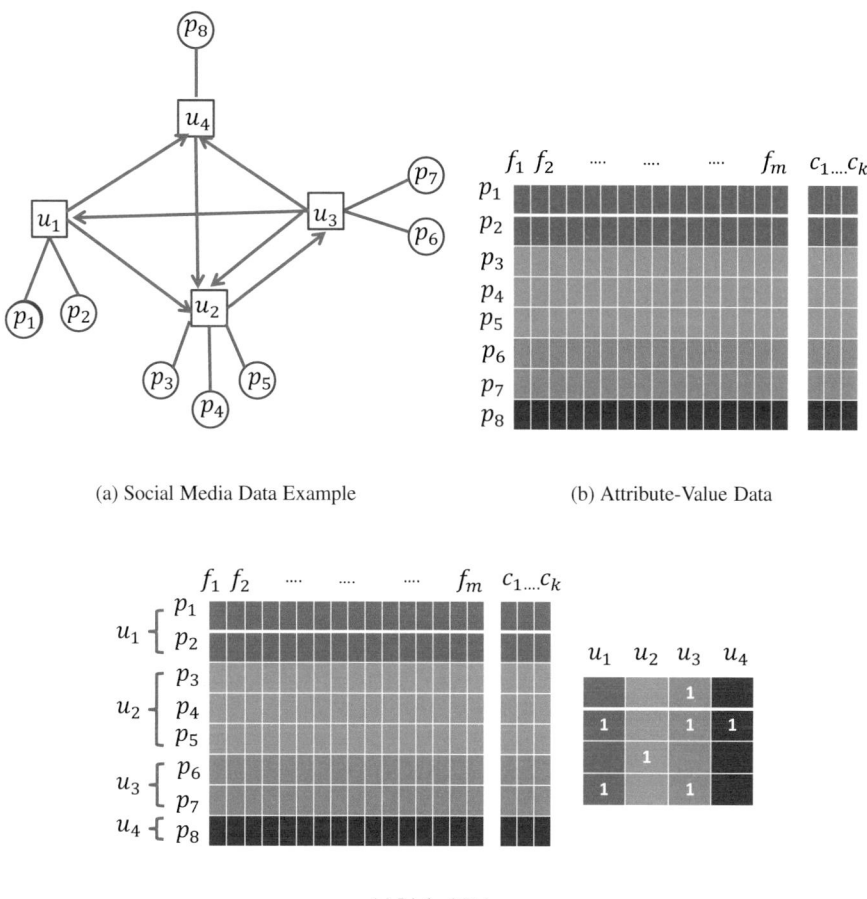

(a) Social Media Data Example   (b) Attribute-Value Data

(c) Linked Data

FIGURE 2.10: Typical linked social media data and its two representations.

social media data with social context and is semi-supervised. In LinkedFS, various relations (co-Post, coFollowing, coFollowed, and Following) are extracted following social correlation theories. LinkedFS significantly improves the performance of feature selection by incorporating these relations into feature selection. There are many issues needing further investigation for linked data, such as handling noise, and incomplete and unlabeled linked social media data.

Acknowledgments

We thank Daria Bazzi for useful comments on the representation, and helpful discussions from Yun Li and Charu Aggarwal about the content of this book chapter. This work is, in part, supported by the National Science Foundation under Grant No. IIS-1217466.

Bibliography

[1] S. Alelyani, J. Tang, and H. Liu. Feature selection for clustering: A review. *Data Clustering: Algorithms and Applications*, Editors: Charu Aggarwal and Chandan Reddy, CRC Press, 2013.

[2] S. Alelyani, L. Wang, and H. Liu. The effect of the characteristics of the dataset on the selection stability. In *Proceedings of the 23rd IEEE International Conference on Tools with Artificial Intelligence*, 2011.

[3] F. R. Bach. Consistency of the group Lasso and multiple kernel learning. *The Journal of Machine Learning Research*, 9:1179–1225, 2008.

[4] H. D. Bondell and B. J. Reich. Simultaneous regression shrinkage, variable selection, and supervised clustering of predictors with oscar. *Biometrics*, 64(1):115–123, 2008.

[5] P. S. Bradley and O. L. Mangasarian. Feature selection via concave minimization and support vector machines. *Proceedings of the Fifteenth International Conference on Machine Learning (ICML '98)*, 82–90. Morgan Kaufmann, 1998.

[6] E. Candes and T. Tao. The Dantzig selector: Statistical estimation when p is much larger than n. *The Annals of Statistics*, 35(6):2313–2351, 2007.

[7] H. Y. Chuang, E. Lee, Y.T. Liu, D. Lee, and T. Ideker. Network-based classification of breast cancer metastasis. *Molecular Systems Biology*, 3(1), 2007.

[8] M. Dash and H. Liu. Feature selection for classification. *Intelligent Data Analysis*, 1(1-4):131–156, 1997.

[9] C. Domeniconi and D. Gunopulos. Local feature selection for classification. *Computational Methods*, page 211, 2008.

[10] R.O. Duda, P.E. Hart, and D.G. Stork. *Pattern Classification*. Wiley-Interscience, 2012.

[11] R.O. Duda, P.E. Hart, and D.G. Stork. *Pattern Classification*. New York: John Wiley & Sons, 2nd edition, 2001.

[12] J.G. Dy and C.E. Brodley. Feature subset selection and order identification for unsupervised learning. In *In Proc. 17th International Conference on Machine Learning*, pages 247–254. Morgan Kaufmann, 2000.

[13] J.G. Dy and C.E. Brodley. Feature selection for unsupervised learning. *The Journal of Machine Learning Research*, 5:845–889, 2004.

[14] J. Fan and R. Li. Variable selection via nonconcave penalized likelihood and its oracle properties. *Journal of the American Statistical Association*, 96(456):1348–1360, 2001.

[15] N. L. C. Talbot G. C. Cawley, and M. Girolami. Sparse multinomial logistic regression via Bayesian l1 regularisation. In *Neural Information Processing Systems*, 2006.

[16] Q. Gu, Z. Li, and J. Han. Towards feature selection in network. In *International Conference on Information and Knowledge Management*, 2011.

[17] Q. Gu, Z. Li, and J. Han. Generalized Fisher score for feature selection. *arXiv preprint arXiv:1202.3725*, 2012.

[18] I. Guyon and A. Elisseeff. An introduction to variable and feature selection. *The Journal of Machine Learning Research*, 3:1157–1182, 2003.

[19] I. Guyon, J. Weston, S. Barnhill, and V. Vapnik. Gene selection for cancer classification using support vector machines. *Machine learning*, 46(1-3):389–422, 2002.

[20] M.A. Hall and L.A. Smith. Feature selection for machine learning: Comparing a correlation-based filter approach to the wrapper. In *Proceedings of the Twelfth International Florida Artificial Intelligence Research Society Conference*, volume 235, page 239, 1999.

[21] T. Hastie, R. Tibshirani, and J. Friedman. *The Elements of Statistical Learning*. Springer, 2001.

[22] T. Hastie and R. Tibshirani. Discriminant adaptive nearest neighbor classification. *IEEE Transactions on Pattern Analysis and Machine Intelligence*, 18(6):607–616, 1996.

[23] J. Huang, J. L. Horowitz, and S. Ma. Asymptotic properties of bridge estimators in sparse high-dimensional regression models. *The Annals of Statistics*, 36(2):587–613, 2008.

[24] J. Huang, S. Ma, and C. Zhang. Adaptive Lasso for sparse high-dimensional regression models. *Statistica Sinica*, 18(4):1603, 2008.

[25] J. Huang, T. Zhang, and D. Metaxas. Learning with structured sparsity. In *Proceedings of the 26th Annual International Conference on Machine Learning*, pages 417–424. ACM, 2009.

[26] I. Inza, P. Larrañaga, R. Blanco, and A. J. Cerrolaza. Filter versus wrapper gene selection approaches in dna microarray domains. *Artificial intelligence in Medicine*, 31(2):91–103, 2004.

[27] L. Jacob, G. Obozinski, and J. Vert. Group Lasso with overlap and graph Lasso. In *Proceedings of the 26th Annual International Conference on Machine Learning*, pages 433–440. ACM, 2009.

[28] G. James, P. Radchenko, and J. Lv. Dasso: Connections between the Dantzig selector and Lasso. *Journal of the Royal Statistical Society: Series B (Statistical Methodology)*, 71(1):127–142, 2009.

[29] R. Jenatton, J. Audibert, and F. Bach. Structured variable selection with sparsity-inducing norms. *Journal of Machine Learning Research*, 12:2777–2824, 2011.

[30] R. Jenatton, J. Mairal, G. Obozinski, and F. Bach. Proximal methods for sparse hierarchical dictionary learning. *Journal of Machine Learning Research*, 12:2297–2334, 2011.

[31] D. Jensen and J. Neville. Linkage and autocorrelation cause feature selection bias in relational learning. In *International Conference on Machine Learning*, pages 259–266, 2002.

[32] S. Kim and E. Xing. Statistical estimation of correlated genome associations to a quantitative trait network. *PLoS Genetics*, 5(8):e1000587, 2009.

[33] S. Kim and E. Xing. Tree-guided group Lasso for multi-task regression with structured sparsity. In *Proceedings of the 27th International Conference on Machine Learing*, Haifa, Israel, 2010.

[34] K. Kira and L. Rendell. A practical approach to feature selection. In *Proceedings of the Ninth International Workshop on Machine Learning*, pages 249–256. Morgan Kaufmann Publishers Inc., 1992.

[35] K. Knight and W. Fu. Asymptotics for Lasso-type estimators. *Annals of Statistics*, 78(5):1356–1378, 2000.

[36] R. Kohavi and G.H. John. Wrappers for feature subset selection. *Artificial Intelligence*, 97(1-2):273–324, 1997.

[37] D. Koller and M. Sahami. Toward optimal feature selection. In *Proceedings of the International Conference on Machine Learning*, pages 284–292, 1996.

[38] H. Liu and H. Motoda. *Feature selection for knowledge discovery and data mining*, volume 454. Springer, 1998.

[39] H. Liu and H. Motoda. *Computational Methods of Feature Selection*. Chapman and Hall/CRC Press, 2007.

[40] H. Liu and L. Yu. Toward integrating feature selection algorithms for classification and clustering. *IEEE Transactions on Knowledge and Data Engineering*, 17(4):491, 2005.

[41] J. Liu, S. Ji, and J. Ye. Multi-task feature learning via efficient l 2, 1-norm minimization. In *Proceedings of the Twenty-Fifth Conference on Uncertainty in Artificial Intelligence*, pages 339–348. AUAI Press, 2009.

[42] J. Liu and J. Ye. Moreau-Yosida regularization for grouped tree structure learning. *Advances in Neural Information Processing Systems*, 187:195–207, 2010.

[43] Z. Liu, F. Jiang, G. Tian, S. Wang, F. Sato, S. Meltzer, and M. Tan. Sparse logistic regression with lp penalty for biomarker identification. *Statistical Applications in Genetics and Molecular Biology*, 6(1), 2007.

[44] B. Long, Z.M. Zhang, X. Wu, and P.S. Yu. Spectral clustering for multi-type relational data. In *Proceedings of the 23rd International Conference on Machine Learning*, pages 585–592. ACM, 2006.

[45] B. Long, Z.M. Zhang, and P.S. Yu. A probabilistic framework for relational clustering. In *Proceedings of the 13th ACM SIGKDD International Conference on Knowledge Discovery and Data Mining*, pages 470–479. ACM, 2007.

[46] S. Ma and J. Huang. Penalized feature selection and classification in bioinformatics. *Briefings in Bioinformatics*, 9(5):392–403, 2008.

[47] S.A. Macskassy and F. Provost. Classification in networked data: A toolkit and a univariate case study. *The Journal of Machine Learning Research*, 8:935–983, 2007.

[48] M. Masaeli, J G Dy, and G. Fung. From transformation-based dimensionality reduction to feature selection. In *Proceedings of the 27th International Conference on Machine Learning*, pages 751–758, 2010.

[49] J. McAuley, J Ming, D. Stewart, and P. Hanna. Subband correlation and robust speech recognition. *IEEE Transactions on Speech and Audio Processing*, 13(5):956–964, 2005.

[50] L. Meier, S. De Geer, and P. Bühlmann. The group Lasso for logistic regression. *Journal of the Royal Statistical Society: Series B (Statistical Methodology)*, 70(1):53–71, 2008.

[51] P. Mitra, C. A. Murthy, and S. Pal. Unsupervised feature selection using feature similarity. *IEEE Transactions on Pattern Analysis and Machine Intelligence*, 24:301–312, 2002.

[52] H. Peng, F. Long, and C. Ding. Feature selection based on mutual information: Criteria of max-dependency, max-relevance, and min-redundancy. *IEEE Transactions on Pattern Analysis and Machine Intelligence*, pages 1226–1238, 2005.

[53] S. Perkins and J. Theiler. Online feature selection using grafting. In *Proceedings of the International Conference on Machine Learning*, pages 592–599, 2003.

[54] J. R. Quinlan. *C4.5: Programs for Machine Learning*. Morgan Kaufmann, 1993.

[55] J. R. Quinlan. Induction of decision trees. *Machine Learning*, 1(1):81–106, 1986.

[56] M. Robnik Sikonja and I. Kononenko. Theoretical and empirical analysis of ReliefF and RReliefF. *Machine Learning*, 53(1-2):23–69, 2003.

[57] Y. Saeys, I. Inza, and P. Larrañaga. A review of feature selection techniques in bioinformatics. *Bioinformatics*, 23(19):2507–2517, 2007.

[58] T. Sandler, P. Talukdar, L. Ungar, and J. Blitzer. Regularized learning with networks of features. *Neural Information Processing Systems*, 2008.

[59] P. Sen, G. Namata, M. Bilgic, L. Getoor, B. Galligher, and T. Eliassi-Rad. Collective classification in network data. *AI Magazine*, 29(3):93, 2008.

[60] M. R. Sikonja and I. Kononenko. Theoretical and empirical analysis of ReliefF and RReliefF. *Machine Learning*, 53:23–69, 2003.

[61] L. Song, A. Smola, A. Gretton, K. Borgwardt, and J. Bedo. Supervised feature selection via dependence estimation. In *Proceedings of the 24th International Conference on Machine Learning*, pages 823–830, 2007.

[62] J. Tang and H. Liu. Feature selection with linked data in social media. In *SDM*, pages 118–128, 2012.

[63] B. Taskar, P. Abbeel, M.F. Wong, and D. Koller. Label and link prediction in relational data. In *Proceedings of the IJCAI Workshop on Learning Statistical Models from Relational Data*. Citeseer, 2003.

[64] R. Tibshirani. Regression shrinkage and selection via the Lasso. *Journal of the Royal Statistical Society. Series B (Methodological)*, pages 267–288, 1996.

[65] R. Tibshirani, M. Saunders, S. Rosset, J. Zhu, and K. Knight. Sparsity and smoothness via the fused Lasso. *Journal of the Royal Statistical Society: Series B (Statistical Methodology)*, 67(1):91–108, 2005.

[66] R. Tibshirani and P. Wang. Spatial smoothing and hot spot detection for cgh data using the fused Lasso. *Biostatistics*, 9(1):18–29, 2008.

[67] J. Wang, P. Zhao, S. Hoi, and R. Jin. Online feature selection and its applications. *IEEE Transactions on Knowledge and Data Engineering*, pages 1–14, 2013.

[68] J. Weston, A. Elisseff, B. Schoelkopf, and M. Tipping. Use of the zero norm with linear models and kernel methods. *Journal of Machine Learning Research*, 3:1439–1461, 2003.

[69] D. Wettschereck, D. Aha, and T. Mohri. A review and empirical evaluation of feature weighting methods for a class of lazy learning algorithms. *Artificial Intelligence Review*, 11:273–314, 1997.

[70] X. Wu, K. Yu, H. Wang, and W. Ding. Online streaming feature selection. In *Proceedings of the 27th International Conference on Machine Learning*, pages 1159–1166, 2010.

[71] Z. Xu, R. Jin, J. Ye, M. Lyu, and I. King. Discriminative semi-supervised feature selection via manifold regularization. In *IJCAI'09: Proceedings of the 21th International Joint Conference on Artificial Intelligence*, 2009.

[72] S. Yang, L. Yuan, Y. Lai, X. Shen, P. Wonka, and J. Ye. Feature grouping and selection over an undirected graph. In *Proceedings of the 18th ACM SIGKDD International Conference on Knowledge Discovery and Data Mining*, pages 922–930. ACM, 2012.

[73] J. Ye and J. Liu. Sparse methods for biomedical data. *ACM SIGKDD Explorations Newsletter*, 14(1):4–15, 2012.

[74] L. Yu and H. Liu. Feature selection for high-dimensional data: A fast correlation-based filter solution. In *International Conference on Machine Learning*, 20:856, 2003.

[75] M. Yuan and Y. Lin. Model selection and estimation in regression with grouped variables. *Journal of the Royal Statistical Society: Series B (Statistical Methodology)*, 68(1):49–67, 2006.

[76] P. Zhao and B. Yu. On model selection consistency of Lasso. *Journal of Machine Learning Research*, 7:2541–2563, 2006.

[77] Z. Zhao and H. Liu. Semi-supervised feature selection via spectral analysis. In *Proceedings of SIAM International Conference on Data Mining*, 2007.

[78] D. Zhou, J. Huang, and B. Schölkopf. Learning from labeled and unlabeled data on a directed graph. In *International Conference on Machine Learning*, pages 1036–1043. ACM, 2005.

[79] J. Zhou, J. Liu, V. Narayan, and J. Ye. Modeling disease progression via fused sparse group Lasso. In *Proceedings of the 18th ACM SIGKDD International Conference on Knowledge Discovery and Data Mining*, pages 1095–1103. ACM, 2012.

[80] H. Zou. The adaptive lasso and its oracle properties. *Journal of the American Statistical Association*, 101(476):1418–1429, 2006.

[81] H. Zou and T. Hastie. Regularization and variable selection via the elastic net. *Journal of the Royal Statistical Society: Series B (Statistical Methodology)*, 67(2):301–320, 2005.

Chapter 3

Probabilistic Models for Classification

Hongbo Deng

Yahoo! Labs
Sunnyvale, CA
hbdeng@yahoo-inc.com

Yizhou Sun

College of Computer and Information Science
Northeastern University
Boston, MA
yzsun@ccs.neu.edu

Yi Chang

Yahoo! Labs
Sunnyvale, CA
yichang@yahoo-inc.com

Jiawei Han

Department of Computer Science
University of Illinois at Urbana-Champaign
Urbana, IL
hanj@uiuc.edu

3.1 Introduction

In machine learning, classification is considered an instance of the supervised learning methods, i.e., inferring a function from labeled training data. The training data consist of a set of training examples, where each example is a pair consisting of an input object (typically a vector) $\mathbf{x} = \langle x_1, x_2, ..., x_d \rangle$ and a desired output value (typically a class label) $y \in \{C_1, C_2, ..., C_K\}$. Given such a set of training data, the task of a classification algorithm is to analyze the training data and produce an inferred function, which can be used to classify new (so far unseen) examples by assigning a correct class label to each of them. An example would be assigning a given email into "spam" or "non-spam" classes.

A common subclass of classification is *probabilistic classification*, and in this chapter we will focus on some probabilistic classification methods. Probabilistic classification algorithms use statistical inference to find the best class for a given example. In addition to simply assigning the best class like other classification algorithms, probabilistic classification algorithms will output a corresponding probability of the example being a member of each of the possible classes. The class with the highest probability is normally then selected as the best class. In general, probabilistic classification algorithms has a few advantages over non-probabilistic classifiers: First, it can output a confidence value (i.e., probability) associated with its selected class label, and therefore it can abstain if its confidence of choosing any particular output is too low. Second, probabilistic classifiers can be more effectively incorporated into larger machine learning tasks, in a way that partially or completely avoids the problem of error propagation.

Within a probabilistic framework, the key point of probabilistic classification is to estimate the posterior class probability $p(C_k|\mathbf{x})$. After obtaining the posterior probabilities, we use decision theory [5] to determine class membership for each new input $\mathbf{x}$. Basically, there are two ways in which we can estimate the posterior probabilities.

In the first case, we focus on determining the class-conditional probabilities $p(\mathbf{x}|C_k)$ for each class C_k individually, and infer the prior class $p(C_k)$ separately. Then using Bayes' theorem, we can obtain the posterior class probabilities $p(C_k|\mathbf{x})$. Equivalently, we can model the joint distribution $p(\mathbf{x}, C_k)$ directly and then normalize to obtain the posterior probabilities. As the class-conditional probabilities define the statistical process that generates the features we measure, these approaches that explicitly or implicitly model the distribution of inputs as well as outputs are known as *generative models*. If the observed data are truly sampled from the generative model, then fitting the parameters of the generative model to maximize the data likelihood is a common method. In this chapter, we will introduce two common examples of probabilistic generative models for classification: Naive Bayes classifier and Hidden Markov model.

Another class of models is to directly model the posterior probabilities $p(C_k|\mathbf{x})$ by learning a discriminative function $f(\mathbf{x}) = p(C_k|\mathbf{x})$ that maps input $\mathbf{x}$ directly onto a class label C_k. This approach is often referred to as the *discriminative model* as all effort is placed on defining the overall discriminative function with the class-conditional probabilities in consideration. For instance, in the case of two-class problems, $f(\mathbf{x}) = p(C_k|\mathbf{x})$ might be continuous value between 0 and 1, such that $f < 0.5$ represents class C_1 and $f > 0.5$ represents class C_2. In this chapter, we will introduce several probabilistic discriminative models, including Logistic Regression, a type of generalized linear models [23], and Conditional Random Fields.

In this chapter, we introduce several fundamental models and algorithms for probabilistic classification, including the following:

- Naive Bayes Classifier. A Naive Bayes classifier is a simple probabilistic classifier based on applying Bayes' theorem with strong (naive) independence assumptions. A good application of Naive Bayes classifier is document classification.

- Logistic Regression. Logistic regression is an approach for predicting the outcome of a categorial dependent variable based on one or more observed variables. The probabilities describing the possible outcomes are modeled as a function of the observed variables using a logistic function.

- Hidden Markov Model. A Hidden Markov model (HMM) is a simple case of dynamic Bayesian network, where the hidden states are forming a chain and only some possible value for each state can be observed. One goal of HMM is to infer the hidden states according to the observed values and their dependency relationships. A very important application of HMM is part-of-speech tagging in NLP.

- Conditional Random Fields. A Conditional Random Field (CRF) is a special case of Markov random field, but each state of node is conditional on some observed values. CRFs can be considered as a type of discriminative classifiers, as they do not model the distribution over observations. Name entity recognition in information extraction is one of CRF's applications.

This chapter is organized as follows. In Section 3.2, we briefly review Bayes' theorem and introduce Naive Bayes classifier, a successful generative probabilistic classification. In Section 3.3, we describe a popular discriminative probabilistic classification logistic regression. We therefore begin our discussion of probabilistic graphical models for classification in Section 3.4, presenting two popular probabilistic classification models, i.e., hidden Markov model and conditional random field. Finally, we give the summary in Section 3.5.

3.2 Naive Bayes Classification

The Naive Bayes classifier is based on the Bayes' theorem, and is particularly suited when the dimensionality of the inputs is high. Despite its simplicity, the Naive Bayes classifier can often achieve comparable performance with some sophisticated classification methods, such as decision tree and selected neural network classifier. Naive Bayes classifiers have also exhibited high accuracy and speed when applied to large datasets. In this section, we will briefly review Bayes' theorem, then give an overview of Naive Bayes classifier and its use in machine learning, especially document classification.

3.2.1 Bayes' Theorem and Preliminary

A widely used framework for classification is provided by a simple theorem of probability [5,11] known as Bayes' theorem or Bayes's rule. Before we introduce Bayes' Theorem, let us first review two fundamental rules of probability theory in the following form:

$$p(X) = \sum_Y p(X,Y) \tag{3.1}$$

$$p(X,Y) = p(Y|X)p(X). \tag{3.2}$$

where the first equation is the *sum rule*, and the second equation is the *product rule*. Here $p(X,Y)$ is a joint probability, the quantity $p(Y|X)$ is a conditional probability, and the quantity $p(X)$ is a marginal probability. These two simple rules form the basis for all of the probabilistic theory that we use throughout this chapter.

Based on the product rule, together with the symmetry property $p(X,Y) = p(Y,X)$, it is easy to obtain the following Bayes' theorem,

$$p(Y|X) = \frac{p(X|Y)p(Y)}{p(X)}, \tag{3.3}$$

which plays a central role in machine learning, especially classification. Using the sum rule, the denominator in Bayes' theorem can be expressed in terms of the quantities appearing in the numerator

$$p(X) = \sum_Y p(X|Y)p(Y).$$

The denominator in Bayes' theorem can be regarded as being the normalization constant required to ensure that the sum of the conditional probability on the left-hand side of Equation (3.3) over all values of Y equals one.

Let us consider a simple example to better understand the basic concepts of probability theory and the Bayes' theorem. Suppose we have two boxes, one red and one white, and in the red box we have two apples, four lemons, and six oranges, and in the white box we have three apples, six lemons, and one orange. Now suppose we randomly pick one of the boxes and from that box we randomly select an item, and have observed which sort of item it is. In the process, we replace the item in the box from which it came, and we could imagine repeating this process many times. Let us suppose that we pick the red box 40% of the time and we pick the white box 60% of the time, and that when we select an item from a box we are equally likely to select any of the items in the box.

Let us define random variable Y to denote the box we choose, then we have

$$p(Y = r) = 4/10$$

and

$$p(Y = w) = 6/10,$$

where $p(Y = r)$ is the marginal probability that we choose the red box, and $p(Y = w)$ is the marginal probability that we choose the white box.

Suppose that we pick a box at random, and then the probability of selecting an item is the fraction of that item given the selected box, which can be written as the following conditional probabilities

$$p(X = a|Y = r) = 2/12 \tag{3.4}$$
$$p(X = l|Y = r) = 4/12 \tag{3.5}$$
$$p(X = o|Y = r) = 6/12 \tag{3.6}$$
$$p(X = a|Y = w) = 3/10 \tag{3.7}$$
$$p(X = l|Y = w) = 6/10 \tag{3.8}$$
$$p(X = o|Y = w) = 1/10. \tag{3.9}$$

Note that these probabilities are normalized so that

$$p(X = a|Y = r) + p(X = l|Y = r) + p(X = o|Y = r) = 1$$

and

$$p(X = a|Y = w) + p(X = l|Y = w) + p(X = o|Y = w) = 1.$$

Now suppose an item has been selected and it is an orange, and we would like to know which box it came from. This requires that we evaluate the probability distribution over boxes conditioned on the identity of the item, whereas the probabilities in Equation (3.4)-(3.9) illustrate the distribution of the item conditioned on the identity of the box. Based on Bayes' theorem, we can calculate the posterior probability by reversing the conditional probability

$$p(Y = r | X = o) = \frac{p(X = o | Y = r) p(Y = r)}{p(X = o)} = \frac{6/12 \times 4/10}{13/50} = \frac{10}{13},$$

where the overall probability of choosing an orange $p(X = o)$ can be calculated by using the sum and product rules

$$p(X = o) = p(X = o | Y = r) p(Y = r) + p(X = o | Y = w) p(Y = w) = \frac{6}{12} \times \frac{4}{10} + \frac{1}{10} \times \frac{6}{10} = \frac{13}{50}.$$

From the sum rule, it then follows that $p(Y = w | X = o) = 1 - 10/13 = 3/13$.

In general cases, we are interested in the probabilities of the classes given the data samples. Suppose we use random variable Y to denote the class label for data samples, and random variable X to represent the feature of data samples. We can interpret $p(Y = C_k)$ as the *prior probability* for the class C_k, which represents the probability that the class label of a data sample is C_k before we observe the data sample. Once we observe the feature X of a data sample, we can then use Bayes' theorem to compute the corresponding *posterior probability* $p(Y|X)$. The quantity $p(X|Y)$ can be expressed as how probable the observed data X is for different classes, which is called the *likelihood*. Note that the likelihood is not a probability distribution over Y, and its integral with respect to Y does not necessarily equal one. Given this definition of likelihood, we can state Bayes' theorem as *posterior* $\propto$ *likelihood* $\times$ *prior*. Now that we have introduced the Bayes' theorem, in the next subsection, we will look at how Bayes' theorem is used in the Naive Bayes classifier.

3.2.2 Naive Bayes Classifier

Naive Bayes classifier is known to be the simplest Bayesian classifier, and it has become an important probabilistic model and has been remarkably successful in practice despite its strong independence assumption [10,20,21,36]. Naive Bayes has proven effective in text classification [25,27], medical diagnosis [16], and computer performance management, among other applications [35]. In the following subsections, we will describe the model of Naive Bayes classifier, and maximum-likelihood estimates as well as its applications.

Problem setting: Let us first define the problem setting as follows: Suppose we have a set of training set $\{(x^{(i)}, y^{(i)})\}$ consisting of N examples, each $x^{(i)}$ is a d-dimensional feature vector, and each $y^{(i)}$ denotes the class label for the example. We assume random variables Y and X with components $X_1, ..., X_d$ corresponding to the label y and the feature vector $x = \langle x_1, x_2, ..., x_d \rangle$. Note that the superscript is used to index training examples for $i = 1, ..., N$, and the subscript is used to refer to each feature or random variable of a vector. In general, Y is a discrete variable that falls into exactly one of K possible classes $\{C_k\}$ for $k \in \{1, ..., K\}$, and the features of $X_1, ..., X_d$ can be any discrete or continuous attributes.

Our task is to train a classifier that will output the posterior probability $p(Y|X)$ for possible values of Y. According to Bayes' theorem, $p(Y = C_k | X = x)$ can be represented as

$$
\begin{aligned}
p(Y = C_k | X = x) \quad &= \quad \frac{p(X = x | Y = C_k) p(Y = C_k)}{p(X = x)} \\
&= \quad \frac{p(X_1 = x_1, X_2 = x_2, ..., X_d = x_d | Y = C_k) p(Y = C_k)}{p(X_1 = x_1, X_2 = x_2, ..., X_d = x_d)}
\end{aligned}
\quad (3.10)
$$

One way to learn $p(Y|X)$ is to use the training data to estimate $p(X|Y)$ and $p(Y)$. We can then use these estimates, together with Bayes' theorem, to determine $p(Y|X = x^{(i)})$ for any new instance $x^{(i)}$.

It is typically intractable to learn exact Bayesian classifiers. Considering the case that Y is boolean and X is a vector of d boolean features, we need to estimate approximately 2^d parameters $p(X_1 = x_1, X_2 = x_2, ..., X_d = x_d | Y = C_k)$. The reason is that, for any particular value C_k, there are 2^d possible values of x, which need to compute $2^d - 1$ independent parameters. Given two possible values for Y, we need to estimate a total of $2(2^d - 1)$ such parameters. Moreover, to obtain reliable estimates of each of these parameters, we will need to observe each of these distinct instances multiple times, which is clearly unrealistic in most practical classification domains. For example, if X is a vector with 20 boolean features, then we will need to estimate more than 1 million parameters.

To handle the intractable sample complexity for learning the Bayesian classifier, the Naive Bayes classifier reduces this complexity by making a conditional independence assumption that the features $X_1, ..., X_d$ are all conditionally independent of one another, given Y. For the previous case, this conditional independence assumption helps to dramatically reduce the number of parameters to be estimated for modeling $p(X|Y)$ from the original $2(2^d - 1)$ to just $2d$. Consider the likelihood $p(X = x | Y = C_k)$ of Equation (3.10), we have

$$
\begin{aligned}
&p(X_1 = x_1, X_2 = x_2, ..., X_d = x_d | Y = C_k) \\
&= \prod_{j=1}^d p(X_j = x_j | X_1 = x_1, X_2 = x_2, ..., X_{j-1} = x_{j-1}, Y = C_k) \\
&= \prod_{j=1}^d p(X_j = x_j | Y = C_k).
\end{aligned}
\tag{3.11}
$$

The second line follows from the chain rule, a general property of probabilities, and the third line follows directly from the above conditional independence, that the value for the random variable X_j is independent of all other feature values, $X_{j'}$ for $j' \neq j$, when conditioned on the identity of the label Y. This is the *Naive Bayes* assumption. It is a relatively strong and very useful assumption. When Y and X_j are boolean variables, we only need $2d$ parameters to define $p(X_j | Y = C_k)$.

After substituting Equation (3.11) in Equation (3.10), we can obtain the fundamental equation for the Naive Bayes classifier

$$
p(Y = C_k | X_1 ... X_d) = \frac{p(Y = C_k) \prod_j p(X_j | Y = C_k)}{\sum_i p(Y = y_i) \prod_j p(X_j | Y = y_i)}.
\tag{3.12}
$$

If we are interested only in the most probable value of Y, then we have the Naive Bayes classification rule:

$$
Y \leftarrow \arg\max_{C_k} \frac{p(Y = C_k) \prod_j p(X_j | Y = C_k)}{\sum_i p(Y = y_i) \prod_i p(X_j | Y = y_i)},
\tag{3.13}
$$

Because the denominator does not depend on C_k, the above formulation can be simplified to the following

$$
Y \leftarrow \arg\max_{C_k} p(Y = C_k) \prod_j p(X_j | Y = C_k).
\tag{3.14}
$$

3.2.3 Maximum-Likelihood Estimates for Naive Bayes Models

In many practical applications, parameter estimation for Naive Bayes models uses the method of maximum likelihood estimates [5, 15]. To summarize, the Naive Bayes model has two types of parameters that must be estimated. The first one is

$$
\pi_k \equiv p(Y = C_k)
$$

for any of the possible values C_k of Y. The parameter can be interpreted as the probability of seeing the label C_k, and we have the constraints $\pi_k \geq 0$ and $\sum_{k=1}^K \pi_k = 1$. Note there are K of these parameters, $(K - 1)$ of which are independent.

For the d input features X_i, suppose each can take on J possible discrete values, and we use $X_i = x_{ij}$ to denote that. The second one is

$$\theta_{ijk} \equiv p(X_i = x_{ij}|Y = C_k)$$

for each input feature X_i, each of its possible values x_{ij}, and each of the possible values C_k of Y. The value for θ_{ijk} can be interpreted as the probability of feature X_i taking value x_{ij}, conditioned on the underlying label being C_k. Note that they must satisfy $\sum_j \theta_{ijk} = 1$ for each pair of i, k values, and there will be dJK such parameters, and note that only $d(J-1)K$ of these are independent.

These parameters can be estimated using maximum likelihood estimates based on calculating the relative frequencies of the different events in the data. Maximum likelihood estimates for θ_{ijk} given a set of training examples are

$$\hat{\theta}_{ijk} = \hat{p}(X_i = x_{ij}|Y = C_k) = \frac{count(X_i = x_{ij} \wedge Y = C_k)}{count(Y = C_k)} \tag{3.15}$$

where $count(x)$ return the number of examples in the training set that satisfy property x, e.g., $count(X_i = x_{ij} \wedge Y = C_k) = \sum_{n=1}^{N}\{X_i^{(n)} = x_{ij} \wedge Y^{(n)} = C_k\}$, and $count(Y = C_k) = \sum_{n=1}^{N}\{Y^{(n)} = C_k\}$. This is a very natural estimate: We simple count the number of times label C_k is seen in conjunction with X_i taking value x_{ij}, and count the number of times the label C_k is seen in total, and then take the ratio of these two terms.

To avoid the case that the data does not happen to contain any training examples satisfying the condition in the numerator, it is common to adapt a smoothed estimate that effectively adds in a number of additional hallucinated examples equally over the possible values of X_i. The smoothed estimate is given by

$$\hat{\theta}_{ijk} = \hat{p}(X_i = x_{ij}|Y = C_k) = \frac{count(X_i = x_{ij} \wedge Y = C_k) + l}{count(Y = C_k) + lJ}, \tag{3.16}$$

where J is the number of distinct values that X_i can take on, and l determines the strength of this smoothing. If l is set to 1, this approach is called Laplace smoothing [6].

Maximum likelihood estimates for π_k take the following form

$$\hat{\pi}_k = \hat{p}(Y = C_k) = \frac{count(Y = C_k)}{N}, \tag{3.17}$$

where $N = \sum_{k=1}^{K} count(Y = C_k)$ is the number of examples in the training set. Similarly, we can obtain a smoothed estimate by using the following form

$$\hat{\pi}_k = \hat{p}(Y = C_k) = \frac{count(Y = C_k) + l}{N + lK}, \tag{3.18}$$

where K is the number of distinct values that Y can take on, and l again determines the strength of the prior assumptions relative to the observed data.

3.2.4 Applications

Naive Bayes classifier has been widely used in many classification problems, especially when the dimensionality of the features is high, such as document classification, spam detection, etc. In this subsection, let us briefly introduce the document classification using Naive Bayes classifier.

Suppose we have a number of documents x, and each document has the occurrence of words w from a dictionary $\mathcal{D}$. Generally, we assume a simple bag-of-words document model, then each document can be modeled as $|\mathcal{D}|$ single draws from a binomial distribution. In that case, for each word w, the probability of the word occurring in the document from class k is p_{kw} (i.e., $p(w|C_k)$), and the probability of it not occurring in the document from class k is obviously $1 - p_{kw}$. If word

w occurs in the document at lease once then we assign the value 1 to the feature corresponding to the word, and if it does not occur in the document we assign the value 0 to the feature. Basically, each document will be represented by a feature vector of ones and zeros with the same length as the size of the dictionary. Therefore, we are dealing with high dimensionality for large dictionaries, and Naive Bayes classifier is particularly suited for solving this problem.

We create a matrix $\mathbf{D}$ with the element D_{xw} to denote the feature (presence or absence) of the word w in document x, where rows correspond to documents and columns represent the dictionary terms. Based on Naive Bayes classifier, we can model the class-conditional probability of a document x coming from class k as

$$p(D_x|C_k) = \prod_{w=1}^{|\mathcal{D}|} p(D_{xw}|C_k) = \prod_{w=1}^{|\mathcal{D}|} p_{kw}^{D_{xw}} (1 - p_{kw})^{1 - D_{xw}}.$$

According to the maximum-likelihood estimate described in Section 3.2.3, we may easily obtain the parameter p_{kw} by

$$\hat{p}_{kw} = \frac{\sum_{x \in C_k} D_{xw}}{N_k},$$

where N_k is the number of documents from class k, and $\sum_{x \in C_k} D_{xw}$ is the number of documents from class k that contain the term w. To handle the case that a term does not occur in the document from class k (i.e., $\hat{p}_{kw} = 0$), the smoothed estimate is given by

$$\hat{p}_{kw} = \frac{\sum_{x \in C_k} D_{xw} + 1}{N_k + 2}.$$

Similarly, we can obtain the estimation for the prior probability $p(C_k)$ based on Equation (3.18). Once the parameters p_{kw} and $p(C_k)$ are estimated, then the estimate of the class conditional likelihood can be plugged into the classification rule Equation (3.14) to make a prediction.

Discussion: Naive Bayes classifier has proven effective in text classification, medical diagnosis, and computer performance management, among many other applications [21, 25, 27, 36]. Various empirical studies of this classifier in comparison to decision tree and neural network classifiers have found it to be comparable in some domains. As the independence assumption on which Naive Bayes classifiers are based almost never holds for natural data sets, it is important to understand why Naive Bayes often works well even though its independence assumption is violated. It has been observed that the optimality in terms of classification error is not necessarily related to the quality of the fit to the probability distribution (i.e., the appropriateness of the independence assumption) [36]. Moreover, Domingos and Pazzani [10] found that an optimal classifier is obtained as long as both the actual and estimated distributions agree on the most-probable class, and they proved Naive Bayes optimality for some problems classes that have a high degree of feature dependencies, such as disjunctive and conductive concepts. In summary, considerable theoretical and experimental evidence has been developed that a training procedure based on the Naive Bayes assumptions can yield an optimal classifier in a variety of situations where the assumptions are wildly violated. However, in practice this is not always the case, owing to inaccuracies in the assumptions made for its use, such as the conditional independence and the lack of available probability data.

3.3 Logistic Regression Classification

We have introduced an example of a generative classifier by fitting class-conditional probabilities and class priors separately and then applying Bayes' theorem to find the parameters. In a direct way, we can maximize a likelihood function defined through the conditional distribution

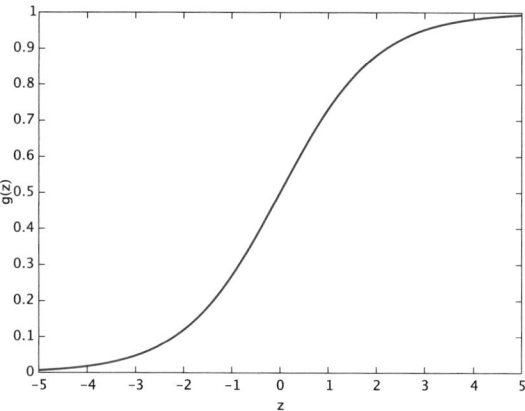

FIGURE 3.1: Illustration of logistic function. In logistic regression, $p(Y = 1|X)$ is defined to follow this form.

$p(C_k|X)$, which represents a form of discriminative model. In this section, we will introduce a popular discriminative classifier, logistic regression, as well as the parameter estimation method and its applications.

3.3.1 Logistic Regression

Logistic regression is an approach to learning $p(Y|X)$ directly in the case where Y is discrete-valued, and $X = \langle X_1, ..., X_d \rangle$ is any vector containing discrete or continuous variables. In this section, we will primarily consider the case where Y is a boolean variable and that $Y \in \{0, 1\}$.

The goal of logistic regression is to directly estimate the distribution $P(Y|X)$ from the training data. Formally, the logistic regression model is defined as

$$p(Y = 1|X) = g(\theta^T X) = \frac{1}{1 + e^{-\theta^T X}}, \tag{3.19}$$

where

$$g(z) = \frac{1}{1 + e^{-z}}$$

is called the logistic function or the sigmoid function, and

$$\theta^T X = \theta_0 + \sum_{i=1}^{d} \theta_i X_i$$

by introducing the convention of letting $X_0 = 1$ ($X = \langle X_0, X_1, ..., X_d \rangle$). Notice that $g(z)$ tends towards 1 as $z \to \infty$, and $g(z)$ tends towards 0 as $z \to -\infty$. Figure 3.1 shows a plot of the logistic function. As we can see from Figure 3.1, $g(z)$, and hence $g(\theta^T X)$ and $p(Y|X)$, are always bounded between 0 and 1.

As the sum of the probabilities must equal 1, $p(Y = 0|X)$ can be estimated using the following equation

$$p(Y = 0|X) = 1 - g(\theta^T X) = \frac{e^{-\theta^T X}}{1 + e^{-\theta^T X}}. \tag{3.20}$$

Because logistic regression predicts probabilities, rather than just classes, we can fit it using likelihood. For each training data point, we have a vector of features, $X = \langle X_0, X_1, ..., X_d \rangle$ ($X_0 = 1$), and an observed class, $Y = y_k$. The probability of that class was either $g(\theta^T X)$ if $y_k = 1$, or

$1 - g(\theta^T X)$ if $y_k = 0$. Note that we can combine both Equation (3.19) and Equation (3.20) as a more compact form

$$p(Y = y_k | X; \theta) = (p(Y = 1|X))^{y_k} (p(Y = 0|X))^{1-y_k} \tag{3.21}$$

$$= \left(g(\theta^T X)\right)^{y_k} \left(1 - g(\theta^T X)\right)^{1-y_k}. \tag{3.22}$$

Assuming that the N training examples were generated independently, the likelihood of the parameters can be written as

$$L(\theta) = p(\overrightarrow{Y} | X; \theta)$$

$$= \prod_{n=1}^{N} p(Y^{(n)} = y_k | X^{(n)}; \theta)$$

$$= \prod_{n=1}^{N} \left(p(Y^{(n)} = 1|X^{(n)})\right)^{Y^{(n)}} \left(p(Y^{(n)} = 0|X^{(n)})\right)^{1-Y^{(n)}}$$

$$= \prod_{n=1}^{N} \left(g(\theta^T X^{(n)})\right)^{Y^{(n)}} \left(1 - g(\theta^T X^{(n)})\right)^{1-Y^{(n)}}$$

where θ is the vector of parameters to be estimated, $Y^{(n)}$ denotes the observed value of Y in the nth training example, and $X^{(n)}$ denotes the observed value of X in the nth training example. To classify any given X, we generally want to assign the value y_k to Y that maximizes the likelihood as discussed in the following subsection.

3.3.2 Parameters Estimation for Logistic Regression

In general, maximum likelihood is used to determine the parameters of the logistic regression. The conditional likelihood is the probability of the observed Y values in the training data, conditioned on their corresponding X values. We choose parameters θ that satisfy

$$\theta \leftarrow \arg\max L(\theta),$$

where $\theta = \langle \theta_0, \theta_1, ..., \theta_d \rangle$ is the vector of parameters to be estimated, and this model has $d + 1$ adjustable parameters for a d-dimensional feature space.

Maximizing the likelihood is equivalent to maximizing the log likelihood:

$$l(\theta) = \log L(\theta) = \sum_{n=1}^{N} \log p(Y^{(n)} = y_k | X^{(n)}; \theta)$$

$$= \sum_{n=1}^{N} Y^{(n)} \log g(\theta^T X^{(n)}) + (1 - Y^{(n)}) \log \left(1 - g(\theta^T X^{(n)})\right) \tag{3.23}$$

$$= \sum_{n=1}^{N} Y^{(n)} (\theta^T X^{(n)}) - \log \left(1 + e^{\theta^T X^{(n)}}\right). \tag{3.24}$$

To estimate the vector of parameters θ, we maximize the log likelihood with the following rule

$$\theta \leftarrow \arg\max_{\theta} l(\theta). \tag{3.25}$$

Since there is no closed form solution to maximizing $l(\theta)$ with respect to θ, we use gradient ascent [37, 38] to solve the problem. Written in vectorial notation, our updates will therefore be given by $\theta \leftarrow \theta + \alpha \Delta_\theta l(\theta)$. After taking partial derivatives, the ith component of the vector gradient has the form

$$\frac{\partial l(\theta)}{\partial \theta_i} = \sum_{n=1}^{N} \left(Y^{(n)} - g(\theta^T X^{(n)}) \right) X_i^{(n)}, \tag{3.26}$$

where $g(\theta^T X^{(n)})$ is the logistic regression prediction using Equation (3.19). The term inside the parentheses can be interpreted as the prediction error, which is the difference between the observed $Y^{(n)}$ and its predicted probability. Note that if $Y^{(n)} = 1$ then we expect $g(\theta^T X^{(n)})$ ($p(Y = 1|X)$) to be 1, whereas if $Y^{(n)} = 0$ then we expect $g(\theta^T X^{(n)})$ to be 0, which makes $p(Y = 0|X)$ equal to 1. This error term is multiplied by the value of $X_i^{(n)}$, so as to take account for the magnitude of the $\theta_i X_i^{(n)}$ term in making this prediction.

According to the standard gradient ascent and the derivative of each θ_i, we can repeatedly update the weights in the direction of the gradient as follows:

$$\theta_i \leftarrow \theta_i + \alpha \sum_{n=1}^{N} \left(Y^{(n)} - g(\theta^T X^{(n)}) \right) X_i^{(n)}, \tag{3.27}$$

where α is a small constant (e.g., 0.01) that is called as the learning step or step size. Because the log likelihood $l(\theta)$ is a concave function in θ, this gradient ascent procedure will converge to a global maximum rather than a local maximum. The above method looks at every example in the entire training set on every step, and it is called *batch gradient ascent*. Another alternative method is called *stochastic gradient ascent*. In that method, we repeatedly run through the training set, and each time we encounter a training example, then we update the parameters according to the gradient of error with respect to that single training example only, which can be expressed as

$$\theta_i \leftarrow \theta_i + \alpha \left(Y^{(n)} - g(\theta^T X^{(n)}) \right) X_i^{(n)}. \tag{3.28}$$

In many cases where the computational efficiency is important, it is common to use the stochastic gradient ascent to estimate the parameters.

Besides maximum likelihood, Newton's method [32, 39] is a different algorithm for maximizing $l(\theta)$. Newton's method typically enjoys faster convergence than (batch) gradient descent, and requires many fewer iterations to get very close to the minimum. For more details about Newton's method, please refer to [32].

3.3.3 Regularization in Logistic Regression

Overfitting generally occurs when a statistical model describes random error or noise instead of the underlying relationship. It is a problem that can arise in logistic regression, especially when the training data is sparse and high dimensional. A model that has been overfit will generally have poor predictive performance, as it can exaggerate minor fluctuations in the data. Regularization is an approach to reduce overfitting by adding a regularization term to penalize the log likelihood function for large values of θ. One popular approach is to add L_2 norm to the log ilkelihood as follows:

$$l(\theta) = \sum_{n=1}^{N} \log p(Y^{(n)} = y_k | X^{(n)}; \theta) - \frac{\lambda}{2} \|\theta\|^2,$$

which constrains the norm of the weight vector to be small. Here λ is a constant that determines the strength of the penalty term.

By adding this penalty term, it is easy to show that maximizing it corresponds to calculating the MAP estimate for θ under the assumption that the prior distribution $p(\theta)$ is a Normal distribution with mean zero, and a variance related to $1/\lambda$. In general, the MAP estimate for θ can be written as

$$\sum_{n=1}^{N} \log p(Y^{(n)} = y_k | X^{(n)}; \theta) + \log p(\theta),$$

and if $p(\theta)$ is a zero mean Gaussian distribution, then $\log p(\theta)$ yields a term proportional to $\|\theta\|^2$. After taking partial derivatives, we can easily obtain the form

$$\frac{\partial l(\theta)}{\partial \theta_i} = \sum_{n=1}^{N} \left(Y^{(n)} - g(\theta^T X^{(n)}) \right) X_i^{(n)} - \lambda \theta_i,$$

and the modified gradient descent rule becomes

$$\theta_i \leftarrow \theta_i + \alpha \sum_{n=1}^{N} \left(Y^{(n)} - g(\theta^T X^{(n)}) \right) X_i^{(n)} - \alpha \lambda \theta_i.$$

3.3.4 Applications

Logistic regression is used extensively in numerous disciplines [26], including the Web, and medical and social science fields. For example, logistic regression might be used to predict whether a patient has a given disease (e.g., diabetes), based on observed characteristics of the patient (age, gender, body mass index, results of various blood tests, etc.) [3,22]. Another example [2] might be to predict whether an American voter will vote Democratic or Republican, based on age, income, gender, race, state of residence, votes in previous elections, etc. The technique can also be used in engineering, especially for predicting the probability of failure of a given process, system, or product. It is also used in marketing applications such as predicting of a customer's propensity for purchasing a product or ceasing a subscription, etc. In economics it can be used to predict the likelihood of a person's choosing to be in the labor force, and in a business application, it would be used to predict the likelihood of a homeowner defaulting on a mortgage.

3.4 Probabilistic Graphical Models for Classification

A graphical model [13, 17] is a probabilistic model for which a graph denotes the conditional independence structure between random variables. Graphical models provide a simple way to visualize the structure of a probabilistic model and can be used to design and motivate new models. In a probabilistic graphical model, each node represents a random variable, and the links express probabilistic relationships between these variables. The graph then captures the way in which the joint distribution over all of the random variables can be decomposed into a product of factors, each depending only on a subset of the variables. There are two branches of graphical representations of distributions that are commonly used: *directed* and *undirected*. In this section, we discuss the key aspects of graphical models for classification. We will mainly cover a directed graphical model hidden Markov model (HMM), and undirected graphical model conditional random fields (CRF). In addition, we will introduce basic Bayesian networks and Markov random fields as the preliminary contents for HMM and CRF, respectively.

3.4.1 Bayesian Networks

Bayesian networks (BNs), also known as *belief networks* (or Bayes nets for short), belong to the *directed graphical models*, in which the links of the graphs have a particular directionality indicated by arrows. Formally, BNs are directed acyclic graphs (DAG) whose nodes represent random variables, and whose edges represent conditional dependencies. For example, a link from random variable X to Y can be informally interpreted as indicating that X "causes" Y. For the classification

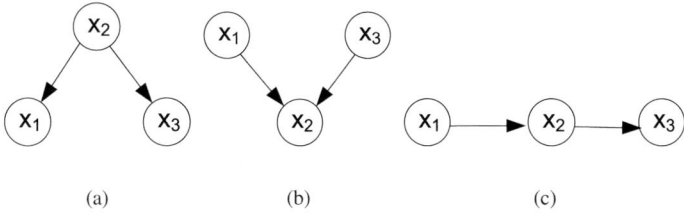

FIGURE 3.2: Examples of directed acyclic graphs describing the dependency relationships among variables.

task, we can create a node for the random variable that is going to be classified, and the goal is to infer the discrete value associated with the random variable. For example, in order to classify whether a patient has lung cancer, a node representing "lung cancer" will be created together with other factors that may directly or indirectly cause lung cancer and outcomes that will be caused by lung cancer. Given the observations for other random variables in BN, we can infer the probability of a patient having lung cancer.

3.4.1.1 Bayesian Network Construction

Once the nodes are created for all the random variables, the graph structure between these nodes needs to be specified. Bayesian network assumes a simple conditional independence assumption between random variables, which can be stated as follows: a node is conditionally independent of its non-descendants given its parents, where the parent relationship is with respect to some fixed topological ordering of the nodes. This is also called *local Markov property*, denoted by $X_v \perp\!\!\!\perp X_{V \setminus de(v)} | X_{pa(v)}$ for all $v \in V$ (the notation $\perp\!\!\!\perp$ represents "is independent of"), where $de(v)$ is the set of descendants of v and $pa(v)$ denotes the parent node set of v. For example, as shown in Figure 3.2(a), we obtain $x_1 \perp\!\!\!\perp x_3 | x_2$.

One key calculation for BN is to compute the joint distribution of an observation for all the random variables, $p(\mathbf{x})$. By using the chain rule of probability, the above joint distribution can be written as a product of conditional distributions, given the topological order of these random variables:

$$p(x_1, ..., x_n) = p(x_1)p(x_2|x_1)...p(x_n|x_{n-1}, ..., x_1). \tag{3.29}$$

By utilizing the conditional independence assumption, the joint probability of all random variables can be factored into a product of density functions for all of the nodes in the graph, conditional on their parent variables. That is, since $X_i \perp\!\!\!\perp \{X_1, ..., X_{i-1}\} \setminus pa_i | pa_i$, $p(x_i|x_1, ..., x_{i-1}) = p(x_i|pa_i)$. More precisely, for a graph with n nodes (denoted as $x_1, ..., x_n$), the joint distribution is given by:

$$p(x_1, ..., x_n) = \Pi_{i=1}^n p(x_i|pa_i), \tag{3.30}$$

where pa_i is the set of parents of node x_i.

Considering the graph shown in Figure 3.2, we can go from this graph to the corresponding representation of the joint distribution written in terms of the product of a set of conditional probability distributions, one for each node in the graph. The joint distributions for Figure 3.2(a)-(c) are therefore $p(x_1, x_2, x_3) = p(x_1|x_2)p(x_2)p(x_3|x_2)$, $p(x_1, x_2, x_3) = p(x_1)p(x_2|x_1, x_3)p(x_3)$, and $p(x_1, x_2, x_3) = p(x_1)p(x_2|x_1)p(x_3|x_2)$, respectively.

In addition to the graph structure, the conditional probability distribution, $p(X_i|Pa_i)$, needs to be given or learned. When the random variable is discrete, this distribution can be represented using conditional probability table (CPT), which lists the probability of each value taken by X_i given all the possible configurations of its parents. For example, assuming X_1, X_2, and X_3 in Figure 3.2(b) are all binary random variables, a possible conditional probability table for $p(X_2|X_1,X_3)$ could be:

	$X_1 = 0; X_3 = 0$	$X_1 = 0; X_3 = 1$	$X_1 = 1; X_3 = 0$	$X_1 = 1; X_3 = 1$
$X_2 = 0$	0.1	0.4	0.8	0.3
$X_2 = 1$	0.9	0.6	0.2	0.7

3.4.1.2 Inference in a Bayesian Network

The classification problem then can be solved by computing the probability of the interested random variable (e.g., lung cancer) taking a certain class label, given the observations of other random variables, $p(x_i|x_1,\ldots,x_{i-1},x_{i+1},\ldots,x_n)$, which can be calculated if the joint probability of all the variables is known:

$$p(x_i|x_1,\ldots,x_{i-1},x_{i+1},\ldots,x_n) = \frac{p(x_1,\ldots,x_n)}{\sum_{x_i'} p(x_1,\ldots,x_i',\ldots,x_n)}. \quad (3.31)$$

Because a BN is a complete model for the variables and their relationships, a complete joint probability distribution (JPD) over all the variables is specified for a model. Given the JPD, we can answer all possible inference queries by summing out (marginalizing) over irrelevant variables. However, the JPD has size $O(2^n)$, where n is the number of nodes, and we have assumed each node can have 2 states. Hence summing over the JPD takes exponential time. The most common *exact inference* method is **Variable Elimination** [8]. The general idea is to perform the summation to eliminate the non-observed non-query variables one by one by distributing the sum over the product. The reader can refer to [8] for more details. Instead of exact inference, a useful *approximate algorithm* called *Belief propagation* [30] is commonly used on general graphs including Bayesian network, which will be introduced in Section 3.4.3.

3.4.1.3 Learning Bayesian Networks

The learning of Bayesian networks is comprised of two components: parameter learning and structure learning. When the graph structure is given, only the parameters such as the probabilities in conditional probability tables need to be learned given the observed data. Typically, MLE (Maximum Likelihood Estimation) or MAP (Maximum a Posterior) estimations will be used to find the best parameters. If the graph structure is also unknown, both the graph structure and the parameters need to be learned from the data.

From the data point of view, it could be either complete or incomplete, depending on whether all the random variables are observable. When hidden random variables are involved, EM algorithms are usually used.

3.4.2 Hidden Markov Models

We now introduce a special case of Bayesian network, the hidden Markov model. In a hidden Markov model, hidden states are linked as a chain and governed by a Markov process; and the observable values are independently generated given the hidden state, which form a sequence. Hidden Markov model is widely used in speech recognition, gesture recognition, and part-of-speech tagging, where the hidden classes are dependent on each other. Therefore, the class labels for an observation are not only dependent on the observation but also dependent on the adjacent states.

In a regular Markov model as Figure 3.3(a), the state x_i is directly visible to the observer, and therefore the state transition probabilities $p(x_i|x_{i-1})$ are the only parameters. Based on the Markov property, the joint distribution for a sequence of n observations under this model is given by

$$p(x_1,\ldots,x_n) = p(x_1)\prod_{i=2}^{n} p(x_i|x_{i-1}). \quad (3.32)$$

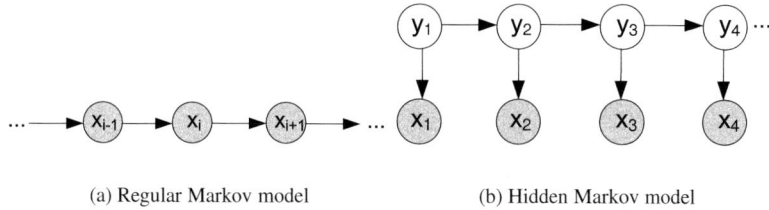

(a) Regular Markov model   (b) Hidden Markov model

FIGURE 3.3: Graphical structures for the regular and hidden Markov model.

Thus, if we use such a model to predict the next observation in a sequence, the distribution of predictions will depend on the value of the immediately preceding observation and will be independent of all earlier observations, conditional on the preceding observation.

A hidden Markov model (HMM) can be considered as the simplest dynamic Bayesian network. In a hidden Markov model, the state y_i is not directly visible, and only the output x_i, dependent on the state, is visible. The hidden state space is discrete, and is assumed to consist of one of N possible values, which is also called latent variable. The observations can be either discrete or continuous, and are typically generated from a categorical distribution or a Gaussian distribution. Generally, an HMM can be considered as a generalization of a *mixture model* where the hidden variables are related through a Markov process rather than independent of each other.

Suppose the latent variables form a first-order Markov chain as shown in Figure 3.3(b). The random variable y_t is the hidden state at time t, and the random variable x_t is the observation at time t. The arrows in the figure denote conditional dependencies. From the diagram, it is clear that y_{t-1} and y_{t+1} are independent given y_t, so that $y_{t+1} \perp\!\!\!\perp y_{t-1}|y_t$. This is the key conditional independence property, which is called the *Markov property*. Similarly, the value of the observed variable x_t only depends on the value of the hidden variable y_t. Then, the joint distribution for this model is given by

$$p(x_1,...,x_n,y_1,...,y_n) = p(y_1) \prod_{t=2}^{n} p(y_t|y_{t-1}) \prod_{t=1}^{n} p(x_t|y_t), \qquad (3.33)$$

where $p(y_t|y_{t-1})$ is the state transition probability, and $p(x_t|y_t)$ is the observation probability.

3.4.2.1 The Inference and Learning Algorithms

Consider the case where we are given a set of possible states $\Omega_Y = \{q_1,...,q_N\}$ and a set of possible observations $\Omega_X = \{o_1,...,o_M\}$. The parameter learning task of HMM is to find the best set of state transition probabilities $A = \{a_{ij}\}$, $a_{ij} = p(y_{t+1} = q_j|y_t = q_i)$ and observation probabilities $B = \{b_i(k)\}$, $b_i(k) = p(x_t = o_k|y_t = q_i)$ as well as the initial state distribution $\Pi = \{\pi_i\}, \pi_i = p(y_0 = q_i)$ for a set of output sequences. Let $\Lambda = \{A,B,\Pi\}$ denote the parameters for a given HMM with fixed Ω_Y and Ω_X. The task is usually to derive the *maximum likelihood estimation* of the parameters of the HMM given the set of output sequences. Usually a local maximum likelihood can be derived efficiently using the *Baum-Welch algorithm* [4], which makes use of *forward-backward algorithm* [33], and is a special case of the generalized EM algorithm [9].

Given the parameters of the model Λ, there are several typical inference problems associated with HMMs, as outlined below. One common task is to compute the probability of a particular output sequence, which requires summation over all possible state sequences: The probability of observing a sequence $X_1^T = o_1,...,o_T$ of length T is given by $P(X_1^T|\Lambda) = \sum_{Y_1^T} P(X_1^T|Y_1^T,\Lambda)P(Y_1^T|\Lambda)$, where the sum runs over all possible hidden-node sequences $Y_1^T = y_1,...,y_T$.

This problem can be handled efficiently using the **forward-backward algorithm**. Before we describe the algorithm, let us define the forward (alpha) values and backward (beta) values as follows: $\alpha_t(i) = P(x_1 = o_1, ..., x_t = o_t, y_t = q_i | \Lambda)$ and $\beta_t(i) = P(x_{t+1} = o_{t+1}, ..., x_T = o_T | y_t = q_i, \Lambda)$. Note the forward values enable us to solve the problem through marginalizing, then we obtain

$$P(X_1^T | \Lambda) = \sum_{i=1}^{N} P(o_1, ..., o_T, y_T = q_i | \Lambda) = \sum_{i=1}^{N} \alpha_T(i).$$

The forward values can be computed efficiently with the principle of *dynamic programming*:

$$\alpha_1(i) = \pi_i b_i(o_1),$$

$$\alpha_{t+1}(j) = \left[\sum_{i=1}^{N} \alpha_t(i) a_{ij}\right] b_j(o_{t+1}).$$

Similarly, the backward values can be computed as

$$\beta_T(i) = 1,$$

$$\beta_t(i) = \sum_{j=1}^{N} a_{ij} b_j(o_{t+1}) \beta_{t+1}(j).$$

The backward values will be used in the Baum-Welch algorithm.

Given the parameters of HMM and a particular sequence of observations, another interesting task is to compute the most likely sequence of states that could have produced the observed sequence. We can find the most likely sequence by evaluating the joint probability of both the state sequence and the observations for each case. For example, in Part-of-speech (POS) tagging [18], we observe a token (word) sequence $X_1^T = o_1, ..., o_T$, and the goal of POS tagging is to find a stochastic optimal tag sequence $Y_1^T = y_1 y_2 ... y_T$ that maximizes $P(Y_1^n, X_1^n)$. In general, finding the most likely explanation for an observation sequence can be solved efficiently using the **Viterbi algorithm** [12] by the recurrence relations:

$$V_1(i) = b_i(o_1)\pi_i,$$
$$V_t(j) = b_i(o_t) \max_i (V_{t-1}(i) a_{ij}).$$

Here $V_t(j)$ is the probability of the most probable state sequence responsible for the first t observations that has q_j as its final state. The Viterbi path can be retrieved by saving back pointers that remember which state $y_t = q_j$ was used in the second equation. Let $Ptr(y_t, q_i)$ be the function that returns the value of y_{t-1} used to compute $V_t(i)$ as follows:

$$y_T = \arg\max_{q_i \in \Omega_Y} V_T(i),$$

$$y_{t-1} = Ptr(y_t, q_i).$$

The complexity of this algorithm is $O(T \times N^2)$, where T is the length of observed sequence and N is the number of possible states.

Now we need a method of adjusting the parameters Λ to maximize the likelihood for a given training set. The **Baum-Welch algorithm** [4] is used to find the unknown parameters of HMMs, which is a particular case of a generalized EM algorithm [9]. We start by choosing arbitrary values for the parameters, then compute the expected frequencies given the model and the observations. The expected frequencies are obtained by weighting the observed transitions by the probabilities specified in the current model. The expected frequencies obtained are then substituted for the old parameters and we iterate until there is no improvement. On each iteration we improve the probability of being observed from the model until some limiting probability is reached. This iterative procedure is guaranteed to converge on a local maximum [34].

3.4.3 Markov Random Fields

Now we turn to another major class of graphical models that are described by undirected graphs and that again specify both a factorization and a set of conditional independence relations. A Markov random field (MRF), also known as an undirected graphical model [14], has a set of nodes, each of which corresponds to a variable or group of variables, as well as a set of links, each of which connects a pair of nodes. The links are undirected, that is, they do not carry arrows.

3.4.3.1 Conditional Independence

Given three sets of nodes, denoted A, B, and C, in an undirected graph G, if A and B are separated in G after removing a set of nodes C from G, then A and B are conditionally independent given the random variables C, denoted as $A \perp\!\!\!\perp B|C$. The conditional independence is determined by simple graph separation. In other words, a variable is conditionally independent of all other variables given its neighbors, denoted as $X_v \perp\!\!\!\perp X_{V \setminus \{v \cup ne(v)\}}|X_{ne(v)}$, where $ne(v)$ is the set of neighbors of v. In general, an MRF is similar to a Bayesian network in its representation of dependencies, and there are some differences. On one hand, an MRF can represent certain dependencies that a Bayesian network cannot (such as cyclic dependencies); on the other hand, MRF cannot represent certain dependencies that a Bayesian network can (such as induced dependencies).

3.4.3.2 Clique Factorization

As the Markov properties of an arbitrary probability distribution can be difficult to establish, a commonly used class of MRFs are those that can be factorized according to the cliques of the graph. A *clique* is defined as a subset of the nodes in a graph such that there exists a link between all pairs of nodes in the subset. In other words, the set of nodes in a clique is fully connected.

We can therefore define the factors in the decomposition of the joint distribution to be functions of the variables in the cliques. Let us denote a clique by C and the set of variables in that clique by x_C. Then the joint distribution is written as a product of *potential functions* $\psi_C(x_C)$ over the maximal cliques of the graph

$$p(x_1, x_2, ..., x_n) = \frac{1}{Z} \Pi_C \psi_C(x_C),$$

where the *partition function* Z is a normalization constant and is given by $Z = \sum_x \Pi_C \psi_C(x_C)$. In contrast to the factors in the joint distribution for a directed graph, the potentials in an undirected graph do not have a specific probabilistic interpretation. Therefore, how to motivate a choice of potential function for a particular application seems to be very important. One popular potential function is defined as $\psi_C(x_C) = \exp(-\varepsilon(x_C))$, where $\varepsilon(x_C) = \ln \psi_C(x_C)$ is an *energy function* [29] derived from statistical physics. The underlying idea is that the probability of a physical state depends inversely on its energy. In the logarithmic representation, we have

$$p(x_1, x_2, ..., x_n) = \frac{1}{Z} \exp\left(-\sum_C \varepsilon(x_C)\right).$$

The joint distribution above is defined as the product of potentials, and so the total energy is obtained by adding the energies of each of the maximal cliques.

A *log-linear model* is a Markov random field with feature functions f_k such that the joint distribution can be written as

$$p(x_1, x_2, ..., x_n) = \frac{1}{Z} \exp\left(\sum_{k=1}^{K} \lambda_k f_k(x_{C_k})\right),$$

where $f_k(x_{C_k})$ is the function of features for the clique C_k, and λ_k is the weight vector of features. The log-linear model provides a much more compact representation for many distributions, especially when variables have large domains such as text.

3.4.3.3   The Inference and Learning Algorithms

In MRF, we may compute the conditional distribution of a set of nodes given values A to another set of nodes B by summing over all possible assignments to $v \notin A$, B, which is called *exact inference*. However, the exact inference is computationally intractable in the general case. Instead, approximation techniques such as MCMC approach [1] and loopy *belief propagation* [5, 30] are often more feasible in practice. In addition, there are some particular subclasses of MRFs that permit efficient Maximum a posterior (MAP) estimation, or more likely, assignment inference, such as associate networks. Here we will briefly describe belief propagation algorithm.

Belief propagation is a message passing algorithm for performing inference on graphical models, including Bayesian networks and MRFs. It calculates the marginal distribution for each unobserved node, conditional on any observed nodes. Generally, belief propagation operates on a factor graph, which is a bipartite graph containing nodes corresponding to variables V and factors U, with edges between variables and the factors in which they appear. Any Bayesian network and MRF can be represented as a factor graph. The algorithm works by passing real valued function called *messages* along the edges between the nodes. Taking pairwise MRF as an example, let $m_{ij}(x_j)$ denote the message from node i to node j, and a high value of $m_{ij}(x_j)$ means that node i "believes" the marginal value $P(x_j)$ to be high. Usually the algorithm first initializes all messages to uniform or random positive values, and then updates message from i to j by considering all messages flowing into i (except for message from j) as follows:

$$m_{ij}(x_j) = \sum_{x_i} f_{ij}(x_i, x_j) \prod_{k \in ne(i) \setminus j} m_{ki}(x_i),$$

where $f_{ij}(x_i, x_j)$ is the potential function of the pairwise clique. After enough iterations, this process is likely to converge to a consensus. Once messages have converged, the marginal probabilities of all the variables can be determined by

$$p(x_i) \propto \prod_{k \in ne(i)} m_{ki}(x_i).$$

The reader can refer to [30] for more details. The main cost is the message update equation, which is $O(N^2)$ for each pair of variables (N is the number of possible states).

Recently, MRF has been widely used in many text mining tasks, such as text categorization [7] and information retrieval [28]. In [28], MRF is used to model the term dependencies using the joint distribution over queries and documents. The model allows for arbitrary text features to be incorporated as evidence. In this model, an MRF is constructed from a graph G, which consists of query nodes q_i and a document node D. The authors explore full independence, sequential dependence, and full dependence variants of the model. Then, a novel approach is developed to train the model that directly maximizes the mean average precision. The results show significant improvements are possible by modeling dependencies, especially on the larger Web collections.

3.4.4   Conditional Random Fields

So far, we have described the Markov network representation as a joint distribution. One notable variant of an MRF is a conditional random field (CRF) [19, 40], in which each variable may also be conditioned upon a set of global observations. More formally, a CRF is an undirected graph whose nodes can be divided into exactly two disjoint sets, the observed variables X and the output variables Y, which can be parameterized as a set of factors in the same way as an ordinary Markov network. The underlying idea is that of defining a conditional probability distribution $p(Y|X)$ over label sequences Y given a particular observation sequence X, rather than a joint distribution over both label and observation sequences $p(Y, X)$. The primary advantage of CRFs over HMMs is their

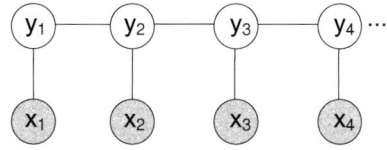

FIGURE 3.4: Graphical structure for the conditional random field model.

conditional nature, resulting in the relaxation of the independence assumptions required by HMMs in order to ensure tractable inference.

Considering a linear-chain CRF with $Y = \{y_1, y_2, ..., y_n\}$ and $X = \{x_1, x_2, ..., x_n\}$ as shown in Figure 3.4, an input sequence of observed variable X represents a sequence of observations and Y represents a sequence of hidden state variables that needs to be inferred given the observations. The y_i's are structured to form a chain, with an edge between each y_i and y_{i+1}. The distribution represented by this network has the form:

$$p(y_1, y_2, ..., y_n | x_1, x_2, ..., x_n) = \frac{1}{Z(X)} \exp\left(\sum_{k=1}^{K} \lambda_k f_k(y_i, y_{i-1}, x_i) \right),$$

where $Z(X) = \sum_{y_i} \exp\left(\sum_{k=1}^{K} \lambda_k f_k(y_i, y_{i-1}, x_i) \right)$.

3.4.4.1 The Learning Algorithms

For general graphs, the problem of exact inference in CRFs is intractable. Basically, the inference problem for a CRF is the same as for an MRF. If the graph is a chain or a tree, as shown in Figure 3.4, message passing algorithms yield exact solutions, which are similar to the forward-backward [4, 33] and Viterbi algorithms [12] for the case of HMMs. If exact inference is not possible, generally the inference problem for a CRF can be derived using approximation techniques such as MCMC [1, 31], loopy *belief propagation* [5, 30], and so on. Similar to HMMs, the parameters are typically learned by maximizing the likelihood of training data. It can be solved using an iterative technique such as iterative scaling [19] and gradient-descent methods [37].

CRF has been applied to a wide variety of problems in natural language processing, including POS tagging [19], shallow parsing [37], and named entity recognition [24], being an alternative to the related HMMs. Based on HMM models, we can determine the sequence of labels by maximizing a joint probability distribution $p(X, \mathcal{Y})$. In contrast, CRMs define a single log-linear distribution, i.e., $p(\mathcal{Y}|X)$, over label sequences given a particular observation sequence. The primary advantage of CRFs over HMMs is their conditional nature, resulting in the relaxation of the independence assumptions required by HMMs in order to ensure tractable inference. As expected, CRFs outperform HMMs on POS tagging and a number of real-word sequence labeling tasks [19, 24].

3.5 Summary

This chapter has introduced the most frequently used probabilistic models for classification, which include generative probabilistic classifiers, such as naive Bayes classifiers and hidden Markov models, and discriminative probabilistic classifiers, such as logistic regression and conditional random fields. Some of the classifiers, like the naive Bayes classifier, may be more appropriate for high-dimensional text data, whereas others such as HMM and CRF are used more commonly for temporal or sequential data. The goal of learning algorithms for these probabilistic models are to find MLE or MAP estimations for parameters in these models. For simple naive Bayes classifier and logistic regression, there are closed form solutions for them. In other cases, iterative algorithms such as the gradient-descent method are good options.

Bibliography

[1] C. Andrieu, N. De Freitas, A. Doucet, and M. Jordan. An introduction to mcmc for machine learning. *Machine Learning*, 50(1):5–43, 2003.

[2] J. Antonakis and O. Dalgas. Predicting elections: Child's play! *Science*, 323(5918):1183–1183, 2009.

[3] S. C. Bagley, H. White, and B. A. Golomb. Logistic regression in the medical literature: Standards for use and reporting, with particular attention to one medical domain. *Journal of Clinical Epidemiology*, 54(10):979–985, 2001.

[4] L. Baum, T. Petrie, G. Soules, and N. Weiss. A maximization technique occurring in the statistical analysis of probabilistic functions of Markov chains. *The Annals of Mathematical Statistics*, 41(1):164–171, 1970.

[5] C. Bishop. *Pattern Recognition and Machine Learning*, volume 4. Springer New York, 2006.

[6] S. F. Chen and J. Goodman. An empirical study of smoothing techniques for language modeling. In *Proceedings of the 34th Annual Meeting on Association for Computational Linguistics*, pages 310–318. Association for Computational Linguistics, 1996.

[7] S. Chhabra, W. Yerazunis, and C. Siefkes. Spam filtering using a Markov random field model with variable weighting schemas. In *Proceedings of the 4th IEEE International Conference on Data Mining, 2004. ICDM'04*, pages 347–350, 2004.

[8] F. Cozman. Generalizing variable elimination in Bayesian networks. In *Workshop on Probabilistic Reasoning in Artificial Intelligence*, pages 27–32. 2000.

[9] A. P. Dempster, N. M. Laird, and D. B. Rubin. Maximum likelihood from incomplete data via the em algorithm. *Journal of the Royal Statistical Society. Series B*, 39(1):1–38, 1977.

[10] P. Domingos and M. Pazzani. On the optimality of the simple Bayesian classifier under zero-one loss. *Machine Learning*, 29(2-3):103–130, 1997.

[11] R. O. Duda and P. E. Hart. *Pattern Classification and Scene Analysis*, volume 3. Wiley New York, 1973.

[12] G. Forney Jr. The viterbi algorithm. *Proceedings of the IEEE*, 61(3):268–278, 1973.

[13] M. I. Jordan. Graphical models. *Statistical Science*, 19(1):140–155, 2004.

[14] R. Kindermann, J. Snell, and American Mathematical Society. *Markov Random Fields and Their Applications*. American Mathematical Society Providence, RI, 1980.

[15] D. Kleinbaum and M. Klein. Maximum likelihood techniques: An overview. *Logistic regression*, pages 103–127, 2010.

[16] R. Kohavi. Scaling up the accuracy of Naive-Bayes classifiers: A decision-tree hybrid. In *KDD*, pages 202–207, 1996.

[17] D. Koller and N. Friedman. *Probabilistic graphical models*. MIT Press, 2009.

[18] J. Kupiec. Robust part-of-speech tagging using a hidden markov model. *Computer Speech & Language*, 6(3):225–242, 1992.

[19] J. D. Lafferty, A. McCallum, and F. C. N. Pereira. Conditional random fields: Probabilistic models for segmenting and labeling sequence data. In *ICML*, pages 282–289, 2001.

[20] P. Langley, W. Iba, and K. Thompson. An analysis of Bayesian classifiers. In *AAAI*, volume 90, pages 223–228, 1992.

[21] D. D. Lewis. Naive (Bayes) at forty: The independence assumption in information retrieval. In *ECML*, pages 4–15, 1998.

[22] J. Liao and K.-V. Chin. Logistic regression for disease classification using microarray data: Model selection in a large p and small n case. *Bioinformatics*, 23(15):1945–1951, 2007.

[23] P. MacCullagh and J. A. Nelder. *Generalized Linear Models*, volume 37. CRC Press, 1989.

[24] A. McCallum and W. Li. Early results for named entity recognition with conditional random fields, feature induction and web-enhanced lexicons. In *Proceedings of the 7th Conference on Natural Language Learning at HLT-NAACL 2003-Volume 4*, pages 188–191, ACL, 2003.

[25] A. McCallum, K. Nigam, et al. A comparison of event models for Naive Bayes text classification. In *AAAI-98 Workshop on Learning for Text Categorization*, volume 752, pages 41–48. 1998.

[26] S. Menard. *Applied Logistic Regression Analysis*, volume 106. Sage, 2002.

[27] V. Metsis, I. Androutsopoulos, and G. Paliouras. Spam filtering with Naive Bayes—Which Naive Bayes? In *Third Conference on Email and Anti-Spam*, pp. 27–28, 2006.

[28] D. Metzler and W. Croft. A Markov random field model for term dependencies. In *Proceedings of the 28th Annual International ACM SIGIR Conference on Research and Development in Information Retrieval*, pages 472–479. ACM, 2005.

[29] T. Minka. Expectation propagation for approximate Bayesian inference. In *Uncertainty in Artificial Intelligence*, volume 17, pages 362–369. 2001.

[30] K. Murphy, Y. Weiss, and M. Jordan. Loopy belief propagation for approximate inference: An empirical study. In *Proceedings of the Fifteenth Conference on Uncertainty in AI*, volume 9, pages 467–475, 1999.

[31] R. M. Neal. Markov chain sampling methods for dirichlet process mixture models. *Journal of Computational and Graphical Statistics*, 9(2):249–265, 2000.

[32] L. Qi and J. Sun. A nonsmooth version of Newton's method. *Mathematical Programming*, 58(1-3):353–367, 1993.

[33] L. Rabiner. A tutorial on hidden Markov models and selected applications in speech recognition. *Proceedings of the IEEE*, 77(2):257–286, 1989.

[34] L. R. Rabiner and B. H. Juang. An introduction to hidden Markov models. *IEEE ASSP Magazine*, 3(1): 4–15, January 1986.

[35] I. Rish. An empirical study of the naive bayes classifier. In *IJCAI 2001 Workshop on Empirical Methods in Artificial Intelligence*, pages 41–46, 2001.

[36] I. Rish, J. Hellerstein, and J. Thathachar. An analysis of data characteristics that affect Naive Bayes performance. In *Proceedings of the Eighteenth Conference on Machine Learning*, 2001.

[37] F. Sha and F. Pereira. Shallow parsing with conditional random fields. In *Proceedings of the 2003 Conference of the North American Chapter of the Association for Computational Linguistics on Human Language Technology*, pages 134–141. 2003.

[38] J. A. Snyman. *Practical Mathematical Optimization: An Introduction to Basic Optimization Theory and Classical and New Gradient-Based Algorithms*, volume 97. Springer, 2005.

[39] J. Stoer, R. Bulirsch, R. Bartels, W. Gautschi, and C. Witzgall. *Introduction to Numerical Analysis*, volume 2. Springer New York, 1993.

[40] C. Sutton and A. McCallum. An introduction to conditional random fields for relational learning. In *Introduction to Statistical Relational Learning*, L. Getoor and B. Taskar(eds.) pages 95–130, MIT Press, 2006.

Chapter 4

Decision Trees: Theory and Algorithms

Victor E. Lee

John Carroll University
University Heights, OH
vlee@jcu.edu

Lin Liu

Kent State University
Kent, OH
lliu@cs.kent.edu

Ruoming Jin

Kent State University
Kent, OH
jin@cs.kent.edu

4.1 Introduction

One of the most intuitive tools for data classification is the decision tree. It hierarchically partitions the input space until it reaches a subspace associated with a class label. Decision trees are appreciated for being easy to interpret and easy to use. They are enthusiastically used in a range of business, scientific, and health care applications [12,15,71] because they provide an intuitive means of solving complex decision-making tasks. For example, in business, decision trees are used for everything from codifying how employees should deal with customer needs to making high-value investments. In medicine, decision trees are used for diagnosing illnesses and making treatment decisions for individuals or for communities.

A decision tree is a rooted, directed tree akin to a flowchart. Each internal node corresponds to a partitioning decision, and each leaf node is mapped to a class label prediction. To classify a data item, we imagine the data item to be traversing the tree, beginning at the root. Each internal node is programmed with a *splitting rule*, which partitions the domain of one (or more) of the data's attributes. Based on the splitting rule, the data item is sent forward to one of the node's children. This testing and forwarding is repeated until the data item reaches a leaf node.

Decision trees are nonparametric in the statistical sense: they are not modeled on a probability distribution for which parameters must be learned. Moreover, decision tree induction is almost always nonparametric in the algorithmic sense: there are no weight parameters which affect the results.

Each directed edge of the tree can be translated to a Boolean expression (e.g., $x_1 > 5$); therefore, a decision tree can easily be converted to a set of *production rules*. Each path from root to leaf generates one rule as follows: form the conjunction (logical AND) of all the decisions from parent to child.

Decision trees can be used with both numerical (ordered) and categorical (unordered) attributes. There are also techniques to deal with missing or uncertain values. Typically, the decision rules are univariate. That is, each partitioning rule considers a single attribute. *Multivariate decision rules* have also been studied [8,9]. They sometimes yield better results, but the added complexity is often not justified. Many decision trees are binary, with each partitioning rule dividing its subspace into two parts. Even binary trees can be used to choose among several class labels. Multiway splits are also common, but if the partitioning is into more than a handful of subdivisions, then both the interpretability and the stability of the tree suffers. *Regression trees* are a generalization of decision trees, where the output is a real value over a continuous range, instead of a categorical value. For the remainder of the chapter, we will assume binary, univariate trees, unless otherwise stated.

Table 4.1 shows a set of training data to answer the classification question, "What sort of contact lenses are suitable for the patient?" This data was derived from a public dataset available from the UCI Machine Learning Repository [3]. In the original data, the age attribute was categorical with three age groups. We have modified it to be a numerical attribute with age in years. The next three attributes are binary-valued. The last attribute is the class label. It is shown with three values (lenses types): {*hard*, *soft*, *no*}. Some decision tree methods support only binary decisions. In this case, we can combine *hard* and *soft* to be simply *yes*.

Next, we show four different decision trees, all induced from the same data. Figure 4.1(a) shows the tree generated by using the Gini index [8] to select split rules when the classifier is targeting all three class values. This tree classifies the training data exactly, with no errors. In the leaf nodes, the number in parentheses indicates how many records from the training dataset were classified into this bin. Some leaf nodes indicate a single data item. In real applications, it may be unwise to permit the tree to branch based on a single training item because we expect the data to have some noise or uncertainty. Figure 4.1(b) is the result of *pruning* the previous tree, in order to achieve a smaller tree while maintaining nearly the same classification accuracy. Some leaf nodes now have a pair of number: (record count, classification errors).

Figure 4.2(a) shows a 2-class classifier (yes, no) and uses the C4.5 algorithm for selecting the splits [66]. A very aggressively pruned tree is shown in Figure 4.2(b). It misclassifies 3 out of 24 training records.

TABLE 4.1: Example: Contact Lens Recommendations

Age	Near-/Far-sightedness	Astigmatic	Tears	Contacts Recommended
13	nearsighted	no	reduced	no
18	nearsighted	no	normal	soft
14	nearsighted	yes	reduced	no
16	nearsighted	yes	normal	hard
11	farsighted	no	reduced	no
18	farsighted	no	normal	soft
8	farsighted	yes	reduced	no
8	farsighted	yes	normal	hard
26	nearsighted	no	reduced	no
35	nearsighted	no	normal	soft
39	nearsighted	yes	reduced	no
23	nearsighted	yes	normal	hard
23	farsighted	no	reduced	no
36	farsighted	no	normal	soft
35	farsighted	yes	reduced	no
32	farsighted	yes	normal	no
55	nearsighted	no	reduced	no
64	nearsighted	no	normal	no
63	nearsighted	yes	reduced	no
51	nearsighted	yes	normal	hard
47	farsighted	no	reduced	no
44	farsighted	no	normal	soft
52	farsighted	yes	reduced	no
46	farsighted	yes	normal	no

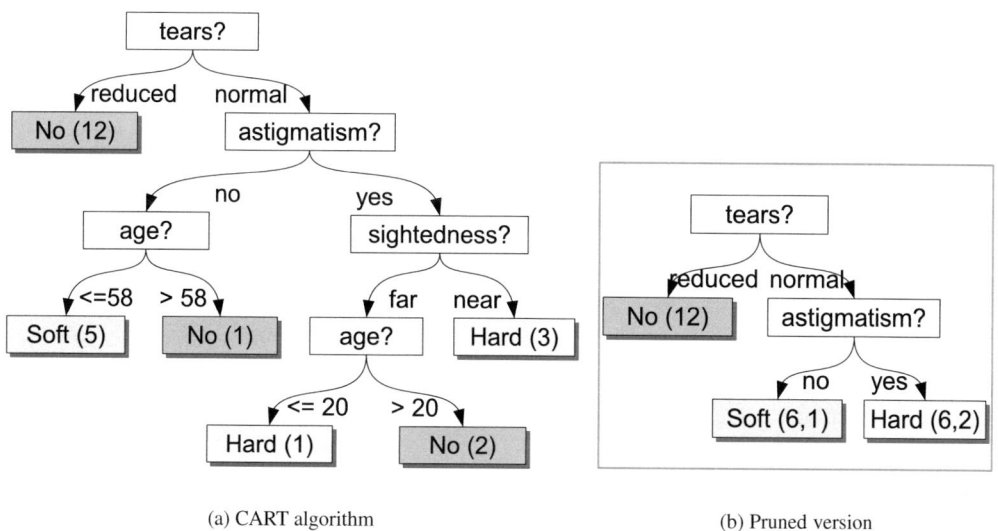

(a) CART algorithm   (b) Pruned version

FIGURE 4.1 (See color insert.): 3-class decision trees for contact lenses recommendation.

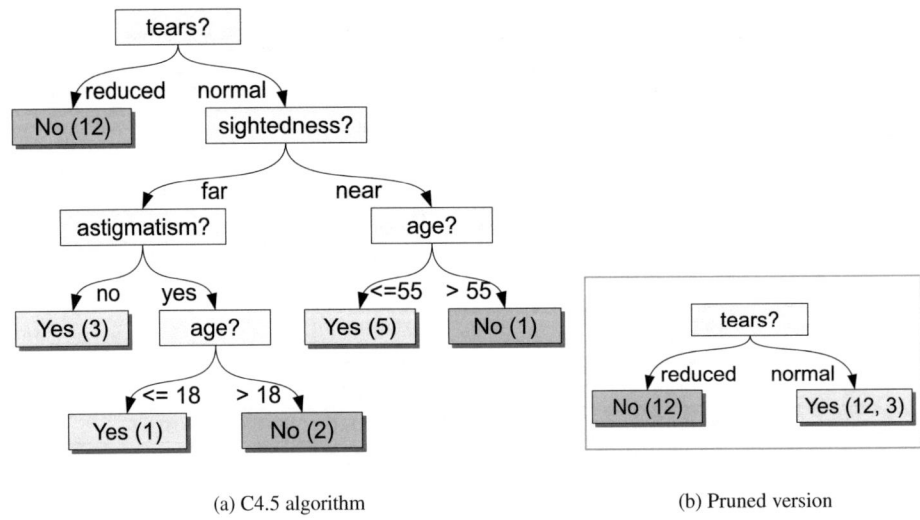

(a) C4.5 algorithm   (b) Pruned version

FIGURE 4.2 (See color insert.): 2-class decision trees for contact lenses recommendation.

Optimality and Complexity

Constructing a decision tree that correctly classifies a consistent[1] data set is not difficult. Our problem, then, is to construct an optimal tree, but what does optimal mean? Ideally, we would like a method with fast tree construction, fast predictions (shallow tree depth), accurate predictions, and robustness with respect to noise, missing values, or concept drift. Should it treat all errors the same, or it is more important to avoid certain types of false positives or false negatives? For example, if this were a medical diagnostic test, it may be better for a screening test to incorrectly predict that a few individuals have an illness than to incorrectly decide that some other individuals are healthy.

Regardless of the chosen measure of goodness, finding a globally optimal result is not feasible for large data sets. The number of possible trees grows exponentially with the number of attributes and with the number of distinct values for each attribute. For example, if one path in the tree from root to leaf tests d attributes, there are $d!$ different ways to order the tests. Hyafil and Rivest [35] proved that constructing a binary decision tree that correctly classifies a set of N data items such that the expected number of steps to classify an item is minimal is *NP*-complete. Even for more esoteric schemes, namely randomized decision trees and quantum decision trees, the complexity is still *NP*-complete [10].

In most real applications, however, we know we cannot make a perfect predictor anyway (due to unknown differences between training data and test data, noise and missing values, and concept drift over time). Instead, we favor a tree that is efficient to construct and/or update and that matches the training set "well enough."

Notation We will use the following notation (further summarized for data and partition in Table 4.2) to describe the data, its attributes, the class labels, and the tree structure. A data item x is a vector of d attribute values with an optional class label y. We denote the set of attributes as $\mathbf{A} = \{A_1, A_2, \ldots, A_d\}$. Thus, we can characterize x as $\{x_1, x_2, \ldots, x_d\}$, where $x_1 \in A_1, x_2 \in A_2, \ldots, x_d \in A_d$. Let $\mathbf{Y} = \{y_1, y_2, \ldots, y_m\}$ be the set of class labels. Each training item x is mapped to a class value y where $y \in \mathbf{Y}$. Together they constitute a data tuple $\langle x, y \rangle$. The complete set of training data is $\mathbf{X}$.

---

[1] A training set is inconsistent if two items have different class values but are identical in all other attribute values.

TABLE 4.2: Summary of Notation for Data and Partitions

Symbol	Definition
$\mathbf{X}$	set of all training data = $\{x_1, \ldots, x_n\}$
$\mathbf{A}$	set of attributes = $\{A_1, \ldots, A_d\}$
$\mathbf{Y}$	domain of class values = $\{y_1, \ldots, y_m\}$
X_i	a subset of X
S	a splitting rule
$\mathbf{X_S}$	a partitioning of $\mathbf{X}$ into $\{X_1, \ldots, X_k\}$

A partitioning rule S subdivides data set $\mathbf{X}$ into a set of subsets collectively known as $\mathbf{X_S}$; that is, $\mathbf{X_S} = \{X_1, X_2, \ldots, X_k\}$ where $\bigcup_i X_i = \mathbf{X}$. A decision tree is a rooted tree in which each set of children of each parent node corresponds to a partitioning ($\mathbf{X_S}$) of the parent's data set, with the full data set associated with the root. The number of items in X_i that belong to class y_j is $|X_{ij}|$. The probability that a randomly selected member of X_i is of class y_j is $p_{ij} = \frac{|X_{ij}|}{|X_i|}$.

The remainder of the chapter is organized as follows. Section 4.2 describes the operation of classical top-down decision tree induction. We break the task down into several subtasks, examining and comparing specific splitting and pruning algorithms that have been proposed. Section 4.3 features case studies of two influential decision tree algorithms, C4.5 [66] and CART [8]. Here we delve into the details of start-to-finish decision tree induction and prediction. In Section 4.4, we describe how data summarization and parallelism can be used to achieve scalability with very large datasets. Then, in Section 4.5, we introduce techniques and algorithms that enable incremental tree induction, especially in the case of streaming data. We conclude with a review of the advantages and disadvantages of decision trees compared to other classification methods.

4.2 Top-Down Decision Tree Induction

The process of learning the structure of a decision tree, to construct the classifier, is called decision tree induction [63]. We first describe the classical approach, top-down selection of partitioning rules on a static dataset. We introduce a high-level generic algorithm to convey the basic idea, and then look deeper at various steps and subfunctions within the algorithm.

The decision tree concept and algorithm can be tracked back to two independent sources: AID (Morgan and Sonquist, 1963) [56] and CLS in *Experiments in Induction* (Hunt, Marin, and Stone, 1966) [34]. The algorithm can be written almost entirely as a single recursive function, as shown in Algorithm 4.1. Given a set of data items, which are each described by their attribute values, the function builds and returns a subtree. The key subfunctions are shown in small capital letters. First, the function checks if it should stop further refinement of this branch of the decision tree (line 3, subfunction STOP). If so, it returns a leaf node, labeled with the class that occurs most frequently in the current data subset X'. Otherwise, it procedes to try all feasible splitting options and selects the best one (line 6, FINDBESTSPLITTINGRULE). A splitting rule partitions the dataset into subsets. What constitutes the "best" rule is perhaps the most distinctive aspect of one tree induction algorithm versus another. The algorithm creates a tree node for the chosen rule (line 8).

If a splitting rule draws all the classification information out of its attribute, then the attribute is exhausted and is ineligible to be used for splitting in any subtree (lines 9–11). For example, if a discrete attribute with k different values is used to create k subsets, then the attribute is "exhausted."

As a final but vital step, for each of the data subsets generated by the splitting rule, we recursively call BUILDSUBTREE (lines 13–15). Each call generates a subtree that is then attached as a child to the principal node. We now have a tree, which is returned as the output of the function.

Algorithm 4.1 Recursive Top-Down Decision Tree Induction

Input: Data set X, Attribute set A

  1: *tree* $\leftarrow$ BUILDSUBTREE$(X, A, 0)$;

  2: **function** BUILDSUBTREE$(X', A', depth)$

  3:     **if** STOP$(X', depth)$ **then**

  4:         return *CreateNode*$(nullRule, majorityClass(X'))$;

  5:     **else**

  6:         *rule* $\leftarrow$ FINDBESTSPLITTINGRULE(X', A');

  7:         *attr* $\leftarrow$ *attributeUsed*$(rule)$;

  8:         *node* $\leftarrow$ *CreateNode*$(rule, nullClass)$;

  9:         **if** *rule* "exhausts" *attr* **then**

10:            remove *attr* from A';

11:         **end if**

12:         *DataSubsets* $\leftarrow$ *ApplyRule*$(X', rule)$;

13:         **for** $X_i \in DataSubsets$ **do**

14:            *child* $\leftarrow$ BUILDSUBTREE$(X_i, A', depth + 1)$;

15:            *node.addChild*$(child)$;

16:         **end for**

17:         return *node*;

18:     **end if**

19: **end function**

4.2.1 Node Splitting

Selecting a splitting rule has two aspects: (1) What possible splits shall be considered, and (2) which of these is the best one? Each attribute contributes a set of candidate splits. This set is determined by the attribute's data type, the actual values present in the training data, and possible restrictions set by the algorithm.

- **Binary attributes**: only one split is possible.

- **Categorical (unordered) attributes**: If an attribute A is unordered, then the domain of A is a mathematical set. Any nonempty proper subset S of A defines a binary split $\{S, A \setminus S\}$. After ruling out redundant and empty partitions, we have $2^{k-1} - 1$ possible binary splits. However, some algorithms make k-way splits instead.

- **Numerical (ordered) attributes**: If there are k different values, then we can make either $(k - 1)$ different binary splits or one single k-way split.

These rules are summarized in Table 4.3. For the data set in Table 4.1, the *age* attribute has 20 distinct values, so there are 19 possible splits. The other three attributes are binary, so they each offer one split. There are a total of 22 ways to split the root node.

We now look at how each splitting rule is evaluated for its goodness. An ideal rule would form subsets that exhibit class purity: each subset would contain members belonging to only one class y. To optimize the decision tree, we seek the splitting rule S which minimizes the impurity function $F(\mathbf{X_S})$. Alternately, we can seek to maximize the amount that the impurity decreases due to the split:

TABLE 4.3: Number of Possible Splits Based on Attribute Type

Attribute Type	Binary Split	Multiway Split
Binary	1	1 (same as binary)
Categorical (unordered)	$\frac{2^k-2}{2} = 2^{k-1} - 1$	one k-way split
Numerical (ordered)	$k - 1$	one k-way split

$\Delta F(S) = F(\mathbf{X}) - F(\mathbf{X_S})$. The authors of CART [8] provide an axiomatic definition of an impurity function F:

Definition 4.1 *An impurity function F for a m-state discrete variable Y is a function defined on the set of all m-tuple discrete probability vectors* $(p_1, p_2, \ldots, p_m)$ *such that*

1. *F is maximum only at* $(\frac{1}{m}, \frac{1}{m}, \ldots, \frac{1}{m})$,

2. *F is minimum only at the "purity points"* $(1, 0, \ldots, 0), (0, 1, \ldots, 0), \ldots, (0, 0, \ldots, 1)$,

3. *F is symmetric with respect to* $p_1, p_2, \ldots, p_m$.

We now consider several basic impurity functions that meet this definition.

1. **Error Rate**

   A simple measure is the percentage of misclassified items. If y_j is the class value that appears most frequently in partition X_i, then the **error rate** for X_i is $\mathcal{E}(X_i) = \frac{|\{y \neq y_j : (x,y) \in X_i\}|}{|X_i|} = 1 - p_{ij}$. The error rate for the entire split X_S is the weighted sum of the error rates for each subset. This equals the total number of misclassifications, normalized by the size of $\mathbf{X}$.

$$\Delta F_{error}(S) = \mathcal{E}(X) - \sum_{i \in S} \frac{|X_i|}{|X|} \mathcal{E}(X_i). \tag{4.1}$$

   This measure does not have good discriminating power. Suppose we have a two-class system, in which y is the majority class not only in X, but in every available partitioning of X. Error rate will consider all such partitions equal in preference.

2. **Entropy and Information Gain (ID3)**

   Quinlan's first decision tree algorithm, ID3 [62], uses information gain, or equivalently, entropy loss, as the goodness function for dataset partitions. In his seminal work on information theory, Shannnon [70] defined information entropy as the degree of uncertainty of a (discrete) random variable Y, formulated as

$$H_X(Y) = -\sum_{y \in Y} p_y \log p_y \tag{4.2}$$

   where p_y is the probability that a random selection would have state y. We add a subscript (e.g., X) when it is necessary to indicate the dataset being measured. Information entropy can be interpreted at the expected amount of information, measured in bits, needed to describe the state of a system. Pure systems require the least information. If all objects in the system are in the same state k, then $p_k = 1$, $\log p_k = 0$, so entropy $H = 0$. There is no randomness in the system; no additional classification information is needed. At the other extreme is maximal uncertainty, when there are an equal number of objects in each of the $|Y|$ states, so $p_y = \frac{1}{|Y|}$, for all y. Then, $H(Y) = -|Y|(\frac{1}{|Y|} \log \frac{1}{|Y|}) = \log |Y|$. To describe the system we have to fully specify all the possible states using $\log |Y|$ bits. If the system is pre-partitioned into subsets according to another variable (or splitting rule) S, then the information entropy of the overall system is the weighted sum of the entropies for each partition, $H_{X_i}(Y)$. This is equivalent to the conditional entropy $H_X(Y|S)$.

$$
\begin{aligned}
\Delta F_{infoGain}(S) &= -\sum_{y \in Y} p_y \log p_y & &+ \sum_{i \in S} \frac{|X_i|}{|X|} \sum_{y \in Y} p_{iy} \log p_{iy} & (4.3)\\
&= H_X(Y) & &- \sum_{i \in S} p_i H_{X_i}(Y)\\
&= H_X(Y) & &- H_X(Y|S).
\end{aligned}
$$

We know $\lim_{p \to 0} p \log p$ goes to 0, so if a particular class value y is not represented in a dataset, then it does not contribute to the system's entropy.

A shortcoming of the information gain criterion is that it is biased towards splits with larger k. Given a candidate split, if subdividing any subset provides additional class differentiation, then the information gain score will always be better. That is, there is no cost to making a split. In practice, making splits into many small subsets increases the sensitivity to individual training data items, leading to overfit. If the split's cardinality k is greater than the number of class values m, then we might be "overclassifying" the data.

For example, suppose we want to decide whether conditions are suitable for holding a seminar, and one of the attributes is day of the week. The "correct" answer is Monday through Friday are candidates for a seminar, while Saturday and Sunday are not. This is naturally a binary split, but ID3 would select a 7-way split.

3. **Gini Criterion (CART)**
   Another influential and popular decision tree program, CART [8], uses the Gini index for a splitting criterion. This can be interpreted as the expected error if each individual item were randomly classified according to the probability distribution of class membership within each subset. Like ID3, the Gini index is biased towards splits with larger k.

$$
Gini(X_i) = \sum_{y \in Y} p_{iy}(1 - p_{iy}) = 1 - \sum_{y \in Y} p_{iy}^2 \qquad (4.4)
$$

$$
\Delta F_{Gini}(S) = Gini(X) - \sum_{i \in S} \frac{|X_i|}{|X|} Gini(X_i). \qquad (4.5)
$$

4. **Normalized Measures — Gain Ratio (C4.5)**
   To remedy ID3's bias towards higher k splits, Quinlan normalized the information gain to create a new measure called **gain ratio** [63]. Gain ratio is featured in Quinlan's well-known decision tree tool, C4.5 [66].

$$
splitInfo(S) = -\sum_{i \in S} \frac{|X_i|}{|X|} \log \frac{|X_i|}{|X|} \qquad\qquad = H(S) \qquad (4.6)
$$

$$
\Delta F_{gainRatio}(S) = \frac{\Delta F_{infoGain}(S)}{splitInfo(S)} \qquad\qquad = \frac{H(Y) - H(Y|S)}{H(S)}. \qquad (4.7)
$$

SplitInfo considers only the number of subdivisions and their relative sizes, not their purity. It is higher when there are more subdivisions and when they are more balanced in size. In fact, *splitInfo* is the entropy of the split where S is the random variable of interest, not Y. Thus, the gain ratio seeks to factor out the information gained from the type of partitioning as opposed to what classes were contained in the partitions.

The gain ratio still has a drawback. A very imbalanced partitioning will yield a low value for $H(S)$ and thus a high value for $\Delta F_{gainRatio}$, even if the information gain is not very good. To overcome this, C4.5 only considers splits whose information gain scores are better than the average value [66].

5. **Normalized Measures — Information Distance**
   López de Mántaras has proposed using normalized information distance as a splitting criterion [51]. The distance between a target attribute Y and a splitting rule S is the sum of the two conditional entropies, which is then normalized by dividing by their joint entropy:

$$d_N(Y,S) = \frac{H(Y|S) + H(S|Y)}{H(Y,S)} \quad (4.8)$$
$$= \frac{\sum_{i \in S} p_i \sum_{y \in Y} p_{iy} \log p_{iy} + \sum_{y \in Y} p_i \sum_{i \in S} p_{yi} \log p_{yi}}{\sum_{i \in S} \sum_{y \in Y} p_{iy} \log p_{iy}}.$$

   This function is a distance metric; that is, it meets the nonnegative, symmetry, and triangle inequality properties, and $d_N(Y,S) = 0$ when $Y = S$. Moreover its range is normalized to $[0,1]$. Due to its construction, it solves both the high-k bias and the imbalanced partition problems of information gain (ID3) and gain ratio (C4.5).

6. **DKM Criterion for Binary Classes**
   In cases where the class attribute is binary, the DKM splitting criterion, named after Dietterich, Kearns, and Mansour [17, 43] offers some advantages. The authors have proven that for a given level of prediction accuracy, the expected size of a DKM-based tree is smaller than for those constructed using C4.5 or Gini.

$$DKM(X_i) = 2\sqrt{p_i(1 - p_i)} \quad (4.9)$$
$$\Delta F_{DKM}(S) = DKM(X) - \sum_{i \in S} \frac{|X_i|}{|X|} DKM(X_i).$$

7. **Binary Splitting Criteria**
   Binary decision trees as often used, for a variety of reasons. Many attributes are naturally binary, binary trees are easy to interpret, and binary architecture avails itself of additional mathematical properties. Below are a few splitting criteria especially for binary splits. The two partitions are designated 0 and 1.

   (a) **Twoing Criterion**
       This criterion is presented in CART [8] as an alternative to the Gini index. Its value coincides with the Gini index value for two-way splits.

$$\Delta F_{twoing}(S) = \frac{p_0 \cdot p_1}{4} \left(\sum_{y \in Y} |p_{y0} - p_{y1}| \right)^2. \quad (4.10)$$

   (b) **Orthogonality Measure**
       Fayyad and Irani measure the cosine angle between the class probability vectors from the two partitions [22]. (Recall the use of probability vectors in Definition 4.1 for an impurity function.) If the split puts all instances of a given class in one partition but

not the other, then the two vectors are orthogonal, and the cosine is 0. The measure is formulated as $(1 - cos\theta)$ so that we seek a maximum value.

$$ORT(S) = 1 - cos(P_0, P_1) = 1 - \frac{\sum\limits_{y \in Y} p_{y0} p_{y1}}{||P_0|| \cdot ||P_1||}. \tag{4.11}$$

(c) **Other Vector Distance Measures**
Once we see that a vector representation and cosine distance can be used, many other distance or divergence criteria come to mind, such as Jensen-Shannon divergence [75], Kolmogorov-Smirmov distance [24], and Hellinger distance [13].

8. **Minimum Description Length**
Quinlan and Rivest [65] use the Minimum Description Length (MDL) approach to simultaneous work towards the best splits and the most compact tree. A description of a solution consists of an encoded classification model (tree) plus discrepancies between the model and the actual data (misclassifications). To find the minimum description, we must find the best balance between a small tree with relatively high errors and a more expansive tree with little or no misclassifications. Improved algorithms are offered in [83] and [82].

9. **Comparing Splitting Criteria**
When a new splitting criterion is introduced, it is compared to others for tree size and accuracy, and sometimes for robustness, training time, or performance with specific types of data. The baselines for comparison are often C4.5(Gain ratio), CART(Gini), and sometimes DKM. For example, Shannon entropy is compared to Rényi and Tsallis entropies in [50]. While one might not be too concerned about tree size, perhaps because one does not have a large number of attributes, a smaller (but equally accurate) tree implies that each decision is accomplishing more classification work. In that sense, a smaller tree is a better tree.

In 1999, Lim and Loh compared 22 decision tree, nine statistical, and two neural network algorithms [49]. A key result was that there was no statistically significant difference in classification accuracy among the top 21 algorithms. However, there were large differences in training time. The most accurate algorithm, POLYCAST, required hours while similarly accurate algorithms took seconds. It is also good to remember that if high quality training data are not available, then algorithms do not have the opportunity to perform at their best. The best strategy may be to pick a fast, robust, and competitively accurate algorithm. In Lim and Loh's tests, C4.5 and an implementation of CART were among the best in terms of balanced speed and accuracy.

Weighted error penalties
All of the above criteria have assumed that all errors have equal significance. However, for many appliations this is not the case. For a binary classifier (class is either *Yes* or *No*), there are four possible prediction results, two correct results and two errors:

1. Actual class $= Yes$; Predicted class $= Yes$: **True Positive**

2. Actual class $= Yes$; Predicted class $= No$: **False Negative**

3. Actual class $= No$; Predicted class $= Yes$: **False Positive**

4. Actual class $= No$; Predicted class $= No$: **True Negative**

If there are k different class values, then there are $k(k-1)$ different types of errors. We can assign each type of error a weight and then modify the split goodness criterion to include these weights. For example, a binary classifier could have w_{FP} and w_{FN} for False Positive and False Negative, respectively. More generally, we can say the weight is $w_{p,t}$, where p is the predicted class and t is the true class. $w_{tt} = 0$ because this represents a correct classification, hence no error.

Let us look at a few examples of weights being incorporated into an impurity function. Let T_i be the class predicted for partition X_i, which would be the most populous class value in X_i.

Weighted Error Rate: Instead of simply counting all the misclassifications, we count how many of each type of classification occurs, multiplied by its weight.

$$F_{error,wt} = \frac{\sum_{i \in S} \sum_{y \in Y} w_{T_i,y} |X_{iy}|}{|X|} = \sum_{i \in S} \sum_{y \in Y} w_{T_i,y} p_{iy}. \tag{4.12}$$

Weighted Entropy: The modified entropy can be incorporated into the information gain, gain ratio, or information distance criterion.

$$H(Y)_{wt} = -\sum_{y \in Y} w_{T_i,y} p_y \log p_y. \tag{4.13}$$

4.2.2 Tree Pruning

Using the splitting rules presented above, the recursive decision tree induction procedure will keep splitting nodes as long as the goodness function indicates some improvement. However, this greedy strategy can lead to *overfitting*, a phenomenon where a more precise model decreases the error rate for the training dataset but increases the error rate for the testing dataset. Additionally, a large tree might offer only a slight accuracy improvement over a smaller tree. Overfit and tree size can be reduced by *pruning*, replacing subtrees with leaf nodes or simpler subtrees that have the same or nearly the same classification accuracy as the unpruned tree. As Breiman et al. have observed, pruning algorithms affect the final tree much more than the splitting rule.

Pruning is basically tree growth in reverse. However, the splitting criteria must be different than what was used during the growth phase; otherwise, the criteria would indicate that no pruning is needed. Mingers (1989) [55] lists five different pruning algorithms. Most of them require using a different data sample for pruning than for the initial tree growth, and most of them introduce a new impurity function. We describe each of them below.

1. **Cost Complexity Pruning, also called Error Complexity Pruning (CART) [8]**
   This multi-step pruning process aims to replace subtrees with a single node. First, we define a cost-benefit ratio for a subtree: the number of misclassification errors it removes divided by the number of leaf nodes it adds. If L_X is the set of leaf node data subsets of X, then

   $$error\_complexity = \frac{\mathcal{E}(X) - \sum_{L_i \in L_X} \mathcal{E}(L_i)}{|L_X| - 1}. \tag{4.14}$$

   Compute *error_complexity* for each internal node, and convert the one with the smallest value (least increase in error per leaf) to a leaf node. Recompute values and repeat pruning until only the root remains, but save each intermediate tree. Now, compute new (estimated) error rates for every pruned tree, using a new test dataset, different than the original training set. Let T_0 be the pruned tree with the lowest error rate. For the final selection, pick the smallest tree T' whose error rate is within one standard error of T_0's error rate, where standard error is defined as $SE = \sqrt{\frac{\mathcal{E}(T_0)(1-\mathcal{E}(T_0))}{N}}$.

   For example, in Figure 4.2(a), the $[age > 55?]$ decision node receives 6 training items: 5 items have $age \leq 55$, and 1 item has $age > 55$. The subtree has no misclassification errors and 2 leaves. If we replace the subtree with a single node, it will have an error rate of $1/6$. Thus, $error\_complexity(age55) = \frac{1/6-0}{2-1} = 0.167$. This is better than the $[age > 18?]$ node, which has an error complexity of 0.333.

2. **Critical Value Pruning [54]**

   This pruning strategy follows the same progressive pruning approach as cost-complexity pruning, but rather than defining a new goodness measure just for pruning, it reuses the same criterion that was used for tree growth. It records the splitting criteria values for each node during the growth phase. After tree growth is finished, it goes back in search of interior nodes whose splitting criteria values fall below some threshold. If all the splitting criteria values in a node's subtree are also below the threshold, then the subtree is replaced with a single leaf node. However, due to the difficulty of selecting the right threshold value, the practical approach is to prune the weakest subtree, remember this pruned tree, prune the next weakest subtree, remember the new tree, and so on, until only the root remains. Finally, among these several pruned trees, the one with the lowest error rate is selected.

3. **Minimum Error Pruning [58]**

   The error rate defined in Equation 4.1 is the actual error rate for the sample (training) data. Assuming that all classes are in fact equally likely, Niblett and Bratko [58] prove that the expected error rate is

$$\mathcal{E}'(X_i) = \frac{|X_i| \cdot \mathcal{E}(X_i) + m - 1}{|X_i| + m}, \tag{4.15}$$

   where m is the number of different classes. Using this as an impurity criterion, this pruning method works just like the tree growth step, except it merges instead of splits. Starting from the parent of a leaf node, it compares its expected error rate with the size-weighted sum of the error rates of its children. If the parent has a lower expected error rate, the subtree is converted to a leaf. The process is repeated for all parents of leaves until the tree has been optimized. Mingers [54] notes a few flaws with this approach: 1) the assumption of equally likely classes is unreasonable, and 2) the number of classes strongly affects the degree of pruning.

   Looking at the $[age > 55?]$ node in Figure 4.2(a) again, the current subtree has a score $\mathcal{E}'(subtree) = (5/6)\frac{5(0)+2-1}{5+2} + (1/6)\frac{1(0)+2-1}{1+2} = (5/6)(1/7) + (1/6)(1/3) = 0.175$. If we change the node to a leaf, we get $\mathcal{E}'(leaf) = \frac{6(1/6)+2-1}{6+2} = 2/8 = 0.250$. The pruned version has a higher expected error, so the subtree is not pruned.

4. **Reduced Error Pruning (ID3) [64]**

   This method uses the same goodness criteria for both tree growth and pruning, but uses different data samples for the two phases. Before the initial tree induction, the training dataset is divided into a growth dataset and a pruning dataset. Initial tree induction is performed using the growth dataset. Then, just as in minimum error pruning, we work bottom-to-top, but this time the pruning dataset is used. We retest each parent of children to see if the split is still advantageous, when a different data sample is used. One weakness of this method is that it requires a larger quantity of training data. Furthermore, using the same criteria for growing and pruning will tend to under-prune.

5. **Pessimistic Error Pruning (ID3, C4.5) [63, 66]**

   This method eliminates the need for a second dataset by estimating the training set bias and compensating for it. A modified error function is created (as in minimum error pruning), which is used for bottom-up retesting of splits. In Quinlan's original version, the adjusted error is estimated to be $\frac{1}{2}$ per leaf in a subtree. Given tree node v, let $T(v)$ be the subtree rooted at v, and $L(v)$ be the leaf nodes under v. Then,

$$\text{not pruned: } \mathcal{E}_{pess}(T(v)) = \sum_{l \in L(v)} \mathcal{E}(l) + \frac{|L(v)|}{2} \tag{4.16}$$

$$\text{if pruned: } \mathcal{E}_{pess}(v) = \mathcal{E}(v) + \frac{1}{2}. \tag{4.17}$$

Because this adjustment alone might still be too optimistic, the actual rule is that a subtree will be pruned if the decrease in error is larger than the Standard Error.

In C4.5, Quinlan modified pessimistic error pruning to be more pessimistic. The new estimated error is the upper bound of the binomial distribution confidence interval, $U_{CF}(\mathcal{E}, |X_i|)$. C4.5 uses 25% confidence by default. Note that the binomial distribution should not be approximated by the normal distribution, because the approximation is not good for small error rates.

For our $[age > 55?]$ example in Figure 4.2(a), C4.5 would assign the subtree an error score of $(5/6)U_{CF}(0,5) + (1/6)U_{CF}(0,1) = (0.833)0.242 + (0.166).750 = 0.327$. If we prune, then the new root has a score of $U_{CF}(1,6) = 0.390$. The original split has a better error score, so we retain the split.

6. **Additional Pruning Methods**
   Mansour [52] has computed a different upper bound for pessimistic error pruning, based on the Chernoff bound. The formula is simpler than Quinlan's but requires setting two parameters. Kearns and Mansour [44] describe an algorithm with good theoretical guarantees for near- optimal tree size. Mehta et al. present an MDL-based method that offers a better combination of tree size and speed than C4.5 or CART on their test data.

Esposito et al. [20] have compared the five earlier pruning techniques. They find that cost-complexity pruning and reduced error pruning tend to overprune, i.e., create smaller but less accurate decision trees. Other methods (error-based pruning, pessimistic error pruning, and minimum error pruning) tend to underprune. However, no method clearly outperforms others on all measures. The wisest strategy for the user seems to be to try several methods, in order to have a choice.

4.3 Case Studies with C4.5 and CART

To illustrate how a complete decision tree classifier works, we look at the two most prominent algorithms: CART and C4.5. Interestingly, both are listed among the top 10 algorithms in all of data mining, as chosen at the 2006 International Conference on Data Mining [85].

CART was developed by Breiman, Friedman, Olshen, and Stone, as a research work. The decision tree induction problem and their solution is described in detail in their 1984 book, *Classification and Regression Trees* [8]. C4.5 gradually evolved out of ID3 during the late 1980s and early 1990s. Both were developed by J. Ross Quinlan. Early on C4.5 was put to industrial use, so there has been a steady stream of enhancements and added features. We focus on the version described in Quinlan's 1993 book, *C4.5: Programs for Machine Learning* [66].

4.3.1 Splitting Criteria

CART: CART was designed to construct binary trees only. It introduced both the Gini and twoing criteria. Numerical attributes are first sorted, and then the $k-1$ midpoints between adjacent numerical values are used as candidate split points. Categorical attributes with large domains must try all $2^{k-1}-1$ possible binary splits. To avoid excessive computation, the authors recommend not having a large number of different categories.

C4.5: In default mode, C4.5 makes binary splits for numerical attributes and k-way splits for categorical attributes. ID3 used the Information Gain criterion. C4.5 normally uses the Gain Ratio, with the caveat that the chosen splitting rule must also have an Information Gain that is stronger than the average Information Gain. Numerical attributes are first sorted. However, instead of selecting the midpoints, C4.5 considers each of the values themselves as the split points. If the sorted values are $(x_1, x_2, \ldots, x_n)$, then the candidate rules are $\{x > x_1, x > x_2, \ldots, x > x_{n-1}\}$. Optionally, instead of splitting categorical attributes into k branches, one branch for each different attribute value, they can be split into b branches, where b is a user-designated number. To implement this, C4.5 first performs the k-way split and then greedily merges the most similar children until there are b children remaining.

4.3.2 Stopping Conditions

In top-down tree induction, each split produces new nodes that recursively become the starting points for new splits. Splitting continues as long as it is possible to continue and to achieve a net improvement, as measured by the particular algorithm. Any algorithm will naturally stop trying to split a node when either the node achieves class purity or the node contain only a single item. The node is then designated a leaf node, and the algorithm chooses a class label for it.

However, stopping only under these absolute conditions tends to form very large trees that are overfit to the training data. Therefore, additional stopping conditions that apply *pre-pruning* may be used. Below are several possible conditions. Each is independent of the other and employs some threshold parameter.

- Data set size reaches a minimum.

- Splitting the data set would make children that are below the minimize size.

- Splitting criteria improvement is too small.

- Tree depth reaches a maximum.

- Number of nodes reaches a maximum.

CART: Earlier, the authors experimented with a minimum improvement rule: $|\Delta F_{Gini}| > \beta$. However, this was abandoned because there was no right value for β. While the immediate benefit of splitting a node may be small, the cumulative benefit from multiple levels of splitting might be substantial. In fact, even if splitting the current node offers only a small reduction of impurity, its chidren could offer a much larger reduction. Consequently, CART's only stopping condition is a minimum node size. Instead, it strives to perform high-quality pruning.

C4.5: In ID3, a Chi-squared test was used as a stopping condition. Seeing that this sometimes caused overpruning, Quinlan removed this stopping condition in C4.5. Like CART, the tree is allowed to grow unfettered, with only one size constraint: any split must have at least two children containing at least n_{min} training items each, where n_{min} defaults to 2.

4.3.3   Pruning Strategy

CART uses cost complexity pruning. This is one of the most theoretically sound pruning methods. To remove data and algorithmic bias from the tree growth phase, it uses a different goodness criterion and a different dataset for pruning. Acknowledging that there is still statistical uncertainty about the computations, a standard error statistic is computed, and the smallest tree that is with the error range is chosen.

C4.5 uses pessimistic error pruning with the binomial confidence interval. Quinlan himself acknowledges that C4.5 may be applying statistical concepts loosely [66]. As a heuristic method, however, it works about as well as any other method. Its major advantage is that it does not require a separate dataset for pruning. Moreover, it allows a subtree to be replaced not only by a single node but also by the most commonly selected child.

4.3.4   Handling Unknown Values: Induction and Prediction

One real-world complication is that some data items have missing attribute values. Using the notation $x = \{x_1, x_2, \ldots, x_d\}$, some of the x_i values may be missing (null). Missing values generate three concerns:

1. Choosing the Best Split: If a candidate splitting criterion uses attribute A_i but some items have no values for A_i, how should we account for this? How do we select the best criteria, when they have different proportions of missing values?

2. Partitioning the Training Set: Once a splitting criteria is selected, to which child node will the imcomplete data items be assigned?

3. Making Predictions: If making class predictions for items with missing attribute values, how will they proceed down the tree?

Recent studies have compared different techniques for handling missing values in decision trees [18, 67]. CART and C4.5 take very different approaches for addressing these concerns.

CART: CART assumes that missing values are sparse. It calculates and compares splitting criteria using only data that contain values for the relevant attributes. However, if the top scoring splitting criteria is on an attribute with some missing values, then CART selects the best *surrogate split* that has no missing attribute values. For any splitting rule S, a surrogate rule generates similar partitioning results, and *the* surrogate S' is the one that is most strongly correlated. For each actual rule selected, CART computes and saves a small ordered list of top surrogate rules. Recall that CART performs binary splits. For dataset X_i, p_{11} is the fraction of items that is classified by both S and S' as state 1; p_{00} is the fraction that is classifed by both as state 0. The probability that a random item is classified the same by both S and S' is $p(S,S') = p_{11}(S,S') + p_{00}(S,S')$. This measure is further refined in light of the discriminating power of S. The final predictive measure of association between S and S' is

$$\lambda(S'|S) = \frac{min(p_0(S), p_1(S)) - (1 - p(S,S'))}{min(p_0(S), p_1(S))}. \tag{4.18}$$

The scaling factor $min(p_0(S), p_1(S))$ estimates the probability that S correctly classifies an item. Due to the use of surrogates, we need not worry about how to partition items with missing attribute values.

When trying to predict the class of a new item, if a missing attribute is encountered, CART looks for the best surrogate rule for which the data item does have an attribute value. This rule is

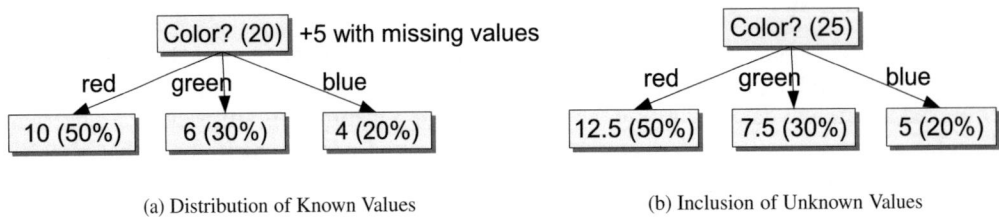

(a) Distribution of Known Values   (b) Inclusion of Unknown Values

FIGURE 4.3: C4.5 distributive classification of items with unknown values.

used instead. So, underneath the primary splitting rules in a CART tree are a set of backup rules. This method seems to depend much on there being highly correlated attributes. In practice, decision trees can have some robustness; even if an item is misdirected at one level, there is some probability that it will be correctly classified in a later level.

C4.5: To compute the splitting criteria, C4.5 computes the information gain using only the items with known attribute values, then weights this result by the fraction of total items that have known values for A. Let X_A be the data subset of X that has known values for attribute A.

$$\Delta F_{infoGain}(S) = \frac{|X_A|}{|X|}(H_{X_A}(Y) - H_{X_A}(Y|S)). \qquad (4.19)$$

Additionally, $splitInfo(S)$, the denominator in C4.5's Gain Ratio, is adjusted so that the set of items with unknown values is considered a separate partition. If S previously made a k-way split, $splitInfo(S)$ is computed as though it were a $(k+1)$-way split.

  To partitioning the training set, C4.5 spreads the items with unknown values according to the same distribution ratios as the items with known attribute values. In the example in Figure 4.3, we have 25 items. Twenty of them have known colors and are partitioned as in Figure 4.3(a). The 5 remaining items are distributed in the same proportions as shown in Figure 4.3(b). This generates *fractional* training items. In subsequent tree levels, we may make fractions of fractions. We now have a probabilistic tree.

  If such a node is encountered while classifying unlabeled items, then all children are selected, not just one, and the probabilities are noted. The prediction process will end at several leaf nodes, which collectively describe a probability distribution. The class with the highest probability can be used for the prediction.

4.3.5 Other Issues: Windowing and Multivariate Criteria

  To conclude our case studies, we take a look at a few notable features that distinguish C4.5 and CART.

Windowing in C4.5: Windowing is the name that Quinlan uses for a sampling technique that was originally intended to speed up C4.5's tree induction process. In short, a small sample, the *window*, of the training set is used to construct an initial decision tree. The initial tree is tested using the remaining training data. A portion of the misclassified items are added to the window, a new tree is inducted, and the non-window training data are again used for testing. This process is repeated until the decision tree's error rate falls below a target threshold or the error rate converges to a constant level.

  In early versions, the initial window was selected uniformly randomly. By the time of this 1993 book, Quinlan had discovered that selecting the window so that the different class values were represented about equally yielded better results. Also by that time, computer memory size and processor speeds had improved enough so that the multiple rounds with windowed data were not always faster than a single round with all the data. However, it was discovered that the multiple rounds improve classification accuracy. This is logical, since the windowing algorithm is a form of boosting.

Multivariate Rules in CART: Breiman et al. investigated the use of multivariate splitting criteria, decision rules that are a function of more than one variable. They considered three different forms: linear combinations, Boolean combinations, and ad hoc combinations. CART considers combining only numerical attributes. For this discussion, assume $A = (A_1, \ldots, A_d)$ are all numerical. In the univariable case, for A_i, we search the $|A_i| - 1$ possible split points for the one that yields the maximal value of C. Using a geometric analogy, if $d = 3$, we have a 3-dimensional data space. A univariable rule, such as $x_i < C$, defines a half-space that is orthogonal to one of the axes. However, if we lift the restriction that the plane is orthogonal to an axis, then we have the more general half-space $\sum_i c_i x_i < C$. Note that a coefficient c_i can be positive or negative. Thus, to find the best multivariable split, we want to find the values of C and $c = (c_1, \ldots, c_d)$, normalized to $\sum_i c_i^2 = 1$, such that ΔF is optimized. This is clearly an expensive search. There are many search heuristics that could accelerate the search, but they cannot guarantee to find the globally best rule. If a rule using all d different attributes is found, it is likely that some of the attributes will not contribute much. The weakest coefficients can be pruned out.

CART also offers to search for Boolean combinations of rules. It is limited to rules containing only conjunction or disjuction. If S_i is a rule on attribute A_i, then candidate rules have the form $S = S_1 \wedge S_2 \wedge \cdots \wedge S_d$ or $S = S_1 \vee S_2 \vee \cdots \vee S_d$. A series of conjunctions is equivalent to walking down a branch of the tree. A series of disjunctions is equivalent to merging children. Unlike linear combinations of rules that offer possible splits that are unavailable with univariate splits, Boolean combinations do not offer a new capability. They simply compress what would otherwise be a large, bushy decision tree.

The ad hoc combination is a manual pre-processing to generate new attributes. Rather than a specific computational technique, this is an acknowledgment that the given attributes might not have good linear correlation with the class variable, but that humans sometimes can study a smal dataset and have helpful intuitions. We might see that a new intermediate function, say the log or square of any existing parameter, might fit better.

None of these features have been aggressively adapted in modern decision trees. In the end, a standard univariate decision tree induction algorithm can always create a tree to classify a training set. The tree might not be as compact or as accurate on new data as we would like, but more often than not, the results are competitive with those of other classification techniques.

4.4 Scalable Decision Tree Construction

The basic decision tree algorithms, such as C4.5 and CART, work well when the dataset can fit into main memory. However, as datasets tend to grow faster than computer memory, decision trees often need to be constructed over disk-resident datasets. There are two major challenges in scaling a decision tree construction algorithm to support large datasets. The first is the very large number of candidate splitting conditions associated with numerical attributes. The second is the recursive nature of the algorithm. For the computer hardware to work efficiently, the data for each node should be stored in sequential blocks. However, each node split conceptually regroups the data. To maintain disk and memory efficiency, we must either relocate the data after every split or maintain special data structures.

One of the first decision tree construction methods for disk-resident datasets was SLIQ [53]. To find splitting points for a numerical attribute, SLIQ requires separation of the input dataset into attribute lists and sorting of attribute lists associated with a numerical attribute. An attribute list in SLIQ has a record-id and attribute value for each training record. To be able to determine the records associated with a non-root node, a data-structure called a *class list* is also maintained. For each training record, the class list stores the class label and a pointer to the current node in the tree. The need for maintaining the class list limits the scalability of this algorithm. Because the class list is accessed randomly and frequently, it must be maintained in main memory. Moreover, in parallelizing the algorithm, it needs to be either replicated, or a high communication overhead is incurred.

A somewhat related approach is SPRINT [69]. SPRINT also requires separation of the dataset into class labels and sorting of attribute lists associated with numerical attributes. The attribute lists in SPRINT store the class label for the record, as well as the record-id and attribute value. SPRINT does not require a class list data structure. However, the attribute lists must be partitioned and written back when a node is partitioned. Thus, there may be a significant overhead for rewriting a disk-resident data set. Efforts have been made to reduce the memory and I/O requirements of SPRINT [41, 72]. However, they do not guarantee the same precision from the resulting decision tree, and do not eliminate the need for writing-back the datasets.

In 1998, Gehrke proposed RainForest [31], a general framework for scaling decision tree construction. It can be used with any splitting criteria. We provide a brief overview below.

4.4.1   RainForest-Based Approach

RainForest scales decision tree construction to larger datasets, while also effectively exploiting the available main memory. This is done by isolating an AVC (Attribute-Value, Classlabel) set for a given attribute and node being processed. An AVC set for an attribute simply records the number of occurrences of each class label for each distinct value the attribute can take. The size of the AVC set for a given node and attribute is proportional to the product of the number of distinct values of the attribute and the number of distinct class labels. The AVC set can be constructed by taking one pass through the training records associated with the node.

Each node has an AVC group, which is the collection of AVC sets for all attributes. The key observation is that though an AVC group does not contain sufficient information to reconstruct the training dataset, it contains all the necessary information for selecting the node's splitting criterion. One can expect the AVC group for a node to easily fit in main memory, though the RainForest framework includes algorithms that do not require this. The algorithm initiates by reading the training dataset once and constructing the AVC group of the root node. Then, the criteria for splitting the root node is selected.

The original RainForest proposal includes a number of algorithms within the RainForest framework to split decision tree nodes at lower levels. In the RF-read algorithm, the dataset is never partitioned. The algorithm progresses level by level. In the first step, the AVC group for the root node is built and a splitting criteria is selected. At any of the lower levels, all nodes at that level are processed in a single pass if the AVC group for all the nodes fit in main memory. If not, multiple passes over the input dataset are made to split nodes at the same level of the tree. Because the training dataset is not partitioned, this can mean reading each record multiple times for one level of the tree.

Another algorithm, RF-write, partitions and rewrites the dataset after each pass. The algorithm RF-hybrid combines the previous two algorithms. Overall, RF-read and RF-hybrid algorithms are able to exploit the available main memory to speed up computations, but without requiring the dataset to be main memory resident.

Figure 4.4(a) and 4.4(b) show the AVC tables for our Contact Lens dataset from Table 4.1. The Age table is largest because it is a numeric attribute with several values. The other three tables are small because their attributes have only two possible values.

Age	Hard	No	Soft
8	1	1	0
11	0	1	0
13	0	1	0
14	0	1	0
16	1	0	0
18	0	0	2
23	1	1	0
26	0	1	0
32	0	1	0
35	0	1	1
36	0	0	1
39	0	1	0
44	0	0	1
46	0	1	0
47	0	1	0
51	1	0	0
52	0	1	0
55	0	1	0
63	0	1	0
64	0	1	0

(a) AVC Table for Age Attribute (RainForest)

Near/Far	Hard	No	Soft
Far	1	8	3
Near	3	7	2

Tears	Hard	No	Soft
Normal	4	3	5
Reduced	0	12	0

Astigmatis	Hard	No	Soft
No	0	7	5
Yes	4	8	0

(b) Other Three AVC Tables (RainForest and SPIES)

Age	Hard	No	Soft
1-10	1	1	0
11-19	1	3	2
20-29	1	2	0
30-39	0	3	2
40-49	0	2	1
50-59	1	2	0
60-69	0	2	0

(c) Concise AVC Table for Age Attribute (SPIES)

FIGURE 4.4: AVC tables for RainForest and SPIES.

4.4.2 SPIES Approach

In [39], a new approach, referred to as SPIES (Statistical Pruning of Intervals for Enhanced Scalability), is developed to make decision tree construction more memory and communication efficient. The algorithm is presented in the procedure *SPIES-Classifier* (Figure 4.2). The SPIES method is based on AVC groups, like the RainForest approach. The key difference is in how the numerical attributes are handled. In SPIES, the AVC group for a node is comprised of three subgroups:

Small AVC group: This is primarily comprised of AVC sets for all categorical attributes. Since the number of distinct elements for a categorical attribute is usually not very large, the size of these AVC sets is small. In addition, SPIES also adds the AVC sets for numerical attributes that only have a small number of distinct elements. These are built and treated in the same fashion as in the RainForest approach.

Concise AVC group: The range of numerical attributes that have a large number of distinct elements in the dataset is divided into *intervals*. The number of intervals and how the intervals are constructed are important parameters to the algorithm. The original SPIES implementation uses equal-width intervals. The concise AVC group records the class histogram (i.e., the frequency of occurrence of each class) for each interval.

Partial AVC group: Based upon the concise AVC group, the algorithm computes a subset of the values in the range of the numerical attributes that are likely to contain the split point. The partial AVC group stores the class histogram for the points in the range of a numerical attribute that has been determined to be a candidate for being the split condition.

SPIES uses two passes to efficiently construct the above AVC groups. The first pass is a quick *Sampling Step*. Here, a sample from the dataset is used to estimate small AVC groups and concise

---

Algorithm 4.2 SPIES-Classifier

1: **function** SPIES-CLASSIFIER(Level **L**, Dataset **X**)
2:     **for** Node $v \in$ Level **L do**
    { *Sampling Step* }
3:         Sample **S** ← Sampling(**X**);
4:         Build_Small_AVCGroup(**S**);
5:         Build_Concise_AVCGroup(**S**);
6:         g' ← Find_Best_Gain(AVCGroup);
7:         Partial_AVCGroup ← Pruning(g', Concise_AVCGroup);

    { *Completion Step* }
8:         Build_Small_AVCGroup(**X**);
9:         Build_Concise_AVCGroup(**X**);
10:         Build_Partial_AVCGroup(**X**);
11:         g ← Find_Best_Gain(AVCGroup);

12:         **if** False_Pruning(g, Concise_AVCGroup)
    {*Additional Step*}
13:             Partial_AVCGroup ← Pruning(g, Concise_AVCGroup);
14:             Build_Partial_AVCGroup(**X**);
15:             g ← Find_Best_Gain(AVCGroup);
16:         **if not** satisfy_stop_condition(v)
17:             Split_Node(v);
18:     **end for**
19: **end function**

---

numerical attributes. Based on these, it obtains an estimate of the best (highest) gain, denoted as g'. Then, using g', the intervals that do not appear likely to include the split point will be pruned. The second pass is the *Completion Step*. Here, the entire dataset is used to construct complete versions of the three AVC subgroups. The partial AVC groups will record the class histogram for all of the points in the unpruned intervals.

After that, the best gain g from these AVC groups can be obtained. Because the pruning is based upon only an estimate of small and concise AVC groups, *false pruning* may occur. However, false pruning can be detected using the updated values of small and concise AVC groups during the completion step. If false pruning has occurred, SPIES can make another pass on the data to construct partial AVC groups for points in falsely pruned intervals. The experimental evaluation shows SPIES significantly reduces the memory requirements, typically by 85% to 95%, and that false pruning rarely happens.

In Figure 4.4(c), we show the concise AVC set for the Age attribute, assuming 10-year ranges. The table size depends on the selected range size. Compare its size to the RainForest AVC in Figure 4.4(a). For discrete attributes and numerical attributes with small distinct values, RainForest and SPIES generate the same small AVC tables, as in Figure 4.4(b).

Other scalable decision tree construction algorithms have been developed over the years; the representatives include BOAT [30] and CLOUDS [2]. BOAT uses a statistical technique called bootstrapping to reduce decision tree construction to as few as two passes over the entire dataset. In addition, BOAT can handle insertions and deletions of the data. CLOUDS is another algorithm that uses intervals to speed up processing of numerical attributes [2]. However, CLOUDS' method does not guarantee the same level of accuracy as one would achieve by considering all possible numerical splitting points (though in their experiments, the difference is usually small). Further, CLOUDS always requires two scans over the dataset for partitioning the nodes at one level of the tree. More recently, SURPASS [47] makes use of linear discriminants during the recursive partitioning process. The summary statistics (like AVC tables) are obtained incrementally. Rather than using summary statistics, [74] samples the training data, with confidence levels determined by PAC learning theory.

4.4.3 Parallel Decision Tree Construction

Several studies have sought to further speed up the decision tree construction using parallel machines. One of the first such studies is by Zaki et al. [88], who develop a shared memory parallelization of the SPRINT algorithm on disk-resident datasets. In parallelizing SPRINT, each attribute list is assigned to a separate processor. Also, Narlikar has used a fine-grained threaded library for parallelizing a decision tree algorithm [57], but the work is limited to memory-resident datasets. A shared memory parallelization has been proposed for the RF-read algorithm [38].

The SPIES approach has been parallelized [39] using a middleware system, FREERIDE (Framework for Rapid Implementation of Datamining Engines) [36, 37], which supports both distributed and shared memory parallelization on disk-resident datasets. FREERIDE was developed in early 2000 and can be considered an early prototype of the popular MapReduce/Hadoop system [16]. It is based on the observation that a number of popular data mining algorithms share a relatively similar structure. Their common processing structure is essentially that of *generalized reductions*. During each phase of the algorithm, the computation involves reading the data instances in an arbitrary order, processing each data instance (similar to *Map* in MapReduce), and updating elements of a *Reduction object* using associative and commutative operators (similar to *Reduce* in MapReduce).

In a distributed memory setting, such algorithms can be parallelized by dividing the data items among the processors and replicating the reduction object. Each node can process the data items it owns to perform a local reduction. After local reduction on all processors, a global reduction is performed. In a shared memory setting, parallelization can be done by assigning different data items to different threads. The main challenge in maintaining the correctness is avoiding race conditions when different threads may be trying to update the same element of the reduction object. FREERIDE has provided a number of techniques for avoiding such race conditions, particularly focusing on the memory hierarchy impact of the use of locking. However, if the size of the reduction object is relatively small, race conditions can be avoided by simply replicating the reduction object.

The key observation in parallelizing the SPIES-based algorithm is that construction of each type of AVC group, i.e., small, concise, and partial, essentially involves a reduction operation. Each data item is read, and the class histograms for appropriate AVC sets are updated. The order in which the data items are read and processed does not impact the final value of AVC groups. Moreover, if separate copies of the AVC groups are initialized and updated by processing different portions of the data set, a final copy can be created by simply adding the corresponding values from the class histograms. Therefore, this algorithm can be easily parallelized using the FREERIDE middleware system.

More recently, a general strategy was proposed in [11] to transform centralized algorithms into algorithms for learning from distributed data. Decision tree induction is demonstrated as an example, and the resulting decision tree learned from distributed data sets is identical to that obtained in the centralized setting. In [4] a distributed hierarchical decision tree algorithm is proposed for a group of computers, each having its own local data set. Similarly, this distributed algorithm induces the same decision tree that would come from a sequential algorithm with full data on each computer. Two univariate decision tree algorithms, C4.5 and univariate linear discriminant tree, are parallelized in [87] in three ways: feature-based, node-based, and data-based. Fisher's linear discriminant function is the basis for a method to generate a multivariate decision tree from distributed data [59]. In [61] MapReduce is employed for massively parallel learning of tree ensembles. Ye et al. [86] take

FIGURE 4.5: Illustration of streaming data.

on the challenging task of combining bootstrapping, which implies sequential improvement, with distributed processing.

4.5 Incremental Decision Tree Induction

Non-incremental decision tree learning methods assume that the training items can be accommodated simultaneously in the main memory or disk. This assumption grieviously limits the power of decision trees when dealing with the following situations: 1) training data sets are too large to fit into the main memory or disk, 2) the entire training data sets are not available at the time the tree is learned, and 3) the underlying distribution of training data sets is changed. Therefore, incremental decision tree learning methods have received much attention from the very beginning [68]. In this section, we examine the techniques for learning decision tree incrementally, especially in a streaming data setting.

Streaming data, represented by an endless sequence of data items, often arrive at high rates. Unlike traditional data available for batch (or off-line) prcessing, the labeled and unlabeled items are mixed together in the stream as shown in Figure 4.5.

In Figure 4.5, the shaded blocks are labeled records. We can see that labeled items can arrive unexpectedly. Therefore, this situation proposes new requirements for learning algorithms from streaming data, such as iterative, single pass, any-time learning [23].

To learn decision trees from streaming data, there are two main strategies: a greedy approach [14, 68, 78–80] and a statistical approach [19, 33]. In this section, we introduce both approaches, which are illustrated by two famous families of decision trees, respectively: *ID3* and *VFDT*.

4.5.1 ID3 Family

Incremental induction has been discussed almost from the start. Schlimmer [68] considers incremental concept induction in general, and develops an incremental algorithm named ID4 with a modification of Quinlan's top-down ID3 as a case study. The basic idea of ID4 is listed in Algorithm 4.3.

In Algorithm 4.3, A_v stands for all the attributes contained in tree node v, and A_v^* for the attribute with the lowest E-score. Meanwhile count $n_{ijy}(v)$ records the number of records observed by node v having value x_{ij} for attribute A_i and being in class y. In [68], the authors only consider positive and negative classes. That means $|\mathbf{Y}| = 2$. v_r stands for the immediate child of v containing item r.

Here, the E-score is the result of computing Quinlan's expected information function E of an attribute at any node. Specifically, at node v,

- n^p: # positive records;

- n^n: # negative records;

- n_{ij}^p: # positive records with value x_{ij} for attribute A_i;

- n_{ij}^n: # negative records with value x_{ij} for attribute A_i;

Algorithm 4.3 ID4(v, r)

Input: v: current decision tree node;
Input: r: data record $r = \langle x, y \rangle$;

 1: **for** each $A_i \in \mathbf{A}_v$, where $\mathbf{A}_v \subseteq \mathbf{A}$ **do**
 2:      Increment count $n_{ijy}(v)$ of class y for A_i;
 3: **end for**
 4: **if** all items observed by v are from class y **then**
 5:      **return**;
 6: **else**
 7:      **if** v is a leaf node **then**
 8:           Change v to be an internal node;
 9:           Update A_v^* with lowest E-score;
10:      **else if** A_v^* does not have the lowest E-score **then**
11:           Remove all children $Child(v)$ of v;
12:           Update A_v^*;
13:      **end if**
14:      **if** $Child(v) = \emptyset$ **then**
15:           Generate the set $Child(v)$ for all values of attribute A_v^*;
16:      **end if**
17:      **ID4**(v_r, r);
18: **end if**

Then

$$E(A_i) = \sum_{j=1}^{|A_i|} \frac{n_{ij}^p + n_{ij}^n}{n^p + n^n} I(n_{ij}^p, n_{ij}^n), \qquad (4.20)$$

with

$$I(x, y) = \begin{cases} 0 & \text{if } x = 0 \text{ or } y = 0 \\ -\frac{x}{x+y} \log \frac{x}{x+y} - \frac{y}{x+y} \log \frac{y}{x+y} & \text{otherwise.} \end{cases}$$

In Algorithm 4.3, we can see that whenever an erroneous splitting attribute is found at v (Line 10), ID4 simply removes all the subtrees rooted at v's immediate children (Line 11), and computes the correct splitting attribute A_v^* (Line 12).

Clearly ID4 is not efficient because it removes the entire subtree when a new A_v^* is found, and this situation could render certain concepts unlearnable by ID4, which could be induced by ID3. Utgoff introduced two improved algorithms: ID5 [77] and ID5R [78]. In particular, ID5R guarantees it will produce the same decision tree that ID3 would have if presented with the same training items.

In Algorithm 4.4, when a splitting test is needed at node v, an arbitrary attribute $A_o \in \mathbf{A}_v$ is chosen; further, according to counts $n_{ijy}(v)$ the optimal splitting attribute A_v^* is calculated based on E-score. If $A_v^* \neq A_o$, the splitting attribute A_v^* is pulled up from all its subtrees (Line 10) to v, and all its subtrees are recursively updated similarly (Line 11 and 13).

The fundamental difference between ID4 (Algorithm 4.3) and ID5R (Algorithm 4.4) is that when ID5R finds a wrong subtree, it restructures the subtree (Line 11 and 13) instead of discarding it and replacing it with a leaf node for the current splitting attribute. The restructuring process in Algorithm 4.4 is called the *pull-up* procedure. The general pull-up procedure is as follows, and illustrated in Figure 4.6. In Figure 4.6, left branches satisfy the splitting tests, and right ones do not.

1. if attribute A_v^* is already at the root, then stop;

2. otherwise,

Algorithm 4.4 ID5R(v, r)

Input: v: the current decision tree node;

Input: r: the data record $r = \langle x, y \rangle$;

1: If v = null, make v a leaf node; **return**;

2: If v is a leaf, and all items observed by v are from class y; **return**;

3: **if** v is a leaf node **then**

4:     Split v by choosing an arbitrary attribute;

5: **end if**

6: **for** $A_i \in \mathbf{A}_v$, where $\mathbf{A}_v \subseteq \mathbf{A}$ **do**

7:     Increment count $n_{ijy}(v)$ of class y for A_i;

8: **end for**

9: **if** A_v^* does not have the lowest E-score **then**

10:     Update A_v^*, and restructure v;

11:     Recursively reestablish $v_c \in Child(v)$ except the branch v_r for r;

12: **end if**

13: Recursively update subtree v_r along the branches with the value occuring in x;

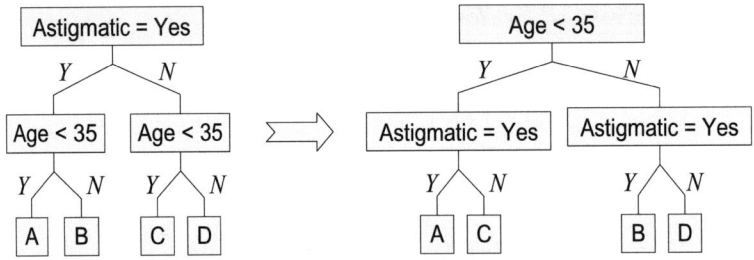

FIGURE 4.6: Subtree restructuring.

(a) Recursively pull the attribute A_v^* to the root of each immediate subtree of v. Convert any leaves to internal nodes as necessary, choosing A_v^* as splitting attribute.

(b) Transpose the subtree rooted at v, resulting in a new subtree with A_v^* at the root, and the old root attribute A_o at the root of each immediate subtree of v.

There are several other works that fall into the ID3 family. A variation for multivariate splits appears in [81], and an improvement of this work appears in [79], which is able to handle numerical attributes. Having achieved an arguably efficient technique for incrementally restructuring a tree, Utgoff applies this technique to develop Direct Metric Tree Induction (DMTI). DMTI leverages fast tree restructuring to fashion an algorithm that can explore more options than traditional greedy top-down induction [80]. Kalles [42] speeds up ID5R by estimating the minimum number of training items for a new attribute to be selected as the splitting attribute.

4.5.2 VFDT Family

In the Big Data era, applications that generate vast streams of data are ever-present. Large retail chain stores like Walmart produce millions of transaction records every day or even every hour, giant telecommunication companies connect millions of calls and text messages in the world, and large banks receive millions of ATM requests throughout the world. These applications need machine learning algorithms that can learn from extremely large (probably infinite) data sets, but spend only a small time with each record. The VFDT learning system was proposed by Domingos [19] to handle this very situation.

VFDT (Very Fast Decision Tree learner) is based on the *Hoeffding tree*, a decision tree learning method. The intuition of the Hoeffding tree is that to find the best splitting attribute it is sufficient to consider only a small portion of the training items available at a node. To acheive this goal, the *Hoeffding bound* is utilized. Basically, given a real-valued random variable r having range R, if we have observed n values for this random variable, and the sample mean is $\bar{r}$, then the Hoeffding bound states that, with probability $1 - \delta$, the true mean of r is at least $\bar{r} - \varepsilon$, where

$$\varepsilon = \sqrt{\frac{R^2 ln(1/\delta)}{2n}}. \tag{4.21}$$

Based on the above analysis, if at one node we find that $\bar{F}(A_i) - \bar{F}(A_j) \geq \varepsilon$, where $\bar{F}$ is the splitting criterion, and A_i and A_j are the two attributes with the best and second best $\bar{F}$ respectively, then A_i is the correct choice with probability $1 - \delta$. Using this novel observation, the Hoeffding tree algorithm is developed (Algorithm 4.5).

Algorithm 4.5 HoeffdingTree(S, $\mathbf{A}$, F, δ)

Input: S: the streaming data;
Input: $\mathbf{A}$: the set of attributes;
Input: F: the split function;
Input: δ: $1 - \delta$ is the probability of choosing the correct attribute to split;
Output: T: decision tree
 1: Let T be a tree with a single leaf v_1;
 2: Let $\mathbf{A}_1 = \mathbf{A}$;
 3: Set count $n_{ijy}(v_1) = 0$ for each $y \in \mathbf{Y}$, each value x_{ij} of each attribute $A_i \in \mathbf{A}$
 4: **for** each training record $\langle x, y \rangle$ in S **do**
 5:     Leaf $v = \text{classify}(\langle x, y \rangle, T)$;
 6:     For each $x_{ij} \in x$: Increment count $n_{ijy}(v)$;
 7:     **if** $n_{ijy}(v)$ does not satisfy any stop conditions **then**
 8:         $A_v^* = F(n_{ijy}(v), \delta)$;
 9:         Replace v by an internal node that splits on A_v^*;
10:         **for** each child v_m of v **do**
11:             Let $\mathbf{A}_m = \mathbf{A}_v - \{A_v^*\}$;
12:             Initialize $n_{ijy}(v_m)$;
13:         **end for**
14:     **end if**
15: **end for**
16: **return** T;

In Algorithm 4.5, the n_{ijy} counts are sufficient to calculate $\bar{F}$. Initially decision tree T only contains a leaf node v_1(Line 1), and v_1 is labeled by predicting the most frequent class. For each item $\langle x, y \rangle$, it is first classified into a leaf node v through T (Line 5). If the items in v are from more than one class, then v is split according to the *Hoeffding* bound (Line 8). The key property of the Hoeffding tree is that under realistic assumptions (see [19] for details), it is possible to guarantee that the generated tree is asymptotically close to the one produced by a batch learner.

When dealing with streaming data, one practical problem that needs considerable attention is concept drift, which does not satisfy the assumption of VFDT: that the sequential data is a random sample drawn from a *stationary* distribution. For example, the behavior of customers of online shopping may change from weekdays to weekends, from season to season. CVFDT [33] has been developed to deal with concept drift.

CVFDT utilizes two strategies: a sliding window W of training items, and alternate subtrees $ALT(v)$ for each internal node v. The decision tree records the statistics for the $|W|$ most recent unique training items. More specifically, instead of learning a new model from scratch when a new training item $\langle x, y \rangle$ comes, CVFDT increments the sufficient statistics n_{ijy} at corresponding nodes for the new item and decrements the counts for the oldest records $\langle x_o, y_o \rangle$ in the window. Periodically, CVFDT reevaluates the classification quality and replaces a subtree with one of the alternate subtrees if needed.

Algorithm 4.6 CVFDT(S, $\mathbf{A}$, F, δ, w, f)

Input: S: the streaming data;
Input: $\mathbf{A}$: the set of attributes;
Input: F: the split function;
Input: δ: $1 - \delta$ is the probability of choosing the correct attribute to split;
Input: w: the size of window;
Input: f: # training records between checks for drift;
Output: T: decision tree;
  1:   let T be a tree with a single leaf v_1;
  2:   let $ALT(v_1)$ be an initially empty set of alternate trees for v_1;
  3:   let $\mathbf{A}_1 = \mathbf{A}$;
  4:   Set count $n_{ijy}(v_1) = 0$ for each $y \in \mathbf{Y}$ and each value x_{ij} for $A_i \in \mathbf{A}$;
  5:   Set record window $W = \emptyset$;
  6:   **for** each training record $\langle x, y \rangle$ in S **do**
  7:      $L = classify(\langle x, y \rangle, T)$, where L contains all nodes that $\langle x, y \rangle$ passes through using T and all trees in ALT;
  8:      $W = W \cup \{\langle x, y \rangle\}$;
  9:      **if** $|W| > w$ **then**
 10:         let $\langle x_o, y_o \rangle$ be the oldest element of W;
 11:         ForgetExample($T, ALT, \langle x_o, y_o \rangle$);
 12:         let $W = W \setminus \{\langle x_o, y_o \rangle\}$;
 13:      **end if**
 14:      CVFDTGrow($T, L, \langle x, y \rangle, \delta$);
 15:      if there have been f examples since the last checking of alternate trees
 16:         CheckSplitValidity(T, δ);
 17: **end for**
 18: **return** T;

An outline of CVFDT is shown in Algorithm 4.6. When a new record $\langle x, y \rangle$ is received, we classify it according to the current tree. We record in a structure L every node in the tree T and in the alternate subtrees ALT that are encountered by $\langle x, y \rangle$ (Line 7). Lines 8 to 14 keep the sliding window up to date. If the tree's number of data items has now exceeded the maximum window size (Line 9), we remove the oldest data item from the statistics (Line 11) and from W (Line 12). *ForgetExample* traverses the decision tree and decrements the corresponding counts n_{ijy} for $\langle x_o, y_o \rangle$ in any node of T or ALT. We then add $\langle x, y \rangle$ to the tree, increasing n_{ijy} statistics according to L (Line 14). Finally, once every f items, we invoke Procedure *CheckSplitValidity*, which scans T and ALT looking for better splitting attributes for each internal node. It revises T and ALT as necessary.

Of course, more recent works can be found following this family. Both VFDT and CVFDT only consider discrete attributes; the VFDTc [26] system extends VFDT in two major directions: 1) VFDTc is equipped with the ability to deal with numerical attributes; and 2) a naïve Bayesian classifier is utilized in each leaf. Jin [40] presents a numerical interval pruning (NIP) approach to efficiently handle numerical attributes, and speeds up the algorithm by reducing the sample size. Further, Bifet [6] proposes a more efficient decision tree learning method than [26] by replacing naïve Bayes with perceptron classifiers, while maintaining competitive accuracy. Hashemi [32] develops a flexible decision tree (FlexDT) based on fuzzy logic to deal with noise and missing values in streaming data. Liang [48] builds a decision tree for uncertain streaming data.

Notice that there are some general works on handling concept drifting for streaming data. Gama [27, 28] detects drifts by tracing the classification errors for the training items based on PAC framework.

4.5.3 Ensemble Method for Streaming Data

An ensemble classifier is a collection of several base classifiers combined together for greater prediction accuracy. There are two well-known ensemble learning approaches: *Bagging* and *Boosting*.

Online bagging and boosting algorithms are introduced in [60]. They rely on the following observation: The probability for each base model to contain each of the original training items k times follows the binomial distribution. As the number of training items goes to infinity, k follows the *Possion(1)* distribution, and this assumption is suitable for streaming data.

In [73], Street et al. propose an ensemble method to read training items sequentially as blocks. When a new training block D comes, a new classifier C_i is learned, and C_i will be evaluated by the next training block. If the ensemble committee E is not full, C_i is inserted into E; otherwise, C_i could replace some member classifier C_j in E if the quality of C_i is better than that of C_j. However, both [73] and [60] fail to explicitly take into consideration the concept drift problem.

Based on Tumer's work [76], Wang et al. [84] prove that ensemble classifier E produces a smaller error than a single classifier $G \in E$, if all the classifiers in E have weights based on their expected classification accuracy on the test data. Accordingly they propose a new ensemble classification method that handles concept drift as follows: When a new chunk D of training items arrives, not only is a new classifier C trained, but also the weights of the previously trained classifiers are recomputed.

They propose the following training method for classifiers on streaming chunks of data, shown in Algorithm 4.7.

Algorithm 4.7 EnsembleTraining(S, K, C)

Input: S: a new chunk of data;
Input: K: the number of classifiers;
Input: C: the set of previously trained classifiers;
 1: Train a new classifier C' based on S;
 2: Compute the weight w' for C';
 3: **for** each $C_i \in C$ **do**
 4:     Recompute the weight w_i for C_i based on S;
 5: **end for**
 6: $C \leftarrow$ top K weighted classifiers from $C \cup \{C'\}$;

In Algorithm 4.7, we can see that when a new chunk S arrives, not only a new classifier C' is trained, but also the weights of the previous trained classifiers are recomputed in this way to handle the concept drifting.

Kolter et al. [45] propose another ensemble classifier to detect concept drift in streaming data. Similar to [84], their method dynamically adjusts the weight of each base classifier according to its accuracy. In contrast, their method has a weight parameter threshold to remove bad classifiers and trains a new classifier for the new item if the existing ensemble classifier fails to identity the correct class.

Fan [21] notices that the previous works did not answer the following questions: When would the old data help detect concept drift and which old data would help? To answer these questions, the author develops a method to sift the old data and proposes a simple cross-validation decision tree ensemble method.

Gama [29] extends the Hoeffding-based Ultra Fast Forest of Trees (UFFT) [25] system to handle concept drifting in streaming data. In a similar vein, Abulsalam [1] extends the random forests ensemble method to run in amortized $O(1)$ time, handles concept drift, and judges whether a sufficient quantity of labeled data has been received to make reasonable predictions. This algorithm also handles multiple class values. Bifet [7] provides a new experimental framework for detecting concept drift and two new variants of bagging methods: ADWIN Bagging and Adaptive-Size Hoeffding Tree (ASHT). In [5], Bifet et al. combine Hoeffding trees using stacking to classify streaming data, in which each Hoeffding tree is built using a subset of item attributes, and ADWIN is utilized both for the perceptron meta-classifier for resetting learning rate and for the ensemble members to detect concept drifting.

4.6 Summary

Compared to other classification methods [46], the following stand out as advantages of decision trees:

- Easy to interpret. A small decision tree can be visualized, used, and understood by a layperson.

- Handling both numerical and categorical attributes. Classification methods that rely on weights or distances (neural networks, k-nearest neighbor, and support vector machines) do not directly handle categorical data.

- Fast. Training time is competitive with other classification methods.

- Tolerant of missing values and irrelevant attributes.

- Can learn incrementally.

The shortcomings tend to be less obvious and require a little more explanation. The following are some weaknesses of decision trees:

- Not well-suited for multivariate partitions. Support vector machines and neural networks are particularly good at making discriminations based on a weighted sum of all the attributes. However, this very feature makes them harder to interpret.

- Not sensitive to relative spacing of numerical values. Earlier, we cited decision trees' ability to work with either categorical or numerical data as an advantage. However, most split criteria do not use the numerical values directly to measure a split's goodness. Instead, they use the values to sort the items, which produces an ordered sequence. The ordering then determines the candidate splits; a set of n ordered items has $n - 1$ splits.

- Greedy approach may focus too strongly on the training data, leading to overfit.

- Sensitivity of induction time to data diversity. To determine the next split, decision tree induction needs to compare every possible split. As the number of different attribute values increases, so does the number of possible splits.

Despite some shortcomings, the decision tree continues to be an attractive choice among classification methods. Improvements continue to be made: more accurate and robust split criteria, ensemble methods for even greater accuracy, incremental methods that handle streaming data and concept drift, and scalability features to handle larger and distributed datasets. A simple concept that began well before the invention of the computer, the decision tree remains a valuable tool in the machine learning toolkit.

Bibliography

[1] Hanady Abdulsalam, David B. Skillicorn, and Patrick Martin. Classification using streaming random forests. *IEEE Transactions on Knowledge and Data Engineering*, 23(1):22–36, 2011.

[2] Khaled Alsabti, Sanjay Ranka, and Vineet Singh. Clouds: A decision tree classifier for large datasets. In *Proceedings of the Fourth International Conference on Knowledge Discovery and Data Mining*, KDD'98, pages 2–8. AAAI, 1998.

[3] K. Bache and M. Lichman. UCI machine learning repository. `http://archive.ics.uci.edu/ml`, 2013.

[4] Amir Bar-Or, Assaf Schuster, Ran Wolff, and Daniel Keren. Decision tree induction in high dimensional, hierarchically distributed databases. In *Proceedings of the Fifth SIAM International Conference on Data Mining*, SDM'05, pages 466–470. SIAM, 2005.

[5] Albert Bifet, Eibe Frank, Geoffrey Holmes, and Bernhard Pfahringer. Accurate ensembles for data streams: Combining restricted hoeffding trees using stacking. *Journal of Machine Learning Research: Workshop and Conference Proceedings*, 13:225–240, 2010.

[6] Albert Bifet, Geoff Holmes, Bernhard Pfahringer, and Eibe Frank. Fast perceptron decision tree learning from evolving data streams. *Advances in Knowledge Discovery and Data Mining*, pages 299–310, 2010.

[7] Albert Bifet, Geoff Holmes, Bernhard Pfahringer, Richard Kirkby, and Ricard Gavaldà. New ensemble methods for evolving data streams. In *Proceedings of the 15th ACM SIGKDD International Conference on Knowledge Discovery and Data Mining*, KDD'09, pages 139–148. ACM, 2009.

[8] Leo Breiman, Jerome Friedman, Charles J. Stone, and Richard A. Olshen. *Classification and Regression Trees*. Chapman & Hall/CRC, 1984.

[9] Carla E. Brodley and Paul E. Utgoff. Multivariate decision trees. *Machine Learning*, 19(1):45–77, 1995.

[10] Harry Buhrman and Ronald De Wolf. Complexity measures and decision tree complexity: A survey. *Theoretical Computer Science*, 288(1):21–43, 2002.

[11] Doina Caragea, Adrian Silvescu, and Vasant Honavar. A framework for learning from distributed data using sufficient statistics and its application to learning decision trees. *International Journal of Hybrid Intelligent Systems*, 1(1):80–89, 2004.

[12] Xiang Chen, Minghui Wang, and Heping Zhang. The use of classification trees for bioinformatics. *Wiley Interdisciplinary Reviews: Data Mining and Knowledge Discovery*, 1(1):55–63, 2011.

[13] David A. Cieslak and Nitesh V. Chawla. Learning decision trees for unbalanced data. In *Proceedings of the 2008 European Conference on Machine Learning and Knowledge Discovery in Databases - Part I*, ECML PKDD'08, pages 241–256. Springer, 2008.

[14] S. L. Crawford. Extensions to the cart algorithm. *International Journal of Man-Machine Studies*, 31(2):197–217, September 1989.

[15] Barry De Ville. *Decision Trees for Business Intelligence and Data Mining: Using SAS Enterprise Miner*. SAS Institute Inc., 2006.

[16] Jeffrey Dean and Sanjay Ghemawat. Mapreduce: Simplified data processing on large clusters. *Communications of the ACM*, 51(1):107–113, 2008. Originally presented at OSDI '04: 6th Symposium on Operating Systems Design and Implementation.

[17] Tom Dietterich, Michael Kearns, and Yishay Mansour. Applying the weak learning framework to understand and improve C4.5. In *Proceedings of the Thirteenth International Conference on Machine Learning*, ICML'96, pages 96–104. Morgan Kaufmann, 1996.

[18] Yufeng Ding and Jeffrey S. Simonoff. An investigation of missing data methods for classification trees applied to binary response data. *The Journal of Machine Learning Research*, 11:131–170, 2010.

[19] Pedro Domingos and Geoff Hulten. Mining high-speed data streams. In *Proceedings of the Sixth ACM SIGKDD International Conference on Knowledge Discovery and Data Mining*, KDD'00, pages 71–80. ACM, 2000.

[20] Floriana Esposito, Donato Malerba, Giovanni Semeraro, and J. Kay. A comparative analysis of methods for pruning decision trees. *IEEE Transactions on Pattern Analysis and Machine Intelligence*, 19(5):476–491, 1997.

[21] Wei Fan. Systematic data selection to mine concept-drifting data streams. In *Proceedings of the Tenth ACM SIGKDD International Conference on Knowledge Discovery and Data Mining*, pages 128–137. ACM, 2004.

[22] Usama M. Fayyad and Keki B. Irani. The attribute selection problem in decision tree generation. In *Proceedings of the Tenth National Conference on Artificial Intelligence*, AAAI'92, pages 104–110. AAAI Press, 1992.

[23] Usama M. Fayyad, Gregory Piatetsky-Shapiro, Padhraic Smyth, and Ramasamy Uthurusamy. *Advances in Knowledge Discovery and Data Mining*. The MIT Press, February 1996.

[24] Jerome H. Friedman. A recursive partitioning decision rule for nonparametric classification. *IEEE Transactions on Computers*, 100(4):404–408, 1977.

[25] João Gama, Pedro Medas, and Ricardo Rocha. Forest trees for on-line data. In *Proceedings of the 2004 ACM Symposium on Applied Computing*, SAC'04, pages 632–636. ACM, 2004.

[26] João Gama, Ricardo Rocha, and Pedro Medas. Accurate decision trees for mining high-speed data streams. In *Proceedings of the Ninth ACM SIGKDD International Conference on Knowledge Discovery and Data Mining*, KDD'03, pages 523–528. ACM, 2003.

[27] João Gama and Gladys Castillo. Learning with local drift detection. In *Advanced Data Mining and Applications*, volume 4093, pages 42–55. Springer-Verlag, 2006.

[28] João Gama, Pedro Medas, Gladys Castillo, and Pedro Rodrigues. Learning with drift detection. *Advances in Artificial Intelligence–SBIA 2004*, pages 66–112, 2004.

[29] João Gama, Pedro Medas, and Pedro Rodrigues. Learning decision trees from dynamic data streams. In *Proceedings of the 2005 ACM Symposium on Applied Computing*, SAC'05, pages 573–577. ACM, 2005.

[30] Johannes Gehrke, Venkatesh Ganti, Raghu Ramakrishnan, and Wei-Yin Loh. Boat– optimistic decision tree construction. *ACM SIGMOD Record*, 28(2):169–180, 1999.

[31] Johannes Gehrke, Raghu Ramakrishnan, and Venkatesh Ganti. Rainforest — A framework for fast decision tree construction of large datasets. In *Proceedings of the International Conference on Very Large Data Bases*, VLDB'98, pages 127–162, 1998.

[32] Sattar Hashemi and Ying Yang. Flexible decision tree for data stream classification in the presence of concept change, noise and missing values. *Data Mining and Knowledge Discovery*, 19(1):95–131, 2009.

[33] Geoff Hulten, Laurie Spencer, and Pedro Domingos. Mining time-changing data streams. In *Proceedings of the Seventh ACM SIGKDD International Conference on Knowledge Discovery and Data Mining*, KDD'01, pages 97–106. ACM, 2001.

[34] Earl Busby Hunt, Janet Marin, and Philip J. Stone. *Experiments in Induction*. Academic Press, New York, London, 1966.

[35] Laurent Hyafil and Ronald L. Rivest. Constructing optimal binary decision trees is np-complete. *Information Processing Letters*, 5(1):15–17, 1976.

[36] Ruoming Jin and Gagan Agrawal. A middleware for developing parallel data mining implementations. In *Proceedings of the 2001 SIAM International Conference on Data Mining*, SDM'01, April 2001.

[37] Ruoming Jin and Gagan Agrawal. Shared memory parallelization of data mining agorithms: Techniques, programming interface, and performance. In *Proceedings of the Second SIAM International Conference on Data Mining*, SDM'02, pages 71–89, April 2002.

[38] Ruoming Jin and Gagan Agrawal. Shared memory parallelization of decision tree construction using a general middleware. In *Proceedings of the 8th International Euro-Par Parallel Processing Conference*, Euro-Par'02, pages 346–354, Aug 2002.

[39] Ruoming Jin and Gagan Agrawal. Communication and memory efficient parallel decision tree construction. In *Proceedings of the Third SIAM International Conference on Data Mining*, SDM'03, pages 119–129, May 2003.

[40] Ruoming Jin and Gagan Agrawal. Efficient decision tree construction on streaming data. In *Proceedings of the Ninth ACM SIGKDD International Conference on Knowledge Discovery and Data Mining*, KDD'03, pages 571–576. ACM, 2003.

[41] Mahesh V. Joshi, George Karypis, and Vipin Kumar. Scalparc: A new scalable and efficient parallel classification algorithm for mining large datasets. In *First Merged Symp. IPPS/SPDP 1998: 12th International Parallel Processing Symposium and 9th Symposium on Parallel and Distributed Processing*, pages 573–579. IEEE, 1998.

[42] Dimitrios Kalles and Tim Morris. Efficient incremental induction of decision trees. *Machine Learning*, 24(3):231–242, 1996.

[43] Michael Kearns and Yishay Mansour. On the boosting ability of top-down decision tree learning algorithms. In *Proceedings of the 28th Annual ACM Symposium on Theory of Computing*, STOC'96, pages 459–468. ACM, 1996.

[44] Michael Kearns and Yishay Mansour. A fast, bottom-up decision tree pruning algorithm with near-optimal generalization. In *Proceedings of the 15th International Conference on Machine Learning*, pages 269–277, 1998.

[45] Jeremy Z. Kolter and Marcus A. Maloof. Dynamic weighted majority: A new ensemble method for tracking concept drift. In *Proceedings of the Third IEEE International Conference on Data Mining, 2003.*, ICDM'03, pages 123–130. IEEE, 2003.

[46] S. B. Kotsiantis. Supervised machine learning: A review of classification techniques. In *Proceedings of the 2007 Conference on Emerging Artifical Intelligence Applications in Computer Engineering*, pages 3–24. IOS Press, 2007.

[47] Xiao-Bai Li. A scalable decision tree system and its application in pattern recognition and intrusion detection. *Decision Support Systems*, 41(1):112–130, 2005.

[48] Chunquan Liang, Yang Zhang, and Qun Song. Decision tree for dynamic and uncertain data streams. In *2nd Asian Conference on Machine Learning*, volume 3, pages 209–224, 2010.

[49] Tjen-Sien Lim, Wei-Yin Loh, and Yu-Shan Shih. A comparison of prediction accuracy, complexity, and training time of thirty-three old and new classification algorithms. *Machine Learning*, 40(3):203–228, 2000.

[50] Christiane Ferreira Lemos Lima, Francisco Marcos de Assis, and Cleonilson Protásio de Souza. Decision tree based on shannon, rényi and tsallis entropies for intrusion tolerant systems. In *Proceedings of the Fifth International Conference on Internet Monitoring and Protection*, ICIMP'10, pages 117–122. IEEE, 2010.

[51] R. López de Mántaras. A distance-based attribute selection measure for decision tree induction. *Machine Learning*, 6(1):81–92, 1991.

[52] Yishay Mansour. Pessimistic decision tree pruning based on tree size. In *Proceedings of the 14th International Conference on Machine Learning*, pages 195–201, 1997.

[53] Manish Mehta, Rakesh Agrawal, and Jorma Rissanen. Sliq: A fast scalable classifier for data mining. In *Proceedings of the Fifth International Conference on Extending Database Technology*, pages 18–32, Avignon, France, 1996.

[54] John Mingers. Expert systems—Rule induction with statistical data. *Journal of the Operational Research Society*, 38(1): 39–47, 1987.

[55] John Mingers. An empirical comparison of pruning methods for decision tree induction. *Machine Learning*, 4(2):227–243, 1989.

[56] James N. Morgan and John A. Sonquist. Problems in the analysis of survey data, and a proposal. *Journal of the American Statistical Association*, 58(302):415–434, 1963.

[57] G. J. Narlikar. A parallel, multithreaded decision tree builder. Technical Report CMU-CS-98-184, School of Computer Science, Carnegie Mellon University, 1998.

[58] Tim Niblett. Constructing decision trees in noisy domains. In I. Bratko and N. Lavrac, editors, *Progress in Machine Learning*. Sigma, 1987.

[59] Jie Ouyang, Nilesh Patel, and Ishwar Sethi. Induction of multiclass multifeature split decision trees from distributed data. *Pattern Recognition*, 42(9):1786–1794, 2009.

[60] Nikunj C. Oza and Stuart Russell. Online bagging and boosting. In *Eighth International Workshop on Artificial Intelligence and Statistics*, pages 105–112. Morgan Kaufmann, 2001.

[61] Biswanath Panda, Joshua S. Herbach, Sugato Basu, and Roberto J. Bayardo. Planet: Massively parallel learning of tree ensembles with mapreduce. *Proceedings of the VLDB Endowment*, 2(2):1426–1437, 2009.

[62] J. Ross Quinlan. Learning efficient classification procedures and their application to chess end-games. In *Machine Learrning: An Artificial Intelligence Approach*. Tioga Publishing Company, 1983.

[63] J. Ross Quinlan. Induction of decision trees. *Machine Learning*, 1(1):81–106, March 1986.

[64] J. Ross Quinlan. Simplifying decision trees. *International Journal of Man-Machine Studies*, 27(3):221–234, 1987.

[65] J. Ross Quinlan and Ronald L. Rivest. Inferring decision trees using the minimum description length principle. *Information and Computing*, 80:227–248, 1989.

[66] John Ross Quinlan. *C4.5: Programs for Machine Learning*. Morgan Kaufmann, 1993.

[67] Maytal Saar-Tsechansky and Foster Provost. Handling missing values when applying classification models. *Journal of Machine Learning Research*, 8:1623–1657, 2007.

[68] Jeffrey C. Schlimmer and Douglas Fisher. A case study of incremental concept induction. In *Proceedings of the Fifth National Conference on Artificial Intelligence*, pages 495–501. Morgan Kaufmann, 1986.

[69] John Shafer, Rakeeh Agrawal, and Manish Mehta. SPRINT: A scalable parallel classifier for data mining. In *Proceedings of the 22nd International Conference on Very Large Databases (VLDB)*, pages 544–555, September 1996.

[70] Claude E. Shannon. A mathematical theory of communication. *Bell System Technical Journal*, 27(3):379–423, July/October 1948.

[71] Harold C. Sox and Michael C. Higgins. *Medical Decision Making*. Royal Society of Medicine, 1988.

[72] Anurag Srivastava, Eui-Hong Han, Vipin Kumar, and Vineet Singh. Parallel formulations of decision-tree classification algorithms. In *Proceedings of the 1998 International Conference on Parallel Processing*, ICPP'98, pages 237–261, 1998.

[73] W. Nick Street and YongSeog Kim. A streaming ensemble algorithm (sea) for large-scale classification. In *Proceedings of the Seventh ACM SIGKDD International Conference on Knowledge Discovery and Data Mining*, KDD'01, pages 377–382. ACM, 2001.

[74] Hyontai Sug. A comprehensively sized decision tree generation method for interactive data mining of very large databases. In *Advanced Data Mining and Applications*, pages 141–148. Springer, 2005.

[75] Umar Syed and Golan Yona. Using a mixture of probabilistic decision trees for direct prediction of protein function. In *Proceedings of the Seventh Annual International Conference on Research in Computational Molecular Biology*, RECOMB'03, pages 289–300. ACM, 2003.

[76] Kagan Tumer and Joydeep Ghosh. Error correlation and error reduction in ensemble classifiers. *Connection Science*, 8(3-4):385–404, 1996.

[77] Paul E. Utgoff. Id5: An incremental id3. In *Proceedings of the Fifth International Conference on Machine Learning*, ICML'88, pages 107–120, 1988.

[78] Paul E. Utgoff. Incremental induction of decision trees. *Machine Learning*, 4(2):161–186, 1989.

[79] Paul E. Utgoff. An improved algorithm for incremental induction of decision trees. In *Proceedings of the Eleventh International Conference on Machine Learning*, ICML'94, pages 318–325, 1994.

[80] Paul E Utgoff, Neil C Berkman, and Jeffery A Clouse. Decision tree induction based on efficient tree restructuring. *Machine Learning*, 29(1):5–44, 1997.

[81] Paul E. Utgoff and Carla E. Brodley. An incremental method for finding multivariate splits for decision trees. In *Proceedings of the Seventh International Conference on Machine Learning*, ICML'90, pages 58–65, 1990.

[82] Paul A. J. Volf and Frans M.J. Willems. Context maximizing: Finding mdl decision trees. In *Symposium on Information Theory in the Benelux*, volume 15, pages 192–200, 1994.

[83] Chris S. Wallace and J. D. Patrick. Coding decision trees. *Machine Learning*, 11(1):7–22, 1993.

[84] Haixun Wang, Wei Fan, Philip S. Yu, and Jiawei Han. Mining concept-drifting data streams using ensemble classifiers. In *Proceedings of the Ninth ACM SIGKDD International Conference on Knowledge Discovery and Data Mining*, pages 226–235. ACM, 2003.

[85] Xindong Wu, Vipin Kumar, J. Ross Quinlan, Joydeep Ghosh, Qiang Yang, Hiroshi Motoda, Geoffrey J. McLachlan, Angus Ng, Bing Liu, Philip S. Yu, Zhi-Hua Zhou, Michael Steinbach, David J. Hand, and Dan Steinberg. Top 10 algorithms in data mining. *Knowledge and Information Systems*, 14(1):1–37, 2008.

[86] Jerry Ye, Jyh-Herng Chow, Jiang Chen, and Zhaohui Zheng. Stochastic gradient boosted distributed decision trees. In *Proceedings of the 18th ACM Conference on Information and Knowledge Management*, CIKM'09, pages 2061–2064. ACM, 2009.

[87] Olcay Taner Yıldız and Onur Dikmen. Parallel univariate decision trees. *Pattern Recognition Letters*, 28(7):825–832, 2007.

[88] M. J. Zaki, Ching-Tien Ho, and Rakesh Agrawal. Parallel classification for data mining on shared-memory multiprocessors. In *Proceedings of the Fifteenth International Conference on Data Engineering*, ICDE'99, pages 198–205, May 1999.

Chapter 5

Rule-Based Classification

Xiao-Li Li

Institute for Infocomm Research
Singapore
xlli@i2r.a-star.edu.sg

Bing Liu

University of Illinois at Chicago (UIC)
Chicago, IL
liub@cs.uic.edu

5.1 Introduction

Classification is an important problem in machine learning and data mining. It has been widely applied in many real-world applications. Traditionally, to build a classifier, a user first needs to collect a set of training examples/instances that are labeled with predefined classes. A classification algorithm is then applied to the training data to build a classifier that is subsequently employed to assign the predefined classes to test instances (for evaluation) or future instances (for application) [1].

In the past three decades, many classification techniques, such as Support Vector Machines (SVM) [2], Neural Network (NN) [3], Rule Learning [9], Naïve Bayesian (NB) [5], K-Nearest Neighbour (KNN) [6], and Decision Tree [4], have been proposed. In this chapter, we focus on rule learning, also called rule-based classification. Rule learning is valuable due to the following advantages.

1. Rules are very natural for knowledge representation, as people can understand and interpret them easily.

2. Classification results are easy to explain. Based on a rule database and input data from the user, we can explain which rule or set of rules is used to infer the class label so that the user is clear about the logic behind the inference.

3. Rule-based classification models can be easily enhanced and complemented by adding new rules from domain experts based on their domain knowledge. This has been successfully implemented in many expert systems.

4. Once rules are learned and stored into a rule database, we can subsequently use them to classify new instances rapidly through building index structures for rules and searching for relevant rules efficiently.

5. Rule based classification systems are competitive with other classification algorithms and in many cases are even better than them.

Now, let us have more detailed discussions about rules. Clearly, rules can represent information or knowledge in a very simple and effective way. They provide a very good data model that human beings can understand very well. Rules are represented in the logic form as **IF-THEN** statements, e.g., a commonly used rule can be expressed as follows:

IF *condition* **THEN** *conclusion*.

where the **IF** part is called the "antecedent" or "condition" and the **THEN** part is called the "consequent" or "conclusion." It basically means that if the *condition* of the rule is satisfied, we can infer or deduct the *conclusion*. As such, we can also write the rule in the following format, namely, *condition* $\rightarrow$ *conclusion*. The *condition* typically consists of one or more feature tests (e.g. $feature_1 > value_2$, $feature_5 = value_3$) connected using logic operators (i.e., "and", "or", "not"). For example, we can have a rule like: If *sex="female" and* ($35 < age < 45$) *and* (*salary="high"* or *creditlimit="high"*), then *potential_customer ="yes"*. In the context of classification, the conclusion can be the class label, e.g., *"yes"* (*potential_customer ="yes"*) or *"no"* (*potential_customer ="no"*). In other words, a rule can be used for classification if its "consequent" will be one of those predefined classes and its "antecedent" or "precondition" contains conditions of various features and their corresponding values.

Many machine learning and data mining techniques have been proposed to automatically learn rules from data. In computer science domain, rule-based systems have been extensively used as an effective way to store knowledge and to do logic inference. Furthermore, based on the given inputs and the rule database, we can manipulate the stored knowledge for interpreting the generated outputs as well as for decision making. Particularly, rules and rule based classification systems have been widely applied in various expert systems, such as fault diagnosis for aerospace and manufacturing, medical diagnosis, highly interactive or conversational Q&A systems, mortgage expert systems, etc.

In this chapter, we will introduce some representative techniques for rule-based classification, which includes two key components, namely 1) rule induction, which learns rules from a given training database/set automatically; and 2) classification, which makes use of the learned rule set for classification. Particularly, we will study two popular rule-based classification approaches: (1) rule induction and (2) classification based on association rule mining.

1. **Rule induction**. Many rule induction/learning algorithms, such as [9], [10], [11], [12], [13], [14], have adopted the sequential covering strategy, whose basic idea is to learn a list of rules from the training data *sequentially*, or one by one. That is, once a new rule has been learned, it will remove the corresponding training examples that it *covers*, i.e., remove those training examples that satisfy the rule antecedent. This learning process, i.e., learn a new rule and remove its covered training data, is repeated until rules can cover the whole training data or no new rule can be learned from the remaining training data.

2. **Classification based on association rule mining**. Association rule mining [16], is perhaps the most important model invented by data mining researchers. Many efficient algorithms have been proposed to detect association rules from large amounts of data. One special type of association rules is called class association rules (CARs). The consequent of a CAR must be a class label, which makes it attractive for classification purposes. We will describe Classification Based on Associations (CBA) — the first system that uses association rules for classification [30], as well as a number of more recent algorithms that perform classification based on mining and applying association rules.

5.2   Rule Induction

The process of learning rules from data directly is called rule induction or rule learning. Most rule induction systems have utilized a learning strategy that is described as *sequential covering*. A rule-based classifier built with this strategy typically consists of a list of discovered rules, which is also called a *decision list* [8]. Note in the *decision list*, the ordering of the rules is very important since it decides which rule should be first used for classification.

The basic idea of sequential covering is to learn a list of rules *sequentially*, one at a time, to cover the whole training data. A rule covering a data instance (either training or test instance), means that the instance satisfies the conditions of the rule. As such, covering the whole training data simply means every training instance in the training data satisfies the conditions of at least one rule in the *decision list* — it is possible that one training/test instance satisfies multiple rules and typically each rule can cover multiple instances. A key learning step using the *sequential covering* strategy is as follows: after each rule is learned, the training examples covered by the rule are removed from the training data and only the remaining training data are used to learn subsequent rules. These key learning steps are performed repeatedly until the remaining training set becomes empty or no new rule can be learned from the training data.

In this section, we study two specific algorithms based on the sequential covering strategy. Both of them are well-known and highly cited. The first algorithm is the CN2 induction algorithm [9] and the second algorithm is based on the ideas from RIPPER algorithm and its variations such as RIPPER [13], FOIL [10], I-REP [11], and REP [12]. Note some ideas are also taken from [14].

5.2.1   Two Algorithms for Rule Induction

We now present these two algorithms, namely CN2 and RIPPER (and its variations), in Section 5.2.1.1 and Section 5.2.1.2, respectively.

5.2.1.1 CN2 Induction Algorithm (Ordered Rules)

CN2 algorithm learns each rule without pre-fixing a class [9]. That is, in each iteration, a rule for any class may be learned. As such, rules for different classes may intermix in the final *decision list RULE_LIST*. As we have mentioned earlier, the ordering of rules is essential for classification, as rules are highly dependent on each other.

CN2 was designed to incorporate ideas from both the AQ algorithm [18] and the ID3 algorithm [4]. Before presenting CN2, we first introduce several basic concepts that were introduced in the AQ algorithm as well as later in CN2 algorithm. Each rule is in the form of "if $< cover >$ then predict $< class >$", where $< cover >$ is a Boolean combination of multiple attribute tests. The basic test on an attribute is called a *selector*, e.g., $< cloudy = yes >$, $< weather = wet \lor stormy >$, and $< Temperature \geq 25 >$. A conjunction of selectors is called a *complex*, e.g., $< cloudy = yes > and < weather = wet \lor stormy >$. A disjunct of multiple complexes is called a *cover*.

The CN2 rule induction algorithm, which is based on ordered rules, is given below, which uses sequential covering.

INPUT

    Let E be a set of classified (training) examples

    Let *SELECTORS* be the set of all possible selectors

CN2 Induction Algorithm CN2(E)

1. Let *RULE_LIST* be the empty list;   // initialize an empty rule set in the beginning
2. Repeat until *Best_CPX* is nil or E is empty;
3.   Let *Best_CPX* be *Find_Best_Complex(E)*
4.   If *Best_CPX* is not nil
5.   Then let E' be the examples covered by *Best_CPX*
6.   Remove from E the examples E' covered by *Best_CPX*
7.   Let C be the most common class of examples in E'
8.   Add the rule "If *Best_CPX* then the class is C" to the end of *RULE_LIST*.
9. **Output *RULE_LIST*.**

In this algorithm, we need two inputs, namely, E and *SELECTORS*. E is the training data and *SELECTORS* is the set of all possible selectors that test each attribute and its corresponding values. Set *RULE_LIST* is the *decision list*, storing the final output list of rules, which is initialized as to empty set in step 1. *Best_CPX* records the best rule detected in each iteration. The function *Find_Best_Complex(E)* learns the *Best_CPX*. We will elaborate the details of this function later in Section 5.2.2. Steps 2 to 8 form a Repeat-loops which learns the best rule and refines the training data. In particular, in each Repeat-loop, once a non-empty rule is learned from the data (steps 3 and 4), all the training examples that are covered by the rule are removed from the data (steps 5 and 6). The rule discovered, consisting of the rule condition and the most common class label of the examples covered by the rule, is added at the end of *RULE_LIST* (steps 7 and 8).

The stopping criteria for the Repeat-loop (from steps 2–8) can be either $E = \emptyset$ (no training examples left for learning) or Rule *Best_CPX* is *nil* (there is no new rule learned from the training data). After the rule learning process completes (i.e., satisfies one of the two stopping criteria), a default class c is inserted at the end of *RULE_LIST*. This step is performed because of the following two reasons: 1) there may still be some training examples that are not covered by any rule as no good rule can be mined from them, and 2) some test instances may not be covered by any rule in the *RULE_LIST* and thus we cannot classify it if we do not have a *default-class*. Clearly, with this *default-class*, we are able to classify any test instance. The *default-class* is typically the majority class among all the classes in the training data, which will be used only if no rule learned from the training data can be used to classify a test example. The final list of rules, together with the *default-class*, is represented as follows:

$$<r_1, r_2, \ldots, r_k, \textit{default-class}>, \text{ where } r_i \text{ is a rule mined from the training data.}$$

Finally, using the list of rules for classification is rather straightforward. For a given test instance, we simply try each rule in the list *sequentially*, starting from r_1, then r_2 (if r_1 cannot cover the test instance), r_3 (if both r_1 and r_2 cannot cover the test instance) and so on. The class consequent of the first rule that covers this test instance is assigned as the class of the test instance. If no rule (from $r_1, r_2, \ldots, r_k$) applies to the test instance, the *default-class* is applied.

5.2.1.2 RIPPER Algorithm and Its Variations (Ordered Classes)

We now introduce the second algorithm (Ordered Classes), which is based on the RIPPER algorithm [13, 51, 71], as well as earlier variations such as FOIL [10], I-REP [11], and REP [12].

RIPPER algorithm and its variations (D, C)

1. *RuleList* $\leftarrow \emptyset$; // initialize *RuleList* as an empty rule set
2. **For** each class $c \in C$ do
3.   Prepare data (Pos, Neg), where *Pos* contains all the examples of class c from D, and *Neg* contains the rest of the examples in D;

4.   **While** $Pos \neq \emptyset$ **do**
5.   *Rule* $\leftarrow$ learn-one-rule(Pos, Neg, c)
6.   **If** *Rule* is NULL **then**
7.   **exit-While-loop**
8.   **Else** *RuleList* $\leftarrow$ insert *Rule* at the end of RuleList;
9.   Remove examples covered by *Rule* from (Pos, Neg)
10.   **EndIf**
11.   **EndWhile**
12. **EndFor**
13. **Output** *RuleList*.

Different from the CN2 algorithm that learns each rule without pre-fixing a class, RIPPLE learns all rules for each class individually. In particular, only after rule learning for one class is completed, it moves on to the next class. As such, all rules for each class appear together in the rule *decision list*. The sequence of rules for each individual class is not important, but the rule subsets for different classes are ordered and still important. The algorithm usually mines rules for the least frequent/minority/rare class first, then the second minority class, and so on. This process ensures that some rules are learned for rare or minority classes. Otherwise, they may be dominated by frequent or majority classes and we will end up with no rules for the minority classes. The RIPPER rule induction algorithm is shown as follows, which is also based on sequential covering:

In this algorithm, the data set D is split into two subsets, namely, *Pos* and *Neg*, where *Pos* contains all the examples of class c from D, and *Neg* the rest of the examples in D (see step 3), i.e., in a one-vs.-others manner. Here $c \in C$ is the current working class of the algorithm, which is initialized as the least frequent class in the first iteration. As we can observe from the algorithm, steps 2 to 12 is a **For-loop**, which goes through all the classes one by one, starting from the minority class. That is why this method is called *Ordered Classes*, from the least frequent class to the most frequent class. For each class c, we have an internal **While-loop** from steps 4 to 11 that includes the rule learning procedure, i.e., perform the Learn-One-Rule() function to learn a rule *Rule* in step

5; insert the learned *Rule* at the end of *RuleList* in step 8; remove examples covered by *Rule* from (*Pos, Neg*) in step 9. Note two stopping conditions for internal rule learning of each class *c* are given in step 4 and step 6, respectively — we stop the **while-loop** for the internal rule learning process for the class *c* when the *Pos* becomes empty or no new rule can be learned by function Learn-One-Rule(*Pos, Neg, c*) from the remaining training data.

The other parts of the algorithm are very similar to those of the CN2 algorithm. The Learn-One-Rule() function will be described later in Section 5.2.2.

Finally, applying the *RuleList* for classification is done in a similar way as for the CN2 algorithm. The only difference is that the order of all rules within each class is not important anymore since they share the same class label, which will lead to the same classification results. Since the rules are now ranked by classes, given a test example, we will try the rules for the least frequent classes first until we can find a single rule that can cover the test example to perform its classification; otherwise, we have to apply the *default-class*.

5.2.2 Learn One Rule in Rule Learning

In the two algorithms described above, we have not elaborated on the two important functions used by them, where the first algorithm uses *Find_Best_Complex*() and the second algorithm uses learn-one-rule(). In this section, we explain the overall idea of the two functions that learn a rule from all or partial training data.

First, the rule starts with an empty set of conditions. In the first iteration, only one condition will be added. In order to find a best condition to add, all possible conditions are explored, which form a set of candidate rules. A condition is an attribute-value pair in the form of A_i *op* v, where A_i is an attribute and v is a value of A_i. Particularly, for a discrete attribute of v, assuming *op* is "=", then a condition will become $A_i = v$. On the other hand, for a continuous attribute, $op \in \{>, \leq\}$ and the condition becomes $A_i > v$ or $A_i \leq v$. The function then evaluates all the possible candidates to detect the *best* one from them (all the remaining candidates are discarded). Note the *best* candidate condition should be the one that can be used to better distinguish different classes, e.g., through an entropy function that has been successfully used in decision tree learning [4].

Next, after the first best condition is added, it further explores adding the second condition and so on in the same manner until some stopping conditions are satisfied. Note that we have omitted the rule class here because it implies the majority class of the training data *covered* by the conditions. In other words, it means that when we apply the rule, the class label that we predict should be correct most of the time.

Obviously, this is a heuristic and greedy algorithm in that after a condition is added, it will not be changed or removed through backtracking [15]. Ideally, we would like to explore all possible *combinations of attributes and values*. However, this is not practical as the number of possibilities grows exponentially with the increased number of conditions in rules. As such, in practice, the above greedy algorithm is used to perform rule induction efficiently.

Nevertheless, instead of keeping only the best set of conditions, we can improve the function a bit by keeping *k* best sets of conditions ($k > 1$) in each iteration. This is called the **beam search** (*k* beams), which ensures that a larger space (more attribute and value combinations) is explored, which may generate better results than the standard greedy algorithm, which only keeps the best set of conditions.

Below, we present the two specific implementations of the functions, namely *Find_Best_Complex*() and learn-one-rule() where *Find_Best_Complex*() is used in the CN2 algorithm, and learn-one-rule() is used in the RIPPER algorithm and its variations.

Find_Best_Complex(D)

The function *Find_Best_Complex(D)* uses beam search with k as its number of beams. The details of the function are given below.

Function *Find_Best_Complex(D)*

1. *BestCond* $\leftarrow \emptyset$;   // rule with no condition.
2. *candidateCondSet* $\leftarrow$ {*BestCond*};
3. *attributeValuePairs* $\leftarrow$ the set of all attribute-value pairs in D of the form $(A_i\ op\ v)$, where A_i is an attribute and v is a value or an interval;
4. **While** *candidateCondSet* $\neq \emptyset$ **do**
5.   *newCandidateCondSet* $\leftarrow \emptyset$;
6.   **For** each candidate *cond* in *candidateCondSet* **do**
7.   **For** each attribute-value pair a in *attributeValuePairs* **do**
8.   *newCond* $\leftarrow$ *cond* $\bigcup$ {a};
9.   *newCandidateCondSet* $\leftarrow$ *newCandidateCondSet* $\bigcup$ {*newCond*};
10.   **EndFor**
11.   **EndFor**
12.   remove duplicates and inconsistencies, e.g., $\{A_i = v_1, A_i = v_2\}$;
13.   **For** each candidate *newCond* in *newCandidateCondSet* **do**
14.   **If** *evaluation(newCond, D)* > *evaluation(BestCond, D)* **then**
15.   *BestCond* $\leftarrow$ *newCond*
16.   **EndIf**
17.   **Endor**
18.   *candidateCondSet* $\leftarrow$ the k best members of *newCandidateCondSet* according to the results of the evaluation function;
19. **EndWhile**
20. **If** *evaluation(BestCond, D)* − *evaluation(∅, D)* > *threshold* **then**
21.   **Output** the rule: "*BestCond* $\rightarrow$ *c*" where is c the majority class of the data covered by *BestCond*
22. **Else Output** NULL
23. **EndIf**

In this function, set *BestCond* stores the conditions of a rule to be returned. The class is omitted here as it refers to the majority class of the training data covered by *BestCond*. Set *candidateCondSet* stores the current best condition sets (which are the frontier beams) and its size is less than or equal to k. Each condition set contains a set of conditions connected by "and" (conjunction). Set *newCandidateCondSet* stores all the new candidate condition sets after adding each attribute-value pair (a possible condition) to every candidate in *candidateCondSet* (steps 5-11). Steps 13-17 update the *BestCond*. Note that an evaluation function is used to assess whether each new candidate condition set is better than the existing best condition set *BestCond* (step 14). If a new candidate condition set has been found better, it then replaces the current *BestCond* (step 15). Step 18 updates *candidateCondSet*, which selects k best condition sets (new beams).

Once the final *BestCond* is found, it is evaluated to check if it is significantly better than without any condition (∅) using a threshold (step 20). If yes, a rule is formed using *BestCond* and the most frequent (or the majority) class of the data covered by *BestCond* (step 21). If not, NULL is returned, indicating that no significant rule is found.

Note the evaluation() function shown below employs the entropy function, the same as in the decision tree learning, to evaluate how good the *BestCond* is.

Function evaluation(*BestCond*, *D*)

1. $D' \leftarrow$ the subset of training examples in D covered by *BestCond*

2. $entropy(D') = -\sum_{j=1}^{|C|} Pr(c_j) \times log_2 Pr(c_j);$

3. **Output** $-entropy(D')$ // since entropy measures impurity.

Specifically, in the first step of the evaluation() function, it obtains an example set D' that consists of a subset of training examples in D covered by *BestCond*. In its second step, it calculates an entropy function $entropy(D')$ based on the probability distribution — $Pr(c_j)$ is the probability of class c_j in the data set D', which is defined as the number of examples of class c_j in D' divided by the total number of examples in D'. In the entropy computation, $0 \times log 0 = 0$. The unit of entropy is **bit**. We now provide some examples to help understand the entropy measure.

Assume the data set D' has only two classes, namely positive class ($c_1=P$) and negative class ($c_2=N$). Based on the following three different combinations of probability distributions, we can compute their entropy values as follows:

1. The data set D' has 50% positive examples (i.e., $Pr(P) = 0.5$) and 50% negative examples (i.e., $Pr(N) = 0.5$). Then, $entropy(D') = -0.5 \times log_2 0.5 - 0.5 \times log_2 0.5 = 1$.

2. The data set D' has 20% positive examples (i.e., $Pr(P) = 0.2$) and 80% negative examples (i.e., $Pr(N) = 0.8$). Then, $entropy(D') = -0.2 \times log_2 0.2 - 0.8 \times log_2 0.8 = 0.722$.

3. The data set D' has 100% positive examples (i.e., $Pr(P) = 1$) and no negative examples (i.e., $Pr(N) = 0$). Then, $entropy(D') = -1 \times log_2 1 - 0 \times log_2 0 = 0$.

From the three scenarios shown above, we can observe that when the data class becomes purer and purer (e.g., all or most of the examples belong to one individual class), the entropy value becomes smaller and smaller. As a matter of fact, it can be shown that for this binary case (only has positive and negative classes), when $Pr(P) = 0.5$ and $Pr(N) = 0.5$, the entropy has the maximum value, i.e., 1 bit. When all the data in D' belong to one class, the entropy has the minimum value, i.e., 0 bit. It is clear that the entropy measures the amount of impurity according to the data class distribution. Obviously, we would like to have a rule that has a low entropy or even 0 bit, since it means that the rule will lead to one major class and we are thus more confident to apply the rule for classification.

In addition to the entropy function, other evaluation functions can also be applied. Note that when *BestCond* = $\emptyset$, it covers every example in D, i.e., $D = D'$.

Learn-One-Rule

In the Learn-One-Rule() function, a rule is first generated and then subjected to a pruning process. This method starts by splitting the positive and negative training data *Pos* and *Neg*, into growing and pruning sets. The growing sets, *GrowPos* and *GrowNeg*, are used to generate a rule, called *BestRule*. The pruning sets, *PrunePos* and *PruneNeg*, are used to prune the rule because *BestRule* may overfit the training data with too many conditions, which could lead to poor predictive performance on the unseen test data. Note that *PrunePos* and *PruneNeg* are actually validation sets, which are used to access the rule's generalization. If a rule has 50% error rate in the validation sets, then it does not generalize well and thus the function does not output it.

Function Learn-One-Rule(*Pos*, *Neg*, *class*)

1. split (*Pos*, *Neg*) into (*GrowPos*, *GrowNeg*), and (*PrunePos*, *PruneNeg*)
2. *BestRule* ← GrowRule(*GrowPos*, *GrowNeg*, *class*)   // grow a new rule
3. *BestRule* ← PruneRule(*BestRule*, *PrunePos*, *PruneNeg*)   // prune the rule
4. **If** the error rate of *BestRule* on (*PrunePos*, *PruneNeg*) exceeds 50% **Then**
5.   return NULL
6. **Endif**
7. **Output** *BestRule*

GrowRule() function: GrowRule() generates a rule (called *BestRule*) by repeatedly adding a condition to its condition set that maximizes an evaluation function until the rule covers only some positive examples in *GrowPos* but no negative examples in *GrowNeg*, i.e., 100% purity. This is basically the same as the Function *Find_Best_Complex(E)*, but without beam search (i.e., only the best rule is kept in each iteration). Let the current partially developed rule be *R*:

$$R: av_1, \ldots, av_k \to class$$

where each av_j (j=1, 2, …k) in rule R is a condition (an attribute-value pair). By adding a new condition av_{k+1}, we obtain the rule R^+: $av_1, \ldots, av_k, av_{k+1} \to class$. The evaluation function for R^+ is the following **information gain** criterion (which is different from the gain function used in decision tree learning):

$$gain(R, R^+) = p_1 \times (log_2 \frac{p_1}{p_1 + n_1} - log_2 \frac{p_0}{p_0 + n_0}) \tag{5.1}$$

where p_0 (respectively, n_0) is the number of positive (or negative) examples covered by R in *Pos* (or *Neg*), and p_1 (or n_1) is the number of positive (or negative) examples covered by R^+ in *Pos* (or *Neg*). R^+ will be better than R if R^+ can cover more proportion of positive examples than R. The GrowRule() function simply returns the rule R^+ that maximizes the gain.

PruneRule() function: To prune a rule, we consider deleting every subset of conditions from the *BestRule*, and choose the deletion that maximizes:

$$v(BestRule, PrunePos, PruneNeg) = \frac{p - n}{p + n}, \tag{5.2}$$

where p (respectively n) is the number of examples in *PrunePos* (or *PruneNeg*) covered by the current rule (after a deletion).

5.3 Classification Based on Association Rule Mining

In the last section, we introduced how to mine rules through rule induction systems. In this section, we discuss *Classification Based on Association Rule Mining*, which makes use of the association rule mining techniques to mine association rules and subsequently to perform classification tasks by applying the discovered rules. Note that *Classification Based on Association Rule Mining* detects *all rules* in data that satisfy the user-specified minimum support (minsup) and minimum confidence (minconf) constraints while a rule induction system detects only *a subset of the rules* for classification. In many real-world applications, rules that are not found by a rule induction system may be of high value for enhancing the classification performance, or for other uses.

The basic idea of *Classification Based on Association Rule Mining* is to first find strong correlations or associations between the frequent itemsets and class labels based on association rule mining techniques. These rules can be subsequently used for classification for test examples. Empirical evaluations have demonstrated that classification based on association rules are competitive with the state-of-the-art classification models, such as decision trees, navie Bayes, and rule induction algorithms.

In Section 5.3.1, we will present the concepts of association rule mining and an algorithm to automatically detect rules from transaction data in an efficient way. Then, in Section 5.3.2, we will introduce mining class association rules, where the class labels or *target attributes*) are on the right-hand side of the rules. Finally, in Section 5.3.3, we describe some techniques for performing classification based on discovered association rules.

5.3.1 Association Rule Mining

Association rule mining, formulated by Agrawal et al. in 1993 [16], is perhaps the most important model invented and extensively studied by the database and data mining communities. Mining association rules is a fundamental and unique data mining task. It aims to discover all co-occurrence relationships (or associations, correlations) among data items, from very large data sets in an efficient way. The discovered associations can also be very useful in data clustering, classification, regression, and many other data mining tasks.

Association rules represent an important class of regularities in data. Over the past two decades, data mining researchers have proposed many efficient association rule mining algorithms, which have been applied across a wide range of real-world application domains, including business, finance, economy, manufacturing, aerospace, biology, and medicine, etc. One interesting and successful example is Amazon book recommendation. Once association rules are detected automatically from the book purchasing history database, they can be applied to recommend those books that are relevant to users based on other peoples/community purchasing experiences.

The classic application of association rule mining is the market basket data analysis, aiming to determine how items purchased by customers in a supermarket (or a store/shop) are associated or co-occurring together. For example, an association rule mined from market basket data could be:

$$Bread \rightarrow Milk \ [support = 10\%, confidence = 80\%].$$

The rule basically means we can use *Bread* to infer *Milk* or those customers who buy *Bread* also frequently buy *Milk*. However, it should be read together with two important quality metrics, namely *support* and *confidence*. Particularly, the *support* of 10% for this rule means that 10% of customers buy Bread and Milk together, or 10% of all the transactions under analysis show that Bread and Milk are purchased together. In addition, a *confidence* of 80% means that those who buy Bread also buy Milk 80% of the time. This rule indicates that item Bread and item Milk are closely associated. Note in this rule, these two metrics are actually used to measure the rule strength, which will be defined in Section 5.3.1.1. Typically, association rules are considered interesting or useful if they satisfy two constraints, namely their *support* is larger than a **minimum support threshold** and their *confidence* is larger than a **minimum confidence threshold**. Both thresholds are typically provided by users and good thresholds may need users to investigate the mining results and vary the threholds multiple times.

Clearly, this association rule mining model is very generic and can be used in many other applications. For example, in the context of the Web and text documents, it can be used to find word co-occurrence relationships and Web usage patterns. It can also be used to find frequent substructures such as subgraphs, subtrees, or sublattices, etc [19], as long as these substructures frequently occurr together in the given dataset.

Note standard association rule mining, however, does not consider the sequence or temporal order in which the items are purchased. Sequential pattern mining takes the sequential information into consideration. An example of a sequential pattern is "5% of customers buy bed first, then mattress and then pillows". The items are not purchased at the same time, but one after another. Such patterns are useful in Web usage mining for analyzing click streams in server logs [20].

5.3.1.1 Definitions of Association Rules, Support, and Confidence

Now we are ready to formally define the problem of mining association rules. Let $I = \{i_1, i_2, \ldots, i_m\}$ be a set of **items**. In the market basket data analysis scenario, for example, set I contains all the items sold in a supermarket. Let $T = (t_1, t_2, \ldots, t_n)$ be a set of **transactions** (the database), where each transaction t_i is a record, consisting of a set of items such that $t_i \subseteq I$. In other words, a transaction is simply a set of items purchased in a basket by a customer and a transaction database includes all the transactions that record all the baskets (or the purchasing history of all customers). An **association rule** is an implication of the following form: $X \rightarrow Y$, where $X \subset I$, $Y \subset I$, and $X \cap Y = \emptyset$. X (or Y) is a set of items, called an **itemset**.

Let us give a concrete example of a *transaction*: $t_i = \{Beef, Onion, Potato\}$, which indicates that a customer purchased three items, i.e. *Beef*, *Onion*, and *Potato*, in his/her basket. An association rule could be in the following form:

$$Beef, Onion \rightarrow Potato,$$

where $\{Beef, Onion\}$ is X and $\{Potato\}$ is Y. Note brackets "{" and "}" are usually not explicitly included in both transactions and rules for simplicity.

As we mentioned before, each rule will be measured by its *support* and *confidence*. Next, we define both of them to evaluate the strength of rules.

A transaction $t_i \in T$ is said to contain an itemset X if X is a subset of t_i.

For example, itemset $\{Beef, Onion, Potato\}$ contains the following seven itemsets: $\{Beef\}$, $\{Onion\}$, $\{Potato\}$, $\{Beef, Onion\}$, $\{Beef, Potato\}$, $\{Onion, Potato\}$, and $\{Beef, Onion, Potato\}$.

Below, we define the support of an itemset and a rule, respectively.

Support of an itemset: The *support* count of an itemset X in T (denoted by $X.count$) is the number of transactions in T that contain X.

Support of a rule: The support of a *rule*, $X \rightarrow Y$ where X and Y are non-overlapping itemsets, is defined as the percentage of transactions in T that contains $X \cup Y$. The rule support thus determines how frequently the rule is applicable in the whole transaction set T. Let n be the number of transactions in T. The support of the rule $X \rightarrow Y$ is computed as follows:

$$support = \frac{(X \cup Y).count}{n}. \tag{5.3}$$

Note that support is a very important measure for filtering out those non-frequent rules that have a very low support since they occur in a very small percentage of the transactions and their occurrences could be simply due to chance.

Next, we define the confidence of a rule.

Confidence of a rule: The confidence of a rule, $X \rightarrow Y$, is the percentage of transactions in T that contain X also contain Y, which is computed as follows:

$$Confidence = \frac{(X \cup Y).count}{X.count}. \tag{5.4}$$

Confidence thus determines the predictability and reliability of a rule. In other words, if the confidence of a rule is too low, then one cannot reliably infer or predict Y given X. Clearly, a rule with low predictability is not very useful in practice.

TABLE 5.1: An Example of a Transaction Database

t_1	Beef, Chicken, Milk
t_2	Beef, Cheese
t_3	Cheese, Boots
t_4	Beef, Chicken, Cheese
t_5	Beef, Chicken, Clothes, Cheese, Milk
t_6	Chicken, Clothes, Milk
t_7	Chicken, Milk, Clothes

Given a transaction set T, the problem of mining association rules from T is to discover *all* association rules in T that have support and confidence greater than or equal to the user-specified **minimum support** (represented by **minsup**) and **minimum confidence** (represented by **minconf**).

Here we emphasize the keyword "all", i.e., association rule mining requires the completeness of rules. The mining algorithms should not miss any rule that satisfies both **minsup** and **minconf** constraints.

Finally, we illustrate the concepts mentioned above using a concrete example, shown in Table 5.1.

We are given a small transaction database, which contains a set of seven **transactions** $T = (t_1, t_2, \ldots, t_7)$. Each transaction t_i (i= 1, 2, ..., 7) is a set of items purchased in a basket in a supermarket by a customer. The set I is the set of all items sold in the supermarket, namely, $\{Beef, Boots, Cheese, Chicken, Clothes, Milk\}$.

Given two user-specified constraints, i.e. minsup = 30% and minconf = 80%, we aim to find all the association rules from the transaction database T. The following is one of the association rules that we can obtain from T, where **sup**= 3/7 is the support of the rule, and **conf**= 3/3 is the confidence of the rule.

$$\text{Chicken, Clothes} \rightarrow Milk \ [\textbf{sup} = 3/7, \textbf{conf} = 3/3]$$

Let us now explain how to calculate the support and confidence for this transaction database. Out of the seven transactions (i.e. $n = 7$ in Equation 5.3), there are three of them, namely, t_5, t_6, t_7 contain itemset {Chicken, Clothes} $\cup$ {Milk} (i.e., $(X \cup Y).count$=3 in Equation 5.3). As such, the support of the rule, **sup**=$(X \cup Y).count/n$=3/7=42.86%, which is larger than the minsup =30% (i.e. 42.86% > 30%).

On the other hand, out of the three transactions t_5, t_6, t_7 containing the condition itemset {Chicken, Clothes} (i.e., $X.count$=3), they also contain the consequent item {Milk}, i.e., {Chicken, Clothes} $\cup$ {Milk}= $(X \cup Y).count = 3$. As such, the confidence of the rule, **conf** = $(X \cup Y).count/X.count = 3/3 = 100\%$, which is larger than the minconf = 80% (100% > 80%). As this rule satisfies both the given minsup and minconf, it is thus valid.

We notice that there are potentially other valid rules. For example, the following one has two items as its consequent, i.e.,

$$\text{Clothes} \rightarrow Milk, Chicken \ [\textbf{sup} = 3/7, \textbf{conf} = 3/3].$$

Over the past 20 years, a large number of association rule mining algorithms have been proposed. They mainly improve the mining efficiency since it is critical to have an efficient algorithm to deal with large scale transaction databases in many real-world applications. Please refer to [49] for detailed comparison across various algorithms in terms of their efficiencies.

Note that no matter which algorithms are applied, the final results, i.e., association rules minded, are all the same based on the definition of association rules. In other words, given a transaction data set T, as well as a minimum support **minsup** and a minimum confidence **minconf**, the set of association rules occurring in T is uniquely determined. All the algorithms should find the same set of rules although their computational efficiencies and memory requirements could be different. In the next session, we introduce the best known mining algorithm, namely the **Apriori** algorithm, proposed by Agrawal in [17].

5.3.1.2 The Introduction of Apriori Algorithm

The well-known Apriori algorithm consists of the following two steps:

1. *Generate all frequent itemsets*: A frequent itemset is an itemset that has a transaction support **sup** above **minsup**, i.e. **sup>=minsup**.

2. *Generate all confident association rules from the frequent itemsets*: A confident association rule is a rule with a confidence **conf** above **minconf**, i.e. **conf >= minconf**.

Note that the size of an itemset is defined as the number of items occurred in it — an itemset of size k (or k-itemset) contains k items. Following the example in Table 5.1, $\{Chicken, Clothes, Milk\}$ is a 3-itemset, containing three items, namely, *Chicken, Clothes*, and *Milk*. It is a *frequent 3-itemset* since its support **sup** = 3/7 is larger than **minsup** = 30%. From the 3-itemset, we can generate the following **three confident association rules** since their confidence **conf** = 100% is greater than **minconf** = 80%:

Rule 1: Chicken, Clothes → *Milk* [sup = 3/7, conf = 3/3]

Rule 2: Clothes, Milk → Chicken [sup = 3/7, conf = 3/3]

Rule 3: Clothes → Milk, Chicken [sup = 3/7, conf = 3/3].

Next, we discuss the two key steps of Apriori Algorithm, namely 1) Frequent Itemset Generation and 2) Association Rule Generation, in detail.

STEP1: Frequent Itemset Generation

In the first step of Apriori algorithm, it generates all frequent itemsets efficiently by taking advantage of the following important property, i.e., *apriori property* or *downward closure property*.

Downward Closure Property: If an itemset has minimum support (or its support **sup** is larger than **minsup**), then its every non-empty subset also has minimum support.

The intuition behind this property is very simple because if a transaction contains a set of itemset X, then it must contain any non-empty subset of X. For example, $\{Chicken, Clothes, Milk\}$ is a frequent three-itemset (sup=3/7). Any non-empty subset of $\{Chicken, Clothes, Milk\}$, say $\{Chicken, Clothes\}$ is also a frequent itemset since out of the three transactions containing $\{Chicken, Clothes, Milk\}$, they all contain its subset $\{Chicken, Clothes\}$. This property and a suitable *minsup* threshold have been exploited to prune a large number of itemsets that cannot be *frequent*. In the Apriori algorithm, it assumes that the items in I as well as all the itemsets are sorted in the **lexicographic order** to ensure efficient frequent itemset generation. For example, suppose we have a k-itemset $w = \{w_1, w_2, \ldots, w_k\}$ that consists of the items $w_1, w_2, \ldots, w_k$, where $w_1 < w_2 < \ldots < w_k$ according to the **lexicographic order**.

Apriori algorithm for frequent itemset generation [16] is a *bottom-up* based approach and uses level-wise search, which starts from 1-itemset and expands to higher-level bigger itemsets, i.e., 2-itemset, 3-itemset, and so on. The overall algorithm is shown in Algorithm Apriori below. It generates all frequent itemsets by making multiple passes over the transaction database. In the first pass, it counts the supports of individual items, i.e., Level 1 items or 1-itemset in C_1 (as shown in step 1, C_1 is candidate 1-itemset) and determines whether each of them is frequent (step 2) where F_1 is the set of frequent 1-itemsets. After this initialization step, each of the subsequent passes, k ($k \geq$ 2), consist of the following three steps:

1. It starts with the seed set of itemsets F_{k-1} found to be frequent in the $(k$-$1)$-th pass. It then uses this seed set to generate candidate itemsets C_k (step 4), which are potential frequent itemsets. This step used the candidate-gen() procedure, as shown in Algorithm 4.

2. The transaction database is then passed over again and the actual support of each candidate itemset c in C_k is counted (steps 5-10). Note that it is not necessary to load the entire data into memory before processing. Instead, at any time point, only one transaction needs to reside in memory. This is a very important feature of the Apriori algorithm as it makes the algorithm scalable to huge data sets that cannot be loaded into memory.

3. At the end of the pass, it determines which of the candidate itemsets are actually frequent (step 11).

Algorithm Apriori for generating frequent itemsets

1. $C_1 \leftarrow$ init-pass(T); // the first pass over T
2. $F_1 \leftarrow \{f | f \in C_1, f.count/n \geq minsup\}$; // n is the no. of transactions in T;
3. **For** ($k = 2$; $F_{k-1} \neq \emptyset$; k++) do
4. $C_k \leftarrow$ candidate-gen(F_{k-1});
5.   **For** each transaction $t \in T$ do
6.   **For** each candidate $c \in C_k$ do
7.   **If** c is contained in t **then**
8. $c.count + +$;
9.   **EndFor**
10.   **EndFor**
11. $F_k \leftarrow \{c \in C_k | c.count/n \geq minsup\}$
12. **EndFor**
13. **Output** $F \leftarrow \bigcup_k F_k$.

The final output of the algorithm is the set F of all frequent itemsets (step 13) where set F contains frequent itemsets with different sizes, i.e., frequent 1-itemsets, frequent 2-itemsets, ..., frequent k-itemsets (k is the highest order of the frequent itemsets).

Next, we elaborate the key candidate-gen() procedure that is called in step 4. Candidate-gen () generates candidate frequent itemsets in two steps, namely the *join step* and the *pruning step*. *Join step* (steps 2-6 in Algorithm 4): This step joins two frequent $(k$-$1)$-itemsets to produce a possible candidate c (step 6). The two frequent itemsets f_1 and f_2 have exactly the same $k - 2$ items (i.e., $i_1, \ldots, i_{k-2}$) except the last one ($i_{k-1} \neq i'_{k-1}$ in steps 3-5). The joined k-itemset c is added to the set of candidates C_k (step 7). *Pruning step* (steps 8–11 in the next algorithm): A candidate c from the join step may not be a final frequent item-set. This step determines whether all the k-1 subsets (there are k of them) of c are in F_{k-1}. If any one of them is not in F_{k-1}, then c cannot be frequent according to the *downward closure property*, and is thus deleted from C_k.

Finally, we will provide an example to illustrate the candidate-gen() procedure.

Given a set of frequent itemsets at level 3, $F_3 = \{\{1,2,3\}, \{1,2,4\}, \{1,3,4\}, \{1,3,5\}, \{2,3,4\}\}$, the join step (which generates candidates C_4 for level 4) produces two candidate itemsets, $\{1,2,3,4\}$ and $\{1,3,4,5\}$. $\{1,2,3,4\}$ is generated by joining the first and the second itemsets in F_3 as their first and second items are the same. $\{1,3,4,5\}$ is generated by joining the third and the fourth itemsets in F_3, i.e., $\{1,3,4\}$ and $\{1,3,5\}$. $\{1,3,4,5\}$ is pruned because $\{3,4,5\}$ is not in F_3.

Procedure candidate-gen() is shown in the following algorithm:

Algorithm Candidate-gen(F_{k-1})

1. $C_k = \emptyset$; // initialize the set of candidates
2. **For** all $f_1, f_2 \in F_{k-1}$ // find all pairs of frequent itemsets
3.   with $f_1 = \{i_1, \ldots, i_{k-2}, i_{k-1}\}$ // that differ only in the last item
4.   with $f_2 = \{i_1, \ldots, i_{k-2}, i'_{k-1}\}$
5.   and $i_{k-1} < i'_{k-1}$ do // according to the lexicographic order
6. $c \leftarrow \{i_1, \ldots, i_{k-1}, i'_{k-1}\}$; // join the two itemsets f_1 and f_2
7. $C_k \leftarrow C_k \cup \{c\}$; // add the new itemset c to the candidates
8.   **For** each $(k-1)$-subset s of c do
9.   **If** $(s \notin F_{k-1})$ **then**
10.   delete c from C_k; // delete c from the candidates
11.   **EndFor**
12. **EndFor**
13. **Output** C_k.

We now provide a running example of the whole Apriori algorithm based on the transactions shown in Table 5.1. In this example, we have used **minsup** = 30%.

Apriori algorithm first scans the transaction data to count the supports of individual items. Those items whose supports are greater than or equal to 30% are regarded as frequent and are stored in set F_1, namely frequent 1-itemsets.

$$F_1: \{\{Beef\}:4, \{Cheese\}:4, \{Chicken\}:5, \{Clothes\}:3, \{Milk\}:4\}$$

In F_1, the number after each frequent itemset is the support count of the corresponding itemset. For example, {Beef}:4 means that the itemset {Beef} has occurred in four transactions, namely t_1, t_2, t_4, and t_5. A minimum support count of 3 is sufficient for being frequent (all the itemsets in F_1 have sufficient supports ≥ 3).

We then perform the Candidate-gen procedure using F_1, which generates the following candidate frequent itemsets C_2:

$$C_2: \{\{Beef, Cheese\}, \{Beef, Chicken\}, \{Beef, Clothes\}, \{Beef, Milk\},$$
$$\{Cheese, Chicken\}, \{Cheese, Clothes\}, \{Cheese, Milk\},$$
$$\{Chicken, Clothes\}, \{Chicken, Milk\}, \{Clothes, Milk\}\}$$

For each itemset in C_2, we need to determine if it is frequent by scanning the database again and storing the frequent 2-itemsets in set F_2:

$$F_2: \{\{Beef, Chicken\}:3, \{Beef, Cheese\}:3, \{Chicken, Clothes\}:3,$$
$$\{Chicken, Milk\}:4, \{Clothes, Milk\}:3\}$$

We now complete the level-2 search (for all 2-itemsets). Similarly, we generate the candidate frequent itemsets C_3 via Candidate-gen procedure:

$$C_3: \{\{Chicken, Clothes, Milk\}\}$$

Note that in C_3, itemset {Beef, Cheese, Chicken}, is also produced in step 6 of the Candidate-gen procedure. However, as its subset {Cheese, Chicken} is not in F_2, it is pruned and not included in C_3, according to **downward closure property**.

Finally, we count the frequency of {Chicken, Clothes, Milk} in database and it is stored in F_3 given that its support is greater than the minimal support.

$$F_3: \{\{\text{Chicken, Clothes, Milk}\}:3\}.$$

Note that since we only have one itemset in F_3, the algorithm stops since we need at least two itemsets to generate a candidate itemset for C_4. Apriori algorithm is just a representative of a large number of association rule mining algorithms that have been developed over the last 20 years. For more algorithms, please see [19].

STEP2: Association Rule Generation As we mentioned earlier, the Apriori algorithm can generate all frequent itemsets as well as all confident association rules. Interestingly, generating association rules is fairly straightforward compared with frequent itemset generation. In fact, we generate all association rules from frequent itemsets. For each frequent itemset f, we use all its non-empty subsets to generate association rules. In particular, for each such subset $\beta, \beta \subseteq f$, we output a Rule 5.5 if the confidence condition in Equation 5.6 is satisfied.

$$(f - \beta) \to \beta, \tag{5.5}$$

$$confidence = \frac{f.count}{(f - \beta).count} \geq minconf \tag{5.6}$$

Note the $f.count$ and $(f - \beta).count$ are the supports count of itemset f and itemset $(f - \beta)$, respectively. According to Equation 5.3, our rule support is $f.count/n$, where n is the total number of transactions in the transaction set T. Clearly, if f is frequent, then any of its non-empty subsets is also frequent according to the downward closure property. In addition, all the support counts needed for confidence computation in Equation 5.6, i.e., $f.count$ and $(f - \beta).count$, are available as we have recorded the supports for all the frequent itemsets during the mining process, e.g., using the Apriori algorithm. As such, there is no additional database scan needed for association rule generation.

5.3.2 Mining Class Association Rules

The association rules mined using Apriori algorithm are generic and flexible. An item can appear as part of the conditions or as part of the consequent in a rule. However, in some real-world applications, users are more interested in those rules with some fixed *target* items (or class labels) on the right-hand side. Obviously, such kinds of rules are very useful for our rule-based classification models.

For example, banks typically maintain a customer database that contains demographic and financial information of individual customers (such as gender, age, ethnicity, address, employment status, salary, home ownership, current loan information, etc) as well as the target features such as whether or not they repaid the loans or defaulted. Using association rule mining technique, we can investigate what kinds of customers are likely to repay (*good* credit risks) or to default (*bad* credit risks) — both of them are target feature values, so that banks can reduce the rate of loan defaults if they can predict those customers who are likely to default in advance based on their personal demographic and financial information. In other words, we are interested in a special set of rules whose consequents are only those target features — these rules are called class association rules (CARs) where we require only target feature values to occur as consequents of rules, although the conditions can be any items or their combinations from financial and demographic information.

Let T be a transaction data set consisting of n transactions. Each transaction in T has been labeled with a class y ($y \in Y$; Y is the set of all class labels or target features/items). Let I be the set of all items in T, and $I \cap Y = \emptyset$. Note here we treat the label set Y differently from the standard items in I and they do not have any overlapping. A class association rule (CAR) is an implication of the following form:

$$X \rightarrow y, where \ X \subseteq I, \ and \ y \in Y. \tag{5.7}$$

The definitions of **support** and **confidence** are the same as those for standard association rules. However, a class association rule is different from a standard association rule in the following two points:

1. The consequent of a CAR has only a *single* item, while the consequent of a standard association rule can have any number of items.

2. The consequent y of a CAR must be only from the class label set Y, i.e., $y \in Y$. No item from I can appear as the consequent, and no class label can appear as a rule condition. In contrast, a standard association rule can have any item as a condition or a consequent.

Clearly, the main objective of mining CARs is to automatically generate a complete set of CARs that satisfy both the user-specified minimum support constraint (minsup) and minimum confidence (minconf) constraint.

Intuitively, we can mine the given transaction data by first applying the Apriori algorithm to get all the rules and then perform a post-processing to select only those class association rules, as CARs are a *special* type of association rules with a target as their consequent. However, this is not efficient due to combinatorial explosion. Now, we present an efficient mining algorithm specifically designed for mining CARs.

This algorithm can mine CARs in a single step. The key operation is to find all *ruleitems* that have support above the given minsup. A ruleitem is a pair that has a *condset* and a class label y, namely, *(condset, y)*, where *condset* $\subseteq I$ is a set of items, and $y \in Y$ is a class label. The support count of a *condset* (called *condsupCount*) is the number of transactions in T that contain the *condset*. The support count of a ruleitem (called *rulesupCount*) is the number of transactions in T that contain the *condset* and are associated with class y. Each ruleitem *(condset, y)* represents a rule:

$$condset \rightarrow y,$$

whose support is *(rulesupCount/n)*, where n is the total number of transactions in T, and whose confidence is *(rulesupCount/condsupCount)*.

Ruleitems that satisfy the minsup are called frequent ruleitems, while the rest are called infrequent ruleitems. Similarly, ruleitems that satisfy the minconf are called confident ruleitems and correspondingly the rules are confident.

The rule generation algorithm, called CAR-Apriori, is given in the pseudocode below. The CAR-Apriori algorithm is based on the Apriori algorithm, which generates all the frequent ruleitems by passing the database multiple times. In particular, it computes the support count in the first pass for each 1-ruleitem that contains only one item in its condset (step 1). All the 1-candidate ruleitems, which pair one item in I and a class label, are stored in set C_1.

$$C_1 = \{(\{i\}, y) | i \in I, and \ y \in Y\} \tag{5.8}$$

Then, step 2 chooses the *frequent* 1-ruleitems (and stores into F_1) whose support count is greater than or equal to the given minsup value. From frequent 1-ruleitems, we generate 1-condition CARs — rules with only one condition in step 3. In a subsequent pass, say k ($k \geq 2$), it starts with the seed set F_{k-1} of $(k-1)$ frequent ruleitems found in the $(k-1)$-th pass, and uses this seed set to generate new possibly frequent k-ruleitems, called candidate k-ruleitems (C_k in step 5). The actual support counts for both condsupCount and rulesupCount are updated during the scan of the data (steps 6-13) for each candidate k-ruleitem. At the end of the data scan, it determines which of the candidate k-ruleitems in C_k are actually frequent (step 14). From the frequent k-ruleitems, step 15 generates k-condition CARs, i.e., class association rules with k conditions.

Algorithm CAR-Apriori(T)

1. $C_1 \leftarrow$ init-pass(T); // the first pass over T
2. $F_1 \leftarrow \{f | f \in C_1, f.rulesupCount / f.condsupCount \geq minsup\}$; // n is the no. of transactions in T;
3. $CAR_1 \leftarrow \{f | f \in F_1, f.rulesupCount / n \geq minconf\}$; // n is the no. of transactions in T;
4. **For** (k = 2; $F_{k-1} \neq \emptyset$; k++) do
5. $C_k \leftarrow$ CARcandidate-gen(F_{k-1});
6.   **For** each transaction $t \in T$ do
7.   **For** each candidate $c \in C_k$ do
8.   **If** $c.condset$ is contained in t **then** // c is a subset of t
9. $c.condsupCount ++$;
10.   if $t.class = c.class$ **then**
11. $c.rulesupCount ++$;
12.   **EndFor**
13.   **EndFor**
14. $F_k \leftarrow \{c \in C_k | c.rulesupCount / n \geq minsup\}$
15. $CAR_k \leftarrow \{f | f \in F_k, f.rulesupCount / f.condsupCount \geq minconf\}$;
16. **EndFor**
17. **Output** $CAR \leftarrow \bigcup_k CAR_k$.

One important observation regarding ruleitem generation is that if a ruleitem/rule has a confidence of 100%, then extending the ruleitem with more conditions, i.e., adding items to its condset, will also result in rules with 100% confidence although their supports may drop with additional items. In some applications, we may consider these subsequent rules with more conditions *redundant* because these additional conditions do not provide any more information for classification. As such, we should not extend such ruleitems in candidate generation for the next level (from $k-1$ to k), which can reduce the number of generated rules significantly. Of course, if desired, redundancy handling procedure can be added in the CAR-Apriori algorithm easily to stop the unnecessary expanding process.

Finally, the CARcandidate-gen() function is very similar to the candidate-gen() function in the Apriori algorithm, and it is thus not included here. The main difference lies in that in CARcandidate-gen(), ruleitems with the same class label are combined together by joining their condsets.

We now give an example to illustrate the usefulness of CARs. Table 5.2 shows a sample loan application dataset from a bank, which has four attributes, namely Age, Has_job, Own_house, and Credit_rating. The first attribute Age has three possible values, i.e., young, middle, and old. The second attribute Has_Job indicates whether an applicant has a job, with binary values: true (has a job) and false (does not have a job). The third attribute Own_house shows whether an applicant owns a house (similarly, it has two values denoted by true and false). The fourth attribute, Credit_rating, has three possible values: fair, good, and excellent. The last column is the class/target attribute, which shows whether each loan application was approved (denoted by Yes) or not (denoted by No) by the bank.

Assuming the user-specified minimal support *minsup* = 2/15 = 13.3% and the minimal confidence *minconf* = 70%, we can mine the above dataset to find the following rules that satisfy the two constraints:

Own_house = *false*, *Has_job* = *true* → *Class* = *Yes* [sup=3/15, conf=3/3]
Own_house = *true* → *Class* = *Yes* [sup=6/15, conf=6/6]
Own_house = *false*, *Has_job* = *true* → *Class* = *Yes*[sup=3/15, conf=3/3]

TABLE 5.2: A Loan Application Data Set

ID	Age	Has_job	Own_house	Credit_rating	Class
1	young	false	false	fair	No
2	young	false	false	good	No
3	young	true	false	good	Yes
4	young	true	true	fair	Yes
5	young	false	false	fair	No
6	middle	false	false	fair	No
7	middle	false	false	good	No
8	middle	true	true	good	Yes
9	middle	false	true	excellent	Yes
10	middle	false	true	excellent	Yes
11	old	false	true	excellent	Yes
12	old	false	true	good	Yes
13	old	true	false	good	Yes
14	old	true	false	excellent	Yes
15	old	false	false	fair	No

$Own\_house = false, Has\_job = false \rightarrow Class = No$ [sup=6/15, conf=6/6]
$Age = young, Has\_job = true \rightarrow Class = Yes$[sup=2/15, conf=2/2]
$Age = young, Has\_job = false \rightarrow Class = No$[sup=3/15, conf=3/3]
$Credit\_rating = fair \rightarrow Class = No$[sup=4/15, conf=4/5] .....

5.3.3 Classification Based on Associations

In this section, we discuss how to employ the discovered class association rules for classification purposes. Since the consequents of CARs are the class labels, it is thus logical to infer the class label of any test transaction, i.e., to do classification. CBA (Classification Based on Associations) is the first system that uses association rules for classification [30]. Note classifiers built using association rules are often called associative classifiers.

Following the above example, after we detect CARs, we intend to use them for learning a classification model to classify or automatically judge future loan applications. In other words, when a new customer visits the bank to apply for a loan, after providing his/her age, whether he/she has a job, whether he/she owns a house, and his/her credit rating, the classification model should predict whether his/her loan application should be approved so that we can use our constructed classification model to automate the loan application approval process.

5.3.3.1 Additional Discussion for CARs Mining

Before introducing how to build a classifier using CARs, we first give some additional discussions about some important points for mining high quality CARs.

Rule Pruning: CARs could be redundant and some of them are not statistically significant, which makes our classifier overfit the training examples. Such a classifier does not have good generalization capability. As such, we need to perform rule pruning to address these issues. Specifically, we can remove some conditions in CARs so that they are shorter, and have higher supports to be statistically significant. In addition, pruning some rules may cause some shortened/revised rules to become redundant — we thus need to remove these repeated rules. Generally speaking, pruning rules could lead to a more concise and accurate rule set as shorter rules are less likely to overfit the training data and potentially perform well on the unseen test data. Pruning is also called gener-alization as it makes rules more general and more applicable to test instances. Of course, we still need to maintain high confidences of CARs during the pruning process so that we can achieve more reliable and accurate classification results once the confident rules are applied. Readers can refer to papers [30, 31] for details of some pruning methods.

Multiple Minimum Class Supports: In many real-life classification problems, the datasets could have uneven or imbalanced class distributions, where majority classes cover a large proportion of the training data, while other minority classes (rare or infrequent classes) only cover a very small portion of the training data. In such a scenario, a single minsup may be inadequate for mining CARs. For example, we have a fraud detection dataset with two classes C_1 (represents "normal class") and C_2 (denotes for "fraud class"). In this dataset, 99% of the data belong to the majority class C_1, and only 1% of the data belong to the minority class C_2, i.e., we do not have many instances from "fraud class." If we set minsup = 1.5%, we may not be able to find any rule for the minority class C_2 as this minsup is still too high for minority class C_2. To address the problem, we need to reduce the minsup, say set minsup = 0.2% so that we can detect some rules for class C_2. However, we may find a huge number of overfitting rules for the majority class C_1 because minsup = 0.2% is too low for class C_1. The solution for addressing this problem is to apply multiple minimum class supports for different classes, depending on their sizes. More specifically, we could assign a different minimum class support $minsup_i$ for each class C_i, i.e., all the rules of class C_i must satisfy corresponding $minsup_i$. Alternatively, we can provide one single total minsup, denoted by $total\_minsup$, which is then distributed to each class according to the class distribution:

$$minsup_i = total\_minsup \times \frac{Number\ of\ Transactions\ in\ C_i}{Total\ Number\ of\ Transactions\ in\ Database}. \qquad (5.9)$$

The equation sets higher minsups for those majority classes while it sets lower minsups for those minority classes.

Parameter Selection: The two parameters used in CARs mining are the minimum support and the minimum confidence. While different minimum confidences may also be used for each class, they do not affect the classification results much because the final classifier tends to use high confi-dence rules. As such, one minimum confidence is usually sufficient. We thus are mainly concerned with how to determine the best support $minsup_i$ for each class C_i. Similar to other classification algorithms, we can apply the standard cross-validation technique to partition the training data into n folds where $n - 1$ folds are used for training and the remaining 1 fold is used for testing (we can repeat this n times so that we have n different combinations of training and testing sets). Then, we can try different values for $minsup_i$ in the training data to mine CARs and finally choose the value for $minsup_i$ that gives the best average classification performance on the test sets.

5.3.3.2 Building a Classifier Using CARs

After all CARs are discovered through the mining algorithm, a classifier is built to exploit the rules for classification. We will introduce five kinds of approaches for classifier building.

Use the Strongest Rule: This is perhaps the simplest strategy. It simply uses the strongest/most powerful CARs directly for classification. For each test instance, it first finds the strongest rule that covers the instance. Note that a rule covers an instance only if the instance satisfies the conditions of the rule. The class label of the strongest rule is then assigned to the test instance. The strength of a rule can be measured in various ways, e.g., based on rule confidence value only, χ^2 test, or a combination of both support and confidence values, etc.

Select a Subset of the Rules to Build a Classifier: This method was used in the CBA system. This method is similar to the sequential covering method, but applied to class association rules with additional enhancements. Formally, let D and S be the training data set and the set of all discovered CARs, respectively. The basic idea of this strategy is to select a subset L ($L \subseteq S$) of high confidence rules to cover the training data D. The set of selected rules, including a default class, is then used as the classifier. The selection of rules is based on a total order defined on the rules in S. Given two rules, r_i and r_j, we say $r_i \succ r_j$ or r_i precedes r_j or r_i has a higher precedence than r_j if

1. the confidence of r_i is greater than that of r_j, or
2. their confidences are the same, but the support of r_i is greater than that of r_j, or
3. both the confidences and supports of r_i and r_j are the same, but r_i is generated earlier than r_j.

A CBA classifier C is of the form:

$$C = < r_1, r_2, \ldots, r_k, default - class > \qquad (5.10)$$

where $r_i \in S$, $r_i \succ r_j$ if $j > i$. When classifying a test case, the first rule that satisfies the case will be used to classify it. If there is not a single rule that can be applied to the test case, it takes the default class, i.e., $default - class$, in Equation 5.10. A simplified version of the algorithm for building such a classifier is given in the following algorithm. The classifier is the *RuleList*.

Algorithm CBA (T)

1. $S = sort(S)$; // sorting is done according to the precedence $\succ$
2. $RuleList = \emptyset$; // the rule list classifier is initialized as empty set
3. **For** each rule $r \in S$ in sequence **do**
4.   **If** $(D \neq \emptyset)$ AND r classifies at least one example in D correctly **Then**
5.   delete from D all training examples covered by r;
6.   add r at the end of *RuleList*
7.   **EndIf**
8. **EndFor**
9. add the majority class as the default class at the end of RuleList

In Algorithm CBA, we first sort all the rules in S according to their precedence defined above. Then we through the rules one by one, from the highest precedence to the lowest precedence, during the for-loop. Particularly, for each rule, we will perform sequential covering from step 3 to 8. Finally, we construct our *RuleList* by appending the majority class so that we can classify any test instance.

Combine Multiple Rules: Like the first method *Use the Strongest Rule*, this method does not take any additional step to build a classifier. Instead, at the classification time, for each test instance, the system first searches a subset of rules that cover the instance.

1. If all the rules in the subset have the same class, then the class is assigned to the test instance.

2. If the rules have different classes, then the system divides the rules into a number of groups according to their classes, i.e., all rules of from the same class are in the same group. The system then compares the aggregated effects of the rule groups and finds the strongest group. Finally, the class label of the strongest group is assigned to the test instance.

To measure the strength of each rule group, there again can be many possible ways. For example, the CMAR system uses a weighted χ^2 measure [31].

Class Association Rules as Features: In this method, rules are used as features to augment the original data or simply form a new data set, which is subsequently fed to a traditional classification algorithm, e.g., Support Vector Machines (SVM), Decision Trees (DT), Naïve Bayesian (NB), K-Nearest Neighbour (KNN), etc.

To make use of CARs as features, only the conditional part of each rule is needed. For each training and test instance, we will construct a feature vector where each dimension corresponds to a specific rule. Specifically, if a training or test instance in the original data satisfies the conditional part of a rule, then the value of the feature/attribute in its vector will be assigned 1; 0 otherwise. The reason that this method is helpful is that CARs capture multi-attribute or multi-item correlations with class labels. Many classification algorithms, like Naïve Bayesian (which assumes the features are independent), do not take such correlations into consideration for classifier building. Clearly, the correlations among the features can provide additional insights on how different feature combinations can better infer the class label and thus they can be quite useful for classification. Several applications of this method have been reported [32–35].

Classification Using Normal Association Rules

Not only can *class association rules* be used for classification, but also *normal association rules*. For example, normal association rules are regularly employed in e-commerce Web sites for product recommendations, which work as follows: When a customer purchases some products, the system will recommend him/her some other related products based on what he/she has already purchased as well as the previous transactions from all the customers.

Recommendation is essentially a classification or prediction problem. It predicts what a customer is likely to buy. Association rules are naturally applicable to such applications. The classification process consists of the following two steps:

1. The system first mines normal association rules using previous purchase transactions (the same as market basket transactions). Note, in this case, there are no fixed classes in the data and mined rules. Any item can appear on the left-hand side or the right-hand side of a rule. For recommendation purposes, usually only one item appears on the right-hand side of a rule.

2. At the prediction (or recommendation) stage, given a transaction (e.g., a set of items already purchased by a given customer), all the rules that cover the transaction are selected. The strongest rule is chosen and the item on the right-hand side of the rule (i.e., the consequent) is then the predicted item and is recommended to the user. If multiple rules are very strong, multiple items can be recommended to the user simultaneously.

This method is basically the same as the "use the strongest rule" method described above. Again, the rule strength can be measured in various ways, e.g., confidence, χ^2 test, or a combination of both support and confidence [42]. Clearly, the other methods, namely, *Select a Subset of the Rules to Build a Classifier*, and *Combine Multiple Rules*, can be applied as well.

The key advantage of using association rules for recommendation is that they can predict any item since any item can be the class item on the right-hand side.

Traditional classification algorithms, on the other hand, only work with a single fixed class attribute, and are not easily applicable to recommendations.

Finally, in recommendation systems, multiple minimum supports can be of significant help. Otherwise, **rare items** will never be recommended, which causes the **coverage problem**. It is shown in [43] that using multiple minimum supports can dramatically increase the coverage. Note that rules from rule induction cannot be used for this recommendation purpose because the rules are not independent of each other.

5.3.4   Other Techniques for Association Rule-Based Classification

Since CBA was proposed to use association rules for classification [30] in 1998, many techniques in this direction have been proposed. We introduce some of the representative ones, including CMAR [31], XRules [43]. Note XRules is specifically designed for classifying semi-structured data, such as XML.

1. CMAR

CMAR stands for classification based on multiple association rules CMAR [31]. Like CBA, CMAR also consists of two phases, namely *rule generation phase* and *classification phase*. In *rule generation phase*, CMAR mines the complete set of rules in the form of $R : P \rightarrow c$, where P is a pattern in the transaction training data set, and c is a class label, i.e., R is a *class association rule*. The support and confidence of the rule R, namely $sup(R)$ and $conf(R)$, satisfy the user pre-defined minimal support and confidence thresholds, respectively.

CMAR used an effective and scalable association rule mining algorithm based on the FP-growth method [21]. As we know, existing association rule mining algorithms typically consist of two steps: 1) detect all the frequent patterns and 2) mine association rules that satisfy the confidence threshold based on the mined frequent patterns. CMAR, on the other hand, has no separated rule generation step. It constructs a class distribution-associated FP-tree and for every pattern, it maintains the distribution of various class labels among examples matching the pattern, without any overhead in the procedure of counting database. As such, once a frequent pattern is detected, rules with regard to the pattern can be generated straightaway. In addition, CMAR makes use of the class label distribution to prune. Given a frequent pattern P, let us assume c is the most dominant/mojority class in the set of examples matching P. If the number of examples having class label c and matching P is less than the support threshold, then there is no need to search any superpattern (superset) P' of P. This is very clear as any rule in the form of $P' \rightarrow c$ cannot satisfy the support threshold either as superset P' will have no larger support than pattern P.

Once rules are mined from the given transaction data, CMAR builds a CR-tree to save space in storing rules as well as to search for rules efficiently. CMAR also performs a rule pruning step to remove redundant and noise rules. In particular, three principles were used for rule pruning, including 1) use more general and high-confidence rules to prune those more specific and lower confidence ones; 2) select only positively correlated rules based on χ^2 testing; 3) prune rules based on database coverage.

Finally, in the *classification* phase, for a given test example, CMAR extracts a subset of rules matching the test example and predicts its class label by analyzing this subset of rules. CMAR first groups rules according to their class labels and then finds the *strongest* group to perform classification. It uses a weighted χ^2 measure [30] to integrate both information of intra-group rule correlation and popularity. In other words, if those rules in a group are highly positively correlated and have good support, then the group has higher *strength*.

2. XRules

Different from CBA and CMAR which are applied to transaction data sets consisting of multidimensional records, XRules [44] on the other hand, build a structural rule-based classifier for semi-structured data, e.g., XML. In the training stage, it constructs *structural rules* that indicate what kind of structural patterns in an XML document are closely related to a particular class label. In the testing stage, it employs these structural rules to perform the structural classification.

Based on the definition of structural rules, XRules performed the following three steps during the training stage: 1) Mine frequent structural rules specific to each class using its proposed XMiner (which extends TreeMiner to find all frequent trees related to some class), with sufficient support and strength. Note that users need to provide a minimum support π_i^{min} for each class c_i. 2) Prioritize or order the rules in decreasing level of precedence as well as removing unpredictive rules. 3) Determine a special class called *default-class*, which will be used to classify those test examples when none of the mined structural rules are applicable. After training, the classification model consists of an ordered rule set, and a *default-class*.

Finally, the testing stage performs classification on the given test examples without class labels. Given a test example S, there are two main steps for its classification, including, i.e., the *rule retrieval* step, which finds all matching rules (stored in set $R(S)$) for S, as well as *class prediction* step, which combines the statistics from each matching rule in $R(S)$ to predict the most likely class for S. Particularly, if $R(S) = \emptyset$, i.e., there are no matching rules, then default class is assigned to S; otherwise, $R(S) \neq \emptyset$. Assume $R_i(S)$ represent the matching rules in $R(S)$ with class c_i as their consequents. XRules used an average confidence method, i.e., for each class c_i, it computes the average rule strength for all the rules in $R_i(S)$. If the average rule strength for class c_i is big enough, the algorithm assigns the class c_i to the test example S. If the average rule strengths for all the classes are all very small, then the *default class* is used again to assign to S.

5.4 Applications

In this section, we briefly introduce some applications of applying rule based classification methods in text categorization [51], intrusion detection [74], diagnostic data mining [25], as well as gene expression data mining [50].

5.4.1 Text Categorization

It is well-known that Support Vector Machines (SVM) [57], Naïve Bayesian (NB) [58], and Rocchio's algorithm [60] are among the most popular techniques for text categorization, also called text classification. Their variations have also been applied to different types of learning tasks, e.g., learning with positive and unlabeled examples (PU learning) [59, 61–64]. However, these existing techniques are typically used as black-boxes. Rule-based classification techniques, on the other hand, can explain their classification results based on rules, and thus have also drawn a lot of attention. RIPPER [13], sleeping-experts [56], and decision-tree-based rule induction systems [52–55], have all been employed for text categorization.

Features used in the standard classification methods (such as SVM, Naïve Bayesian (NB), and Rocchio) are usually the individual terms in the form of words or word stems. Given a single word w in a document d, w's influence on d's predicted class is assumed to be independent of other words in d [60].

This assumption does not hold since w's *context*, encoded by the other words present in the document d, typically can provide more specific meanings and better indications on the d's classification, than w itself. As such, rule-based systems, such as RIPPER [13] and sleeping-experts [56], have exploited *context* information of the words for text categorization [51]. Both techniques performed very well across different data sets, such as AP title corpus, TREC-AP corpus, and Reuters etc, outperforming classification methods, like decision tree [4] and Rocchio algorithm [60].

Next, we will introduce how RIPPER and sleeping-experts (or specifically sleeping-experts for phrases) make use of context information for text categorization, respectively.

RIPPER for text categorization

In RIPPER, the context of a word w_1 is a conjunction of the form

$$w_1 \in d \ \text{ and } \ w_2 \in d \ \ldots \ \text{ and } \ w_k \in d.$$

Note that the context of a word w_1 consists of a number of other words $w_2, \ldots,$ and w_k, that need to co-occur with w_1, but they may occur in any order, and in any location in document d.

The standard RIPPER algorithm was extended in the following two ways so that it can be better used for text categorization.

1. Allow users to specify a *loss ratio* [65]. A loss ratio is defined as the ratio of the cost of a false negative to the cost of a false positive. The objective of the learning is to minimize misclassification cost on the unseen or test data. RIPPER can balance the recall and precision for a given class by setting a suitable loss ratio. Specifically, during the RIPPER's pruning and optimization stages, suitable weights are assigned to false positive errors and false negative errors, respectively.

2. In text classification, while a large corpus or a document collection contains many different words, a particular document will usually only contain quite limited words. To save space for representation, a document is represented as a *single attribute a*, with its *value* as the set of words that appear in the document or a word list of the document, i.e., $a = \{w_1, w_2, ..., w_n\}$. The primitive tests (conditions) on a set-valued attribute a are in the form of $w_i \in a$.

For a rule construction, RIPPER will repeatedly add conditions to rule r_0, which is initialized as an empty antecedent. Specifically, at each iteration i, a single condition is added to the rule r_i, producing an expanded rule r_{i+1}. The condition added to r_{i+1} is the one that maximizes information gain with regards to r_i. Given the set-valued attributes, RIPPER will carry out the following two steps to find a best condition to add:

1. For the current rule r_i, RIPPER will iterate over the set of examples/documents S that are covered by r_i and record a word list W where each word $w_i \in W$ appears as an element/value of attribute a in S. For each $w_i \in W$, RIPPER also computes two statistics, namely p_i and n_i, which represent the number of positive and negative examples in S that contain w_i, respectively.

2. RIPPER will go over all the words $w_i \in W$, and use p_i and n_i to calculate the information gain for its condition $w_i \in a$. We can then choose the condition that yields the largest information gain and add it to r_i to form rule r_{i+1}.

The above process of adding new literals/conditions continues until the rule does not cover negative examples or until no condition has a positive information gain. Note the process only requires time linear in the size of S, facilitating its applications to handle large text corpora.

RIPPER has been used to classify or filter personal emails [69] based on a relatively small sets of labeled messages.

Sleeping-experts for phrases for text categorization

Sleeping-experts [56] is an ensemble framework that builds a *master algorithm* to integrate the "advice" of different "experts" or classifiers [51, 76]. Given a test example, the master algorithm uses a weighted combination of the predictions of the experts. One efficient weighted assignment algorithm is the *multiplicative* update method where weights for each individual experts are updated by multiplying them by a constant. Particularly, those "correct" experts that make right classification will be able to keep their weights unchanged (i.e., multiplying 1) while those "bad" experts that make wrong classification have to multiply a constant (less than 1) so that their weights will become smaller.

In the context of text classification, the experts correspond to all length-k phrases that occur in a corpus. Given a document that needs to be classified, those experts are "awake" and make predictions if they appear in the document; the remaining experts are said to be "sleeping" on the document. Different from the context information used in the RIPPER, the context information in sleeping-experts (or sleeping-experts for phrases), is defined in the following phrase form:

$$w_{i_1}, w_{i_2} \ldots w_{i_j}$$

where $i_1 < i_2 < \ldots < i_{j-1} < i_j$ and $i_j - i_1 < n$.

Note that there could be some "holes" or "gaps" between any two words in the context /phrase. The detailed sleeping-experts for phrases algorithm is as follows.

The sleeping-experts algorithm for phrases

Input Parameters: $\beta \in (0,1), \theta_C \in (0,1)$, number of labeled documents T **Initialize**: $Pool \leftarrow \emptyset$

Do for $t=1, 2, \ldots, T$

1. Receive a new document $w_1^t, w_2^t, \ldots, w_l^t$, and its classification c^t

2. Define the set of active phrases:
$$W^t = \{\bar{w} | \bar{w} = w_{i_1}^t, w_{i_2}^t, \ldots, w_{i_j}^t, \ 1 \le i_1 < i_2 < \ldots < i_{j-1} < i_j < l, \ i_j - i_1 < n\}$$

3. Define the set of active mini-experts:
$$E^t = \{\bar{w}_k | \bar{w} \in W^t, k \in \{0, 1\}\}$$

4. Initialize the weights of new mini-experts:
$$\forall \bar{w}_k \in E^t \ s.t. \ \bar{w}_k \notin Pool : \ p_{\bar{w}_k}^t = 1$$

5. Classify the document as positive if
$$y^t = \frac{\sum_{\bar{w} \in W^t} p_{\bar{w}_1}^t}{\sum_{\bar{w} \in W^t} \sum_{k=0,1} p_{\bar{w}_k}^t} > \theta_C$$

6. Update weights:
$$l(\bar{w}_k) = \begin{cases} 0, & \text{if } c^t = k \\ 1, & \text{if } c^t \neq k \end{cases} \Rightarrow p_{\bar{w}_k}^{t+1} = p_{\bar{w}_k}^t \times \beta^{l(\bar{w}_k)} = \begin{cases} p_{\bar{w}_k}^t, & \text{if } c^t = k \\ \beta \times p_{\bar{w}_k}^t, & \text{if } c^t \neq k \end{cases}$$

7. Renormalize weights:

    (a) $Z_t = \sum_{\bar{w}_k' \in E^t} p_{\bar{w}_k'}^t$

    (b) $Z_{t+1} = \sum_{\bar{w}_k' \in E^t} p_{\bar{w}_k'}^{t+1}$

    (c) $p_{\bar{w}_k'}^{t+1} = \frac{Z_t}{Z_{t+1}} p_{\bar{w}_k'}^{t+1}$

8. Update: $Pool \leftarrow Pool \cup E^t$.

In this algorithm, the master algorithm maintains a *pool*, recording the sparse phrases that appeared in the previous documents and a set **p**, containing one weight for each sparse phrase in the pool.

This algorithm iterates over all the T labeled examples to update the weight set **p**. Particularly, at each time step t, we have a document $w_1^t, w_2^t, \ldots, w_l^t$ with length l, and its classification label c^t (step 1). In step 2, we search for a set of active phrases, denoted by W^t from the given document. Step 3 defines two active mini-experts $\bar{w}_1$ and $\bar{w}_0$ for each phrase $\bar{w}$ where $\bar{w}_1$ ($\bar{w}_0$) predicts the document belongs to the class (does not belong to the class). Obviously, given the actual class label, only one of them is correct. In step 4, this algorithm initializes the weights of new mini-experts (not in the pool) as 1. Step 5 classifies the document by calculating the weighted sum of the min-experts and storing the sum into the variable y^t — the document is classified as positive (class 1) if $y^t > \theta_C$; otherwise the negative (class 0). $\theta_C = \frac{1}{2}$ has been set to minimize the errors and get a balanced precision and recall. After performing classification, Step 6 updated weights to reflect the correlation between the classification results and the actual class label. It first computes the *loss* $l(\bar{w}_k)$ of each mini-expert $\bar{w}_k$ — if the predicted label is equal to the actual label, then the loss $l(\bar{w}_k)$ is zero; 1 otherwise. The weight of each expert is then multiplied by a factor $\beta^{l(\bar{w}_k)}$ where $\beta < 1$ is called the learning rate, which controls how quickly the weights are updated. The value for β is in the range [0.1,0.5]. Basically, this algorithm keeps the weight of the correctly classified mini-expert unchanged but lowers the weight of the wrongly classified mini-expert by multiplying β. Finally, step 7 normalizes the active mini-experts so that the total weight of the active mini-experts does not change. In effect, this re-normalization is to increase the weights of the mini-experts that were correct in classification.

5.4.2 Intrusion Detection

Nowadays, network-based computer systems play crucial roles in society. However, criminals have attempted to intrude into and compromise these systems in various manners. According to Heady [72], an intrusion is defined as any set of actions that attempt to compromise the integrity, confidentiality, or availability of a resource, e.g., illegally accessing administrator or superuser privilege, attacking and rendering a system out of services, etc. While some intrusion prevention techniques, such as user authentication by using passwords or biometrics as well as information protection by encryption, have been applied, they are not sufficient to address this problem as these systems typically have weaknesses due to their designs and programming errors [73]. As such, intrusion detection systems are thus imperative to serve as an additional shield to protect these computer systems from malicious activities or policy violations by closely monitoring the network and system activities.

There are some existing intrusion detection systems, which are manually constructed to protect a computer system based on some prior knowledge, such as known intrusion behaviors and the current computer system information. However, when facing new computer system environments and newly designed attacking/intruding methods, these types of *manual* and *ad hoc* intrusion detection systems are not flexible enough and will not be effective any more due to their limited adaptability.

We introduce a generic framework for building an intrusion detection system by analyzing the audit data [74], which refers to time-stamped data streams that can be used for detecting intrusions. The system first mines the audit data to detect the frequent activity patterns, which are in turn used to guide the selection of system features as well as construction of additional temporal and statistical features. Classifiers can then be built based on these features and served as intrusion detection models to classify whether an observed system activity is legitimate or intrusive. Compared with those methods with hand-crafted intrusion signatures to represent the intrusive activities, the approach has more generalized detection capabilities.

In general, there are two types of intrusion detection techniques, namely, anomaly detection and misuse detection. Anomaly detection determines whether deviation from an established normal behavior profile is an intrusion. In particular, a profile typically comprises a few statistical measures on system activities, e.g., frequency of system commands during a user login session and CPU usage. Deviation from a profile can then be calculated as the weighted sum of the deviations of the constituent statistical measures. Essentially, this is an unsupervised method as it does not need users to provide known specific intrusions to learn from, and it can detect unknown, abnormal, and suspicious activities. The challenging issue for anomaly detection is how to define and maintain normal profiles — improper definition, such as lack of sufficient examples to represent different types of normal activities, could lead to high level false alarms, i.e., some non-intrusion activities are flagged as intrusions.

Misuse detection, on the other hand, exploits known intrusion activities/patterns (e.g., more than three consecutive failed logins within a few minutes is a penetration attempt) or weak spots of a system (e.g., system utilities that have the "buffer overflow" vulnerabilities) as training data to identify intrusions. Compared with anomaly detection, misuse detection is a supervised learning method, which can be used to identify those known intrusions effectively and efficiently as long as they are similar to the training intrusions. However, it cannot detect unknown or newly invented attacks that could lead to unacceptable false negative error rates, i.e., some real intrusions are not able to be detected.

In order to perform intrusion predictions, we need to access those rich audit data that record system activities/events, the evidence of legitimate and intrusive users, as well as program activities. Anomaly detection searches for the normal usage patterns from the audit data while misuse detection encodes and matches intrusion patterns using the audit data [74].

For example, anomaly detection was performed for system programs [74], such as *sendmail*, as intruders use them to perform additional malicious activities. From the sequence of run-time system calls (e.g., open, read, etc), the audit data were segmented into a list of records, each of which had 11 consecutive system calls. RIPPER has been employed to detect rules that serve as normal (execution) profiles. In total, 252 rules are mined to characterize the normal co-occurrences of these system calls and to identify the intrusions that deviate from the normal system calls.

In addition, another type of intrusions, where intruders aim to disrupt network services by attacking the weakness in TCP/IP protocols, has also been identified [74]. By processing the raw packet-level data, it is possible to create a time series of connection-level records that capture the connection information such as duration, number of bytes transferred in each direction, and the flag that specifies whether there is an error according to the protocol etc. Once again, RIPPER has been applied to mine 20 rules that serve as normal network profile, characterizing the normal traffic patterns for each network service. Given the temporal nature of activity sequences [75], the temporal measures over features and the sequential correlation of features are particularly useful for accurate identification. Note the above anomaly detection methods need sufficient data which can cover as much variation of the normal behaviors as possible. Otherwise, given insufficient audit data, the anomaly detection will not be successful as some normal activities will be flagged as intrusions.

5.4.3 Using Class Association Rules for Diagnostic Data Mining

Liu et al [25] reported a deployed data mining system for Motorola, called Opportunity Map, that is based on class association rules mined from CBA [30]. The original objective of the system was to identify causes of cellular phone call failures. Since its deployment in 2006, it has been used for all kinds of applications.

The original data set contained cellular phone call records, and has more than 600 attributes and millions of records. After some pre-processing by domain experts, about 200 attributes are regarded as relevant to call failures. The data set is like any classification data set. Some of the attributes are continuous and some are discrete. One attribute indicates the final disposition of the call such as failed during setup, dropped while in progress, and ended successfully. This attribute is the class attribute in classification with discrete values. Two types of mining are usually performed with this kind of data:

1. Predictive data mining: The objective is to build predictive or classification models that can be used to classify future cases or to predict the classes of future cases. This has been the focus of research of the machine learning community.

2. Diagnostic data mining: The objective here is usually to understand the data and to find causes of some problems in order to solve the problems. No prediction or classification is needed.

In the above example, the problems are failed during setup and dropped while in progress. A large number of data mining applications in engineering domains are of this type because product improvement is the key task. The above application falls into the second type. The objective is not prediction, but to better understand the data and to find causes of call failures or to identify situations in which calls are more likely to fail. That is, the user wants interesting and actionable knowledge. Clearly, the discovered knowledge has to be understandable. Class association rules are suitable for this application.

It is easy to see that such kinds of rules can be produced by classification algorithms such as decision trees and rule induction (e.g., CN2 and RIPPER), but they are not suitable for the task due to three main reasons:

1. A typical classification algorithm only finds a very small subset of the rules that exist in data. Most of the rules are not discovered because their objective is to find only enough rules for classification. However, the subset of discovered rules may not be useful in the application. Those useful rules are left undiscovered. We call this the completeness problem.

2. Due to the completeness problem, the context information of rules are lost, which makes rule analysis later very difficult as the user does not see the complete information.

3. Since the rules are for classification purposes, they usually contain many conditions in order to achieve high accuracy. Long rules are, however, of limited use according to our experience because the engineers can hardly take any action based on them. Furthermore, the data coverage of long rules is often so small that it is not worth doing anything about them.

Class association rule mining [30] is found to be more suitable as it generates all rules. The Opportunity Map system basically enables the user to visualize class association rules in all kinds of ways through OLAP operations in order to find those interesting rules that meet the user needs.

5.4.4 Gene Expression Data Analysis

In recent years, association rule mining techniques have been applied in the bioinformatics domain, e.g., detecting patterns, and clustering or classifying gene expression data [39, 50, 66]. Microarray technology enables us to measure the expression levels of tens of thousands of genes in cells simultaneously [66] and has been applied in various clinical research [39]. The gene expression datasets generated by microarray technology typically contain a large number of columns (corresponding to tens of thousands of human genes) but a much smaller number of rows (corresponding to only tens or hundreds of conditions), which can be considered as tens or hundreds of very high-dimensional data. This is in contrast to those typical transaction databases that have many more rows (e.g., millions of transactions) than columns (tens or hundreds of features).

The objective of microarray dataset analysis is to detect important correlations between gene expression patterns (genes and their corresponding expression value ranges) and disease outcomes (certain cancer or normal status), which are very useful biomedical knowledge and can be utilized for clinical diagnostic purposes [67, 68].

The rules that can be detected from gene expression data are in the following form:

$$gene_1[a_1, b_1], \ldots, gene_n[a_n, b_n] \rightarrow class \tag{5.11}$$

where $gene_i$ is the name of a gene and $[a_i, b_i]$ is its expression value range or interval. In other words, the antecedent of the rule in Equation 5.11 consists of a set of conjunctive gene expression level intervals and the consequent is a single class label. For example, $X95735[-\infty, 994] \rightarrow ALL$ is a rule that was discovered from the gene expression profiles of ALL/AML tissues [50]. It has only one condition for gene $X95735$, whose expression value is less than 994. We have two classes for the dataset where class ALL stands for *Acute Lymphocytic Leukemia* cancer and AML stands for *Acute Myelogenous Leukemia* cancer. Obviously, association rules are very useful in analyzing gene expression data. The discovered rules, due to their simplicity, can be easily interpreted by clinicians and biologists, which provides direct insights and potential knowledge that could be used for medical diagnostic purposes. This is quite different from other machine learning methods, such as Support Vector Machines (SVM), which typically serve as a black box in many applications. Although they could be more accurate in certain datasets for classification purposes, it is almost impossible to convince clinicians to adopt their predictions for diagnostics in practice, as the logic behind the predictions is hard to explain compared with rule-based methods.

RCBT [50], Refined Classification Based on Top-k covering rule *groups* (TopkRGS), was proposed to address two challenging issues in mining the gene expression data. First, a huge number of rules can be mined from the high-dimensional gene expression dataset, even with rather high minimum support and confidence thresholds. It will be extremely difficult for biologists/clinicians to dig out clinically useful rules or diagnostic knowledge from a large amount of rules. Secondly, the high dimensionality (tens of thousands of genes) and the huge number of rules leads to an extremely long mining process.

TABLE 5.3: Example of Gene Expression Data and Rule Groups

Rows/conditions	Discretized gene expression data	Class label
r1	a, b, c, d, e	C
r2	a, b, c, o, p	C
r3	c, d, e, f, g	C
r4	c, d, e, f, g	¬C
r5	e, f, g, h, o	¬C

To address the above challenging problems, RCBT discovers the most significant TopkRGS for *each row* of a gene expression dataset. Note that TopkRGS can provide a more complete description for each row, which is different from existing interestingness measures that may fail to discover any interesting rules to cover some of the rows if given a higher support threshold. As such, the information in those rows that are not covered will not be captured in the set of rules. Given that gene expression datasets have a small number of rows, RCBT will not lose important knowledge.

Particularly, the rule group conceptually clusters rules from the same set of rows. We use the example in Table 5.3 to illustrate the concept of a rule group [50]. Note the gene expression data in Table 5.3 have been discretized. They consist of 5 rows, namely, $r1$, $r2$, ..., $r5$ where the first three rows have class label C while the last two have label $\neg C$. Given an item set I, its Item Support Set, denoted $R(I)$, is defined as the largest set of rows that contain I. For example, given item set $I = \{a, b\}$, its Item Support Set, $R(I) = \{r1, r2\}$. In fact, we observe that $R(a) = R(b) = R(ab) = R(ac) = R(bc) = R(abc) = \{r1, r2\}$. As such, they make up a rule group $\{a \rightarrow C, b \rightarrow C, \ldots, abc \rightarrow C\}$ of consequent C, with the upper bound $abc \rightarrow C$ and the lower bounds $a \rightarrow C$, and $b \rightarrow C$.

Obviously all rules in the same rule group have the exactly same support and confidence since they are essentially derived from the same subset of rows [50], i.e. $\{r1, r2\}$ in the above example. We can easily identify the remaining rule members based on the upper bound and all the lower bounds of a rule group. In addition, the significance of different rule groups can be evaluated based on both their confidence and support scores.

In addition, RCBT has designed a row enumeration technique as well as several pruning strategies that make the rule mining process very efficient. A classifier has been constructed from the top-k covering rule groups. Given a test instance, RCBT also aims to reduce the chance of classifying it based on the *default class* by building additional stand-by classifiers. Specifically, given k sets of rule groups $RG1, \ldots, RGk$, k classifiers $CL1, \ldots, CLk$ are built where $CL1$ is the main classifier and $CL2, \ldots, CLk$ are stand-by classifiers. It makes a final classification decision by aggregating voting scores from all the classifiers.

A number of experiments have been carried out on real bioinformatics datasets, showing that the RCBT algorithm is orders of magnitude faster than previous association rule mining algorithms.

5.5 Discussion and Conclusion

In this chapter, we discussed two types of popular rule-based classification approaches, i.e., rule induction and classification based on association rules. Rule induction algorithms generate a small set of rules directly from the data. Well-known systems include AQ by Michalski et al. [36], CN2 by Clark and Niblett [9], FOIL by Quinlan [10], FOCL by Pazzani et al. [37], I-REP by Furnkranz and Widmer [11], and RIPPER by Cohen [13]. Using association rules to build classifiers was proposed by Liu et al. in [30], which also reported the CBA system. CBA selects a small subset of class association rules as the classifier. Other classifier building techniques include combining multiple rules by Li et al. [31], using rules as features by Meretakis and Wuthrich [38], Antonie and Zaiane [32], Deshpande and Karypis [33], and Lesh et al. [35], generating a subset of rules by Cong et al. [39], Wang et al. [40], Yin and Han [41], and Zaki and Aggarwal [44]. Additional systems include those by Li et al. [45], Yang et al. [46], etc.

Note that well-known decision tree methods [4], such as ID3 and C4.5, build a tree structure for classification. The tree has two different types of nodes, namely decision nodes (internal nodes) and leaf nodes. A decision node specifies a test based on a single attribute while a leaf node indicates a class label. A decision tree can also be converted to a set of IF-THEN rules. Specifically, each path from the root to a leaf forms a rule where all the decision nodes along the path form the conditions of the rule and the leaf node forms the consequent of the rule. The main differences between decision tree and rule induction are in their *learning strategy* and *rule understandability*. Decision tree learning uses the divide-and-conquer strategy. In particular, at each step, all attributes are evaluated and one is selected to partition/divide the data into m disjoint subsets, where m is the number of values of the attribute. Rule induction, however, uses the separate-and-conquer strategy, which evaluates all attribute-value pairs (conditions) and selects only one. Thus, each step of divide-and-conquer expands m rules, while each step of separate-and-conquer expands only one rule. On top of that, the number of attribute-value pairs are much larger than the number of attributes. Due to these two effects, the separate-and-conquer strategy is much slower than the divide-and-conquer strategy. In terms of rule understandability, while if-then rules are easy to understand by human beings, we should be cautious about rules generated by rule induction (e.g., using the sequential covering strategy) since they are generated in order. Such rules can be misleading because the covered data are removed after each rule is generated. Thus the rules in the rule list are not independent of each other. In addition, a rule r may be of high quality in the context of the data D' from which r was generated. However, it may be a very weak rule with a very low accuracy (confidence) in the context of the whole data set D ($D' \subseteq D$) because many training examples that can be covered by r have already been removed by rules generated before r. If you want to understand the rules generated by rule induction and possibly use them in some real-world applications, you should be aware of this fact. The rules from decision trees, on the other hand, are independent of each other and are also mutually exclusive. The main differences between decision tree (or a rule induction system) and class association rules (CARs) are in their mining algorithms and the final rule sets. CARs mining detects all rules in data that satisfy the user-specified minimum support (minsup) and minimum confidence (minconf) constraints while a decision tree or a rule induction system detects only a small subset of the rules for classification. In many real-world applications, rules that are not found in the decision tree (or a rule list) may be able to perform classification more accurately. Empirical comparisons have demonstrated that in many cases, classification based on CARs performs more accurately than decision trees and rule induction systems.

The complete set of rules from CARs mining could also be beneficial from a rule usage point of view. For example, in a real-world application for finding causes of product problems (e.g., for diagnostic purposes), more rules are preferred to fewer rules because with more rules, the user is more likely to find rules that indicate the causes of the problems. Such rules may not be generated by a decision tree or a rule induction system. A deployed data mining system based on CARs is reported in [25]. Finally, CARs mining, like standard association rule mining, can only take discrete attributes for its rule mining, while decision trees can deal with continuous attributes naturally. Similarly, rule induction can also use continuous attributes. But for CARs mining, we first need to apply an attribute discretization algorithm to automatically discretize the value range of a continuous attribute into suitable intervals [47, 48], which are then considered as discrete values to be used for CARs mining algorithms. This is not a problem as there are many discretization algorithms available.

Bibliography

[1] Li, X. L., Liu, B., and Ng, S.K. Learning to identify unexpected instances in the test set. In *Proceedings of Twentieth International Joint Conference on Artificial Intelligence*, pages 2802–2807, India, 2007.

[2] Cortes, Corinna and Vapnik, Vladimir N. Support-vector networks. *Machine Learning*, 20 (3):273–297, 1995.

[3] Hopfield, J. J. Neural networks and physical systems with emergent collective computational abilities. In *Proceedings of the National Academy of Sciences USA*, 79 (8):2554–2558, 1982.

[4] Quinlan, J. *C4.5: Programs for machine learning*. Morgan Kaufmann Publishers, 1993.

[5] George H. John and Pat Langley. Estimating continuous distributions in Bayesian classifiers. In *Proceedings of the Eleventh Conference on Uncertainty in Artificial Intelligence*, pages 338–345, San Mateo, 1995.

[6] Bremner, D., Demaine, E., Erickson, J., Iacono, J., Langerman, S., Morin, P., and Toussaint, G. Output-sensitive algorithms for computing nearest-neighbor decision boundaries. *Discrete and Computational Geometry*, 33 (4):593–604, 2005.

[7] Hosmer, David W. and Lemeshow, Stanley. *Applied Logistic Regression*. Wiley, 2000.

[8] Rivest, R. Learning decision lists. *Machine Learning*, 2(3):229–246, 1987.

[9] Clark, P. and Niblett, T. The CN2 induction algorithm. *Machine Learning*, 3(4):261–283, 1989.

[10] Quinlan, J. Learning logical definitions from relations. *Machine Learning*, 5(3):239–266, 1990.

[11] Furnkranz, J. and Widmer, G. Incremental reduced error pruning. In *Proceedings of International Conference on Machine Learning (ICML-1994)*, pages 70–77, 1994.

[12] Brunk, C. and Pazzani, M. An investigation of noise-tolerant relational concept learning algorithms. In *Proceedings of International Workshop on Machine Learning*, pages 389–393, 1991.

[13] Cohen, W. W. Fast effective rule induction. In *Proceedings of the Twelfth International Conference on Machine Learning*, pages 115–123, 1995.

[14] Mitchell, T. *Machine Learning*. McGraw Hill. 1997.

[15] Donald E. K. *The Art of Computer Programming*. Addison-Wesley, 1968.

[16] Agrawal, R., Imieliski, T., and Swami, A. Mining association rules between sets of items in large databases. In *Proceedings of ACM SIGMOD International Conference on Management of Data (SIGMOD-1993)*, pages 207–216, 1993.

[17] Agrawal, R. and Srikant, R. Fast algorithms for mining association rules in large databases. In *Proceedings of International Conference on Very Large Data Bases (VLDB-1994)*, pages 487–499, 1994.

[18] Michalski, R. S. On the quasi-minimal solution of the general covering problem. In *Proceedings of the Fifth International Symposium on Information Processing*, pages 125–128, 1969.

[19] Han, J. W., Kamber, M., and Pei, J. *Data Mining: Concepts and Technqiues.* 3rd edition, Morgan Kaufmann, 20011.

[20] Liu, B. *Web Data Mining: Exploring Hyperlinks, Contents, and Usage Data.* Springer, 2006.

[21] Han, J. W., Pei, J., and Yin, Y. Mining frequent patterns without candidate generation. In *Proceedings of ACM SIGMOD Conference on Management of Data (SIGMOD-2000)*, pages 1–12, 2000.

[22] Bayardo, Jr., R. and Agrawal, R. Mining the most interesting rules. In *Proceedings of ACM SIGKDD International Conference on Knowledge Discovery and Data Mining (KDD-1999)*, pages 145–154, 1999.

[23] Klemettinen, M., Mannila, H., Ronkainen, P., Toivonen, H., and Verkamo, A. Finding interesting rules from large sets of discovered association rules. In *Proceedings of ACM International Conference on Information and Knowledge Management (CIKM-1994)*, pages 401–407, 1994.

[24] Liu, B., Hsu, W., and Ma, Y. Pruning and summarizing the discovered associations. In *Proceedings of ACM SIGKDD International Conference on Knowledge Discovery and Data Mining (KDD-1999)*, pages 125–134, 1999.

[25] Liu, B., Zhao, K., Benkler, J., and Xiao, W. Rule interestingness analysis using OLAP operations. In *Proceedings of ACM SIGKDD International Conference on Knowledge Discovery and Data Mining (KDD-2006)*, pages 297–306, 2006.

[26] Padmanabhan, B. and Tuzhilin, A. Small is beautiful: discovering the minimal set of unexpected patterns. In *Proceedings of ACM SIGKDD International Conference on Knowledge Discovery and Data Mining (KDD-2000)*, pages 54–63, 2000.

[27] Piatetsky-Shapiro, G. Discovery, analysis, and presentation of strong rules. In *Knowledge discovery in databases*, pages 229–248, 1991.

[28] Silberschatz, A. and Tuzhilin, A. What makes patterns interesting in knowledge discovery systems. *IEEE Transactions on Knowledge and Data Engineering*, 8 (6):970–974, 1996.

[29] Tan, P., Kumar, V., and Srivastava, J. Selecting the right interestingness measure for association patterns. In *Proceedings of ACM SIGKDD International Conference on Knowledge Discovery and Data Mining (KDD-2002)*, pages 32-41, 2002.

[30] Liu, B., Hsu, W., and Ma, Y. Integrating classification and association rule mining. In *Proceedings of ACM SIGKDD International Conference on Knowledge Discovery and Data Mining (KDD-1998)*, pages 80–86, 1998.

[31] Li, W., Han, J., and Pei, J. *CMAR:* Accurate and efficient classification based on multiple class-association rules. In *Proceedings of IEEE International Conference on Data Mining (ICDM-2001)*, pages 369–376, 2001.

[32] Antonie, M. and Zaïane, O. Text document categorization by term association. In *Proceedings of IEEE International Conference on Data Minig (ICDM-2002)*, Pages, 19–26, 2002.

[33] Deshpande, M. and Karypis, G. Using conjunction of attribute values for classification. In *Proceedings of ACM International Conference on Information and Knowledge Management (CIKM-2002)*, pages 356–364, 2002.

[34] Jindal, N. and Liu, B. Identifying comparative sentences in text documents. In *Proceedings of ACM SIGIR Conference on Research and Development in Information Retrieval (SIGIR-2006)*, pages 244–251, 2006.

[35] Lesh, N., Zaki, M., and Ogihara, M. Mining features for sequence classification. In *Proceedings of ACM SIGKDD International Conference on Knowledge Discovery and Data Mining (KDD-1999)*, pages 342-346, 1999.

[36] Michalski, R., Mozetic, I., Hong, J., and Lavrac, N. The multi-purpose incremental learning system AQ15 and its testing application to three medical domains. In *Proceedings of National Conference on Artificial Intelligence (AAAI-86)*, pages 1041–1045, 1986.

[37] Pazzani, M., Brunk, C., and Silverstein, G. A knowledge-intensive approach to learning relational concepts. In *Proceedings of International Workshop on Machine Learning (ML-1991)*, pages 432–436, 1991.

[38] Meretakis, D. and Wuthrich, B. Extending naïve Bayes classifiers using long itemsets. In *Proceedings of ACM SIGKDD International Conference on Knowledge Discovery and Data Mining (KDD-1999)*, pages 165–174, 1999.

[39] Cong, G., Tung, A.K.H., Xu, X., Pan, F., and Yang, J. Farmer: Finding interesting rule groups in microarray datasets. In *Proceedings of ACM SIGMOD Conference on Management of Data (SIGMOD-2004)*, pages 143–154, 2004.

[40] Wang, K., Zhou, S., and He, Y. Growing decision trees on support-less association rules. In *Proceedings of ACM SIGKDD International Conference on Knowledge Discovery and Data Mining (KDD-2000)*, pages 265-269, 2000.

[41] Yin, X. and Han, J. CPAR: Classification based on predictive association rules. In *Proceedings of SIAM International Conference on Data Mining (SDM-2003)*, pages 331-335, 2003.

[42] Lin, W., Alvarez, S., and Ruiz, C. Efficient adaptive-support association rule mining for recommender systems. *Data Mining and Knowledge Discovery*, 6(1):83-105, 2002.

[43] Mobasher, B., Dai, H., Luo, T., and Nakagawa, M. Effective personalization based on association rule discovery from web usage data. In *Proceedings of ACM Workshop on Web Information and Data Management*, pages 9–15, 2001.

[44] Zaki, M. and Aggarwal, C. XRules: an effective structural classifier for XML data. In *Proceedings of ACM SIGKDD International Conference on Knowledge Discovery and Data Mining (KDD-2003)*, pages 316–325, 2003.

[45] Li, J., Dong, G., Ramamohanarao, K., and Wong, L. DeEPs: A new instance-based lazy discovery and classification system. *Machine Learning*, 54(2):99-124, 2004.

[46] Yang, Q., Li, T., and Wang, K. Building association-rule based sequential classifiers for web-document prediction. *Data Mining and Knowledge Discovery*, 8(3):253–273, 2004.

[47] Dougherty, J., Kohavi, R., and Sahami, M. Supervised and unsupervised discretization of continuous features. In *Proceedings of International Conference on Machine Learning (ICML-1995)*, pages 194–202, 1995.

[48] Fayyad, U. and Irani, K. Multi-interval discretization of continuous-valued attributes for classification learning. In *Proceedings of the International Joint Conference on Artificial Intelligence (IJCAI-1993)*, pages 1022–1028, 1993.

[49] Zheng, Z., Kohavi, R., and Mason, L. Real world performance of association rule algorithms. In *Proceedings of ACM SIGKDD International Conference on Knowledge Discovery and Data Mining (KDD-2001)*, pages 401-406, 2001.

[50] Cong, G., Tan, K.-L., Tung A.K.H., and Xu, X. Mining top-k covering rule groups for gene expression data. In *Proceedings of the 2005 ACM-SIGMOD International Conference on Management of Data (SIGMOD–05)*, pages 670–681, 2005.

[51] Cohen, W.W., and. Yoram, S. Context-sensitive learning methods for text categorization. *ACM Transactions on Information Systems*, 17(2):141–173, 1999.

[52] Johnson, D., Oles. F., Zhang T., and Goetz, T. A decision tree-based symbolic rule induction system for text categorization. *IBM Systems Journal*, 41(3):428–437, 2002.

[53] Apte, C., Damerau, F., and Weiss, S. Automated learning of decision rules for text categorization. *ACM Transactions on Information Systems*, 12(3):233–251, 1994.

[54] Weiss, S. M., Apte C., Damerau, F., Johnson, D., Oles, F., Goetz, T., and Hampp, T. Maximizing text-mining performance. *IEEE Intelligent Systems*, 14(4):63–69, 1999.

[55] Weiss, S. M. and Indurkhya, N. Optimized rule induction. *IEEE Expert*, 8(6):61–69, 1993.

[56] Freund, Y., Schapire, R., Singer, Y., and Warmuth, M. Using and combining predictors that specialize. In *Proceedings of the 29th Annual ACM Symposium on Theory of Computing*, pp. 334–343, 1997.

[57] Joachims, T. Text categorization with support vector machines: learning with many relevant features. In *Proceedings of the European Conference on Machine Learning (ECML)*, pages 137–142, 1998.

[58] Andrew, M. and Nigam, K. A comparison of event models for Naïve Bayes text classification. In *Proceedings of AAAI-98 workshop on learning for text categorization*. Vol. 752. 1998.

[59] Liu, B., Lee, W. S., Yu, P. S. and Li, X. L. Partially supervised classification of text documents. In *Proceedings of the Nineteenth International Conference on Machine Learning (ICML-2002)*, pages 387–394, Australia, 2002.

[60] Rocchio, J. Relevance feedback in information retrieval. In G. Salton (ed.). *The Smart Retrieval System: Experiments in Automatic Document Processing*, Prentice-Hall, Upper Saddle River, NJ, 1971.

[61] Li, X. L. and Liu, B. Learning to classify texts using positive and unlabeled data. In *Proceedings of Eighteenth International Joint Conference on Artificial Intelligence*, pages 587–592, Mexico, 1993.

[62] Li, X. L., Liu, B., Yu, P. S., and Ng, S. K. Positive unlabeled learning for data stream classification. In *Proceedings of the Ninth SIAM International Conference on Data Mining*, pages 257–268, 2009.

[63] Li, X. L., Liu, B., Yu, P. S., and Ng, S. K. Negative training data can be harmful to text classification. In *Proceedings of the 2010 Conference on Empirical Methods in Natural Language Processing*, pages 218–228, USA, 2010.

[64] Liu, B., Dai, Y., Li, X. L., Lee, W. S., and Yu, P. S. Building text classifiers using positive and unlabeled examples. In *Proceedings of Third IEEE International Conference on Data Mining*, pages 179–186, 2003.

[65] Lewis, D. and Catlett, J. Heterogeneous uncertainty sampling for supervised learning. In *Proceedings of the Eleventh Annual Conference on Machine Learning*, pages 148–156, 1994.

[66] Li, X. L., Tan, Y. C., and Ng, S. K. Systematic gene function prediction from gene expression data by using a fuzzy nearest-cluster method *BMC Bioinformatics*, 7(Suppl 4):S23, 2006.

[67] Han, X.X. and Li, X.L. Multi-resolution independent component analysis for high-performance tumor classification and biomarker discovery, *BMC Bioinformatics*, 12(Suppl 1): S7, 2011

[68] Yang, P., Li, X. L., Mei, J. P., Kwoh, C. K., and Ng, S. K. Positive-unlabeled learning for disease gene identification, *Bioinformatics*, Vol 28(20):2640–2647, 2012

[69] Cohen, W.W. Learning rules that classify e-mail. In *Proceedings of the AAAI Spring Symposium on Machine Learning in Information Access*, pages 18–25, 1996.

[70] Liu, B., Dai, Y., Li, X. L., Lee, W. S., and Yu, P. S. Text classification by labeling words. In *Proceedings of the National Conference on Artificial Intelligence*, pages 425–430, USA, 2004.

[71] Cohen, W.W. Learning trees and rules with set-valued features In *Proceedings of the Thirteenth National Conference on Artificial Intelligence*, pages 709–716, 1996.

[72] Heady, R., Luger, G., Maccabe, A., and Servilla, M. The Architecture of a Network Level Intrusion Detection System. Technical report, University of New Mexico, 1990.

[73] Lee, W., Stolfo, S. J., and Mok, K. W. Adaptive intrusion detection: A data mining approach. *Artificial Intelligence Review - Issues on the Application of Data Mining Archive*, 14(6):533–567, 2000.

[74] Lee, W. and Stolfo, S. J. Data mining approaches for intrusion detection. In *Proceedings of the 7th USENIX Security Symposium*, San Antonio, TX, 1998.

[75] Mannila, H. and Toivonen, H. Discovering generalized episodes using minimal occurrences. In *Proceedings of the 2nd International Conference on Knowledge Discovery in Databases and Data Mining*. Portland, Oregon, pages 146–151,1996.

[76] Friedman, J.H. and Popescu, B.E. Predictive learning via rule ensembles *The Annals of Applied Statistics*, 2(3):916–954, 2008.

Chapter 6

Instance-Based Learning: A Survey

Charu C. Aggarwal

IBM T. J. Watson Research Center
Yorktown Heights, NY
charu@us.ibm.com

6.1 Introduction

Most classification methods are based on building a model in the training phase, and then using this model for specific test instances, during the actual classification phase. Thus, the classification process is usually a two-phase approach that is cleanly separated between processing training and test instances. As discussed in the introduction chapter of this book, these two phases are as follows:

- *Training Phase:* In this phase, a model is constructed from the training instances.

- *Testing Phase:* In this phase, the model is used to assign a label to an unlabeled test instance.

Examples of models that are created during the first phase of training are decision trees, rule-based methods, neural networks, and support vector machines. Thus, the first phase creates *pre-compiled abstractions* or *models* for learning tasks. This is also referred to as *eager* learning, because the models are constructed in an eager way, without waiting for the test instance. In instance-based learning, this clean separation between the training and testing phase is usually not present. The specific instance, which needs to be classified, is used to create a model that is *local* to a specific test instance. The classical example of an instance-based learning algorithm is the k-nearest neighbor classification algorithm, in which the k nearest neighbors of a classifier are used in order to create a local model for the test instance. An example of a local model using the k nearest neighbors could be that the majority class in this set of k instances is reported as the corresponding label, though more complex models are also possible. Instance-based learning is also sometimes referred to as *lazy* learning, since most of the computational work is not done upfront, and one waits to obtain the test instance, before creating a model for it [9]. Clearly, instance-based learning has a different set of tradeoffs, in that it requires very little or no processing for creating a *global* abstraction of the training data, but can sometimes be expensive at classification time. This is because instance-based learning typically has to determine the relevant local instances, and create a local model from these instances at classification time. While the obvious way to create a local model is to use a k-nearest neighbor classifier, numerous other kinds of lazy solutions are possible, which combine the power of lazy learning with other models such as locally-weighted regression, decision trees, rule-based methods, and SVM classifiers [15,36,40,77]. This chapter will discuss all these different scenarios. It is possible to use the traditional "eager" learning methods such as Bayes methods [38], SVM methods [40], decision trees [62], or neural networks [64] in order to improve the effectiveness of local learning algorithms, by applying them only on the local neighborhood of the test instance at classification time.

It should also be pointed out that many instance-based algorithms may require a pre-processing phase in order to improve the efficiency of the approach. For example, the efficiency of a nearest neighbor classifier can be improved by building a similarity index on the training instances. In spite of this pre-processing phase, such an approach is still considered lazy learning or instance-based learning since the pre-processing phase is not really a classification model, but a data structure that enables efficient implementation of the run-time modeling for a given test instance.

Instance-based learning is related to but not quite the same as case-based reasoning [1,60,67], in which previous examples may be used in order to make predictions about specific test instances. Such systems can modify cases or use parts of cases in order to make predictions. Instance-based methods can be viewed as a particular kind of case-based approach, which uses specific kinds of algorithms for instance-based classification. The framework of instance-based algorithms is more amenable for reducing the computational and storage requirements, noise and irrelevant attributes. However, these terminologies are not clearly distinct from one another, because many authors use the term "case-based learning" in order to refer to instance-based learning algorithms. Instance-specific learning can even be extended to distance function learning, where instance-specific distance functions are learned, which are local to the query instance [76].

Instance-based learning methods have several advantages and disadvantages over traditional learning methods. The lazy aspect of instance-based learning is its greatest advantage. The global pre-processing approach of eager learning algorithms is inherently myopic to the characteristics of specific test instances, and may create a model, which is often not optimized towards specific instances. The advantage of instance-based learning methods is that they can be used in order to create models that are optimized to specific test instances. On the other hand, this can come at a cost, since the computational load of performing the classification can be high. As a result, it may often not be possible to create complex models because of the computational requirements. In some cases, this may lead to oversimplification. Clearly, the usefulness of instance-based learning (as in all other class of methods) depends highly upon the data domain, size of the data, data noisiness and dimensionality. These aspects will be covered in some detail in this chapter.

This chapter will provide an overview of the basic framework for instance-based learning, and the many algorithms that are commonly used in this domain. Some of the important methods such as nearest neighbor classification will be discussed in more detail, whereas others will be covered at a much higher level. This chapter is organized as follows. Section 6.2 introduces the basic framework for instance-based learning. The most well-known instance-based method is the nearest neighbor classifier. This is discussed in Section 6.3. Lazy SVM classifiers are discussed in Section 6.4. Locally weighted methods for regression are discussed in section 6.5. Locally weighted naive Bayes methods are introduced in Section 6.6. Methods for constructing lazy decision trees are discussed in Section 6.7. Lazy rule-based classifiers are discussed in Section 6.8. Methods for using neural networks in the form of radial basis functions are discussed in Section 6.9. The advantages of lazy learning for diagnostic classification are discussed in Section 6.10. The conclusions and summary are discussed in Section 6.11.

6.2 Instance-Based Learning Framework

The earliest instance-based algorithms were synonymous with nearest neighbor pattern classification [31, 33], though the field has now progressed well beyond the use of such algorithms. These algorithms were often criticized for a number of shortcomings, especially when the data is high dimensional, and distance-function design is too challenging [45]. In particular, they were seen to be computationally expensive, intolerant of attribute noise, and sensitive to the choice of distance function [24]. Many of these shortcomings have subsequently been addressed, and these will be discussed in detail in this chapter.

The principle of instance-based methods was often understood in the earliest literature as follows:

"... *similar instances have similar classification.*" (Page 41, [11])

However, a broader and more powerful principle to characterize such methods would be:

Similar instances are easier to model with a learning algorithm, because of the simplification of the class distribution within the locality of a test instance.

Note that the latter principle is a bit more general than the former, in that the former principle seems to advocate the use of a nearest neighbor classifier, whereas the latter principle seems to suggest that *locally optimized* models to the test instance are usually more effective. Thus, according to the latter philosophy, a vanilla nearest neighbor approach may not always obtain the most accurate results, but a locally optimized regression classifier, Bayes method, SVM or decision tree may sometimes obtain better results because of the simplified modeling process [18, 28, 38, 77, 79]. This class of methods is often referred to as *lazy learning*, and often treated differently from traditional instance-based learning methods, which correspond to nearest neighbor classifiers. Nevertheless, the two classes of methods are closely related enough to merit a unified treatment. Therefore, this chapter will study both the traditional instance-based learning methods and lazy learning methods within a single generalized umbrella of instance-based learning methods.

The primary output of an instance-based algorithm is a concept description. As in the case of a classification model, this is a function that maps instances to category values. However, unlike traditional classifiers, which use extensional concept descriptions, instance-based concept descriptions may typically contain a set of stored instances, and optionally some information about how the stored instances may have performed in the past during classification. The set of stored instances can change as more instances are classified over time. This, however, is dependent upon the underlying classification scenario being temporal in nature. There are three primary components in all instance-based learning algorithms.

1. *Similarity or Distance Function:* This computes the similarities between the training instances, or between the test instance and the training instances. This is used to identify a locality around the test instance.

2. *Classification Function:* This yields a classification for a particular test instance with the use of the locality identified with the use of the distance function. In the earliest descriptions of instance-based learning, a nearest neighbor classifier was assumed, though this was later expanded to the use of any kind of locally optimized model.

3. *Concept Description Updater:* This typically tracks the classification performance, and makes decisions on the choice of instances to include in the concept description.

Traditional classification algorithms construct explicit abstractions and generalizations (e.g., decision trees or rules), which are constructed *in an eager way in a pre-processing phase*, and are independent of the choice of the test instance. These models are then used in order to classify test instances. This is different from instance-based learning algorithms, where instances are used along with the training data to construct the concept descriptions. Thus, the approach is *lazy* in the sense that knowledge of the test instance is required before model construction. Clearly the tradeoffs are different in the sense that "eager" algorithms avoid too much work at classification time, but are myopic in their ability to create a specific model for a test instance in the most accurate way. Instance-based algorithms face many challenges involving efficiency, attribute noise, and significant storage requirements. A work that analyzes the last aspect of storage requirements is discussed in [72].

While nearest neighbor methods are almost always used as an intermediate step for identifying data locality, a variety of techniques have been explored in the literature beyond a majority vote on the identified locality. Traditional modeling techniques such as decision trees, regression modeling, Bayes, or rule-based methods are commonly used to create an optimized classification model around the test instance. *It is the optimization inherent in this localization that provides the greatest advantages of instance-based learning.* In some cases, these methods are also combined with some level of global pre-processing so as to create a combination of instance-based and model-based algorithms [55]. In any case, many instance-based methods combine typical classification generalizations such as regression-based methods [15], SVMs [54, 77], rule-based methods [36], or decision trees [40] with instance-based methods. Even in the case of pure distance-based methods, some amount of model building may be required at an early phase for learning the underlying distance functions [75]. This chapter will also discuss such techniques within the broader category of instance-based methods.

6.3 The Nearest Neighbor Classifier

The most commonly used instance-based classification method is the nearest neighbor method. In this method, the nearest k instances to the test instance are determined. Then, a simple model is constructed on this set of k nearest neighbors in order to determine the class label. For example, the majority class among the k nearest neighbors may be reported as the relevant labels. For the purpose of this paper, we always use a binary classification (two label) assumption, in which case the use of the majority class is relevant. However, the method can be easily extended to the multi-class scenario very easily by using the class with the largest presence, rather than the majority class. Since the different attributes may be defined along different scales (e.g., age versus salary), a common approach is to scale the attributes either by their respective standard deviations or the observed range of that attribute. The former is generally a more sound approach from a statistical point of view. It has been shown in [31] that the nearest neighbor rule provides at most twice the error as that provided by the local Bayes probability.

Such an approach may sometimes not be appropriate for imbalanced data sets, in which the rare class may not be present to a significant degree among the nearest neighbors, even when the test instance belongs to the rare class. In the case of cost-sensitive classification or rare-class learning the majority class is determined after weighting the instances with the relevant costs. These methods will be discussed in detail in Chapter 17 on rare class learning. In cases where the class label is continuous (regression modeling problem), one may use the weighted average numeric values of the target class. Numerous variations on this broad approach are possible, both in terms of the distance function used or the local model used for the classification process.

- The choice of the distance function clearly affects the behavior of the underlying classifier. In fact, the problem of distance function learning [75] is closely related to that of instance-based learning since nearest neighbor classifiers are often used to validate distance-function learning algorithms. For example, for numerical data, the use of the euclidian distance assumes a spherical shape of the clusters created by different classes. On the other hand, the true clusters may be ellipsoidal and arbitrarily oriented with respect to the axis system. Different distance functions may work better in different scenarios. The use of feature-weighting [69] can also change the distance function, since the weighting can change the contour of the distance function to match the patterns in the underlying data more closely.

- The final step of selecting the model from the local test instances may vary with the application. For example, one may use the majority class as the relevant one for classification, a cost-weighted majority vote, or a more complex classifier within the locality such as a Bayes technique [38, 78].

One of the nice characteristics of the nearest neighbor classification approach is that it can be used for practically any data type, as long as a distance function is available to quantify the distances between objects. Distance functions are often designed with a specific focus on the classification task [21]. Distance function design is a widely studied topic in many domains such as time-series data [42], categorical data [22], text data [56], and multimedia data [58] or biological data [14]. Entropy-based measures [29] are more appropriate for domains such as strings, in which the distances are measured in terms of the amount of effort required to transform one instance to the other. Therefore, the simple nearest neighbor approach can be easily adapted to virtually every data domain. This is a clear advantage in terms of usability. A detailed discussion of different aspects of distance function design may be found in [75].

A key issue with the use of nearest neighbor classifiers is the *efficiency* of the approach in the classification process. This is because the retrieval of the k nearest neighbors may require a running time that is linear in the size of the data set. With the increase in typical data sizes over the last few years, this continues to be a significant problem [13]. Therefore, it is useful to create indexes, which can efficiently retrieve the k nearest neighbors of the underlying data. This is generally possible for many data domains, but may not be true of all data domains in general. Therefore, scalability is often a challenge in the use of such algorithms. A common strategy is to use either indexing of the underlying instances [57], sampling of the data, or aggregations of some of the data points into smaller clustered pseudo-points in order to improve accuracy. While the indexing strategy seems to be the most natural, it rarely works well in the high dimensional case, because of the curse of dimensionality. Many data sets are also very high dimensional, in which case a nearest neighbor index fails to prune out a significant fraction of the data points, and may in fact do worse than a sequential scan, because of the additional overhead of indexing computations.

Such issues are particularly challenging in the streaming scenario. A common strategy is to use very fine grained clustering [5, 7] in order to replace multiple local instances within a small cluster (belonging to the same class) with a pseudo-point of that class. Typically, this pseudo-point is the centroid of a small cluster. Then, it is possible to apply a nearest neighbor method on these pseudo-points in order to obtain the results more efficiently. Such a method is desirable, when scalability is of great concern and the data has very high volume. Such a method may also reduce the noise that is associated with the use of individual instances for classification. An example of such an approach is provided in [7], where classification is performed on a fast data stream, by summarizing the stream into micro-clusters. Each micro-cluster is constrained to contain data points only belonging to a particular class. The class label of the closest micro-cluster to a particular instance is reported as the relevant label. Typically, the clustering is performed with respect to different time-horizons, and a cross-validation approach is used in order to determine the time-horizon that is most relevant at a given time. Thus, the model is instance-based in a dual sense, since it is not only a nearest neighbor classifier, but it also determines the relevant time horizon in a lazy way, which is specific to the time-stamp of the instance. Picking a smaller time horizon for selecting the training data may often be desirable when the data evolves significantly over time. The streaming scenario also benefits from laziness in the temporal dimension, since the most appropriate model to use for the same test instance may vary with time, as the data evolves. It has been shown in [7], that such an "on demand" approach to modeling provides more effective results than eager classifiers, because of its ability to optimize for the test instance from a temporal perspective. Another method that is based on the nearest neighbor approach is proposed in [17]. This approach detects the changes in the distribution of the data stream on the past window of instances and accordingly re-adjusts the classifier. The approach can handle symbolic attributes, and it uses the Value Distance Metric (VDM) [60] in order to measure distances. This metric will be discussed in some detail in Section 6.3.1 on symbolic attributes.

A second approach that is commonly used to speed up the approach is the concept of *instance selection* or *prototype selection* [27, 41, 72, 73, 81]. In these methods, a subset of instances may be pre-selected from the data, and the model is constructed with the use of these pre-selected instances. It has been shown that a good choice of pre-selected instances can often lead to *improvement* in accuracy, in addition to the better efficiency [72, 81]. This is because a careful pre-selection of instances reduces the noise from the underlying training data, and therefore results in better classification. The pre-selection issue is an important research issue in its own right, and we refer the reader to [41] for a detailed discussion of this important aspect of instance-based classification. An empirical comparison of the different instance selection algorithms may be found in [47].

In many rare class or cost-sensitive applications, the instances may need to be weighted differently corresponding to their importance. For example, consider an application in which it is desirable to use medical data in order to diagnose a specific condition. The vast majority of results may be normal, and yet it may be costly to miss a case where an example is abnormal. Furthermore, a nearest neighbor classifier (which does not weight instances) will be naturally biased towards identifying instances as normal, especially when they lie at the border of the decision region. In such cases, costs are associated with instances, where the cost associated with an abnormal instance is the same as the relative cost of misclassifying it (false negative), as compared to the cost of misclassifying a normal instance (false positive). The weights on the instances are then used for the classification process.

Another issue with the use of nearest neighbor methods is that it does not work very well when the dimensionality of the underlying data increases. This is because the quality of the nearest neighbor decreases with an increasing number of irrelevant attributes [45]. The noise effects associated with the irrelevant attributes can clearly degrade the quality of the nearest neighbors found, especially when the dimensionality of the underlying data is high. This is because the cumulative effect of irrelevant attributes often becomes more pronounced with increasing dimensionality. For the case of numeric attributes, it has been shown [2], that the use of fractional norms (i.e. L_p-norms for $p < 1$) provides superior quality results for nearest neighbor classifiers, whereas L_∞ norms provide the poorest behavior. Greater improvements may be obtained by designing the distance function more carefully, and weighting more relevant ones. This is an issue that will be discussed in detail in later subsections.

In this context, the issue of distance-function design is an important one [50]. In fact, an entire area of machine learning has been focussed on distance function design. Chapter 18 of this book has been devoted entirely to distance function design, and an excellent survey on the topic may be found in [75]. A discussion of the applications of different similarity methods for instance-based classification may be found in [32]. In this section, we will discuss some of the key aspects of instance-function design, which are important in the context of nearest neighbor classification.

6.3.1   Handling Symbolic Attributes

Since most natural distance functions such as the L_p-norms are defined for numeric attributes, a natural question arises as to how the distance function should be computed in data sets in which some attributes are symbolic. While it is always possible to use a distance-function learning approach [75] for an arbitrary data type, a simpler and more efficient solution may sometimes be desirable. A discussion of several unsupervised symbolic distance functions is provided in [22], though it is sometimes desirable to use the class label in order to improve the effectiveness of the distance function.

A simple, but effective supervised approach is to use the value-difference-metric (VDM), which is based on the class-distribution conditional on the attribute values [60]. The intuition here is that similar symbolic attribute values will show similar class distribution behavior, and the distances should be computed on this basis. Thus, this distance function is clearly a supervised one (unlike the euclidian metric), since it explicitly uses the class distributions in the training data.

Let x_1 and x_2 be two possible symbolic values for an attribute, and $P(C_i|x_1)$ and $P(C_i|x_2)$ be the conditional probabilities of class C_i for these values, respectively. These conditional probabilities can be estimated in a data-driven manner. Then, the value different metric $VDM(x_1,x_2)$ is defined as follows:

$$VDM(x_1,x_2) = \sum_{i=1}^{k} (P(C_i|x_1) - P(C_i|x_2))^q \tag{6.1}$$

Here, the parameter q can be chosen either on an ad hoc basis, or in a data-driven manner. This choice of metric has been shown to be quite effective in a variety of instance-centered scenarios [36, 60]. Detailed discussions of different kinds of similarity functions for symbolic attributes may be found in [22, 30].

6.3.2   Distance-Weighted Nearest Neighbor Methods

The simplest form of the nearest neighbor method is when the the majority label among the k-nearest neighbor distances is used. In the case of distance-weighted neighbors, it is assumed that all nearest neighbors are not equally important for classification. Rather, an instance i, whose distance d_i to the test instance is smaller, is more important. Then, if c_i is the label for instance i, then the number of votes $V(j)$ for class label j from the k-nearest neighbor set S_k is as follows:

$$V(j) = \sum_{i:i\in S_k, c_i=j} f(d_i) \tag{6.2}$$

Here $f(\cdot)$ is either an increasing or decreasing function of its argument, depending upon when d_i represents similarity or distance, respectively. It should be pointed out that if the appropriate weight is used, then it is not necessary to use the k nearest neighbors, but simply to perform this average over the entire collection.

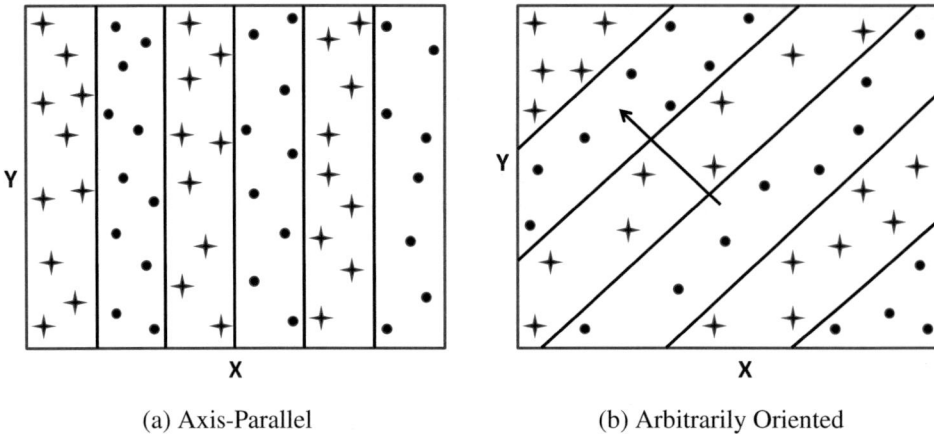

| (a) Axis-Parallel | (b) Arbitrarily Oriented |

FIGURE 6.1: Illustration of importance of feature weighting for nearest neighbor classification.

6.3.3 Local Distance Scaling

A different way to improve the quality of the k-nearest neighbors is by using a technique, that is referred to as *local distance scaling* [65]. In local distance scaling, the distance of the test instance to each training example $\overline{X_i}$ is scaled by a weight r_i, which is specific to the training example $\overline{X_i}$. The weight r_i for the training example $\overline{X_i}$ is the largest distance from $\overline{X_i}$ such that it does not contain a training example from a class that is different from $\overline{X_i}$. Then, the new scaled distance $d(\hat{\overline{X}}, \overline{X_i})$ between the test example $\overline{X}$ and the training example $\overline{X_i}$ is given by the following scaled value:

$$d(\hat{\overline{X}}, \overline{X_i}) = d(\overline{X}, \overline{X_i})/r_i \qquad (6.3)$$

The nearest neighbors are computed on this set of distances, and the majority vote among the k nearest neighbors is reported as the class labels. This approach tends to work well because it picks the k nearest neighbor group in a noise-resistant way. For test instances that lie on the decision boundaries, it tends to discount the instances that lie on the noisy parts of a decision boundary, and instead picks the k nearest neighbors that lie away from the noisy boundary. For example, consider a data point $\overline{X_i}$ that lies reasonably close to the test instance, but even closer to the decision boundary. Furthermore, since $\overline{X_i}$ lies almost on the decision boundary, it lies extremely close to another training example in a different class. In such a case the example $\overline{X_i}$ should not be included in the k-nearest neighbors, because it is not very informative. The small value of r_i will often ensure that such an example is not picked among the k-nearest neighbors. Furthermore, such an approach also ensures that the distances are scaled and normalized by the varying nature of the patterns of the different classes in different regions. Because of these factors, it has been shown in [65] that this modification often yields more robust results for the quality of classification.

6.3.4 Attribute-Weighted Nearest Neighbor Methods

Attribute-weighting is the simplest method for modifying the distance function in nearest neighbor classification. This is closely related to the use of the Mahalanobis distance, in which arbitrary *directions* in the data may be weighted differently, as opposed to the actual attributes. The Mahalanobis distance is equivalent to the Euclidian distance computed on a space in which the different directions along an arbitrarily oriented axis system are "stretched" differently, according to the co-variance matrix of the data set. Attribute weighting can be considered a simpler approach, in which the directions of stretching are parallel to the original axis system. By picking a particular weight to be zero, that attribute is eliminated completely. This can be considered an implicit form of feature selection. Thus, for two d-dimensional records $\overline{X} = (x_1 \ldots x_d)$ and $\overline{Y} = (y_1 \ldots y_d)$, the feature weighted distance $d(\overline{X}, \overline{Y}, \overline{W})$ with respect to a d-dimensional vector of weights $\overline{W} = (w_1 \ldots w_d)$ is defined as follows:

$$d(\overline{X}, \overline{Y}, \overline{W}) = \sqrt{\sum_{i=1}^{d} w_i \cdot (x_i - y_i)^2}. \tag{6.4}$$

For example, consider the data distribution, illustrated in Figure 6.1(a). In this case, it is evident that the feature X is much more discriminative than feature Y, and should therefore be weighted to a greater degree in the final classification. In this case, almost perfect classification can be obtained with the use of feature X, though this will usually not be the case, since the decision boundary is noisy in nature. It should be pointed out that the Euclidian metric has a spherical decision boundary, whereas the decision boundary in this case is linear. This results in a bias in the classification process because of the difference between the shape of the model boundary and the true boundaries in the data. The importance of weighting the features becomes significantly greater, when the classes are not cleanly separated by a decision boundary. In such cases, the natural noise at the decision boundary may combine with the significant bias introduced by an unweighted Euclidian distance, and result in even more inaccurate classification. By weighting the features in the Euclidian distance, it is possible to elongate the model boundaries to a shape that aligns more closely with the class-separation boundaries in the data. The simplest possible weight to use would be to normalize each dimension by its standard deviation, though in practice, the class label is used in order to determine the best feature weighting [48]. A detailed discussion of different aspects of feature weighting schemes is provided in [70].

In some cases, the class distribution may not be aligned neatly along the axis system, but may be arbitrarily oriented along different directions in the data, as in Figure 6.1(b). A more general distance metric is defined with respect to a $d \times d$ matrix A rather than a vector of weights $\overline{W}$.

$$d(\overline{X}, \overline{Y}, A) = \sqrt{(\overline{X} - \overline{Y})^T \cdot A \cdot (\overline{X} - \overline{Y})}. \tag{6.5}$$

The matrix A is also sometimes referred to as a metric. The value of A is assumed to be the inverse of the covariance matrix of the data in the standard definition of the Mahalanobis distance for *unsupervised* applications. Generally, the Mahalanobis distance is more sensitive to the global data distribution and provides more effective results.

The Mahalanobis distance does not, however take the class distribution into account. In supervised applications, it makes much more sense to pick A based on the *class* distribution of the underlying data. The core idea is to "elongate" the neighborhoods along less discriminative directions, and to shrink the neighborhoods along more discriminative dimensions. Thus, in the modified metric, a small (unweighted) step along a discriminative direction, would result in relatively greater distance. This naturally provides greater importance to more discriminative directions. Numerous methods such as the linear discriminant [51] can be used in order to determine the most discriminative dimensions in the underlying data. However, the key here is to use a *soft* weighting of the different directions, rather than selecting specific dimensions in a hard way. The goal of the matrix A is to accomplish this. How can A be determined by using the distribution of the classes? Clearly, the matrix A should somehow depend on the within-class variance and between-class variance, in the context of linear discriminant analysis. The matrix A defines the shape of the neighborhood within a threshold distance, to a given test instance. The neighborhood directions with low ratio of inter-class variance to intra-class variance should be elongated, whereas the directions with high ratio of the inter-class to intra-class variance should be shrunk. Note that the "elongation" of a neighborhood direction is achieved by scaling that component of the distance by a larger factor, and therefore de-emphasizing that direction.

Let $\mathcal{D}$ be the full database, and $\mathcal{D}_i$ be the portion of the data set belonging to class i. Let $\overline{\mu}$ represent the mean of the entire data set. Let $p_i = |\mathcal{D}_i|/|\mathcal{D}|$ be the fraction of records belonging to class i, $\overline{\mu_i}$ be the d-dimensional row vector of means of $\mathcal{D}_i$, and Σ_i be the $d \times d$ covariance matrix of $\mathcal{D}_i$. Then, the scaled[1] within class scatter matrix S_w is defined as follows:

$$S_w = \sum_{i=1}^{k} p_i \cdot \Sigma_i. \tag{6.6}$$

The between-class scatter matrix S_b may be computed as follows:

$$S_b = \sum_{i=1}^{k} p_i (\overline{\mu_i} - \overline{\mu})^T (\overline{\mu_i} - \overline{\mu}). \tag{6.7}$$

Note that the matrix S_b is a $d \times d$ matrix, since it results from the product of a $d \times 1$ matrix with a $1 \times d$ matrix. Then, the matrix A (of Equation 6.5), which provides the desired distortion of the distances on the basis of class-distribution, can be shown to be the following:

$$A = S_w^{-1} \cdot S_b \cdot S_w^{-1}. \tag{6.8}$$

It can be shown that this choice of the metric A provides an excellent discrimination between the different classes, where the elongation in each direction depends inversely on ratio of the between-class variance to within-class variance along the different directions. The aforementioned description is based on the discussion in [44]. The reader may find more details of implementing the approach in an effective way in that work.

A few special cases of the metric of Equation 6.5 are noteworthy. Setting A to the identity matrix corresponds to the use of the Euclidian distance. Setting the non-diagonal entries of A entries to zero results in a similar situation to a d-dimensional vector of weights for individual dimensions. Therefore, the non-diagonal entries contribute to a rotation of the axis-system before the stretching process. For example, in Figure 6.1(b), the optimal choice of the matrix A will result in greater importance being shown to the direction illustrated by the arrow in the figure in the resulting metric. In order to avoid ill-conditioned matrices, especially in the case when the number of training data points is small, a parameter ε can be used in order to perform smoothing.

$$A = S_w^{-1/2} \cdot (S_w^{-1/2} \cdot S_b \cdot S_w^{-1/2} + \varepsilon \cdot I) \cdot S_w^{-1/2}. \tag{6.9}$$

Here ε is a small parameter that can be tuned, and the identity matrix is represented by I. The use of this modification assumes that any particular direction does not get infinite weight. This is quite possible, when the number of data points is small. The use of this parameter ε is analogous to *Laplacian smoothing* methods, and is designed to avoid overfitting.

Other heuristic methodologies are also used in the literature for learning feature relevance. One common methodology is to use cross-validation in which the weights are trained using the original instances in order to minimize the error rate. Details of such a methodology are provided in [10,48]. It is possible to also use different kinds of search methods such as Tabu search [61] in order to improve the process of learning weights. This kind of approach has also been used in the context of text classification, by learning the relative importance of the different words, for computing the similarity function [43]. Feature relevance has also been shown to be important for other domains such as image retrieval [53]. In cases where domain knowledge can be used, some features can be eliminated very easily with tremendous performance gains. The importance of using domain knowledge for feature weighting has been discussed in [26].

---

[1]The unscaled version may be obtained by multiplying S_w with the number of data points. There is no difference from the final result, whether the scaled or unscaled version is used, within a constant of proportionality.

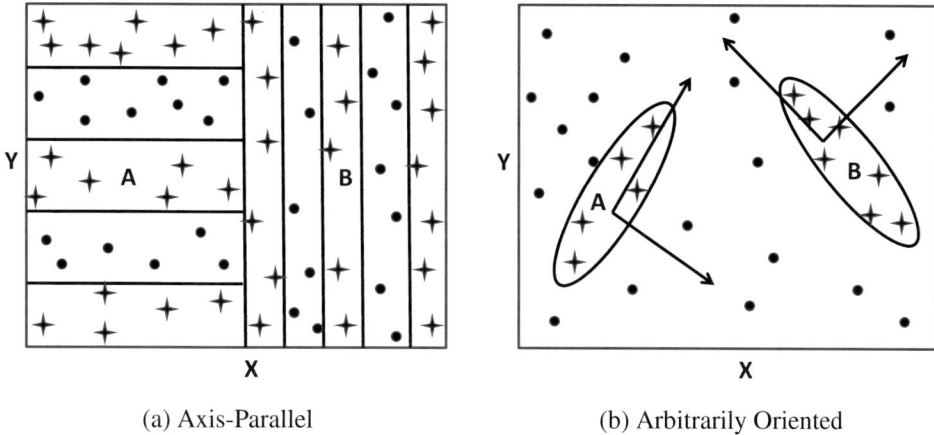

(a) Axis-Parallel   (b) Arbitrarily Oriented

FIGURE 6.2: Illustration of importance of local adaptivity for nearest neighbor classification.

The most effective distance function design may be performed at query-time, by using a weighting that is *specific* to the test instance [34, 44]. This can be done by learning the weights only on the basis of the instances that are near a specific test instance. While such an approach is more expensive than global weighting, it is likely to be more effective because of local optimization specific to the test instance, and is better aligned with the spirit and advantages of lazy learning. This approach will be discussed in detail in later sections of this survey.

It should be pointed out that many of these algorithms can be considered rudimentary forms of distance function learning. The problem of distance function learning [3, 21, 75] is fundamental to the design of a wide variety of data mining algorithms including nearest neighbor classification [68], and a significant amount of research has been performed in the literature in the context of the classification task. A detailed survey on this topic may be found in [75]. The nearest neighbor classifier is often used as the prototypical task in order to evaluate the quality of distance functions [2, 3, 21, 45, 68, 75], which are constructed using distance function learning techniques.

6.3.5 Locally Adaptive Nearest Neighbor Classifier

Most nearest neighbor methods are designed with the use of a *global* distance function such as the Euclidian distance, the Mahalanobis distance, or a discriminant-based weighted distance, in which a particular set of weights is used over the entire data set. In practice, however, the importance of different features (or data directions) is not local, but global. This is especially true for high dimensional data, where the relative importance, relevance, and noise in the different dimensions may vary significantly with data locality. Even in the unsupervised scenarios, where no labels are available, it has been shown [45] that the relative importance of different dimensions for finding nearest neighbors may be very different for different localities. In the work described in [45], the relative importance of the different dimensions is determined with the use of a contrast measure, which is independent of class labels. Furthermore, the weights of different dimensions vary with the data locality in which they are measured, and these weights can be determined with the use of genetic algorithm discussed in [45]. Even for the case of such an unsupervised approach, it has been shown that the effectiveness of a nearest neighbor classifier improves significantly. The reason for this behavior is that different features are noisy in different data localities, and the local feature selection process sharpens the quality of the nearest neighbors.

Since local distance functions are more effective in unsupervised scenarios, it is natural to explore whether this is also true of supervised scenarios, where even greater information in available in the form of different labels in order to make inferences about the relevance of the different dimensions. A recent method for distance function learning [76] also constructs instance-specific distances with supervision, and shows that the use of locality provides superior results. However, the supervision in this case is not specifically focussed on the traditional classification problem, since it is defined in terms of similarity or dissimilarity constraints between instances, rather than labels attached to instances. Nevertheless, such an approach can also be used in order to construct instance-specific distances for classification, by transforming the class labels into similarity or dissimilarity constraints. This general principle is used frequently in many works such as those discussed by [34, 39, 44] for locally adaptive nearest neighbor classification.

Labels provide a rich level of information about the relative importance of different dimensions in different localities. For example, consider the illustration of Figure 6.2(a). In this case, the feature Y is more important in locality A of the data, whereas feature X is more important in locality B of the data. Correspondingly, the feature Y should be weighted more, when using test instances in locality A of the data, whereas feature X should be weighted more in locality B of the data. In the case of Figure 6.2(b), we have shown a similar scenario as in Figure 6.2(a), except that different *directions* in the data should be considered more or less important. Thus, the case in Figure 6.2(b) is simply an arbitrarily oriented generalization of the challenges faced in Figure 6.2(a).

One of the earliest methods for performing locally adaptive nearest neighbor classification was proposed in [44], in which the matrix A (of Equation 6.5) and the neighborhood of the test data point $\overline{X}$ are learned iteratively from one another in a local way. The major difference here is that the matrix A will depend upon the choice of test instance X, since the metric is designed to be locally adaptive. The algorithm starts off by setting the $d \times d$ matrix A to the identity matrix, and then determines the k-nearest neighbors N_k, using the generalized distance described by Equation 6.5. Then, the value of A is updated using Equation 6.8 *only on the neighborhood points N_k found in the last iteration*. This procedure is repeated till convergence. Thus, the overall iterative procedure may be summarized as follows:

1. Determine N_k as the set of the k nearest neighbors of the test instance, using Equation 6.5 in conjunction with the current value of A.

2. Update A using Equation 6.8 on the between-class and within-class scatter matrices of N_k.

At completion, the matrix A is used in Equation 6.8 for k-nearest neighbor classification. In practice, Equation 6.9 is used in order to avoid giving infinite weight to any particular direction. It should be pointed out that the approach is [44] is quite similar in principle to an unpublished approach proposed previously in [39], though it varies somewhat on the specific details of the implementation, and is also somewhat more robust. Subsequently, a similar approach to [44] was proposed in [34], which works well with limited data. It should be pointed out that linear discriminant analysis is not the only method that may be used for "deforming" the shape of the metric in an instance-specific way to conform better to the decision boundary. Any other model such as an SVM or neural network that can find important directions of discrimination in the data may be used in order to determine the most relevant metric. For example, an interesting work discussed in [63] determines a small number of k-nearest neighbors for each class, in order to span a linear subspace for that class. The classification is done based not on distances to prototypes, but the distances to subspaces. The intuition is that the linear subspaces essentially generate pseudo training examples for that class. A number of methods that use support-vector machines will be discussed in the section on Support Vector Machines. A review and discussion of different kinds of feature selection and feature weighting schemes are provided in [8]. It has also been suggested in [48] that feature weighting may be superior to feature selection, in cases where the different features have different levels of relevance.

6.3.6 Combining with Ensemble Methods

It is evident from the aforementioned discussion that lazy learning methods have the advantage of being able to optimize more effectively to the locality of a specific instance. In fact, one issue that is encountered sometimes in lazy learning methods is that the quest for local optimization can result in overfitting, especially when some of the features are noisy within a specific locality. In many cases, it may be possible to create *multiple models* that are more effective for different instances, but it may be hard to know which model is more effective without causing overfitting. An effective method to achieve the goal of simultaneous local optimization and robustness is to use ensemble methods that combine the results from multiple models for the classification process. Some recent methods [4,16,80] have been designed for using ensemble methods in order to improve the effectiveness of lazy learning methods.

An approach discussed in [16] samples random subsets of features, which are used for nearest neighbor classification. These different classifications are then combined together in order to yield a more robust classification. One disadvantage of this approach is that it can be rather slow, especially for larger data sets. This is because the nearest neighbor classification approach is inherently computationally intensive because of the large number of distance computations *for every instance-based classification*. This is in fact one of the major issues for all lazy learning methods. A recent approach *LOCUST* discussed in [4] proposes a more efficient technique for ensemble lazy learning, which can also be applied to the streaming scenario. The approach discussed in [4] builds inverted indices on discretized attributes of each instance in real time, and then uses efficient intersection of randomly selected inverted lists in order to perform real time classification. Furthermore, a limited level of feature bias is used in order to reduce the number of ensemble components for effective classification. It has been shown in [4] that this approach can be used for resource-sensitive lazy learning, where the number of feature samples can be tailored to the resources available for classification. Thus, this approach can also be considered an *anytime* lazy learning algorithm, which is particularly suited to the streaming scenario. This is useful in scenarios where one cannot control the input rate of the data stream, and one needs to continuously adjust the processing rate, in order to account for the changing input rates of the data stream.

6.3.7 Multi-Label Learning

In traditional classification problems, each class is associated with exactly one label. This is not the case for multi-label learning, in which each class may be associated with more than one class. The number of classes with which an instance is associated is unknown a-priori. This kind of scenario is quite common in many scenarios such as document classification, where a single document may belong to one of several possible categories.

The unknown number of classes associated with a test instance creates a challenge, because one now needs to determine not just the most relevant classes, but also the number of relevant ones for a given test instance. One way of solving the problem is to decompose it into multiple independent binary classification problems. However, such an approach does not account for the correlations among the labels of the data instances.

An approach known as ML-KNN is proposed in [78], where the Bayes method is used in order to estimate the probability that a label belongs to a particular class. The broad approach used in this process is as follows. For each test instance, its k nearest neighbors are identified. The statistical information in the label sets of these neighborhood instances is then used to determine the label set. The maximum a-posteriori principle is applied to determine the label set of the test instance.

Let $\mathcal{Y}$ denote the set of labels for the multi-label classification problem, and let n be the total number of labels. For a given test instance T, the first step is to compute its k nearest neighbors. Once the k nearest neighbors have been computed, the number of occurrences $\overline{C} = (C(1)\dots C(n))$ of each of the n different labels is computed. Let $E^1(T,i)$ be the event that the test instance T contains the label i, and $E^0(T,i)$ be the event that the test instance T does not contain the label i. Then, in order to determine whether or not the label i is included in test instance T, the maximum posterior principle is used:

$$b = \text{argmax}_{b \in \{0,1\}} \{ P(E^b(T,i)|\overline{C}) \}. \tag{6.10}$$

In other words, we wish to maximize between the probability of the events of label i being included or not. Therefore, the Bayes rule can be used in order to obtain the following:

$$b = \text{argmax}_{b \in \{0,1\}} \left\{ \frac{P(E^b(T,i)) \cdot P(\overline{C}|E^b(T,i))}{P(\overline{C})} \right\}. \tag{6.11}$$

Since the value of $P(\overline{C})$ is independent of b, it is possible to remove it from the denominator, without affecting the maximum argument. This is a standard approach used in all Bayes methods. Therefore, the best matching label may be expressed as follows:

$$b = \text{argmax}_{b \in \{0,1\}} \left\{ P(E^b(T,i)) \cdot P(\overline{C}|E^b(T,i)) \right\}. \tag{6.12}$$

The prior probability $P(E^b(T,i))$ can be estimated as the fraction of the labels belonging to a particular class. The value of $P(\overline{C}|E^b(T,i))$ can be estimated by using the naive Bayes rule.

$$P(\overline{C}|E^b(T,i)) = \prod_{j=1}^{n} P(C(j)|E^b(T,i)). \tag{6.13}$$

Each of the terms $P(C(j)|E^b(T,i))$ can be estimated in a data driven manner by examining among the instances satisfying the value b for class label i, the fraction that contains exactly the count $C(j)$ for the label j. Laplacian smoothing is also performed in order to avoid ill conditioned probabilities. Thus, the correlation between the labels is accounted for by the use of this approach, since each of the terms $P(C(j)|E^b(T,i))$ indirectly measures the correlation between the labels i and j.

This approach is often popularly understood in the literature as a nearest neighbor approach, and has therefore been discussed in the section on nearest neighbor methods. However, it is more similar to a local naive Bayes approach (discussed in Section 6.6) rather than a distance-based approach. This is because the statistical frequencies of the neighborhood labels are used for local Bayes modeling. Such an approach can also be used for the standard version of the classification problem (when each instance is associated with exactly one label) by using the statistical behavior of the neighborhood features (rather than label frequencies). This yields a lazy Bayes approach for classification [38]. However, the work in [38] also estimates the Bayes probabilities *locally* only over the neighborhood in a data driven manner. Thus, the approach in [38] sharpens the use of locality even further for classification. This is of course a tradeoff, depending upon the amount of training data available. If more training data is available, then local sharpening is likely to be effective. On the other hand, if less training data is available, then local sharpening is not advisable, because it will lead to difficulties in robust estimations of conditional probabilities from a small amount of data. This approach will be discussed in some detail in Section 6.6.

If desired, it is possible to combine the two methods discussed in [38] and [78] for multi-label learning in order to learn the information in both the features and labels. This can be done by using both feature *and* label frequencies for the modeling process, and the product of the label-based and feature-based Bayes probabilities may be used for classification. The extension to that case is straightforward, since it requires the multiplication of Bayes probabilities derived form two different methods. An experimental study of several variations of nearest neighbor algorithms for classification in the multi-label scenario is provided in [59].

6.4 Lazy SVM Classification

In an earlier section, it was shown that linear discriminant analysis can be very useful in adapting the behavior of the nearest neighbor metric for more effective classification. It should be pointed out that SVM classifiers are also linear models that provide directions discriminating between the different classes. Therefore, it is natural to explore the connections between SVM and nearest neighbor classification.

The work in [35] proposes methods for computing the relevance of each feature with the use of an SVM approach. The idea is to use a support vector machine in order to compute the gradient vector (local to test instance) of the rate change in class label in each direction, which is parallel to the axis. It has been shown in [35] that the support vector machine provides a natural way to estimate this gradient. The component along any direction provides the relevance R_j for that feature j. The idea is that when the class labels change at a significantly higher rate along a given direction j, then that feature should be weighted more in the distance metric. The actual weight w_j for the feature j is assumed to be proportional to $e^{\lambda \cdot R_j}$, where λ is a constant that determines the level of importance to be given to the different relevance values. A similar method for using SVM to modify the distance metric is proposed in [54], and the connections of the approach to linear discriminant analysis have also been shown in this work.

Note that the approaches in [35, 54] use a nearest neighbor classifier, and the SVM is only used to learn the distance function more accurately by setting the relevance weights. Therefore, the approach is still quite similar to that discussed in the last section. A completely different approach would be to use the nearest neighbors in order to isolate the locality around the test instance, and then *build an SVM that is optimized to that data locality*. For example, in the case of Figures 6.3(a) and (b) the data in the two classes is distributed in such a way that a global SVM cannot separate the classes very well. However, in the locality around the test instances A and B, it is easy to create a linear SVM that separates the two classes well. Such an approach has been proposed in [23, 77]. The main difference between the work in [23] and [77] is that the former is designed for the L_2-distances, whereas the latter can work for arbitrary and complex distance functions. Furthermore, the latter also proposes a number of optimizations for the purpose of efficiency. Therefore, we will describe the work in [77], since it is a bit more general. The specific steps that are used in this approach are as follows:

1. Determine the k-nearest neighbors of the test instance.

2. If all the neighbors have the same label, then that label is reported, otherwise all pairwise distances between the k-nearest neighbors are computed.

3. The distance matrix is converted into a kernel matrix using the kernel trick and a local SVM classifier is constructed on this matrix.

4. The test instance is classified with this local SVM classifier.

A number of optimizations such as caching have been proposed in [77] in order to improve the efficiency of the approach. Local SVM classifiers have been used quite successfully for a variety of applications such as spam filtering [20].

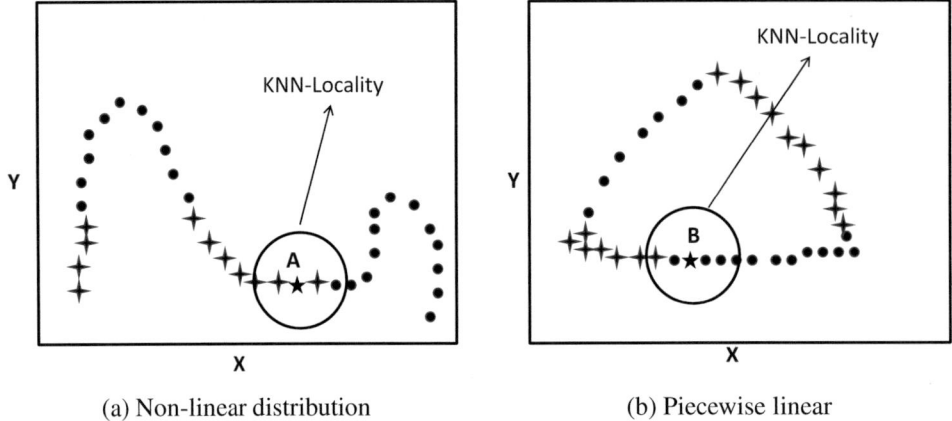

(a) Non-linear distribution   (b) Piecewise linear

FIGURE 6.3: Advantages of instance-centered local linear modeling.

6.5 Locally Weighted Regression

Regression models are linear classifiers, which use a regression model in order to relate the attribute values to the class label. The key idea in locally weighted regression is that global regression planes are not well optimized to the local distributions in the data, as is evident from the illustrations in Figure 6.2. Correspondingly, it is desirable to create a test-instance specific local regression model. Such a method has been discussed in [15, 28]. The latter work [28] is also applicable to the multi-label case.

The core idea in locally weighted regression is to create a regression model that relates the feature values to the class label, as in all traditional regression models. The only difference is that the k-nearest neighbors are used for constructing the regression model. Note that this approach differs from the local discriminant-based models (which are also linear models), in that the votes among the k nearest neighbors are not used for classification purposes. Rather, a linear regression model is constructed in a lazy way, which is specific to the test instance. Several choices may arise in the construction of the model:

- The nature of the fit may be different for different kinds of data sets. For example, linear models may be more appropriate for some kinds of data, whereas quadratic models may be more appropriate for others.

- The choice of the error function may be different for different kinds of data. Two common error functions that are used are the least-squares error and the distance-weighted least squares error.

The primary advantage of locally weighted regression is when the data shows non-linearity in the underlying class distribution on a global scale, but can be approximated by linear models at the local scale. For example, in Figures 6.3(a) and (b) there are no linear functions that can approximate the discrimination between the two classes well, but it is easy to find local linear models that discriminate well within the locality of the test instances A and B, respectively. Since instance-based learning provides knowledge about the appropriate locality (by virtue of the laziness in the model construction process), it is possible to build a regression model that is well optimized to the corresponding data locality. An interesting local linear model with the use of the recursive least squares algorithm is proposed in [18]. In this approach, the number of nearest neighbors is also selected in a data-driven manner.

6.6 Lazy Naive Bayes

Lazy learning has also been extended to the case of the naive Bayes classifiers [38,79]. The work in [38] proposes a locally weighted naive Bayes classifier. The naive Bayes classifier is used in a similar way as the local SVM method [77], or the locally weighted regression method [15] discussed in previous sections. A local Bayes model is constructed with the use of a subset of the data in the neighborhood of the test instance. Furthermore, the training instances in this neighborhood are weighted, so that training instances that are closer to the test instance are weighted more in the model. This local model is then used for classification. The subset of data in the neighborhood of the test instance are determined by using the k-nearest neighbors of the test instance. In practice, the subset of k neighbors is selected by setting the weight of any instance beyond the k-th nearest neighbor to 0. Let D represent the distance of the test instance to the kth nearest neighbor, and d_i represent the distance of the test instance to the ith training instance. Then, the weight w_i of the ith instance is set to the following:

$$w_i = f(d_i/D). \tag{6.14}$$

The function $f(\cdot)$ is a monotonically non-increasing function, which is defined as follows:

$$f(y) = \max\{1 - y, 0\}. \tag{6.15}$$

Therefore, the weight decreases linearly with the distance, but cannot decrease beyond 0, once the k-nearest neighbor has been reached. The naive Bayes method is applied in a standard way, *except that the instance weights are used in estimating all the probabilities for the Bayes classifier*. Higher values of k will result in models that do not fluctuate much with variations in the data, whereas very small values of k will result in models that fit the noise in the data. It has been shown in [38] that the approach is not too sensitive to the choice of k within a reasonably modest range of values of k. Other schemes have been proposed in the literature, which use the advantages of lazy learning in conjunction with the naive Bayes classifier. The work in [79] fuses a standard rule-based learner with naive Bayes models. A technique discussed in [74] lazily learns multiple Naive Bayes classifiers, and uses the classifier with the highest estimated accuracy in order to decide the label for the test instance.

The work discussed in [78] can also be viewed as an unweighted local Bayes classifier, where the weights used for all instances are 1, rather than the weighted approach discussed above. Furthermore, the work in [78] uses the frequencies of the *labels* for Bayes modeling, rather than the actual *features* themselves. The idea in [78] is that the other labels themselves serve as features, since the same instance may contain multiple labels, and sufficient correlation information is available for learning. This approach is discussed in detail in Section 6.3.7.

6.7 Lazy Decision Trees

The work in [40] proposes a method for constructing lazy decision trees, which are specific to a test instance, which is referred to as *LazyDT*. In practice, only a *path* needs to be constructed, because a test instance follows only one path along the decision tree, and other paths are irrelevant from the perspective of lazy learning. As discussed in a later section, such paths can also be interactively explored with the use of visual methods.

One of the major problems with decision trees is that they are *myopic*, since a split at a higher level of the tree may not be optimal for a specific test instance, and in cases where the data contains many relevant features, only a small subset of them may be used for splitting. When a data set contains N data points, a decision tree is allowed only $O(\log(N))$ (approximately balanced) splits, and this may be too small in order to use the best set of features for a particular test instance. Clearly, the knowledge of the test instance allows the use of more relevant features for construction of the appropriate decision path at a higher level of the tree construction.

The *additional knowledge of the test instance* helps in the recursive construction of a path in which only relevant features are used. One method proposed in [40] is to use a split criterion, which successively reduces the size of the training associated with the test instance, until either all instances have the same class label, or the same set of features. In both cases, the majority class label is reported as the relevant class. In order to discard a set of irrelevant instances in a particular iteration, any standard decision tree split criterion is used, and only those training instances, that satisfy the predicate in the same way as the test instance will be selected for the next level of the decision path. The split criterion is decided using any standard decision tree methodology such as the normalized entropy or the gini-index. The main difference from the split process in the traditional decision tree is that *only the node containing the test instance is relevant in the split*, and the information gain or gini-index is computed on the basis of this node. One challenge with the use of such an approach is that the information gain in a single node can actually be negative if the original data is imbalanced. In order to avoid this problem, the training examples are re-weighted, so that the aggregate weight of each class is the same. It is also relatively easy to deal with missing attributes in test instances, since the split only needs to be performed on attributes that are present in the test instance. It has been shown [40] that such an approach yields better classification results, because of the additional knowledge associated with the test instance during the decision path construction process.

A particular observation here is noteworthy, since such decision paths can also be used to construct a robust any-time classifier, with the use of principles associated with a random forest approach [25]. It should be pointed out that a random forest translates to a random path created by a random set of splits from the instance-centered perspective. Therefore a natural way to implement the instance-centered random forest approach would be to discretize the data into ranges. A test instance will be relevant to exactly one range from each attribute. A random set of attributes is selected, and the intersection of the ranges provides one possible classification of the test instance. This approach can be repeated in order to provide a very efficient lazy ensemble, and the number of samples provides the tradeoff between running time and accuracy. Such an approach can be used in the context of an any-time approach in resource-constrained scenarios. It has been shown in [4] how such an approach can be used for efficient any-time classification of data streams.

6.8   Rule-Based Classification

A method for lazy rule-based classification has been proposed in [36], in which a unification has been proposed between instance-based methods and rule-based methods. This system is referred to as *Rule Induction from a Set of Exemplars*, or *RISE* for short. All rules contain at most one condition for each attribute on the left-hand side. In this approach, no distinction is assumed between instances and rules. An instance is simply treated as a rule in which all interval values on the left hand side of the rule are degenerate. In other words, an instance can be treated as a rule, by choosing appropriate conditions for the antecedent, and the class variable of the instance as the consequent. The conditions in the antecedent are determined by using the values in the corresponding instance. For example, if an attribute value for the instance for x_1 is numeric valued at 4, then the condition is assumed to be

$4 \leq x_i \leq 4$. If x_1 is symbolic and its value is a, then the corresponding condition in the antecedent is $x_1 = a$.

As in the case of instance-centered methods, a distance is defined between a test instance and a rule. Let $R = (A_1 \ldots A_m, C)$ be a rule with the m conditions $A_1 \ldots A_m$ in the antecedent, and the class C in the consequent. Let $\overline{X} = (x_1 \ldots x_d)$ be a d-dimensional example. Then, the distance $\Delta(\overline{X}, R)$ between the instance $\overline{X}$ and the rule R is defined as follows.

$$\Delta(\overline{X}, R) = \sum_{j=1}^{m} \delta(i)^s. \tag{6.16}$$

Here s is a real valued parameter such as 1,2,3, etc., and $\delta(i)$ represents the distance on the ith conditional. The value of $\delta(i)$ is equal to the distance of the instance to the nearest end of the range for the case of a numeric attribute and the value difference metric (VDM) of Equation 6.1 for the case of a symbolic attribute. This value of $\delta(i)$ is zero, if the corresponding attribute value is a match for the antecedent condition. The class label for a test instance is defined by the label of the nearest rule to the test instance. If two or more rules have the same accuracy, then the one with the greatest accuracy on the training data is used.

The set of rules in the RISE system are constructed as follows. RISE constructs good rules by using successive generalizations on the original set of instances in the data. Thus, the algorithm starts off with the training set of examples. RISE examines each rule one by one, and finds the nearest example of the same class that is not already covered by the rule. The rule is then generalized in order to cover this example, by expanding the corresponding antecedent condition. For the case of numeric attributes, the ranges of the attributes are increased minimally so as to include the new example, and for the case of symbolic attributes, a corresponding condition on the symbolic attribute is included. If the effect of this generalization on the global accuracy of the rule is non-negative, then the rule is retained. Otherwise, the generalization is not used and the original rule is retained. It should be pointed out that even when generalization does not improve accuracy, it is desirable to retain the more general rule because of the desirable bias towards simplicity of the model. The procedure is repeated until no rule can be generalized in a given iteration. It should be pointed out that some instances may not be generalized at all, and may remain in their original state in the rule set. In the worst case, no instance is generalized, and the resulting model is a nearest neighbor classifier.

A system called *DeEPs* has been proposed in [49], which combines the power of rules and lazy learning for classification purposes. This approach examines how the frequency of an instance's subset of features varies among the training classes. In other words, patterns that sharply differentiate between the different classes for a particular test instance are leveraged and used for classification. Thus, the specificity to the instance plays an important role in this discovery process. Another system, *HARMONY*, has been proposed in [66], which determines rules that are optimized to the different training instances. Strictly speaking, this is not a lazy learning approach, since the rules are optimized to training instances (rather than test instances) in a pre-processing phase. Nevertheless, the effectiveness of the approach relies on the same general principle, and it can also be generalized for lazy learning if required.

6.9 Radial Basis Function Networks: Leveraging Neural Networks for Instance-Based Learning

Radial-basis function networks (RBF) are designed in a similar way to regular nearest neighbor classifiers, except that a set of N *centers* are learned from the training data. In order to classify a test instance, a distance is computed from the test instance to each of these centers $x_1 \ldots x_N$, and a density function is computed at the instance using these centers. The combination of functions computed from each of these centers is computed with the use of a neural network. Radial basis functions can be considered three-layer feed-forward networks, in which each hidden unit computes a function of the form:

$$f_i(x) = e^{-||x-x_i||^2/2\cdot\sigma_i^2}. \tag{6.17}$$

Here σ_i^2 represents the local variance at center x_i. Note that the function $f_i(x)$ has a very similar form to that commonly used in kernel density estimation. For ease in discussion, we assume that this is a binary classification problem, with labels drawn from $\{+1, -1\}$, though this general approach extends much further, even to the extent of regression modeling. The final function is a weighted combination of these values with weights c_i.

$$f^*(x) = \sum_{i=1}^{N} w_i \cdot f_i(x). \tag{6.18}$$

Here $x_1 \ldots x_N$ represent the N different centers, and w_i denotes the weight of center i, which is learned in the neural network. In classical instance-based methods, each data point x_i is an individual training instance, and the weight w_i is set to $+1$ or -1, depending upon its label. However, in RBF methods, the weights are *learned* with a neural network approach, since the centers are derived from the underlying training data, and do not have a label directly attached to them.

The N centers $x_1 \ldots x_N$ are typically constructed with the use of an unsupervised approach [19, 37, 46], though some recent methods also use supervised techniques for constructing the centers [71]. The unsupervised methods [19, 37, 46] typically use a clustering algorithm in order to generate the different centers. A smaller number of centers typically results in smaller complexity, and greater efficiency of the classification process. Radial-basis function networks are related to sigmoidal function networks (SGF), which have one unit for each instance in the training data. In this sense sigmoidal networks are somewhat closer to classical instance-based methods, since they do not have a first phase of cluster-based summarization. While radial-basis function networks are generally more efficient, the points at the different cluster boundaries may often be misclassified. It has been shown in [52] that RBF networks may sometimes require ten times as much training data as SGF in order to achieve the same level of accuracy. Some recent work [71] has shown how supervised methods even at the stage of center determination can significantly improve the accuracy of these classifiers. Radial-basis function networks can therefore be considered an evolution of nearest neighbor classifiers, where more sophisticated (clustering) methods are used for prototype selection (or re-construction in the form of cluster centers), and neural network methods are used for combining the density values obtained from each of the centers.

6.10 Lazy Methods for Diagnostic and Visual Classification

Instance-centered methods are particularly useful for diagnostic and visual classification, since the role of a diagnostic method is to find diagnostic characteristics that *are specific to that test-instance*. When an eager model is fully constructed as a pre-processing step, it is often not optimized to finding the best diagnostic characteristics that are specific to a particular test instance. Therefore, instance-centered methods are a natural approach in such scenarios. The *subspace decision path* method [6] can be considered a lazy and interactive version of the decision tree, which provides insights into the specific characteristics of a particular test instance.

Kernel density estimation is used in order to create a visual profile of the underlying data. It is assumed that the data set $\mathcal{D}$ contains N points and d dimensions. The set of points in $\mathcal{D}$ are denoted

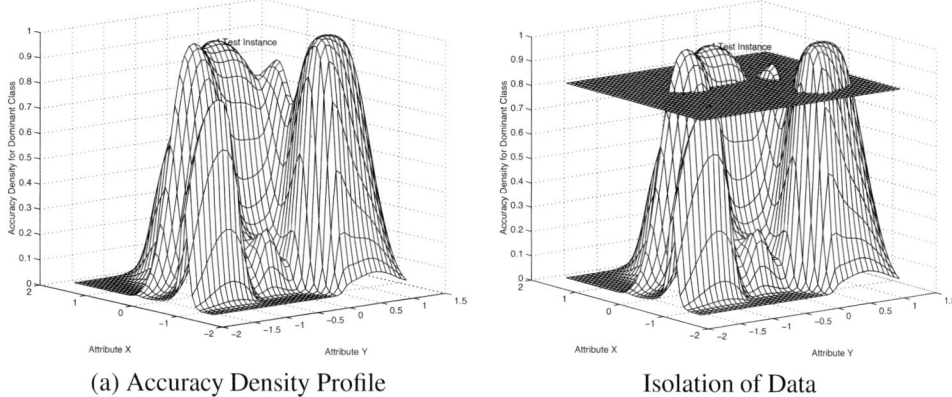

(a) Accuracy Density Profile   Isolation of Data

FIGURE 6.4: Density profiles and isolation of data.

by $X_1 \ldots X_N$. Let us further assume that the k classes in the data are denoted by $C_1 \ldots C_k$. The number of points belonging to the class C_i is n_i, so that $\sum_{i=1}^{k} n_i = N$. The data set associated with the class i is denoted by $\mathcal{D}_i$. This means that $\cup_{i=1}^{k} \mathcal{D}_i = \mathcal{D}$. The probability density at a given point is determined by the sum of the smoothed values of the kernel functions $K_h(\cdot)$ associated with each point in the data set. Thus, the density estimate of the data set $\mathcal{D}$ at the point x is defined as follows:

$$f(x, \mathcal{D}) = \frac{1}{n} \cdot \sum_{X_i \in \mathcal{D}} K_h(x - X_i). \tag{6.19}$$

The kernel function is a smooth unimodal distribution such as the Gaussian function:

$$K_h(x - X_i) = \frac{1}{\sqrt{2\pi} \cdot h} \cdot e^{-\frac{||x - X_i||^2}{2h^2}}. \tag{6.20}$$

The kernel function is dependent on the use of a parameter h, which is the level of smoothing. The accuracy of the density estimate depends upon this width h, and several heuristic rules have been proposed for estimating the bandwidth [6].

The value of the density $f(x, \mathcal{D})$ may differ considerably from $f(x, \mathcal{D}_i)$ because of the difference in distributions of the different classes. Correspondingly, the *accuracy density* $\mathcal{A}(x, C_i, \mathcal{D})$ for the class C_i is defined as follows:

$$\mathcal{A}(x, C_i, \mathcal{D}) = \frac{n_i \cdot f(x, \mathcal{D}_i)}{\sum_{i=1}^{k} n_i \cdot f(x, \mathcal{D}_i)}. \tag{6.21}$$

The above expression always lies between 0 and 1, and is simply an estimation of the Bayesian posterior probability of the test instance belonging to class C_i. It is assumed that the a-priori probability of the test instance belonging to class C_i, (without knowing any of its feature values) is simply equal to its fractional presence n_i/N in class C_i. The conditional probability density after knowing the feature values x is equal to $f(x, \mathcal{D}_i)$. Then, by applying the Bayesian formula for posterior probabilities, the condition of Equation 6.21 is obtained. The higher the value of the accuracy density, the greater the relative density of C_i compared to the other classes.

Another related measure is the *interest density* $I(x, C_i, \mathcal{D})$, which is the *ratio* of the density of the class C_i to the overall density of the data.

$$I(x, C_i, \mathcal{D}) = \frac{f(x, \mathcal{D}_i)}{f(x, \mathcal{D})}. \tag{6.22}$$

The class C_i is over-represented at x, when the interest density is larger than one. The *dominant class* at the coordinate x is denoted by $CM(x, \mathcal{D})$, and is equal to $\text{argmax}_{i \in \{1,...k\}} I(x, C_i, \mathcal{D})$. Correspondingly, the maximum interest density at x is denoted by $IM(x, \mathcal{D}) = \max_{i \in \{1,...k\}} I(x, C_i, \mathcal{D})$. Both the interest and accuracy density are valuable quantifications of the level of dominance of the different classes. The interest density is more effective at comparing among the different classes at a given point, whereas the accuracy density is more effective at providing an idea of the absolute accuracy at a given point.

So far, it has been assumed that all of the above computations are performed in the full dimensional space. However, it is also possible to project the data onto the subspace E in order to perform this computation. Such a calculation would quantify the discriminatory power of the subspace E at x. In order to denote the use of the subspace E in any computation, the corresponding expression will be superscripted with E. Thus the density in a given subspace E is denoted by $f^E(\cdot, \cdot)$, the accuracy density by $\mathcal{A}^E(\cdot, \cdot, \cdot)$, and the interest density by $I^E(\cdot, \cdot, \cdot)$. Similarly, the dominant class is defined using the subspace-specific interest density at that point, and the accuracy density profile is defined for that particular subspace. An example of the accuracy density profile (of the dominant class) in a 2-dimensional subspace is illustrated in Figure 6.4(a). The test instance is also labeled in the same figure in order to illustrate the relationship between the density profile and test instance.

It is desired to find those projections of the data in which the interest density value $IM^E(t, \mathcal{D})$ is the maximum. It is quite possible that in some cases, different subspaces may provide different information about the class behavior of the data; these are the difficult cases in which a test instance may be difficult to classify accurately. In such cases, the user may need to isolate particular data localities in which the class distribution is further examined by a hierarchical exploratory process. While the density values are naturally defined over the continuous space of quantitative attributes, it has been shown in [6] that intuitively analogous values can be defined for the interest and accuracy densities even when categorical attributes are present.

For a given test example, the end user is provided with unique options in exploring various characteristics that are indicative of its classification. A subspace determination process is used on the basis of the highest interest densities at a given test instance. Thus, the subspace determination process finds the appropriate *local* discriminative subspaces for a given test example. These are the various possibilities (or branches) of the decision path, which can be utilized in order to explore the regions in the locality of the test instance. In each of these subspaces, the user is provided with a visual profile of the accuracy density. This profile provides the user with an idea of which branch is likely to lead to a region of high accuracy for that test instance. This visual profile can also be utilized in order to determine which of the various branches are most suitable for further exploration. Once such a branch has been chosen, the user has the option to further explore into a particular region of the data that has high accuracy density. This process of data localization can quickly isolate an arbitrarily shaped region in the data containing the test instance. This sequence of data localizations creates a path (and a locally discriminatory combination of dimensions) that reveals the underlying classification causality to the user.

In the event that a decision path is chosen that is not strongly indicative of any class, the user has the option to backtrack to a higher level node and explore a different path of the tree. In some cases, different branches may be indicative of the test example belonging to different classes. These are the "ambiguous cases" in which a test example could share characteristics from multiple classes. Many standard modeling methods may classify such an example incorrectly, though the subspace decision path method is much more effective at providing the user with an intensional knowledge of the test example because of its exploratory approach. This can be used in order to understand the causality behind the ambiguous classification behavior of that instance.

The overall algorithm for decision path construction is illustrated in Figure 6.5. The details of the subroutines in the procedure are described in [6], though a summary description is provided here. The input to the system is the data set $\mathcal{D}$, the test instance t for which one wishes to find the

Algorithm *SubspaceDecisionPath(Test Example: t, Data Set: $\mathcal{D}$,*
      MaxDimensionality: l, MaxBranchFactor: b_{max}, Minimum Interest Ratio: ir_{min})
begin
  PATH$= \{\mathcal{D}\}$;
  while *not(termination)* **do**
  begin
    Pick the last data set $\mathcal{L}$ indicated in PATH;
    $\mathcal{E} = \{E_1 \ldots E_q\} = ComputeClassificationSubspaces(\mathcal{L}, t, l, b_{max}, ir_{min})$;
    for each E_i *ConstructDensityProfile(E_i, $\mathcal{L}$, t)*;
    { " Zooming in" refers to further subspace exploration }
    if (*zoom-in (user-specified)*) **then**
    begin
      User specifies choice of branch E_i;
      User specifies accuracy density threshold λ for zoom-in;
      { $p(\mathcal{L}', C_i)$ is accuracy significance of class
        C_i in $\mathcal{L}'$ with respect to $\mathcal{L}$ }
      $(\mathcal{L}', p(\mathcal{L}', C_1) \ldots p(\mathcal{L}', C_k)) = IsolateData(\mathcal{L}, t, \lambda)$;
      Add $\mathcal{L}'$ to the end of PATH;
    end;
    else begin (retreat))
      User specifies data set $\mathcal{L}'$ on PATH to backtrack to;
      Delete all data set pointers occuring after $\mathcal{L}'$ on PATH; (backtrack)
    end;
    { Calculate cumulative dominance of each class C_i along PATH
    in order to provide the user a measure of its significance }
    for each class C_i **do**
      Output $CD(PATH, C_i) = 1 - \pi_{(\mathcal{L} \in PATH, \mathcal{L} \neq \mathcal{D})}(1 - p(\mathcal{L}, C_i))$ to user;
  end;
end

FIGURE 6.5: The subspace decision path method.

diagnostic characteristics, a maximum branch factor b_{max}, and a minimum interest density ir_{min}. In addition, the maximum dimensionality l of any subspace utilized in data exploration is utilized as user input. The value of $l = 2$ is especially interesting because it allows for the use of visual profile of the accuracy density. Even though it is natural to use 2-dimensional projections because of their visual interpretability, the data exploration process along a given path reveals a higher dimensional combination of dimensions, which is most suitable for the test instance. The branch factor b_{max} is the maximum number of possibilities presented to the user, whereas the value of ir_{min} is the corresponding minimum interest density of the test instance in any subspace presented to the user. The value of ir_{min} is chosen to be 1, which is the break-even value for interest density computation. This break-even value is one at which the interest density neither under-estimates nor over-estimates the accuracy of the test instance with respect to a class. The variable PATH consists of the pointers to the sequence of successively reduced training data sets, which are obtained in the process of interactive decision tree construction. The list PATH is initialized to a single element, which is the pointer to the original data set $\mathcal{D}$. At each point in the decision path construction process, the subspaces $E_1 \ldots E_q$ are determined, which have the greatest interest density (of the dominant class) in the locality of the test instance t. This process is accomplished by the procedure *ComputeClassificationSubspaces*. Once these subspaces have been determined, the density profile is constructed for each of them by the procedure *ConstructDensityProfile*. Even though one subspace may have higher interest density at the test instance than another, the true value of a subspace in separating the data locality around the test instance is often a subjective judgement that depends both upon the interest density of the test instance and the spatial separation of the classes. Such a judgement requires human intuition, which can be harnessed with the use of the visual profile of the accuracy density profiles of the various possibilities. These profiles provide the user with an intuitive idea of the class behavior of the data set in various projections. If the class behavior across different projections is not very consistent (different projections are indicative of different classes), then such a node is not very revealing of valuable information. In such a case, the user may choose to backtrack by specifying an earlier node on PATH from which to start further exploration.

On the other hand, if the different projections provide a consistent idea of the class behavior, then the user utilizes the density profile in order to isolate a small region of the data in which the accuracy density of the data in the locality of the test instance is significantly higher for a particular class. This is achieved by the procedure *IsolateData*. This isolated region may be a cluster of arbitrary shape depending upon the region covered by the dominating class. However, the use of the visual profile helps to maintain the interpretability of the isolation process in spite of the arbitrary contour of separation. An example of such an isolation is illustrated in Figure 6.4(b), and is facilitated by the construction of the visual density profile. The procedure returns the isolated data set L' along with a number of called the *accuracy significance* $p(L', C_i)$ of the class C_i. The pointer to this new data set L' is added to the end of PATH. At that point, the user decides whether further exploration into that isolated data set is necessary. If so, the same process of subspace analysis is repeated on this node. Otherwise, the process is terminated and the most relevant class label is reported. The overall exploration process also provides the user with a good diagnostic idea of how the local data distributions along different dimensions relate to the class label.

6.11   Conclusions and Summary

In this chapter, we presented algorithms for instance-based learning and lazy-learning. These methods are usually computationally more expensive than traditional models, but have the advantage of using information about the test instance at the time of model construction. In many scenarios, such methods can provide more accurate results, when a pre-processed model cannot fully capture the underlying complexities of the data. The other advantage of instance-based methods is that they are typically very simple and easy to implement for arbitrary data types, because the complexity is usually abstracted in determining an appropriate distance function between objects. Since distance function design has been widely studied in the literature for different data types, this allows the use of this methodology for arbitrary data types. Furthermore, existing model-based algorithms can be generalized to local variations in order to transform them to instance-learning methods. Such methods have great diagnostic value for specific test instances, since the model is constructed local to the specific test instance, and also provides knowledge to the analyst that is test-instance specific. Lazy methods are highly flexible and can be generalized to streaming and other temporal scenarios, where the laziness of the approach can be used in order to learn the appropriate temporal horizon, in which lazy learning should be performed.

Bibliography

[1] A. Aamodt and E. Plaza. Case-based reasoning: Foundational issues, methodological variations, and system approaches, *AI communications*, 7(1):39–59, 1994.

[2] C. Aggarwal, A. Hinneburg, and D. Keim. On the surprising behavior of distance metrics in high dimensional space, *Data Theory—ICDT Conference*, 2001; *Lecture Notes in Computer Science*, 1973:420–434, 2001.

[3] C. Aggarwal. Towards systematic design of distance functions for data mining applications, *Proceedings of the KDD Conference*, pages 9–18, 2003.

[4] C. Aggarwal and P. Yu. Locust: An online analytical processing framework for high dimensional classification of data streams, *Proceedings of the IEEE International Conference on Data Engineering*, pages 426–435, 2008.

[5] C. Aggarwal and C. Reddy. *Data Clustering: Algorithms and Applications*, CRC Press, 2013.

[6] C. Aggarwal. Toward exploratory test-instance-centered diagnosis in high-dimensional classification, *IEEE Transactions on Knowledge and Data Engineering*, 19(8):1001–1015, 2007.

[7] C. Aggarwal, J. Han, J. Wang, and P. Yu. A framework for on demand classification of evolving data streams. In *IEEE TKDE Journal*, 18(5):577–589, 2006.

[8] D. Aha, P. Clark, S. Salzberg, and G. Blix. Incremental constructive induction: An instance-based approach, *Proceedings of the Eighth International Workshop on Machine Learning*, pages 117–121, 1991.

[9] D. Aha. *Lazy learning*. Kluwer Academic Publishers, 1997.

[10] D. Aha. Tolerating noisy, irrelevant and novel attributes in instance-based learning algorithms, *International Journal of Man-Machine Studies*, 36(2):267–287, 1992.

[11] D. Aha, D. Kibler, and M. Albert. Instance-based learning algorithms, *Machine Learning*, 6(1):37–66, 1991.

[12] D. Aha. Lazy learning: Special issue editorial, *Artificial Intelligence Review*, 11:7–10, 1997.

[13] F. Angiulli. Fast nearest neighbor condensation for large data sets classification, *IEEE Transactions on Knowledge and Data Engineering*, 19(11):1450–1464, 2007.

[14] M. Ankerst, G. Kastenmuller, H.-P. Kriegel, and T. Seidl. Nearest Neighbor Classification in 3D Protein Databases, *ISMB-99 Proceedings*, pages 34–43, 1999.

[15] C. Atkeson, A. Moore, and S. Schaal. Locally weighted learning, *Artificial Intelligence Review*, 11(1–5):11–73, 1997.

[16] S. D. Bay. Combining nearest neighbor classifiers through multiple feature subsets. *Proceedings of ICML Conference*, pages 37–45, 1998.

[17] J. Beringer and E Hullermeier. Efficient instance-based learning on data streams, *Intelligent Data Analysis*, 11(6):627–650, 2007.

[18] M. Birattari, G. Bontempi, and M. Bersini. Lazy learning meets the recursive least squares algorithm, *Advances in neural information processing systems*, 11:375–381, 1999.

[19] C. M. Bishop. Improving the generalization properties of radial basis function neural networks, *Neural Computation*, 3(4):579–588, 1991.

[20] E. Blanzieri and A. Bryl. Instance-based spam filtering using SVM nearest neighbor classifier, *FLAIRS Conference*, pages 441–442, 2007.

[21] J. Blitzer, K. Weinberger, and L. Saul. Distance metric learning for large margin nearest neighbor classification, *Advances in neural information processing systems*: 1473–1480, 2005.

[22] S. Boriah, V. Chandola, and V. Kumar. Similarity measures for categorical data: A comparative evaluation. *Proceedings of the SIAM Conference on Data Mining*, pages 243–254, 2008.

[23] L. Bottou and V. Vapnik. Local learning algorithms, *Neural COmputation*, 4(6):888–900, 1992.

[24] L. Breiman, J. Friedman, R. Olshen, and C. Stone. *Classification and Regression Trees*, Wadsworth, 1984.

[25] L. Breiman. Random forests, *Machine Learning*, 48(1):5–32, 2001.

[26] T. Cain, M. Pazzani, and G. Silverstein. Using domain knowledge to influence similarity judgements, *Proceedings of the Case-Based Reasoning Workshop*, pages 191–198, 1991.

[27] C. L. Chang. Finding prototypes for nearest neighbor classifiers: *IEEE Transactions on Computers*, 100(11):1179–184, 1974.

[28] W. Cheng and E. Hullermeier. Combining instance-based learning and logistic regression for multilabel classification, *Machine Learning*, 76 (2–3):211–225, 2009.

[29] J. G. Cleary and L. E. Trigg. K^*: An instance-based learner using an entropic distance measure, *Proceedings of ICML Conference*, pages 108–114, 1995.

[30] S. Cost and S. Salzberg. A weighted nearest neighbor algorithm for learning with symbolic features, *Machine Learning*, 10(1):57–78, 1993.

[31] T. Cover and P. Hart. Nearest neighbor pattern classification, *IEEE Transactions on Information Theory*, 13(1):21–27, 1967.

[32] P. Cunningham, A taxonomy of similarity mechanisms for case-based reasoning, *IEEE Transactuions on Knowledge and Data Engineering*, 21(11):1532–1543, 2009.

[33] B. V. Dasarathy. *Nearest Neighbor (NN) Norms: NN Pattern Classification Techniques*, IEEE Computer Society Press, 1990.

[34] C. Domeniconi, J. Peng, and D. Gunopulos. Locally adaptive metric nearest-neighbor classification, *IEEE Transactions on Pattern Analysis and Machine Intelligence*, 24(9):1281–1285, 2002.

[35] C. Domeniconi and D. Gunopulos. Adaptive nearest neighbor classification using support vector machines, *Advances in Neural Information Processing Systems*, 1: 665–672, 2002.

[36] P. Domingos. Unifying instance-based and rule-based induction, *Machine Learning*, 24(2):141–168, 1996.

[37] G. W. Flake. Square unit augmented, radially extended, multilayer perceptrons, *In Neural Networks: Tricks of the Trade*, pages 145–163, 1998.

[38] E. Frank, M. Hall, and B. Pfahringer. Locally weighted Naive Bayes, *Proceedings of the Nineteenth Conference on Uncertainty in Artificial Intelligence*, pages 249–256, 2002.

[39] J. Friedman. *Flexible Nearest Neighbor Classification*, Technical Report, Stanford University, 1994.

[40] J. Friedman, R. Kohavi, and Y. Yun. Lazy decision trees, *Proceedings of the National Conference on Artificial Intelligence*, pages 717–724, 1996.

[41] S. Garcia, J. Derrac, J. Cano, and F. Herrera. Prototype selection for nearest neighbor classification: Taxonomy and empirical study, *IEEE Transactions on Pattern Analysis and Machine Intelligence*, 34(3):417–436, 2012.

[42] D. Gunopulos and G. Das. Time series similarity measures. In *Tutorial notes of the sixth ACM SIGKDD International Conference on Knowledge Discovery and Data Mining*, pages 243–307, 2000.

[43] E. Han, G. Karypis, and V. Kumar. Text categorization using weight adjusted k-nearest neighbor classification, *Proceedings of the Pacific-Asia Conference on Knowledge Discovery and Data Mining*, pages 53–65, 2001.

[44] T. Hastie and R. Tibshirani. Discriminant adaptive nearest neighbor classification, *IEEE Transactions on Pattern Analysis and Machine Intelligence*, 18(6):607–616, 1996.

[45] A. Hinneburg, C. Aggarwal, and D. Keim. What is the nearest neighbor in high dimensional spaces? *Proceedings of VLDB Conference*, pages 506–515, 2000.

[46] Y. S. Hwang and S. Y. Bang. An efficient method to construct a radial basis function neural network classifier, *Neural Networks*, 10(8):1495–1503, 1997.

[47] N. Jankowski and M. Grochowski, Comparison of instance selection algorithms, I algorithms survey, *Lecture Notes in Computer Science*, 3070:598–603, 2004.

[48] R. Kohavi, P. Langley, and Y. Yun. The utility of feature weighting in nearest-neighbor algorithms, *Proceedings of the Ninth European Conference on Machine Learning*, pages 85–92, 1997.

[49] J. Li, G. Dong, K. Ramamohanarao, and L. Wong. Deeps: A new instance-based lazy discovery and classification system, *Machine Learning*, 54(2):99–124, 2004.

[50] D. G. Lowe. Similarity metric learning for a variable-kernel classifier, *Neural computation*, 7(1):72–85, 1995.

[51] G. J. McLachlan. *Discriminant analysis and statistical pattern recognition*, Wiley-Interscience, Vol. 544, 2004.

[52] J. Moody and C. Darken. Fast learning in networks of locally-tuned processing units, *Neural computation*, 1(2), 281–294, 1989.

[53] J. Peng, B. Bhanu, and S. Qing. Probabilistic feature relevance learning for content-based image retrieval, *Computer Vision and Image Understanding*, 75(1):150–164, 1999.

[54] J. Peng, D. Heisterkamp, and H. Dai. LDA/SVM driven nearest neighbor classification, *Computer Vision and Pattern Recognition Conference*, 1–58, 2001.

[55] J. R. Quinlan. Combining instance-based and model-based learning. *Proceedings of ICML Conference*, pages 236–243, 1993.

[56] G. Salton and M. J. McGill. *Introduction to modern information retrieval.* McGraw-Hill, 1983.

[57] H. Samet. *Foundations of multidimensional and metric data structures*, Morgan Kaufmann, 2006.

[58] M. Sonka, H. Vaclav, and R. Boyle. *Image Processing, Analysis, and Machine Vision.* Thomson Learning, 1999.

[59] E. Spyromitros, G. Tsoumakas, and I. Vlahavas. An empirical study of lazy multilabel classification algorithms, In *Artificial Intelligence: Theories, Models and Applications*, pages 401–406, 2008.

[60] C. Stanfil and D. Waltz. Toward memory-based reasoning, *Communications of the ACM*, 29(12):1213–1228, 1986.

[61] M. Tahir, A. Bouridane, and F. Kurugollu. Simultaneous feature selection and feature weighting using Hybrid Tabu Search in K-nearest neighbor classifier, *Pattern Recognition Letters*, 28(4):483–446, 2007.

[62] P. Utgoff. Incremental induction of decision trees, *Machine Learning*, 4(2):161–186, 1989.

[63] P. Vincent and Y. Bengio. K-local hyperplane and convex distance nearest algorithms, *Neural Information Processing Systems*, pages 985–992, 2001.

[64] D. Volper and S. Hampson. Learning and using specific instances, *Biological Cybernetics*, 57(1–2):57–71, 1987.

[65] J. Wang, P. Neskovic, and L. Cooper. Improving nearest neighbor rule with a simple adaptive distance measure, *Pattern Recognition Letters*, 28(2):207–213, 2007.

[66] J. Wang and G. Karypis. On mining instance-centric classification rules, *IEEE Transactions on Knowledge and Data Engineering*, 18(11):1497–1511, 2006.

[67] I. Watson and F. Marir. Case-based reasoning: A review, *Knowledge Engineering Review*, 9(4):327–354, 1994.

[68] K. Weinberger, J. Blitzer, and L. Saul. Distance metric learning for large margin nearest neighbor classification, *NIPS Conference*, MIT Press, 2006.

[69] D. Wettschereck, D. Aha, and T. Mohri. A review and empirical evaluation of feature weighting methods for a class of lazy learning algorithms, *Artificial Intelligence Review*, 11(1–5):273–314, 1997.

[70] D. Wettschereck and D. Aha. Weighting features. *Case-Based Reasoning Research and Development*, pp. 347–358, Springer Berlin, Heidelberg, 1995.

[71] D. Wettschereck and T. Dietterich. Improving the performance of radial basis function networks by learning center locations, *NIPS*, Vol. 4:1133–1140, 1991.

[72] D. Wilson and T. Martinez. Reduction techniques for instance-based learning algorithms, *Machine Learning* 38(3):257–286, 2000.

[73] D. Wilson and T. Martinez. An integrated instance-based learning algorithm, *Computers and Intelligence*, 16(1):28–48, 2000.

[74] Z. Xie, W. Hsu, Z. Liu, and M. L. Lee. Snnb: A selective neighborhood based naive Bayes for lazy learning, *Advances in Knowledge Discovery and Data Mining*, pages 104–114, Springer, 2002.

[75] L. Yang. Distance Metric Learning: A Comprehensive Survey, 2006. `http://www.cs.cmu.edu/~liuy/frame_survey_v2.pdf`

[76] D.-C. Zhan, M. Li, Y.-F. Li, and Z.-H. Zhou. Learning instance specific distances using metric propagation, *Proceedings of ICML Conference*, pages 1225–1232, 2009.

[77] H. Zhang, A. Berg, M. Maire, and J. Malik. SVM-KNN: Discriminative nearest neighbor classification for visual category recognition, *Computer Vision and Pattern Recognition*, pages 2126–2136, 2006.

[78] M. Zhang and Z. H. Zhou. ML-kNN: A lazy learning approach to multi-label learning, *Pattern Recognition*, 40(7): 2038–2045, 2007.

[79] Z. Zheng and G. Webb. Lazy learning of Bayesian rules. *Machine Learning*, 41(1):53–84, 2000.

[80] Z. H. Zhou and Y. Yu. Ensembling local learners through multimodal perturbation, *IEEE Transactions on Systems, Man, and Cybernetics, Part B: Cybernetics*, 35(4):725–735, 2005.

[81] X. Zhu and X. Wu. Scalable representative instance selection and ranking, *Proceedings of IEEE International Conference on Pattern Recognition*, Vol 3:352–355, 2006.

Chapter 7

Support Vector Machines

Po-Wei Wang

National Taiwan University
Taipei, Taiwan
b97058@csie.ntu.edu.tw

Chih-Jen Lin

National Taiwan University
Taipei, Taiwan
cjlin@csie.ntu.edu.tw

7.1 Introduction

Machine learning algorithms have a tendency to over-fit. It is possible to achieve an arbitrarily low training error with some complex models, but the testing error may be high, because of poor generalization to unseen test instances. This is problematic, because the goal of classification is not to obtain good accuracy on known training data, but to predict unseen test instances correctly. Vapnik's work [34] was motivated by this issue. His work started from a statistical derivation on linearly separable scenarios, and found that classifiers with maximum margins are less likely to overfit. This concept of maximum margin classifiers eventually evolved into support vector machines (SVMs).

SVM is a theoretically sound approach for controlling model complexity. It picks important instances to construct the separating surface between data instances. When the data is not linearly separable, it can either penalize violations with loss terms, or leverage kernel tricks to construct non-linear separating surfaces. SVMs can also perform multiclass classifications in various ways, either by an ensemble of binary classifiers or by extending margin concepts. The optimization techniques of SVMs are mature, and SVMs have been used widely in many application domains.

This chapter will introduce SVMs from several perspectives. We introduce maximum margin classifiers in Section 7.2 and the concept of regularization in Section 7.3. The dual problems of SVMs are derived in Section 7.4. Then, to construct nonlinear separating surfaces, feature transformations and kernel tricks are discussed in Section 7.5. To solve the kernelized SVM problem,

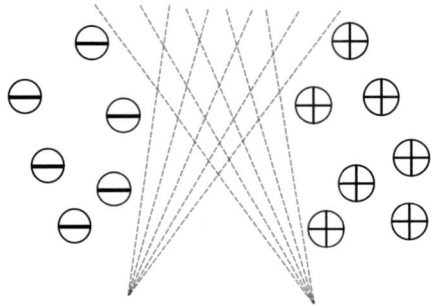

FIGURE 7.1: Infinitely many classifiers could separate the positive and negative instances.

decomposition methods are introduced in Section 7.6. Further, we discuss the multiclass strategies in Section 7.7 and make a summary in Section 7.8.

This chapter is not the only existing introduction of SVM. A number of surveys and books may be found in [6, 12, 33]. While this chapter does not discuss applications of SVMs, the reader is referred to [16] for a guide to such applications.

7.2 The Maximum Margin Perspective

Given the training data in the form of label-feature pairs,

$$S = \{(y_i, \boldsymbol{x}_i) \mid y_i \in \{+1, -1\}, \, \boldsymbol{x}_i \in \mathbb{R}^n, \, \forall i = 1 \ldots l \},$$

the goal of binary classification is to find a decision function $f(\boldsymbol{x})$ that could correctly predict the label y of an input feature $\boldsymbol{x}$. The problem could be transformed to the determination of a function $h(\boldsymbol{x})$, such that

$$f(\boldsymbol{x}) = \begin{cases} +1 & \text{when } h(\boldsymbol{x}) > 0, \\ -1 & \text{when } h(\boldsymbol{x}) < 0. \end{cases}$$

The separating surface, which represents the boundary between positive and negative regions, is defined as follows:

$$\mathcal{H} = \{\boldsymbol{x} \mid h(\boldsymbol{x}) = 0\}.$$

Now consider a linearly separable scenario, where some linear functions

$$h(\boldsymbol{x}) \equiv \boldsymbol{w}^\top \boldsymbol{x} + b$$

could separate the two regions. As shown in Figure 7.1, there may be infinitely many hyperplanes that separate the positive and negative instances correctly. A reasonable choice is the one with the largest gap between both classes. This setting, called the **Maximum Margin Classifer**, may be more resistant to any perturbation of the training data. Since the margin could be measured in different ways, we formally define the distances of a feature $\boldsymbol{x}$ to the set $\mathcal{H}$.

$$\text{Distance}(\boldsymbol{x}, \mathcal{H}) \equiv \min_{\boldsymbol{p} \in \mathcal{H}} \|\boldsymbol{x} - \boldsymbol{p}\|.$$

Note that we may use different norms in the definition of distance. By [24, Theorem 2.2], there is a closed-form solution when $\mathcal{H}$ is a hyperplane:

$$\text{Distance}(\boldsymbol{x}, \mathcal{H}) = \frac{|\boldsymbol{w}^\top \boldsymbol{x} + b|}{\|\boldsymbol{w}\|^*},$$

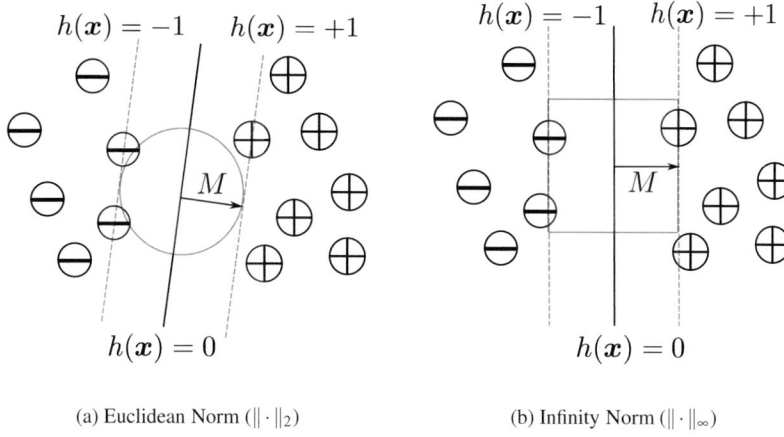

(a) Euclidean Norm ($\|\cdot\|_2$) (b) Infinity Norm ($\|\cdot\|_\infty$)

FIGURE 7.2: Maximum margin classifiers of different norms.

in which $\|\cdot\|^*$ is the dual norm of the norm we choose in the definition of distance. The margin, which measures the gap between the separating surface $\mathcal{H}$ and the nearest instance, is defined as follows:

$$M \equiv \min_i \frac{|\boldsymbol{w}^\top \boldsymbol{x}_i + b|}{\|\boldsymbol{w}\|^*}.$$

When the instance $\boldsymbol{x}_i$ lies on the correct side of the hyperplane, the numerator of margin M could be simplified as follows:

$$|\boldsymbol{w}^\top \boldsymbol{x}_i + b| = y_i(\boldsymbol{w}^\top \boldsymbol{x}_i + b) > 0.$$

Thus, to find a maximum margin classifier, we can instead solve the following problem:

$$\max_{\boldsymbol{w},\,b,\,M} M, \quad \text{s.t.} \quad \frac{y_i(\boldsymbol{w}^\top \boldsymbol{x}_i + b)}{\|\boldsymbol{w}\|^*} \geq M.$$

It could be verified that any non-zero multiple of the optimal solution $(\bar{\boldsymbol{w}}, \bar{b})$ is still an optimal solution. Therefore, we could set

$$M\|\boldsymbol{w}\|^* = 1,$$

so that the problem could be rewritten in a simpler form

$$\min_{\boldsymbol{w},\,b} \|\boldsymbol{w}\|^*, \quad \text{s.t.} \quad y_i(\boldsymbol{w}^\top \boldsymbol{x}_i + b) \geq 1, \ \forall i = 1\ldots l.$$

If the Euclidean norm $\|\boldsymbol{x}\|_2 \equiv \sqrt{\sum_i x_i^2}$ is chosen, this formulation reverts to the original SVM formulation in [1]. If an infinity norm $\|\boldsymbol{x}\|_\infty \equiv \max_i(|x_i|)$ is chosen, it instead leads to the L1 regularized SVM. An illustration is provided in Figures 7.2(a) and 7.2(b).

The concept of maximum margin could also be explained by statistical learning theory. From [35], we know that if training data and testing data are sampled i.i.d., under probability $1 - \eta$ we have

$$\text{testing error} \leq \text{training error} + \sqrt{\frac{\text{VC} \cdot (\ln \frac{2l}{\text{VC}} + 1) - \ln \frac{\eta}{4}}{l}},$$

in which VC is the Vapnik-Chervonenkis dimension of our model. In our separable scenario, the training error is zero. Therefore, the minimization of the testing error could be achieved by controlling the VC dimension of our model. However, optimizing the VC dimension is a difficult task.

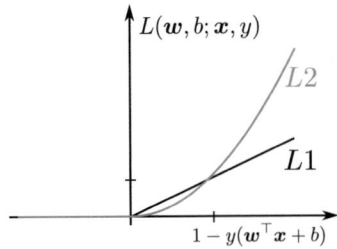

FIGURE 7.3: L1 and L2 hinge losses.

Instead, we may optimize a simpler upper-bound for VC. Assume that all data lie in an n-dimension sphere of radius R. It is proven in [35, Theorem 10.3] that the VC dimension of the linear classifier is upper bounded,

$$\text{VC} \le \min(\lceil \|\boldsymbol{w}\|_2^2 R^2 \rceil, n) + 1.$$

As a result, $\|\boldsymbol{w}\|_2^2$ bounds the VC dimension. The latter bounds the testing error. The maximum margin classifier, which has the smallest $\|\boldsymbol{w}\|_2^2$, minimizes an upper bound of the testing error. Thus, SVM is likely to have a smaller generalization error than other linear classifiers [28, 35].

7.3 The Regularization Perspective

In the previous section, we assume that data are linearly separable. However, in real world scenarios, data are less likely to satisfy this neat assumption. Therefore, the model should also be capable of handling nonseparable data. Recall that the separable condition is equivalent to the following constraints,

$$y_i(\boldsymbol{w}^\top \boldsymbol{x}_i + b) - 1 \ge 0, \quad \forall i = 1 \ldots l.$$

To penalize nonseparable instances, a **Loss Function** $\xi(\cdot)$ is used to measure the violation of these inequalities. For example, we can use an absolute value on the violation, and the loss function is referred to as the L1 hinge loss,

$$\xi_{\text{L1}}(\boldsymbol{w}, b; y_i, \boldsymbol{x}_i) = \max(0, \ 1 - y_i(\boldsymbol{w}^\top \boldsymbol{x}_i + b)).$$

If we square the violation, the loss function is referred to as the L2 hinge loss,

$$\xi_{\text{L2}}(\boldsymbol{w}, b; y_i, \boldsymbol{x}_i) = \max(0, \ 1 - y_i(\boldsymbol{w}^\top \boldsymbol{x}_i + b))^2.$$

An illustration is provided in Figure 7.3.

In contrast to maximizing *only* the margin in the separable scenario, we also need to penalize the loss terms $\xi(\boldsymbol{w}, b; y_i, \boldsymbol{x}_i)$ for each i, while maintaining a large margin. Thus, there are two objectives, and there may be multiple ways of balancing them. For example, one can have a smaller margin with one violated instance, as illustrated in Figure 7.4(a). Alternatively, one can have a larger margin with more violated instances, as illustrated in Figure 7.4(b). Since it is more difficult to minimize multiple objectives simultaneously, some tradeoffs between $\|\boldsymbol{w}\|^*$ and the loss terms are introduced. If the distance is defined by the Euclidean norm, and the L1 hinge loss is chosen, we have the classic SVM formulation in [1],

$$\min_{\boldsymbol{w}, b} \quad \frac{1}{2} \boldsymbol{w}^\top \boldsymbol{w} + C \sum_{i=1}^{l} \xi_{\text{L1}}(\boldsymbol{w}, b; y_i, \boldsymbol{x}_i). \tag{7.1}$$

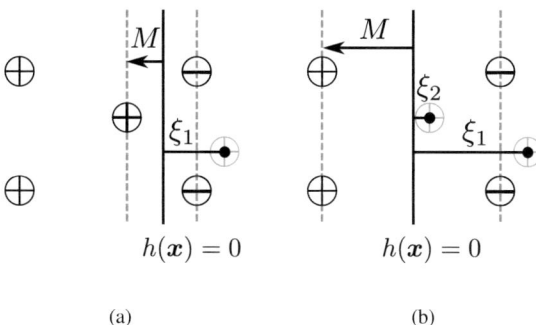

FIGURE 7.4: We can take different tradeoffs between margin size and violations.

If L2 hinge loss is selected, it would lead to another formulation

$$\min_{\boldsymbol{w}, b} \quad \frac{1}{2}\boldsymbol{w}^\top \boldsymbol{w} + C\sum_{i=1}^{l} \xi_{\mathrm{L2}}(\boldsymbol{w}, b; y_i, \boldsymbol{x}_i).$$

On the other hand, when the infinity norm is chosen to define the distance, we have the L1-regularized SVM problem in [5],

$$\min_{\boldsymbol{w}, b} \quad \|\boldsymbol{w}\|_1 + C\sum_{i=1}^{l} \xi_{\mathrm{L1}}(\boldsymbol{w}, b; y_i, \boldsymbol{x}_i).$$

As we can see, all these formulations are composed of two parts: a norm from the maximum margin objective, and the loss terms from the violation of the models. The maximum margin part is also called the **Regularization Term**, which is used to control the model complexity with respect to the violation of the models. From the perspective of numeric stability, the regularization term also plays a role in the balance between the accuracy of the model and the numeric range of $\boldsymbol{w}$. It ensures that the value of $\boldsymbol{w}$ does not take in extreme values.

Because of the tradeoffs between regularization and loss terms, the loss might not be zero even if the data are linearly separable. However, for L2 regularized SVM with L1 hinge loss, it is proved that when C is large enough, the solution $(\bar{\boldsymbol{w}}, \bar{b})$ of the SVM with loss terms are identical to the solution of the SVM without loss terms under the separable scenario. A proof of this is available in [23].

7.4 The Support Vector Perspective

We will consider an equivalent formulation of (7.1). By introducing a non-negative scalar $\xi_i = \xi_{\mathrm{L1}}(\boldsymbol{w}, b; y_i, \boldsymbol{x}_i)$, the L2 regularized L1 hinge loss SVM can be rewritten as follows:

$$\min_{\boldsymbol{w}, b, \boldsymbol{\xi}} \quad \frac{1}{2}\boldsymbol{w}^\top \boldsymbol{w} + C\sum_{i=1}^{l} \xi_i, \quad \text{s.t.} \quad y_i(\boldsymbol{w}^\top \boldsymbol{x}_i + b) \geq 1 - \xi_i, \ \xi_i \geq 0, \ \forall i = 1\ldots l. \tag{7.2}$$

Different from the original non-differentiable problem, this formulation is a convex quadratic programming problem over some linear constraints. Since this problem is a constrained optimization problem, a Lagrangian relaxation approach may be used to derive the optimality condition. The Lagrangian is defined to be the sum of the objective function, and a constraint-based penalty term with non-negative Lagrange multipliers $\boldsymbol{\alpha} \geq \mathbf{0}$ and $\boldsymbol{\beta} \geq \mathbf{0}$,

$$L(\boldsymbol{w}, b, \boldsymbol{\xi}, \boldsymbol{\alpha}, \boldsymbol{\beta}) \equiv \frac{1}{2} \boldsymbol{w}^\top \boldsymbol{w} + C \sum_{i=1}^{l} \xi_i - \sum_{i=1}^{l} \alpha_i (y_i (\boldsymbol{w}^\top \boldsymbol{x}_i + b) - 1 + \xi_i) - \sum_{i=1}^{l} \beta_i \xi_i.$$

Consider the problem $\min_{\boldsymbol{w}, b, \boldsymbol{\xi}} P(\boldsymbol{w}, b, \boldsymbol{\xi})$ and $\max_{\boldsymbol{\alpha} \geq 0, \boldsymbol{\beta} \geq 0} D(\boldsymbol{\alpha}, \boldsymbol{\beta})$, in which

$$P(\boldsymbol{w}, b, \boldsymbol{\xi}) \equiv \max_{\boldsymbol{\alpha} \geq 0, \boldsymbol{\beta} \geq 0} L(\boldsymbol{w}, b, \boldsymbol{\xi}, \boldsymbol{\alpha}, \boldsymbol{\beta}) \quad \text{and} \quad D(\boldsymbol{\alpha}, \boldsymbol{\beta}) \equiv \min_{\boldsymbol{w}, b, \boldsymbol{\xi}} L(\boldsymbol{w}, b, \boldsymbol{\xi}, \boldsymbol{\alpha}, \boldsymbol{\beta}).$$

We point out the following property of the function P. When $(\boldsymbol{w}, b, \boldsymbol{\xi})$ is a feasible solution of (7.2), $P(\boldsymbol{w}, b, \boldsymbol{\xi})$ equals the objective of (7.2) by setting the Lagrangian multipliers to zero. Otherwise,

$$P(\boldsymbol{w}, b, \boldsymbol{\xi}) \to \infty \quad \text{when} \quad (\boldsymbol{w}, b, \boldsymbol{\xi}) \text{ is not feasible.}$$

As a result, the problem $\min_{\boldsymbol{w}, b, \boldsymbol{\xi}} P(\boldsymbol{w}, b, \boldsymbol{\xi})$ is equivalent to (7.2). Furthermore, because $L(\boldsymbol{w}, b, \boldsymbol{\xi}, \boldsymbol{\alpha}, \boldsymbol{\beta})$ is convex in $(\boldsymbol{w}, b, \boldsymbol{\xi})$, and concave in $(\boldsymbol{\alpha}, \boldsymbol{\beta})$, we have the following equality from the saddle point property of concave-convex functions,

$$\min_{\boldsymbol{w}, b, \boldsymbol{\xi}} \overbrace{\max_{\boldsymbol{\alpha} \geq 0, \boldsymbol{\beta} \geq 0} L(\boldsymbol{w}, b, \boldsymbol{\xi}, \boldsymbol{\alpha}, \boldsymbol{\beta})}^{P(\boldsymbol{w}, b, \boldsymbol{\xi})} = \max_{\boldsymbol{\alpha} \geq 0, \boldsymbol{\beta} \geq 0} \overbrace{\min_{\boldsymbol{w}, b, \boldsymbol{\xi}} L(\boldsymbol{w}, b, \boldsymbol{\xi}, \boldsymbol{\alpha}, \boldsymbol{\beta})}^{D(\boldsymbol{\alpha}, \boldsymbol{\beta})}.$$

This property is popularly referred to as the **Minimax Theorem** [32] of concave-convex functions. The left-hand side, $\min_{\boldsymbol{w}, b, \boldsymbol{\xi}} P(\boldsymbol{w}, b, \boldsymbol{\xi})$, is called the **Primal Problem** of SVM, and the right-hand side, $\max_{\boldsymbol{\alpha} \geq 0, \boldsymbol{\beta} \geq 0} D(\boldsymbol{\alpha}, \boldsymbol{\beta})$, is the **Dual Problem** of SVM. Similar to the primal objective function,

$$D(\boldsymbol{\alpha}, \boldsymbol{\beta}) \to -\infty \quad \text{when} \quad \sum_{i=1}^{l} \alpha_i y_i \neq 0 \text{ or } C\boldsymbol{e} - \boldsymbol{\alpha} + \boldsymbol{\beta} \neq \mathbf{0},$$

where $\boldsymbol{e}$ is the vector of ones. However, such an $(\boldsymbol{\alpha}, \boldsymbol{\beta})$ does not need to be considered, because the dual problem maximizes its objective function. Under any given $(\boldsymbol{\alpha}, \boldsymbol{\beta})$ with $D(\boldsymbol{\alpha}, \boldsymbol{\beta}) \neq -\infty$, the minimum $(\hat{\boldsymbol{w}}, \hat{b}, \hat{\boldsymbol{\xi}})$ is achieved in the unconstrained problem $\min_{\boldsymbol{w}, b, \boldsymbol{\xi}} L(\boldsymbol{w}, b, \boldsymbol{\xi}, \boldsymbol{\alpha}, \boldsymbol{\beta})$, if and only if $\nabla_{\boldsymbol{w}, b, \boldsymbol{\xi}} L(\boldsymbol{w}, b, \boldsymbol{\xi}, \boldsymbol{\alpha}, \boldsymbol{\beta}) = \mathbf{0}$. In other words,

$$\frac{\partial L}{\partial \boldsymbol{w}} = 0 \Rightarrow \hat{\boldsymbol{w}} - \sum_{i=1}^{l} \alpha_i y_i \boldsymbol{x}_i = \mathbf{0}, \quad \frac{\partial L}{\partial b} = 0 \Rightarrow \sum_{i=1}^{l} y_i \alpha_i = 0, \quad \frac{\partial L}{\partial \boldsymbol{\xi}} = 0 \Rightarrow C\boldsymbol{e} - \boldsymbol{\alpha} + \boldsymbol{\beta} = \mathbf{0}. \quad (7.3)$$

Thus, the dual objective, $D(\boldsymbol{\alpha}, \boldsymbol{\beta}) \equiv \min_{\boldsymbol{w}, b, \boldsymbol{\xi}} L(\boldsymbol{w}, b, \boldsymbol{\xi}, \boldsymbol{\alpha}, \boldsymbol{\beta})$, could be expressed as follows:

$$D(\boldsymbol{\alpha}, \boldsymbol{\beta}) = L(\hat{\boldsymbol{w}}, \hat{b}, \hat{\boldsymbol{\xi}}, \boldsymbol{\alpha}, \boldsymbol{\beta}) = -\frac{1}{2} \sum_{i=1}^{l} \sum_{j=1}^{l} \alpha_i \alpha_j y_i y_j \boldsymbol{x}_i^\top \boldsymbol{x}_j + \sum_{i=1}^{l} \alpha_i.$$

As a result, the corresponding dual problem is as follows:

$$\max_{\boldsymbol{\alpha}, \boldsymbol{\beta}} \quad -\frac{1}{2} \sum_{i=1}^{l} \sum_{j=1}^{l} \alpha_i \alpha_j y_i y_j \boldsymbol{x}_i^\top \boldsymbol{x}_j + \sum_{i=1}^{l} \alpha_i,$$

$$\text{s.t.} \quad \sum_{i=1}^{l} y_i \alpha_i = 0, \ C - \alpha_i - \beta_i = 0, \ \alpha_i \geq 0, \ \beta_i \geq 0, \ \forall i = 1 \ldots l.$$

Moreover, by $C - \alpha_i = \beta_i \geq 0$, we can remove β_i by imposing an additional inequality constraint on α_i. The dual problem could be simplified as follows:

$$\min_{\boldsymbol{\alpha}} \quad \frac{1}{2} \boldsymbol{\alpha}^\top Q \boldsymbol{\alpha} - \boldsymbol{e}^\top \boldsymbol{\alpha}, \quad \text{s.t.} \quad 0 = \boldsymbol{y}^\top \boldsymbol{\alpha}, \, \boldsymbol{0} \leq \boldsymbol{\alpha} \leq C\boldsymbol{e}, \, Q_{ij} = y_i y_j \boldsymbol{x}_i^\top \boldsymbol{x}_j. \tag{7.4}$$

Consider the solutions $(\bar{\boldsymbol{w}}, \bar{b}, \bar{\boldsymbol{\xi}})$ and $(\bar{\boldsymbol{\alpha}}, \bar{\boldsymbol{\beta}})$ for the primal and dual problems, respectively. By the minimax theorem and the definition of P and D, we have the following:

$$L(\bar{\boldsymbol{w}}, \bar{b}, \bar{\boldsymbol{\xi}}, \bar{\boldsymbol{\alpha}}, \bar{\boldsymbol{\beta}}) \leq \overbrace{\max_{\boldsymbol{\alpha}, \boldsymbol{\beta}} L(\bar{\boldsymbol{w}}, \bar{b}, \bar{\boldsymbol{\xi}}, \boldsymbol{\alpha}, \boldsymbol{\beta})}^{P(\bar{\boldsymbol{w}}, \bar{b}, \bar{\boldsymbol{\xi}})} = \overbrace{\min_{\boldsymbol{w}, b, \boldsymbol{\xi}} L(\boldsymbol{w}, b, \boldsymbol{\xi}, \bar{\boldsymbol{\alpha}}, \bar{\boldsymbol{\beta}})}^{D(\bar{\boldsymbol{\alpha}}, \bar{\boldsymbol{\beta}})} \leq L(\bar{\boldsymbol{w}}, \bar{b}, \bar{\boldsymbol{\xi}}, \bar{\boldsymbol{\alpha}}, \bar{\boldsymbol{\beta}}).$$

That is, $P(\bar{\boldsymbol{w}}, \bar{b}, \bar{\boldsymbol{\xi}}) = D(\bar{\boldsymbol{\alpha}}, \bar{\boldsymbol{\beta}}) = L(\bar{\boldsymbol{w}}, \bar{b}, \bar{\boldsymbol{\xi}}, \bar{\boldsymbol{\alpha}}, \bar{\boldsymbol{\beta}})$. As a result, (7.3) should also hold for $(\bar{\boldsymbol{w}}, \bar{b}, \bar{\boldsymbol{\xi}}, \bar{\boldsymbol{\alpha}}, \bar{\boldsymbol{\beta}})$.

$$\bar{\boldsymbol{w}} - \sum_{i=1}^{l} \bar{\alpha}_i y_i \boldsymbol{x}_i = \boldsymbol{0}, \quad \sum_{i=1}^{l} y_i \bar{\alpha}_i = 0, \quad C\boldsymbol{e} - \bar{\boldsymbol{\alpha}} + \bar{\boldsymbol{\beta}} = \boldsymbol{0}. \tag{7.5}$$

Further, by $P(\bar{\boldsymbol{w}}, \bar{b}, \bar{\boldsymbol{\xi}}) = L(\bar{\boldsymbol{w}}, \bar{b}, \bar{\boldsymbol{\xi}}, \bar{\boldsymbol{\alpha}}, \bar{\boldsymbol{\beta}})$,

$$0 = \sum_{i=1}^{l} \bar{\alpha}_i (y_i(\bar{\boldsymbol{w}}^\top \boldsymbol{x}_i + \bar{b}) - 1 + \bar{\xi}_i) + \sum_{i=1}^{l} \bar{\beta}_i \bar{\xi}_i.$$

With the feasibility constraints on solutions,

$$0 \leq \bar{\alpha}_i, \quad 0 \leq \bar{\beta}_i, \quad 0 \leq \bar{\xi}_i, \quad y_i(\bar{\boldsymbol{w}}^\top \boldsymbol{x}_i + \bar{b}) \geq 1 - \bar{\xi}_i, \quad \forall i = 1 \ldots l, \tag{7.6}$$

we have

$$\bar{\alpha}_i (y_i(\bar{\boldsymbol{w}}^\top \boldsymbol{x}_i + \bar{b}) - 1 + \bar{\xi}_i) = 0, \quad \bar{\beta}_i \bar{\xi}_i = 0, \quad \forall i = 1 \ldots l. \tag{7.7}$$

The conditions (7.5), (7.6), and (7.7) are referred to as the **Karush-Kuhn-Tucher (KKT) optimality conditions** [4]. We have shown that the optimal solution should satisfy the KKT conditions. In fact, for a convex optimization problem with linear constraints, the KKT conditions are both necessary and sufficient for optimality.

Further, the KKT conditions link the primal and dual solutions. We have $\bar{\boldsymbol{w}} = \sum_{i=1}^{l} \bar{\alpha}_i y_i \boldsymbol{x}_i$ from (7.5), and $\bar{b}$ could be inferred from (7.7).[1] Once $(\bar{\boldsymbol{\alpha}}, \bar{b})$ has been calculated, we could evaluate the decision value as follows:

$$h(\boldsymbol{x}) = \bar{\boldsymbol{w}}^\top \boldsymbol{x} + \bar{b} = \sum_{i=1}^{l} \bar{\alpha}_i y_i \boldsymbol{x}_i^\top \boldsymbol{x} + \bar{b}. \tag{7.8}$$

One can observe that only non-zero $\bar{\alpha}_i$'s affect the decision value. We call instances with non-zero $\bar{\alpha}_i$ as the support vectors. Further, from the KKT condition,

$$y_i(\bar{\boldsymbol{w}}^\top \boldsymbol{x}_i + \bar{b}) \begin{cases} < 1 & \implies \bar{\alpha}_i = C, \\ > 1 & \implies \bar{\alpha}_i = 0, \\ = 1 & \implies 0 \leq \bar{\alpha}_i \leq C. \end{cases}$$

The result shows that all the misclassified instances have $\bar{\alpha}_i = C$, and all correctly classified instances off the margin have $\bar{\alpha}_i = 0$. Under the separable scenario, only a few instances are mis-classified or fall within the margin, while all other $\bar{\alpha}_i$ are zero. This means that the set of support vectors should be sparse.

---

[1] If $0 < \bar{\alpha}_i < C$, then (7.5), (7.6), and (7.7) imply that $\bar{\xi}_i = 0$ and $\bar{b} = y_i - \bar{\boldsymbol{w}}^\top \boldsymbol{x}_i$. If all $\bar{\alpha}_i$ are bounded (i.e., $\bar{\alpha}_i = 0$ or $\bar{\alpha}_i = C$), the calculation of $\bar{b}$ can be done by a slightly different setting; see, for example, [7].

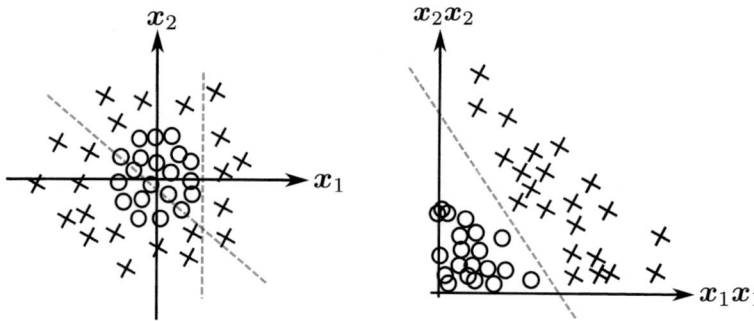

(a) No linear classifier performs well on the data.

(b) The transformed data are linearly separable.

FIGURE 7.5: Feature transformation.

7.5 Kernel Tricks

A linear model may not always be appropriate for every kind of data. For example, consider the scenario in Figure 7.5(a). For any linear model, one can obtain only about 50% accuracy. This is not satisfactory, because it is only as good as random guesses. However, if we apply a feature transformation $\phi(x)$ to all the data by the following:

$$\phi(x) = \begin{bmatrix} x_1 x_1 & x_2 x_2 \end{bmatrix}^\top.$$

In this case, the data set becomes linearly separable in the mapped space, as in Figure 7.5(b). In general, when the decision surface is decided by a smooth function $h(x)$, it could be expressed and approximated by Taylor expansions. For example, the second-order Taylor expansion of $h(x)$ is as follows:

$$h(x) \approx w_1 + \begin{bmatrix} w_2 & w_3 \end{bmatrix} \begin{bmatrix} x_1 \\ x_2 \end{bmatrix} + \frac{1}{2} \begin{bmatrix} x_1 & x_2 \end{bmatrix} \begin{bmatrix} w_4 & w_5 \\ w_5 & w_6 \end{bmatrix} \begin{bmatrix} x_1 \\ x_2 \end{bmatrix}. \tag{7.9}$$

It should be pointed out that although $h(x)$ is nonlinear in x, it is linear in w. Consequently, it could be rewritten as $h(x) = w^\top \phi(x)$, in which

$$\phi(x) = \begin{bmatrix} 1 & x_1 & x_2 & \frac{1}{2}x_1 x_1 & x_1 x_2 & \frac{1}{2}x_2 x_2 \end{bmatrix}^\top. \tag{7.10}$$

The primal problem with transformed instances is quite similar to the original one, as the change is to replace x_i with $\phi(x_i)$.

$$\min_{w,b,\xi} \frac{1}{2}w^\top w + C \sum_{i=1}^{l} \xi_i, \quad \text{s.t.} \quad y_i(w^\top \phi(x_i) + b) \geq 1 - \xi_i, \; \xi_i \geq 0, \; \forall i = 1 \ldots l.$$

However, the transformation has a serious problem. It is known that the dth-order Taylor expansion for an n-dimensional input has $\frac{(d+n)!}{d!n!}$ coefficients, because of all the combinations of coordinates of x less than or equal to order d. Thus, when the dimensionality of features and the degree of Taylor expansion grow, the dimension of w, which is identical to the number of coefficients in the Taylor expansion, may become very large. By replacing each x_i in (7.4) with the mapped vector $\phi(x_i)$, the dual formulation with transformed instances is as follows:

$$\min_{\alpha} \frac{1}{2}\alpha^\top Q\alpha - e^\top \alpha, \quad \text{s.t.} \quad 0 = y^\top \alpha, \; 0 \leq \alpha \leq Ce, \; Q_{ij} = y_i y_j \phi(x_i)^\top \phi(x_j).$$

The dual problem only has l variables, and the calculation of Q_{ij} may not even require expanding $\phi(\boldsymbol{x}_i)$ and $\phi(\boldsymbol{x}_j)$. For example, suppose that we carefully scale the feature mapping of (7.10) as follows:

$$\phi(\boldsymbol{x}) = \begin{bmatrix} 1 & \sqrt{2}x_1 & \sqrt{2}x_2 & x_1x_1 & \sqrt{2}x_1x_2 & x_2x_2 \end{bmatrix}^\top,$$

In this case, $\phi(\boldsymbol{x}_i)^\top \phi(\boldsymbol{x}_j)$ could be obtained as follows:

$$\phi(\boldsymbol{x}_i)^\top \phi(\boldsymbol{x}_j) = (1 + \boldsymbol{x}_i^\top \boldsymbol{x}_j)^2.$$

As a result, it is not necessary to explicitly expand the feature mapping. In the same manner, one could apply such a scaling to the dth order Taylor expansion, and the dot product $\phi(\boldsymbol{x}_i)^\top \phi(\boldsymbol{x}_j)$ could be simplified as follows:

$$\phi(\boldsymbol{x}_i)^\top \phi(\boldsymbol{x}_j) = (1 + \boldsymbol{x}_i^\top \boldsymbol{x}_j)^d.$$

In fact, such a method can even handle infinite dimensional expansions. Consider the following feature mapping of a one-dimensional instance x,

$$\phi(x) = \exp(-x^2) \begin{bmatrix} 1 & \sqrt{\frac{2}{1!}}x & \sqrt{\frac{2^2}{2!}}x^2 & \sqrt{\frac{2^3}{3!}}x^3 & \cdots \end{bmatrix}^\top. \tag{7.11}$$

The transformation is constructed by all degrees of polynomials of x, and $\boldsymbol{w}^\top \phi(\boldsymbol{x})$ can be seen as a scaled Taylor expansion with respect to the scalar x. Such a feature transformation could not even be performed *explicitly*, because of the infinite dimensionality. However, the dot product of the transformed feature vectors has a closed form, as follows:

$$\begin{aligned} \phi(x)^\top \phi(z) &= \exp(-x^2)\exp(-z^2)\textstyle\sum_{i=0}^{\infty}\frac{1}{i!}(2xz)^i \\ &= \exp(-x^2)\exp(-z^2)\exp(2xz) \\ &= \exp(-(x-z)^2). \end{aligned}$$

As a result, we could implicitly apply this infinite dimensional feature mapping in the dual formulation, which is expressed only in terms of the dot product. This means that the transformation does not need to be performed explicitly, because the dot products in the dual formulation can be substituted with the kernel function. Such a method is referred to as the kernel trick, and the product is referred to as the **Kernel Function**, $k(\boldsymbol{x},\boldsymbol{z}) \equiv \phi(\boldsymbol{x})^\top \phi(\boldsymbol{z})$. Similar to (7.8), we can predict an input $\boldsymbol{x}$ via

$$h(\boldsymbol{x}) = \sum_{i=1}^{l} \bar{\alpha}_i y_i \phi(\boldsymbol{x}_i)^\top \phi(\boldsymbol{x}) + \bar{b} = \sum_{i=1}^{l} \bar{\alpha}_i y_i k(\boldsymbol{x}_i, \boldsymbol{x}) + \bar{b}. \tag{7.12}$$

The matrix K, in which $K_{ij} \equiv k(\boldsymbol{x}_i, \boldsymbol{x}_j) = \phi(\boldsymbol{x}_i)^\top \phi(\boldsymbol{x}_j)$, is called the **Kernel Matrix**. With the kernel matrix, the dual SVM could be written as the following form in [9].

$$\min_{\boldsymbol{\alpha}} \quad \frac{1}{2}\boldsymbol{\alpha}^\top Q\boldsymbol{\alpha} - \boldsymbol{e}^\top \boldsymbol{\alpha}, \quad \text{s.t.} \quad 0 = \boldsymbol{y}^\top \boldsymbol{\alpha}, \, \boldsymbol{0} \le \boldsymbol{\alpha} \le C\boldsymbol{e}, \, Q_{ij} = y_i y_j K_{ij}.$$

As the definition shows, the kernel matrix is simply a dot map between transformed instances. Solving dual SVMs would only require the kernel matrix K, and we do not even need to know the mapping function $\phi(\cdot)$. In other words, classification could be done in an unknown high dimensional space, if the labels and a kernel matrix of training instances are given.[2] However, not all matrices K are valid kernel matrices, because K must be generated from a dot map. The following theorem discusses when K is a valid kernel matrix.

---

[2]Dual SVM is not the only formulation to apply the kernel trick. In fact, we can directly modify the primal problem to obtain a formulation involving kernels. The reader is referred to [8, 25].

Theorem 1 *A matrix $K \in \mathbb{R}^{l \times l}$ is a valid kernel matrix if and only if it is symmetric positive semidefinite (SPSD). In other words, the following needs to be true:*

$$K_{ij} = K_{ji}, \ \forall i, j \quad and \quad \boldsymbol{u}^\top K \boldsymbol{u} \geq 0, \ \forall \boldsymbol{u} \in \mathbb{R}^l.$$

We will prove both directions of this statement. For the necessary direction, if K is SPSD, then there exists an A such that $K = A^\top A$ by the Spectral Theorem. As a result, there is a corresponding feature mapping $\phi(\boldsymbol{x}_i) = A_i$, where A_i is the ith column of A. Then, by definition, K is a valid kernel matrix. For the sufficient direction, if K is a valid kernel matrix, then it is generated by a corresponding feature mapping $\phi(\boldsymbol{x})$. By that $\phi(\boldsymbol{x}_i)^\top \phi(\boldsymbol{x}_j) = \phi(\boldsymbol{x}_j)^\top \phi(\boldsymbol{x}_i)$, K is symmetric. K is also positive semidefinite because for an arbitrary $\boldsymbol{u} \in \mathbb{R}^l$,

$$\boldsymbol{u}^\top K \boldsymbol{u} = \sum_{i=1}^{l} \sum_{j=1}^{l} u_i u_j \phi(\boldsymbol{x}_i)^\top \phi(\boldsymbol{x}_j) = \| \sum_{i=1}^{l} u_i \phi(\boldsymbol{x}_i) \|^2 \geq 0.$$

For the prediction of an arbitrary point, the kernel function $k(\cdot)$ is needed as shown in (7.12). In the same way, not all functions might be valid kernel functions. The following **Mercer's Theorem** [26, 27] gives a condition for a valid kernel function.

Theorem 2 $k : \mathbb{R}^n \times \mathbb{R}^n \to \mathbb{R}$ *is a valid kernel function if and only if the kernel matrix it generated for any finite sequence $\boldsymbol{x}_1, \boldsymbol{x}_2, \ldots, \boldsymbol{x}_l$ is SPSD.*

This theorem is slightly different from Theorem 1. It ensures a consistent feature mapping $\phi(\boldsymbol{x})$, possibly in an infinite dimensional space, that generates all the kernel matrices from the kernel function. In practice, we do not need to specify what the feature mapping is. For example, a Gaussian kernel function, also called the **Radial Basis Function** (RBF), is a common choice for such an implicit feature transform. It is a multivariate version of (7.11), with the following kernel function

$$k(\boldsymbol{x}, \boldsymbol{z}) = \exp(-\gamma \|\boldsymbol{x} - \boldsymbol{z}\|^2),$$

where γ is the kernel parameter decided by users. Similar to (7.11), the feature mapping of the RBF kernel corresponds to all polynomial expansions of the input instance vector. We conceptually explain that the mapping function of the RBF kernel transforms the input data to become linearly separable in a high-dimensional space. When the input data are separable by any smooth surface, as illustrated in Figure 7.6, a smooth $h(\boldsymbol{x})$ corresponds to the separating surface. Note that smooth functions could be analyzed by Taylor expansions, which are a series of polynomial terms. That is, similar to (7.9), $h(\boldsymbol{x})$ can be expressed as $h(\boldsymbol{x}) = \bar{\boldsymbol{w}}^\top \bar{\phi}(\boldsymbol{x})$ for some $\bar{\boldsymbol{w}}$ and $\bar{\phi}(\boldsymbol{x})$, which includes polynomial terms. Because $\phi(\boldsymbol{x})$ of the RBF kernels is a scaled vector of $\bar{\phi}(\boldsymbol{x})$, we can adjust $\bar{\boldsymbol{w}}$ to another vector $\boldsymbol{w}$. Then $h(\boldsymbol{x}) = \boldsymbol{w}^\top \phi(\boldsymbol{x})$ with RBF's mapping function $\phi(\boldsymbol{x})$. Therefore, $h(\boldsymbol{x})$ is a linear separating hyperplane for $\phi(\boldsymbol{x}_i), \forall i$ in the space of $\phi(\boldsymbol{x})$. However, although highly nonlinear mappings can make training data separable, overfitting may occur to cause poor prediction results.

Note that the kernel trick works even if we do not know which separating surface it corresponds to in the infinite dimensional space. The implicit transformation gives SVMs the power to construct nonlinear separating surfaces, but its lack of explanation is also a disadvantage.

7.6 Solvers and Algorithms

As a mature machine learning model, SVMs have been trained by a wide variety of optimization techniques. We will discuss how to solve the following kernelized dual SVM.

$$\min_{\boldsymbol{\alpha}} \ \frac{1}{2} \boldsymbol{\alpha}^\top Q \boldsymbol{\alpha} - \boldsymbol{e}^\top \boldsymbol{\alpha}, \quad \text{s.t.} \quad 0 = \boldsymbol{y}^\top \boldsymbol{\alpha}, \ \boldsymbol{0} \leq \boldsymbol{\alpha} \leq C\boldsymbol{e}, \ Q_{ij} = y_i y_j K_{ij}.$$

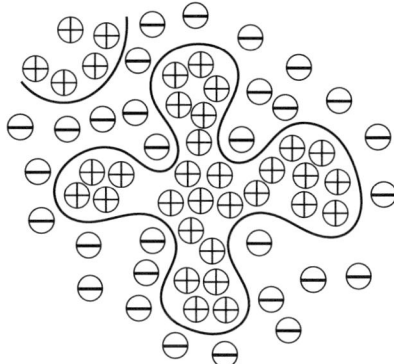

FIGURE 7.6: A smooth separating surface for the data corresponds to a linear separating hyperplane in the mapped space used by the RBF kernel. That is, the RBF kernel transforms data to be linear separable in a higher-dimensional space.

The kernelized dual SVM is a quadratic programming problem with linear constraints, which could be solved by typical quadratic solvers. However, these solvers usually assume full access to the matrix Q. This may be too large to be memory-resident, when the value of l is large. Thus, **Dual Decomposition Methods** [18, 29] are proposed to overcome these challenges. These methods decompose the full scale dual problem into subproblems with less variables, which could be solved efficiently without full access to Q. The **Sequential Minimal Optimization** (SMO) approach, as discussed in [30], is an extreme case of the dual decomposition method by using only two variables in each subproblem.

We now demonstrate how a decomposition method works. We denote $\bar{D}(\boldsymbol{\alpha}) \equiv \frac{1}{2}\boldsymbol{\alpha}^\top Q\boldsymbol{\alpha} - \boldsymbol{e}^\top \boldsymbol{\alpha}$ as the dual objective function, and $\boldsymbol{e}_i$ as the unit vector with respect to the ith coordinate. At each step of SMO, only two variables α_i and α_j in the current $\boldsymbol{\alpha}$ are changed for the next solution $\boldsymbol{\alpha}'$. To make each step feasible, $\boldsymbol{\alpha}'$ must satisfy the equality constraint $0 = \boldsymbol{y}^\top \boldsymbol{\alpha}'$. It can be validated that by choosing

$$\boldsymbol{\alpha}' = \boldsymbol{\alpha} + dy_i\boldsymbol{e}_i - dy_j\boldsymbol{e}_j,$$

$\boldsymbol{\alpha}'$ satisfies the equality for any scalar d. By substituting the new $\boldsymbol{\alpha}'$ into the dual objective function, the subproblem with regards to the variable d becomes the following:

$$\min_{d\in\mathbb{R}} \quad \frac{1}{2}(K_{ii} + K_{jj} - 2K_{ij}) \cdot d^2 + (y_i\nabla_i\bar{D}(\boldsymbol{\alpha}) - y_j\nabla_j\bar{D}(\boldsymbol{\alpha})) \cdot d + \bar{D}(\boldsymbol{\alpha})$$
$$\text{s.t.} \quad 0 \le \alpha_i + y_id \le C, \ 0 \le \alpha_j - y_jd \le C. \tag{7.13}$$

This is a single-variable quadratic optimization problem, with box constraints. It is known to have a closed-form solution. An illustration is provided in Figure 7.7, and more details are provided in [7, Section 6]. In addition, $\nabla_i\bar{D}(\boldsymbol{\alpha})$ and $\nabla_j\bar{D}(\boldsymbol{\alpha})$ could be obtained by

$$\nabla_i\bar{D}(\boldsymbol{\alpha}) = y_i\sum_{t=1}^{l} y_t\alpha_t K_{it} - 1 \quad \text{and} \quad \nabla_j\bar{D}(\boldsymbol{\alpha}) = y_j\sum_{t=1}^{l} y_t\alpha_t K_{jt} - 1.$$

The operation evaluates $2l$ entries of the kernel. Alternatively, by the fact that we only change two variables each time, we could maintain the full gradient at the same cost:

$$\text{For } t = 1,\ldots,l, \quad \nabla_t\bar{D}(\boldsymbol{\alpha}') = \nabla_t\bar{D}(\boldsymbol{\alpha}) + y_t(K_{it} - K_{jt})d.$$

Then, at the next iteration, the two needed gradient elements are directly available. One reason to

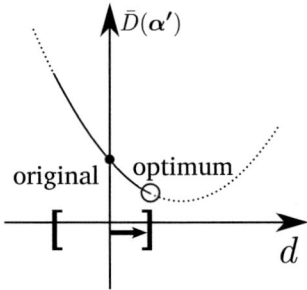

FIGURE 7.7: A single-variable quadratic problem with box constraints.

maintain the full gradient is that it is useful for choosing coordinates i and j. As a result, solving the subproblem costs $O(l)$ kernel evaluations, and $O(l)$ space for storing the full gradient.

The selection of coordinates i and j is an important issue. If they are not not carefully chosen, the objective value might not even decrease. One naive approach is to try all pairs of coordinates and pick the one with the largest reduction. However, it is expensive to examine all $l(l-1)/2$ pairs. Instead, [19] chooses the pair that has the maximum violation in the optimality condition. In another approach, the second-order selection rule [14] used in LIBSVM [7], fixes the coordinate i via the maximum violation, and determines the value of j that could produce the largest decrease in objective value.[3] Both methods require the gradient $\nabla \bar{D}(\boldsymbol{\alpha})$, which is available if we choose to maintain the gradient. In addition, these two methods ensure that all other operations require no more than $O(l)$ cost. Therefore, if each kernel evaluation needs $O(n)$ operations, an SMO step of updating two variables requires $O(l \cdot n)$ cost. These types of decomposition methods effectively solve the challenges associated with memory requirements, because they require only the space for two kernel columns, rather than the full kernel matrix. A review of algorithms and solvers for kernelized SVMs is provided in [3].

7.7 Multiclass Strategies

So far, we have discussed binary SVMs. However, it is common to classify data in more than two classes. In this section, we show that multiclass classification could be achieved by either training several binary subproblems or solving an extension of the SVM formulation. A comparison of different multiclass strategies for SVM is provided in [17].

First, we discuss the use of several binary problems. One of the most straightforward methods is to create a binary classifier for each class to decide whether an instance belongs to that class. This is called the **One-Against-Rest** strategy [2] because each binary subproblem classifies one class against all the others. Consider an example containing four classes. We denote f_m as the mth classifier deciding class c_m, and $+, -$ as the corresponding positive and negative label of classes. The one-against-rest strategy could be visualized by the following table.

---

[3]Note that in selecting j, constraints in (7.13) are relaxed. See details in [14].

Class	Classifiers			
	f_1	f_2	f_3	f_4
c_1	$+$	$-$	$-$	$-$
c_2	$-$	$+$	$-$	$-$
c_3	$-$	$-$	$+$	$-$
c_4	$-$	$-$	$-$	$+$

Each column of the table stands for a binary classifier trained on the specified positive and negative classes. To predict an input x, we apply all the binary classifiers to obtain binary predictions. If all of them are correct, there is exactly one positive value $f_m(x)$, and we choose the corresponding class c_m to be the multiclass prediction. However, the binary predictions could be wrong, and there may be multiple positive values, or even no positive value. To handle such situations, we could compare the decision values $h_m(x)$ of classifiers f_m, $\forall m$ among the k classes, and choose the largest one.

$$f(x) = \arg \max_{m=1\ldots k} h_m(x). \tag{7.14}$$

If $\|w_m\|$, $\forall m$ are the same, a larger $h_m(x)$ implies a larger distance of the instance from the decision surface. In other words, we are more confident about the binary predictions. However, the setting in (7.14) is an ad-hoc solution for resolving conflicting predictions, because $\|w_m\|$, $\forall m$ are likely to be different.

On the other hand, it may not be necessary to train binary classifiers over all the training data. We could construct one binary classifier for each pair of classes. In other words, for k classes, $k(k-1)/2$ binary classifiers are trained on each subset of distinct labels c_r and c_s. This method is called the **One-Against-One** strategy [20]. This is used for SVMs in [7, 15, 21]. The notation f_{rs} denotes the classifier with respect to classes c_r and c_s, and the notation 0 indicates the irrelevant classes. The one-against-one strategy is visualized in the following table.

Class	Classifiers					
	f_{12}	f_{13}	f_{14}	f_{23}	f_{24}	f_{34}
c_1	$+$	$+$	$+$	0	0	0
c_2	$-$	0	0	$+$	$+$	0
c_3	0	$-$	0	$-$	0	$+$
c_4	0	0	$-$	0	$-$	$-$

In prediction, a voting procedure based on predictions of all $k(k-1)/2$ binary classifiers is conducted. A total of k bins are prepared for the k classes. For each classifier, with respect to classes c_r and c_s, we cast a vote to the corresponding bin according to the predicted class. The output of the voting process is the class with maximum votes. If all predictions are correct, the winner's bin should contain $k-1$ votes, while all others should have less than $k-1$ votes.

When kernelized SVMs are used as binary classifiers, the one-against-one strategy usually spends less time in training than the one-against-rest strategy [17]. Although there are more subproblems in the one-against-one strategy, each of the subproblems involves fewer instances and could be trained faster. Assume it costs $O(l^d)$ time to train a dual SVM. If each class has about the same number of instances, the time complexity for the one-against-one strategy is $O((l/k)^d k^2) = O(l^d k^{2-d})$. In contrast, the one-against-rest strategy needs $O(l^d k)$ cost. As a result, the one-against-one strategy has a lower complexity when $d > 1$.

In kernelized SVMs, recall that we use the following equation to evaluate the decision value of a binary classifier,

$$h(x) = \sum_{i=1}^{l} \bar{\alpha}_i y_i k(x_i, x) + \bar{b}.$$

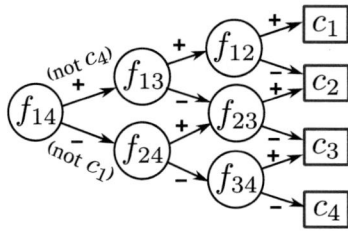

FIGURE 7.8: A direct acyclic graph for SVM prediction.

The kernel evaluations for $k(\boldsymbol{x}_i, \boldsymbol{x}), \forall \bar{\alpha}_i \neq 0$ account for most of the time in prediction. For multiclass methods such as one-against-one and one-against-rest, the values $k(\boldsymbol{x}_i, \boldsymbol{x})$ are the same across different binary subproblems. Therefore, we only need to pre-calculate the kernel entries with nonzero $\bar{\alpha}_i$ in all the binary classifiers once. In practice, the number of "total" support vectors across binary problems of one-against-rest and one-against-one strategies do not differ much. Thus, the required times for prediction are similar for both multiclass strategies.

To improve the one-against-one strategy, we could avoid applying all $k(k-1)/2$ classifiers for prediction. Instead of making decisions on all classifiers, we can apply one classifier at a time to exclude one class, and after $k - 1$ runs there is only one class left. This policy is referred to as the **DAGSVM** [31], because the procedure is similar to walking through a direct acyclic graph (DAG) in Figure 7.8. Though such a policy may not be fair to all the classes, it has the merit of being fast.

The binary reduction strategies may ensemble inconsistent classifiers. If we would like to solve multiclass classification as a whole, we must redefine the separability and the margin of multiclass data. Recall that in the one-against-rest strategy, we predict the class with the largest decision value. For the simplicity of our formulation, we consider a linear separating hyperplane without the bias term, $h_m(\boldsymbol{x}) = \boldsymbol{w}_m^\top \boldsymbol{x}$. As a result, the decision function is

$$f(\boldsymbol{x}) = \arg \max_{m=1\ldots k} \boldsymbol{w}_m^\top \boldsymbol{x}.$$

When all instances are classified correctly under the setting, there are $\boldsymbol{w}_1, \ldots, \boldsymbol{w}_k$ such that $\forall i = 1, \ldots, l$,

$$\boldsymbol{w}_{y_i}^\top \boldsymbol{x}_i > \boldsymbol{w}_m^\top \boldsymbol{x}_i, \quad \forall m = 1 \ldots k, \, m \neq y_i.$$

With the same technique in Section 7.2, the linearly separable condition could be formulated as $\forall i = 1, \ldots, l$,

$$\boldsymbol{w}_{y_i}^\top \boldsymbol{x}_i - \boldsymbol{w}_m^\top \boldsymbol{x}_i \geq 1, \quad \forall m = 1 \ldots k, \, m \neq y_i.$$

Thus, the margin between classes c_r and c_s should be defined as follows:

$$M_{r,s} \equiv \min_{i \text{ s.t.} y_i \in \{r,s\}} \frac{|(\boldsymbol{w}_r - \boldsymbol{w}_s)^\top \boldsymbol{x}_i|}{\|\boldsymbol{w}_r - \boldsymbol{w}_s\|^*}.$$

An illustration is provided in Figure 7.9. To maximize the margin $M_{r,s}$ for every pair r, s, we instead minimize the sum of its inverse $\sum_{r \neq s} 1/M_{r,s}^2$. The work [10] further shows that

$$\sum_{r,s=1,\ldots,k \text{ s.t. } r \neq s} \frac{1}{M_{r,s}^2} \leq k \sum_{m=1\ldots k} \boldsymbol{w}_m^\top \boldsymbol{w}_m.$$

Therefore, a maximum margin classifier for multiclass problems could be formulated as the follow-

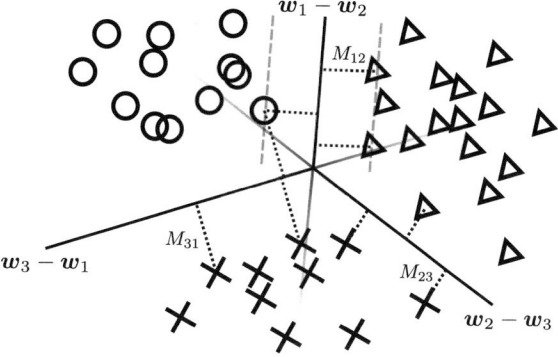

FIGURE 7.9: The margins of multiclass SVM.

ing problem,

$$\min_{\boldsymbol{w}_1,\ldots,\boldsymbol{w}_k} \quad \sum_{m=1}^{k} \boldsymbol{w}_m^\top \boldsymbol{w}_m$$

$$\text{s.t.} \quad \boldsymbol{w}_{y_i}^\top \boldsymbol{x}_i - \boldsymbol{w}_m^\top \boldsymbol{x}_i \geq 1,$$

$$\forall m = 1,\ldots,k, \ m \neq y_i, \ \forall i = 1,\ldots,l.$$

For the nonseparable scenario, the formulation with a loss term $\boldsymbol{\xi}$ is

$$\min_{\boldsymbol{w}_1,\ldots,\boldsymbol{w}_k,\boldsymbol{\xi}} \quad \frac{1}{2}\sum_{m=1}^{k} \boldsymbol{w}_m^\top \boldsymbol{w}_m + C\sum_{i=1}^{l} \xi_i$$

$$\text{subject to} \quad \boldsymbol{w}_{y_i}^\top \boldsymbol{x}_i - \boldsymbol{w}_m^\top \boldsymbol{x}_i \geq (1 - \delta_{y,m}) - \xi_i,$$

$$\forall i = 1\ldots l, \ \forall m = 1\ldots k,$$

where $\delta_{y_i,m}$ is defined to be 1 when y_i equals m, and 0 otherwise. We could validate that when y_i equals m, the corresponding inequality becomes $\xi_i \geq 0$. This formulation is referred to as the **Crammer and Singer's SVM** (CSSVM) [10,11]. Similarly, we can derive its dual problem and the kernelized form. Compared to the binary reduction strategies, the work in [17] shows that CSSVM usually gives similar classification accuracy, but needs more training time. Other similar, but slightly different strategies, are provided in [22,36].

7.8 Conclusion

SVM is now considered a mature machine learning method. It has been widely applied in different applications. Furthermore, SVM has been well-studied, both from theoretical and practical perspectives. In this chapter we focus on the kernelized SVMs, which construct nonlinear separating surfaces. However, on sparse data with many features and instances, a linear SVM may achieve similar accuracy with a kernelized SVM, but is significantly faster. Readers may refer to [13] and [37] for details.

The topic of kernelized SVM is broader than one can hope to cover in a single chapter. For example, we have not discussed topics such as kernelized support vector regression (SVR) and kernel approximations to improve the training speed. Our aim was to introduce the basic concepts of SVM. Readers can use the pointers provided in this chapter, to further study the rich literature of SVM, and learn about recent advancements.

Bibliography

[1] B. E. Boser, I. Guyon, and V. Vapnik. A training algorithm for optimal margin classifiers. In *Proceedings of the Fifth Annual Workshop on Computational Learning Theory*, pages 144–152. ACM Press, 1992.

[2] L. Bottou, C. Cortes, J. S. Denker, H. Drucker, I. Guyon, L. Jackel, Y. LeCun, U. A. Müller, E. Säckinger, P. Simard, and V. Vapnik. Comparison of classifier methods: A case study in handwriting digit recognition. In *International Conference on Pattern Recognition*, pages 77–87. IEEE Computer Society Press, 1994.

[3] L. Bottou and C.-J. Lin. Support vector machine solvers. In L. Bottou, O. Chapelle, D. DeCoste, and J. Weston, editors, *Large Scale Kernel Machines*, pages 301–320. MIT Press, Cambridge, MA, 2007.

[4] S. Boyd and L. Vandenberghe. *Convex Optimization*. Cambridge University Press, 2004.

[5] P. S. Bradley and O. L. Mangasarian. Feature selection via concave minimization and support vector machines. In *Proceedings of the Twenty-Fifth International Conference on Machine Learning (ICML)*, 1998.

[6] C. J. C. Burges. A tutorial on support vector machines for pattern recognition. *Data Mining and Knowledge Discovery*, 2(2):121–167, 1998.

[7] C.-C. Chang and C.-J. Lin. LIBSVM: A library for support vector machines. *ACM Transactions on Intelligent Systems and Technology*, 2:27:1–27:27, 2011. Software available at http://www.csie.ntu.edu.tw/~cjlin/libsvm.

[8] O. Chapelle. Training a support vector machine in the primal. *Neural Computation*, 19(5):1155–1178, 2007.

[9] C. Cortes and V. Vapnik. Support-vector network. *Machine Learning*, 20:273–297, 1995.

[10] K. Crammer and Y. Singer. On the learnability and design of output codes for multiclass problems. In *Computational Learning Theory*, pages 35–46, 2000.

[11] K. Crammer and Y. Singer. On the algorithmic implementation of multiclass kernel-based vector machines. *Journal of Machine Learning Research*, 2:265–292, 2001.

[12] N. Cristianini and J. Shawe-Taylor. *An Introduction to Support Vector Machines*. Cambridge University Press, Cambridge, UK, 2000.

[13] R.-E. Fan, K.-W. Chang, C.-J. Hsieh, X.-R. Wang, and C.-J. Lin. LIBLINEAR: A library for large linear classification. *Journal of Machine Learning Research*, 9:1871–1874, 2008.

[14] R.-E. Fan, P.-H. Chen, and C.-J. Lin. Working set selection using second order information for training SVM. *Journal of Machine Learning Research*, 6:1889–1918, 2005.

[15] J. H. Friedman. Another approach to polychotomous classification. Technical report, Department of Statistics, Stanford University, 1996.

[16] C.-W. Hsu, C.-C. Chang, and C.-J. Lin. A practical guide to support vector classification. Technical report, Department of Computer Science, National Taiwan University, 2003.

[17] C.-W. Hsu and C.-J. Lin. A comparison of methods for multi-class support vector machines. *IEEE Transactions on Neural Networks*, 13(2):415–425, 2002.

[18] T. Joachims. Making large-scale SVM learning practical. In B. Schölkopf, C. J. C. Burges, and A. J. Smola, editors, *Advances in Kernel Methods – Support Vector Learning*, pages 169–184, MIT Press, Cambridge, MA, 1998.

[19] S. S. Keerthi, S. K. Shevade, C. Bhattacharyya, and K. R. K. Murthy. Improvements to Platt's SMO algorithm for SVM classifier design. *Neural Computation*, 13:637–649, 2001.

[20] S. Knerr, L. Personnaz, and G. Dreyfus. Single-layer learning revisited: a stepwise procedure for building and training a neural network. In J. Fogelman, editor, *Neurocomputing: Algorithms, Architectures and Applications*. Springer-Verlag, 1990.

[21] U. H.-G. Kressel. Pairwise classification and support vector machines. In B. Schölkopf, C. J. C. Burges, and A. J. Smola, editors, *Advances in Kernel Methods – Support Vector Learning*, pages 255–268, MIT Press, Cambridge, MA, 1998.

[22] Y. Lee, Y. Lin, and G. Wahba. Multicategory support vector machines. *Journal of the American Statistical Association*, 99(465):67–81, 2004.

[23] C.-J. Lin. Formulations of support vector machines: a note from an optimization point of view. *Neural Computation*, 13(2):307–317, 2001.

[24] O. L. Mangasarian. Arbitrary-norm separating plane. *Operations Research Letters*, 24(1):15–23, 1999.

[25] O. L. Mangasarian. A finite Newton method for classification. *Optimization Methods and Software*, 17(5):913–929, 2002.

[26] J. Mercer. Functions of positive and negative type, and their connection with the theory of integral equations. *Philosophical Transactions of the Royal Society of London. Series*, 209:415–446, 1909.

[27] H. Q. Minh, P. Niyogi, and Y. Yao. Mercer's theorem, feature maps, and smoothing. In *Proceedings of the Nineteenth Annual Workshop on Computational Learning Theory (COLT)*, pages 154–168. 2006.

[28] E. Osuna, R. Freund, and F. Girosi. Support vector machines: Training and applications. AI Memo 1602, Massachusetts Institute of Technology, 1997.

[29] E. Osuna, R. Freund, and F. Girosi. Training support vector machines: An application to face detection. In *Proceedings of IEEE Computer Society Conference on Computer Vision and Pattern Recognition (CVPR)*, pages 130–136, 1997.

[30] J. C. Platt. Fast training of support vector machines using sequential minimal optimization. In B. Schölkopf, C. J. C. Burges, and A. J. Smola, editors, *Advances in Kernel Methods - Support Vector Learning*, MIT Press, Cambridge, MA, 1998.

[31] J. C. Platt, N. Cristianini, and J. Shawe-Taylor. Large margin DAGs for multiclass classification. In *Advances in Neural Information Processing Systems*, volume 12, pages 547–553. MIT Press, Cambridge, MA, 2000.

[32] R. T. Rockafellar. *Convex Analysis*. Princeton University Press, Princeton, NJ, 1970.

[33] B. Schölkopf and A. J. Smola. *Learning with Kernels*. MIT Press, Cambridge, MA, 2002.

[34] V. Vapnik. *Estimation of Dependences Based on Empirical Data*. Springer Verlag, New York, NY, 1982.

[35] V. Vapnik. *Statistical Learning Theory*. Wiley, New York, NY, 1998.

[36] J. Weston and C. Watkins. Multi-class support vector machines. In M. Verleysen, editor, *Proceedings of ESANN99*, pages 219–224, Brussels, 1999. D. Facto Press.

[37] G.-X. Yuan, C.-H. Ho, and C.-J. Lin. Recent advances of large-scale linear classification. *Proceedings of the IEEE*, 100(9):2584–2603, 2012.

Chapter 8

Neural Networks: A Review

Alain Biem

IBM T. J. Watson Research Center
New York, NY
biem@us.ibm.com

8.1 Introduction

Neural networks have recently been rediscovered as an important alternative to various standard classification methods. This is due to a solid theoretical foundation underlying neural network research, along with recently-achieved strong practical results on challenging real-world problems. Early work established neural networks as universal functional approximators [33, 59, 60], able to approximate any given vector space mapping. As classification is merely a mapping from a vector space to a nominal space, in theory, a neural network is capable of performing any given classification task, provided that a judicious choice of the model is made and an adequate training method in implemented. In addition, neural networks are able to directly estimate posterior probabilities, which provides a clear link to classification performance and makes them a reliable estimator of the optimal Bayes classifier [14, 111, 131]. Neural networks are also referred to as connectionist models.

Theories of neural networks have been developed over many years. Since the late 19th century, there have been attempts to create mathematical models that mimic the functioning of the human nervous system. The discovery by Cajal in 1892 [29] that the nervous system is comprised of neurons communicating with each other by sending electrical signals down their axons, was a breakthrough. A neuron receives various signals from other linked neurons, processes them, and then decides to inhibit or send a signal based on some internal logic, thus acting as an electrical gate. This distributed neuronal process is the mechanism the brain uses to perform its various tasks. This finding led to an increase in research on mathematical models that attempt to simulate such behavior, with the hope, yet to be substantiated, of achieving human-like classification performance.

The first broadly-publicized computational model of the neuron was proposed by McCulloch and Pitts in 1943 [91]. It was a binary threshold unit performing a weighted sum of binary input values, and producing a 0 or 1 output depending on whether the weighted sum exceeded a given threshold. This model generated great hope for neural networks, as it was shown that with a sufficient number of such neurons, any arbitrary function could be simulated, provided that the weights were accurately selected [96]. In 1962, Rosenblatt [114] proposed the *perceptron learning rule*, an iterative learning procedure for single-layer linear threshold units in which the unit can take scalar inputs to produce a binary output. The perceptron learning rule was guaranteed to converge to the optimal set of weights, provided that the target function was computable by the network.

Regrettably, research into neural networks slowed down for almost 15 years, largely due to an influential paper by Minsky and Papert [78] which proved that a single-layer perceptron is incapable of simulating an XOR gate, which severely limits its capabilities. In the 1980s, interest in neural networks was revived, based on a proposal that they may be seen as a memory encoding system. Based on this view, associative networks were proposed utilizing some energy measure to optimize the network. This framework led to the Hopfield network [58] and Boltzmann Machine [1]. In terms of usage for classification, the development of the *back-propagation algorithm* [90, 116], as a way to train a multi-layer perceptron, provided a fundamental breakthrough. The multi-layer perceptron (MLP) and its variants do not have the limitation of the earlier single-layer perceptron models, and were proven to be able to approximate any given vector space mapping. Consequently, MLPs have been the most widely-used neural network architecture. The 1990s witnessed the emergence of various neural network applications, mostly around small or medium-size architectures. Networks with one or two hidden layers were common, as networks with more layers suffered training issues such as slow convergence, local minima problems, and complex learning parameter tuning.

Since the mid-2000s, the focus has been on the ability of neural networks to discover internal representations within their multi-layered architecture [15]. The *deep learning* concept [56] was proposed as a way to exploit deep neural networks with many layers and large numbers of units that maximize the extraction of unseen non-linear features within their layers, and thus be able to accomplish complex classification tasks. With the use of high-performance computing hardware and the utilization of modular and selective learning techniques, deep neural networks with millions of weights have been able to achieve breakthrough classification performance.

In terms of practical use, neural networks have been successfully applied to a variety of real world classification problems in various domains [132] including handwriting recognition [32, 47, 73, 89], speech recognition [25, 87], fault detection [6, 61], medical diagnosis [9, 28], financial markets [124, 126], and more.

In this chapter, we provide a review of the fundamental concepts of neural networks, with a focus on their usage in classification tasks. The goal of this chapter is not to provide an exhaustive coverage of all neural network architectures, but to provide the reader with an overview of widely used architectures for classification. In particular, we focus on layered feedforward architectures.

The chapter is organized as follows. In Section 8.2, we introduce the fundamental concepts underlying neural networks. In Section 8.3 we introduce single-layer architectures. We review Radial Basis Function network as an example of a kernel neural network in Section 8.4. We discuss multi-layer architectures in Section 8.5, and deep neural networks in Section 8.6. We summarize in Section 8.7.

8.2 Fundamental Concepts

An artificial neural network (ANN) or neural net is a graph of connected units representing a mathematical model of biological neurons. Those units are sometimes referred to as *processing units*, nodes, or simply neurons. The units are connected through unidirectional or bidirectional arcs with weights representing the strength of the connections between units. This is inspired from the biological model in which the connection weights represent the strength of the synapses between the neurons, inhibiting or facilitating the passage of signals.

The neural network takes input data from a set of dedicated *input units* and delivers its output via a set of dedicated *output units*. A unit can act as both an input unit as well as an output unit. The remaining units are responsible for the computing logic, as represented by the mathematical model of the biological neuron. For classification problems, a neural network is characterized by the following:

- **The neuron model** or the mathematical model of a neuron that describes how a unit in the network produces an output from its inputs and the role it plays in the network (input unit, output unit, or computing unit)

- **The architecture** or the topology that outlines the connections between units, including a well-defined set of input and output units.

- **The data encoding policy** describing how input data or class labels are represented in the network.

- **The training algorithm** used to estimate the optimal set of weights associated with each unit.

In the remainder of this section, we will review each of these characteristics in more detail.

8.2.1 Mathematical Model of a Neuron

The artificial neuron, as illustrated in Figure 8.1, is a computational engine that transforms a set of inputs $\mathbf{x} = (x_1, \cdots, x_d)$ into a single output o using a composition of two functions as follows:

1. *A net value function* ξ, which utilizes the unit's parameters or weights $\mathbf{w}$ to summarize input data into a *net value*, v, as

$$v = \xi(\mathbf{x}, \mathbf{w}). \tag{8.1}$$

The net value function mimics the behavior of a biological neuron as it aggregates signals from linked neurons into an internal representation. Typically, it takes the form of a weighted sum, a distance, or a kernel.

2. An *activation function*, or *squashing function*, ϕ, that transforms net value into the unit's output value o as

$$o = \phi(v). \tag{8.2}$$

The activation function simulates the behaviour of a biological neuron as it decides to fire or inhibit signals, depending on its internal logic. The output value is then dispatched to all receiving units as determined by the underlying topology. Various activations have been proposed. The most widely-used ones include the linear function, the step or threshold function, and the sigmoid and hyperbolic tangent function, as shown in Figure 8.2.

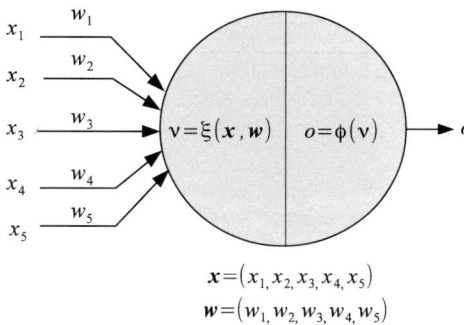

FIGURE 8.1: Mathematical model of an ANN unit. The unit computes a net value as a scalar function of the inputs using local parameters, and then passes the net value to an activation function that produces the unit output.

8.2.2 Types of Units

Given the variety of net value functions and activation functions, various type of neural network units can be derived. However, not all combinations are adequate since the choice of a particular combination depends on the topology of the network and the underlying theoretical justification. The most widely used units include the linear threshold unit, the linear unit, the sigmoidal unit, the Gaussian unit, and the distance unit (Table 8.1).

TABLE 8.1: Type of Units Widely Used in Neural Network

Type of unit	net function	activation function
linear threshold unit	weighted sum	step or sign function
linear unit	weighted sum	linear or piecewise linear
sigmoidal unit	weighted sum	sigmoid or tanh
distance unit	distance	linear or piecewise linear
gaussian unit	distance	gaussian kernel

8.2.2.1   McCullough Pitts Binary Threshold Unit

We start by introducing the McCulloch and Pitts neuron. The McCullough Pitts or binary threshold unit was the first artificial neuron [91] directly inspired from research in neuroscience. The model was a direct attempt to simulate the behavior of a biological neuron in term of its logical functions. It has limited use for data classification tasks, but is a useful historical reference. The input data is binary and binary weights are selected to implement blocking and pass-through effects of nervous signal impulses.

The binary threshold unit uses a weighted sum of inputs as the net value function and a threshold function as the activation function. Hence:

$$\xi(\mathbf{x}, \mathbf{w}) \quad = \quad \mathbf{w}^t \mathbf{x} \tag{8.3}$$

$$= \quad \sum_{i=1}^{d} x_i w_i = v \tag{8.4}$$

$$\phi(v) \quad = \quad \mathbf{1}(v > \theta) \tag{8.5}$$

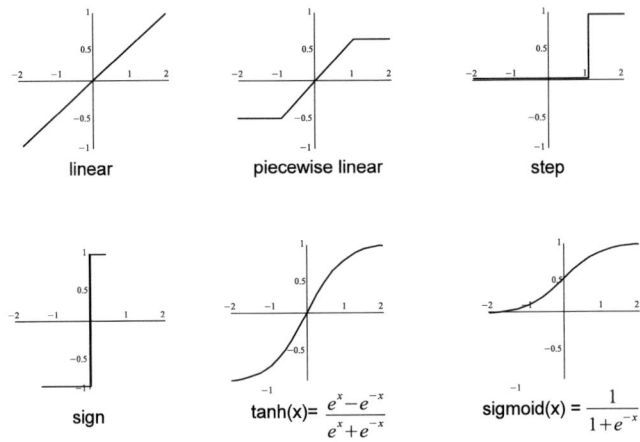

FIGURE 8.2: Widely-used activation functions.

where $\mathbf{w} = (w_1, \cdots, w_d)$ is the binary weight vector representing the unit's parameters, with $w_i \in \{0, 1\}$. The activation function is the step function $\mathbf{1}(b)$ with value equal to 1 when the Boolean value b is true and zero otherwise.

The binary threshold unit performs a two-class classification on a space of binary inputs with a set of binary weights. The net value, as a weighted sum, is similar to a linear regression (notwithstanding the thresholding) and is able to accurately classify a two-class linear problem in this space. It was demonstrated that the binary threshold unit can realize a variety of logic functions including the AND, OR, and NAND gates [91].

However, its inability to perform the XOR gate [114] severely limits its classification capabilities. Another major drawback of the binary threshold unit is the lack of learning algorithms to estimate the weights, given the classification task.

8.2.2.2 Linear Unit

The linear unit uses a weighted sum as the net value function and an affine function as activation function, yielding

$$\xi(\mathbf{x}, \mathbf{w}) = \mathbf{w}^t \mathbf{x} \tag{8.6}$$

$$= \sum_{i=1}^{d} x_i w_i + w_o \tag{8.7}$$

$$\phi(v) = \alpha v$$

where w_o is a bias with $\mathbf{x} = (1, x_1, \cdots, x_d)$ and $\mathbf{w} = (w_o, w_1, \cdots, w_d)$. The activation is a simple linear multiplier with $\alpha = 1$ in most cases. The linear unit is essentially a two-class linear regression model, and inherits linear regression capabilities. In particular, it is a good fit for linearly separable two-class problems. However, the limitation to linear classification tasks can be overcome in the context of a neural network, where various linear units work in concert to implement a distributed processing scheme in a manner similar to linear discrimination functions [35].

The linear unit is widely used across a variety of neural network architectures: there is no restriction a priori on the type of weights, and its smoothness makes it suitable for various gradient-based optimizations.

8.2.2.3 Linear Threshold Unit

The linear threshold unit uses a weighted sum as the net value function with the 0-1 step function as an activation function

$$\phi(v; \theta) = \mathbf{1}(v > \theta). \tag{8.8}$$

The step function provides a degree of non-linearity that was not available with the linear unit. This can potentially increase classification performance beyond linear problems, in particular when used within a neural network architecture. The main issue with the 0-1 step function is its lack of smoothness, which makes that unit non-amenable to optimization via standard gradient descent techniques.

8.2.2.4 Sigmoidal Unit

The sigmoidal unit uses a weighted sum with bias (Equation 8.6) as the value function, but can be implemented using two types of activation functions: the sigmoid function and the hyperbolic tangent (tanh) function. The sigmoid function, defined as

$$\phi(v) = \frac{1}{1 + \exp(-v)} \tag{8.9}$$

maps the net value into interval $[0, 1]$ while the tanh function,

$$\phi(v) = \tanh(v) = \frac{\exp(v) - \exp(-v)}{\exp(v) + \exp(-v)}. \tag{8.10}$$

maps the net value into the interval $[-1, 1]$. The step function is a smooth approximation of the step function while the tanh function is a smooth approximation of the sign function. Smoothness in an activation function is a useful feature as it allows gradient descent optimization techniques and provides a gradual response to various strengths of the net value stimuli. Furthermore, it has been proven that a neural network with sigmoidal units can achieve universal approximation [33, 59, 60]. This important property explains why the sigmoidal unit is the most widely used unit and the predominant unit in feedforward architectures.

8.2.2.5 Distance Unit

The distance unit uses a distance measure as the net value function and a linear function as the activation function, yielding

$$\xi(\mathbf{x}, \mathbf{w}) \quad = \quad ||\mathbf{x} - \mathbf{w}|| \tag{8.11}$$

$$\phi(v) \quad = \quad \alpha v \tag{8.12}$$

with a typical value of $\alpha = 1$, where $\mathbf{x} = (x_1, \cdots, x_d)$ and $\mathbf{w} = (w_1, \cdots, w_d)$. This unit computes the distance between the inputs $\mathbf{x}$ and some prototype vector $\mathbf{w}$ [75]. A commonly used distance is the Euclidean distance, i.e. $\xi(\mathbf{x}; \mathbf{w}) = \sum_i (x_i - w_i)^2$.

8.2.2.6 Radial Basis Unit

The radial basis unit uses a distance function as the net value function and a kernel function as the activation function, producing

$$\xi(\mathbf{x}, \mathbf{w}) \quad = \quad ||\mathbf{x} - \mathbf{w}|| \tag{8.13}$$

$$\phi(v) \quad = \quad \exp(-v) \tag{8.14}$$

for a parameter value $\mathbf{w}$ characterizing the unit's parameters. In Equation 8.14, we use a Gaussian kernel as activation function, which is the primary choice for classification tasks. The net value typically uses a scaled Euclidean distance or the Mahalanobis distance. The radial basis provides a local response activation similar to local receptive patterns in brain cells, centred at the unit's parameters. Hence, the unit is activated when the input data is close in value to the its parameters. The Gaussian kernel enables the unit to achieve non-linear classification capabilities, while enabling better local pattern matching [18].

8.2.2.7 Polynomial Unit

The polynomial unit [116] computes the net value as a polynomial transform of the input data components:

$$\xi(\mathbf{x};\mathbf{w}) = \sum_i w_i x_i + \sum_{i,j} w_{ij} x_i x_j + w_o. \tag{8.15}$$

The associated activation function is typically a linear or a sigmoid function. Polynomial units are infrequently used for classification tasks and we do not discuss them in more detail.

8.2.3 Network Topology

The network topology refers to the underlying graph of units, representing the connectivity between different types of units. Connections are often unidirectional, with the arcs representing flow of data in the network. A unit typically has many successor units to which it passes its outputs. Neural network architectures for classification require at least one input layer, whose role is to bring data to the network, and one output layer, whose role is to provide the network's outputs and the classification decision. The topology can be free-style, layered, modular, or with feedback, depending on the particular assumptions.

8.2.3.1 Layered Network

A network with a layered architecture organizes its units into hierarchical layers, where each layer provides a specific functionality such as receiving input data, performing internal transformations, or providing the network's output. Units in the input layer do not perform any computation; their role is merely to ingest data into the network, and as such, are not included in the layer count. So, a single-layer network comprises an input layer and output layer. A two-layer network is made of an input layer, a single hidden layer, and output layer. Typical layered architectures for classification include:

- Feedforward networks or multi-layer perceptron (MLP), which features an input layer, one or several hidden layers, and an output layer. The units in the feedforward networks are linear or sigmoidal units. Data move from the input layer, through the hidden layers, and up to the output layer. Feedback is not allowed.

- Radial Basis Function (RBF) networks [115] made of a single hidden layer with kernel units and output layer with linear units.

- The Leaning Vector Quantization (LVQ) network [75], featuring a single layer network where input units are connected directly to output units made of distance units.

8.2.3.2 Networks with Feedback

Networks with feedback implement autoregressive computations, either at the unit-level or at the layer level. Standard examples of networks with feedbacks include recurrent feedforward neural networks [113] or recurrent radial basis function networks [17].

8.2.3.3 Modular Networks

Modular networks create a modular representation of the classification task. A module is a neural network designed for a specific task such as performing classification within a subset of categories; for example, classifying only consonants in the phoneme classification task. Modules can be organized hierarchically as done with the large-phonetic Time Delay Neural Network (TDNN-LR) [120] or integrated as mixture of experts [65].

8.2.4 Computation and Knowledge Representation

At the unit level, computation is a two-stage process, in which the unit computes the net value using the net value function, and then computes its output using the activation function. At the network level, computation starts with a presentation of the data to the input units in the network. The input units then transmit the data (without modifications) to the connected successor units. These units are then activated, each one computing its net value and output, and in turn transmitting that output to the connected recipient units. This process is called *spreading activations* or *forward propagation*, as activation values spread from the input layer up to the output layer in a sequential fashion. Depending on the network connectivity, the spreading activation process can be implemented in parallel distributed system or a PDP [117].

In networks without feedback, the process of activation and propagation ends with the production of outputs from the output layer. In networks with feedback, such as recurrent networks, the activation propagation never ends, instead following a dynamic trajectory through the state space, as units and layers are continuously updated.

The *state* of the network is represented by the values of all weights for all units. The values of the weights, the connections between units, and the type of net value functions and activation functions are the mechanism used by the network to encode knowledge learned from the data — similar to the way the brain operates.

8.2.5 Learning

Neural network learning for classification is the estimation of the set of weights that optimally achieve a classification task. Learning is done at the network level to minimize a classification error metric. This includes schemes for each unit to update its weight, to increase the overall performance of the network. Each class of networks provides a learning scheme suited for its particular topology. In principle, the update of weight w_{ij} between neuron j getting input x_i from neuron i and producing an output o_j should reflect the degree to which the two neurons frequently interact in producing the network's output. The weight update Δw_{ij} is thus of the form

$$\Delta w_{ij} = F(x_i, o_j, w_{ij}, \eta) \tag{8.16}$$

where η is some learning rate parameters and $F()$ some multiplicative function. Various update rules are proposed to reflect this broad principle.

8.2.5.1 Hebbian Rule

The Hebbian learning rule [34] states that a weight between two units should be strengthened according to how frequently they jointly participate in the classification task. This led to a formula of the type:

$$\Delta w_{ij} = \eta x_i o_j \tag{8.17}$$

where the update is proportional to some function of input and output values. The Hebbian rule is unsupervised since it does not make use of a targeted output. Also, since the activation function is not explicitly used, it is applicable to all types of units including threshold units.

8.2.5.2 The Delta Rule

The delta rule, also called the Widrow-Hoff learning rule (see Section 8.3.2), is a gradient descent on a square error between the net value v_j and a desired output t_j, yielding

$$\Delta w_{ij} = \eta(t_j - v_j)x_i. \tag{8.18}$$

The above form of the delta can be used on all types of units, provided the availability of a desired output. However, this form of the delta rule does not use the unit's output, which results in a discrepancy between the system used for classification and the one used for training, especially for non-linear units. The generalized form of the delta rule provides a remedy to this issue by embedding the derivative of the activation function, producing an update in the form

$$\Delta w_{ij} = \eta(t_j - o_j)\phi'(v_j)x_i \tag{8.19}$$

where $\phi'()$ is the derivative of the activation function. The generalized delta rule requires the use of a smooth activation function and is the form widely used for classification tasks in many network architectures. It is a supervised training scheme directly derived from a gradient descent on square error between a target function and the output. Training starts with weights initialized to small values.

8.3 Single-Layer Neural Network

A single-layer neural network features one input layer and one output layer. The simplicity of these networks makes them suitable for introducing the inner workings of neural networks. In this section, we review a few standard single-layer networks widely used for classification problems. This architecture includes the perceptron, the adaline network, and the Linear Vector Quantization (LVQ).

8.3.1 The Single-Layer Perceptron

The single-layer perceptron (SLP), as introduced by Rosenblatt [114], is one of the earlier and simplest neural network architecture. The output layer is made of a single linear threshold unit with weights **w** that receives input **x** from a set of input neurons. That output unit uses the sign function as an activation function, thus implementing a two-class classification task onto the space $\{-1, 1\}$, where the class decision is based on the sign of the dot product $\mathbf{w}^t\mathbf{x}$.

8.3.1.1 Perceptron Criterion

We are given the training data $X = \{\mathbf{x}_1, \cdots, \mathbf{x}_N\}$ with associated desired outputs $\{t_1, \cdots, t_N\}$ ($t_n \in \{-1, 1\}$) where -1 represents class C_1 and 1 represents C_2. For each input data $\mathbf{x}_n$, the perceptron generates an output $o_n \in \{-1, 1\}$. Training minimizes the perceptron criterion $\mathcal{E}_p$ [19, 35] defined over the training set as

$$\mathcal{E}_p = -\sum_{n, o_n \neq t_n} t_n \mathbf{w}^t \mathbf{x}_n \tag{8.20}$$

where the sum is performed over misclassified examples ($o_n \neq t_n$). This criterion attempts to find the linear hyperplane in the **x**-space that separates the two classes at the origin. To see this, observe that the dot product $-t_n\mathbf{w}^t\mathbf{x}_n$ is always positive for a misclassified $\mathbf{x}_n$. Therefore, minimizing the

---

Algorithm 8.1 Single-layer perceptron learning

Input: Training data $X = \{\mathbf{x}_1, \cdots, \mathbf{x}_N\}$ and associated labels $\{t_1, \cdots, t_N\}$; ($\mathbf{x}_n \in \mathbb{R}^D$, $t_n \in \{0,1\}$); learning rate $\eta > 0$.
Output: A trained single-layer perceptron with optimal set of weights $\mathbf{w}$.

1: Randomly initialize the set of weight $\mathbf{w} = (w_1, \cdots, w_d)$ to zero.
2: **repeat**
3:     **for** n in $\{1, ..., N\}$ **do**
4:         Compute the neuron output $o_n = \text{sign}(\mathbf{w}^t \mathbf{x}_n)$
5:         **if** $o_n \neq t_n$ **then**
6:             $\mathbf{w} \leftarrow \mathbf{w} + \eta \mathbf{x}_n$ # misclassified data
7:         **end if**
8:     **end for**
9: **until** All training data are correctly satisfied or a specified minimum number of iterations are executed

---

perceptron criterion is equivalent to solving the linear inequality $t_n \mathbf{w}^t \mathbf{x}_n > 0$ for all $\mathbf{x}_n$, which can be done either incrementally, one sample at a time, or in batch fashion.

The SLP learning is shown in Algorithm 8.1. It is a equivalent to stochastic gradient descent on the perceptron criterion. Some variants of the algorithm use a linearly decreasing learning rate $\eta(n)$ as function of n. For each misclassified sample $\mathbf{x}_n$, the training process adds the data to the weight when data belong to class C_2 and subtracts the input data from the weights, when the data belong to C_1. The effect of this label-based selective update is to incrementally re-align the separating hyperplane to the direction that optimally separates the two classes [19,35]. The batch version of the perceptron algorithm performs a similar update on the accumulated sum of misclassified samples:

$$\mathbf{w} \leftarrow \mathbf{w} + \eta(t) \sum_{n, o_n \neq t_n} \mathbf{x}_n t_n \tag{8.21}$$

for a linearly decreasing learning $\eta(t)$ as a function of training time t. The update is done up to a specified number of epochs or when all data have been correctly classified. It is a direct application of the delta rule since weights are updated according to their contribution to the error.

The *perceptron convergence theorem* guarantees the optimality of the process: The learning algorithm, as illustrated in Algorithm 8.1, will find the optimal set of weights within a finite number of iterations, provided that the classes are linearly separable [35,52,114].

For non-linear problems, there is no guarantee of success. In that scenario, learning is aimed at achieving classification performance within a tolerable margin of error γ, such that

$$t_n \mathbf{w}^t \mathbf{x}_n > \gamma \tag{8.22}$$

for all n. The perceptron learning then minimizes misclassification error relative to that margin, which gives

$$\mathbf{w} \leftarrow \mathbf{w} + \eta \mathbf{x}_n \quad \text{if} \ -t_n \mathbf{w}^t \mathbf{x}_n > \gamma. \tag{8.23}$$

For a $\gamma > 0$ satisfying the condition in Equation 8.22, the *perceptron margin theorem* provides a bound on the number of achievable errors. Suppose there exists a normalized set of weights $\mathbf{w}$ such that $\| \mathbf{w} \| = 1$ and that the entire training set is contained in a ball of diameter D ($\| \mathbf{x}_n \| < D$ for all n), then the number of mistakes M on the training set X made by the perceptron learning algorithm is at most $\frac{D^2}{\gamma^2}$. However, a priori knowledge of the optimal γ is hardly available. In practice, γ can be set to

$$\gamma = \min_n t_n \mathbf{w}^t \mathbf{x}_n \tag{8.24}$$

after completion of the learning scheme, and that value can be used to provide a bound on the error rate. Note that the bound does depend on the dimensionality of the input data, making the perceptron applicable to large-dimensional data. To remove dependency on D, it is common to normalize each training data by $\frac{\mathbf{x}_n}{||\mathbf{x}_n||}$, which produces the bound of the form $\frac{1}{\gamma^2}$.

8.3.1.2 Multi-Class Perceptrons

The original perceptron algorithm deals with a two-class problem and uses a single output unit with a threshold activation. For classification problems with $K > 2$ classes, $C_1, \cdots, C_K$, the perceptron features K output units in the output layer, with one output unit per class, and all output units getting the same input data. Each output j has its own set of weights $\mathbf{w}_j$. This is equivalent to a multi-class linear machine [35, 104], where the j-th output unit implements a linear discrimination function for class C_j. For an input data $\mathbf{x} = (1, x_1, \cdots, x_d)$, each output unit j computes a net value $v_j = \mathbf{w}_j^t \mathbf{x}$ and an output value $o_j = v_j$ where column-vector $\mathbf{w}_j = (w_{oj}, w_{1j}, \cdots, w_{dj})$ is the set of weights with w_{ij} representing the connection between the i-th input unit, and the j-th output unit and w_{oj} is a bias. In contrast to the two-class perceptron, output units are linear units; threshold activations are not used. The use of linear activation allows for a gradual comparison of the scores produced by the outputs, which is not possible with threshold units producing either -1 or 1.

Classification is done by selecting the class of the output unit that yields the highest output value:

$$\mathbf{x} \in C_k \qquad \text{if } k = \arg\max_j o_j. \tag{8.25}$$

Supervised training of the multi-class perceptron is aimed at generating a positive value o_k if the input data belongs to the representative category C_k and a negative value if the input data belongs to a category other than C_k. To achieve this outcome, the multiclass perceptron learning implements a competitive learning scheme whenever there is misclassification. Given training data $X = \{\mathbf{x}_1, \cdots, \mathbf{x}_N\}$ with associated desired outputs $\mathbf{t}_n = (t_{n1}, \cdots, t_{nK})$ with t_{nj} encoded using the principle of 1-of-K in the set $\{-1, 1\}$ as

$$t_{nk} = \begin{cases} 1 & \text{if } \mathbf{x}_n \in C_k \\ -1 & \text{otherwise,} \end{cases} \tag{8.26}$$

then, for an input data $\mathbf{x}_n$ of the class C_k that generates an output o_k the multi-class perceptron performs an update whenever o_k is misclassified as a class C_q:

$$\mathbf{w}_j \leftarrow \mathbf{w}_j + \eta t_{nj} \mathbf{x}_n \qquad \text{for } j \in \{k, q\}. \tag{8.27}$$

Similar to the two-class case, the update occurs only when the input data have been misclasssified. The effect is to increase the weight of the output vector of the right category ($\mathbf{w}_k \leftarrow \mathbf{w}_k + \eta \mathbf{x}_n$) and decrease the weight of competing category $\mathbf{w}_q \leftarrow \mathbf{w}_q - \eta \mathbf{x}_n$. The multi-class perceptron learning is shown in Algorithm 8.2.

8.3.1.3 Perceptron Enhancements

The perceptron is essentially a linear discrimination machine. Performance is guaranteed only for linearly separable classification problems within a finite number of iterative steps [35, 114]. Also, the perceptron is very sensitive to noise, which severely limits its performance in real-world scenarios.

The use of a margin γ as a tradeoff in performance enables us to adapt the perceptron to challenging non-linear scenarios. To overcome some learning issues encountered by the original perceptron in noisy settings, more robust variants of the perceptron learning have been proposed. The pocket algorithm [44] keeps a buffer of good weights in a "pocket" during the learning process and uses

---

Algorithm 8.2 Multi-class perceptron learning

Input: Training data $X = \{\mathbf{x}_1, \cdots, \mathbf{x}_N\}$ and learning rate $\eta > 0$.
Output: A trained single-layer multi-class perceptron with optimal set of weights $\mathbf{w}$.

1: initialize the set of weight $\mathbf{W} = (\mathbf{w}_1, \cdots, \mathbf{w}_K)$ to zero; K number of classes.
2: **repeat**
3:     **for** n in $\{1, \cdots, N\}$ **do**
4:         **for** j in $\{1, \cdots, K\}$ **do**
5:             Compute the j-th neuron output $o_{nj} = \mathbf{w}_j^t \mathbf{x}_n$
6:         **end for**
7:         Get $q = \arg\max_j o_{nj}$
8:         Get k index of correct category of $\mathbf{x}_n$
9:         **if** $k \neq q$ **then**   # misclassified data
10:             $\mathbf{w}_k \leftarrow \mathbf{w}_k + \eta\mathbf{x}_n$ # adjust right category
11:             $\mathbf{w}_q \leftarrow \mathbf{w}_q - \eta\mathbf{x}_n$ # adjust wrong category
12:         **end if**
13:     **end for**
14: **until** All training data are correctly satisfied or a specified minimum number of iterations are executed

---

the good weight estimate from the pocket, whenever the standard perceptron is suboptimal. The Adatron [3] proposes an efficient adaptive learning scheme and the voted-perceptron [41] embeds a margin maximization in the learning process. In fact, it has been demonstrated that the voted-perceptron can achieve similar performance as a kernel-based Support Vector Machine while being a much simpler and much easier model to train.

8.3.2 Adaline

The Adaptive Linear Neuron (ADALINE) was proposed by Widrow and Hoff [5] in 1960 as an adaptation of the binary threshold gate of McCullough and Pitts [91]. Like the perceptron, the adaline is a single-layer neural network with an input layer made of multiple units, and a single output unit. And like the perceptron, the output unit is a linear threshold unit. However, while the perceptron focuses on optimizing the sign of the net value function, the adaline focuses on minimizing the square error between the net value and the target. We start by describing the original adaline network aimed at a two-class problem and then describe extensions for multi-class settings.

8.3.2.1 Two-Class Adaline

For a two-class classification problem, C_1 and C_2 with label -1 and -1, the adaline features a single threshold linear neuron of weight $\mathbf{w} = (w_1, \cdots, w_d)$ that received data from the input nodes and produced an ouput $o \in \{-1, 1\}$ similar to the perceptron. Unlike the perceptron the adaline optimizes the cost function $\mathcal{E}_a$, defined as the sum of local mean square errors:

$$\mathcal{E}_a = \sum_n (t_n - v_n)^2 \tag{8.28}$$

with t_n as the desired output of sample $\mathbf{x}_n$, and with $t_n \in \{-1, 1\}$ and v_n as the net value of the output unit. This error is defined using the net value instead of the output value. Minimization of the adaline criterion is an adaptive (iterative) process implemented as the Widrow-Hoff learning rule, also called the delta rule:

$$\mathbf{w} \leftarrow \mathbf{w} + \eta(t_n - v_n)\mathbf{x}_n \tag{8.29}$$

Algorithm 8.3 Single-layer adaline

Input: Training data $X = \{\mathbf{x}_1, \cdots, \mathbf{x}_N\}$ and associated labels $\{t_1, \cdots, t_N\}$; ($\mathbf{x}_n \in \mathbb{R}^d$, $t_n \in \{-1, 1\}$); learning rate $\eta > 0$.
Output: A trained adaline with optimal set of weights $\mathbf{w}$.

1: Randomly initialize the set of weights $\mathbf{w} = (w_1, \cdots, w_d)$ to zero
2: **repeat**
3:     **for** n in $\{1, \cdots, N\}$ **do**
4:         Compute the output unit net value $v_n = \mathbf{w}_n^t \mathbf{x}$
5:         Compute the output $o_n = \text{sign}(v_n)$
6:         **if** $o_n \neq t_n$ **then**   # misclassified data
7:             $\mathbf{w} \leftarrow \mathbf{w} + \eta(t_n - v_n)\mathbf{x}_n$
8:         **end if**
9:     **end for**
10: **until** All training data are correctly satisfied or a specified minimum number of iterations are executed

---

for incoming data $\mathbf{x}_n$ and with a learning rate $\eta > 0$ as shown in Algorithm 8.3.

The learning update in Equation 8.29 converges in the L_2-norm sense to the plane that separates the data (for linearly separable data). Widrow-Hoff also proposed an alternative to Equation 8.29 to make the update insensitive to the size the input data :

$$\mathbf{w} \leftarrow \mathbf{w} + \eta(t_n - v_n)\frac{\mathbf{x}_n}{\|\mathbf{x}_n\|} \tag{8.30}$$

where $\|\|$ refers to the Euclidean norm. The Widrow-Hoff learning method has the following characteristics. The direction of the weight update $(t_n - v_n)\mathbf{x}_n$ has same effect as performing a gradient descent on E_a. The adaline error is defined at the net value level and not at the output level, which means that the Widrow-Hoff rule behaves like a linear regression estimation [19, 35]. If the unit is non-linear, the output value of the unit is not used during training, yielding a clear discrepancy between the optimization done during training and the system used for classification.

8.3.2.2 Multi-Class Adaline

The adaline network for $K > 2$-class classification, $C_1, \cdots, C_K$, uses K output units, with one output unit per class in a manner similar to that of the multi-class perceptron. All output units get the same input data and each output unit j is a linear unit with weight $\mathbf{w}_j$. The error is defined as the mean squared error between a vector of net values and a vector of target values:

$$E_a = \sum_n \sum_{k=1}^{K} (t_{nk} - v_{nk})^2. \tag{8.31}$$

The predicted class corresponds to the output unit with the highest output function:

$$\mathbf{x} \in C_j \qquad \text{if } j = \arg\max_{k=1}^{K} o_k \tag{8.32}$$

where $o_k = v_k$ for all k, given a sequence of training data $\{\mathbf{x}_1, \cdots, \mathbf{x}_N\}$ and corresponding target vector $\mathbf{t}_n = (t_{n1}, \cdots, t_{nK})$ with t_{nj} encoded using the principle of 1-of-K in the set $\{-1, 1\}$.

Multi-class learning applies the adaline learning for each weight $\mathbf{w}_j$:

$$\mathbf{w}_j \leftarrow \mathbf{w}_j + \eta(t_{nj} - v_{nj})\mathbf{x}_n. \tag{8.33}$$

Again, this is equivalent to a stochatic gradient descent on the objection function.

8.3.3 Learning Vector Quantization (LVQ)

The LVQ [75] is a single-layer neural network in which the output layer is made of distance units, referred to as prototypes or references [93]. LVQ's attractiveness is derived from its simple yet powerful classification capabilities. Its architecture provides a natural fit to multi-class problems, and unlike alternative classification schemes such as support vector machines or multi-layer perceptrons, the optimized system is not a black box and is easy to interpret.

For a K-class problem, each C_j has a dedicated set of Q output units with parameters $\mathbf{w}_{jq}$ ($q \in \{1,..,Q\}$), of the same dimension as the input data, producing the output

$$o_{jq} = ||\mathbf{x} - \mathbf{w}_{jq}|| \tag{8.34}$$

where the distance is typically the Euclidean distance. The output layer is made of $K \times Q$ output units. The classification decision selects the class label of the closest prototype:

$$\mathbf{x} \in C_k \quad \text{if} \quad k = \arg\min_j \left\{ \arg\min_q o_{jq} \right\}. \tag{8.35}$$

The classification decision is the class label of the prototype that yields the smallest output, or equivalently, is the closest prototype to input data.

LVQ learning attempts to find the best prototypes that compress the data while achieving the best classification performance. Training falls in the family of discriminative training since prototypes in the output layer are selectively updated based on the class label of the input data. Prototypes of the same class as the input data are moved closer to the data point and prototype of competing classes are moved further. This is done selectively and iteratively. Three training algorithms are available: LVQ1, LVQ2, and LVQ3 [74]. We review LVQ1 and LVQ2 as the most widely used training schemes for LVQ, LVQ3 being very similar to LVQ2.

8.3.3.1 LVQ1 Training

LVQ1 selectively pulls or pushes the output unit's parameters depending on the label of the incoming data. The process is as follows. For an input data at training time t, first select the "winner" prototype, which is the output unit of parameters $\mathbf{w}_{jq}$ that is the closest to the input. Then, update the weight of that prototype as

$$\mathbf{w}_{jq}(t) = \begin{cases} \mathbf{w}_{jq}(t-1) + \eta(t)(\mathbf{x} - \mathbf{w}_{jq}) & \text{if } \mathbf{x} \text{ belongs to } C_j \\ \mathbf{w}_{jq}(t-1) - \eta(t)(\mathbf{x} - \mathbf{w}_{jq}) & \text{otherwise} \end{cases} \tag{8.36}$$

where $\eta(t) > 0$ is a monotically decreasing learning rate as a function of t [92].

At each training iteration, LVQ1 only updates the winning prototype, moving it closer or further from the input data depending on the input data being correctly classified or not. For a correct classification, the winning prototype is moved closer to the input data. For a misclassified sample, the winning prototype is moved farther from the data. Intuitively, LVQ1 re-arranges the prototypes to generate cluster of data that matches the categories of data.

8.3.3.2 LVQ2 Training

LVQ1 training only adjusts the winning prototype, which may not be the optimal process in terms of margin maximization and better generalization. In contrast, LVQ2 focuses on discriminating data at the class boundaries, by jointly updating the closest two winning prototypes of different categories. For an incoming data of class C_k, training only occurs when the following conditions are met: 1) the winning prototype, $\mathbf{w}_{jq}$, has an incorrect label C_j; 2) the second winner, $\mathbf{w}_{kp}$, in the remaining classes has the correct label; 3) the input data is inside a sphere centred in the middle of the two selected winning prototypes $\mathbf{w}_{jq}$ and $\mathbf{w}_{kp}$, ensuring that the two winning prototypes are close to the boundary between the two competing categories. The LVQ2 update is as follows:

$$\mathbf{w}_{jq}(t) = \mathbf{w}_{jq}(t-1) - \eta(t)(\mathbf{x} - \mathbf{w}_{jq}) \quad \text{for the first winner but incorrect category} \quad (8.37)$$

$$\mathbf{w}_{kq}(t) = \mathbf{w}_{kq}(t-1) + \eta(t)(\mathbf{x} - \mathbf{w}_{kq}) \quad \text{for the second winner and correct category} \quad (8.38)$$

LVQ2 training selects and updates the two winning prototypes, whenever misclassification occurs and when these prototypes are directly competing. This is slightly similar to the multi-class perceptron model except that the goal of the LVQ2 is to minimize the distance to the correct prototype while the perceptron minimizes the dot product between the input and the weights.

LVQ can be seen as a supervised extension of vector quantization (VQ), Self-Organization Map (SOM) [76], or k-means clustering. Those algorithms compress data into a set of representative prototypes and generate clusters of data centered at prototypes. These unsupervised techniques are generally used as a starting point for LVQ training, which is then used to readjust the clusters, boundaries based on the data labels. Clustering can generate initial prototypes using the entire data set but it is more effective to perform clustering within each category, thus generating initial prototypes per category, and then use LVQ competitive training to readjust the boundaries [93].

8.3.3.3 Application and Limitations

LVQ inherits the benefits of prototype-based learning algorithms in terms of theoretically proven generalization ability for high dimensional data and are characterized by dimensionality independent large-margin generalization bounds [30]. LVQ has been successfully applied to a wide range of data classification tasks. In particular, an LVQ system was able to achieve similar performance as a more complex neural network in a speech sounds classification task [93]. However, the original LVQ formulation suffers from several drawbacks. One drawback is the lack of an objective function being optimized; LVQ training was derived intuitively and used heuristics that may lead to instability or sensitivity to learning parameters [13]. Another drawback is the use of the Euclidean distance, which limits the performance capabilities of the system.

Various generalizations of LVQ have been proposed to overcome some of these issues. The generalized LVQ (GLVQ) [119] proposes an objective function that quantifies the difference between the closest but wrong category and the second closest but correct category. Minimization, based on gradient descent techniques, is more stable than the original LVQ version. Similarly, LVQ can be seen as a particular instance of the prototype-based minimum classification error (PBMEC) [94, 106]. PBMEC implements an Lp-norm of distances to prototypes with a discriminative objective function whose optimization yields an update similar to LVQ2. Also, alternative distance measures have been used, including the Mahalanobis distance, and the use of kernels with better success [50, 121]

8.4 Kernel Neural Network

We used the term kernel-based neural network for a class of neural network that use a kernel within their architecture. The most prominent example of this type is the Radial Basis Function Network, which is the focus of this section.

8.4.1 Radial Basis Function Network

A radial basis function network (RBFN) [108] implements a two-layer architecture, featuring kernel units in the hidden layer and linear units in the output layer. Although RBFNs were derived

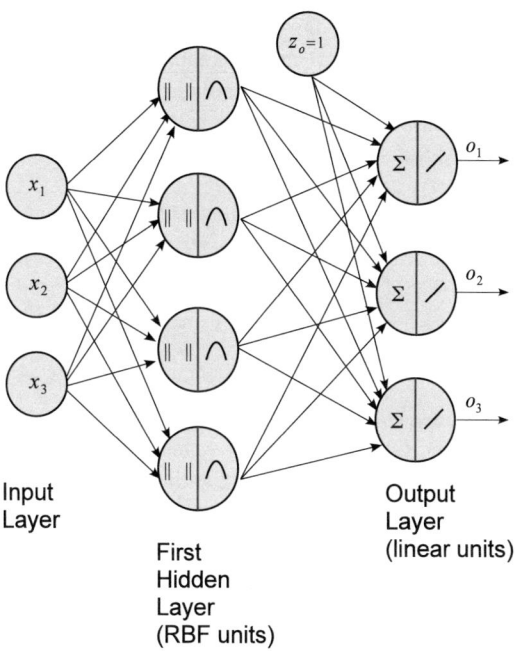

Input Layer

First Hidden Layer (RBF units)

Output Layer (linear units)

FIGURE 8.3: An RBF network with radial basis unit in the hidden layer and linear units in the output layer.

from function approximation and regularization theory with regression as targeted applications [98], they have been successfully adapted to classification problems [101, 102]. Theoretically, RBFNs are universal approximators, making them powerful neural network candidates for any data classification task [45]. In practice, a judicious choice of the kernel should be done with an adequate training scheme. For classification problems, a Gaussian activation is the most widely used kernel and is the one illustrated throughout this section.

An RBF network is illustrated in Figure 8.3. Let us assume a K-class classification problem with an RBFN of Q hidden units (called radial basis functions) and K output units. Each q-th radial basis unit in the hidden layer produces an output z_q from the input units:

$$v_q = ||\mathbf{x} - \mu_q|| \tag{8.39}$$

$$z_q = \exp\left(-\frac{v_q}{2}\right) \tag{8.40}$$

where a common choice for the norm is the quadratic function

$$||\mathbf{x} - \mu_q|| = (\mathbf{x} - \mu_q)^t \Sigma_q (\mathbf{x} - \mu_q) \tag{8.41}$$

and where Σ_q is a positive definite matrix. μ_q and Σ_q are the radial basis center and the width or radii of hidden unit q for $q \in \{1, \cdots, Q\}$. The outputs of the hidden layer are passed to the output layer, where each k-th output unit generates

$$o_k = \mathbf{w}_k^t \mathbf{z} \tag{8.42}$$

$$\tag{8.43}$$

with $\mathbf{z} = (1, z_1, \cdots, z_q)$ and $\mathbf{w}_k = (w_{ok}, w_{1k}, \cdots, w_{Qk})$ as the set of weights characterizing the k-th output unit. An RBFN is fully determined by the set of μ_q, Σ_q, w_{qk} for $q \in \{0, 1, \cdots, Q\}$ and $k \in \{1, \cdots, K\}$.

Similar to the output unit in the multi-class single-layer perceptron, the output units have a set of dedicated weights and are linear units implementing a linear transform on the features generated by the hidden units. The use of a kernel in the hidden layer introduces needed non-linearity capabilities and is equivalent to a projection into a high-dimensional feature space, a feature shared by all kernel machines. Clearly, an o_k is a linear combination of kernels in a manner similar to that of SVMs. A key difference with SVMs is the training scheme and the selection of kernels' parameters.

The classification decision uses the maximum score rule, yielding:

$$\mathbf{x} \in C_j \qquad \text{if } j = \arg\max_k o_k \tag{8.44}$$

for all $k = \{1, \cdots, K\}$.

8.4.2 RBFN Training

RBFN training minimizes the mean square error between the output vector and the target vector, presented as 1-of-K encoded vectors, typically in $\{0, 1\}$ space. RBFN training is done in two steps. The first step determines the set of centers and radiis in the hidden layer. The second step optimizes a classification criterion on the set of weights connecting the hidden layer to the output layer.

8.4.2.1 Using Training Samples as Centers

Earlier formulations of RBFN training advocated the use of all training data as centers [100]; RBFNs are then viewed as a linear combination of radial basis functions centered at each sample in the data set. The Euclidean distance is used with a standard deviation σ used as radii and shared by all hidden units. While this approach is fast as it does not require explicit training and is useful for interpolation problems, which were the earlier motivations of the RBF theory, it may not be convenient for classification problems since the size of the network increases with the size of the training data set. Furthermore, this approach is prone to overfitting, leading to poor performance in deployed settings.

8.4.2.2 Random Selection of Centers

Instead of using the entire training set as centers, an alternative is to randomly select a few samples as centers. That approach is a valid alternative, provided that the training data are uniformly distributed across categories [88]. The radii σ is estimated from the spread of the selected centers and is shared by all units. One such choice is $\sigma = \frac{d}{\sqrt{2M}}$ where d is the maximum distance between centers and M is the number of centers [51].

8.4.2.3 Unsupervised Selection of Centers

An alternative to random selection of centers from the training data is to use clustering techniques to compress the data into a set of prototypes. The prototypes are then used as centers in the hidden layers and the radiis or covariance matrices are estimated from the clusters. This also enables us to estimate a covariance matrix for each hidden unit, thus enhancing modeling capabilities. Unsupervised clustering techniques, such as k-means [21] or Self-Organizing Map (SOM) [40, 77], can be used in this approach. The clustering process can be done on the entire training set. However, a better way is to do clustering within each class [101, 102], generating class-specific centers. The width is derived with some estimate of the within-cluster variance [101] or optimized by gradient descent techniques [102].

8.4.2.4 Supervised Estimation of Centers

Supervised clustering techniques, such as LVQ, can be used to further enhance the set of initial prototypes, thus making use of class labels in the estimation of centers. The use of supervised center estimation has been proven to provide performance superior to a multilayer perceptron in some scenarios [130].

8.4.2.5 Linear Optimization of Weights

The second step in RBFN training estimates the parameter of the output units assuming that hidden units have been previously optimized in the first step. For an input data $\mathbf{x}_n$, generating the set of hidden activations $\mathbf{z}_n$ with target vector $\mathbf{t}_n = (t_{n1}, \cdots, t_{nK})$ with t_{nj} encoded using the principle of 1-of-K in the set $\{0, 1\}$, the goal is to solve the set of linear equations:

$$\mathbf{t}_n = \mathbf{W}^t \mathbf{z}_n \tag{8.45}$$

for all n, where $\mathbf{W} = (\mathbf{w}_1, \cdots, \mathbf{w}_K)$ is the matrix of weight (a column k of $\mathbf{W}$ is the weight-vector of k-th output unit). In matrix form, Equation 8.45 can be rewritten as

$$\mathbf{T} = \mathbf{W}^t \mathbf{Z} \tag{8.46}$$

where $\mathbf{T}$ and $\mathbf{Z}$ correspond to the matrix made of target-vectors and hidden activation vectors, respectively. The general solution to this equation is given by

$$\mathbf{W} = \mathbf{Z}^\dagger \mathbf{T} \tag{8.47}$$

where $\mathbf{Z}^\dagger$ is the pseudo-inverse of $\mathbf{Z}$ defined as $\mathbf{Z}^\dagger = (\mathbf{Z}^t \mathbf{Z})^{-1} \mathbf{Z}$. The pseudo-inverse can be computed via linear solvers such as Cholesky decomposition, singular value decomposition, orthogonal least squares, quasi Newton techniques, or Levenberg-Marquardt [46].

8.4.2.6 Gradient Descent and Enhancements

A global training of the entire networks including centers, radiis, and weights is possible using gradient descent techniques [27, 64]. The objective function is the square error E_{rbf} between the outputs and target vectors:

$$E_{rbf} = \sum_n \sum_{k=1}^K (t_{nk} - o_{nk})^2 \tag{8.48}$$

given a sequence of training data and corresponding target vector $\mathbf{t}_n$. The initial set of centers and weights could be derived from one of the center estimation methods outlined earlier. Weights are usually randomly initialized to small values of zero means and unit variance.

Alternatively, the gradient descent could be performed selectively only on weights while keeping the centers and radii fixed. This has the advantage to avoid matrix inversion as required by the linear solver techniques. It was demonstrated that a wide class of RBFNs feature only a unique minima in the error surface, which leads to one single solution for most training techniques [98].

RBFs can be sensitive to noise and outliers due to the use of cluster centroids as centers. One alternative to this problem is to use a robust statistics learning scheme, as proposed by the mean and median RBF [22, 23].

8.4.3 RBF Applications

The radial basis function capabilities are derived from the use of non-linear activation in the hidden unit. In particular, the use of the kernel is equivalent to the transformation of the original input data into a higher dimensional space, where data are linearly separable. A proof of this result is provided by the Cover theorem [51]. Some RBF variants use a sigmoid at the output layer, further enhancing classification capabilities.

RBFNs are capable of achieving strong classification capabilities with a relatively smaller number of units. Consequently, RBFNs have been successfully applied to a variety of data classification tasks [103] including the recognition of handwritten numerals [83], image recognition [23, 107], and speech recognition [105, 110, 128], process faults detection [85], and various pattern recognition tasks [24, 86].

8.5   Multi-Layer Feedforward Network

In earlier sections, we introduced the perceptron and adaline as examples of single-layer feedforward neural networks. Although these architectures were pioneering utilizations of neural networks, they had limited classification performance, partly to due to the absence of hidden layers that are able to infuse more discriminative capabilities in the network.

The multi-layer feedforward network, also called Multi-Layer Perceptron (MLP), was proposed as an enhancement of the perceptron model, endowed with the following characteristics:

- The use of at least one hidden layer in the network between the input layer and the output layer, which allows internal representation and adds more modeling capabilities.

- The use of sigmoidal units in the hidden layers and the output layer. The smoothness of the sigmoid enables the training of the entire network directly from the output down to the hidden units, which was not possible either with perceptron or the adaline.

- The availability of a learning scheme called backpropagation, which trains the entire network hierarchically.

These characteristics, together with a solid theoretical framework backed by concrete results, have made the MLPs the most widely used neural network architecture for classification tasks. In terms of foundational theory, the work by Cybenko [33], who proved that an MLP with a single hidden layer of sigmoidal units can approximate any mapping at any desired accuracy, helped in establishing MLPs as universal approximator, alongside the RBFNs. This result was extended by Hornik [59, 60] who showed that a neural network with a hidden layer of sigmoidal units can in fact approximate any function and its derivatives. In theory, MLPs can then perform any classification tasks. In practice, a sub-optimal classification performance may arise due to factors such as an inadequate training, an insufficient number of hidden units, or a lack of a sufficient or representative number of data.

8.5.1   MLP Architecture for Classification

The architecture of an MLP enables a rich and powerful representation as illustrated in Figure 8.4. Units are arranged in layers comprising an input layer, one or several hidden layers, and an output layer. Layers are organized sequentially following the flow of data from the input layer to the hidden layers, and ending at the output layer. Input data are collected and processed by the input layer's units, then propagated through the hidden layers' units to the output layer. Each unit in a layer receives data from the preceding layer's units, and the output units provide the final classification decision. This process is known as the *forward propagation* of the network. Units in hidden layers and output layers are typically sigmoidal units. The choice of the number and type of output units is tailored to the classification task.

We start by reviewing MLPs for binary classification problems. We then follow with a more generic model for multi-class problems.

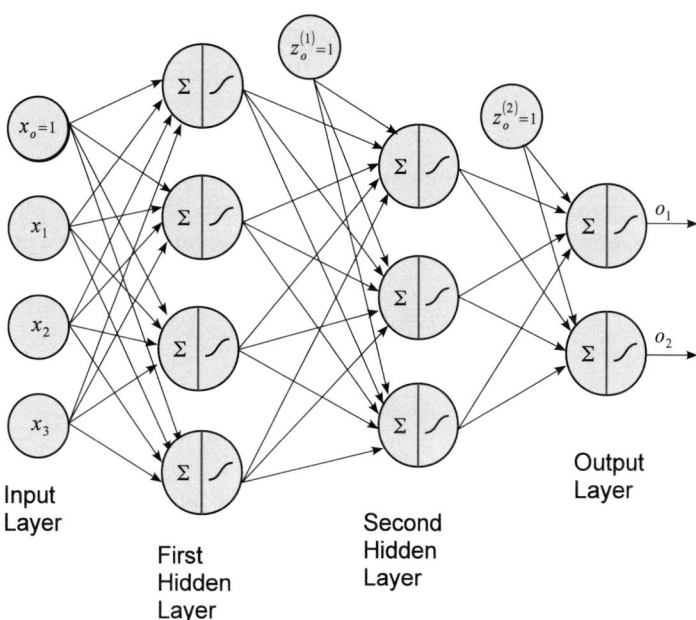

FIGURE 8.4: An MLP network made of three layers: two hidden layers and one output layer. Units in the network are sigmoid units. Bias are represented as input units.

8.5.1.1 Two-Class Problems

For two-class problems, C_1, C_2, the MLP contains that one or more hidden layers and a single sigmoidal output unit which produces a single output value o. Assuming that the target value is 0 for class C_1 and 1 for class C_2, the classification rule is the rule:

$$\mathbf{x} \in C_1 \quad \text{if } o > 0.5 \tag{8.49}$$
$$\mathbf{x} \in C_2 \quad \text{otherwise} \tag{8.50}$$

with the output o in the range of $[0,1]$ (or $[-1,1]$ if the tanh function is used and therefore the decision threshold is zero). Clearly, the output unit implements a logistic regression on the space of features generated by the last hidden layer. An MLP for a two-class problem is similar to the perceptron or adaline in its use of a single output unit. However, its capabilities are greatly enhanced by the use of hidden layers.

8.5.1.2 Multi-Class Problems

For a number of classes $K > 2$, the MLP has one or more hidden layers and K output units, each corresponding to a particular class. The target function is encoded with a 1-of-K encoding scheme with values in $\{0,1\}$ and the output o_k is viewed as a discriminant function of the class C_k, enabling the classification decision as

$$\mathbf{x} \in C_k \quad \text{if } k = \arg\max_j o_j \tag{8.51}$$

for $j \in \{1, \cdots, K\}$ with $o_k \in [0,1]$ when the sigmoid is used as activation function. The use of a 1-of-K encoding scheme for the target function is standard. The MLPs training is equivalent to approximating a mapping $f : \mathbb{R}^d \to \mathbb{R}^K$, where target values are the natural basis of $\mathbb{R}^K$. It has been proven that with the use of the 1-of-K encoding scheme and an adequate error criterion, such as the mean square error or cross-entropy, an MLP output can approximate posterior probabilities [14, 111, 131]:

$$o_k \approx P(C_k|\mathbf{x}) \tag{8.52}$$

for all k. This is, however, an asymptotic result. In practice, the sigmoidal unit's output values do not allow for a probabilistic interpretation since they do not sum to one. For a probabilistic interpretation of the network's outputs, a *softmap transformation* on the output unit's net value or activation can be used [26]:

$$o_k = \frac{e^{v_k}}{\sum_{j=1}^{K} e^{v_j}} \tag{8.53}$$

where clearly $\sum_{k=1}^{K} o_k = 1$.

8.5.1.3 Forward Propagation

The forward propagation refers to the transfer of data from the input layer to the output layer. Let us assume that an MLP is made of L layers where indices $1, \cdots, L-1$ refer to set of hidden layers and the index L represents the output layer. Per convention, we will use index 0 to represent the input layer. For a K-class problem, the MLP features K output units, where each output unit produces an output o_k, given an input example $\mathbf{x}$. As illustrated in Figure 8.4 the input data is propagated from the input layer to the output layer as follows:

From input layer to hidden layer: The input layer passes the d-dimensional input example $\mathbf{x}$ to each unit in the first hidden layer, $\mathcal{H}_1$, where the h-th unit computes a net value $v_h^{(1)}$ and output $z_h^{(1)}$ as:

$$v_h^{(1)} = \sum_{i=0}^{d} w_{ih}^{(1)} x_i \tag{8.54}$$

$$z_h^{(1)} = \phi(v_h^{(1)}) \tag{8.55}$$

where $w_{ih}^{(1)}$ denotes the weights of the synapse connecting the i-th input unit to the h-th hidden unit in the first layer; we assume that $w_{oh}^{(1)}$ refers to bias and $x_0 = 1$. For simplicity, we also assume that all units in the network use the same activation function $\phi()$.

From hidden layer to hidden layer: A unit h in the l-th hidden layer, $\mathcal{H}_l$, gets its data from units in the previous layer, $\mathcal{H}_{l-1}$, to produce an internal output as

$$v_h^{(l)} = \sum_{h' \in \mathcal{H}_{l-1}} w_{h'h}^{(l)} z_{h'}^{(l-1)} \tag{8.56}$$

$$z_h^{(l)} = \phi(v_h^{(l)}) \tag{8.57}$$

where $w_{h'h}^{(l)}$ denotes the synapse weights between the h'-th hidden unit in the layer $\mathcal{H}_{l-1}$ and the h-th hidden unit in the layer $\mathcal{H}_l$; $l \in \{2, \cdots, L-1\}$.

From hidden layer to output layer: Finally, a k-th output unit produces a network's response:

$$v_k^{(L)} = \sum_{h \in \mathcal{H}_{L-1}} w_{hk}^{L} z_h^{(L-1)} \tag{8.58}$$

$$o_k = \phi(v_k^{(L)}) \tag{8.59}$$

for all $k \in \{1, \cdots, K\}$. The forward pass propagates the activations $z_h^{(l)}$ from the first hidden layer to the outputs. Each hidden layer operates a transformation of the data, creating internal representations. Thus, the set of $z_h^{(l)}$ can be seen as a hierarchical sequence of non-linear representations of the input data **x** generated by the MLPs. The output acts as a classification machine on the feature set, $z_h^{(L-1)}$, generated by the last hidden layer. By performing both feature extration and classification in a single architecture, the MLPs can automatically generate representation that best suits the classification task data [69].

8.5.2   Error Metrics

Training an MLP consists of minimizing an objective function $\mathcal{E}(\mathbf{w})$ as a function of the network parameter set **w** in order to minimize the probability of error. The objective function is the expectation of an error metric $E(\mathbf{x}; \mathbf{w})$ that measures the discrepancy of the network output $\mathbf{o} = (o_1, \cdots, o_K)$ and desired response $\mathbf{t} = (t_1, \cdots, t_K)$ for a given input data **x**:

$$\mathcal{E}(\mathbf{w}) = \int_{\mathbf{x}} E(\mathbf{x}; \mathbf{w}) p(\mathbf{x}) d\mathbf{x}. \tag{8.60}$$

In practice, training uses an empirical approximation of the objective function defined on a training set $\mathcal{X}$:

$$\mathcal{E}(\mathbf{w}) = \frac{1}{N} \sum_{\mathbf{x} \in \mathcal{X}} E(\mathbf{x}; \mathbf{w}) \tag{8.61}$$

and the minimization process is monitored on the values of that empirical approximation.

Various error metrics are available for supervised training of the MLP and are defined using the target output function that describes the desirable output vector for a given input data example. A 1-of-K encoding scheme for the target output vector is the norm. We will review a few widely used error metrics used to train MLPs.

8.5.2.1   Mean Square Error (MSE)

The MSE is Euclidean distance between the MLP's output and the target output:

$$E_{mse}(\mathbf{x}; \mathbf{w}) = \sum_{k=0}^{K} (o_k - t_k)^2 \tag{8.62}$$

MSE is a widely used metric, thanks to its quadratic properties. Variants can include more general L_p norms.

8.5.2.2   Cross-Entropy (CE)

The cross entropy between the target and MLP output is derived using information-theoretic principles [8, 53, 67]. The CE criterion is defined as

$$E_{ent}(\mathbf{x}; \mathbf{w}) = -\sum_{k=1}^{K} t_k \log o_k. \tag{8.63}$$

Training a network to minimize the CE objective function can be interpreted as minimizing the Kullback-Liebler information distance [131] or maximizing the mutual information. Faster learning has frequently been reported for information theoretic error metrics compared to the quadratic error. Learning with logarithmic error metrics has also been shown to be less prone to local minima convergence.

8.5.2.3 Minimum Classification Error (MCE)

The MCE criterion is a smoothed approximation of the 0-1 error count [68, 70] that has been used for neural network optimization [69]. The MCE criterion is defined for $\mathbf{x}$ belonging to C_k

$$E_{mce}(\mathbf{x}; \mathbf{w}) = \text{sigmoid}(d_k(\mathbf{x}; \mathbf{w})) \tag{8.64}$$

$$d_k(\mathbf{x}; \mathbf{w}) = -o_k + \left[\frac{1}{K} \sum_{j, j \neq k}^{K} o_j^{\upsilon} \right]^{1/\upsilon} \tag{8.65}$$

with $\upsilon > 0$. The term d_k is called the misclassification measure. A negative value indicates a correct classification decision ($E_{mce}(\mathbf{x}; \mathbf{w}) \approx 0$) and a positive value indicates an incorrect classification decision ($E_{mce}(\mathbf{x}; \mathbf{w}) \approx 1$).

The MCE objection function

$$\mathcal{E}_{mce}(\mathbf{w}) = \frac{1}{N} \sum_{\mathbf{x} \in \mathcal{X}} E_{mce}(\mathbf{x}; \mathbf{w}) \tag{8.66}$$

is a smoothed approximation of the error rate on the training set, and therefore minimizes misclassification errors in a more direct manner than MSE and CE [16].

8.5.3 Learning by Backpropagation

The backpropagation training algorithm is an iterative gradient descent algorithm designed to minimize an error criterion of a feed-forward network. The term backpropagation comes from the propagation of error derivatives from output units, down to hidden units. At its core, the error backprogation performs the following update

$$w \leftarrow w + \nabla w \tag{8.67}$$

with

$$\nabla w = -\eta \frac{\partial E}{\partial w} \tag{8.68}$$

for any given weight w in the network where E is one of the error metrics previously defined. Here, we are considering the stochastic gradient descent that uses the local error, E, in lieu of the empirical objective function.

Due to the interdependence between units in the network, the computation of Equation 8.68 is performed hierarchically in the network, starting from the output units, down to the hidden units and ending at the input units. To illustrate, let w_{jk} be the weight connecting the j-th unit in the last hidden layer and the k-th output unit with net value $\mathbf{v}_k$ and output $o_k = \phi(\mathbf{v}_k^{(L)})$:

$$\frac{\partial E}{\partial w_{jk}} = \frac{\partial E}{\partial \mathbf{v}_k^{(L)}} \frac{\partial \mathbf{v}_k^{(L)}}{\partial w_{jk}} \tag{8.69}$$

$$= \delta_k z_j^{(L-1)} \tag{8.70}$$

where $z_j^{(L-1)}$ is the output of the j-th unit in the last hidden layer $\mathcal{H}_{L-1}$ and where we define the $\delta_k = \frac{\partial E}{\partial \mathbf{v}_k^{(L)}}$. Similarly, the error gradient relative to weight $w_{hj}^{(l)}$ between the h-th unit in hidden layer $\mathcal{H}_{l-1}$ and the j-th unit in hidden layer $\mathcal{H}_l$ is computed as:

$$\frac{\partial E}{\partial w_{hj}^{(l)}} = \frac{E}{\partial \mathbf{v}_j^{(l)}} \frac{\partial \mathbf{v}_j^{(l)}}{\partial w_{hj}^{(l)}} . \tag{8.71}$$

$$= \delta_j^{(l)} z_h^{(l-1)} \tag{8.72}$$

By referencing the output layer as $\mathcal{H}_L$ and input later as $\mathcal{H}_0$, the weight update is simply

$$\nabla w_{ij}^{(l)} = -\eta \delta_j^{(l)} z_i^{(l-1)} \tag{8.73}$$

for any given weight $w_{ij}^{(l)}$ and $l \in \{1, \cdots, L\}$, where $\delta_j^{(l)}$ is typically called the "credit" of unit j in layer $\mathcal{H}_l$. We remark that the MLP's weight update in Equation 8.73 implements the Hebbian learning rule.

To update a unit's weight requires the computation of the unit's credit. Computing the credit of each unit depends on its position within the network. The credit δ_k that corresponds to the k-th output unit (without a softmap transformation, for simplicity) is

$$\delta_k \quad = \quad \frac{\partial E}{\partial o_k} \frac{o_k}{\partial v_k^{(L)}} \tag{8.74}$$

$$= \quad \phi'(v_k^{(L)}) \frac{\partial E}{\partial o_k} \tag{8.75}$$

where $\phi'()$ is the derivative of the activation function. When using the sigmoid function with the MSE criterion, the credit of an output unit is

$$\delta_k \quad = \quad \phi(v_k)(1 - \phi(v_k))(o_k - t_k) \tag{8.76}$$

$$= \quad o_k(1 - o_k)(o_k - t_k). \tag{8.77}$$

For a hidden unit in layer $\mathcal{H}_l$, the credit $\delta_h^{(l)}$ depends on its recipient units in layer $\mathcal{H}_{l+1}$:

$$\delta_h^{(l)} \quad = \quad \frac{\partial E}{\partial z_h^{(l)}} \frac{\partial z_h^{(l)}}{\partial v_h^{(l)}} \tag{8.78}$$

$$= \quad \phi'(v_h^{(l)}) \sum_{h' \in \mathcal{H}_{l+1}} \frac{\partial E}{\partial v_{h'}^{(l+1)}} \frac{\partial v_{h'}^{(l+1)}}{\partial z_h^{(l)}} \tag{8.79}$$

$$= \quad z_h^{(l)}(1 - z_h^{(l)}) \sum_{h' \in \mathcal{H}_{l+1}} \delta_{h'}^{(l+1)} w_{hh'}^{(l+1)} \tag{8.80}$$

where we assumed the use of the sigmoid function in deriving Equation 8.80.

The credit is computed from the output unit using Equation 8.77 and then propagated to the hidden layer using Equation 8.80. These two equations are the foundation of the backpropagation algorithm enabling the computation of the gradient descent for each weight by propagating the credit from the output layer to hidden layers. The backpropagation algorithm is illustrated in Algorithm 8.4.

8.5.4 Enhancing Backpropagation

The standard backpropagation algorithm is a first-order optimization. Its convergence is highly sensitive to the size of the network, the noise level in the data, and the choice of the learning rate: convergence can be slow and may lead to a bad local minima of the error surface [7] . Enhancing the performance of backpropagation algorithms can be done by data normalization and the use of enhanced alternatives to the standard gradient descent approach. Consequently, data are typically normalized to lie between [-1, 1], prior to being presented to the MLP. Similarly, the network weights are typically initialized with zero means and variance of 1, prior to backpropagation.

Algorithm 8.4 Backpropagation algorithm

Input: Training data $X = \{\mathbf{x}_1, \cdots, \mathbf{x}_N\}$ and associated target $\{\mathbf{t}_1, \cdots, \mathbf{t}_K\}$; ($\mathbf{x}_n \in \mathbb{R}^d$,
$\mathbf{t}_n = (t_{n1}, t_{n2}, \cdots, t_{nK})$ with $t_{nj} = \delta_{jk}$ the Kronecker notation for $\mathbf{x}_n \in C_j$,; learning rate $\eta > 0$,
Output: A trained multi-layer perceptron with optimal set of weights $\mathbf{w}$.

1: Randomly initialize the set of weights $\mathbf{w} = \{w_{ij}\}$ with zero means and unit variance, preferably
2: **for** t in $\{1, \cdots, T\}$ **do**   # for each epoch up to T
3:     **for** n in $\{1, \cdots, N\}$ **do**   # for each sample
4:         Forward propagation to get output $\mathbf{o} = (o_1, \cdots, o_K)$ and activations $\{z_j^{(l)}\}$
5:
6:         **for** l in $\{L, L-1, \cdots, 1\}$ **do**
7:             **for** j in $\mathcal{H}_l$ **do**   # for all units in layer $\mathcal{H}_l$
8:                 **if** unit j is an *output* unit **then**
9:
$$\delta_j = \delta_j^{(L)} = o_j(1 - o_j)(o_j - t_j)$$

10:                 **end if**
11:                 **if** unit j is an *internal hidden unit* in layer $\mathcal{H}_l$ **then**
12:
$$\delta_j^{(l)} = z_j^{(l)}(1 - z_j^{(l)}) \sum_{h \in \mathcal{H}_{l+1}} \delta_h^{(l+1)} w_{jh}^{(l+1)}$$

13:                 **end if**
14:
15:                 **for** i in $\mathcal{H}_{l-1}$ **do**
16:                     $w_{ij}^{(l)} \leftarrow w_{ij}^{(l)} - \eta(t) z_i^{(l-1)} \delta_j^{(l)}$ # weight update
17:                 **end for**
18:
19:             **end for**
20:         **end for**
21:     **end for**
22: **end for**
23: $w_{ij}^{(l)}$: weight from unit i to unit j in layer $\mathcal{H}_l$
24: $z_i^{(l)}$: output of unit i in layer $\mathcal{H}_l$
25: $\eta(t)$: learning at iteration t
26: $\delta_j^{(l)}$: error term (credit) of unit j in layer $\mathcal{H}_l$

Enhancements to the gradient descent include, among other techniques, the use of a momentum smoother [116], the use of a dynamic learning rate [7, 129], the resilient backpropagation algorithm, and the Quickprop algorithm.

8.5.4.1   Backpropagation with Momentum

The use of a momentum factor $\alpha > 0$ as

$$\Delta w = -\eta \frac{\partial \mathcal{E}}{\partial w}(t) + \alpha \Delta w(t-1) \tag{8.81}$$

helps alleviate the noisy oscillations during gradient descent. This assumes that gradient descent is done in batch fashion on the objective function $\mathcal{E}$.

8.5.4.2 Delta-Bar-Delta

The optimal choice of the learning rate depends on the complexity of the error surface, and the network architecture. To circumvent this problem, the delta-bar-delta algorithm [66] assigns an individual learning rate η_w for each weight w in the network. This weight is updated during learning, depending on the direction of the gradient. The delta-bar-delta algorithm operates on the objective function $\mathcal{E}$ as follows.

First, it estimates a moving average of the error derivative, $\widehat{\frac{\partial \mathcal{E}}{\partial w}}(t)$, at each iteration t to reduce gradient fluctuation as

$$\widehat{\frac{\partial \mathcal{E}}{\partial w}}(t) = \beta \frac{\partial \mathcal{E}}{\partial w}(t) + (1 - \beta) \widehat{\frac{\partial \mathcal{E}}{\partial w}}(t - 1) \tag{8.82}$$

with $0 < \beta < 1$. Then, the individual learning rate is updated as follows:

$$\eta_w(t+1) = \begin{cases} \eta_w(t) + \kappa & \text{if } \widehat{\frac{\partial \mathcal{E}}{\partial w}}(t) \frac{\partial \mathcal{E}}{\partial w}(t-1) > 0 \\ (1-\rho)\eta_w(t) & \text{if } \widehat{\frac{\partial \mathcal{E}}{\partial w}}(t) \frac{\partial \mathcal{E}}{\partial w}(t-1) < 0 \\ \eta_w(t) & \text{otherwise} \end{cases} \tag{8.83}$$

The learning rate increases when consecutive gradients point in the same direction, and decreases otherwise. The choice of meta-learning parameters κ, ρ, β is driven by various considerations, in particular the type of data. The use of a momentum has been suggested to reduce gradient oscillations [95]. The update is in batch mode. An incremental version of the delta-bar-delta algorithm that can be used in online settings is available [125].

8.5.4.3 Rprop Algorithm

The search for a noise-resilient backpropagation algorithm has been the subject of various investigations [63]. The resilient back propogation (Rprop) algorithm [112] is one solution to this problem as it aims at eliminating the influence of the norm of the error derivatives in the update step, and instead, focuses on the direction of the gradient by monitoring its sign. Rprop operates on the objective function $\mathcal{E}$ and updates each weight at time t as

$$\Delta w(t) = -\gamma_w(t) \text{sign}\left(\frac{\partial \mathcal{E}}{\partial w}(t)\right) \tag{8.84}$$

where γ_w is the update step that is tracked at each training iteration t and updated as

$$\gamma_w(t+1) = \begin{cases} \min\{\eta_+\gamma_w(t), \gamma_{max}\} & \text{if } \frac{\partial \mathcal{E}}{\partial w}(t) \frac{\partial \mathcal{E}}{\partial w}(t-1) > 0 \\ \max\{\eta_-\gamma_w(t), \gamma_{min}\} & \text{if } \frac{\partial \mathcal{E}}{\partial w}(t) \frac{\partial \mathcal{E}}{\partial w}(t-1) < 0 \\ \gamma_w(t) & \text{otherwise} \end{cases} \tag{8.85}$$

where $0 < \eta_- < 1 < \eta_+$ with typical values 0.5 and 1.2 respectively, with the initial update-value set to $\delta_w = 0.1$.

The effect of Rprop is to speed up learning in flat regions of the error surface, and in areas that are close to a local minimum. The size of the update is bounded by the use of the γ_{min} and γ_{max}, to moderate the speed-up.

8.5.4.4 Quick-Prop

Second-order algorithms consider more information about the shape of the error surface than the mere value of its gradient. A more efficient optimization can be done if the curvature of the error surface is considered at each step. However, those methods require the estimation of the Hessian matrix, which is computationally expensive for a feedforward network [7]. Pseudo-Newton methods can be used to simplify the form of the Hessian matrix [10] where only the diagonal components are computed, yielding,

$$\Delta w(t) = \frac{\frac{\partial \mathcal{E}}{\partial w}(t)}{\frac{\partial^2 \mathcal{E}}{\partial^2 w}(t)}. \tag{8.86}$$

The Quick prop algorithm further approximates the second derivative $\frac{\partial^2 E}{\partial^2 w}$ as the difference between consecutive values of the first derivative, producing

$$\Delta w(t) = \Delta w(t-1) \frac{\frac{\partial \mathcal{E}}{\partial w}(t)}{\frac{\partial \mathcal{E}}{\partial w}(t) - \frac{\partial \mathcal{E}}{\partial w}(t-1)}. \tag{8.87}$$

No matrix inversion is necessary and the computational effort involved in finding the required second partial derivatives is limited. Quick-prop relies on numerous assumptions on the error surface, including a quadratic form and a non-multivariate nature of the Hessian matrix. Various heuristics are therefore necessary to make it work in practice. This includes limiting the update $\Delta w(t)$ to some maximum value and switching to a simple gradient update whenever the sign of the approximated Hessian does not change for multiple consecutive iterations [36, 37].

8.5.5   Generalization Issues

The issue of generalization is well known in machine learning and in particular neural networks. Given a training set, the network may achieve a good classification performance on the training that does not translate well to the testing environment. This may be due to multiple factors. One factor is over-training, in which the network learns a data set perfectly instead of learning the underlying mapping. Another factor is the unbalance ratio between the training size and the number of parameters, in which the numbers of weights is too large with respect to the number of data.

Early stopping is a method widely used to prevent overfitting. The idea is to periodically estimate the error using a separate validation set, and then stop training when the error in the validation set starts to increase. Cross validation is as an alternative method, but requires training multiple networks, which could be time-consuming. Another useful method to improve generalization is *training with noise* [2, 20]. Adding small-variance white noise to the training set has the same effect as adding more training data, which may prevent the network from overfitting a particular data set.

8.5.6   Model Selection

Model selection, that is, the appropriate choice of the number of units, is a difficult problem. There are several model selection criteria available for MLPs. In addition to standard model selection criteria such as AIC, BIC, and MDL, criteria-specific to neural networks include the generalized final prediction error (GPE) [99] and the Vapnik and Chernenski (VC) dimension [127], which expresses the capacity of the network to achieve optimal classification tasks and provide bounds on the number of training sets required to achieve a targeted probability of error. However, estimating the VC bound, given a network, is a challenging endeavor that leads, at times, to unreasonably large networks. Instead, recent advances use the concept of deep learning to selectively train a large network for accurate performance, as we shall see in the next section.

8.6   Deep Neural Networks

Deep learning methods aim at learning and exploiting the abilities of neural networks to generate internal feature representations of the data within their hidden layers. As was previously said, for a multi-layer architecture, the output of each hidden layer is a set of features to be processed by the next layer, meaning that the network generates composed and unseen internal representations of the data. The output layer can be viewed as a classification machine on the space spanned by the last hidden layer's outputs. More hidden layers and more units in those layers generate a more powerful network with the ability to generate more meaningful internal representations of the data.

However, training large and deep neural networks (DNN) with a large number of hidden layers and units is a difficult enterprise. With a number of parameters in the order of millions, the error surface is rather complex and has numerous local minima with many of them leading to poor performance [4, 42]. Also, during backpropagation, there is the issue of the *vanishing gradient*, where the norm of the gradient gets smaller as the chain rule of the gradient computation gets expanded across many layers [12, 57]. Lastly, a neural network with a large number of weights requires tremendous computational power. Deep learning techniques come with various strategies and heuristics to overcome these problems. Those strategies include the use of prior knowledge on the domain of application, the application of a layer-wise selective training technique, and the use of a high-performance computing system.

8.6.1 Use of Prior Knowledge

One way to successfully implement and train a deep layer architecture is to exploit prior knowledge about the domain of application that can guide the selection of the architecture, the encoding scheme, and the optimization techniques. Prior knowledge may help define not only the numbers of needed layers and their functions, but also provide information that enables weight-sharing, where units can share weights in the network if they do have similar functionality in the network or are operating on similar types of data. This balancing act between having numerous units but having a fewer number of parameters due to weight-sharing enabled successful training of earlier deep neural network architectures in the 1980s and the mid-1990s with application to speech and handwriting recognition. Prototyping examples include the large-phonetic time delay network (TDNN-LR) [120] and the convolution neural networks [80].

The original time-delay neural network (TDNN) was proposed in the 1980s with a focus on designing a neural network architecture that is time-invariant on the speech signal. The basic TDNN unit features a sigmoidal neuron with delay weights for current and past stimuli that are aggregated together to produce the final output. Weights-sharing is done to ensure that same time-events occuring over time use the same weights. While the original TDNN only used one hidden layer on a small classification task, more complex and layered versions were proposed, using the tied-connection and shift-invariant principles. This includes the TDNN-LR, a modular TDNN designed for the recognition of the entire Japanese phonemic set [97, 120] and the MS-TDNN, targeted at word recognition and embedding a time-alignment component in its optimization criterion [48, 49].

While the TDNN provides a good example of the influence of prior knowledge in the design of an architecture, the first successful implementation of a deep architecture that exploits prior knowledge is the convolutional neural network (CNN) [31, 80]. A CNN and its variants can exhibit more than four or five hidden layers and has achieved impressive results on the NIST handwritten digit recognition task. Its design mimics the human visual cortex. In the visual cortex, system information is processed locally through contiguous receptive fields and then aggregated into higher concepts [62]. This principle has inspired a few neural networks in the past, in particular the neocognitron architecture of Fukushima [43, 71].

The CNN is a multi-layer perceptron made of numerous layers where each layer is a non-linear feature detector performing local processing of contiguous features within each layer, leading to higher conceptual representation as information moves up to the output layer. The CNN's weight space is sparse as each unit in a layer only selectively takes input from a subset of contiguous neurons in the lower layers; this is the local connectivity that mimics the visual cortex. In addition, each layer is made of replicated neurons taking similar types of input from the lower layer and then generating "feature maps" of units with the similar receiving fields. Neurons within a map share the same weights enabling spatial-translation invariance capabilities. Lastly, some layers are designed to perform maxPooling [80] by selecting the maximum value of non-overlapping regions generated by lower layers. The end result is a powerful system that generates patterns in a hierarchical manner similar to that of the visual cortex. Weight-sharing constraints and local connectivity yield a sparse weight set that significantly reduces the degrees of freedom in the system and enables relatively faster training with increased chances of generalization [79].

The CNN has been applied very successfully to numerous classification tasks including hand-written character recognition [32, 118] and face recognition [39, 79].

8.6.2 Layer-Wise Greedy Training

In the absence of prior knowledge, selective or greedy training of layers has been suggested as a way to train deep networks in a generic manner with the goal of overcoming the vanishing gradient issue and helping to avoid attraction to low-performing local minima of the error surface. An earlier suggestion to implement this process was to individually train each neuron or each layer using a supervised training scheme based on the network's classification performance [38, 84]. While this approach can efficiently train large networks, it suffers from the over-fitting problem. Cross-validation or regularization can be used to mitigate that problem but are time-consuming approaches for very large networks.

The deep learning methodology proposes a two-step approach as a solution to this problem. The first step is an unsupervised learning scheme that trains each hidden layer individually, starting from the first hidden layer and up to the last hidden layer. The training problem is broken down into easier steps with fewer parameters to train. The use of unsupervised learning provides a form of regularization to prevent over-fitting but is also a powerful mechanism to let the network learn the inherent characteristics of the data without using class labels. This step is merely for weight initialization in preparation for the next step.

The second step trains the entire network using backpropagation. This step refines and readjusts the network for the targeted classification problem. Since the weigths have previously been optimized to better represent unseen characteristics of the data, they provide a good starting point for backpropagation.

The deep learning methodology combines unsupervised learning of initial weights aimed at feature representation and supervised training of the entire network with a focus on classification performance [56]. It is a design mechanism that uses two different learning criteria. This contrasts with earlier proposals that emphasized the dual learning of features and classification using a unique criterion [16, 69]. The use of the restricted Boltzmann machine (RBM) or stacked auto-encoder are the techniques widely used to perform unsupervised layer-wise network initiatialization [56].

8.6.2.1 Deep Belief Networks (DBNs)

The Restricted Boltzmann Machine [122] is a single-layer network where the output layer is made of binary latent variables encoding the data. It consists of a single visible layer and single hidden layer, where connections are formed between the layers. Given an input data, the RBM uses a latent representation to generate a probability distribution of the data; it is a generative model. It has been argued and shown that RBMs can successfully be used to learn the internal weights of a deep neural network architecture [56].

A deep belief network (DBN) is a neural network whose hidden layers are Restricted Boltzmann Machines (RBMs) and the output layer is an associative network. The hidden layers are trained in a layer-wise greedy fashion, one layer at a time, to provide an internal encoding of the data. Each RBM layer is trained to generate the data in its visible layer, forcing the hidden layer to encode that information. Once an RBM is trained, its output is used as input to the next layer and the process repeated [56]. The overall process starts by training the first RBM, then training another RBM using the first RBM's hidden layer outputs as the second RBM's visible layer, and then continuing up to the last hidden layer.

8.6.2.2 Stack Auto-Encoder

Stacked auto-encoders are MLPs with an architecture designed to regenerate input data [11,55]. The MLP is trained to reproduce an estimate $\hat{\mathbf{x}} = (\hat{x}_1, \cdots, \hat{x}_d)$ of the input $\mathbf{x} = (x_1, \cdots, x_d)$, thus forcing the MLP to encode an internal representation of $\mathbf{x}$ in the hidden units. It generally features a single or two hidden layers that are responsible for creating a hidden representation of the data. The network tries to minimize the mean square error E between input and output as

$$E(\hat{\mathbf{x}}, \mathbf{x}) = \sum_{i=1}^{d} (x_i - \hat{x}_i)^2 \tag{8.88}$$

and by doing so, the MLP encodes an internal presentation in the hidden layers that can be used to regenerate the data in the output layer. Auto-encoding can be likened to a non-linear principal component estimation. When used within a deep-learning system, each hidden layer of the DNN is used as a hidden layer in an encoder and trained accordingly. After training is done, the weights are fixed and used to generate data for the next layer, which is also trained within an auto-encoder using the outputs of the previously trained layer. The process is repeated up to the last hidden layer. A deep CNN (DCNN) design was proposed based on those principles and achieved an improved performance compared to the original CNN version [81, 109]; it was a good showcase of the dual-use of domain knowledge in the design of the CNN architecture and layer-wise training for network optimization.

8.6.3 Limits and Applications

Deep learning is a powerful methodology well-suited to training deep and large networks. However, it does require the availability of powerful computing capabilities, which justifies the use of special accelerator hardware such as GPU in training deep neural networks.

Successful applications of deep networks have already been presented on a large variety of data. We have pointed out the impressive results obtained by the CNN and the DCNN in image recognition [31, 81, 109], speech detection recognition [123], and sound classification [82]. Deep learning has had an important impact on speech recognition where it has helped achieve better performance than state-of-the-art systems. For speech recognition, it is used as a probability estimator within a hidden Markov model architecture, which demonstrates an example of a hybrid system-based deep learning technique [54, 72].

8.7 Summary

We have seen in this chapter that neural networks are powerful universal mappers for data classification endowed with certain properties such as non-linearity, the capacity to represent information in a hierarchical fashion, and the use of parallelism in performing the classification tasks. The resurging interest of neural networks coincides with the availability of high-performance computing to perform computationally demanding tasks. The emergence of training methodologies to exploit these computing resources and the concrete results achieved in selected classification tasks make neural networks an attractive choice for classification problems. The choice of a particular network is still linked to the task to be accomplished. Prior knowledge is an important element to consider and the emerging dual use of unsupervised layer-wise training and supervised training has been proven robust to provide good initialization of the weights and avoid local minima.

Acknowledgements

The author would like to thank Deepak Turaga of IBM T. J. Watson Research Center for his support and encouragement for this review.

Bibliography

[1] D. H. Ackley, G. E. Hinton, and T. J. Sejnowski. A learning algorithm for Boltzmann machines. *Cognitive Science*, 9(1):147–169, 1985.

[2] G An. The effects of adding noise during backpropagation training on a generalization performance. *Neural Computation*, 8:643–674, 1996.

[3] J. K. Anlauf and M. Biehl. The AdaTron: an Adaptive Perceptron Algorithm. *Europhysics Letters*, 10(7):687–692, 1989.

[4] P. Auer, M. Herbster, and M. K. Warmuth. Exponentially many local minima for single neurons. In M. Mozer, D. S. Touretzky, and M. Perrone, editors, *Advances in Neural Information Processing Systems*, volume 8, pages 315–322. MIT Press, Cambridge, MA, 1996.

[5] B. Widrow and M. E. Hoff Adaptive Switching Circuits. In *IRE WESCON Convention Record*, pages 96–104, 1960.

[6] E. B. Barlett and R. E. Uhrig. Nuclear power plant status diagnostics using artificial neural networks, *Nuclear Technology*, 97:272–281, 1992.

[7] T. Battiti. First- and second-order methods for learning: Between steepest descent and Newton's method. *Neural Computation*, 4(2):141–166, 1992.

[8] E. B. Baum and F. Wilczek. Supervised learning of probability distribution by neural network. In A. Andeson, editor, *Neural Information Processing Systems*, pages 52–61. American Institute of Physics, 1988.

[9] W. G. Baxt. Use of an artificial neural network for data analysis in clinical decision-making: The diagnosis of acute coronary occlusion. *Neural Computation*, 2(4):480–489, 1990.

[10] S. Becker and Y. Le Cun. Improving the convergence of back-propagation learning with second order methods. In D. S. Touretzky, G. E. Hinton, and T. J. Sejnowski, editors, *Proceedings of the 1988 Connectionist Models Summer School*, pages 29–37. San Francisco, CA: Morgan Kaufmann, 1989.

[11] Y. Bengio, P. Lamblin, D. Popovici, and H. Larochelle. Greedy layer-wise training of deep networks. In *Proceedings of NIPS*, pages 153–160, MIT Press, Cambridge, MA, 2006.

[12] Y. Bengio, P. Simard, and P. Frasconi. Learning long-term dependencies with gradient descent is difficult. *IEEE Transactions on Neural Networks*, 5(2):157–166, 1994.

[13] M. Biehl, A. Ghosh, and B. Hammer. Dynamics and generalization ability of LVQ algorithms. *Journal of Machine Learning Research*, 8:323–360, 2007.

[14] A. Biem. Discriminative Feature Extraction Applied to Speech Recognition. PhD thesis, Université Paris 6, 1997.

[15] A. Biem and S. Katagiri. Cepstrum liftering based on minimum classification error. In *Technical Meeting of Institute of Electrical Information Communcation Engineering of Japan (IEICE)*, volume 92-126, pages 17–24, July 1992.

[16] A. Biem, S. Katagiri, and B.-H. Juang. Pattern recognition based on discriminative feature extraction. *IEEE Transactions on Signal Processing*, 45(02):500–504, 1997.

[17] S. A. Billings and C. F. Fung. Recurrent radial basis function networks for adaptive noise cancellation. *Neural Networks*, 8(2):273–290, 1995.

[18] C. Bishop. Improving the generalization properties of radial basis function neural networks. *Neural Computation*, 3(4):579–588, 1991.

[19] C. M. Bishop. *Neural Network for Pattern Recognition*. Oxford University Press, 1995.

[20] C. M. Bishop. Training with noise is equivalent to Tikhonov regularization. *Neural Computation*, 7(1):108–116, 1995.

[21] C. M. Bishop. *Neural Networks for Pattern Recognition*. Clarendon Press, Oxford, 1995.

[22] A. G. Bors and I. Pitas. Median radial basis function neural network. *IEEE Transactions on Neural Networks*, 7(6):1351–1364, 1996.

[23] A. G. Bors and I. Pitas. Object classification in 3-D images using alpha-trimmed mean radial basis function network. *IEEE Transactions on Image Processing*, 8(12):1744–1756, 1999.

[24] D. Bounds, P. Lloyd, and B. Mathew. A comparison of neural network and other pattern recognition approaches to the diagnosis of low back disorders. *Neural Networks*, 3(5):583–91, 1990.

[25] H. Bourlard and N. Morgan. Continuous speech recognition by connectionist statistical methods. *IEEE Transactions on Neural Networks*, 4:893–909, 1993.

[26] J. S. Bridle. Probabilistic interpretation of feedforward classification network outputs, with relationships to statistical pattern recognition. In Fogelman-Soulie and Herault, editors, *Neurocomputing: Algorithms, Architectures and Applications*, NATO ASI Series, pages 227–236. Springer, 1990.

[27] O. Buchtala, P. Neumann, and B. Sick. A strategy for an efficient training of radial basis function networks for classification applications. In *Proceedings of the International Joint Conference on Neural Networks*, volume 2, pages 1025–1030, 2003.

[28] H. B. Burke, P. H. Goodman, D. B. Rosen, D. E. Henson, J. N. Weinstein, F. E. Harrell, J. R. Marks, D. P. Winchester, and D. G. Bostwick. Artificial neural networks improve the accuracy of cancer survival prediction. *Cancer*, 79:857–862, 1997.

[29] S. Cajal. A new concept of the histology of the central nervous system. In Rottenberg and Hochberg, editors, *Neurological Classics in Modern Translation*. New York: Hafner, 1977.

[30] K. Crammer, R. Gilad-Bachrach, A. Navot, and A. Tishby. Margin analysis of the LVQ algorithm. In *Advances in Neural Information Processing Systems*, volume 15, pages 462–469. MIT Press, Cambridge, MA, 2003.

[31] Y. Le Cun and Y. Bengio. Convolutional networks for images, speech, and time series. In M.A. Arbib, editor, *Handbook of Brain Theory and Neural Networks*, pages 255–258. MIT Press, Cambridge, MA, 1995.

[32] Y. Le Cun, B. Boser, J. S. Denker, D. Henderson, R. E. Howard, W. Hubberd, and L. D. Jackel. Handwritten digit recognition with a back-propagation network. *Advances in Neural Information Processing Systems*, 2:396–404, 1990.

[33] G. Cybenko. Approximation by superpositions of a sigmoidal function. *Mathematics of Control Signals and Systems*, 2(303-314), 1989.

[34] D. Hebb *Organization of Behavior*. Science Edition, 1961.

[35] R. O. Duda and P. E. Hart. *Pattern Classification and Scene Analysis*. Wiley Interscience Publications, 1973.

[36] S. Fahlman. Faster learning variations on back-propagation: An empirical study. In D. S. Touretzky, G. E. Hinton, and T. J. Sejnowski, editors, *Proceedings of the 1988 Connectionist Models Summer School*, pages 38–51. Morgan Kaufmann, San Francisco, CA, 1989.

[37] S. E. Fahlman. An empirical study of learning speech in back-propagation networks. Technical report, Canergie Mellon University, 1988.

[38] Scott E. Fahlman and Christian Lebiere. The cascade-correlation learning architecture. In D.S. Touretzky, editor, *Advances in Neural Information Processing Systems 2*, pages 524–532. Morgan Kaufmann, San Mateo, 1990.

[39] B. Fasel. Robust face analysis using convolutional neural networks. In *Proceedings of the International Conference on Pattern Recognition*, 2002.

[40] J. A. Flanagan. Self-organisation in Kohonen's SOM. *Neural Networks*, 9(7):1185–1197, 1996.

[41] Y. Freund and R Schapire. Large margin classification using the perceptron algorithm. *Machine Learning*, 37(3):277–296., 1999.

[42] K. Fukumizu and S. Amari. Local minima and plateaus in hierarchical structures of multilayer perceptrons. *Neural Networks*, 13(3):317–327, 2000.

[43] K. Fukushima. Neocognitron: A self-organizing neural network model for a mechanism of pattern recognition unaffected by shift in position. *Biological Cybernetics*, 36:193–202, 1980.

[44] S.I. Gallant. Perceptron-based learning algorithms. *IEEE Transactions on Neural Networks*, 1(2):179–191, 1990.

[45] F. Girosi, M. Jones, and T. Poggio. Regularization theory and neural networks architectures. *Neural Computation*, 7(2):219–269, 1995.

[46] G. H. Golub and C. F. Van Loan. *Matrix Computations*. The Johns Hopkins University Press, 3rd edition, 1996.

[47] I. Guyon. Applications of neural networks to character recognition. *International Journal of Pattern Recognition Artificial Intelligence*, 5:353–382, 1991.

[48] P. Haffner, M. Franzini, and A. Waibel. Integrating time alignment and neural networks for high performance continuous speech recognition. In *Proceedings of IEEE International Conference of Acoustic, Speech, and Signal Processing (ICASSP)*, pages 105–108, May 1991.

[49] P. Haffner and A. Waibel. Multi-state time delay neural networks for continuous speech recognition. In *NIPS*, pages 135–142, 1992.

[50] B. Hammer and T. Villmann. Generalized relevance learning vector quantization. *Neural Networks*, 15(8-9):1059–1068, 2002.

[51] S. Haykin. *Neural Networks. A Comprehensive Foundation*. Macmillan College Publishing, New York, 1994.

[52] J. Hertz, A. Krogh, and R. Palmer. *Introduction to the Theory of Neural Computation*. Addison-Wesley, 1991.

[53] G. Hinton. Connectionist learning procedures. *Artificial Intelligence*, 40(1-3): 185–234, 1989.

[54] G. Hinton, L. Deng, D. Yu, and G. Dahl. Deep neural networks for acoustic modeling in speech recognition, *IEEE Signal Processing Magazine*, 29(6):82–97, 2012.

[55] G. Hinton and R. Salakhutdinov. Reducing the dimensionality of data with neural networks. *Science*, 313(5786):504– 507, 2006.

[56] G. E. Hinton, S. Osindero, and Y. W. Teh. A fast learning algorithm for deep belief nets. *Neural Computation*, 18(7):1527–1554, 2006.

[57] S. Hochreiter. The vanishing gradient problem during learning recurrent neural nets and problem solutions. *International Journal of Uncertainty, Fuzziness and Knowledge-Based Systems*, 06(02):107–116, 1998.

[58] J. Hopfield. Neural networks and physical systems with emergent collective computational abilities. Reprinted in Anderson and Rosenfeld (1988), editors, In *Proceedings of National Academy of Sciences USA*, volume 79, pages 2554–58, April 1982.

[59] K. Hornik. Approximation capabilities of multilayer feedforward networks. *Neural Networks*, 4:251–257, 1991.

[60] K. Hornik, M. Stinchcombe, and H. White. Multilayer feedforward networks are universal approximators. *Neural Networks*, 2:359–366, 1989.

[61] J. C. Hoskins, K. M. Kaliyur, and D. M. Himmelblau. Incipient fault detection and diagnosis using artificial neural networks. *Proceedings of the International Joint Conference on Neural Networks*, pages 81–86, 1990.

[62] D. Hubel and T. Wiesel. Receptive fields and functional architecture of monkey striate cortex. *Journal of Physiology*, 195:215-243, 1968.

[63] C. Igel and M Husken. Empirical evaluation of the improved rprop learning algorithms. *Neurocomputing*, 50:105–123, 2003.

[64] B. Igelnik and Y. H. Pao. Stochastic choice of basis functions in adaptive function approximation and the functional-link net. *IEEE Transactions on Neural Networks*, 6(6):1320–1329, 1995.

[65] R. A. Jacobs, M. I. Jordan, S. J. Nowlan, and G. E. Hinton. Adaptive mixtures of local experts. *Neural Computation*, 3(1):79–87, 1991.

[66] R. A. Jacobs. Increased rate of convergence through learning rate adaptation. *Neural Networks*, 1(4):295–307, 1988.

[67] J. J. Hopfield. Learning algorithms and probability distributions in feed-forward and feed-back networks. In *National Academy of Science, USA*, volume 84, pages 8429–8433, December 1987.

[68] B.-H. Juang and S. Katagiri. Discriminative learning for minimum error classification. *IEEE Transactions on Acoustics, Speech, and Signal Processing*, 40(12):3043–3054, December 1992.

[69] S. Katagiri, B.-H. Juang, and A. Biem. Discriminative Feature Extraction. In R. J. Mammone, editor, *Artificial Neural Networks for Speech and Vision*. Chapman and Hall, 1993.

[70] S. Katagiri, C.-H. Lee, and B.-H. Juang. Discriminative multilayer feed-forward networks. In *Proceedings of the IEEE Worshop on Neural Networks for Signal Processing*, pages 309–318, 1991.

[71] M. Kikuchi and K. Fukushima. Neural network model of the visual system: Binding form and motion. *Neural Networks*, 9(8):1417–1427, 1996.

[72] B. Kingsbury, T. Sainath, and H. Soltau. Scalable minimum Bayes risk training of DNN acoustic models using distributed Hessian-free optimization. In *Interspeech*, 2012.

[73] S. Knerr, L. Personnaz, and G. Dreyfus. Handwritten digit recognition by neural networks with single-layer training. *IEEE Transactions on Neural Networks*, 3:962–968, 1992.

[74] T. Kohonen. The Self-Organizing Map. *Proceedings of the IEEE*, 78(9):1464–1480, 1990.

[75] T. Kohonen, G. Barma, and T. Charles. Statistical pattern recognition with neural networks: Benchmarking studies. In *IEEE Proceedings of ICNN*, volume 1, pages 66–68, 1988.

[76] T. Kohonen. *Self-Organization and Associative Memory*. Springer, Berlin, third edition, 1989.

[77] T. Kohonen. *Self-Organizing Maps*. Springer, Berlin, 1995.

[78] M. L. Minsky and S. A. Papert. *Perceptrons*. MIT Press, Cambridge, MA, 1969.

[79] S. Lawrence, C. L. Giles, A. C. Tsoi, and A. D. Back. Face recognition: A convolutional neural network approach. *IEEE Transactions on Neural Networks*, 8(1):98–113, 1997.

[80] Y. LeCun, L. Bottou, Y. Bengio, and P. Haffner. Gradient-based learning applied to document recognition. *Proceedings of the IEEE*, 86(11):2278–2324, 1998.

[81] H. Lee, R. Grosse, R. Ranganath, and A. Ng. Convolutional deep belief networks for scalable unsupervised learning of hierarchical representations. In *26th International Conference on Machine Learning*, pages 609–616, 2009.

[82] H. Lee, Y. Largman, P. Pham, and A. Ng. Unsupervised feature learning for audio classification using convolutional deep belief networks. In *Advances in Neural Information Processing Systems*, 22:1096–1104, 2009.

[83] Y. Lee. Handwritten digit recognition using k nearest-neighbor, radial-basis function, and backpropagation neural networks. *Neural Computation*, 3(3):440–449, 1991.

[84] R. Lengelle and T. Denoeux. Training MLPs layer by layer using an objective function for internal representations. *Neural Networks*, 9:83–97, 1996.

[85] J. A. Leonard and M. A. Kramer. Radial basis function networks for classifying process faults. *IEEE Control Systems Magazine*, 11(3):31–38, 1991.

[86] R. P Lippmann. Pattern classification using neural networks. *IEEE Communications Magazine*, 27(11):47–50, 59–64, 1989.

[87] R. P. Lippmann. Review of neural networks for speech recognition. *Neural Computation*, 1(1):1–38, 1989.

[88] D. Lowe. Adaptive radial basis function nonlinearities and the problem of generalisation. In *1st IEEE International Conference on Artificial Neural Networks*, pages 171–175, 1989.

[89] G. L. Martin and J. A. Pittman. Recognizing hand-printed letters and digits using backpropagation learning. *Neural Computation*, 3(2):258–267, 1991.

[90] J. L. McClelland, D. E. Rumelhart, and the PDP Research Group. *Parallel Distributed Processing: Explorations in the Microstructure of Cognition*, volume 2. MIT Press, Cambridge, MA, 1986.

[91] W.S. McCulloch and W. Pitts. A logical calculus of the ideas in immanent nervous system. *Bulletin of Mathematical BioPhysics*, 5:115–33, 1943.

[92] E. McDermott and S. Katagiri. Shift-invariant, multi-category phoneme recognition using Kohonen's LVQ2. In *Proceedings of IEEE ICASSP*, 1:81–84, 1989.

[93] E. McDermott and S. Katagiri. LVQ-based shift-tolerant phoneme recognition. *IEEE Transactions on Acoustics, Speech, and Signal Processing*, 39:1398–1411, 1991.

[94] E. McDermott and S. Katagiri. Prototype-based minimum classification error/generalized probabilistic descent for various speech units. *Computer Speech and Language*, 8(8):351–368, Oct. 1994.

[95] A. Minai and R.D. Williams. Acceleration of backpropagation through learning rate and momentum adaptation. In *International Joint Conference on Neural Networks*, pages 676–679, 1990.

[96] M. Minsky. *Computation: Finite and Infinite Machines*. Prentice-Hall, Englewood Cliffs, NJ, 1967.

[97] M. Miyatake, H. Sawai, Y. Minami, and K. Shikano. Integrated training for spotting Japanese phonemes using large phonemic time-delay neural networks. In *Proceedings of IEEE ICASSP*, 1: pages 449–452, 1990.

[98] P. F. M. Bianchini and M. Gori. Learning without local minima in radial basis function networks. *IEEE Transactions on Neural Networks*, 6(3):749–756, 1995.

[99] J. Moody. The effective number of parameters: An analysis of generalization and regularization in nonlinear learning systems. In J. Moody, S. J. Hanson, and R. P Lippmann, editors, *Advances in Neural Information Processing Systems*, 4:847–854, Morgan Kaufmann, San Mateo, CA, 1992.

[100] J. Moody and C. Darken. Learning with localized receptive fields. In D. Touretzky, G. Hinton, and T. Sejnowski, editors, *Proceedings of the 1988 Connectionist Summer School*, pages 133–143. San Mateo, CA: Morgan Kaufmann., 1988.

[101] J. Moody and C. J. Darken. Fast learning in networks of locally-tuned processing units. *Neural Computation*, 1(2):281–294, 1989.

[102] M. Musavi, W. Ahmed, K. Chan, K. Faris, and D Hummels. On the training of radial basis function classifiers. *Neural Networks*, 5:595–603, 1992.

[103] K. Ng. A Comparative Study of the Practical Characteristics of Neural Network and Conventional Pattern Classifiers. M.S. Thesis, Massachusetts Institute of Technology. Dept. of Electrical Engineering and Computer Science, 1990.

[104] N. J. Nilsson. *Learning Machines*. New York, NY: McGraw-Hill, 1965.

[105] M. Niranjan and F. Fallside. Neural networks and radial basis functions in classifying static speech patterns. *Computer Speech and Language*, 4(3):275–289, 1990.

[106] R. Nopsuwanchai and A. Biem. Prototype-based minimum error classifier for handwritten digits recognition. In *IEEE International Conference on Acoustics, Speech and Signal Processing (ICASSP'04)*, volume 5, pages 845–848, 2004.

[107] T. Poggio and S. Edelman. A network that learns to recognize three dimensional objects. *Letters to Nature*, 343:263–266, 1990.

[108] T. Poggio and F. Girosi. Networks for approximation and learning. *Proceedings of the IEEE*, 78:1481–1497, 1990.

[109] M.-A. Ranzato, C. Poultney, S. Chopra, and Y. LeCun. Efficient learning of sparse representations with an energy-based model. In B. Scholkopf, J. Platt, and T. Hoffman, editors, *Advances in Neural Information Processing Systems*, volume 19. MIT Press, 2007.

[110] S. Renals and R. Rohwer. Phoneme classification experiments using radial basis function. In *Proceedings of the International Joint Conference on Neural Networks*, volume 1, pages 461–467, 1989.

[111] M. D. Richard and R. P. Lippmann. Neural network classifiers estimate Bayesian *a posteriori* probabilities. *Neural Computation*, 3(4):461–483, 1991.

[112] M. Riedmiller and H. Braun. A direct adaptive method for faster backpropagation learning: the rprop algorithm. In *IEEE International Conference on Neural Networks*, pages 586–591, San Francisco, CA, 1993.

[113] A.J. Robinson. Application of recurrent nets to phone probability estimation. *IEEE Transactions on Neural Networks*, 5(2):298–305, March 1994.

[114] Frank Rosenblatt. *Principles of Neurodynamics*. Spartan, New York, 1962.

[115] A. Roy, S. Govil, and R. Miranda. An algorithm to generate radial basis function (RBF)-like nets for classification problems. *Neural Networks*, 8(2):179–201, 1995.

[116] D. Rumelhart, G.E. Hinton, and R. Williams. Learning internal representation by error propagation. In D. Rumelhart and J. McClelland, editors, *Parallel Distributed Processing*, volume 1 of *Exploring the Microstructure of Cognition*. MIT Press, 1988.

[117] David E. Rumelhart, G. E. Hinton, and R. J. Williams. *Parallel Distributed Processing*. MIT Press, Cambridge, MA, 1986.

[118] E. Sackinger, B. Boser, J. Bromley, and Y. LeCun. Application of the ANNA neural network chip to high-speed character recognition. *IEEE Transactions on Neural Networks*, 3:498–505, 1992.

[119] A. Sato and K Yamada. Generalized learning vector quantization. In D. S. Touretzky and M. E. Hasselmo, editors, *Advances in neural information processing systems*, volume 8. MIT Press, Cambridge, MA., 1996.

[120] H. Sawai. TDNN-LR continuous speech recognition system using adaptive incremental TDNN training. In *Proceedings of IEEE ICASSP*, volume 1, pages 53–55, 1991.

[121] P. Schneider, M. Biehl, and B. Hammer. Adaptive relevance matrices in learning vector quantization. *Neural Computation*, 21(12):3532–3561, December 2009.

[122] P. Smolensky. Information processing in dynamical systems: Foundations of harmony theory. In D. E. Rumelhart and J. L. McClelland, editors, *Parallel Distributed Processing*, volume 1, chapter 6, pages 194–281. MIT Press, Cambridge, MA, 1986.

[123] S. Sukittanon, A. C. Surendran, J. C. Platt, and C. J. C. Burges. Convolutional networks for speech detection. In *Interspeech*, pages 1077–1080, 2004.

[124] A. J. Surkan and J. C. Singleton. Neural networks for bond rating improved by multiple hidden layers. In *IEEE International Joint Conference on Neural Networks*, volume 2, pages 157–162, 1990.

[125] R. S. Sutton. Adapting bias by gradient-descent: An incremental version of delta-bar-delta. In *Proceedings of AAAI-92*, 1992.

[126] J. Utans and J. Moody. Selecting neural network architecture via the prediction risk: Application to corporate bond rating prediction. In *Proceedings of the 1st International Conference on Artificial Intelligence Applications*, pages 35–41, 1991.

[127] V. N. Vapnik. *The Nature of Statistical Learning Theory*. Springer, New York, 1995.

[128] R. L K Venkateswarlu, R.V. Kumari, and G.V. Jayasri. Speech recognition using radial basis function neural network. In *Electronics Computer Technology (ICECT), 2011 3rd International Conference on*, volume 3, pages 441–445, 2011.

[129] T. P. Vogl, J. K. Mangis, J. K. Rigler, W. T. Zink, and D. L. Alkon. Accelerating the convergence of the backpropagation method. *Biological Cybernetics*, 59:257–263, 1988.

[130] D. Wettschereck and T. Dietterich. Improving the performance of radial basis function networks by learning center locations. In *Advances in Neural Information Processing Systems*, volume 4, pages 1133–1140, Morgan Kaufmann, San Mateo, CA. 1992.

[131] H. White. Learning in artificial neural networks: A statistical perspective. *Neural Computation*, 1(4):425–464, 1989.

[132] B. Widrow, D. E. Rumelhard, and M. A. Lehr. Neural networks: Applications in industry, business and science. *Communications of ACM*, 37:93–105, 1994.

Chapter 9

A Survey of Stream Classification Algorithms

Charu C. Aggarwal

IBM T. J. Watson Research Center
Yorktown Heights, NY 10598
`charu@us.ibm.com`

9.1 Introduction

Advances in hardware technology have led to the increasing popularity of data streams [1]. Many simple operations of everyday life, such as using a credit card or the phone, often lead to automated creation of data. Since these operations often scale over large numbers of participants, they lead to massive data streams. Similarly, telecommunications and social networks often contain large amounts of network or text data streams. The problem of learning from such data streams presents unprecedented challenges, especially in resource-constrained scenarios.

An important problem in the streaming scenario is that of data classification. In this problem, the data instances are associated with labels, and it is desirable to determine the labels on the test instances. Typically, it is assumed that both the training data and the test data may arrive in the form of a stream. In the most general case, the two types of instances are mixed with one another. Since the test instances are classified independently of one another, it is usually not difficult to perform the real-time classification of the test stream. The problems arise because of the fact that the training model is always based on the aggregate properties of multiple records. These aggregate properties may change over time. This phenomenon is referred to as *concept drift*. All streaming models need to account for concept drift in the model construction process. Therefore, the construction of a training model in a streaming and evolving scenarios can often be very challenging.

Aside from the issue of concept drift, many other issues arise in data stream classification, which are specific to the problem at hand. These issues are dependent on the nature of the application in which streaming classification is used. A single approach may not work well in all scenarios. This diversity in problem scenarios is discussed below:

- The classification may need to be performed in a resource-adaptive way. In particular, the stream needs to be classified *on demand*, since it may not be possible to control the rate at which test instances arrive.

- In many scenarios, some of the classes may be rare, and may arrive only occasionally in the data stream. In such cases, the classification of the stream becomes extremely challenging because of the fact that it is often more important to detect the rare class rather than the normal class. This is typical of cost sensitive scenarios. Furthermore, in some cases, previously unseen classes may be mixed with classes for which training data is available.

- Streaming classifiers can be designed to provide either *incremental* or *decremental* learning. In incremental learning, only the impact of incoming data points is accounted for. In decremental sampling, a sliding window model is used, and expiring items at the other end of the window are accounted for, the model update process. Not all classes of models can be used effectively for decremental learning, though incremental learning is usually much easier to address.

- In the context of discrete (categorical) attributes, the massive domain case is particularly important. In these cases, the attributes are drawn on massive domain of categorical values. For example, if the IP-address is an attribute, then the number of possible values may be in the millions. Therefore, it often becomes difficult to store the relevant summary information for classification purposes.

- Many other data domains such as text [6] and graphs [9] have been studied in the context of classification. Such scenarios may require dedicated techniques, because of the difference in the underlying data format.

- In many cases, the entire stream may not be available at a single processor or location. In such cases, distributed mining of data streams becomes extremely important.

Clearly, different scenarios and data domains present different challenges for the stream classification process. This chapter will discuss the different types of stream classification algorithms that are commonly used in the literature. We will also study different data-centric scenarios, corresponding to different domains or levels of difficulty.

This chapter is organized as follows. The next section presents a number of general algorithms for classification of quantitative data. This includes methods such as decision trees, nearest neighbor classifiers, and ensemble methods. Most of these algorithms have been developed for quantitative data, but can easily be extended to the categorical scenario. Section 9.3 discusses the problem of rare class classification in data streams. In Section 9.4, we will study the problem of massive domain stream classification. Different data domains such as text and graphs are studied in Section 9.5. Section 9.6 contains the conclusions and summary.

9.2 Generic Stream Classification Algorithms

In this section, a variety of generic stream classification algorithms will be studied. This represents the most general case, in which the stream is presented as an incoming stream of multidimensional records. In particular, the extension of different kinds of classifiers such as rule-based methods, decision trees, and nearest neighbor methods to the streaming scenario will be studied. Ensemble methods will also be discussed, which have the advantage of robustness in the context of data streams. It should be pointed out that many of these methods use sampling in order to improve the classification accuracy.

Reservoir sampling [86] is a generic technique, which can be used in order to improve the effectiveness of many stream classifiers. The idea is to maintain a continuous sample of the training data, of modest size. At any given time, a learning algorithm can be applied to this sample in order to create the model. This is, of course, a general meta-algorithm, and it does have the disadvantage that the accuracy of the approach is limited by the size of the sample. Nevertheless, its merit is that it can be used in conjunction with any offline classifier of arbitrary complexity, as long as a reasonable training sample size is used for analysis. Such algorithms can handle concept drift, since reservoir methods for creating time-decayed samples [12] are available. When a learning algorithm is applied to such a model, it will be well adjusted to the changes in the underlying data distribution, because of the time-decay bias of the sample. Since reservoirs can also be collected over sliding windows, such methods can also implicitly support decremental learning.

9.2.1 Decision Trees for Data Streams

While many scalable decision tree methods such as *SLIQ* [68], *RainForest* [44], and *BOAT* [43] have been proposed earlier, these methods are not designed for the streaming scenario.

While relatively little known, a large number of classical methods were designed for incremental decision tree induction, on the basis of the ID3 family in the classical literature [49, 83–85]. Most of these methods preceded the advent of the data stream paradigm, and are therefore not well known, at least within the streaming context. One of the earliest methods based on Quinlan's ID3 algorithm was designed by Schlimmer [78] in 1986. Other methods were designed for multivariate induction as well [84, 85], as early as 1990 and 1994, respectively. Kalles [49] proposed one of the most efficient versions by estimating the minimum number of training items for a new attribute to be selected as a split attribute. These methods were generally understood as purely incremental methods, though many of them do have the potential to be applied to data streams. The major downside of these methods is that they are typically not designed to handle concept-drift in the underlying data stream.

The earliest method *specifically* designed for decision-tree-based stream classification was proposed in [26]. This method is known as the *Very Fast Decision Tree (VFDT)* method. The VFDT method is based on the principle of the Hoeffding tree. This method essentially subsamples the data in order to achieve scalability in the construction of the decision tree. This is related to *BOAT* [43], which uses bootstrapping (rather than subsamples) in order to create more accurate trees. However, the *BOAT* method was designed for scalable decision tree construction, and not for the streaming scenario. Bootstrapping is also not an appropriate methodology for analyzing the streaming scenario.

This argument used in the Hoeffding tree is quite similar to *BOAT*. The idea is to show that the entire decision tree constructed would be the same as the one built on sub-sampled data with high probability. First, we describe a Hoeffding tree algorithm, which is constructed with the use of random sampling on static data. The idea is to determine a random sample of sufficient size so that the tree constructed on the sample is the same as that constructed on the entire data set. The

Hoeffding bound is used to show that the decision tree on the sub-sampled tree would make the same split as on the full stream with high probability. This approach can be used with a variety of criteria such as the gini-index, or information gain. For example, consider the case of the gini-index. For two attributes i and j, we would like to pick the attribute i, for which its gini-index G_i is smaller than G_j. While dealing with sampled data, the problem is that an error could be caused by the sampling process, and the order of the gini-index might be reversed. Therefore, for some threshold level ε, if $G_i - G_j < -\varepsilon$ is true for the sampled data, it is desired that $G_i - G_j < 0$ on the original data with high probability. This would result in the same split at that node in the sampled and original data, if i and j correspond to the best and second-best, attributes, respectively. The number of examples required to produce the same split as the original data (with high probability) is determined. The Hoeffding bound is used to determine the number of relevant examples, so that this probabilistic guarantee may be achieved. If all splits in the decision tree are the same, then the same decision tree will be created. The probabilistic guarantees on each split can be converted to probabilistic guarantees on the construction of the entire decision tree, by aggregating the probabilities of error over the individual nodes. The Hoeffding tree can also be applied to data streams, by building the tree incrementally, as more examples stream in, from the higher levels to the lower levels. At any given node, one needs to wait until enough tuples are available in order to make decisions about lower levels. The memory requirements are modest, because only the counts of the different discrete values of the attributes (over different classes) need to be maintained in order to make split decisions. The VFDT algorithm is also based on the Hoeffding tree algorithm, though it makes a number of modifications. Specifically, it is more aggressive about making choices in the tie breaking of different attributes for splits. It also allows the deactivation of less promising leaf nodes. It is generally more memory efficient, because of these optimizations.

The original *VFDT* method is not designed for cases where the stream is evolving over time. The work in [47] extends this method to the case of concept-drifting data streams. This method is referred to as *CVFDT*. *CVFDT* incorporates two main ideas in order to address the additional challenges of drift:

1. A sliding window of training items is used to limit the impact of historical behavior.

2. Alternate subtrees at each internal node i are constructed.

Because of the sliding window approach, the main issue here is the update of the attribute frequency statistics at the nodes, as the sliding window moves forward. For the incoming items, their statistics are added to the attribute frequencies in the current window, and the statistics of the expiring items at the other end of the window are decreased. Therefore, when these statistics are updated, some nodes may no longer meet the Hoeffding bound, and somehow need to be replaced. CVFDT associates each internal node i with a list of alternate subtrees split on different attributes. For each internal node i, a periodic testing mode is used, in which it is decided whether the current state of i should be replaced with one of the alternate subtrees. This is achieved by using the next set of incoming training records as a test set to evaluate the impact of the replacement. A detailed discussion of the CVFDT method may be found in the Chapter 4 on decision trees.

One of the major challenges of the VFDT family of methods is that it is naturally suited to categorical data. This is a natural derivative of the fact that decision tree splits implicitly assume categorical data. Furthermore, discretization of numerical to categorical data is often done offline in order to ensure good distribution of records into the discretized intervals. Of course, it is always possible to test all possible split points, while dealing with numerical data, but the number of possible split points may be very large, when the data is numerical. The bounds used for ensuring that the split is the same for the sampled and the original data may also not apply in this case. The technique in [48] uses numerical interval pruning in order to reduce the number of possible split points, and thereby make the approach more effective. Furthermore, the work uses the properties of the gain function on the entropy in order to achieve the same bound as the VFDT method with the use of a smaller number of samples. The work in [40] also extends the VFDT method for continuous data and drift and applies Bayes classifiers at the leaf nodes.

It was pointed out in [27] that it is often assumed that old data is very valuable in improving the accuracy of stream mining. While old data does provide greater robustness in cases where patterns in the stream are stable, this is not always the case. In cases where the stream has evolved, the old data may not represent the *currently* relevant patterns for classification. Therefore, the work in [27] proposes a method that is able to sensibly select the correct choice from the data with the use of a little extra cost. The technique in [27] uses a decision tree ensemble method, in which each component of the ensemble mixture is essentially a decision tree. While many models achieve the same goal using a sliding window model, the merit of this approach is the ability to perform systematic data selection.

Bifet et al [19] proposed a method, that shares similarities with the work in [40]. However, the work in [19] replaces the naive Bayes with perceptron classifiers. The idea is to gain greater efficiency, while maintaining competitive accuracy. Hashemi et al [46] developed a flexible decision tree, known as *FlexDT*, based on fuzzy logic. This is done in order to address noise and missing values in streaming data. The problem of decision tree construction has also been extended to the uncertain scenario by Liang [59]. A detailed discussion of some of the streaming decision-tree methods, together with pseudocodes, are provided in Chapter 4.

9.2.2   Rule-Based Methods for Data Streams

In rule-based methods, different combinations of attributes are associated with the class label. For the case of quantitative data, different intervals of the attributes are used in order to relate the attribute values to the class labels. Typically, rule-based methods have a very high level of interpretability for the classification process. The major disadvantage of rule-based methods is that if the data evolves significantly, then these methods cannot be used effectively. There are two kinds of rule-based methods that are primarily used in the literature for the static case:

1. *Rule-based methods with sequential covering:* These methods typically use the sequential covering paradigm in order to grow rules in the same way as a decision tree. Examples include *C4.5Rules* [73], *CN2* [23], and *RIPPER* [24]. Some of the classifiers such as *C4.5Rules* are almost directly rule-based versions of tree classifiers.

2. *Leveraging association rules:* In these cases, association patterns are mined in which the consequent of the rule represents a class label. An example is CBA [61].

Among the aforementioned methods, the second is easier to generalize to the streaming scenario, because online methods exist for frequent pattern mining in data streams. Once the frequent patterns have been determined, rule-sets can be constructed from them using any offline algorithm. Since frequent patterns can also be efficiently determined over sliding windows, such methods can also be used for decremental learning. The reader is referred to [1] for a primer on the streaming methods for frequent pattern mining.

Since decision trees can be extended to streaming data, the corresponding rule-based classifiers can also be extended to the streaming scenario. As discussed in the introduction chapter, the rule-growth phase of sequential covering algorithms shares a number of conceptual similarities with decision tree construction. Therefore, many of the methods for streaming decision tree construction can be extended to rule growth. The major problem is that the sequential covering algorithm assumes the availability of all the training examples at one time. These issues can however be addressed by sampling recent portions of the stream.

An interesting rule-based method, which is based on *C4.5Rules*, is proposed in [90]. This method is able to adapt to concept drift. This work distinguishes between proactive and reactive models. The idea in a proactive model is to try to anticipate the concept drift that will take place, and to make adjustments to the model on this basis. A reactive model is one in which the additional concept drift that has already occurred is used in order to modify the model. The work in [90] uses *C4.5Rules* [73] as the base leaner in order to create the triggers for the different scenarios. The section on text stream classification in this chapter also discusses a number of other rule-based methods for streaming classification.

Another recent method, known as *LOCUST* [4], uses a lazy learning method in order to improve the effectiveness of the classification process in an evolving data stream. The reader is advised to refer to Chapter 6 on instance-based learning. Thus, the training phase is completely dispensed with, except for the fact that the training data is organized in the form of inverted lists on the discretized attributes for efficient retrieval. These lists are maintained using an online approach, as more data arrives. Furthermore, the approach is resource-adaptive, because it can adjust to varying speeds of the underlying data stream. The way in which the algorithm is made resource-adaptive is by structuring it as an *any-time* algorithm. The work in [4] defines an online analytical processing framework for real-time classification. In this technique, the data is received continuously over time, and the classification is performed in real time, by sampling local subsets of attributes, which are relevant to a particular data record. This is achieved by sampling the inverted lists on the discretized attributes. The intersection of these inverted lists represents a subspace local to the test instance. Thus, each of the sampled subsets of attributes represents an *instance-specific* rule in the locality of the test instance. The majority class label of the local instances in that subset of attributes is reported as the relevant class label for a particular record for that sample. Since multiple attribute samples are used, this provides greater robustness to the estimation process. Since the approach is structured as an *any-time* method, it can vary on the number of samples used, during periods of very high load.

9.2.3 Nearest Neighbor Methods for Data Streams

In nearest neighbor methods a sample of the data stream is maintained. The class label is defined as the majority (or largest) label presence among the k-nearest neighbors of the test instance. Therefore, the approach is quite simple, in that it only needs to maintain a dynamic sample from the data stream. The process of finding a dynamic sample from a data stream is referred to as *reservoir sampling*. A variety of unbiased [86] and biased [12] sampling methods are available for this purpose. In biased methods, more recent data points are heavily weighted as compared to earlier data points. Such an approach can account for concept drift in the data stream.

One of the earliest nearest neighbor classification methods that was specifically designed for data streams was the *ANNCAD* algorithm [54]. The core idea in this work is that when data is non-uniform, it is difficult to predetermine the number of nearest neighbors that should be used. Therefore, the approach allows an adaptive choice of the number of k nearest neighbors. Specifically, one adaptively expands the nearby area of a test point until a satisfactory classification is obtained. In order to achieve this efficiently, the feature space of the training set is decomposed with the use of a multi-resolution data representation. The information from different resolution levels can then be used for adaptively preassigning a class to every cell. A test point is classified on the basis of the label of its cell. A number of efficient data structures are proposed in [54] in order to speed up the approach.

Another method proposed in [14] uses micro-clustering in order to perform nearest neighbor classification. While vanilla nearest neighbor methods use the original data points for the classification process, the method proposed in [14] uses supervised *micro-clusters* for this purpose. A cluster can be viewed as a pseudo-point, which is represented by its centroid. The advantage of using clusters rather than the individual data points is that there are fewer clusters than data points. Therefore, the classification can be performed much more efficiently. Of course, since each cluster must contain data points from the same class, the micro-clustering approach needs to be modified in order to create class specific clusters. The clusters must also not be too coarse, in order to ensure that sufficient accuracy is retained.

A micro-cluster is a statistical data structure that maintains the zero-th order, first order, and second order moments from the data stream. It can be shown that these moments are sufficient for maintaining most of the required cluster statistics. In this approach, the algorithm dynamically maintains a set of micro-clusters [11], such that each micro-cluster is constrained to contain data points of the same class. Furthermore, snapshots of the micro-clusters are maintained (indirectly) over different historical horizons. This is done by maintaining micro-cluster statistics since the beginning of the stream and additively updating them. Then, the statistics are stored either uniformly or over pyramidally stored intervals. By storing the statistics over pyramidal intervals, better space efficiency can be achieved. The statistics for a particular horizon $(t_c - h, t_c)$ of length h may be inferred by subtracting the statistics at time $t_c - h$ from the statistics at time t_c. Note that statistics over multiple time horizons can be constructed using such an approach.

In order to perform the classification, a key issue is the choice of the horizon to be used for a particular test instance. For this purpose, the method in [14] separates out a portion of the training stream as the *hold out* stream. The classification accuracy is tested over different time horizons over the hold out streams, and the best accuracy horizon is used in each case. This approach is similar to the standard "bucket of models" approach used for parameter-tuning in data classification [92]. The major difference in this case is that the parameter being tuned is temporal time-horizon, since it is a data stream, that is being considered. This provides more effective results because smaller horizons are automatically used in highly evolving streams, whereas larger horizons are used in less evolving streams. It has been shown in [14] that such an approach is highly adaptive to different levels of evolution of the data stream. The microclustering approach also has the advantage that it can be naturally generalized to positive-unlabeled data stream classification [58].

9.2.4 SVM Methods for Data Streams

SVM methods have rarely been used for data stream classification because of the challenges associated with incrementally updating the SVM classification model with an increasing number of records. Some methods such as *SVMLight* [50] and *SVMPerf* [51] have been proposed for scaling up SVM classification. These methods are discussed in the next chapter on big data. However, these methods are not designed for the streaming model, which is much more restrictive. SVM classifiers are particularly important to generalize to the streaming scenario, because of their widespread applicability. These methods use a quadratic programming formulation with as many constraints as the number of data points. This makes the problem computationally intensive. In the case of kernel methods, the size of the kernel matrix scales with the *square* of the number of data points. In a stream, the number of data points constantly increases with time. Clearly, it is not reasonable to assume that such methods can be used directly in the streaming context.

In the context of support vector machines, the main thrust of the work is in three different directions, either separately or in combination:

1. It is desired to make support vector machines incremental, so that the model can be adjusted efficiently as new examples come in, without having to learn from scratch. Because of the large size of the SVM optimization problem, it is particularly important to avoid learning from scratch.

2. In many cases, incremental learning is combined with learning on a *window* of instances. Learning on a window of instances is more difficult, because one has to adjust not only for the incoming data points, but also points that fall off at the expiring end of the window. The process of removing instances from a trained model is referred to as *decremental learning*.

3. In the context of concept drift detection, it is often difficult to estimate the correct window size. During periods of fast drift, the window size should be small. During periods of slow drift, the window size should be larger, in order to minimize generalization error. This ensures

that a larger number of training data points are used during stable periods. Therefore, the window sizes should be adapted according to the varying trends in the data stream. This can often be quite challenging in the streaming scenario.

The work in streaming SVM-classification is quite significant, but relatively less known than the more popular techniques on streaming decision trees and ensemble analysis. In fact, some of the earliest work [52, 82] in streaming SVM learning precedes the earliest streaming work in decision tree construction. The work in [52] is particularly notable, because it is one of the earliest works that uses a dynamic window-based framework for adjusting to concept drift. This precedes most of the work on streaming concept drift performed subsequently by the data mining community. Other significant works on incremental support vector machines are discussed in [30, 36, 38, 74, 76, 80].

The key idea in most of these methods is that SVM classification is a non-linear programming problem, which can be solved with the use of Kuhn-Tucker conditions. Therefore, as long as it is possible to account for the impact of the addition or removal of test instances on these conditions, such an approach is likely to work well. Therefore, these methods show how to efficiently maintain the optimality conditions while adding or removing instances. This is much more efficient than retraining the SVM from scratch, and is also very useful for the streaming scenario. In terms of popularity, the work in [30] is used most frequently as a representative of support vector learning. The notable feature of this work is that it shows that decremental learning provides insights about the relationship between the generalization performance and the geometry of the data.

9.2.5 Neural Network Classifiers for Data Streams

The basic unit in a neural network is a *neuron* or *unit*. Each unit receives a set of inputs, which are denoted by the vector $\overline{X_i}$, which in this case corresponds to the term frequencies in the ith document. Each neuron is also associated with a set of weights A, which are used in order to compute a function $f(\cdot)$ of its inputs. A typical function that is often used in the neural network is the linear function as follows:

$$p_i = A \cdot \overline{X_i} + b. \tag{9.1}$$

Thus, for a vector $\overline{X_i}$ drawn from a lexicon of d words, the weight vector A should also contain d elements. Here b denotes the bias. Now consider a binary classification problem, in which all labels are drawn from $\{+1, -1\}$. We assume that the class label of $\overline{X_i}$ is denoted by y_i. In that case, the sign of the predicted function p_i yields the class label.

Since the sign of the function $A \cdot \overline{X_i} + b$ yields the class label, the goal of the approach is to *learn* the set of weights A with the use of the training data. The idea is that we start off with random weights and gradually update them when a mistake is made by applying the current function on the training example. The magnitude of the update is regulated by a learning rate μ. This forms the core idea of the *perceptron algorithm*, which is as follows:

Perceptron Algorithm
Inputs: Learning Rate: μ
  Training Data $(\overline{X_i}, y_i) \; \forall i \in \{1 \ldots n\}$
Initialize weight vectors in A and b to 0 or small random numbers
repeat
Apply each training data to the neural network to check if the
  sign of $A \cdot \overline{X_i} + b$ matches y_i;
if sign of $A \cdot \overline{X_i} + b$ does **not** match y_i, then
  update weights A based on learning rate μ
until weights in A converge

The weights in A are typically updated (increased or decreased) proportionally to $\mu \cdot \overline{X_i}$, so as to reduce the direction of the error of the neuron. We further note that many different update rules have been proposed in the literature. For example, one may simply update each weight by μ, rather than by $\mu \cdot \overline{X_i}$. This is particularly possible in domains such as text, in which all feature values take on small non-negative values of relatively similar magnitude. Most neural network methods iterate through the training data several times in order to learn the vector of weights A. This is evident from the pseudocode description provided above. This is however, not possible in the streaming scenario. However, in the streaming scenario, the number of training examples are so large that it is not necessary to iterate through the data multiple times. It is possible to continuously train the neural network using incoming examples, as long as the number of examples are large enough. If desired, segments of the data stream may be isolated for multiple iterations, where needed. A number of implementations of neural network methods for text data have been studied in [35, 60, 70, 79, 89]. Neural networks are particularly suitable for data stream scenarios because of their incremental approach to the model update process. The extension of these methods to the stream scenario is straightforward, and is discussed in the section on text stream classification. These methods are not specific to text, and can be extended to any scenario, though they are used very commonly for text data.

9.2.6 Ensemble Methods for Data Streams

Ensemble methods are a particularly popular method for stream classification, because of the fact that classification models can often not be built robustly in a fast data stream. Therefore, the models can be made more robust by using a combination of classifiers.

The two most common ensemble methods include bagging and boosting [71]. These methods model the number of times that a data item occurs in one of the base classifiers, and show that it is suitable for streaming data. The work in [81] proposes an ensemble method to read training items sequentially as blocks. The classifier is learned on one block and evaluated on the next. The ensemble components are selected on the basis of this qualitative evaluation. The major weakness of these methods is that they fail to take the concept drift into account.

Such a method was proposed in [87]. The method is also designed to handle concept drift, since it can effectively capture the evolution in the underlying data. The data stream is partitioned into chunks, and multiple classifiers are utilized on each of these chunks. The final classification score is computed as a function of the score on each of these chunks. In particular, ensembles of classification models are scored, such as C4.5, RIPPER, and naive Bayesian, from sequential chunks of the data stream. The classifiers in the ensemble are weighted based on their expected classification accuracy on the test data under the time-evolving environment. This ensures that the approach is able to achieve a higher degree of accuracy, since the classifiers are dynamically tuned in order to optimize the accuracy for that part of the data stream. The work in [87] shows that an ensemble classifier produces a smaller error than a single classifier, if the weights of all classifiers are assigned based on their expected classification accuracy.

A significant body of work [53, 95, 96] uses the approach of example and model weighting or selection in order to obtain the best accuracy over different portions of the concept drifting data stream. Typically, the idea in all of these methods is to pick the best classifier which works for that test instance or a portion of the training data stream. The work in [88] uses a stochastic optimization model in order to reduce the risk of overfitting and minimize classification bias. The diversity arising from the use of the ensemble approach has a significant role in the improvement of classification accuracy [69].

Both homogeneous and heterogeneous ensembles can be very useful in improving the accuracy of classification, as long as the right subset of data is selected for training in the different ensemblar components. An example of a homogeneous decision tree is presented in [27], in which the individual components of the ensemble are decision trees. The work in [27] carefully selects the subset of data on which to apply the method in order to achieve the most accurate results. It was pointed out in [42] that the appropriate assumptions for mining concept drifting data streams are not always easy to infer from the underlying data. It has been shown in [42] that a simple voting-based ensemblar framework can sometimes perform more effectively than relatively complex models.

Ensemble methods have also been extended to the rare class scenario. For example, the work in [41] is able to achieve effective classification in the context of skewed distributions. This scenario is discussed in detail in the next section.

9.3 Rare Class Stream Classification

In the rare class scenario, one or more classes are presented to a much lower degree than the other classes. Such distributions are known as skewed distributions. The rare class scenario is closely related to supervised outlier detection, and is discussed in detail in Chapter 17 of this book, for the static case. An overview on the topic may be found in [31]. In many cases, costs may be associated with examples, which provide an idea of the relative cost of incorrect classification of the different classes. In the static scenario, most classification methods can be extended to the rare class scenario by using a variety of simple methods:

- *Example Re-Weighting:* In this case, the examples are re-weighted using the misclassification costs.

- *Example Over-sampling or Under-sampling:* In this case, the examples are re-weighted implicitly or explicitly by oversampling or undersampling the different classes.

The supervised scenario is a a very rich one in the temporal domain, because *different kinds of temporal and frequency-based aspects of the classes could correspond to rare class*. In many cases, full supervision is not possible, and therefore semi-supervised outliers may need to be found. Such semi-supervised cases will be discussed in this chapter, because of their importance to the classification process. The different scenarios could correspond to novel class outliers, rare class outliers, or infrequently recurring class outliers. Thus, a combination of methods for concept drift analysis, outlier detection, and rare class detection may need to be used in order to accurately identify such cases in the streaming scenario. Such scenarios could arise quite often in applications such as intrusion detection, in which some known intrusions may be labeled, but new intrusions may also arise over time. Therefore, it is critical for the anomaly detection algorithm to use a combination of supervised and unsupervised methods in order to perform the labeling and detect outliers both in a semi-supervised and fully supervised way. Therefore, this problem may be viewed as a hybrid between outlier detection and imbalanced class learning.

The problem of rare class detection is also related to that of *cost-sensitive learning* in which costs are associated with the records. This is discussed in detail in Chapter 17 on rare class detection. It should be pointed out that the ensemble method discussed in [87] is able to perform cost-sensitive learning, though the method is not particularly focussed on different variations of the rare class detection problem in the streaming scenario. In particular, the method in [41] shows how the ensemble method can be extended to the rare class scenario. The work in [41] uses implicit oversampling of the rare class and undersampling of the copious class in order to achieve effective classification in the rare class scenario. The rare class detection problem has numerous variations in the streaming scenario, which are not quite as evident in the static scenario. This is because of the presence of novel classes and recurring classes in the streaming scenario, which need to be detected.

The problem of rare class detection is related to, but distinct from, the problem of classifying data streams with a limited number of training labels [66]. In the latter case, the labels are limited but may not necessarily be restricted to a particular class, However, in practice, the problem of limited labels arises mostly in binary classification problems where the rare class has a relatively small number of labels, and the remaining records are unlabeled.

The different kinds of scenarios that could arise in these contexts are as follows:

- *Rare Classes:* The determination of such classes is similar to the static supervised scenario, except that it needs to be done efficiently in the streaming scenario. In such cases, a small fraction of the records may belong to a rare class, but they may not necessarily be distributed in a non-homogenous way from a temporal perspective. While some concept drift may also need to be accounted for, the modification to standard stream classification algorithms remains quite analogous to the modifications of the static classification problem to the rare-class scenario.

- *Novel Classes:* These are classes, that were not encountered before in the data stream. Therefore, they may not be reflected in the training model at all. Eventually, such classes may become a normal part of the data over time. This scenario is somewhat similar to semi-supervised outlier detection in the static scenario, though the addition of the temporal component brings a number of challenges associated with it.

- *Infrequently Recurring Classes:* These are classes, that have not been encountered for a while, but may re-appear in the stream. Such classes are different from the first type of outliers, because they arrive in *temporally rare bursts*. Since most data stream classification algorithms use some form of discounting in order to address concept drift, they may sometimes completely age out information about old classes. Such classes cannot be distinguished from novel classes, if the infrequently recurring classes are not reflected in the training model. Therefore, issues of model update and discounting are important in the detection of such classes. The third kind of outlier was first proposed in [64].

We discuss each of the above cases below.

9.3.1 Detecting Rare Classes

Numerous classifiers are available for the streaming scenario [1], especially in the presence of concept drift. For detecting *rare classes*, the only change to be made to these classifiers is to add methods that are intended to handle the *class imbalance*. Such changes are not very different from those of addressing class imbalance in the static context. In general, the idea is that classes are often associated with costs, which enhance the importance of the rare class relative to the normal class. Since the broad principles of detecting rare classes do not change very much between the static and dynamic scenario, the discussion in this section will focus on the other two kinds of outliers.

9.3.2 Detecting Novel Classes

Novel class detection is the problem of detecting classes that have not occurred earlier either in the training or test data. Therefore, no ground truth is available for the problem of novel class detection. The problem of novel class detection will be studied in the online setting, which is the most common scenario in which it is encountered. Detecting novel classes is also a form of *semi-supervision*, because models are available about many of the other classes, but not the class that is being detected. The fully unsupervised setting corresponds to the clustering problem [11,13,94,97], in which newly created clusters are reported as novel classes. In the latter case, the models are created in an unsupervised way with the use of clustering or other unsupervised methods. The work in [64,65] addresses the problem of novel class detection in the streaming scenario.

Much of the traditional work on novel class detection [67] is focussed only on finding novel classes that are different from the current models. However, this approach does not distinguish between the different novel classes that may be encountered over time. A more general way of understanding the novel class detection problem is to view it as a combination of supervised (classification) and unsupervised (clustering) models. Thus, as in unsupervised novel class detection models such as first story detection [13, 97], the cohesion between the test instances of a novel class is important in determining whether they belong to the same novel class or not. The work in [64, 65] combines both supervised and semi-supervised models by:

- Maintaining a supervised model of the classes available in the training data as an ensemble of classification models.

- Maintaining an unsupervised model of the (unlabeled) novel classes received so far as cohesive groups of tightly knit clusters.

When a new test instance is received, the classification model is first applied to it to test whether it belongs to a currently existing (labeled) class. If this is not the case, it is tested whether it naturally belongs to one of the novel classes. The relationship of the test instance to a statistical boundary of the clusters representing the novel classes is used for this purpose. If neither of these conditions hold, it is assumed that the new data point should be in a novel class of its own. Thus, this approach creates a flexible scenario that combines supervised and unsupervised methods for novel class detection.

9.3.3 Detecting Infrequently Recurring Classes

Many outliers in real applications often arrive in *infrequent temporal bursts*. Many classifiers cannot distinguish between novel classes and rare classes, especially if the old models have aged out in the data stream. Therefore, one solution is to simply report a recurring outlier as a novel outlier.

This is however not a very desirable solution because novel-class detection is a semi-supervised problem, whereas the detection of recurring classes is a fully supervised problem. Therefore, by remembering the distribution of the recurring class over time, it is possible to improve the classification accuracy. The second issue is related to computational and resource efficiency. Since novel class detection is much more computationally and memory intensive than the problem of identifying an existing class, it is inefficient to treat recurring classes as novel classes. The work in [64] is able to identify the recurring classes in the data, and is also able to distinguish between the different kinds of recurring classes, when they are distinct from one another, by examining their relationships in the feature space. For example, two kinds of recurring intrusions in a network intrusion detection application may form distinct clusters in the stream. The work in [64] is able to distinguish between the different kinds of recurring classes as well. Another recent method that improves upon the basic technique of detecting rare and recurring classes is proposed in [16].

9.4 Discrete Attributes: The Massive Domain Scenario

In this section, we will study the problem of stream classification with discrete or categorical attributes. Stream classification with discrete attributes is not very different from continuous attributes, and many of the aforementioned methods can be extended easily to categorical data with minor modifications. A particularly difficult case in the space of discrete attributes is the massive domain scenario.

A massive-domain stream is defined as one in which each attribute takes on an extremely large number of possible values. Some examples are as follows:

1. In internet applications, the number of possible source and destination addresses can be very large. For example, there may be well over 10^8 possible IP-addresses.

2. The individual items in supermarket transactions are often drawn from millions of possibilities.

3. In general, when the attributes contain very detailed information such as locations, addresses, names, phone numbers or other such information, the domain size is very large. Recent years have seen a considerable increase in such applications because of advances in data collection techniques.

Many synopsis techniques such as sketches and distinct-element counting are motivated by the massive-domain scenario [34]. Note that these data structures would not be required for streams with a small number of possible values. Therefore, while the importance of this scenario is well understood in the context of synopsis data structures, it is rarely studied in the context of core mining problems. Recent work has also addressed this scenario in the context of core mining problems such as clustering [10] and classification [5].

The one-pass restrictions of data stream computation create further restrictions on the *computational approach*, which may be used for discriminatory analysis. Thus, the massive-domain size creates challenges in terms of space requirements, whereas the stream model further restricts the classes of algorithms that may be used in order to create space-efficient methods. For example, consider the following types of classification models:

- Techniques such as decision trees [20,73] require the computation of the discriminatory power of each possible attribute value in order to determine how the splits should be constructed. In order to compute the relative behavior of different attribute values, the discriminatory power of different attribute values (or combinations of values) need to be maintained. This becomes difficult in the context of massive data streams.

- Techniques such as rule-based classifiers [24] require the determination of combinations of attributes that are relevant to classification. With increasing domain size, it is no longer possible to compute this efficiently either in terms of space or running time.

The stream scenario presents additional challenges for classifiers. This is because the one-pass constraint dictates the choice of data structures and algorithms that can be used for the classification problem. All stream classifiers such as that discussed in [26] implicitly assume that the underlying domain size can be handled with modest main memory or storage limitations.

In order to discuss further, some notations and definitions will be introduced. The data stream $\mathcal{D}$ contains d-dimensional records that are denoted by $\overline{X_1} \ldots \overline{X_N} \ldots$. Associated with each record is a class that is drawn from the index $\{1 \ldots k\}$. The attributes of record $\overline{X_i}$ are denoted by $(x_i^1 \ldots x_i^d)$. It is assumed that the attribute value x_i^k is drawn from the unordered domain set $J_k = \{v_1^k \ldots v_{M^k}^k\}$. The value of M^k denotes the domain size for the kth attribute. The value of M^k can be very large, and may range on the order of millions or billions. When the discriminatory power is defined in terms of subspaces of higher dimensionality, this number multiples rapidly to very large values. Such intermediate computations will be difficult to perform on even high-end machines.

Even though an extremely large number of attribute-value combinations may be possible over the different dimensions and domain sizes, only a limited number of these possibilities are usually relevant for classification purposes. Unfortunately, the intermediate computations required to effectively compare these combinations may not be easily feasible. The one-pass constraint of the data stream model creates an additional challenge in the computation process. In order to perform the classification, it is not necessary to explicitly determine the combinations of attributes that are related to a given class label. The more relevant question is the determination of *whether some combinations of attributes exist* that are strongly related to some class label. As it turns out, a sketch-based approach is very effective in such a scenario.

Sketch-based approaches [34] were designed for enumeration of different kinds of frequency statistics of data sets. The work in [5] extends the well-known count-min sketch [34] to the problem of classification of data streams. In this sketch, a total of $w = \lceil \ln(1/\delta) \rceil$ pairwise independent hash functions are used, each of which map onto uniformly random integers in the range $h = [0, e/\varepsilon]$, where e is the base of the natural logarithm. The data structure itself consists of a two dimensional array with $w \cdot h$ cells with a length of h and width of w. Each hash function corresponds to one of w 1-dimensional arrays with h cells each. In standard applications of the count-min sketch, the hash functions are used in order to update the counts of the different cells in this 2-dimensional data structure. For example, consider a 1-dimensional data stream with elements drawn from a massive set of domain values. When a new element of the data stream is received, each of the w hash functions are applied, in order to map onto a number in $[0 \ldots h-1]$. The count of each of the set of w cells is incremented by 1. In order to *estimate* the count of an item, the set of w cells to which each of the w hash-functions map are determined. The minimum value among all these cells is determined. Let c_t be the true value of the count being estimated. The estimated count is at least equal to c_t, since all counts are non-negative, and there may be an over-estimation because of collisions among hash cells. As it turns out, a probabilistic upper bound to the estimate may also be determined. It has been shown in [34] that for a data stream with T arrivals, the estimate is at most $c_t + \varepsilon \cdot T$ with probability at least $1 - \delta$.

In typical subspace classifiers such as rule-based classifiers, low dimensional projections such as 2-dimensional or 3-dimensional combinations of attributes in the antecedents of the rule are used. In the case of data sets with massive domain sizes, the number of possible combinations of attributes (even for such low-dimensional combinations) can be so large that the corresponding statistics cannot be maintained explicitly during intermediate computations. However, the sketch-based method provides a unique technique for maintaining counts by creating *super-items* from different combinations of attribute values. Each super-item V containing a concatenation of the attribute value strings along with the dimension indices to which these strings belong. Let the actual value-string corresponding to value i_r be $S(i_r)$, and let the dimension index corresponding to the item i_r be $dim(i_r)$. In order to represent the dimension-value combinations corresponding to items $i_1 \ldots i_p$, a new string is created by concatenating the strings $S(i_1) \ldots S(i_p)$ and the dimension indices $dim(i_1) \ldots dim(i_p)$.

This new super-string is then hashed into the sketch table as if it is the attribute value for the special super-item V. For each of the k-classes, a separate sketch of size $w \cdot h$ is maintained, and the sketch cells for a given class are updated only when a data stream item of the corresponding class is received. *It is important to note that the same set of w hash functions is used for updating the sketch corresponding to each of the k classes in the data.* Then, the sketch is updated once for each 1-dimensional attribute value for the d different attributes, and once for each of the super-items created by attribute combinations. For example, consider the case where it is desired to determine discriminatory combinations or 1- or 2-dimensional attributes. There are a total of $d + d \cdot (d-1)/2 = d \cdot (d+1)/2$ such combinations. Then, the sketch for the corresponding class is updated $L = d \cdot (d+1)/2$ times for each of the attribute-values or combinations of attribute-values. In general, L may be larger if even higher dimensional combinations are used, though for cases of massive domain sizes, even a low-dimensional subspace would have a high enough level of specificity for classification purposes. This is because of the extremely large number of combinations of possibilities, most of which would have very little frequency with respect to the data stream. For all practical purposes, one can assume that the use of 2-dimensional or 3-dimensional combinations provides sufficient discrimination in the massive-domain case. The value of L is dependent only on the dimensionality

Algorithm *SketchUpdate*(Labeled Data Stream: $\mathcal{D}$,
      NumClasses: k, MaxDim: r)
begin
  Initialize k sketch tables of size $w \cdot h$ each
  with zero counts in each entry;
  repeat
    Receive next data point $\overline{X}$ from $\mathcal{D}$;
    Add 1 to each of the sketch counts in the
    (class-specific) table for all L value-combinations
    in $\overline{X}$ with dimensionality less than r;
  until(all points in $\mathcal{D}$ have been processed);
end

FIGURE 9.1: Sketch updates for classification (training).

and is independent of the domain size along any of the dimensions. For modest values of d, the value of L is typically much lower than the number of possible combinations of attribute values.

The sketch-based classification algorithm has the advantage of being simple to implement. For each of the classes, a separate sketch table with $w \cdot d$ values is maintained. Thus, there are a total of $w \cdot d \cdot k$ cells that need to be maintained. When a new item from the data stream arrives, a total of $L \cdot w$ cells of the ith sketch table are updated. Specifically, for each item or super-item we update the count of the corresponding w cells in the sketch table by one unit. The overall approach for updating the sketch table is illustrated in Figure 9.1. The input to the algorithm is the data stream $\mathcal{D}$, the maximum dimensionality of the subspace combinations that are tracked, and the number of classes in the data set.

The key to using the sketch-based approach effectively is to be able to efficiently determine discriminative combinations of attributes. While one does not need to determine such combinations *explicitly* in closed form, it suffices to be able to *test* whether a *given* combination of attributes is discriminative. This suffices to perform effective classification of a given test instance. Consider the state of the data stream, when N records have arrived so far. The number of data streams records received from the k different classes are denoted by $N_1 \ldots N_k$, so that we have $\sum_{i=1}^{k} N_i = N$.

Most combinations of attribute values have very low frequency of presence in the data stream. Here, one is interested in those combinations of attribute values that have high relative presence in one class compared to the other classes. Here we are referring to high *relative presence for a given class* in combination with a moderate amount of absolute presence. For example, if a particular combination of values occurs in 0.5% of the records corresponding to the class i, but it occurs in less than 0.1% of the records belonging to the other classes, then the relative presence of the combination in that particular class is high enough to be considered significant. Therefore, the discriminative power of a given combination of values (or super-item) V will be defined. Let $f_i(V)$ denote the fractional presence of the super-item V in class i, and $g_i(V)$ be the fractional presence of the super-item V in all classes other than i. In order to identify classification behavior specific to class i, the super-item V is of interest, if $f_i(V)$ is significantly greater than $g_i(V)$.

Definition 9.4.1 *The discriminatory power $\theta_i(V)$ of the super-item V is defined as the fractional difference in the relative frequency of the attribute-value combination V in class i versus the relative presence in classes other than i. Formally, the value of $\theta_i(V)$ is defined as follows:*

$$\theta_i(V) = \frac{f_i(V) - g_i(V)}{f_i(V)}. \tag{9.2}$$

Since one is interested only in items from which $f_i(V)$ is greater than $g_i(V)$, the value of $\theta_i(V)$

in super-items of interest will lie between 0 and 1. The larger the value of $\theta_i(V)$, the greater the correlation between the attribute-combination V and the class i. A value of $\theta_i(V) = 0$ indicates no correlation, whereas the value of $\theta_i(V) = 1$ indicates perfect correlation. In addition, it is interesting to determine those combinations of attribute values that occur in at least a fraction s of the records belonging to any class i. Such attribute value combinations are referred to as *discriminatory*. The concept of (θ, s)-discriminatory combinations is defined as follows:

Definition 9.4.2 *An attribute value-combination V is defined as (θ, s)-discriminatory with respect to class i, if $f_i(V) \geq s$ and $\theta_i(V) \geq \theta$.*

In other words, the attribute-value combination V occurs in at least a fraction s of the records belonging to class i, and has a discriminatory power of at least θ with respect to class i. Next, an approach for the *decision problem* of testing whether or not a given attribute-value combination is discriminatory will be discussed. This is required to perform classification of test instances.

In order to estimate the value of $\theta_i(V)$, the values of $f_i(V)$ and $g_i(V)$ also need to be estimated. The value of N_i is simply the sum of the counts in the table corresponding to the ith class divided by L. This is because exactly L sketch-table entries for each incoming record are updated. In order to estimate $f_i(V)$ we take the minimum of the w hash cells to which V maps for various hash functions in the sketch table corresponding to class i. This is the approach used for estimating the counts as discussed in [34]. The value of $g_i(V)$ can be estimated similarly, except that we compute a sketch table that contains the sum of the (corresponding cell-specific) counts in the $(k-1)$ classes other than i. This composite sketch table represents the other $(k-1)$ classes, since the same hash-function on each table is used. As in the previous case, the value of $g_i(V)$ is estimated by using the minimum cell value from all the w cells to which V maps. The value of $\theta_i(V)$ can then be estimated from these values of $f_i(V)$ and $g_i(V)$. The following result can be shown [5].

Lemma 9.4.1 *With probability at least $(1 - \delta)$, the values of $f_i(V)$ and $g_i(V)$ are respectively overestimated to within $L \cdot \varepsilon$ of their true values when we use sketch tables with size $w = \lceil ln(1/\delta) \rceil$ and $h = \lceil e/\varepsilon \rceil$.*

Next, the accuracy of estimation of $\theta_i(V)$ is estimated. Note that one is only interested in those attribute-combinations V for which $f_i(V) \geq s$ and $f_i(V) \geq g_i(V)$, since such patterns have sufficient statistical counts and are also discriminatory with $\theta_i(V) \geq 0$.

Lemma 9.4.2 *Let $\beta_i(V)$ be the estimated value of $\theta_i(V)$ for an attribute-combination V with fractional selectivity at least s and $f_i(V) \geq g_i(V)$. Let ε be chosen such that $\varepsilon' = \varepsilon \cdot L/s << 1$. With probability at least $1 - \delta$, it is the case that $\beta_i(V) \leq \theta_i(V) + \varepsilon'$.*

Next, we will examine the case when the value of $\theta_i(V)$ is under-estimated [5].

Lemma 9.4.3 *Let $\beta_i(V)$ be the estimated value of $\theta_i(V)$ for an attribute-combination V with fractional selectivity at least s and $f_i(V) \geq g_i(V)$. Let ε be chosen such that $\varepsilon' = \varepsilon \cdot L/s << 1$. With probability at least $1 - \delta$, it is the case that $\beta_i(V) \geq \theta_i(V) - \varepsilon'$.*

The results of Lemma 9.4.2 and 9.4.3 can be combined to conclude the following:

Lemma 9.4.4 *Let $\beta_i(V)$ be the estimated value of $\theta_i(V)$ for an attribute-combination V with fractional selectivity at least s and $f_i(V) \geq g_i(V)$. Let ε be chosen such that $\varepsilon' = \varepsilon \cdot L/s << 1$. With probability at least $1 - 2 \cdot \delta$, it is the case that $\beta_i(V) \in (\theta_i(V) - \varepsilon', \theta_i(V) + \varepsilon')$.*

This result follows from the fact each of the inequalities in Lemmas 9.4.2 and 9.4.3 are true with probability at least $1 - \delta$. Therefore, both inequalities are true with probability at least $(1 - 2 \cdot \delta)$. Another natural corollary of this result is that any pattern that is truly (θ, s)-discriminatory will be

TABLE 9.1: Storage Requirement of Sketch Table for Different Data and Pattern Dimensionalities ($\varepsilon' = 0.01$, $\delta = 0.01$, $s = 0.01$)

Data Dimensionality	Max. Pattern Dimensionality	L	$\varepsilon = \varepsilon' \cdot s/L$	# Cells = $\lceil e \cdot \ln(1/\delta)/\varepsilon \rceil$	Storage Required	Storage $\varepsilon' X 20$
10	2	55	$10^{-4}/55$	6884350	26.3 MB	1.315 MB
10	3	175	$10^{-4}/175$	$21.9 * 10^6$	83.6 MB	4.18 MB
20	2	210	$10^{-4}/210$	$26.3 * 10^6$	100.27 MB	5.01 MB
20	3	1920	$10^{-4}/1920$	$24 * 10^7$	916.85 MB	45.8 MB
50	2	1275	$10^{-4}/1275$	$15.96 * 10^6$	608.85 MB	30.5 MB
100	2	5050	$10^{-4}/5050$	$63.21 * 10^7$	2.41 GB	120.5 MB
200	2	20100	$10^{-4}/20100$	$25.16 * 10^8$	9.6 GB	480 MB

discovered with probability at least $(1 - 2 \cdot \delta)$ by using the sketch based approach in order to determine all patterns that are at least $(\theta - \varepsilon', s \cdot (1 - \varepsilon'))$-discriminatory. Furthermore, the discriminatory power of such patterns will not be over- or under-estimated by an inaccuracy greater than ε'.

The process of determining whether a super-item V is (θ, s) requires us to determine $f_i(V)$ and $g_i(V)$ only. The value of $f_i(V)$ may be determined in a straightforward way by using the sketch based technique of [34] in order to determine the estimated frequency of the item. We note that $f_i(V)$ can be determined quite accurately since we are only considering those patterns that have a certain minimum support. The value of $g_i(V)$ may be determined by adding up the sketch tables for all the other different classes, and then using the same technique.

What are the space requirements of the approach for classifying large data streams under a variety of practical parameter settings? Consider the case where we have a 10-dimensional massive domain data set that has at least 10^7 values over each attribute. Then, the number of possible 2-dimensional and 3-dimensional value combinations are $10^{14} * 10 * (10 - 1)/2$ and $10^{21} * 10 * (10 - 1) * (10 - 2)/6$, respectively. We note that even the former requirement translates to an order of magnitude of about 4500 tera-bytes. Clearly, the intermediate-space requirements for aggregation-based computations of many standard classifiers are beyond most modern computers. On the other hand, for the sketch-based technique, the requirements continue to be quite modest. For example, let us consider the case where we use the sketch-based approach with $\varepsilon' = 0.01$ and $\delta = 0.01$. Also, let us assume that we wish to have the ability to perform discriminatory classification on patterns with specificity at least $s = 0.01$. We have illustrated the storage requirements for data sets of different dimensionalities in Table 9.1. While the table is constructed for the case of $\varepsilon' = 0.01$, the storage numbers in the last column of the table are illustrated for the case of $\varepsilon' = 0.2$. We will see later that the use of much large values of ε' can provide effective and accurate results. It is clear that the storage requirements are quite modest, and can be held in main memory in most cases. For the case of stringent accuracy requirements of $\varepsilon' = 0.01$, the high dimensional data sets may require a modest amount of disk storage in order to capture the sketch table. However, a practical choice of $\varepsilon = 0.2$ always results in a table that can be stored in main memory. These results are especially useful in light of the fact that even data sets of small dimensionalities cannot be effectively processed with the use of traditional methods.

The sketch table may be leveraged to perform the classification for a given test instance $\overline{Y}$. The first step is to extract all the L patterns in the test instance with dimensionality less than r. Then, we determine those patterns that are (θ, s)-discriminatory with respect to at least one class. The process of finding (θ, s)-discriminatory patterns has already been discussed in the last section. We use a voting scheme in order to determine the final class label. Each pattern that is (θ, s)-discriminatory constitutes a vote for that class. The class with the highest number of votes is reported as the relevant class label. The overall procedure is reported in Figure 9.2.

Algorithm *SketchClassify*(Test Data Point: $\overline{Y}$,
      NumClasses: k, MaxDim: r)
begin
  Determine all L value-combinations specific
  to test instance;
  Use sketch-based approach to determine which
  value-combinations are (θ, s)-discriminatory;
  Add 1 vote to each class for which a pattern is
    (θ, s)-discriminatory;
  Report class with largest number of votes;
end

FIGURE 9.2: Classification (testing phase).

9.5 Other Data Domains

The problem of stream classification is also relevant to other data domains such as text and graphs. In this section, a number of algorithms for these data domains will be studied in detail.

9.5.1 Text Streams

The problem of classification of data streams has been widely studied by the data mining community in the context of different kinds of data [1, 14, 87, 91]. Many of the methods for classifying numerical data can be easily extended to the case of text data, just as the numerical clustering method in [11] has been extended to a text clustering method in [13]. In particular, the classification method of [14] can be extended to the text domain, by adapting the concept of numerical micro-clusters to condensed droplets as discussed in [13]. With the use of the condensed droplet data structure, it is possible to extend all the methods discussed in [14] to the text stream scenario. Similarly, the core idea in [87] uses an ensemble-based approach on chunks of the data stream. This broad approach is essentially a *meta-algorithm* that is not too dependent upon the specifics of the underlying data format. Therefore, the broad method can also be easily extended to the case of text streams.

The problem of text stream classification arises in the context of a number of different IR tasks. The most important of these is *news filtering* [55], in which it is desirable to automatically assign incoming documents to pre-defined categories. A second application is that of email spam filtering [17], in which it is desirable to determine whether incoming email messages are spam or not. As discussed earlier, the problem of text stream classification can arise in two different contexts, depending upon whether the training or the test data arrives in the form of a stream:

- In the first case, the training data may be available for batch learning, but the test data may arrive in the form of a stream.

- In the second case, both the training and the test data may arrive in the form of a stream. The patterns in the training data may continuously change over time, as a result of which the models need to be updated dynamically.

The first scenario is usually easy to handle, because most classifier models are compact and classify individual test instances efficiently. On the other hand, in the second scenario, the training model needs to be constantly updated in order to account for changes in the patterns of the underlying training data. The easiest approach to such a problem is to incorporate temporal decay factors into model construction algorithms, so as to age out the old data. This ensures that the new (and more timely) data is weighted more significantly in the classification model. An interesting technique along this direction has been proposed in [77], in which a temporal weighting factor is introduced in order to modify the classification algorithms. Specifically, the approach has been applied to the Naive Bayes, Rocchio, and k-nearest neighbor classification algorithms. It has been shown that the incorporation of temporal weighting factors is useful in improving the classification accuracy, when the underlying data is evolving over time.

A number of methods have also been designed specifically for the case of text streams. In particular, the method discussed in [39] studies methods for classifying text streams in which the classification model may evolve over time. This problem has been studied extensively in the literature in the context of multi-dimensional data streams [14, 87]. For example, in a spam filtering application, a user may generally delete the spam emails for a particular topic, such as those corresponding to political advertisements. However, in a particular period such as the presidential elections, the user may be interested in the emails for that topic, and so it may not be appropriate to continue to classify that email as spam.

The work in [39] looks at the particular problem of classification in the context of user-interests. In this problem, the label of a document is considered either *interesting* or *non-interesting*. In order to achieve this goal, the work in [39] maintains the interesting and non-interesting topics in a text stream together with the evolution of the theme of the interesting topics. A document collection is classified into multiple topics, each of which is labeled either interesting or non-interesting at a given point. A concept refers to the main theme of interesting topics. A concept drift refers to the fact that the main theme of the interesting topic has changed.

The main goals of the work are to maximize the accuracy of classification and minimize the cost of re-classification. In order to achieve this goal, the method in [39] designs methods for detecting both *concept drift* as well as *model adaptation*. The former refers to the change in the theme of user-interests, whereas the latter refers to the detection of brand new concepts. In order to detect concept drifts, the method in [39] measures the classification error-rates in the data stream in terms of true and false positives. When the stream evolves, these error rates will increase, if the change in the concepts are not detection. In order to determine the change in concepts, techniques from statistical quality control are used, in which we determine the mean μ and standard deviation σ of the error rates, and determine whether this error rate remains within a particular tolerance, which is $[\mu - k \cdot \sigma, \mu + k \cdot \sigma]$. Here the tolerance is regulated by the parameter k. When the error rate changes, we determine when the concepts should be dropped or included. In addition, the drift rate is measured in order to determine the rate at which the concepts should be changed for classification purposes. In addition, methods for dynamic construction and removal of sketches are discussed in [39].

Another related work is that of one-class classification of text streams [93], in which only training data for the positive class is available, but there is no training data available for the negative class. This is quite common in many real applications in which it is easy to find representative documents for a particular topic, but it is hard to find the representative documents in order to model the background collection. The method works by designing an ensemble of classifiers in which some of the classifiers corresponds to a recent model, whereas others correspond to a long-term model. This is done in order to incorporate the fact that the classification should be performed with a combination of short-term and long-term trends. Another method for positive-unlabeled stream classification is discussed in [62].

A rule-based technique, which can learn classifiers incrementally from data streams, is the *sleeping-experts systems* [29, 33]. One characteristic of this rule-based system is that it uses the position of the words in generating the classification rules. Typically, the rules correspond to sets of words that are placed close together in a given text document. These sets of words are related to a class label. For a given test document, it is determined whether these sets of words occur in the document, and are used for classification. This system essentially learns a set of rules (or sleeping experts), which can be updated incrementally by the system. While the technique was proposed prior to the advent of data stream technology, its online nature ensures that it can be effectively used for the stream scenario.

One of the classes of methods that can be easily adapted to stream classification is the broad category of *neural networks* [79, 89]. This is because neural networks are essentially designed as a classification model with a network of perceptrons and corresponding weights associated with the term-class pairs. Such an incremental update process can be naturally adapted to the streaming context. These weights are incrementally learned as new examples arrive. The first neural network methods for online learning were proposed in [79, 89]. In these methods, the classifier starts off by setting all the weights in the neural network to the same value. The incoming training example is classified with the neural network. In the event that the result of the classification process is correct, then the weights are not modified. On the other hand, if the classification is incorrect, then the weights for the terms are either increased or decreased depending upon which class the training example belongs to. Specifically, if the class to which the training example belongs is a positive instance, the weights of the corresponding terms (in the training document) are increased by α. Otherwise, the weights of these terms are reduced by α. The value of α is also known as the *learning rate*. Many other variations are possible in terms of how the weights may be modified. For example, the method in [25] uses a multiplicative update rule, in which two multiplicative constants $\alpha_1 > 1$ and $\alpha_2 < 1$ are used for the classification process. The weights are multiplied by α_1, when the example belongs to the positive class, and is multiplied by α_2 otherwise. Another variation [56] also allows the modification of weights, when the classification process is correct. A number of other online neural network methods for text classification (together with background on the topic) may be found in [21, 28, 70, 75]. A Bayesian method for classification of text streams is discussed in [22]. The method in [22] constructs a Bayesian model of the text, which can be used for online classification. The key components of this approach are the design of a Bayesian online perceptron and a Bayesian online Gaussian process, which can be used effectively for online learning.

9.5.2 Graph Streams

Recently, the emergence of Internet applications has lead to increasing interest in *dynamic graph streaming applications* [7, 8, 37]. Such network applications are defined on a *massive underlying universe* of nodes. Some examples are as follows:

- The communication pattern of users in a social network in a modest time window can be decomposed into a group of disconnected graphs, which are defined over a massive-domain of user nodes.

- The browsing pattern of a single user (constructed from a proxy log) is typically a small sub-graph of the Web graph. The browsing pattern over all users can be interpreted as a continuous stream of such graphs.

- The intrusion traffic on a communication network is a stream of localized graphs on the massive IP-network.

The most challenging case is one in which the data is available only in the form of a stream of graphs (or edge streams with graph identifiers). Furthermore, it is assumed that the *number of distinct edges is so large, that it is impossible to hold summary information about the underlying structures explicitly*. For example, a graph with more than 10^7 nodes may contains as many as 10^{13} edges. Clearly, it may not be possible to store information about such a large number of edges explicitly. Thus, the complexity of the graph stream classification problem arises in two separate ways:

1. Since the graphs are available only in the form of a stream, this restrains the kinds of algorithms that can be used to mine structural information for future analysis. An additional

complication is caused by the fact the the edges for a particular graph may arrive *out of order* (i.e., not contiguously) in the data stream. Since re-processing is not possible in a data stream, such out-of-order edges create challenges for algorithms that extract structural characteristics from the graphs.

2. The massive size of the graph creates a challenge for effective extraction of information that is relevant to classification. For example, it is difficult to even store summary information about the large number of distinct edges in the data. In such cases, the determination of frequent discriminative subgraphs may be computationally and space-inefficient to the point of being impractical.

A *discriminative subgraph mining approach* was proposed in [9] for graph stream classification. We will define discriminative subgraphs both in terms of edge pattern co-occurrence and the class label distributions. The presence of such subgraphs in test instances can be used in order to infer their class labels. Such discriminative subgraphs are difficult to determine with enumeration-based algorithms because of the massive domain of the underlying data. A probabilistic algorithm was proposed for determining the discriminative patterns. The probabilistic algorithm uses a hash-based summarization approach to capture the discriminative behavior of the underlying massive graphs. This compressed representation can be leveraged for effective classification of test (graph) instances.

The graph is defined over the node set N. This node set is assumed to be drawn from a massive domain of identifiers. The individual graphs in the stream are denoted by $G_1 \ldots G_n \ldots$. Each graph G_i is associated with the class label C_i, which is drawn from $\{1 \ldots m\}$. It is assumed that the data is *sparse*. The sparsity property implies that even though the node and edge domain may be very large, the number of edges in the individual graphs may be relatively modest. This is generally true over a wide variety of real applications such as those discussed in the introduction section. In the streaming case, the edges of each graph G_i may not be neatly received at a given moment in time. Rather, the edges of different graphs may appear *out of order* in the data stream. This means that the edges for a given graph may not appear contiguously in the stream. This is definitely the case for many applications such as social networks and communication networks in which one cannot control the ordering of messages and communications across different edges. This is a particularly difficult case. It is assumed that the edges are received in the form $< GraphId >< Edge >< ClassLabel >$. Note that the class label is the same across all edges for a particular graph identifier. For notational convenience, we can assume that the class label is appended to the graph identifier, and therefore we can assume (without loss of generality or class information) that the incoming stream is of the form $< GraphId >< Edge >$. The value of the variable $< Edge >$ is defined by its two constituent nodes.

A rule-based approach was designed in [9], which associates discriminative subgraphs to specific class labels. Therefore, a way of quantifying the significance of a particular subset of edges P for the purposes of classification is needed. Ideally, one would like the subgraph P to have significant statistical presence in terms of the *relative frequency* of its constituent edges. At the same time, one would like P to be discriminative in terms of the class distribution.

The first criterion retains subgraphs with high relative frequency of presence, whereas the second criterion retains only subgraphs with high discriminative behavior. It is important to design effective techniques for determining patterns that satisfy both of these characteristics. First, the concept of a significant subgraph in the data will be defined. A significant subgraph P is defined as a subgraph (set of edges), for which the edges are correlated with one another in terms of absolute presence. This is also referred to as *edge coherence*. This concept is formally defined as follows:

Definition 9.5.1 *Let $f_\cap(P)$ be the fraction of graphs in $G_1 \ldots G_n$ in which **all** edges of P are present. Let $f_\cup(P)$ be the fraction of graphs in which **at least one or more** of the edges of P are present. Then, the edge coherence $C(P)$ of the subgraph P is denoted by $f_\cap(P)/f_\cup(P)$.*

We note that the above definition of edge coherence is focussed on *relative presence* of subgraph patterns rather than the absolute presence. This ensures that only significant patterns are found. Therefore, large numbers of irrelevant patterns with high frequency but low significance are not considered. While the coherence definition is more effective, it is computationally quite challenging because of the size of the search space that may need to be explored. The randomized scheme discussed here is specifically designed in order to handle this challenge.

Next, the class discrimination power of the different subgraphs is defined. For this purpose, the *class confidence* of the edge set P with respect to the class label $r \in \{1 \ldots m\}$ is defined as follows.

Definition 9.5.2 *Among all graphs containing subgraph P, let $s(P,r)$ be the fraction belonging to class label r. We denote this fraction as the confidence of the pattern P with respect to the class r.*

Correspondingly, the concept of the dominant class confidence for a particular subgraph is defined.

Definition 9.5.3 *The dominant class confidence $DI(P)$ or subgraph P is defined as the maximum class confidence across all the different classes $\{1 \ldots m\}$.*

A significantly large value of $DI(P)$ for a particular test instance indicates that the pattern P is very relevant to classification, and the corresponding dominant class label may be an attractive candidate for the test instance label.

In general, one would like to determine patterns that are interesting in terms of absolute presence, and are also discriminative for a particular class. Therefore, the parameter-pair (α, θ) is defined, which corresponds to threshold parameters on the edge coherence and class interest ratio.

Definition 9.5.4 *A subgraph P is said to be be (α, θ)-significant, if it satisfies the following two* edge-coherence *and* class discrimination *constraints:*
(a) Edge Coherence Constraint: *The edge-coherence $C(P)$ of subgraph P is at least α. In other words, it is the case that $C(P) \geq \alpha$.*
(b) Class Discrimination Constraint: *The dominant class confidence $DI(P)$ is at least θ. In other words, it is the case that $DI(P) \geq \theta$.*

The above constraints are quite challenging because of the size of the search space that needs to be explored in order to determine relevant patterns. This approach is used, because it is well suited to massive graphs in which it is important to prune out as many irrelevant patterns as possible. The edge coherence constraint is designed to prune out many patterns that are abundantly present, but are not very significant from a relative perspective. This helps in more effective classification.

A probabilistic *min-hash approach* is used for determining discriminative subgraphs. The min-hash technique is an elegant probabilistic method, which has been used for the problem of finding interesting 2-itemsets [32]. This technique cannot be easily adapted to the graph classification problem, because of the large number of distinct edges in the graph. Therefore, w a 2-dimensional compression technique was used, in which a min-hash function will be used in combination with a more straightforward randomized hashing technique. We will see that this combination approach is extremely effective for graph stream classification.

The aim of constructing this synopsis is to be able to design a continuously updatable data structure, which can determine the most relevant discriminative subgraphs for classification. Since the size of the synopsis is small, it can be maintained in main memory and be used at any point during the arrival of the data stream. The ability to continuously update an in-memory data structure is a natural and efficient design for the stream scenario. At the same time, this *structural synopsis* maintains sufficient information, which is necessary for classification purposes.

For ease in further discussion, a tabular binary representation of the graph data set with N rows and n columns will be utilized. This table is only *conceptually* used for description purposes, but it is *not explicitly maintained* in the algorithms or synopsis construction process. The N rows correspond to the N different graphs present in the data. While columns represent the different edges in the data, this is not a one-to-one mapping. This is because the number of possible edges is so large that it is necessary to use a uniform random hash function in order to map the edges onto the n columns. The choice of n depends upon the space available to hold the synopsis effectively, and affects the quality of the final results obtained. Since many edges are mapped onto the same column by the hash function, the support counts of subgraph patterns are over-estimated with this approach. Since the edge-coherence $C(P)$ is represented as a ratio of supports, the edge coherence may either be over- or under-estimated. The details of this approach are discussed in [9]. A number of other models and algorithms for graph stream classification have recently been proposed [45,57,72]. The work in [57] uses discriminative hash kernels for graph stream classification. A method that uses a combination of hashing and factorization is proposed in [45]. Finally, the work in [72] addresses the problem of graph stream classification in the presence of imbalanced distributions and noise.

Social streams can be viewed as graph streams that contain a combination of text and structural information. An example is the *Twitter* stream, which contains both structural information (in the form of sender-recipient information), and text in the form of the actual tweet. The problem of event detection is closely related to that of classification. The main difference is that event labels are associated with time-instants rather than individual records. A method for performing supervised event detection in social streams is discussed in [15]. The work in [15] also discusses unsupervised methods for event detection.

9.5.3 Uncertain Data Streams

Probabilistic data has become increasingly popular because of the presence of numerous scenarios in which the data is recorded probabilistically. In such cases, the information about the probabilistic distribution of the data can be used in order to make better predictions. The work in [2] creates a density transform that can be used in order to create a kernel density estimate of the data. The kernel density estimate can be computed more accurately by using the information about the uncertainty behavior of the data. The work in [2] uses a relatively simple estimate of the uncertainty in terms of the standard deviation. It has been shown in [2] that the approach can be extended to real-time classification of data streams by using a micro-clustering approach. The micro-clusters can be maintained in real time, and the corresponding density estimates can be also be computed in real time. These can be used in order to create rules that are specific to the test instance in lazy fashion and perform the classification. The decision tree method has also been extended to the case of uncertain data streams [59].

9.6 Conclusions and Summary

This chapter provides a survey of data stream classification algorithms. We present the different types of methods for stream classification, which include the extension of decision trees, rule-based classifiers, nearest neighbor methods, and Bayes methods. In addition, ensemble methods were discussed for classification. A number of different scenarios were also discussed in this chapter, such as rare class methods and categorical data. The massive domain scenario is particularly challenging, because of the large number of distinct attribute values, that need to be considered. Different data domains such as text and graph data were also studied in the chapter. Domains such as graph data are likely to be especially important in the near future, because of the increasing importance of social streams and social networks in which such data sets arise. The related topic of big data classification is discussed in the next chapter.

Bibliography

[1] C. Aggarwal. *Data Streams: Models and Algorithms*, Springer, 2007.

[2] C. Aggarwal. On density-based transforms for uncertain data mining, *Proceedings of the ICDE Conference*, pages 866–875, 2007.

[3] C. Aggarwal. *Outlier Analysis*, Springer, 2013.

[4] C. Aggarwal and P. Yu. LOCUST: An online analytical processing framework for high dimensional classification of data streams, *Proceedings of the ICDE Conference*, pages 426–435, 2008.

[5] C. Aggarwal and P. Yu. On classification of high cardinality data streams. *Proceedings of the SDM Conference*, pages 802–813, 2010.

[6] C. Aggarwal and C. X. Zhai, *Mining Text Data*, Springer, 2012.

[7] C. Aggarwal, Y. Li, P. Yu, and R. Jin. On dense pattern mining in graph streams, *Proceedings of the VLDB Conference*, 3(1-2):975–984, 2010.

[8] C. Aggarwal, Y. Zhao, and P. Yu. On clustering graph streams, *Proceedings of the SDM Conference*, 2010.

[9] C. Aggarwal, On classification of graph streams, *Proceedings of the SDM Conference*, 2010.

[10] C. Aggarwal. A framework for clustering massive-domain data streams, *Proceedings of the ICDE Conference*, pages 102–113, 2009.

[11] C. C. Aggarwal, J. Han. J. Wang, and P. Yu. A framework for clustering evolving data streams, *Proceedings of the VLDB Conference*, pages 81–92, 2003.

[12] C. C. Aggarwal. On biased reservoir sampling in the presence of stream evolution, *Proceedings of the VLDB Conference*, pages 607–618, 2006.

[13] C. C. Aggarwal and P. Yu. On clustering massive text and categorical data streams, *Knowledge and Information Systems*, 24(2), pp. 171–196, 2010.

[14] C. Aggarwal, J. Han, J. Wang, and P. Yu. A framework for classification of evolving data streams. In *IEEE Transactions on Knowledge and Data Engineering*, 18(5):577–589, 2006.

[15] C. Aggarwal and K. Subbian. Event detection in social streams, *Proceedings of the SDM Conference*, 2012.

[16] T. Al-Khateeb, M. Masud, L. Khan, C. Aggarwal, J. Han, and B. Thuraisingham. Recurring and novel class detection using class-based ensemble, *Proceedings of the ICDM Conference*, 2012.

[17] I. Androutsopoulos, J. Koutsias, K. V. Chandrinos, and C. D. Spyropoulos. An experimental comparison of naive Bayesian and keyword-based anti-spam filtering with personal e-mail messages. *Proceedings of the ACM SIGIR Conference*, pages 160–167, 2000.

[18] A. Banerjee and J. Ghosh. Competitive learning mechanisms for scalable, balanced and incremental clustering of streaming texts, *Neural Networks*, pages 2697–2702, 2003.

[19] A. Bifet, G. Holmes, B. Pfahringer, and E. Frank. Fast perceptron decision tree learning from evolving data streams. *Advances in Knowledge Discovery and Data Mining*, pages 299–310, 2010.

[20] L. Breiman, J. Friedman, and C. Stone, *Classification and Regrssion Trees*, Chapman and Hall, 1984.

[21] K. Crammer and Y. Singer. A new family of online algorithms for category ranking, *ACM SIGIR Conference*, pages 151–158, 2002.

[22] K. Chai, H. Ng, and H. Chiu. Bayesian online classifiers for text classification and filtering, *ACM SIGIR Conference*, pages 97–104, 2002.

[23] P. Clark and T. Niblett. The CN2 induction algorithm, *Machine Learning*, 3(4): 261–283, 1989.

[24] W. Cohen. Fast effective rule induction, *ICML Conference*, pages 115–123, 1995.

[25] I. Dagan, Y. Karov, and D. Roth. Mistake-driven learning in text categorization. *Proceedings of Conference Empirical Methods in Natural Language Processing*, pages 55–63, 1997.

[26] P. Domingos and G. Hulten, Mining high-speed data streams, *ACM KDD Conference*, pages 71–80, 2000.

[27] W. Fan. Systematic data selection to mining concept drifting data streams, *ACM KDD Conference*, pages 128–137, 2004.

[28] Y. Freund and R. Schapire. Large margin classification using the perceptron algorithm, *Machine Learning*, 37(3):277–296, 1998.

[29] Y. Freund, R. Schapire, Y. Singer, M. Warmuth. Using and combining predictors that specialize. *Proceedings of the 29th Annual ACM Symposium on Theory of Computing*, pages 334–343, 1997.

[30] G. Cauwenberghs and T. Poggio. Incremental and decremental support vector machine learning, *NIPS Conference*, pages 409–415, 2000.

[31] N. Chawla, N. Japkowicz, and A. Kotcz. Editorial: special issue on learning from imbalanced data sets. *ACM SIGKDD Explorations Newsletter*, 6(1):1–6, 2004.

[32] E. Cohen, M. Datar, S. Fujiwara, A. Gionis, P. Indyk, R. Motwani, J. Ullman, and C. Yang, Finding interesting associations without support pruning, *IEEE TKDE*, 13(1): 64–78, 2001.

[33] W. Cohen and Y. Singer. Context-sensitive learning methods for text categorization. *ACM Transactions on Information Systems*, 17(2): 141–173, 1999.

[34] G. Cormode and S. Muthukrishnan. An improved data-stream summary: The count-min sketch and its applications, *Journal of Algorithms*, 55(1):58–75, 2005.

[35] I. Dagan, Y. Karov, and D. Roth. Mistake-driven learning in text categorization, *Proceedings of EMNLP*, pages 55-63, 1997.

[36] C. Domeniconi and D. Gunopulos. Incremental support vector machine construction. *ICDM Conference*, pages 589–592, 2001.

[37] J. Feigenbaum, S. Kannan, A. McGregor, S. Suri, and J. Zhang, Graph distances in the data stream model. *SIAM Jour. on Comp.*, 38(5):1709–1727, 2005.

[38] G. Fung and O. L. Mangasarian. Incremental support vector machine classification. *SIAM Conference on Data Mining*, pages 77–86, 2002.

[39] G. P. C. Fung, J. X. Yu, and H. Lu. Classifying text streams in the presence of concept drifts. *Advances in Knowledge Discovery and Data Mining*, 3056:373–383, 2004.

[40] J. Gama, R. Fernandes, and R. Rocha. Decision trees for mining data streams. *Intelligent Data Analysis*, 10:23–45, 2006.

[41] J. Gao, W. Fan, J. Han, and P. Yu. A general framework for mining concept drifting data stream with skewed distributions, *SDM Conference*, 2007.

[42] J. Gao, W. Fan, and J. Han. On appropriate assumptions to mine data streams: Analysis and practice, *ICDM Conference*, pages 143–152, 2007.

[43] J. Gehrke, V. Ganti, R. Ramakrishnan, and W.-Y. Loh. BOAT: Optimistic decision tree construction, *ACM SIGMOD Conference*, pages 169–180, 1999.

[44] J. Gehrke, R. Ramakrishnan, and V. Ganti. Rainforest—A framework for fast decision tree construction of large datasets, *VLDB Conference*, pages 416–427, 1998.

[45] T. Guo, L. Chi, and X. Zhu. Graph hashing and factorization for fast graph stream classification. *ACM CIKM Conference*, pages 1607–1612, 2013.

[46] S. Hashemi and Y. Yang. Flexible decision tree for data stream classification in the presence of concept change, noise and missing values. *Data Mining and Knowledge Discovery*, 19(1):95–131, 2009.

[47] G. Hulten, L. Spencer, and P. Domingos. Mining time-changing data streams. *ACM KDD Conference*, pages 97–106, 2001.

[48] R. Jin and G. Agrawal. Efficient Decision Tree Construction on Streaming Data, *ACM KDD Conference*, pages 571–576, 2003.

[49] D. Kalles and T. Morris. Efficient incremental induction of decision trees. *Machine Learning*, 24(3):231–242, 1996.

[50] T. Joachims. Making large scale SVMs practical. *Advances in Kernel Methods, Support Vector Learning*, pp. 169–184, MIT Press, Cambridge, 1998.

[51] T. Joachims. Training linear SVMs in linear time. *KDD*, pages 217–226, 2006.

[52] R. Klinkenberg and T. Joachims. Detecting concept drift with support vector machines. *ICML Conference*, pages 487–494, 2000.

[53] J. Kolter and M. Maloof. Dynamic weighted majority: A new ensemble method for tracking concept drift, *ICDM Conference*, pages 123–130, 2003.

[54] Y.-N. Law and C. Zaniolo. An adaptive nearest neighbor classification algorithm for data streams, *PKDD Conference*, pages 108–120, 2005.

[55] D. Lewis. The TREC-4 filtering track: description and analysis. *Proceedings of TREC-4, 4th Text Retrieval Conference*, pages 165–180, 1995.

[56] D. Lewis, R. E. Schapire, J. P. Callan, and R. Papka. Training algorithms for linear text classifiers. *ACM SIGIR Conference*, pages 298–306, 1996.

[57] B. Li, X. Zhu, L. Chi, and C. Zhang. Nested subtree hash kernels for large-scale graph classification over streams. *ICDM Conference*, pages 399–408, 2012.

[58] X. Li, P. Yu, B. Liu, and S. K. Ng. Positive-unlabeled learning for data stream classification, *SDM Conference*, pages 257–268, 2009.

[59] C. Liang, Y. Zhang, and Q. Song. Decision tree for dynamic and uncertain data streams. *2nd Asian Conference on Machine Learning*, volume 3, pages 209–224, 2010.

[60] N. Littlestone. Learning quickly when irrelevant attributes abound: A new linear-threshold algorithm. *Machine Learning*, 2: pages 285–318, 1988.

[61] B. Liu, W. Hsu, and Y. Ma, Integrating classification and association rule mining, *ACM KDD Conference*, pages 80–86, 1998.

[62] S. Pan, Y. Zhang, and X. Li. Dynamic classifier ensemble for positive unlabeled text stream classification. *Knowledge and Information Systems*, 33(2):267–287, 2012.

[63] J. R. Quinlan. *C4.5: Programs in Machine Learning*, Morgan-Kaufmann Inc, 1993.

[64] M. Masud, T. Al-Khateeb, L. Khan, C. Aggarwal, J. Gao, J. Han, and B. Thuraisingham. Detecting recurring and novel classes in concept-drifting data streams. *ICDM Conference*, pages 1176–1181, 2011.

[65] M. Masud, Q. Chen, L. Khan, C. Aggarwal, J. Gao, J. Han, A. Srivastava, and N. Oza. Classification and adaptive novel class detection of feature-evolving data streams, *IEEE Transactions on Knowledge and Data Engineering*, 25(7):1484–1487, 2013. Available at `http://doi.ieeecomputersociety.org/10.1109/TKDE.2012.109`

[66] M. Masud, J. Gao, L. Khan, J. Han, and B. Thuraisingham: A practical approach to classify evolving data streams: Training with limited amount of labeled data. *ICDM Conference*, pages 929–934, 2008.

[67] M. Markou and S. Singh. Novelty detection: A review, Part 1: Statistical approaches, *Signal Processing*, 83(12):2481–2497, 2003.

[68] M. Mehta, R. Agrawal, and J. Rissanen. SLIQ: A fast scalable classifier for data mining, *EDBT Conference*, pages 18–32, 1996.

[69] L. Minku, A. White, and X. Yao. The impact of diversity on online ensemble learning in the presence of concept drift, *IEEE Transactions on Knowledge and Data Engineering*, 22(6):730–742, 2010.

[70] H. T. Ng, W. B. Goh, and K. L. Low. Feature selection, perceptron learning, and a usability case study for text categorization. *SIGIR Conference*, pages 67–73, 1997.

[71] N. Oza and S. Russell. Online bagging and boosting. In *Eighth Int. Workshop on Artificial Intelligence and Statistics*, pages 105–112. Morgan Kaufmann, 2001.

[72] S. Pan and X. Zhu. Graph classification with imbalanced class distributions and noise. *AAAI Conference*, pages 1586–1592, 2013.

[73] R. Quinlan. *C4.5: Programs for Machine Learning*. Morgan Kaufmann Publishers, 1993.

[74] L. Ralaivola and F. d'Alché-Buc. Incremental support vector machine learning: A local approach. *Artificial Neural Network*, pages 322–330, 2001.

[75] F. Rosenblatt. The perceptron: A probabilistic model for information and storage organization in the brain, *Psychological Review*, 65: pages 386–407, 1958.

[76] S. Ruping. Incremental learning with support vector machines. *IEEE ICDM Conference*, pp. 641–642, 2001.

[77] T. Salles, L. Rocha, G. Pappa, G. Mourao, W. Meira Jr., and M. Goncalves. Temporally-aware algorithms for document classification. *ACM SIGIR Conference*, pages 307–314, 2010.

[78] J. Schlimmer and D. Fisher. A case study of incremental concept induction. *Proceedings of the Fifth National Conference on Artificial Intelligence*, pages 495–501. Morgan Kaufmann, 1986.

[79] H. Schutze, D. Hull, and J. Pedersen. A comparison of classifiers and document representations for the routing problem. *ACM SIGIR Conference*, pages 229–237, 1995.

[80] A. Shilton, M. Palaniswami, D. Ralph, and A. Tsoi. Incremental training of support vector machines. *IEEE Transactions on Neural Networks*, 16(1):114–131, 2005.

[81] W. N. Street and Y. Kim. A streaming ensemble algorithm (sea) for large-scale classification. *ACM KDD Conference*, pages 377–382, 2001.

[82] N. Syed, H. Liu, and K. Sung. Handling concept drifts in incremental learning with support vector machines. *ACM KDD Conference*, pages 317–321, 1999.

[83] P. Utgoff. Incremental induction of decision trees. *Machine Learning*, 4(2):161–186, 1989.

[84] P. Utgoff and C. Brodley. An incremental method for finding multivariate splits for decision trees. *Proceedings of the Seventh International Conference on Machine Learning*, pages 58–65, 1990.

[85] P. Utgoff. An improved algorithm for incremental induction of decision trees. *Proceedings of the Eleventh International Conference on Machine Learning*, pages 318–325, 1994.

[86] J. Vitter. Random sampling with a reservoir, *ACM Transactions on Mathematical Software (TOMS)*, 11(1):37–57, 1985.

[87] H. Wang, W. Fan, P. Yu, J. Han. Mining concept-drifting data streams with ensemble classifiers, *KDD Conference*, pages 226–235, 2003.

[88] H. Wang, J. Yin, J. Pei, P. Yu, and J. X. Yu. Suppressing model overfitting in concept drifting data streams, *ACM KDD Conference*, pages 736–741, 2006.

[89] E. Wiener, J. O. Pedersen, and A. S. Weigend. A neural network approach to topic spotting, *SDAIR*, pages 317–332, 1995.

[90] Y. Yang, X. Wu, and X. Zhu. Combining proactive and reactive predictions for data streams, *ACM KDD Conference*, pages 710–715, 2005.

[91] K. L. Yu and W. Lam. A new on-line learning algorithm for adaptive text filtering, *ACM CIKM Conference*, pages 156–160, 1998.

[92] S. Dzeroski and B. Zenko. Is combining classifiers better than selecting the best one? *Machine Learning*, 54(3):255–273, 2004.

[93] Y. Zhang, X. Li, and M. Orlowska. One class classification of text streams with concept drift, *ICDMW Workshop*, pages 116–125, 2008.

[94] J. Zhang, Z. Ghahramani, and Y. Yang. A probabilistic model for online document clustering with application to novelty detection, *NIPS*, 2005.

[95] P. Zhang, Y. Zhu, and Y. Shi. Categorizing and mining concept drifting data streams, *ACM KDD Conference*, pages 812–820, 2008.

[96] X. Zhu, X. Wu, and Y. Zhang. Dynamic classifier selection for effective mining from noisy data streams, *ICDM Conference*, pages 305–312, 2004.

[97] http://www.itl.nist.gov/iad/mig/tests/tdt/tasks/fsd.html

Chapter 10

Big Data Classification

Hanghang Tong

City College
City University of New York
New York, New York
hanghang.tong@gmail.com

10.1 Introduction

We are in the age of 'big data.' Big Data has the potential to revolutionize many scientific disciplines, ranging from astronomy, biology, and education, to economics, social science, etc [16]. From an algorithmic point of view, the main challenges that 'big data' brings to classification can be summarized by three characteristics, that are often referred to as the "three Vs," namely *volume*, *variety*, and *velocity*. The first characteristic is volume, which corresponds to the fact that the data is being generated at unprecedented scale. For example, it is estimated that [16] there were more than 13 *exabytes* of new data stored by enterprises and users in 2010. The second characteristic is variety, according to which, real data is often heterogeneous, comprised of multiple different types and/or coming from different sources. The third characteristic is that the data is not only large and complex, but also generated at a very high rate.

These new characteristics of big data have brought new challenges and opportunities for classification algorithms. For example, to address the challenge of *velocity*, on-line streaming classification algorithms have been proposed (please refer to the previous chapter for details); in order to address the challenge of *variety*, the so-called heterogeneous machine learning has been emerging, including multi-view classification for data heterogeneity, transfer learning and multi-task classification for classification task heterogeneity, multi-instance learning for instance heterogeneity, and classification with crowd-sourcing for oracle heterogeneity; in order to address the challenge of *volume*, many efforts have been devoted to scaling up classification algorithms.

In this chapter, we will summarize some recent efforts in classification algorithms to address the scalability issue in response to the volume challenge in big data. For discussion on the streaming scenario, we refer to the previous chapter. We will start by introducing how to scale up classification algorithms on a single machine. Then we will introduce how to further scale up classification algorithms by parallelism.

10.2   Scale-Up on a Single Machine

In this section, we introduce some representative work on how to speed up classification algorithms on a single machine. While some methods such as associative classifiers and nearest neighbors can be scaled up with better choice of subroutines, this is not the case for other methods such as decision trees and support vector machines. Some of the earliest scalable methods for decision trees include *SLIQ* [22], *RainForest* [11], and *BOAT* [10]. Recently, decision trees have also been scaled up to data streams, along with other ensemble-based methods [6, 14, 15, 28]. Some of these methods are discussed in Chapter 4 on decision trees and Chapter 9 on data streams, and will not be addressed in this chapter. The focus in this chapter will be on support vector machines. We start with SVMs and then generalize our discussion to other classification algorithms that fit in the so-called regularized risk minimization framework.

10.2.1   Background

Given a training set $(\mathbf{x}_i, y_i)(i = 1, \dots, n)$, where $\mathbf{x}$ is a d-dimensional feature vector and $y_i = \pm 1$ is the class label, linear support vector machines (SVM) aim to find a classifier in the form of $h_{\mathbf{w},b}(\mathbf{x}) = \text{sign}(\mathbf{w}^T\mathbf{x} + b)$, where $\mathbf{w}$ and b are the parameters.

By introducing a dummy feature of constant value $1s$, we ignore the parameter b. Then, the parameter $\mathbf{w}$ can be learnt from the training set by solving the following optimization problem:

$$\min_{\mathbf{w}, \xi_i \geq 0} \frac{1}{2}\mathbf{w}\mathbf{w}^T + \frac{C}{n}\sum_{i=1}^{n}\xi_i$$
$$s.t. \forall i \in \{1, \dots, n\} : y_i(\mathbf{w}^T\mathbf{x}_i) \geq 1 - \xi_i. \tag{10.1}$$

Traditionally, this optimization problem can be solved in its dual form, which can be solved in turn by quadratic programming, e.g., SMO [23], LIBSVM [3], SVMLight [18], etc. All of these methods scale linearly with respect to the dimensionality d of the feature vector. However, they usually require a *super-linear* time in the number of training examples n. On the other hand, many algorithms have been proposed to speed up the training process of SVMs in the past decade to achieve linear scalability with respect to the number of the training examples n, such as the interior-point based approach [7], the proximal-based method [9], the newton-based approach [21], etc. However, all these methods still scale *quadratically* with the dimensionality d of the feature space.

10.2.2   SVMPerf

For many big data applications, it is often the case that we have a large number of training examples n *and* a high dimensional feature space d. For example, for the text classification problem, we could have millions of training examples and hundreds of thousands of features. For such applications, it is highly desirable to have a classification algorithm that scales well with respect to both n and d.

To this end, SVMPerf was proposed in [20], whose time complexity is $O(ns)$, where s is the average non-zero feature values per training example. Note that such a time-complexity is very attractive for many real applications where the features are sparse (small s) even though its dimensionality d is high.

The key idea of SVMPerf is to re-formulate the original SVM in Equation (10.1) as the following structural SVM:

$$\min_{\mathbf{w}, \xi \geq 0} \frac{1}{2} \mathbf{w} \mathbf{w}^T + C\xi$$
$$s.t. \forall \mathbf{c} \in \{0,1\}^n : \frac{1}{n} \mathbf{w}^T \sum_{i=1}^{n} (\mathbf{c}_i y_i \mathbf{x}_i) \geq \frac{1}{n} \sum_{i=1}^{n} \mathbf{c}_i - \xi. \tag{10.2}$$

Compared with the original formulation in Equation (10.1), the above formulation has a much simpler objective function, where we only have a single slack variable ξ. On the other hand, there are 2^n constraints in the structural SVM, each of which corresponds to the sum of a subset of the constraints in the original formulation and such a subset is specified by the vector $\mathbf{c} = (\mathbf{c}_1, ..., \mathbf{c}_n) \in \{0,1\}^n$. Thus, the reduction in the number of slack variables is achieved at the expense of having many more constraints. So how is this better?

Despite the fact that we have an exponential number of constraints in Equation (10.2), it turns out we can resort to the cutting-plane algorithm to efficiently solve it. The basic idea is as follows. Instead of working on the entire 2^n constraints directly, we keep an active current work set of constrains $\mathcal{W}$. The algorithm starts with an empty set of constraint set $\mathcal{W}$, and then iteratively (1) solves Equation (10.2) by considering the constraints that are only in the current set $\mathcal{W}$; and (2) expands the current working set $\mathcal{W}$ of constraints, based on the current classifier, until a pre-defined precision is reached.

The key component in this algorithm is the approach for expanding the current working set $\mathcal{W}$. The work in [20] suggests using the constraint in Equation (10.2) that requires the biggest ξ to make it feasible. To be specific, if $\mathbf{w}$ is the currently learnt weighted vector, we define $\mathbf{c}$ as $\mathbf{c}_i = 1$ if $y_i(\mathbf{w}^T \mathbf{x}_i) < 1$; and $\mathbf{c}_i = 0$ otherwise. Then, we will add this constraint $\mathbf{c}$ in the current work set $\mathcal{W}$.

Note that the above algorithm is designed for discrete binary classification. The work in [20] further generalizes it to the ordinal regression for the task of ranking (as opposed to binary classification), where $y_i \in \{1, ..., R\}$ indicates an ordinal scale. The algorithm for this case has the complexity of $O(sn\log(n))$.

In SVMPerf, there is a parameter ε that controls the accuracy in the iterative process. The overall time complexity is also related to this parameter. To be specific, the number of required iterations in SVMPerf is $O(\frac{1}{\varepsilon^2})$. This yields an overall complexity of $O(\frac{ns}{\varepsilon^2})$.

10.2.3 Pegasos

Note that we still have a linear term with respect to the number of training examples $O(n)$ in SVMPerf. For the applications from big data, this term is often on the order of millions or even billions. In order to further speed up the training process, Pegasos was proposed in [25] based on a stochastic sub-gradient descent method.

Unlike SVMPerf and many other methods, which formulate SVM as a *constrained* optimization problem, Pegasos relies on the the following *un-constrained* optimization formulation of SVM:

$$\min_{\mathbf{w}} \frac{\lambda}{2} \mathbf{w} \mathbf{w}^T + \frac{1}{n} \sum_{i=1}^{n} l(\mathbf{w}, (\mathbf{x}_i, y_i))$$
$$\text{where} \quad l(\mathbf{w}, (\mathbf{x}_i, y_i)) = \max\{0, 1 - y_i \mathbf{w}^T \mathbf{x}_i\}. \tag{10.3}$$

Basic Pegasos. Pegasos is an on-line algorithm. In other words, in each iteration, it aims to update the weight vector $\mathbf{w}$ by using only *one* randomly sampled example $\mathbf{x}_{i_t}$ through the following approximate objective function:

$$f(\mathbf{w}; i_t) = \frac{\lambda}{2} \mathbf{w} \mathbf{w}^T + l(\mathbf{w}, (\mathbf{x}_{i_t}, y_{i_t})) \tag{10.4}$$

where $t = 1, 2, \ldots$ is the iteration number and $(i_t \in \{1, \ldots, n\}$ is the index of the training example that is sampled in the tth iteration.

Note that $f(\mathbf{w}; i_t)$ has one non-differential point. Therefore, we use its sub-gradient instead to update the weight vector $\mathbf{w}$ as

$$\mathbf{w}_{t+1} = \mathbf{w}_t - \eta_t \nabla_t. \tag{10.5}$$

In the above update process, ∇_t is the *sub-gradient* of $f(\mathbf{w}; i_t)$ in the tth iteration: $\nabla_t = \lambda \mathbf{w}_t - \mathbf{1}[y_{i_t} \mathbf{w}^T \mathbf{x}_{i_t} < 1] y_{i_t} \mathbf{x}_{i_t}$, where $\mathbf{1}[.]$ is the indicator function. In Equation (10.5), the learning rate η_t is set as $\eta_t = \frac{1}{\lambda t}$. Finally, the training pair $(\mathbf{x}_{i_t}, y_{i_t})$ is chosen uniformly at random.

The authors in [25] also introduced an optional projection step to limit the admissible solution of $\mathbf{w}$ to a ball of radius $\frac{1}{\sqrt{\lambda}}$. In other words, after updating $\mathbf{w}$ using Equation (10.5), we further update it as follows:

$$\mathbf{w}_{t+1} \leftarrow \min\{1, \frac{1/\sqrt{\lambda}}{\|\mathbf{w}_{t+1}\|}\} \mathbf{w}_{t+1}.$$

It turns out that the time complexity of the above algorithm is $O(s/(\lambda \varepsilon))$. Compared with the complexity of SVMPerf, we can see that it is *independent* of the number of training examples n. This makes it especially suitable for large training data set sizes.

Mini-Batch Pegasos. The authors in [25] also propose several variants of the above basic version of the Pegasos algorithm. The first variant is the so-called *Mini-Batch* update, where in each iteration, we use k training examples (as opposed to a *single* example) in order to update the weight vector $\mathbf{w}$:

$$f(\mathbf{w}; i_t) = \frac{\lambda}{2} \mathbf{w} \mathbf{w}^T + \frac{1}{k} \sum_{i \in A_t} l(\mathbf{w}, (\mathbf{x}_i, y_i)) \tag{10.6}$$

where $1 \leq k \leq n$ and $A_t \subset \{1, \ldots, n\}$, $\|A_t\| = k$. In this *Mini-Batch* mode, in order to update the weight vector $\mathbf{w}_t$ at each iteration t, we first need to find out which of these k selected examples violate the margin rule based on the current weight vector $\mathbf{w}_t$: $A_t^+ = \{i \in A_t : y_i \mathbf{w}_t' \mathbf{x}_i < 1\}$. Then, we will leverage these training examples in A_t^+ to update the weight vector $\mathbf{w}$:

$$\mathbf{w}_{t+1} \leftarrow (1 - \frac{1}{t}) \mathbf{w}_t + \frac{1}{\lambda t k} \sum_{i \in A_t^+} y_i \mathbf{x}_i.$$

Kernelized Pegasos. The second variant is the kernelized Pegasos, which can address cases where the decision boundary is nonlinear. This variant may be formulated as the following optimization problem :

$$\min_{\mathbf{w}} \frac{\lambda}{2} \mathbf{w} \mathbf{w}^T + \frac{1}{n} \sum_{i=1}^{n} l(\mathbf{w}, \phi(\mathbf{x}_i, y_i))$$

$$\text{where } l(\mathbf{w}, (\mathbf{x}_i, y_i)) = \max\{0, 1 - y_i \mathbf{w}^T \phi(\mathbf{x}_i)\} \tag{10.7}$$

where the mapping $\phi(.)$ is implicitly specified by the kernel operator $\mathcal{K}(\mathbf{x}_1, \mathbf{x}_2) = \phi(\mathbf{x}_1)' \phi(\mathbf{x}_2)$. The authors in [25] show that we can still solve the above optimization problem in its primal form using stochastic gradient descent, whose iteration number is still $O(\frac{s}{\lambda \varepsilon})$. Nonetheless, in the kernelized case, we might need $\min(t, n)$ kernel evaluations in the tth evaluations. This makes the overall complexity $O(\frac{sn}{\lambda \varepsilon})$.

10.2.4 Bundle Methods

Regularized Risk Minimization. So far, we have focused on how to scale up the SVM classifier. The work in [26] generalizes the above ideas and provides a generic solution for the so-called *regularized risk minimization* problem, which includes many machine learning formulations (e.g., SVM, regression, Gaussian process, etc.) as its specialized cases.

In a *regularized risk minimization* problem, we aim to find a parameter $\mathbf{w}$ (e.g., the weight vector in SVM) from the training set by minimizing the following objective function:

$$J(\mathbf{w}) = R_{emp}(\mathbf{w}) + \lambda\Omega(\mathbf{w}) \tag{10.8}$$

where the second term is the regularization term for the parameter $\mathbf{w}$, which is often chosen to be a smooth, convex function, e.g., $\Omega(\mathbf{w}) = \frac{1}{2}\mathbf{w}'\mathbf{w}$ in the case of SVM and regression; and $\lambda > 0$ is the regularization weight. The first term R_{emp}, which can often be written as the summation over all the training examples: $R_{emp} = \frac{1}{n}\sum_{i=1}^{n} l(\mathbf{x}_i, y_i, \mathbf{w})$ and $l(\mathbf{x}_i, y_i, \mathbf{w})$, is the loss function over the ith training pair $(\mathbf{x}_i, y_i)$ (e.g., the hinge loss in SVM, etc.). The challenge of regularized risk minimization problems mostly comes from R_{emp} as (1) it involves all the training examples so that the computation of this term is costly for large-scale problems; and (2) R_{emp} itself is often not smooth over the entire parameter space.

Convex Bundle Methods. To address these challenges, the basic idea of bundle method [26] is to iteratively approximate R_{emp} by the first-order Taylor expansion, and then update the parameter $\mathbf{w}$ by solving such an approximation. Note that R_{emp} itself may not be differential (i.e., not smooth). In this case, we will use the so-called *sub-differential* to do the Taylor expansion. We summarize its basic ideas below.

Recall that μ is called subgradient of a convex, finite function F at a given point $\mathbf{w}$ if for any $\tilde{\mathbf{w}}$, we have $F(\tilde{\mathbf{w}}) \geq F(\mathbf{w}) + <\tilde{\mathbf{w}} - \mathbf{w}, \mu>$. The subdifferential is the set of all subgradients. Therefore, if we approximate R_{emp} by the first-order Taylor expansion at the current estimation of the parameter $\mathbf{w}_t$ using its subdifferential, each approximation by one of its subgradients provides a lower bound of R_{emp}.

In the convex bundle method, the goal is to minimize the maximum of all such lower bounds (plus the regularization term) to update the parameter $\mathbf{w}$. The convergence analysis shows that this method converges in $O(1/\varepsilon)$ iterations for a general convex function. Its convergence rate can be further improved as $O(log(1/\varepsilon))$ if R_{emp} is continuously differential.

Non-Convex Bundle Methods. Note that in [26], we require that the function R_{emp} be convex. The reason is that if the function R_{emp} is convex, its Taylor approximation at the current estimation always provides a lower bound (i.e., under-estimator) of R_{emp}. While empirical risk function R_{emp} for some classification problems (e.g., standard SVM, logistic regression, etc.) is indeed convex, this is not the case for some other more complicated classification problems, including the so-called transductive SVM (TSVM) [4, 19], ramp loss SVM [29], convolutional nets [17], etc.

A classic technique for such non-connex optimization problems is to use convex relaxation so that we can transfer the original non-convex function into a convex function. Nonetheless, it is not an easy task to find such a transformation. In [5], the authors generalize the standard bundle methods to handle such a non-convex function. The overall procedure is similar to that in [26] in the sense that we still repeatedly use the first-order Taylor expansion of the R_{emp} at the current estimation to approximate the true R_{emp}. The difference is that we need to *off-set* the Taylor approximation so that the approximation function is still an under-estimator of the true empirical loss function. Note that in the non-convex case, it is not clear if we can still have $O(1/\varepsilon)$ or a similar convergence rate.

10.3 Scale-Up by Parallelism

Another way to scale up large-scale classification problems is through parallelism. In this section, we first give some examples of how to speed up decision tree classifiers and how to solve the optimization problem of SVM in parallel, respectively; and then present some generic frameworks to parallel classification algorithms (or machine learning algorithms in general) using multi-core and/or MapReduce types of computational infrastructures.

10.3.1 Parallel Decision Trees

A widely used classifier is decision trees. For large scale data, the computational challenges mainly lie in the training stage. In this subsection, we review two representative works of parallel decision trees. Some discussions on scalable methods may be found in Chapter 4 on decision trees and streaming data. Here, a discussion of parallel methods is provided.

SPRINT. In order to speed up the decision trees training, we need to answer two key questions, including (Q1) how to find split points to generate tree nodes; and (Q2) given a split point, how to partition the data set. SRPINT [24], which stands for <u>S</u>calable <u>Pa</u><u>R</u>allelizable <u>IN</u>dution of decision <u>T</u>rees, addresses these two challenges mainly through two important data structures.

The first data structure is *attribute lists*. Each attribute list contains an attribute value, a class label, and the index of the data record, where these attribute values are from. The initial list is sorted according to the attribute value and sits at the root of the decision tree. As the tree grows, the attribute list is split accordingly and passed down to the corresponding children nodes in the decision tree. Here, the important point to notice is that the partition on each attribute list keeps its ordering. In other words, we do not need to re-sort the attribute list after partition. The second important data structure is *histogram*. For categorical attribute, it is called count matrix, which consists of the class distribution for each attribute value. On the other hand, if the attribute is continuous, for each tree node that is about to split, we would need two histograms, i.e., each of them for the class distribution associated with the attribute value below and above the splitting point, respectively.

Compared with its earlier version, e.g., SLIQ [22], a unique feature of SPRINT is that it removes the constraint that the algorithm needs to keep all or a subset of the input data set in the main memory. Like SLIQ, SPRINT only requires one-time sorting for the attribute list. On the other hand, the memory requirement of SPRINT (i.e., mainly for the histograms) is *independent* of the input data size. The data structure in SPRINT is specially designed to remove the dependence among data points, which makes it easily parallelized on the shared-nothing multiprocessor infrastructure. The experimental evaluations in [24] show that SPRINT has comparable performance as SLIQ when the class list can be fit in the main memory. On the other hand, it achieves a much better performance in terms of sizeup, scaleup, and speedup simultaneously, even when no other parallel decision tree algorithms (by the time SPINT was proposed) can achieve a comparable performance.

MReC4.5 is designed to scale up the C4.5 ensemble classifier with MapReduce infrastructure [30]. It works as follows. At the beginning of the algorithm (the partition stage), the master generates *bootstrap* partition of the input data set; then (the base-classifier building stage), each mapper builds a C4.5 decision tree based on a bootstrap sampling, respectively; finally (ensemble stage), the reducer is responsible for merging all the base-classifiers from the mapper stage using the bagging ensemble algorithm. Overall, MReC4.5 provides an interesting example of leveraging the strength of MapReduce infrastructure to scale up classification algorithms that use bagging-type ensembles.

10.3.2 Parallel SVMs

Unlike in the previous section, Parallel SVM is usually solved in its dual form, which is a standard quadratic program problem.

$$\min \ \mathcal{D}(\alpha) = \frac{1}{2}\alpha'\mathbf{Q}\alpha - \alpha'\mathbf{1}$$
$$s.t. \mathbf{0} \leq \alpha \leq C, \mathbf{y}'\alpha = 0 \tag{10.9}$$

where $\mathbf{Q}(i,j) = K(\mathbf{x}_i, \mathbf{x}_j)$ and $\alpha \in R^n$. The final classifier is defined by $y = \text{sign}(\sum_{i=1}^n \mathbf{y}_i \alpha_i K(\mathbf{x}, \mathbf{x}_i + b))$.

Bayesian Committee Machine. In [27], the authors proposed first partitioning the training set into a few subsets, and then training a separate SVM on each subset independently. In the test stage, the final prediction is a specific combination of the prediction from each SVM.

Here, the key is in the method for combining these prediction results. In Bayesian Committee Machine, the authors suggested a combination scheme that is based on the variance of the test data from each individual SVM.

Cascade SVM. In [13], the authors propose the following procedure to solve the QP in (10.9). Its basic idea is to 'chunk' and to eliminate the non-support vectors as early as possible. To be specific, we first partition the entire training set into small chunks, and then train a separate SVM for each chunk (layer 1); the output of each chunk is a (hopefully) small number of support vectors. These support vectors are treated as the new training set. We repeat the above process — partition the support vectors from layer 1 into chunks, train a separate SVM for each chunk, and output the new support vectors from each chunk. This process will be repeated again until we only have one single chunk. The output (i.e., the support vectors) from that chunk will be fed back to layer 1 to test the global convergence.

It can be seen that with this approach, each layer except the first layer will only operate on a small number of training examples (i.e., the support vectors from the previous layers); and for each chunk at a given layer, we can train an SVM by solving its corresponding QP independently and thus it can be parallelized easily.

Parallel IPM-SVM. On a single machine, one of the most powerful methods to solve the QP in Equation (10.9) is the so-called prime-dual Interior-Point Method (IPM) [8]. Please refer to [1] for the details of using IPM to solve the QP problem in SVM. The key and most expensive step in IPM involves the multiplication between a matrix inverse $\boldsymbol{\Sigma}$ and a vector $\mathbf{q}$: $\boldsymbol{\Sigma}^{-1}\mathbf{q}$, where $\boldsymbol{\Sigma} = \mathbf{Q} + \mathbf{D}$ and $\mathbf{D}$ is some diagonal matrix. It is very expensive ($O(n^3)$) to compute $\boldsymbol{\Sigma}^{-1}\mathbf{q}$.

However, if we can approximate $\mathbf{Q}$ by its low-rank approximation, we can largely speed up this step. To be specific, if $\mathbf{Q} = \mathbf{H}'\mathbf{H}$, where $\mathbf{H}$ is a rank-k ($k \ll n$) matrix (e.g., by the incomplete Cholesky decomposition), by the Sharman-Morrison-Woodury theorem, we have

$$\begin{aligned} \boldsymbol{\Sigma}^{-1}\mathbf{q} &= (\mathbf{D} + \mathbf{H}'\mathbf{H})^{-1}\mathbf{q} \\ &= \mathbf{D}^{-1}\mathbf{q} - \mathbf{D}^{-1}\mathbf{H}(\mathbf{I} + \mathbf{H}'\mathbf{D}^{-1}\mathbf{H})^{-1}\mathbf{H}'\mathbf{D}^{-1}\mathbf{q} \end{aligned} \tag{10.10}$$

In other words, instead of computing the inverse of an $n \times n$ matrix, we only need to compute the inverse of a $k \times k$ matrix. In [1], the authors further propose to parallel this step as well as factorizing the original $\mathbf{Q}$ matrix by the incomplete Cholesky decomposition.

10.3.3 MRM-ML

A recent, remarkable work to parallel classification algorithms (machine learning algorithms in general) was done in [2]. Unlike most of the previous work, which aims to speed up a *single* machine learning algorithm, in that work, the authors identified a *family* of machine learning algorithms that can be easily sped up in the parallel computation setting. Specially, they showed that any algorithm that fit in the so-called *statistical query model* can be parallelized easily under the multi-core MapReduce environment.

The key is that such an algorithm can be re-formulated in some "summation form." The classification algorithms that fall in this category include linear regression, locally weighted linear regression, logistic regression, naive Bayes, SVM with the quadratic loss, etc. For all these algorithms (and many more un-supervised learning algorithms), they achieve a linear speedup with respect to the number of processors in the cluster.

Let us use linear regression to explain its basic ideas. In linear regression, we look for a parameter vector θ by minimizing $\sum_{i=1}^{n}(\theta'\mathbf{x}_i - y_i)^2$. If we stack all the feature vectors $\mathbf{x}_i(i = 1,...,n)$ into an $n \times d$ matrix $\mathbf{X}$, and all the labels $y_i(i = 1,...,n)$ into an $n \times 1$ vector $\mathbf{y}$, in the most simple case (i.e., without any regularization terms), we can solve θ as $\theta = (\mathbf{X}'\mathbf{X})^{-1}\mathbf{X}'\mathbf{y}$. In the case we have a large number of training examples, the main computational bottleneck is to compute $\mathbf{A} = \mathbf{X}'\mathbf{X}$ and $\mathbf{b} = \mathbf{X}'\mathbf{y}$. Each of these two terms can be re-written in the "summation" form as follows: $\mathbf{A} = \sum_{i=1}^{n}(\mathbf{x}_i\mathbf{x}_i')$ and $\mathbf{b} = \sum_{i=1}^{n}(\mathbf{x}_i y_i)$. Note that the summation is taken over different training examples. Therefore, the computation can be divided into equal size partitions of the data, each of which can be done by a mapper job, and the final summation can be done by a reducer.

Another example that fits in this family is logistic regression. Recall that in logistic regression, the classifier is in the form of $h_\theta(\mathbf{x}) = g(\theta'\mathbf{x}) = \frac{1}{1+exp(-\theta'\mathbf{x})}$. Here, the weight vector θ can be learnt in a couple of different ways, one of which is by the Newton-Raphson approach as follows: after some initialization, we will iteratively update the weighted vector θ by: $\theta \leftarrow \theta - H^{-1}\nabla_\theta$. For a large training data set, the computational bottleneck is to compute the Hessian matrix H and the gradient ∇_θ. Like in the linear regression case, both terms can be re-written as the "summation" forms at each Newton-Raphson step t: the Hessian matrix $H(i,j) := H(i,j) + h_\theta(\mathbf{x}^{(t)})(h_\theta(\mathbf{x}^{(t)}) - 1)\mathbf{x}_i^{(t)}\mathbf{x}_j^{(t)}$; and the gradient $\nabla_\theta = \sum_{i=1}^{n}(y^{(t)} - h_\theta(\mathbf{x}^{(t)}))\mathbf{x}_i^{(t)}$. Therefore, if we divide the training data into equal size partitions, both terms can be parallelized: each mapper will do the summation over the corresponding partition and the final summation can be done by a reducer.

10.3.4 SystemML

MRM-ML answers the question of what kind of classification (or machine learning in general) algorithms *can* be parallelized in a MapReduce environment. However, it is still costly and even prohibitive to implement a large class of machine learning algorithms as low-level MapReduce jobs. To be specific, in MRM-ML, we still need to hand-code each individual MapReduce job for a given classification algorithm. What is more, in order to improve the performance, the programmer still needs to manually schedule the execution plan based on the size of the input data set as well as the size of the MapReduce cluster.

To address these issues, SystemML was proposed in [12]. The main advantage of SystemML is that it provides a high-level language called Declarative Machine Learning Language (DML). DML encodes two common and important building blocks shared by many machine learning algorithms, including linear algebra and iterative optimization. Thus, it frees the programmer from lower-level implementation details. By automatic program analysis, SystemML further breaks an input DML script into smaller so-called statement blocks. So, in this way, it also frees the users from manual scheduling of the execution plan.

Let us illustrate this functionality by the matrix-matrix multiplication — a common operation in many machine learning algorithms. To be specific, if we partition the two input matrices ($\mathbf{A}$ and $\mathbf{B}$) in the block forms, the multiplication between them can be written in the form block as $\mathbf{C}_{i,j} = \sum_k \mathbf{A}_{i,k}\mathbf{B}_{k,j}$, where $\mathbf{C}_{i,j}$ is the corresponding block in the resulting matrix $\mathbf{C}$. In the parallel environment, we have different choices to implement such a block-based matrix multiplication. In the first strategy (referred to as "replication based matrix multiplication" in [12]), we only need one single map-reduce job, but some matrix (like $\mathbf{A}$) might be replicated and sent to multiple reducers. In contrast, in the second strategy (referred to as "cross product based matrix multiplication" in [12]), each block of the two input matrices will only be sent to a single reducer, with the cost of an additional map-reduce job. Here, which strategy we should choose is highly dependent on the characteristic of the two input matrices. For example, if the number of rows in **A** is much smaller than its columns (say **A** is a topic-document matrix), the replication based strategy might be much more efficient than the cross-product based strategy. In another example, in many iterative matrix-based machine learning algorithms (e.g., NMF, etc), we might need the product between *three* input matrices **ABC**. Here, the ordering of such a matrix multiplication (e.g., (**AB**)**C** vs. **A**(**BC**)) also depends on the characteristic (i.e., the size) of the input matrices.

SystemML covers a large family of machine learning algorithms including linear models, matrix factorization, principal component analysis, pagerank, etc. Its empirical evaluations show that it scales to very large data sets and its performance is comparable to those hand-coded implementations.

10.4   Conclusion

Given that (1) the data size keeps growing explosively and (2) the intrinsic complexity of a classification algorithm is often high, the scalability seems to be a 'never-ending' challenge in classification. In this chapter, we have briefly reviewed two basic techniques to speed up and scale up a classification algorithm. This includes (a) how to scale up classification algorithms on a single machine by carefully solving its associated optimization problem; and (b) how to further scale up classification algorithms by parallelism. A future trend of big data classification is to address the scalability in conjunction with other challenges of big data (e.g., variety, velocity, etc).

Bibliography

[1] Edward Y. Chang, Kaihua Zhu, Hao Wang, Hongjie Bai, Jian Li, Zhihuan Qiu, and Hang Cui. PSVM: Parallelizing support vector machines on distributed computers. In *NIPS*, 2007.

[2] Cheng-Tao Chu, Sang Kyun Kim, Yi-An Lin, YuanYuan Yu, Gary R. Bradski, Andrew Y. Ng, and Kunle Olukotun. Map-reduce for machine learning on multicore. In *NIPS*, pages 281–288, 2006.

[3] Chih Chung Chang and Chih-Jen Lin. LIBSVM: A library for support vector machines, 2001.

[4] Ronan Collobert, Fabian H. Sinz, Jason Weston, and Léon Bottou. Trading convexity for scalability. In *ICML*, pages 201–208, 2006.

[5] Trinh-Minh-Tri Do and Thierry Artieres. Regularized bundle methods for convex and non-convex risks. *Journal of Machine Learning Research*, 13(Dec.): pages 3539–3583, MIT Press, 2012.

[6] Pedro Domingos and Geoff Hulten, Mining high-speed data streams, *ACM KDD Conference*, pages 71–80, 2000.

[7] Michael C. Ferris and Todd S. Munson. Interior-point methods for massive support vector machines. *SIAM Journal on Optimization*, 13(3):783–804, 2002.

[8] Katsuki Fujisawa. The implementation of the primal-dual interior-point method for the semidefinite programs and its engineering applications (Unpublished manuscript), 1992.

[9] Glenn Fung and Olvi L. Mangasarian. Proximal support vector machine classifiers. In *KDD*, pages 77–86, 2001.

[10] Johannes Gehrke, Venkatesh Ganti, Raghu Ramakrishnan, and Wei-Yin Loh. BOAT: Optimistic Decision Tree Construction, *ACM SIGMOD Conference*, 1999.

[11] Johannes Gehrke, Raghu Ramakrishnan, and Venkatesh Ganti. RainForest—A framework for fast decision tree construction of large datasets, *VLDB Conference*, pages 416–427, 1998.

[12] Amol Ghoting, Rajasekar Krishnamurthy, Edwin P. D. Pednault, Berthold Reinwald, Vikas Sindhwani, Shirish Tatikonda, Yuanyuan Tian, and Shivakumar Vaithyanathan. Systemml: Declarative machine learning on Map-Reduce. In *ICDE*, pages 231–242, 2011.

[13] Hans Peter Graf, Eric Cosatto, Léon Bottou, Igor Durdanovic, and Vladimir Vapnik. Parallel support vector machines: The cascade svm. In *NIPS*, 2004.

[14] Geoff Hulten, Laurie Spencer, and Pedro Domingos. Mining time-changing data streams. *ACM KDD Conference*, 2001.

[15] Ruoming Jin and Gagan Agrawal. Efficient Decision Tree Construction on Streaming Data, *ACM KDD Conference*, 2003.

[16] Community white paper. Challenges and opportunities with big data. Technical Report available online at http://www.cra.org/ccc/files/docs/init/bigdatawhitepaper.pdf/

[17] Kevin Jarrett, Koray Kavukcuoglu, Marc'Aurelio Ranzato, and Yann LeCun. What is the best multi-stage architecture for object recognition? In *ICCV*, pages 2146–2153, 2009.

[18] Thorsten Joachims. Making large-scale support vector machine learning practical, In *Advances in Kernel Methods*, pages 169–184, MIT Press, Cambridge, MA, 1999.

[19] Thorsten Joachims. Transductive inference for text classification using support vector machines. In *ICML*, pages 200–209, 1999.

[20] Thorsten Joachims. Training linear SVMs in linear time. In *KDD*, pages 217–226, 2006.

[21] S. Sathiya Keerthi and Dennis DeCoste. A modified finite Newton method for fast solution of large scale linear SVMs. *Journal of Machine Learning Research*, 6:341–361, 2005.

[22] Manish Mehta, Rakesh Agrawal, and Jorma Rissanen. SLIQ: A fast scalable classifier for data mining. In *Extending Database Technology*, pages 18–32, 1996.

[23] John C. Platt. Sequential minimal optimization: A fast algorithm for training support vector machines. In *Advances in Kernel Methods – Suport Vector Learning*, 1998.

[24] John C. Shafer, Rakesh Agrawal, and Manish Mehta. SPRINT: A scalable parallel classifier for data mining. In *VLDB*, pages 544–555, 1996.

[25] Shai Shalev-Shwartz, Yoram Singer, Nathan Srebro, and Andrew Cotter. Pegasos: primal estimated sub-gradient solver for SVM. *Mathematical Programming*, 127(1):3–30, 2011.

[26] Choon Hui Teo, S. V. N. Vishwanathan, Alex J. Smola, and Quoc V. Le. Bundle methods for regularized risk minimization. *Journal of Machine Learning Research*, 11:311–365, 2010.

[27] Volker Tresp. A Bayesian committee machine. *Neural Computation*, 12(11):2719–2741, 2000.

[28] Haixun Wang, Wei Fan, Philip Yu, and Jiawei Han, Mining concept-drifting data streams using ensemble classifiers, *KDD Conference*, pages 226–235, 2003.

[29] Zhuang Wang and Slobodan Vucetic. Fast online training of ramp loss support vector machines. In *ICDM*, pages 569–577, 2009.

[30] Gongqing Wu, Haiguang Li, Xuegang Hu, Yuanjun Bi, Jing Zhang, and Xindong Wu. MReC4.5: C4.5 ensemble classification with MapReduce. In *Fourth ChinaGrid Annual Conference*, pages 249–255, 2009.

Chapter 11

Text Classification

Charu C. Aggarwal

IBM T. J. Watson Research Center
Yorktown Heights, NY
charu@us.ibm.com

ChengXiang Zhai

University of Illinois at Urbana-Champaign
Urbana, IL
czhai@cs.uiuc.edu

11.1 Introduction

The problem of classification has been widely studied in the database, data mining, and information retrieval communities. The problem of classification is defined as follows. Given a set of records $\mathcal{D} = \{X_1, \ldots, X_N\}$ and a set of k different discrete values indexed by $\{1 \ldots k\}$, each representing a category, the task is to assign one category (equivalently the corresponding index value) to each record X_i. The problem is usually solved by using a supervised learning approach where a set of training data records (i.e., records with known category labels) are used to construct a *classification model*, which relates the features in the underlying record to one of the class labels. For a given *test instance* for which the class is unknown, the training model is used to predict a class label for this instance. The problem may also be solved by using unsupervised approaches that do not require labeled training data, in which case keyword queries characterizing each class are often manually created, and bootstrapping may be used to heuristically obtain pseudo training data. Our review focuses on supervised learning approaches.

There are some variations of the basic problem formulation given above for text classifcation. In the *hard version* of the classification problem, a particular label is explicitly assigned to the instance, whereas in the *soft version* of the classification problem, a probability value is assigned to the test instance. Other variations of the classification problem allow ranking of different class choices for a test instance, or allow the assignment of multiple labels [60] to a test instance. The classification problem assumes categorical values for the labels, though it is also possible to use continuous values as labels. The latter is referred to as the regression modeling problem. The problem of text classification is closely related to that of classification of records with set-valued features [34]; however, this model assumes that only information about the presence or absence of words is used in a document. In reality, the frequency of words also plays a helpful role in the classification process, and the typical domain-size of text data (the entire lexicon size) is much greater than a typical set-valued classification problem.

A broad survey of a wide variety of classification methods may be found in [50,72], and a survey that is specific to the text domain may be found in [127]. A relative evaluation of different kinds of text classification methods may be found in [153]. A number of the techniques discussed in this chapter have also been converted into software and are publicly available through multiple toolkits such as the *BOW* toolkit [107], Mallot [110], WEKA,[1] and LingPipe.[2]

The problem of text classification finds applications in a wide variety of domains in text mining. Some examples of domains in which text classification is commonly used are as follows:

- **News Filtering and Organization:** Most of the news services today are electronic in nature in which a large volume of news articles are created every single day by the organizations. In such cases, it is difficult to organize the news articles manually. Therefore, automated methods can be very useful for news categorization in a variety of Web portals [90]. In the special case of binary categorization where the goal is to distinguish news articles interesting to a user from those that are not, the application is also referred to as *text filtering*.

- **Document Organization and Retrieval:** The above application is generally useful for many applications beyond news filtering and organization. A variety of supervised methods may be used for document organization in many domains. These include large digital libraries of doc-

---

[1] http://www.cs.waikato.ac.nz/ml/weka/
[2] http://alias-i.com/lingpipe/

uments, Web collections, scientific literature, or even social feeds. Hierarchically organized document collections can be particularly useful for browsing and retrieval [24].

- **Opinion Mining:** Customer reviews or opinions are often short text documents that can be mined to determine useful information from the review. Details on how classification can be used in order to perform opinion mining are discussed in [101]. A common classification task is to classify an opinionated text object (e.g., a product review) into positive or negative sentiment categories.

- **Email Classification and Spam Filtering:** It is often desirable to classify email [29, 33, 97] in order to determine either the subject or to determine junk email [129] in an automated way. This is also referred to as *spam filtering* or *email filtering*.

A wide variety of techniques have been designed for text classification. In this chapter, we will discuss the broad classes of techniques, and their uses for classification tasks. We note that these classes of techniques also generally exist for other data domains such as quantitative or categorical data. Since text may be modeled as quantitative data with frequencies on the word attributes, it is possible to use most of the methods for quantitative data directly on text. However, text is a particular kind of data in which the word attributes are sparse, and high dimensional, with low frequencies on most of the words. Therefore, it is critical to design classification methods that effectively account for these characteristics of text. In this chapter, we will focus on the specific changes that are applicable to the text domain. Some key methods, that are commonly used for text classification are as follows:

- **Decision Trees:** Decision trees are designed with the use of a hierarchical division of the underlying data space with the use of different text features. The hierarchical division of the data space is designed in order to create class partitions that are more skewed in terms of their class distribution. For a given text instance, we determine the partition that it is most likely to belong to, and use it for the purposes of classification.

- **Pattern (Rule)-Based Classifiers:** In rule-based classifiers we determine the word patterns that are most likely to be related to the different classes. We construct a set of rules, in which the left-hand side corresponds to a word pattern, and the right-hand side corresponds to a class label. These rules are used for the purposes of classification.

- **SVM Classifiers:** SVM Classifiers attempt to partition the data space with the use of linear or non-linear delineations between the different classes. The key in such classifiers is to determine the optimal boundaries between the different classes and use them for the purposes of classification.

- **Neural Network Classifiers:** Neural networks are used in a wide variety of domains for the purposes of classification. In the context of text data, the main difference for neural network classifiers is to adapt these classifiers with the use of word features. We note that neural network classifiers are related to SVM classifiers; indeed, they both are in the category of discriminative classifiers, which are in contrast with the *generative classifiers* [116].

- **Bayesian (Generative) Classifiers:** In Bayesian classifiers (also called generative classifiers), we attempt to build a probabilistic classifier based on modeling the underlying word features in different classes. The idea is then to classify text based on the posterior probability of the documents belonging to the different classes on the basis of the word presence in the documents.

- **Other Classifiers:** Almost all classifiers can be adapted to the case of text data. Some of the other classifiers include nearest neighbor classifiers, and genetic algorithm-based classifiers. We will discuss some of these different classifiers in some detail and their use for the case of text data.

The area of text categorization is so vast that it is impossible to cover all the different algorithms in detail in a single chapter. Therefore, our goal is to provide the reader with an overview of the most important techniques, and also the pointers to the different variations of these techniques.

Feature selection is an important problem for text classification. In feature selection, we attempt to determine the features that are most relevant to the classification process. This is because some of the words are much more likely to be correlated to the class distribution than others. Therefore, a wide variety of methods have been proposed in the literature in order to determine the most important features for the purpose of classification. These include measures such as the gini-index or the entropy, which determine the level at which the presence of a particular feature skews the class distribution in the underlying data. We will discuss the different feature selection methods that are commonly used for text classification.

The rest of this chapter[3] is organized as follows. In the next section, we will discuss methods for feature selection in text classification. In Section 11.3, we will describe decision tree methods for text classification. Rule-based classifiers are described in detail in Section 11.4. We discuss naive Bayes classifiers in Section 11.5. The nearest neighbor classifier is discussed in Section 11.7. In Section 11.6, we will discuss a number of linear classifiers, such as the SVM classifier, direct regression modeling, and the neural network classifier. A discussion of how the classification methods can be adapted to text and Web data containing hyperlinks is discussed in Section 11.8. In Section 11.9, we discuss a number of different meta-algorithms for classification such as boosting, bagging, and ensemble learning. Methods for enhancing classification methods with additional training data are discussed in Section 11.10. Section 11.11 contains the conclusions and summary.

11.2   Feature Selection for Text Classification

Before any classification task, one of the most fundamental tasks that needs to be accomplished is that of document representation and feature selection. While feature selection is also desirable in other classification tasks, it is especially important in text classification due to the high dimensionality of text features and the existence of irrelevant (noisy) features. In general, text can be represented in two separate ways. The first is as a bag of words, in which a document is represented as a set of words, together with their associated frequency in the document. Such a representation is essentially independent of the sequence of words in the collection. The second method is to represent text directly as *strings*, in which each document is a sequence of words. Most text classification methods use the bag-of-words representation because of its simplicity for classification purposes. In this section, we will discuss some of the methods that are used for feature selection in text classification.

The most common feature selection that is used in both supervised and unsupervised applications is that of stop-word removal and stemming. In stop-word removal, we determine the common words in the documents that are not specific or discriminatory to the different classes. In stemming, different forms of the same word are consolidated into a single word. For example, singular, plural, and different tenses are consolidated into a single word. We note that these methods are not specific

---

[3]Another survey on text classification may be found in [1]. This chapter is an up-to-date survey, and contains material on topics such as SVM, Neural Networks, active learning, semisupervised learning, and transfer learning.

to the case of the classification problem, and are often used in a variety of unsupervised applications such as clustering and indexing. In the case of the classification problem, it makes sense to supervise the feature selection process with the use of the class labels. This kind of selection process ensures that those features that are highly skewed towards the presence of a particular class label are picked for the learning process. A wide variety of feature selection methods are discussed in [154, 156]. Many of these feature selection methods have been compared with one another, and the experimental results are presented in [154]. We will discuss each of these feature selection methods in this section.

11.2.1 Gini Index

One of the most common methods for quantifying the discrimination level of a feature is the use of a measure known as the *gini-index*. Let $p_1(w) \ldots p_k(w)$ be the fraction of class-label presence of the k different classes for the word w. In other words, $p_i(w)$ is the conditional probability that a document belongs to class i, given the fact that it contains the word w. Therefore, we have:

$$\sum_{i=1}^{k} p_i(w) = 1. \tag{11.1}$$

Then, the gini-index for the word w, denoted by $G(w)$, is defined[4] as follows:

$$G(w) = \sum_{i=1}^{k} p_i(w)^2 \tag{11.2}$$

The value of the gini-index $G(w)$ always lies in the range $(1/k, 1)$. Higher values of the gini-index $G(w)$ represent a greater discriminative power of the word w. For example, when all documents that contain word w belong to a particular class, the value of $G(w)$ is 1. On the other hand, when documents containing word w are evenly distributed among the k different classes, the value of $G(w)$ is $1/k$.

One criticism with this approach is that the global class distribution may be skewed to begin with, and therefore the above measure may sometimes not accurately reflect the discriminative power of the underlying attributes. Therefore, it is possible to construct a normalized gini-index in order to reflect the discriminative power of the attributes more accurately. Let $P_1 \ldots P_k$ represent the global distributions of the documents in the different classes. Then, we determine the normalized probability value $p_i'(w)$ as follows:

$$p_i'(w) = \frac{p_i(w)/P_i}{\sum_{j=1}^{k} p_j(w)/P_j}. \tag{11.3}$$

Then, the gini-index is computed in terms of these normalized probability values.

$$G(w) = \sum_{i=1}^{k} p_i'(w)^2. \tag{11.4}$$

The use of the global probabilities P_i ensures that the gini-index more accurately reflects class-discrimination in the case of biased class distributions in the whole document collection. For a document corpus containing n documents, d words, and k classes, the complexity of the information gain computation is $O(n \cdot d \cdot k)$. This is because the computation of the term $p_i(w)$ for all the different words and the classes requires $O(n \cdot d \cdot k)$ time.

---

[4]The gini-index is also sometimes defined as $1 - \sum_{i=1}^{k} p_i(w)^2$, with lower values indicating greater discriminative power of the feature w.

11.2.2 Information Gain

Another related measure that is commonly used for text feature selection is that of information gain or entropy. Let P_i be the global probability of class i, and $p_i(w)$ be the probability of class i, given that the document contains the word w. Let $F(w)$ be the fraction of the documents containing the word w. The information gain measure $I(w)$ for a given word w is defined as follows:

$$I(w) = -\sum_{i=1}^{k} P_i \cdot \log(P_i) + F(w) \cdot \sum_{i=1}^{k} p_i(w) \cdot \log(p_i(w)) +$$

$$+ (1 - F(w)) \cdot \sum_{i=1}^{k} (1 - p_i(w)) \cdot \log(1 - p_i(w)).$$

The greater the value of the information gain $I(w)$, the greater the discriminatory power of the word w. For a document corpus containing n documents and d words, the complexity of the information gain computation is $O(n \cdot d \cdot k)$.

11.2.3 Mutual Information

This *mutual information measure* is derived from information theory [37], and provides a formal way to model the mutual information between the features and the classes. The pointwise mutual information $M_i(w)$ between the word w and the class i is defined on the basis of the level of co-occurrence between the class i and word w. We note that the expected co-occurrence of class i and word w on the basis of mutual independence is given by $P_i \cdot F(w)$. The true co-occurrence is of course given by $F(w) \cdot p_i(w)$. In practice, the value of $F(w) \cdot p_i(w)$ may be much larger or smaller than $P_i \cdot F(w)$, depending upon the level of correlation between the class i and word w. The mutual information is defined in terms of the ratio between these two values. Specifically, we have:

$$M_i(w) = \log\left(\frac{F(w) \cdot p_i(w)}{F(w) \cdot P_i}\right) = \log\left(\frac{p_i(w)}{P_i}\right). \tag{11.5}$$

Clearly, the word w is positively correlated to the class i, when $M_i(w) > 0$, and the word w is negatively correlated to class i, when $M_i(w) < 0$. We note that $M_i(w)$ is specific to a particular class i. We need to compute the overall mutual information as a function of the mutual information of the word w with the different classes. These are defined with the use of the average and maximum values of $M_i(w)$ over the different classes.

$$M_{avg}(w) = \sum_{i=1}^{k} P_i \cdot M_i(w)$$

$$M_{max}(w) = \max_i\{M_i(w)\}.$$

Either of these measures may be used in order to determine the relevance of the word w. The second measure is particularly useful, when it is more important to determine high levels of positive correlation of the word w with any of the classes.

11.2.4 χ^2-Statistic

The χ^2 statistic is a different way to compute the lack of independence between the word w and a particular class i. Let n be the total number of documents in the collection, $p_i(w)$ be the conditional probability of class i for documents that contain w, P_i be the global fraction of documents containing the class i, and $F(w)$ be the global fraction of documents that contain the word w. The χ^2-statistic of the word between word w and class i is defined as follows:

$$\chi_i^2(w) = \frac{n \cdot F(w)^2 \cdot (p_i(w) - P_i)^2}{F(w) \cdot (1 - F(w)) \cdot P_i \cdot (1 - P_i))} \tag{11.6}$$

As in the case of the mutual information, we can compute a global χ^2 statistic from the class-specific values. We can use either the average or maximum values in order to create the composite value:

$$\chi_{avg}^2(w) = \sum_{i=1}^{k} P_i \cdot \chi_i^2(w)$$

$$\chi_{max}^2(w) = \max_i \chi_i^2(w)$$

We note that the χ^2-statistic and mutual information are different ways of measuring the correlation between terms and categories. One major advantage of the χ^2-statistic over the mutual information measure is that it is a normalized value, and therefore these values are more comparable across terms in the same category.

11.2.5 Feature Transformation Methods: Unsupervised and Supervised LSI

While feature selection attempts to reduce the dimensionality of the data by picking from the original set of attributes, feature transformation methods create a new (and smaller) set of features as a function of the original set of features. The motivation of these feature transformation methods is to capture semantic associations of words so that words related to the same latent concept would be grouped together into one latent concept feature. A typical example of such a feature transformation method is Latent Semantic Indexing (LSI) [46], and its probabilistic variant PLSA [67]. The LSI method transforms the text space of a few hundred thousand word features to a new axis system (of size about a few hundred), which is a linear combination of the original word features. In order to achieve this goal, Principal Component Analysis techniques [81] are used to determine the axis-system that retains the greatest level of information about the variations in the underlying attribute values. The main disadvantage of using techniques such as LSI is that these are unsupervised techniques that are blind to the underlying class-distribution. Thus, the features found by LSI are not necessarily the directions along which the *class-distribution* of the underlying documents can best be separated. A modest level of success has been obtained in improving classification accuracy by using boosting techniques in conjunction with the conceptual features obtained from unsupervised pLSA method [22]. A more recent study has systematically compared pLSA and LDA (which is a Bayesian version of pLSA) in terms of their effectiveness in transforming features for text categorization, and has drawn a similar conclusion and found that pLSA and LDA tend to perform similarly [104]. In the case of classification tasks that require fine-grained discrimination, such as search engine applications where the task is to distinguish relevant from non-relevant documents to a query, the latent low-dimensional features obtained by LSI or PLSA alone are often insufficient, and a more discriminative representation based on keywords may have to be used as well to obtain a multiscale representation [161].

A number of techniques have also been proposed to perform the feature transformation methods by using the class labels for effective supervision. The most natural method is to adapt LSI in order to make it work more effectively for the supervised case. A number of different methods have been proposed in this direction. One common approach is to perform local LSI on the subsets of data representing the individual classes, and identify the discriminative eigenvectors from the different reductions with the use of an iterative approach [142]. This method is known as SLSI (Supervised Latent Semantic Indexing), and the advantages of the method seem to be relatively limited, because the experiments in [142] show that the improvements over a standard SVM classifier, which did not use a dimensionality reduction process, are relatively limited. The work in [149] uses a combination of class-specific LSI and global analysis. As in the case of [142], class-specific LSI representations are created. Test documents are compared against each LSI representation in order to create the most discriminative reduced space. One problem with this approach is that the different local LSI representations use a different subspace, and therefore it is difficult to compare the similarities of the different documents across the different subspaces. Furthermore, both methods in [142, 149] tend to be computationally expensive. Similar ideas were also implemented in a probabilistic topic model, called supervised LDA [16], where class labels are used to "supervise" the discovery of topics in text data [16].

A method called *sprinkling* is proposed in [26], in which artificial terms are added to (or "sprinkled" into) the documents, which correspond to the class labels. In other words, we create a term corresponding to the class label, and add it to the document. LSI is then performed on the document collection with these added terms. The sprinkled terms can then be removed from the representation, once the eigenvectors have been determined. The sprinkled terms help in making the LSI more sensitive to the class distribution during the reduction process. It has also been proposed in [26] that it can be generally useful to make the sprinkling process *adaptive*, in which all classes are not necessarily treated equally, but the relationships between the classes are used in order to regulate the sprinkling process. Furthermore, methods have also been proposed in [26] to make the sprinkling process adaptive to the use of a particular kind of classifier.

11.2.6 Supervised Clustering for Dimensionality Reduction

One technique that is commonly used for feature transformation is that of text clustering [8, 83, 95, 140]. In these techniques, the clusters are constructed from the underlying text collection, with the use of supervision from the class distribution. The exception is [95] in which supervision is not used. In the simplest case, each class can be treated as a separate cluster, though better results may be obtained by using the classes for supervision of the clustering process. The frequently occurring words in these supervised clusters can be used in order to create the new set of dimensions. The classification can be performed with respect to this new feature representation. One advantage of such an approach is that it retains interpretability with respect to the original words of the document collection. The disadvantage is that the optimum directions of separation may not necessarily be represented in the form of clusters of words. Furthermore, the underlying axes are not necessarily orthonormal to one another. The use of supervised methods [2, 8, 83, 140] has generally led to good results either in terms of improved classification accuracy, or significant performance gains at the expense of a small reduction in accuracy. The results with the use of unsupervised clustering [95, 99] are mixed. For example, the work in [95] suggests that the use of unsupervised term-clusters and phrases is generally not helpful [95] for the classification process. The key observation in [95] is that the loss of granularity associated with the use of phrases and term clusters is not necessarily advantageous for the classification process. The loss of discrimination due to dimension reduction is also recently discussed in the context of using probabilistic topic models to obtain a low-dimensional feature representation for text data [161]. The work in [9] has shown that the use of the information bottleneck method for feature distributional clustering can create clustered pseudo-word representations that are quite effective for text classification.

11.2.7 Linear Discriminant Analysis

Another method for feature transformation is the use of linear discriminants, which explicitly try to construct directions in the feature space, along which there is best separation of the different classes. A common method is the *Fisher's linear discriminant* [54]. The main idea in the Fisher's discriminant method is to determine the directions in the data along which the points are as well separated as possible. The subspace of lower dimensionality is constructed by iteratively finding such unit vectors α_i in the data, where α_i is determined in the ith iteration. We would also like to ensure that the different values of α_i are orthonormal to one another. In each step, we determine this vector α_i by discriminant analysis, and project the data onto the remaining orthonormal subspace. The next vector α_{i+1} is found in this orthonormal subspace. The quality of vector α_i is measured by an objective function that measures the separation of the different classes. This objective function reduces in each iteration, since the value of α_i in a given iteration is the optimum discriminant in that subspace, and the vector found in the next iteration is the optimal one from a smaller search space. The process of finding linear discriminants is continued until the class separation, as measured by an objective function, reduces below a given threshold for the vector determined in the current iteration. The power of such a dimensionality reduction approach has been illustrated in [23], in which it has been shown that a simple decision tree classifier can perform much more effectively on this transformed data, as compared to more sophisticated classifiers.

Next, we discuss how the Fisher's discriminant is actually constructed. First, we will set up the objective function $J(\overline{\alpha})$, which determines the level of separation of the different classes along a given direction (unit-vector) $\overline{\alpha}$. This sets up the crisp optimization problem of determining the value of $\overline{\alpha}$, which maximizes $J(\overline{\alpha})$. For simplicity, let us assume the case of binary classes. Let D_1 and D_2 be the two sets of documents belonging to the two classes. Then, the projection of a document $\overline{X} \in D_1 \cup D_2$ along $\overline{\alpha}$ is given by $\overline{X} \cdot \overline{\alpha}$. Therefore, the squared class separation $S(D_1, D_2, \overline{\alpha})$ along the direction $\overline{\alpha}$ is given by:

$$S(D_1, D_2, \overline{\alpha}) = \left(\frac{\sum_{\overline{X} \in D_1} \overline{\alpha} \cdot \overline{X}}{|D_1|} - \frac{\sum_{\overline{X} \in D_2} \overline{\alpha} \cdot \overline{X}}{|D_2|} \right)^2. \tag{11.7}$$

In addition, we need to normalize this absolute class separation with the use of the underlying class variances. Let $Var(D_1, \overline{\alpha})$ and $Var(D_2, \overline{\alpha})$ be the individual class variances along the direction α. In other words, we have:

$$Var(D_1, \overline{\alpha}) = \frac{\sum_{\overline{X} \in D_1} (\overline{X} \cdot \overline{\alpha})^2}{|D_1|} - \left(\frac{\sum_{\overline{X} \in D_1} \overline{X} \cdot \overline{\alpha}}{|D_1|} \right)^2. \tag{11.8}$$

The value of $Var(D_2, \overline{\alpha})$ can be defined in a similar way. Then, the normalized class-separation $J(\overline{\alpha})$ is defined as follows:

$$J(\overline{\alpha}) = \frac{S(D_1, D_2, \overline{\alpha})}{Var(D_1, \overline{\alpha}) + Var(D_2, \overline{\alpha})} \tag{11.9}$$

The optimal value of α needs to be determined subject to the constraint that $\overline{\alpha}$ is a unit vector. Let μ_1 and μ_2 be the means of the two data sets D_1 and D_2, and C_1 and C_2 be the corresponding covariance matrices. It can be shown that the optimal (unscaled) direction $\overline{\alpha} = \overline{\alpha^*}$ can be expressed in closed form, and is given by the following:

$$\overline{\alpha^*} = \left(\frac{C_1 + C_2}{2} \right)^{-1} (\mu_1 - \mu_2). \tag{11.10}$$

The main difficulty in computing the above equation is that this computation requires the inversion of the covariance matrix, which is sparse and computationally difficult in the high-dimensional text domain. Therefore, a gradient descent approach can be used in order to determine the value of $\overline{\alpha}$ in a more computationally effective way. Details of the approach are presented in [23].

Another related method that attempts to determine projection directions that maximize the topical differences between different classes is the *Topical Difference Factor Analysis* method proposed in [84]. The problem has been shown to be solvable as a generalized eigenvalue problem. The method was used in conjunction with a k-nearest neighbor classifier, and it was shown that the use of this approach significantly improves the accuracy over a classifier that uses the original set of features.

11.2.8 Generalized Singular Value Decomposition

While the method discussed above finds one vector $\overline{\alpha_i}$ at a time in order to determine the relevant dimension transformation, it is possible to be much more direct in finding the optimal subspaces simultaneously by using a generalized version of dimensionality reduction [68, 69]. It is important to note that this method has really been proposed in [68,69] as an *unsupervised* method that preserves the underlying *clustering* structure, assuming the data has already been clustered in a pre-processing phase. Thus, the generalized dimensionality reduction method has been proposed as a much more aggressive dimensionality reduction technique, which preserves the underlying clustering structure rather than the individual points. This method can however also be used as a *supervised* technique in which the different classes are used as input to the dimensionality reduction algorithm, instead of the clusters constructed in the pre-processing phase [152]. This method is known as the *Optimal Orthogonal Centroid Feature Selection Algorithm (OCFS)*, and it directly targets at the maximization of inter-class scatter. The algorithm is shown to have promising results for supervised feature selection in [152].

11.2.9   Interaction of Feature Selection with Classification

Since the classification and feature selection processes are dependent upon one another, it is interesting to test how the feature selection process interacts with the underlying classification algorithms. In this context, two questions are relevant:

- Can the feature-specific insights obtained from the intermediate results of some of the classification algorithms be used for creating feature selection methods that can be used more generally by other classification algorithms?

- Do the different feature selection methods work better or worse with different kinds of classifiers?

Both these issues were explored in some detail in [113]. In regard to the first question, it was shown in [113] that feature selection, which was derived from *linear* classifiers, provided very effective results. In regard to the second question, it was shown in [113] that the sophistication of the feature selection process itself was more important than the specific pairing between the feature selection process and the classifier.

Linear Classifiers are those for which the output of the linear predictor is defined to be $p = \overline{A} \cdot \overline{X} + b$, where $\overline{X} = (x_1 \ldots x_n)$ is the normalized document word frequency vector, $\overline{A} = (a_1 \ldots a_n)$ is a vector of linear coefficients with the same dimensionality as the feature space, and b is a scalar. Both the basic neural network and basic SVM classifiers [75] (which will be discussed later in this chapter) belong to this category. The idea here is that if the coefficient a_i is close to zero, then the corresponding feature does not have a significant effect on the classification process. On the other hand, since large absolute values of a_j may significantly influence the classification process, such features should be selected for classification. In the context of the SVM method, which attempts to determine linear planes of separation between the different classes, the vector $\overline{A}$ is essentially the normal vector to the corresponding plane of separation between the different classes. This intuitively explains the choice of selecting features with large values of $|a_j|$. It was shown in [113] that this class of feature selection methods was quite robust, and performed well even for classifiers such as the Naive Bayes method, which were unrelated to the linear classifiers from which these features were derived. Further discussions on how SVM and maximum margin techniques can be used for feature selection may be found in [59,65].

11.3   Decision Tree Classifiers

A decision tree [122] is essentially a hierarchical decomposition of the (training) data space, in which a *predicate* or a condition on the attribute value is used in order to divide the data space hierarchically. In the context of text data, such predicates are typically conditions on the presence or absence of one or more words in the document. The division of the data space is performed recursively in the decision tree, until the leaf nodes contain a certain minimum number of records, or some conditions on class purity. The majority class label (or cost-weighted majority label) in the leaf node is used for the purposes of classification. For a given test instance, we apply the sequence of predicates at the nodes, in order to traverse a path of the tree in top-down fashion and determine the relevant leaf node. In order to further reduce the overfitting, some of the nodes may be pruned by holding out a part of the data, which are not used to construct the tree. The portion of the data which is held out is used in order to determine whether or not the constructed leaf node should be pruned or not. In particular, if the class distribution in the training data (for decision tree construction) is very different from the class distribution in the training data that is used for pruning, then it is assumed that the node overfits the training data. Such a node can be pruned. A detailed discussion of decision tree methods may be found in [20, 50, 72, 122].

In the particular case of text data, the predicates for the decision tree nodes are typically defined in terms of the terms in the underlying text collection. For example, a node may be partitioned into its children nodes depending upon the presence or absence of a particular term in the document. We note that different nodes at the same level of the tree may use different terms for the partitioning process.

Many other kinds of predicates are possible. It may not be necessary to use individual terms for partitioning, but one may measure the similarity of documents to correlated sets of terms. These correlated sets of terms may be used to further partition the document collection, based on the similarity of the document to them. The different kinds of splits are as follows:

- **Single Attribute Splits:** In this case, we use the presence or absence of particular words (or even phrases) at a particular node in the tree in order to perform the split. At any given level, we pick the word that provides the maximum discrimination between the different classes. Measures such as the gini-index or information gain can be used in order to determine the level of entropy. For example, the DT-min10 algorithm [93] is based on this approach.

- **Similarity-Based Multi-Attribute Split:** In this case, we use documents (or meta-documents such as frequent word clusters), and use the similarity of the documents to these word clusters in order to perform the split. For the selected word cluster, the documents are further partitioned into groups by rank ordering the documents by similarity value, and splitting at a particular threshold. We select the word-cluster for which rank-ordering by similarity provides the best separation between the different classes.

- **Discriminant-Based Multi-Attribute Split:** For the multi-attribute case, a natural choice for performing the split is to use discriminants such as the Fisher discriminant for performing the split. Such discriminants provide the directions in the data along which the classes are best separated. The documents are projected on this discriminant vector for rank ordering, and then split at a particular coordinate. The choice of split point is picked in order to maximize the discrimination between the different classes. The work in [23] uses a discriminant-based split, though this is done indirectly because of the use of a feature transformation to the discriminant representation, before building the classifier.

Some of the earliest implementation of classifiers may be found in [92,93,99,147]. The last of these is really a rule-based classifier, which can be interpreted either as a decision tree or a rule-based classifier. Most of the decision tree implementations in the text literature tend to be small variations on standard packages such as ID3 and C4.5, in order to adapt the model to text classification. Many of these classifiers are typically designed as baselines for comparison with other learning models [75].

A well known implementation of the decision tree classifier based on the C4.5 taxonomy of algorithms [122] is presented in [99]. More specifically, the work in [99] uses the successor to the C4.5 algorithm, which is also known as the C5 algorithm. This algorithm uses single-attribute splits at each node, where the feature with the highest information gain [37] is used for the purpose of the split. Decision trees have also been used in conjunction with boosting techniques. An adaptive boosting technique [56] is used in order to improve the accuracy of classification. In this technique, we use n different classifiers. The ith classifier is constructed by examining the errors of the $(i-1)$th classifier. A voting scheme is applied among these classifiers in order to report the final label. Other boosting techniques for improving decision tree classification accuracy are proposed in [134].

The work in [51] presents a decision tree algorithm based on the Bayesian approach developed in [28]. In this classifier, the decision tree is grown by recursive greedy splits, where the splits are chosen using Bayesian posterior probability of model structure. The structural prior penalizes additional model parameters at each node. The output of the process is a class probability rather than a deterministic class label for the test instance.

11.4 Rule-Based Classifiers

Decision trees are also generally related to *rule-based classifiers*. In rule-based classifiers, the data space is modeled with a set of rules, in which the left-hand side is a condition on the underlying feature set, and the right-hand side is the class label. The rule set is essentially the model that is generated from the training data. For a given test instance, we determine the set of rules for which the test instance satisfies the condition on the left-hand side of the rule. We determine the predicted class label as a function of the class labels of the rules that are satisfied by the test instance. We will discuss more on this issue slightly later.

In its most general form, the left-hand side of the rule is a boolean condition, which is expressed in Disjunctive Normal Form (DNF). However, in most cases, the condition on the left-hand side is much simpler and represents a set of terms, all of which must be present in the document for the condition to be satisfied. The *absence* of terms is rarely used, because such rules are not likely to be very informative for sparse text data, in which most words in the lexicon will typically not be present in it by default (sparseness property). Also, while the *set intersection* of conditions on term presence is used often, the union of such conditions is rarely used in a single rule. This is because such rules can be split into two separate rules, each of which is more informative on its own. For example, the rule *Honda* ∪ *Toyota* ⇒ *Cars* can be replaced by two separate rules *Honda* ⇒ *Cars* and *Toyota* ⇒ *Cars* without any loss of information. In fact, since the confidence of each of the two rules can now be measured separately, this can be more useful. On the other hand, the rule *Honda* ∩ *Toyota* ⇒ *Cars* is certainly much more informative than the individual rules. Thus, in practice, for sparse data sets such as text, rules are much more likely to be expressed as a simple conjunction of conditions on term presence.

We note that decision trees and decision rules both tend to encode rules on the feature space, except that the decision tree tends to achieve this goal with a hierarchical approach. In fact, the original work on decision tree construction in C4.5 [122] studied the decision tree problem and decision rule problem within a single framework. This is because a particular path in the decision tree can be considered a rule for classification of the text instance. The main difference is that the decision tree framework is a strict hierarchical partitioning of the data space, whereas rule-based classifiers allow for overlaps in the decision space. The general principle is to create a rule set, such that all points in the decision space are covered by *at least* one rule. In most cases, this is achieved by generating a set of targeted rules that are related to the different classes, and one default *catch-all* rule, which can cover all the remaining instances.

A number of criteria can be used in order to generate the rules from the training data. Two of the most common conditions that are used for rule generation are those of *support* and *confidence*. These conditions are common to all rule-based pattern classifiers [100] and may be defined as follows:

- **Support:** This quantifies the **absolute number** of instances in the training data set that are relevant to the rule. For example, in a corpus containing 100,000 documents, a rule in which **both** the left-hand set and right-hand side are satisfied by 50,000 documents is more important than a rule that is satisfied by 20 documents. Essentially, this quantifies the statistical *volume* that is associated with the rule. However, it does not encode the strength of the rule.

- **Confidence:** This quantifies the **conditional probability** that the right-hand side of the rule is satisfied, if the left-hand side is satisfied. This is a more direct measure of the strength of the underlying rule.

We note that the aforementioned measures are not the only measures that are possible, but are widely used in the data mining and machine learning literature [100] for both textual and non-textual data, because of their intuitive nature and simplicity of interpretation. One criticism of the above measures is that they do not normalize for the a-priori presence of different terms and features, and are therefore prone to misinterpretation, when the feature distribution or class-distribution in the underlying data set is skewed.

The training phase constructs all the rules, which are based on measures such as the above. For a given test instance, we determine all the rules that are relevant to the test instance. Since we allow overlaps, it is possible that more than one rule may be relevant to the test instance. If the class labels on the right-hand sides of all these rules are the same, then it is easy to pick this class as the relevant label for the test instance. On the other hand, the problem becomes more challenging when there are conflicts between these different rules. A variety of different methods are used to rank-order the different rules [100], and report the most relevant rule as a function of these different rules. For example, a common approach is to rank-order the rules by their confidence, and pick the top-*k* rules as the most relevant. The class label on the right-hand side of the greatest number of these rules is reported as the relevant one.

An interesting rule-based classifier for the case of text data has been proposed in [6]. This technique uses an iterative methodology, which was first proposed in [148] for generating rules. Specifically, the method determines the single best rule related to any particular class in the training data. The best rule is defined in terms of the confidence of the rule, as defined above. This rule along with its corresponding instances are removed from the training data set. This approach is continuously repeated, until it is no longer possible to find strong rules in the training data, and complete predictive value is achieved.

The transformation of decision trees to rule-based classifiers is discussed generally in [122], and for the particular case of text data in [80]. For each path in the decision tree a rule can be generated, which represents the conjunction of the predicates along that path. One advantage of the rule-based classifier over a decision tree is that it is not restricted to a strict hierarchical partitioning of the feature space, and it allows for overlaps and inconsistencies among the different rules. Therefore, if a new set of training examples are encountered, which are related to a new class or new part of the feature space, then it is relatively easy to modify the rule set for these new examples. Furthermore, rule-based classifiers also allow for a tremendous interpretability of the underlying decision space. In cases in which domain-specific expert knowledge is known, it is possible to encode this into the classification process by manual addition of rules. In many practical scenarios, rule-based techniques are more commonly used because of their ease of maintenance and interpretability.

One of the most common rule-based techniques is the *RIPPER* technique discussed in [32–34]. The *RIPPER* technique uses the sequential covering paradigm to determine the combinations of words that are related to a particular class. The *RIPPER* method has been shown to be especially effective in scenarios where the number of training examples is relatively small [31]. Another method called *sleeping experts* [32,57] generates rules that take the placement of the words in the documents into account. Most of the classifiers such as *RIPPER* [32–34] treat documents as set-valued objects, and generate rules based on the co-presence of the words in the documents. The rules in *sleeping experts* are different from most of the other classifiers in this respect. In this case [32,57], the left-hand side of the rule consists of a *sparse phrase*, which is a group of words close to one another in the document (though not necessarily completely sequential). Each such rule has a weight, which depends upon its classification specificity in the training data. For a given test example, we determine the sparse phrases that are present in it, and perform the classification by combining the weights of the different rules that are fired. The *sleeping experts* and *RIPPER* systems have been compared in [32], and have been shown to have excellent performance on a variety of text collections.

11.5 Probabilistic and Naive Bayes Classifiers

Probabilistic classifiers are designed to use an implicit mixture model for generation of the underlying documents. This mixture model typically assumes that each class is a component of the mixture. Each mixture component is essentially a generative model, which provides the probability of sampling a particular term for that component or class. This is why this kind of classifiers are often also called generative classifiers. The naive Bayes classifier is perhaps the simplest and also the most commonly used generative classifier. It models the distribution of the documents in each class using a probabilistic model with independence assumptions about the distributions of different terms. Two classes of models are commonly used for naive Bayes classification. Both models essentially compute the posterior probability of a class, based on the distribution of the words in the document. These models ignore the actual position of the words in the document, and work with the "bag of words" assumption. The major difference between these two models is the assumption in terms of taking (or not taking) word frequencies into account, and the corresponding approach for sampling the probability space:

- **Multivariate Bernoulli Model:** In this model, we use the presence or absence of words in a text document as features to represent a document. Thus, the frequencies of the words are not used for the modeling a document, and the word features in the text are assumed to be binary, with the two values indicating presence or absence of a word in text. Since the features to be modeled are binary, the model for documents in each class is a multivariate Bernoulli model.

- **Multinomial Model:** In this model, we capture the frequencies of terms in a document by representing a document with a bag of words. The documents in each class can then be modeled as samples drawn from a multinomial word distribution. As a result, the conditional probability of a document given a class is simply a product of the probability of each observed word in the corresponding class.

No matter how we model the documents in each class (be it a multivariate Bernoulli model or a multinomial model), the component class models (i.e., generative models for documents in each class) can be used in conjunction with the Bayes rule to compute the posterior probability of the class for a given document, and the class with the highest posterior probability can then be assigned to the document.

There has been considerable confusion in the literature on the differences between the multivariate Bernoulli model and the multinomial model. A good exposition of the differences between these two models may be found in [108]. In the following, we describe these two models in more detail.

11.5.1 Bernoulli Multivariate Model

This class of techniques treats a document as a set of distinct words with no frequency information, in which an element (term) may be either present or absent. The seminal work on this approach may be found in [94].

Let us assume that the lexicon from which the terms are drawn are denoted by $V = \{t_1 \ldots t_n\}$. Let us assume that the bag-of-words (or text document) in question contains the terms $Q = \{t_{i_1} \ldots t_{i_m}\}$, and the class is drawn from $\{1 \ldots k\}$. Then, our goal is to model the posterior probability that the document (which is assumed to be generated from the term distributions of one of the classes) belongs to class i, given that it contains the terms $Q = \{t_{i_1} \ldots t_{i_m}\}$. The best way to understand the Bayes method is by understanding it as a sampling/generative process from the underlying mixture model of classes. The Bayes probability of class i can be modeled by sampling a set of terms T from the term distribution of the classes:

If we sampled a term set T of any size **from the term distribution of one of the randomly chosen classes**, *and the final outcome is the set Q, then what is the posterior probability that we had originally picked class i for sampling? The a-priori probability of picking class i is equal to its fractional presence in the collection.*

We denote the class of the sampled set T by C^T and the corresponding posterior probability by $P(C^T = i | T = Q)$. This is essentially what we are trying to find. It is important to note that since we do not allow replacement, we are essentially picking a subset of terms from V with no frequencies attached to the picked terms. Therefore, the set Q may not contain duplicate elements. Under the naive Bayes assumption of independence between terms, this is essentially equivalent to either selecting or not selecting each term with a probability that depends upon the underlying term distribution. Furthermore, it is also important to note that this model has no restriction on the number of terms picked. As we will see later, these assumptions are the key differences with the multinomial Bayes model. The Bayes approach classifies a given set Q based on the posterior probability that Q is a sample from the data distribution of class i, i.e., $P(C^T = i | T = Q)$, and it requires us to compute the following two probabilities in order to achieve this:

1. What is the prior probability that a set T is a sample from the term distribution of class i? This probability is denoted by $P(C^T = i)$.

2. If we sampled a set T of any size *from the term distribution of class i*, then what is the probability that our sample is the set Q? This probability is denoted by $P(T = Q | C^T = i)$.

We will now provide a more mathematical description of Bayes modeling. In other words, we wish to model $P(C^T = i | Q$ is sampled$)$. We can use the Bayes rule in order to write this conditional probability in a way that can be *estimated* more easily from the underlying corpus. In other words, we can simplify as follows:

$$P(C^T = i | T = Q) = \frac{P(C^T = i) \cdot P(T = Q | C^T = i)}{P(T = Q)}$$

$$= \frac{P(C^T = i) \cdot \prod_{t_j \in Q} P(t_j \in T | C^T = i) \cdot \prod_{t_j \notin Q} (1 - P(t_j \in T | C^T = i))}{P(T = Q)}.$$

We note that the last condition of the above sequence uses the *naive independence assumption*, because we are assuming that the probabilities of occurrence of the different terms are independent of one another. This is practically necessary, in order to transform the probability equations to a form that can be estimated from the underlying data.

The class assigned to Q is the one with the highest posterior probability given Q. It is easy to see that this decision is not affected by the denominator, which is the marginal probability of observing Q. That is, we will assign the following class to Q:

$$\hat{i} = \arg\max_i P(C^T = i | T = Q)$$

$$= \arg\max_i P(C^T = i) \cdot$$

$$\prod_{t_j \in Q} P(t_j \in T | C^T = i) \cdot \prod_{t_j \notin Q} (1 - P(t_j \in T | C^T = i)).$$

It is important to note that all terms in the right hand-side of the last equation can be estimated from the training corpus. The value of $P(C^T = i)$ is estimated as the global fraction of documents belonging to class i, the value of $P(t_j \in T | C^T = i)$ is the fraction of documents in the ith class that contain term t_j. We note that all of the above are maximum likelihood estimates of the corresponding probabilities. In practice, Laplacian smoothing [144] is used, in which small values are added to the frequencies of terms in order to avoid zero probabilities of sparsely present terms.

In most applications of the Bayes classifier, we only care about the *identity* of the class with the highest probability value, rather than the actual probability value associated with it, which is why we do not need to compute the normalizer $P(T = Q)$. In fact, in the case of **binary** classes, a number of simplifications are possible in computing these Bayes "probability" values by using the logarithm of the Bayes expression, and removing a number of terms that do not affect the ordering of class probabilities. We refer the reader to [124] for details.

Although for classification, we do not need to compute $P(T = Q)$, some applications necessitate the exact computation of the posterior probability $P(C^T = i | T = Q)$. For example, in the case of supervised anomaly detection (or rare class detection), the exact posterior probability value $P(C^T = i | T = Q)$ is needed in order to fairly compare the probability value over different test instances, and rank them for their anomalous nature. In such cases, we would need to compute $P(T = Q)$. One way to achieve this is simply to take a sum over all the classes:

$$P(T = Q) = \sum_i P(T = Q | C^T = i) P(C^T = i).$$

This is based on the conditional independence of features for each class. Since the parameter values are estimated for each class separately, we may face the problem of data sparseness. An alternative way of computing it, which may alleviate the data sparseness problem, is to further make the assumption of (global) independence of terms, and compute it as:

$$P(T = Q) = \prod_{j \in Q} P(t_j \in T) \cdot \prod_{t_j \notin Q} (1 - P(t_j \in T))$$

where the term probabilities are based on global term distributions in *all* the classes.

A natural question arises, as to whether it is possible to design a Bayes classifier that does not use the naive assumption, and models the dependencies between the terms during the classification process. Methods that generalize the naive Bayes classifier by not using the independence assumption do not work well because of the higher computational costs and the inability to estimate the parameters accurately and robustly in the presence of limited data. The most interesting line of work in relaxing the independence assumption is provided in [128]. In this work, the tradeoffs in spectrum of allowing different levels of dependence among the terms have been explored. On the one extreme, an assumption of complete dependence results in a Bayesian network model that turns out to be computationally very expensive. On the other hand, it has been shown that allowing limited levels of dependence can provide good tradeoffs between accuracy and computational costs. We note that while the independence assumption is a practical approximation, it has been shown in [35,47] that the approach does have some theoretical merit. Indeed, extensive experimental tests have tended to show that the naive classifier works quite well in practice.

A number of papers [24,74,86,91,124,129] have used the naive Bayes approach for classification in a number of different application domains. The classifier has also been extended to modeling temporally aware training data, in which the importance of a document may decay with time [130]. As in the case of other statistical classifiers, the naive Bayes classifier [129] can easily incorporate *domain-specific knowledge* into the classification process. The particular domain that the work in [129] addresses is that of filtering junk email. Thus, for such a problem, we often have a lot of additional domain knowledge that helps us determine whether a particular email message is junk or not. For example, some common characteristics of the email that would make an email to be more or less likely to be junk are as follows:

- The domain of the sender such as *.edu* or *.com* can make an email to be more or less likely to be junk.

- Phrases such as *"Free Money"* or over emphasized punctuation such as "!!!" can make an email more likely to be junk.

- Whether the recipient of the message was a particular user, or a mailing list.

The Bayes method provides a natural way to incorporate such additional information into the classification process, by creating new features for each of these characteristics. The standard Bayes technique is then used in conjunction with this augmented representation for classification. The Bayes technique has also been used in conjunction with the incorporation of other kinds of domain knowledge, such as the incorporation of hyperlink information into the classification process [25,118].

The Bayes method is also suited to hierarchical classification, when the training data is arranged in a taxonomy of topics. For example, the Open Directory Project (ODP), *Yahoo!* Taxonomy, and a variety of news sites have vast collections of documents that are arranged into hierarchical groups. The hierarchical structure of the topics can be exploited to perform more effective classification [24, 86], because it has been observed that context-sensitive feature selection can provide more useful classification results. In hierarchical classification, a Bayes classifier is built at *each node*, which then provides us with the next branch to follow for classification purposes. Two such methods are proposed in [24, 86], in which node specific features are used for the classification process. Clearly, much fewer features are required at a particular node in the hierarchy, because the features that are picked are relevant to that branch. An example in [86] suggests that a branch of the taxonomy that is related to *Computer* may have no relationship with the word "cow." These node-specific features are referred to as *signatures* in [24]. Furthermore, it has been observed in [24] that in a given node, the most discriminative features for a given class may be different from their parent nodes. For example, the word "health" may be discriminative for the *Yahoo!* category @*Health*, but the word "baby" may be much more discriminative for the category @*Health@Nursing*. Thus, it is critical to have an appropriate feature selection process at each node of the classification tree. The methods in [24, 86] use different methods for this purpose.

- The work in [86] uses an information-theoretic approach [37] for feature selection, which takes into account the dependencies between the attributes [128]. The algorithm greedily eliminates the features one-by-one so as to least disrupt the conditional class distribution at that node.

- The node-specific features are referred to as *signatures* in [24]. These node-specific signatures are computed by calculating the ratio of intra-class variance to inter-class variance for the different words at each node of the tree. We note that this measure is the same as that optimized by the Fisher's discriminant, except that it is applied to the original set of words, rather than solved as a general optimization problem in which arbitrary directions in the data are picked.

A Bayesian classifier is constructed at each node in order to determine the appropriate branch. A small number of context-sensitive features provide one advantage of these methods, i.e., Bayesian classifiers work much more effectively with a much smaller number of features. Another major difference between the two methods is that the work in [86] uses the Bernoulli model, whereas that in [24] uses the multinomial model, which will be discussed in the next subsection. This approach in [86] is referred to as the *Pachinko Machine classifier* and that in [24] is known as *TAPER* (*Taxonomy and Path Enhanced Retrieval System*).

Other noteworthy methods for hierarchical classification are proposed in [13, 109, 151]. The work [13] addresses two common problems associated with hierarchical text classification: (1) error propagation; (2) non-linear decision surfaces. The problem of error propagation occurs when the classification mistakes made at a parent node are propagated to its children nodes. This problem was solved in [13] by using cross validation to obtain a training data set for a child node that is more similar to the actual test data passed to the child node from its parent node than the training data set normally used for training a classifier at the child node. The problem of non-linear decision surfaces refers to the fact that the decision boundary of a category at a higher level is often non-linear (since its members are the union of the members of its children nodes). This problem is addressed by using the tentative class labels obtained at the children nodes as features for use at a parent node. These are general strategies that can be applied to any base classifier, and the experimental results in [13] show that both strategies are effective.

11.5.2 Multinomial Distribution

This class of techniques treats a document as a set of words with frequencies attached to each word. Thus, the set of words is allowed to have duplicate elements.

As in the previous case, we assume that the set of words in document is denoted by Q, drawn from the vocabulary set V. The set Q contains the distinct terms $\{t_{i_1} \ldots t_{i_m}\}$ with associated frequencies $F = \{F_{i_1} \ldots F_{i_m}\}$. We denote the terms and their frequencies by $[Q, F]$. The total number of terms in the document (or document length) is denoted by $L = \sum_{j=1}^{m} F(i_j)$. Then, our goal is to model the posterior probability that the document T belongs to class i, given that it contains the terms in Q *with the associated frequencies* F. The Bayes probability of class i can be modeled by using the following sampling process:

If we sampled L terms sequentially **from the term distribution of one of the randomly chosen classes** *(allowing repetitions) to create the term set T, and the final outcome for sampled set T is the set Q with the corresponding frequencies F, then what is the posterior probability that we had originally picked class i for sampling? The a-priori probability of picking class i is equal to its fractional presence in the collection.*

The aforementioned probability is denoted by $P(C^T = i | T = [Q, F])$. An assumption that is commonly used in these models is that the length of the document is independent of the class label. While it is easily possible to generalize the method, so that the document length is used as a prior, independence is usually assumed for simplicity. As in the previous case, we need to estimate two values in order to compute the Bayes posterior.

1. What is the prior probability that a set T is a sample from the term distribution of class i? This probability is denoted by $P(C^T = i)$.

2. If we sampled L terms *from the term distribution of class i* (with repetitions), then what is the probability that our sampled set T is the set Q with associated frequencies F? This probability is denoted by $P(T = [Q, F] | C^T = i)$.

Then, the Bayes rule can be applied to this case as follows:

$$P(C^T = i | T = [Q, F]) = \frac{P(C^T = i) \cdot P(T = [Q, F] | C^T = i)}{P(T = [Q, F])}$$

$$\propto P(C^T = i) \cdot P(T = [Q, F] | C^T = i). \tag{11.11}$$

As in the previous case, it is not necessary to compute the denominator, $P(T = [Q, F])$, for the purpose of deciding the class label for Q. The value of the probability $P(C^T = i)$ can be estimated as the fraction of documents belonging to class i. The computation of $P([Q, F] | C^T = i)$ is more complicated. When we consider the sequential order of the L different samples, the number of possible ways to sample the different terms so as to result in the outcome $[Q, F]$ is given by $\frac{L!}{\prod_{i=1}^{m} F_i!}$. The probability of *each* of these sequences is given by $\prod_{t_j \in Q} P(t_j \in T)^{F_j}$, by using the naive independence assumption. Therefore, we have:

$$P(T = [Q, F] | C^T = i) = \frac{L!}{\prod_{i=1}^{m} F_i!} \cdot \prod_{t_j \in Q} P(t_j \in T | C^T = i)^{F_j}. \tag{11.12}$$

We can substitute Equation 11.12 in Equation 11.11 to obtain the class with the highest Bayes posterior probability, where the class priors are computed as in the previous case, and the probabilities $P(t_j \in T | C^T = i)$ can also be easily estimated as previously with Laplacian smoothing [144]. Note that for the purpose of choosing the class with the highest posterior probability, we do not really have to compute $\frac{L!}{\prod_{i=1}^{m} F_i!}$, as it is a constant not depending on the class label (i.e., the same for all the classes). We also note that the probabilities of class absence are not present in the above equations because of the way in which the sampling is performed.

A number of different variations of the multinomial model have been proposed in [61, 82, 96, 109, 111, 117]. In the work [109], it is shown that a category hierarchy can be leveraged to improve the estimate of multinomial parameters in the naive Bayes classifier to significantly improve classification accuracy. The key idea is to apply shrinkage techniques to smooth the parameters for data-sparse child categories with their common parent nodes. As a result, the training data of related categories are essentially "shared" with each other in a weighted manner, which helps improve the robustness and accuracy of parameter estimation when there are insufficient training data for each individual child category. The work in [108] has performed an extensive comparison between the Bernoulli and the multinomial models on different corpora, and the following conclusions were presented:

- The multi-variate Bernoulli model can sometimes perform better than the multinomial model at small vocabulary sizes.

- The multinomial model outperforms the multi-variate Bernoulli model for large vocabulary sizes, and almost always beats the multi-variate Bernoulli when vocabulary size is chosen optimally for both. On the average a 27% reduction in error was reported in [108].

The afore-mentioned results seem to suggest that the two models may have different strengths, and may therefore be useful in different scenarios.

11.5.3 Mixture Modeling for Text Classification

We note that the afore-mentioned Bayes methods simply assume that each component of the mixture corresponds to the documents belonging to a class. A more general interpretation is one in which the components of the mixture are created by a clustering process, and the class membership probabilities are modeled in terms of this mixture. Mixture modeling is typically used for unsupervised (probabilistic) clustering or topic modeling, though the use of clustering can also help in enhancing the effectiveness of probabilistic classifiers [98, 117]. These methods are particularly useful in cases where the amount of training data is limited. In particular, clustering can help in the following ways:

- The Bayes method implicitly estimates the word probabilities $P(t_i \in T | C^T = i)$ of a large number of terms in terms of their fractional presence in the corresponding component. This is clearly noisy. By treating the clusters as separate entities from the classes, we now only need to relate (a much smaller number of) cluster membership probabilities to class probabilities. This reduces the number of parameters and greatly improves classification accuracy [98].

- The use of clustering can help in incorporating unlabeled documents into the training data for classification. The premise is that unlabeled data is much more copiously available than labeled data, and when labeled data is sparse, it should be used in order to assist the classification process. While such unlabeled documents do not contain class-specific information, they do contain a lot of information about the clustering behavior of the underlying data. This can be very useful for more robust modeling [117], when the amount of training data is low. This general approach is also referred to as *co-training* [10, 17, 45].

The common characteristic of both methods [98, 117] is that they both use a form of supervised clustering for the classification process. While the goal is quite similar (limited training data), the approach used for this purpose is quite different. We will discuss both of these methods in this section.

In the method discussed in [98], the document corpus is modeled with the use of supervised word clusters. In this case, the k mixture components are clusters that are correlated to, but are distinct from the k groups of documents belonging to the different classes. The main difference from the Bayes method is that the term probabilities are computed indirectly by using clustering as an intermediate step. For a sampled document T, we denote its class label by $C^T \in \{1 \ldots k\}$, and its mixture component by $M^T \in \{1 \ldots k\}$. The k different mixture components are essentially word-clusters whose frequencies are generated by using the frequencies of the terms in the k different classes. This ensures that the word clusters for the mixture components are correlated to the classes, but they are not assumed to be drawn from the same distribution. As in the previous case, let us assume that the A document contains the set of words Q. Then, we would like to estimate the probability $P(T = Q | C^T = i)$ for each class i. An interesting variation of the work in [98] from the Bayes approach is that it does not attempt to determine the posterior probability $P(C^T = i | T = Q)$. Rather, it simply reports the class with the highest likelihood $P(T = Q | C^T = i)$. This is essentially equivalent to assuming, in the Bayes approach, that the prior distribution of each class is the same.

The other difference of the approach is in terms of how the value of $P(T = Q | C^T = i)$ is computed. As before, we need to estimate the value of $P(t_j \in T | C^T = i)$, according to the naive Bayes rule. However, unlike the standard Bayes classifier, this is done very indirectly with the use of mixture modeling. Since the mixture components do not directly correspond to the class, this term can only be estimated by summing up the expected value over all the mixture components:

$$P(t_j \in T | C^T = i) = \sum_{s=1}^{k} P(t_j \in T | M^T = s) \cdot P(M^T = s | C^T = i). \tag{11.13}$$

The value of $P(t_j \in T | M^T = s)$ is easy to estimate by using the fractional presence of term t_j in the sth mixture component. The main unknown here is the set of model parameters $P(M^T = s | C^T = i)$. Since a total of k classes and k mixture-components are used, this requires the estimation of only k^2 model parameters, which is typically quite modest for a small number of classes. An EM-approach has been used in [98] in order to estimate this small number of model parameters in a robust way. It is important to understand that the work in [98] is an interesting combination of supervised topic modeling (dimensionality reduction) and Bayes classification after reducing the effective dimensionality of the feature space to a much smaller value by clustering. The scheme works well because of the use of supervision in the topic modeling process, which ensures that the use of an intermediate clustering approach does not lose information for classification. We also note that in this model, the number of mixtures can be made to vary from the number of classes. While the work in [98] does not explore this direction, there is really no reason to assume that the number of mixture components is the same as the number of classes. Such an assumption can be particularly useful for data sets in which the classes may not be contiguous in the feature space, and a natural clustering may contain far more components than the number of classes.

Next, we will discuss the second method [117], which uses unlabeled data. The approach in [117] uses the unlabeled data in order to improve the training model. Why should unlabeled data help in classification at all? In order to understand this point, recall that the Bayes classification process effectively uses k mixture components, which are assumed to be the k different classes. If we had an infinite amount of training data, it would be possible to create the mixture components, but it would not be possible to assign labels to these components. However, the most data-intensive part of modeling the mixture is that of determining the shape of the mixture components. The actual assignment of mixture components to class labels can be achieved with a relatively small number of class labels. It has been shown in [30] that the accuracy of assigning components to classes increases exponentially with the number of labeled samples available. Therefore, the work in [117] designs an EM-approach [44] to simultaneously determine the relevant mixture model and its class assignment.

It turns out that the EM-approach, as applied to this problem, is quite simple to implement. It has been shown in [117] that the EM-approach is equivalent to the following iterative methodology. First, a naive Bayes classifier is constructed by estimating the model parameters from the labeled documents only. This is used in order to assign probabilistically weighted class labels to the unlabeled documents. Then, the Bayes classifier is reconstructed, except that we also use the newly labeled documents in the estimation of the underlying model parameters. We again use this classifier to reclassify the (originally unlabeled) documents. The process is continually repeated till convergence is achieved. This process is in many ways similar to pseudo-relevance feedback in information retrieval where a portion of top-ranked documents returned from an initial round of retrieval would be assumed to be relevant document examples (may be weighted), which can then be used to learn an improved query representation for improving ranking of documents in the next round of retrieval [150].

The ability to significantly improve the quality of text classification with a small amount of labeled data, and the use of clustering on a large amount of unlabeled data, has been a recurring theme in the text mining literature. For example, the method in [141] performs purely unsupervised clustering (with no knowledge of class labels), and then as a final step assigns all documents in the cluster to the *dominant* class label of that cluster (as an evaluation step for the unsupervised clustering process in terms of its ability in matching clusters to known topics).[5] It has been shown that this approach is able to achieve a comparable accuracy of matching clusters to topics as a supervised naive Bayes classifier trained over a small data set of about 1000 documents. Similar results were obtained in [55] where the quality of the unsupervised clustering process was shown to be comparable to an SVM classifier that was trained over a small data set.

---

[5]In a supervised application, the last step would require only a small number of class labels in the cluster to be known to determine the dominant label very accurately.

11.6 Linear Classifiers

Linear Classifiers are those for which the output of the linear predictor is defined to be $p = \overline{A} \cdot \overline{X} + b$, where $\overline{X} = (x_1 \ldots x_n)$ is the normalized document word frequency vector, $\overline{A} = (a_1 \ldots a_n)$ is a vector of linear coefficients with the same dimensionality as the feature space, and b is a scalar. A natural interpretation of the predictor $p = \overline{A} \cdot \overline{X} + b$ in the *discrete scenario* (categorical class labels) would be as a *separating hyperplane* between the different classes. *Support Vector Machines* [36, 145] are a form of classifiers that attempt to determine "good" linear separators between the different classes. One characteristic of linear classifiers is that they are closely related to many *feature transformation methods* (such as the Fisher discriminant), which attempt to use these directions in order to transform the feature space, and then use other classifiers on this transformed feature space [59, 65, 113]. Thus, linear classifiers are intimately related to linear feature transformation methods as well.

Regression modeling (such as the least squares method) is a more direct and traditional statistical method for text classification. However, it is generally used in cases where the target variable to be learned is numerical rather than categorical, though logistic regression [116] is commonly used for text classification (indeed, classification in general). A number of methods have been proposed in the literature for adapting such methods to the case of text data classification [155]. A comparison of different linear regression techniques for classification, including SVM, may be found in [159].

Finally, simple neural networks are also a form of linear classifiers, since the function computed by a set of neurons is essentially linear. The simplest form of neural network, known as the *perceptron* (or single layer network) is essentially designed for linear separation, and works well for text. However, by using multiple layers of neurons, it is also possible to generalize the approach for non-linear separation. In this section, we will discuss the different linear methods for text classification.

11.6.1 SVM Classifiers

Support vector machines are generally defined for binary classification problems. Therefore, the class variable y_i for the ith training instance $\overline{X_i}$ is assumed to be drawn from $\{-1, +1\}$. One advantage of the SVM method is that since it attempts to determine the optimum direction of discrimination in the feature space by examining the appropriate combination of features, it is quite robust to high dimensionality. It has been noted in [74] that text data is ideally suited for SVM classification because of the sparse high-dimensional nature of text, in which few features are irrelevant, but they tend to be correlated with one another and generally organized into linearly separable categories. We note that it is not necessary to use a linear function for the SVM classifier. Rather, with the kernel trick [7], SVM can construct a non-linear *decision surface* in the original feature space by mapping the data instances non-linearly to an inner product space where the classes can be separated linearly with a hyperplane.

The most important criterion, which is commonly used for SVM classification, is that of the *maximum margin hyperplane*. In order to understand this point, consider the case of linearly separable data illustrated in Figure 11.1(a). Two possible separating hyperplanes, with their corresponding *support vectors* and *margins* have been illustrated in the figure. It is evident that one of the separating hyperplanes has a much larger margin than the other, and is therefore more desirable because of its greater generality for unseen test examples. Therefore, one of the important criteria for support vector machines is to achieve maximum margin separation of the hyperplanes.

In general, it is assumed for d dimensional data that the separating hyperplane is of the form $\overline{W} \cdot \overline{X} + b = 0$. Here $\overline{W}$ is a d-dimensional vector representing the coefficients of the hyperplane of separation, and b is a constant. Without loss of generality, it may be assumed (because of appropriate coefficient scaling) that the two symmetric support vectors have the form $\overline{W} \cdot \overline{X} + b = 1$ and $\overline{W} \cdot$

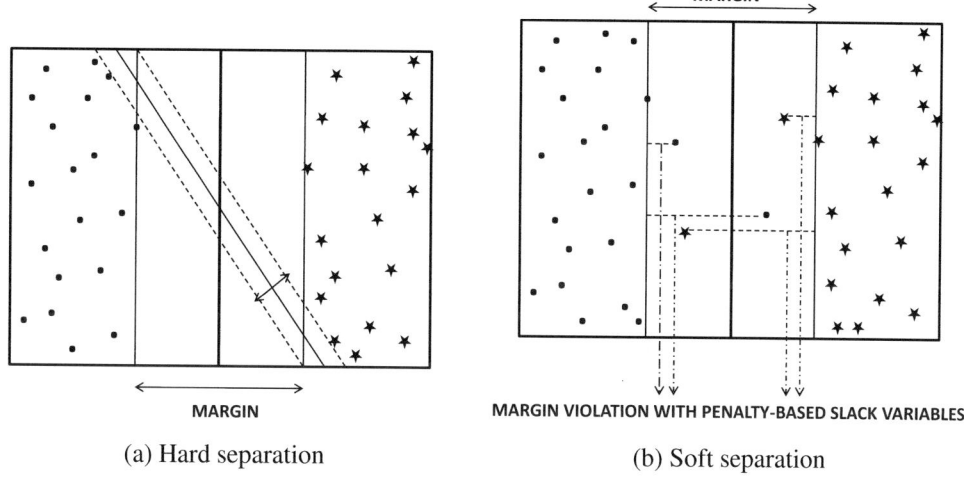

(a) Hard separation (b) Soft separation

FIGURE 11.1: Hard and soft support vector machines.

$\overline{X} + b = -1$. The coefficients $\overline{W}$ and b need to be learned from the training data $\mathcal{D}$ in order to maximize the margin of separation between these two parallel hyperplanes. It can be shown from elementary linear algebra that the distance between these two hyperplanes is $2/||\overline{W}||$. Maximizing this objective function is equivalent to minimizing $||\overline{W}||^2/2$. The constraints are defined by the fact that the training data points for each class are on one side of the support vector. Therefore, these constraints are as follows:

$$\overline{W} \cdot \overline{X_i} + b \geq +1 \quad \forall i : y_i = +1 \tag{11.14}$$

$$\overline{W} \cdot \overline{X_i} + b \leq -1 \quad \forall i : y_i = -1. \tag{11.15}$$

This is a constrained convex quadratic optimization problem, which can be solved using Lagrangian methods. In practice, an off-the-shelf optimization solver may be used to achieve the same goal.

In practice, the data may not be linearly separable. In such cases, soft-margin methods may be used. A slack $\xi_i \geq 0$ is introduced for training instance, and a training instance is allowed to violate the support vector constraint, for a penalty, which is dependent on the slack. This situation is illustrated in Figure 1.2(b). Therefore, the new set of constraints are now as follows:

$$\overline{W} \cdot \overline{X} + b \geq +1 - \xi_i \quad \forall i : y_i = +1 \tag{11.16}$$

$$\overline{W} \cdot \overline{X} + b \leq -1 + \xi_i \quad \forall i : y_i = -1 \tag{11.17}$$

$$\xi_i \geq 0. \tag{11.18}$$

Note that additional non-negativity constraints also need to be imposed in the slack variables. The objective function is now $||\overline{W}||^2/2 + C \cdot \sum_{i=1}^{n} \xi_i$. The constant C regulates the importance of the margin and the slack requirements. In other words, small values of C make the approach closer to soft-margin SVM, whereas large values of C make the approach more of the hard-margin SVM. It is also possible to solve this problem using off-the-shelf optimization solvers.

It is also possible to use transformations on the feature variables in order to design non-linear SVM methods. In practice, non-linear SVM methods are learned using kernel methods. The key idea here is that SVM formulations can be solved using only pairwise dot products (similarity values) between objects. In other words, the optimal decision about the class label of a test instance, from the solution to the quadratic optimization problem in this section, can be expressed in terms of the following:

1. Pairwise dot products of different training instances.

2. Pairwise dot product of the test instance and different training instances.

The reader is advised to refer to [143] for the specific details of the solution to the optimization formulation. The dot product between a pair of instances can be viewed as notion of similarity among them. Therefore, the aforementioned observations imply that it is possible to perform SVM classification, with pairwise similarity information between training data pairs and training-test data pairs. The actual feature values are not required.

This opens the door for using transformations, which are represented by their similarity values. These similarities can be viewed as kernel functions $K(\overline{X}, \overline{Y})$, which measure similarities between the points $\overline{X}$ and $\overline{Y}$. Conceptually, the kernel function may be viewed as dot product between the pair of points in a newly transformed space (denoted by mapping function $\Phi(\cdot)$). However, this transformation does not need to be explicitly computed, as long as the kernel function (dot product) $K(\overline{X}, \overline{Y})$ is already available:

$$K(\overline{X}, \overline{Y}) = \Phi(\overline{X}) \cdot \Phi(\overline{Y}). \tag{11.19}$$

Therefore, all computations can be performed in the original space using the dot products implied by the kernel function. Some interesting examples of kernel functions include the Gaussian radial basis function, polynomial kernel, and hyperbolic tangent, which are listed below in the same order.

$$K(\overline{X_i}, \overline{X_j}) = e^{-||\overline{X_i} - \overline{X_j}||^2 / 2\sigma^2}. \tag{11.20}$$

$$K(\overline{X_i}, \overline{X_j}) = (\overline{X_i} \cdot \overline{X_j} + 1)^h. \tag{11.21}$$

$$K(\overline{X_i}, \overline{X_j}) = \tanh(\kappa \overline{X_i} \cdot \overline{X_j} - \delta). \tag{11.22}$$

These different functions result in different kinds of nonlinear decision boundaries in the original space, but they correspond to a linear separator in the transformed space. The performance of a classifier can be sensitive to the choice of the kernel used for the transformation. One advantage of kernel methods is that they can also be extended to arbitrary data types, as long as appropriate pairwise similarities can be defined.

The first set of SVM classifiers, as adapted to the text domain, were proposed in [74–76]. A deeper theoretical study of the SVM method has been provided in [77]. In particular, it has been shown why the SVM classifier is expected to work well under a wide variety of circumstances. This has also been demonstrated experimentally in a few different scenarios. For example, the work in [49] applied the method to email data for classifying it as spam or non-spam data. It was shown that the SVM method provides much more robust performance as compared to many other techniques such as boosting decision trees, the rule based RIPPER method, and the Rocchio method. The SVM method is flexible and can easily be combined with interactive user-feedback methods [123].

The major downside of SVM methods is that they are slow. Our discussion in this section shows that the problem of finding the best separator is a Quadratic Programming problem. The number of constraints is proportional to the number of data points. This translates directly into the number of Lagrangian relaxation variables in the optimization problem. This can sometimes be slow, especially for high dimensional domains such as text. It has been shown [51] that by breaking a large Quadratic Programming problem (QP problem) into a set of smaller problems, an efficient solution can be derived for the task.

A number of other methods have been proposed to scale up the SVM method for the special structure of text. A key characteristic of text is that the data is high dimensional, but an individual document contains very few features from the full lexicon. In other words, the number of *non-zero* features is small. A number of different approaches have been proposed to address these issues. The first approach [78], referred to as *SVMLight*, shares a number of similarities with [51]. The first approach also breaks down the quadratic programming problem into smaller subproblems. This achieved by using a working set of Lagrangian variables, which are optimized, while keeping the other variables fixed. The choice of variables to select is based on the gradient of the objective function with respect to these variables. This approach does not, however, fully leverage the sparsity of text. A second approach, known as *SVMPerf* [79] also leverages the sparsity of text data. This approach reduces the number of slack variables in the quadratic programming formulation, while increasing the constraints. A cutting plane algorithm is used to solve the optimization problem efficiently. The approach is shown to require $O(n \cdot s)$ time, where n is the number of training examples, and s is the average number of non-zero features per training document. The reader is referred to Chapter 10 on big-data classification, for a detailed discussion of this approach. The SVM approach has also been used successfully [52] in the context of a hierarchical organization of the classes, as often occurs in Web data. In this approach, a different classifier is built at different positions of the hierarchy.

SVM methods are very popular and tend to have high accuracy in the text domain. Even the linear SVM works rather well for text in general, though the specific accuracy is obviously data-set dependent. An introduction to SVM methods may be found in [36, 40, 62, 132, 133, 145]. Kernel methods for support vector machines are discussed in [132].

11.6.2 Regression-Based Classifiers

Regression modeling is a method that is commonly used in order to learn the relationships between real-valued attributes. Typically, these methods are designed for real valued attributes, as opposed to binary attributes. This, however, is not an impediment to its use in classification, because the binary value of a class may be treated as a rudimentary special case of a real value, and some regression methods such as logistic regression can also naturally model discrete response variables.

An early application of regression to text classification is the Linear Least Squares Fit (LLSF) method [155], which works as follows. Suppose the predicted class label is $p_i = \overline{A} \cdot \overline{X_i} + b$, and y_i is known to be the true class label, then our aim is to learn the values of A and b, such that the *Linear Least Squares Fit (LLSF)* $\sum_{i=1}^{n} (p_i - y_i)^2$ is minimized. In practice, the value of b is set to 0 for the learning process. Let P be $1 \times n$ vector of binary values indicating the binary class to which the corresponding class belongs. Thus, if X is the $n \times d$ term-matrix, then we wish to determine the $1 \times d$ vector of regression coefficients A for which $||A \cdot X^T - P||$ is minimized, where $|| \cdot ||$ represents the Froebinus norm. The problem can be easily generalized from the binary class scenario to the multi-class scenario with k classes, by using P as a $k \times n$ matrix of binary values. In this matrix, exactly one value in each column is 1, and the corresponding row identifier represents the class to which that instance belongs. Similarly, the set A is a $k \times d$ vector in the multi-class scenario. The LLSF method has been compared to a variety of other methods [153, 155, 159], and has been shown to be very robust in practice.

A more natural way of modeling the classification problem with regression is the logistic regression classifier [116], which differs from the LLSF method in that the objective function to be optimized is the likelihood function. Specifically, instead of using $p_i = \overline{A} \cdot \overline{X_i} + b$ directly to fit the true label y_i, we assume that the probability of observing label y_i is:

$$p(C = y_i | X_i) = \frac{exp(\overline{A} \cdot \overline{X_i} + b)}{1 + exp(\overline{A} \cdot \overline{X_i} + b).}$$

This gives us a conditional generative model for y_i given X_i. Putting it in another way, we assume that the logit transformation of $p(C = y_i | X_i)$ can be modeled by the linear combination of features of the instance X_i, i.e.,

$$\log \frac{p(C = y_i | X_i)}{1 - p(C = y_i | X_i)} = \overline{A} \cdot \overline{X_i} + b.$$

Thus logistic regression is also a linear classifier as the decision boundary is determined by a linear function of the features. In the case of binary classification, $p(C = y_i | X_i)$ can be used to determine the class label (e.g., using a threshold of 0.5). In the case of multi-class classification, we have $p(C = y_i|X_i) \propto exp(\overline{A} \cdot \overline{X_i} + b)$, and the class label with the highest value according to $p(C = y_i|X_i)$ would be assigned to X_i. Given a set of training data points $\{(X_1, y_i), ...(X_n, y_n)\}$, the logistic regression classifier can be trained by choosing parameters $\overline{A}$ to maximize the conditional likelihood $\prod_{i=1}^{n} p(y_i|X_i)$.

In some cases, the domain knowledge may be of the form where some sets of words are more important than others for a classification problem. For example, in a classification application, we may know that certain domain-words (*Knowledge Words (KW)*) may be more important to classification of a particular target category than other words. In such cases, it has been shown [43] that it may be possible to encode such domain knowledge into the logistic regression model in the form of prior on the model parameters and use Bayesian estimation of model parameters.

It is clear that the regression classifiers are extremely similar to the SVM model for classification. Indeed, since LLSF, Logistic Regression, and SVM are all linear classifiers, they are thus identical at a conceptual level; the main difference among them lies in the details of the optimization formulation and implementation. As in the case of SVM classifiers, training a regression classifier also requires an expensive optimization process. For example, fitting LLSF requires expensive matrix computations in the form of a singular value decomposition process.

11.6.3 Neural Network Classifiers

Neural networks attempt to simulate biological systems, corresponding to the human brain. In the human brain, neurons are connected to one another via points, which are referred to as *synapses*. In biological systems, learning is performed by changing the strength of the synaptic connections, in response to impulses.

This biological analogy is retained in an artificial neural network. The basic computation unit in an artificial neural network is a *neuron* or *unit*. These units can be arranged in different kinds of architectures by connections between them. The most basic architecture of the neural network is a perceptron, which contains a set of input nodes and an output node. The output unit receives a set of inputs from the input units. There are d different input units, which is exactly equal to the dimensionality of the underlying data. The data is assumed to be numerical. Categorical data may need to be transformed to binary representations, and therefore the number of inputs may be larger. The output node is associated with a set of weights $\overline{W}$, which are used in order to compute a function $f(\cdot)$ of its inputs. Each component of the weight vector is associated with a connection from the input unit to the output unit. The weights can be viewed as the analogue of the synaptic strengths in biological systems. In the case of a perceptron architecture, the input nodes do not perform any computations. They simply transmit the input attribute forward. Computations are performed only at the output nodes in the basic perceptron architecture. The output node uses its weight vector along with the input attribute values in order to compute a function of the inputs. A typical function, which is computed at the output nodes, is the signed linear function:

$$z_i = \text{sign}\{\overline{W} \cdot \overline{X_i} + b\}. \tag{11.23}$$

The output is a predicted value of the binary class variable, which is assumed to be drawn from $\{-1, +1\}$. The notation b denotes the bias. Thus, for a vector $\overline{X_i}$ drawn from a dimensionality of d, the weight vector $\overline{W}$ should also contain d elements. Now consider a binary classification problem, in which all labels are drawn from $\{+1, -1\}$. We assume that the class label of $\overline{X_i}$ is denoted by y_i. In that case, the sign of the predicted function z_i yields the class label. Thus, the goal of the approach is to *learn* the set of weights $\overline{W}$ with the use of the training data, so as to minimize the least squares error $(y_i - z_i)^2$. The idea is that we start off with random weights and gradually update them, when a mistake is made by applying the current function on the training example. The magnitude of the update is regulated by a learning rate λ. This update is similar to the updates in gradient descent,

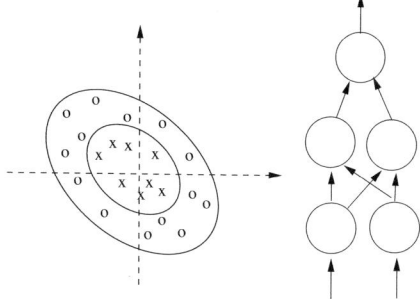

FIGURE 11.2: Multi-layered neural networks for nonlinear separation.

which are made for least-squares optimization. In the case of neural networks, the update function is as follows.

$$\overline{W}^{t+1} = \overline{W}^t + \lambda(y_i - z_i)\overline{X_i} \qquad (11.24)$$

Here, $\overline{W}^t$ is the value of the weight vector in the tth iteration. It is not difficult to show that the incremental update vector is related to the negative gradient of $(y_i - z_i)^2$ with respect to $\overline{W}$. It is also easy to see that updates are made to the weights only when mistakes are made in classification. When the outputs are correct, the incremental change to the weights is zero. The overall perceptron algorithm is illustrated below.

Perceptron Algorithm
Inputs: Learning Rate: λ
      Training Data $(\overline{X_i}, y_i) \; \forall i \in \{1 \ldots n\}$
Initialize weight vectors in $\overline{W}$ and b to small random numbers
repeat
Apply each training data to the neural network to check if the
      sign of $\overline{W} \cdot \overline{X_i} + b$ matches y_i;
if sign of $\overline{W} \cdot \overline{X_i} + b$ does **not** match y_i, then
      update weights $\overline{W}$ based on learning rate λ
until weights in $\overline{W}$ converge

The similarity to support vector machines is quite striking, in the sense that a linear function is also learned in this case, and the sign of the linear function predicts the class label. In fact, the perceptron model and support vector machines are closely related, in that both are linear function approximators. In the case of support vector machines, this is achieved with the use of maximum margin optimization. In the case of neural networks, this is achieved with the use of an incremental learning algorithm, which is approximately equivalent to least squares error optimization of the prediction.

The constant λ regulates the learning rate. The choice of learning rate is sometimes important, because learning rates that are too small will result in very slow training. On the other hand, if the learning rates are too fast, this will result in oscillation between suboptimal solutions. In practice, the learning rates are fast initially, and then allowed to gradually slow down over time. The idea here is that initially large steps are likely to be helpful, but are then reduced in size to prevent oscillation between suboptimal solutions. For example, after t iterations, the learning rate may be chosen to be proportional to $1/t$.

The aforementioned discussion was based on the simple perceptron architecture, which can model only linear relationships. A natural question arises as to how a neural network may be used, if all the classes may not be neatly separated from one another with a linear separator. For example, in Figure 11.2, we have illustrated an example in which the classes may not be separated with the use of a single linear separator. The use of *multiple layers of neurons* can be used in order to induce such non-linear classification boundaries. The effect of such multiple layers is to induce multiple piece-wise linear boundaries, which can be used to approximate enclosed regions belonging to a particular class. In such a network, the outputs of the neurons in the earlier layers feed into the neurons in the later layers. The training process of such networks is more complex, as the errors need to be back-propagated over different layers. Some examples of such classifiers include those discussed in [87, 126, 146, 153]. However, the general observation [135, 149] for text has been that linear classifiers generally provide comparable results to non-linear data, and the improvements of non-linear classification methods are relatively small. This suggests that the additional complexity of building more involved non-linear models does not pay for itself in terms of significantly better classification.

In practice, the neural network is arranged in three layers, referred to as the *input layer*, *hidden layer*, and the *output layer*. The input layer only transmits the inputs forward, and therefore, there are really only two layers to the neural network, which can perform computations. Within the hidden layer, there can be any number of layers of neurons. In such cases, there can be an arbitrary number of layers in the neural network. In practice, there is only one hidden layer, which leads to a two-layer network. The perceptron can be viewed as a very special kind of neural network, which contains only a single layer of neurons (corresponding to the output node). Multilayer neural networks allow the approximation of nonlinear functions, and complex decision boundaries, by an appropriate choice of the network topology, and non-linear functions at the nodes. In these cases, a logistic or sigmoid function, known as a *squashing function*, is also applied to the inputs of neurons in order to model non-linear characteristics. It is possible to use different non-linear functions at different nodes. Such general architectures are very powerful in approximating arbitrary functions in a neural network, given enough training data and training time. This is the reason that neural networks are sometimes referred to as *universal function approximators*.

In the case of single layer perceptron algorithms, the training process is easy to perform by using a gradient descent approach. The major challenge in training multilayer networks is that it is no longer known for intermediate (hidden layer) nodes what their "expected" output should be. This is only known for the final output node. Therefore, some kind of "error feedback" is required, in order to determine the changes in the weights at the intermediate nodes. The training process proceeds in two phases, one of which is in the forward direction, and the other is in the backward direction.

1. *Forward Phase:* In the forward phase, the activation function is repeatedly applied to propagate the inputs from the neural network in the forward direction. Since the final output is supposed to match the class label, the final output at the output layer provides an error value, depending on the training label value. This error is then used to update the weights of the output layer, and propagate the weight updates backwards in the next phase.

2. *Backpropagation Phase:* In the backward phase, the errors are propagated backwards through the neural network layers. This leads to the updating of the weights in the neurons of the different layers. The gradients at the previous layers are learned as a function of the errors and weights in the layer ahead of it. The learning rate λ plays an important role in regulating the rate of learning.

In practice, any arbitrary function can be approximated well by a neural network. The price of this generality is that neural networks are often quite slow in practice. They are also sensitive to noise, and can sometimes overfit the training data.

The previous discussion assumed only binary labels. It is possible to create a k-label neural network, by either using a multiclass "one-versus-all" meta-algorithm, or by creating a neural network architecture in which the number of output nodes is equal to the number of class labels. Each output represents prediction to a particular label value. A number of implementations of neural network methods have been studied in [41, 102, 115, 135, 149], and many of these implementations are designed in the context of text data. It should be pointed out that both neural networks and SVM classifiers use a linear model that is quite similar. The main difference between the two is in how the optimal linear hyperplane is determined. Rather than using a direct optimization methodology, neural networks use a *mistake-driven* approach to data classification [41]. Neural networks are described in detail in [15, 66].

11.6.4 Some Observations about Linear Classifiers

While the different linear classifiers have been developed independently from one another in the research literature, they are surprisingly similar at a basic conceptual level. Interestingly, these different lines of work have also resulted in a number of similar conclusions in terms of the effectiveness of the different classifiers. We note that the main difference between the different classifiers is in terms of the details of the objective function that is optimized, and the iterative approach used in order to determine the optimum direction of separation. For example, the SVM method uses a Quadratic Programming (QP) formulation, whereas the LLSF method uses a closed-form least-squares formulation. On the other hand, the perceptron method does not try to formulate a closed-form objective function, but works with a softer iterative hill climbing approach. This technique is essentially inherited from the iterative learning approach used by neural network algorithms. However, its goal remains quite similar to the other two methods. Thus, the differences between these methods are really at a detailed level, rather than a conceptual level, in spite of their very different research origins.

Another general observation about these methods is that all of them can be implemented with non-linear versions of their classifiers. For example, it is possible to create non-linear decision surfaces with the SVM classifier, just as it is possible to create non-linear separation boundaries by using layered neurons in a neural network [153]. However, the general consensus has been that the linear versions of these methods work very well, and the additional complexity of non-linear classification does not tend to pay for itself, except for some special data sets. The reason for this is perhaps because text is a high dimensional domain with highly correlated features and small non-negative values on sparse features. For example, it is hard to easily create class structures such as that indicated in Figure 11.2 for a sparse domain such as text containing only small non-negative values on the features. On the other hand, the high dimensional nature of correlated text dimensions is especially suited to classifiers that can exploit the redundancies and relationships between the different features in separating out the different classes. Common text applications have generally resulted in class structures that are linearly separable over this high dimensional domain of data. This is one of the reasons that linear classifiers have shown an unprecedented success in text classification.

11.7 Proximity-Based Classifiers

Proximity-based classifiers essentially use distance-based measures in order to perform the classification. The main thesis is that documents which belong to the same class are likely to be close to one another based on similarity measures such as the dot product or the cosine metric [131]. In order to perform the classification for a given test instance, two possible methods can be used:

- We determine the k-nearest neighbors in the training data to the test instance. The majority (or most abundant) class from these k neighbors are reported as the class label. Some examples of such methods are discussed in [31, 63, 155]. The choice of k typically ranges between 20 and

40 in most of the afore-mentioned work, depending upon the size of the underlying corpus. In practice, it is often set empirically using cross validation.

- We perform training data aggregation during pre-processing, in which clusters or groups of documents belonging to the same class are created. A representative meta-document is created from each group. The same k-nearest neighbor approach is applied as discussed above, except that it is applied to this new set of meta-documents (or *generalized instances* [88]) rather than to the original documents in the collection. A pre-processing phase of summarization is useful in improving the efficiency of the classifier, because it significantly reduces the number of distance computations. In some cases, it may also boost the accuracy of the technique, especially when the data set contains a large number of outliers. Some examples of such methods are discussed in [64, 88, 125].

A method for performing nearest neighbor classification in text data is the *WHIRL* method discussed in [31]. The *WHIRL* method is essentially a method for performing soft similarity joins on the basis of text attributes. By *soft* similarity joins, we refer to the fact that the two records may not be exactly the same on the joined attribute, but may be approximately similar based on a pre-defined notion of similarity. It has been observed in [31] that any method for performing a similarity-join can be adapted as a nearest neighbor classifier, by using the relevant text documents as the joined attributes.

One observation in [155] about nearest neighbor classifiers was that feature selection and document representation play an important part in the effectiveness of the classification process. This is because most terms in large corpora may not be related to the category of interest. Therefore, a number of techniques were proposed in [155] in order to learn the associations between the words and the categories. These are then used to create a feature representation of the document, so that the nearest neighbor classifier is more sensitive to the classes in the document collection. A similar observation has been made in [63], in which it has been shown that the addition of weights to the terms (based on their class-sensitivity) significantly improves the underlying classifier performance. The nearest neighbor classifier has also been extended to the temporally-aware scenario [130], in which the timeliness of a training document plays a role in the model construction process. In order to incorporate such factors, a temporal weighting function has been introduced in [130], which allows the importance of a document to gracefully decay with time.

For the case of classifiers that use grouping techniques, the most basic among such methods is that proposed by Rocchio in [125]. In this method, a *single* representative meta-document is constructed from each of the representative classes. For a given class, the weight of the term t_k is the normalized frequency of the term t_k in documents belonging to that class, minus the normalized frequency of the term in documents which do not belong to that class. Specifically, let f_p^k be the expected weight of term t_k in a randomly picked document belonging to the positive class, and f_n^k be the expected weight of term t_k in a randomly picked document belonging to the negative class. Then, for weighting parameters α_p and α_n, the weight $f_{rocchio}^k$ is defined as follows:

$$f_{rocchio}^k = \alpha_p \cdot f_p^k - \alpha_n \cdot f_n^k \tag{11.25}$$

The weighting parameters α_p and α_n are picked so that the positive class has much greater weight as compared to the negative class. For the relevant class, we now have a vector representation of the terms $(f_{rocchio}^1, f_{rocchio}^2 \cdots f_{rocchio}^n)$. This approach is applied separately to each of the classes, in order to create a separate meta-document for each class. For a given test document, the closest meta-document to the test document can be determined by using a vector-based dot product or other similarity metric. The corresponding class is then reported as the relevant label. The main distinguishing characteristic of the Rocchio method is that it creates a single profile of the entire class. This class of methods is also referred to as the *Rocchio framework*. The main disadvantage of this method is that if a single class occurs in multiple disjoint clusters that are not very well connected in the data, then the centroid of these examples may not represent the class behavior very well. This is likely to be a source of inaccuracy for the classifier. The main advantage of this method is its extreme simplicity and efficiency; the training phase is linear in the corpus size, and the number of computations in the testing phase are linear to the number of classes, since all the documents have already been aggregated into a small number of classes. An analysis of the Rocchio algorithm, along with a number of different variations may be found in [74].

In order to handle the shortcomings of the Rocchio method, a number of classifiers have also been proposed [2, 19, 64, 88], which explicitly perform the clustering of each of the classes in the document collection. These clusters are used in order to generate class-specific profiles. These profiles are also referred to as *generalized instances* in [88]. For a given test instance, the label of the closest generalized instance is reported by the algorithm. The method in [19] is also a centroid-based classifier, but is specifically designed for the case of text documents. The work in [64] shows that there are some advantages in designing schemes in which the similarity computations take account of the dependencies between the terms of the different classes.

We note that the nearest neighbor classifier can be used in order to generate a ranked list of categories for each document. In cases where a document is related to multiple categories, these can be reported for the document, as long as a thresholding method is available. The work in [157] studies a number of thresholding strategies for the k-nearest neighbor classifier. It has also been suggested in [157] that these thresholding strategies can be used to understand the thresholding strategies of other classifiers that use ranking classifiers.

11.8   Classification of Linked and Web Data

In recent years, the proliferation of the Web and social network technologies has led to a tremendous amount of document data, which are expressed in the form of linked networks. The simplest example of this is the Web, in which the documents are linked to one another with the use of hyperlinks. Social networks can also be considered a noisy example of such data, because the comments and text profiles of different users are connected to one another through a variety of links. Linkage information is quite relevant to the classification process, because documents of similar subjects are often linked together. This observation has been used widely in the *collective classification literature* [14], in which a subset of network nodes are labeled, and the remaining nodes are classified on the basis of the linkages among the nodes.

In general, a content-based network may be denoted by $G = (N, A, C)$, where N is the set of nodes, A is the set of edges between the nodes, and C is a set of text documents. Each node in N corresponds to a text document in C, and it is possible for a document to be empty, when the corresponding node does not contain any content. A subset of the nodes in N are labeled. This corresponds to the training data. The classification problem in this scenario is to determine the labels of the remaining nodes with the use of the training data. It is clear that both the content and structure can play a useful and complementary role in the classification process.

An interesting method for combining linkage and content information for classification was discussed in [25]. In this paper, a hypertext categorization method was proposed, which uses the content and labels of neighboring Web pages for the classification process. When the labels of all the nearest neighbors are available, a Bayesian method can be adapted easily for classification purposes. Just as the presence of a word in a document can be considered a Bayesian feature for a text classifier, the presence of a link between the target page and a page for which the label is known can be considered a feature for the classifier. The real challenge arises when the labels of all the nearest neighbors are not available. In such cases, a relaxation labeling method was proposed in order to perform the classification. Two methods have been proposed in this work:

- **Fully Supervised Case of Radius One Enhanced Linkage Analysis:** In this case, it is assumed that all the neighboring class labels are known. In such a case, a Bayesian approach is utilized in order to treat the labels on the nearest neighbors as features for classification purposes. In this case, the linkage information is the sole information that is used for classification purposes.

- **When the class labels of the nearest neighbors are not known:** In this case, an iterative approach is used for combining text and linkage based classification. Rather than using the pre-defined labels (which are not available), we perform a first labeling of the neighboring documents with the use of document content. These labels are then used to classify the label of the target document, with the use of *both* the local text and the class labels of the neighbors. This approach is used iteratively for re-defining the labels of both the target document and its neighbors until convergence is achieved.

The conclusion from the work in [25] is that a combination of text and linkage based classification always improves the accuracy of a text classifier. Even when none of the neighbors of the document have known classes, it always seemed to be beneficial to add link information to the classification process. When the class labels of all the neighbors are known, the advantages of using the scheme seem to be quite significant.

An additional idea in the paper is that of the use of *bridges* in order to further improve the classification accuracy. The core idea in the use of a bridge is the use of *2-hop* propagation for link-based classification. The results with the use of such an approach are somewhat mixed, as the accuracy seems to reduce with an increasing number of hops. The work in [25] shows results on a number of different kinds of data sets such as the *Reuters database*, *US patent database*, and *Yahoo!* Since the *Reuters database* contains the least amount of noise, pure text classifiers were able to do a good job. On the other hand, the *US patent database* and the *Yahoo! database* contain an increasing amount of noise, which reduces the accuracy of text classifiers. An interesting observation in [25] was that a scheme that simply absorbed the neighbor text into the current document performed *significantly worse* than a scheme that was based on pure text-based classification. This is because there are often significant cross-boundary linkages between topics, and such linkages are able to confuse the classifier. A publicly available implementation of this algorithm may be found in the *NetKit* tool kit available in [106].

Another relaxation labeling method for graph-based document classification is proposed in [5]. In this technique, the probability that the end points of a link take on a particular pair of class labels is quantified. We refer to this as the *link-class pair probability*. The posterior probability of classification of a node T into class i is expressed as the sum of the probabilities of pairing all possible class labels of the neighbors of T with class label i. We note a significant percentage of these (exponential number of) possibilities are pruned, since only the currently most probable[6] labelings are used in this approach. For this purpose, it is assumed that the class labels of the different neighbors of T (while dependent on T) are independent of each other. This is similar to the naive assumption, which is often used in Bayes classifiers. Therefore, the probability for a particular combination of labels on the neighbors can be expressed as the product of the corresponding link-class pair probabilities. The approach starts off with the use of a standard content-based Bayes or SVM classifier in order to assign the initial labels to the nodes. Then, an iterative approach is used to refine the labels, by using the most probable label estimations from the previous iteration in order to refine the labels in the current iteration. We note that the link-class pair probabilities can

---

[6]In the case of *hard labeling*, the single most likely labeling is used, whereas in the case of *soft labeling*, a small set of possibilities is used.

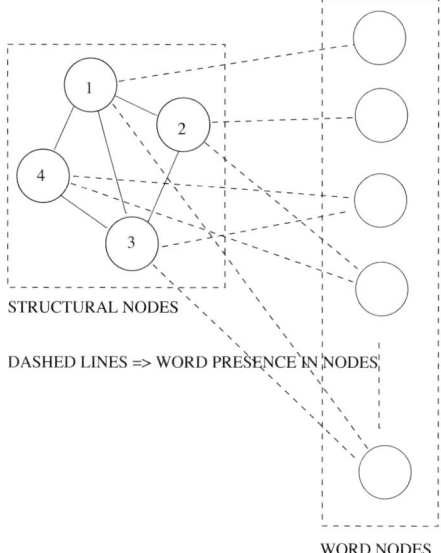

STRUCTURAL NODES

DASHED LINES => WORD PRESENCE IN NODES

WORD NODES

FIGURE 11.3: The Semi-bipartite transformation.

be estimated as the smoothed fraction of edges in the last iteration that contains a particular pair of classes as the end points (hard labeling), or it can also be estimated as the average product of node probabilities over all edges which take on that particular class pair (soft labeling). This approach is repeated to convergence.

Another method that uses a naive Bayes classifier to enhance link-based classification is proposed in [118]. This method incrementally assigns class labels, starting off with a temporary assignment and then gradually making them permanent. The initial class assignment is based on a simple Bayes expression based on both the terms and links in the document. In the final categorization, the method changes the term weights for Bayesian classification of the target document with the terms in the neighbor of the current document. This method uses a broad framework which is similar to that in [25], except that it differentiates between the classes in the neighborhood of a document in terms of their influence on the class label of the current document. For example, documents for which the class label was either already available in the training data, or for which the algorithm has performed a final assignment, have a different confidence weighting factor than those documents for which the class label is currently temporarily assigned. Similarly, documents that belong to a completely different subject (based on content) are also removed from consideration from the assignment. Then, the Bayesian classification is performed with the re-computed weights, so that the document can be assigned a final class label. By using this approach the technique is able to compensate for the noise and inconsistencies in the link structures among different documents.

One major difference between the work in [25] and [118] is that the former is focussed on using link information in order to propagate the labels, whereas the latter attempts to use the content of the neighboring pages. Another work along this direction, which uses the content of the neighboring pages more explicitly, is proposed in [121]. In this case, the content of the neighboring pages is broken up into different fields such as titles, anchor text, and general text. The different fields are given different levels of importance, which is learned during the classification process. It was shown in [121] that the use of title fields and anchor fields is much more relevant than the general text. This accounts for much of the accuracy improvements demonstrated in [121].

The work in [3] proposes a method for dynamic classification in text networks with the use of a random-walk method. The key idea in the work is to *transform* the combination of structure and content in the network into a pure network containing only content. Thus, we transform the original network $G = (N, A, C)$ into an *augmented* network $G^A = (N \cup N_c, A \cup A_c)$, where N_c and A_c are an additional set of nodes and edges added to the original network. Each node in N_c corresponds to a distinct word in the lexicon. Thus, the augmented network contains the original structural nodes N, and a new set of word nodes N_c. The added edges in A_c are undirected edges added between the structural nodes N and the word nodes N_c. Specifically, an edge (i, j) is added to A_c, if the word $i \in N_c$ occurs in the text content corresponding to the node $j \in N$. Thus, this network is *semi-bipartite*, in that there are no edges between the different word nodes. An illustration of the semi-bipartite content-structure transformation is provided in Figure 11.3.

It is important to note that once such a transformation has been performed, any of the collective classification methods [14] can be applied to the structural nodes. In the work in [3], a random-walk method has been used in order to perform the collective classification of the underlying nodes. In this method, repeated random walks are performed starting at the unlabeled nodes that need to be classified. The random walks are defined only on the structural nodes, and each hop may either be a *structural* hop or a *content* hop. We perform l different random walks, each of which contains h nodes. Thus, a total of $l \cdot h$ nodes are encountered in the different walks. The class label of this node is predicted to be the label with the highest frequency of presence in the different $l \cdot h$ nodes encountered in the different walks. The error of this random walk-based sampling process has been bounded in [14]. In addition, the method in [14] can be adapted to dynamic content-based networks, in which the nodes, edges, and their underlying content continuously evolve over time. The method in [3] has been compared to that proposed in [23] (based on the implementation in [106]), and it has been shown that the classification methods of [14] are significantly superior.

Another method for classification of linked text data is discussed in [160]. This method designs two separate regularization conditions; one is for the text-only classifier (also referred to as the *local* classifier), and the other is for the link information in the network structure. These regularizers are expressed in the terms of the underlying kernels; the link regularizer is related to the standard graph regularizer used in the machine learning literature, and the text regularizer is expressed in terms of the kernel gram matrix. These two regularization conditions are combined in two possible ways. One can either use linear combinations of the regularizers, or linear combinations of the associated kernels. It was shown in [160] that both combination methods perform better than either pure structure-based or pure text-based methods. The method using a linear combination of regularizers was slightly more accurate and robust than the method that used a linear combination of the kernels.

A method in [38] designs a classifier that combines a naive Bayes classifier (on the text domain), and a rule-based classifier (on the structural domain). The idea is to invent a set of predicates, which are defined in the space of links, pages, and words. A variety of predicates (or relations) are defined depending upon the presence of the word in a page, linkages of pages to each other, the nature of the anchor text of the hyperlink, and the neighborhood words of the hyperlink. These essentially encode the graph structure of the documents in the form of boolean predicates, and can also be used to construct relational learners. The main contribution in [38] is to combine the relational learners on the structural domain with the naive Bayes approach in the text domain. We refer the reader to [38, 39] for the details of the algorithm, and the general philosophy of such relational learners.

One of the interesting methods for collective classification in the context of email networks was proposed in [29]. The technique in [29] is designed to classify *speech acts* in email. Speech acts essentially characterize whether an email refers to a particular kind of action (such as scheduling a meeting). It has been shown in [29] that the use of sequential thread-based information from the email is very useful for the classification process. An email system can be modeled as a network in several ways, one of which is to treat an email as a node, and the edges as the thread relationships between the different emails. In this sense, the work in [29] devises a network-based mining procedure that uses both the content and the structure of the email network. However, this work is rather specific to the case of email networks, and it is not clear whether the technique can be adapted (effectively) to more general networks.

A different line of solutions to such problems, which are defined on a heterogeneous feature space, is to use latent space methods in order to simultaneously homogenize the feature space, and also determine the latent factors in the underlying data. The resulting representation can be used in conjunction with any of the text classifiers that are designed for latent space representations. A method in [162] uses a matrix factorization approach in order to construct a latent space from the underlying data. Both supervised and unsupervised methods were proposed for constructing the latent space from the underlying data. It was then shown in [162] that this feature representation provides more accurate results, when used in conjunction with an SVM-classifier.

Finally, a method for Web page classification is proposed in [138]. This method is designed for using intelligent agents in Web page categorization. The overall approach relies on the design of two functions that correspond to scoring Web pages and links respectively. An advice language is created, and a method is proposed for mapping advice to neural networks. It is has been shown in [138] how this general purpose system may be used in order to find home pages on the Web.

11.9 Meta-Algorithms for Text Classification

Meta-algorithms play an important role in classification strategies because of their ability to enhance the accuracy of existing classification algorithms by combining them, or making a general change in the different algorithms to achieve a specific goal. Typical examples of classifier meta-algorithms include *bagging*, *stacking*, and *boosting* [50]. Some of these methods change the underlying distribution of the training data, others combine classifiers, and yet others change the algorithms in order to satisfy specific classification criteria. We will discuss these different classes of methods in this section.

11.9.1 Classifier Ensemble Learning

In this method, we use *combinations* of classifiers in conjunction with a voting mechanism in order to perform the classification. The idea is that since different classifiers are susceptible to different kinds of overtraining and errors, a combination classifier is likely to yield much more robust results. This technique is also sometimes referred to as *stacking* or *classifier committee construction*.

Ensemble learning has been used quite frequently in text categorization. Most methods simply use weighted combinations of classifier outputs (either in terms of scores or ranks) in order to provide the final classification result. For example, the work by Larkey and Croft [91] used weighted linear combinations of the classifier scores or ranks. The work by Hull [70] used linear combinations of probabilities for the same goal. A linear combination of the normalized scores was used for classification [158]. The work in [99] used classifier selection techniques and voting in order to provide the final classification result. Some examples of such voting and selection techniques are as follows:

- In a binary-class application, the class label that obtains the majority vote is reported as the final result.

- For a given test instance, a specific classifier is selected, depending upon the performance of the classifiers that are closest to that test instance.

- A weighted combination of the results from the different classifiers are used, where the weight is regulated by the performance of the classifier on validation instances that are most similar to the current test instance.

The last two methods above try to select the final classification in a smarter way by discriminating between the performances of the classifiers in different scenarios. The work by [89] used category-averaged features in order to construct a different classifier for each category.

The major challenge in ensemble learning is to provide the appropriate combination of classifiers for a particular scenario. Clearly, this combination can significantly vary with the scenario and the data set. In order to achieve this goal, the method in [12] proposes a method for probabilistic combination of text classifiers. The work introduces a number of variables known as *reliability variables* in order to regulate the importance of the different classifiers. These reliability variables are learned dynamically for each situation, so as to provide the best classification.

11.9.2   Data Centered Methods: Boosting and Bagging

While ensemble techniques focus on combining different classifiers, data-centered methods such as boosting and bagging typically focus on training the same classifier on different parts of the training data in order to create different models. For a given test instance, a combination of the results obtained from the use of these different models is reported. Another major difference between ensemble-methods and boosting methods is that the training models in a boosting method are not constructed independently, but are constructed sequentially. Specifically, after i classifiers are constructed, the $(i+1)$th classifier is constructed on those parts of the training data that the first i classifiers are unable to accurately classify. The results of these different classifiers are combined together carefully, where the weight of each classifier is typically a function of its error rate. The most well known meta-algorithm for boosting is the *AdaBoost* algorithm [56]. Such boosting algorithms have been applied to a variety of scenarios such as decision tree learners, rule-based systems, and Bayesian classifiers [57, 71, 85, 114, 134, 137].

We note that boosting is also a kind of ensemble learning methodology, except that we train the same model on different subsets of the data in order to create the ensemble. One major criticism of boosting is that in many data sets, some of the training records are noisy, and a classification model should be resistant to overtraining on the data. Since the boosting model tends to weight the error-prone examples more heavily in successive rounds, this can cause the classification process to be more prone to overfitting. This is particularly noticeable in the case of noisy data sets. Some recent results have suggested that all convex boosting algorithms may perform poorly in the presence of noise [103]. These results tend to suggest that the choice of boosting algorithm may be critical for a successful outcome, depending upon the underlying data set.

Bagging methods [21] are generally designed to reduce the model overfitting error that arises during the learning process. The idea in bagging is to pick *bootstrap samples* (samples with replacement) from the underlying collection, and train the classifiers in these samples. The classification results from these different samples are then combined together in order to yield the final result. Bagging methods are generally used in conjunction with decision trees, though these methods can be used in principle with any kind of classifier. The main criticism of the bagging method is that it can sometimes lead to a reduction in accuracy because of the smaller size of each individual training sample. Bagging is useful only if the model is unstable to small details of the training algorithm, because it reduces the overfitting error. An example of such an algorithm would be the decision tree model, which is highly sensitive to how the higher levels of the tree are constructed in a high dimensional feature space such as text. The main goal in bagging methods is to reduce the variance component of the underlying classifier.

11.9.3   Optimizing Specific Measures of Accuracy

We note that the use of the absolute classification accuracy is not the only measure that is relevant to classification algorithms. For example, in skewed-class scenarios, as often arise in the context of applications such as fraud detection, and spam filtering, it is more costly to misclassify examples of one class than another. For example, while it may be tolerable to misclassify a few spam emails (thereby allowing them into the inbox), it is much more undesirable to incorrectly mark a legitimate email as spam. Cost-sensitive classification problems also naturally arise in cases in which one class is more rare than the other, and it is therefore more desirable to identify the rare examples. In such cases, it is desirable to optimize the *cost-weighted accuracy* of the classification process. We note that many of the broad techniques that have been designed for non-textual data [48, 50, 53] are also applicable to text data, because the specific feature representation is not material to how standard algorithms are modified to the cost-sensitive case. A good understanding of cost-sensitive classification both for the textual and non-textual case may be found in [4, 48, 53]. Some examples of how classification algorithms may be modified in straightforward ways to incorporate cost-sensitivity are as follows:

- In a decision-tree, the split condition at a given node tries to maximize the accuracy of its children nodes. In the cost-sensitive case, the split is engineered to maximize the cost-sensitive accuracy.

- In rule-based classifiers, the rules are typically quantified and ordered by measures corresponding to their predictive accuracy. In the cost-sensitive case, the rules are quantified and ordered by their *cost-weighted accuracy*.

- In Bayesian classifiers, the posterior probabilities are weighted by the cost of the class for which the prediction is made.

- In linear classifiers, the optimum hyperplane separating the classes is determined in a cost-weighted sense. Such costs can typically be incorporated in the underlying objective function. For example, the least-square error in the objective function of the LLSF method can be weighted by the underlying costs of the different classes.

- In a k-nearest neighbor classifier, we report the cost-weighted majority class among the k nearest neighbors of the test instance.

We note that the use of a cost-sensitive approach is essentially a change of the objective function of classification, which can also be formulated as an optimization problem. While the standard classification problem generally tries to optimize accuracy, the cost-sensitive version tries to optimize a cost-weighted objective function. A more general approach was proposed in [58] in which a meta-algorithm was proposed for optimizing a specific figure of merit such as the accuracy, precision, recall, or F_1-measure. Thus, this approach generalizes this class of methods to *any arbitrary objective function*, making it essentially an *objective-centered classification method*. A generalized probabilistic descent algorithm (with the desired objective function) is used in conjunction with the classifier of interest in order to derive the class labels of the test instance. The work in [58] shows the advantages of using the technique over a standard SVM-based classifier.

11.10 Leveraging Additional Training Data

In this class of methods, *additional labeled or unlabeled data* are used to enhance classification. While the use of additional unlabeled training data was discussed briefly in the section on Bayes classification, this section will discuss it in more detail. Both these methods are used when there is a direct paucity of the underlying training data. In the case of transfer learning (to be discussed later), additional training (labeled) data from a different domain or problem are used to supervise the classification process. On the other hand, in the case of semi-supervised learning, unlabeled data

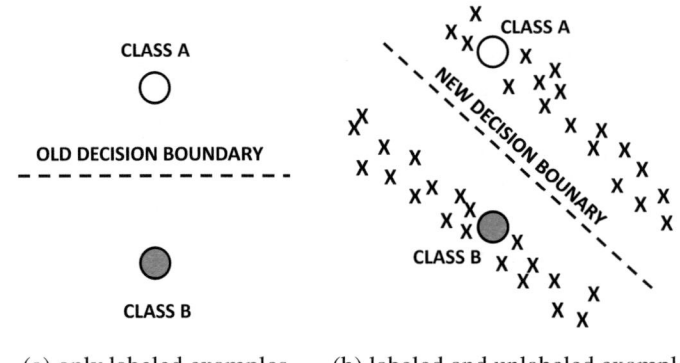

(a) only labeled examples   (b) labeled and unlabeled examples

FIGURE 11.4: Impact of unsupervised examples on classification process.

are used to enhance the classification process. These methods are briefly described in this section. A survey on transfer learning methods may be found in [68].

11.10.1   Semi-Supervised Learning

Co-training and semi-supervised learning methods improve the effectiveness of learning methods with the use of *unlabeled* data [112]. In domains such as text, unlabeled data is copious and freely available from many sources such as the Web. On the other hand, it is much harder to obtain labeled data. The main difference of semi-supervised learning from transfer learning methods is that *unlabeled* data with the same features is used in the former, whereas external labeled data (possibly from a different source) is used in the latter. A key question arises as to why unlabeled data should improve the effectiveness of classification in any way, when it does not provide any additional labeling knowledge. The reason for this is that unlabeled data provides a good idea of the manifolds in which the data is embedded, as well as the density structure of the data in terms of the clusters and sparse regions. The key assumption is that the classification labels exhibit a smooth variation over different parts of the manifold structure of the underlying data. This manifold structure can be used to determine feature correlations, and joint feature distributions, which are very helpful for classification. The semi-supervised setting is also sometimes referred to as the *transductive* setting, when all the unlabeled examples need to be specified together with the labeled training examples.

The motivation of semisupervised learning is that knowledge of the dense regions in the space and correlated regions of the space are helpful for classification. Consider the two-class example illustrated in Figure 11.4(a), in which only a single training example is available for each class. In such a case, the decision boundary between the two classes is the straight line perpendicular to the one joining the two classes. However, suppose that some additional unsupervised examples are available, as illustrated in Figure 11.4(b). These unsupervised examples are denoted by "x". In such a case, the decision boundary changes from Figure 11.4(a). The major assumption here is that the classes vary *less* in dense regions of the training data, because of the *smoothness* assumption. As a result, even though the added examples do not have labels, they contribute significantly to improvements in classification accuracy.

In this example, the *correlations* between feature values were estimated with unlabeled training data. This has an intuitive interpretation in the context of text data, where *joint* feature distributions can be estimated with unlabeled data. For example, consider a scenario where training data is available about predicting whether a document is in the "*politics*" category. It may be possible that the word "*Obama*" (or some of the less common words) may not occur in any of the (small number of) training documents. However, the word "*Obama*" may often co-occur with many features of the "*politics*" category in the unlabeled instances. Thus, the unlabeled instances can be used to learn the relevance of these less common features to the classification process, especially when the amount of available training data is small.

Similarly, when the data are clustered, each cluster in the data is likely to predominantly contain data records of one class or the other. The identification of these clusters only requires unsupervised data rather than labeled data. Once the clusters have been identified from unlabeled data, only a small number of labeled examples are required in order to determine confidently which label corresponds to which cluster. Therefore, when a test example is classified, its clustering structure provides critical information for its classification process, even when a smaller number of labeled examples are available. It has been argued in [117] that the accuracy of the approach may increase exponentially with the number of labeled examples, as long as the assumption of smoothness in label structure variation holds true. Of course, in real life, this may not be true. Nevertheless, it has been shown repeatedly in many domains that the addition of unlabeled data provides significant advantages for the classification process. An argument for the effectiveness of semi-supervised learning, which uses the spectral clustering structure of the data, may be found in [18]. In some domains such as graph data, semisupervised learning is the only way in which classification may be performed. This is because a given node may have very few neighbors of a specific class.

Text classification from labeled and unlabeled documents uses EM. Semi-supervised methods are implemented in a wide variety of ways. Some of these methods directly try to label the unlabeled data in order to increase the size of the training set. The idea is to incrementally add the most confidently predicted label to the training data. This is referred to as *self training*. Such methods have the downside that they run the risk of overfitting. For example, when an unlabeled example is added to the training data with a specific label, the label might be incorrect because of the specific characteristics of the feature space, or the classifier. This might result in further propagation of the errors. The results can be quite severe in many scenarios.

Therefore, semisupervised methods need to be carefully designed in order to avoid overfitting. An example of such a method is *co-training* [17], which partitions the attribute set into two subsets, on which classifier models are independently constructed. The top label predictions of one classifier are used to augment the training data of the other, and vice-versa. Specifically, the steps of co-training are as follows:

1. Divide the feature space into two disjoint subsets f_1 and f_2.

2. Train two independent classifier models $\mathcal{M}_1$ and $\mathcal{M}_2$, which use the disjoint feature sets f_1 and f_2, respectively.

3. Add the unlabeled instance with the most confidently predicted label from $\mathcal{M}_1$ to the training data for $\mathcal{M}_2$ and vice-versa.

4. Repeat all the above steps.

Since the two classifiers are independently constructed on different feature sets, such an approach avoids overfitting. The partitioning of the feature set into f_1 and f_2 can be performed in a variety of ways. While it is possible to perform random partitioning of features, it is generally advisable to leverage redundancy in the feature set to construct f_1 and f_2. Specifically, each feature set f_i should be picked so that the features in f_j (for $j \neq i$) are redundant with respect to it. Therefore, each feature set represents a different view of the data, which is sufficient for classification. This ensures that the "confident" labels assigned to the other classifier are of high quality. At the same time, overfitting is avoided to at least some degree, because of the disjoint nature of the feature set used by the two classifiers. Typically, an erroneously assigned class label will be more easily detected by the disjoint feature set of the other classifier, which was not used to assign the erroneous label. For a test instance, each of the classifiers is used to make a prediction, and the combination score from the two classifiers may be used. For example, if the naive Bayes method is used as the base classifier, then the product of the two classifier scores may be used.

The aforementioned methods are generic meta-algorithms for semi-supervised leaning. It is also possible to design variations of existing classification algorithms such as the EM-method, or transductive SVM classifiers. EM-based methods [117] are very popular for text data. These methods attempt to model the joint probability distributions of the features and the labels with the use of partially supervised clustering methods. This allows the estimation of the conditional probabilities in the Bayes classifier to be treated as missing data, for which the EM-algorithm is very effective. This approach shows a connection between the partially supervised clustering and partially supervised classification problems. The results show that partially supervised classification is most effective when the clusters in the data correspond to the different classes. In transductive SVMs, the labels of the unlabeled examples are also treated as integer decision variables. The SVM formulation is modified in order to determine the maximum margin SVM, with the best possible label assignment of unlabeled examples. The SVM classifier has also been shown to be useful in large scale scenarios in which a large amount of unlabeled data and a small amount of labeled data is available [139]. This is essentially a semi-supervised approach because of its use of unlabeled data in the classification process. This technique is also quite scalable because of its use of a number of modified quasi-Newton techniques, which tend to be efficient in practice. Surveys on semi-supervised methods may be found in [27, 163].

11.10.2 Transfer Learning

As in the case of semi-supervised learning, transfer learning methods are used when there is a direct paucity of the underlying training data. However, the difference from semi-supervised learning is that, instead of using unlabeled data, labeled data from a different domain is used to enhance the learning process. For example, consider the case of learning the class label of Chinese documents, where enough training data is not available about the documents. However, similar English documents may be available, that contain training labels. In such cases, the knowledge in training data for the English documents can be *transferred* to the Chinese document scenario for more effective classification. Typically, this process requires some kind of "bridge" in order to relate the Chinese documents to the English documents. An example of such a "bridge" could be pairs of similar Chinese and English documents, though many other models are possible. In many cases, a small amount of auxiliary training data, in the form of labeled Chinese training documents, may also be available in order to further enhance the effectiveness of the transfer process. This general principle can also be applied to cross-category or cross-domain scenarios where knowledge from one classification category is used to enhance the learning of another category [120], or the knowledge from one data domain (e.g., text) is used to enhance the learning of another data domain (e.g., images) [42, 120, 121]. In the context of text data, transfer learning is generally applied in two different ways:

1. *Crosslingual Learning:* In this case, the documents from one language are transferred to the other. An example of such an approach is discussed in [11].

2. *Crossdomain learning:* In this case, knowledge from the text domain is typically transferred to multimedia, and vice-versa. An example of such an approach, which works between the text and image domain, is discussed in [119, 121].

Broadly speaking, transfer learning methods fall into one of the following four categories:

1. *Instance-based Transfer:* In this case, the feature space of the two domains are highly overlapping; even the class labels may be the same. Therefore, it is possible to transfer knowledge from one domain to the other by simply re-weighting the features.

2. *Feature-based Transfer:* In this case, there may be some overlaps among the features, but a significant portion of the feature space may be different. Often, the goal is to perform a transformation of each feature set into a new low dimensional space, which can be shared across related tasks.

3. *Parameter-Based Transfer:* In this case, the motivation is that a good training model has typically learned a lot of structure. Therefore, if two tasks are related, then the structure can be transferred to learn the target task.

4. *Relational-Transfer Learning:* The idea here is that if two domains are related, they may share some similarity relations among objects. These similarity relations can be used for transfer learning across domains.

The major challenge in such transfer learning methods is that *negative* transfer can be caused in some cases when the side information used is very noisy or irrelevant to the learning process. Therefore, it is critical to use the transfer learning process in a careful and judicious way in order to truly improve the quality of the results. A survey on transfer learning methods may be found in [105], and a detailed discussion on this topic may be found in Chapter 21.

11.10.3 Active Learning

A different way of enhancing the classification process is to focus on *label acquisition* actively during the learning process, so as to enhance the training data. Most classification algorithms assume that the learner is a passive recipient of the data set, which is then used to create the training model. Thus, the data collection phase is cleanly separated out from modeling, and is generally not addressed in the context of model construction. However, data collection is costly, and is often the (cost) bottleneck for many classification algorithms. In active learning, the goal is to collect more labels *during the learning process* in order to improve the effectiveness of the classification process at a low cost. Therefore, the learning process and data collection process are tightly integrated with one another and enhance each other. Typically, the classification is performed in an interactive way with the learner providing well chosen examples to the user, for which the user may then provide labels.

In general, the examples are typically chosen for which the learner has the greatest level of uncertainty based on the current training knowledge and labels. This choice evidently provides the greatest additional information to the learner in cases where the greatest uncertainty exists about the current label. As in the case of semi-supervised learning, the assumption is that unlabeled data are copious, but acquiring labels for them is expensive. Therefore, by using the help of the learner in choosing the appropriate examples to label, it is possible to greatly reduce the effort involved in the classification process. Active learning algorithms often use support vector machines, because the latter is particularly good at determining the boundaries between the different classes. Examples that lie on these boundaries are good candidates to query the user, because the greatest level of uncertainty exists for these examples. Numerous criteria exist for training example choice in active learning algorithms, most of which try to either reduce the uncertainty in classification or reduce the error associated with the classification process. A survey on active learning methods may be found in [136]. Active learning is discussed in detail in Chapter 22.

11.11 Conclusions and Summary

The classification problem is one of the most fundamental problems in the machine learning and data mining literature. In the context of text data, the problem can also be considered similar to that of classification of *discrete set-valued* attributes, when the frequencies of the words are ignored.

The domains of these sets are rather large, as it comprises the entire lexicon. Therefore, text mining techniques need to be designed to effectively manage large numbers of elements with varying frequencies. Almost all the known techniques for classification such as decision trees, rules, Bayes methods, nearest neighbor classifiers, SVM classifiers, and neural networks have been extended to the case of text data. Recently, a considerable amount of emphasis has been placed on linear classifiers such as neural networks and SVM classifiers, with the latter being particularly suited to the characteristics of text data. In recent years, the advancement of Web and social network technologies have led to a tremendous interest in the classification of text documents containing links or other meta-information. Recent research has shown that the incorporation of linkage information into the classification process can significantly improve the quality of the underlying results.

Bibliography

[1] C. C. Aggarwal, C. Zhai. A survey of text classification algorithms, In *Mining Text Data*, pages 163–222, Springer, 2012.

[2] C. C. Aggarwal, S. C. Gates, and P. S. Yu. On using partial supervision for text categorization, *IEEE Transactions on Knowledge and Data Engineering*, 16(2):245–255, 2004.

[3] C. C. Aggarwal and N. Li. On node classification in dynamic content-based networks, *SDM Conference*, 2011.

[4] I. Androutsopoulos, J. Koutsias, K. Chandrinos, G. Paliouras, and C. Spyropoulos. An evaluation of naive Bayesian anti-spam filtering. *Proceedings of the* Workshop on *Machine Learning in the New Information Age*, in conjunction with *ECML Conference*, 2000. http://arxiv.org/PS_cache/cs/pdf/0006/0006013v1.pdf

[5] R. Angelova and G. Weikum. Graph-based text classification: Learn from your neighbors, *ACM SIGIR Conference*, pages 485–492, 2006.

[6] C. Apte, F. Damerau, and S. Weiss. Automated learning of decision rules for text categorization, *ACM Transactions on Information Systems*, 12(3):233–251, 1994.

[7] M. Aizerman, E. Braverman, and L. Rozonoer. Theoretical foundations of the potential function method in pattern recognition learning, *Automation and Remote Control*, 25: 821–837, 1964.

[8] L. Baker and A. McCallum. Distributional clustering of words for text classification, *ACM SIGIR Conference*, pages 96–103, 1998.

[9] R. Bekkerman, R. El-Yaniv, Y. Winter, and N. Tishby. On feature distributional clustering for text categorization, *ACM SIGIR Conference*, pages 146–153, 2001.

[10] S. Basu, A. Banerjee, and R. J. Mooney. Semi-supervised clustering by seeding, *ICML Conference*, pages 27–34, 2002.

[11] N. Bel, C. Koster, and M. Villegas. Cross-lingual text categorization. In *Research and advanced technology for digital libraries*, Springer, Berlin Heidelberg, pages 126–139, 2003.

[12] P. Bennett, S. Dumais, and E. Horvitz. Probabilistic combination of text classifiers using reliability indicators: Models and results, *ACM SIGIR Conference*, pages 207, 214, 2002.

[13] P. Bennett and N. Nguyen. Refined experts: Improving classification in large taxonomies. *Proceedings of the 32nd ACM SIGIR Conference*, pages 11–18, 2009.

[14] S. Bhagat, G. Cormode, and S. Muthukrishnan. Node classification in social networks, In *Social Network Data Analytics*, Ed. Charu Aggarwal, Springer, 2011.

[15] C. Bishop. *Neural Networks for Pattern Recognition*, Oxford University Press, 1996.

[16] D. M. Blei and J. D. McAuliffe. Supervised topic models, *NIPS 2007*.

[17] A. Blum and T. Mitchell. Combining labeled and unlabeled data with co-training, *COLT*, pages 92–100, 1998.

[18] M. Belkin and P. Niyogi. Semi-supervised learning on Riemannian manifolds, *Machine Learning*, 56:209–239, 2004.

[19] D. Boley, M. Gini, R. Gross, E.-H. Han, K. Hastings, G. Karypis, V. Kumar, B. Mobasher, and J. Moore. Partitioning-based clustering for web document categorization, *Decision Support Systems*, 27(3):329–341, 1999.

[20] L. Brieman, J. Friedman, R. Olshen, and C. Stone. *Classification and Regression Trees*, CRC Press, Boca Raton, FL, 1984.

[21] L. Breiman. Bagging predictors, *Machine Learning*, 24(2):123–140, 1996.

[22] L. Cai, T. Hofmann. Text categorization by boosting automatically extracted concepts, *ACM SIGIR Conference*, pages 182–189, 2003.

[23] S. Chakrabarti, S. Roy, and M. Soundalgekar. Fast and accurate text classification via multiple linear discriminant projections, *VLDB Journal*, 12(2):172–185, 2003.

[24] S. Chakrabarti, B. Dom. R. Agrawal, and P. Raghavan. Using taxonomy, discriminants and signatures for navigating in text databases, *VLDB Conference*, pages 446–455, 1997.

[25] S. Chakrabarti, B. Dom, and P. Indyk. Enhanced hypertext categorization using hyperlinks, *ACM SIGMOD Conference*, pages 307–318, 1998.

[26] S. Chakraborti, R. Mukras, R. Lothian, N. Wiratunga, S. Watt, and D. Harper. Supervised latent semantic indexing using adaptive sprinkling, *IJCAI*, pages 1582–1587, 2007.

[27] O. Chapelle, B. Scholkopf, and A. Zien. *Semi-Supervised Learning*, Vol. 2, MIT Press, Cambridge, MA, 2006.

[28] D. Chickering, D. Heckerman, and C. Meek. A Bayesian approach for learning Bayesian networks with local structure, *Thirteenth Conference on Uncertainty in Artificial Intelligence*, pages 80–89, 1997.

[29] V. R. de Carvalho and W. Cohen. On the collective classification of email "speech acts", *ACM SIGIR Conference*, pages 345–352, 2005.

[30] V. Castelli and T. M. Cover. On the exponential value of labeled samples, *Pattern Recognition Letters*, 16(1):105–111, 1995.

[31] W. Cohen and H. Hirsh. Joins that generalize: text classification using WHIRL, *ACM KDD Conference*, pages 169–173, 1998.

[32] W. Cohen and Y. Singer. Context-sensitive learning methods for text categorization, *ACM Transactions on Information Systems*, 17(2):141–173, 1999.

[33] W. Cohen. Learning rules that classify e-mail, *AAAI Conference*, pages 18–25, 1996.

[34] W. Cohen. Learning trees and rules with set-valued features, *AAAI Conference*, pages 709–716, 1996.

[35] W. Cooper. Some inconsistencies and misnomers in probabilistic information retrieval, *ACM Transactions on Information Systems*, 13(1):100–111, 1995.

[36] C. Cortes and V. Vapnik. Support-vector networks, *Machine Learning*, 20(3):273–297, 1995.

[37] T. M. Cover and J. A. Thomas. *Elements of Information Theory*, New York: John Wiley and Sons, 1991.

[38] M. Craven and S. Slattery. Relational learning with statistical predicate invention: Better models for hypertext. *Machine Learning*, 43(1-2):97–119, 2001.

[39] M. Craven, D. DiPasquo, D. Freitag, A. McCallum, T. Mitchell, K. Nigam, and S. Slattery. Learning to extract symbolic knowledge from the worldwide web, *AAAI Conference*, pages 509–516, 1998.

[40] N. Cristianini and J. Shawe-Taylor. *An Introduction to Support Vector Machines and Other Kernel-based Learning Methods*, Cambridge University Press, 2000.

[41] I. Dagan, Y. Karov, and D. Roth. Mistake-driven learning in text categorization, *Proceedings of EMNLP*, pages 55–63, 1997.

[42] W. Dai, Y. Chen, G.-R. Xue, Q. Yang, and Y. Yu. Translated learning: Transfer learning across different feature spaces, *Proceedings of Advances in Neural Information Processing Systems*, 2008.

[43] A. Dayanik, D. Lewis, D. Madigan, V. Menkov, and A. Genkin. Constructing informative prior distributions from domain knowledge in text classification, *ACM SIGIR Conference*, pages 493–500, 2006.

[44] A. P. Dempster, N.M. Laird, and D.B. Rubin. Maximum likelihood from incomplete data via the em algorithm, *Journal of the Royal Statistical Society, Series B*, 39(1): pp. 1–38, 1977.

[45] F. Denis and A. Laurent. Text Classification and Co-Training from Positive and Unlabeled Examples, *ICML 2003 Workshop: The Continuum from Labeled to Unlabeled Data.* http://www.grappa.univ-lille3.fr/ftp/reports/icmlws03.pdf.

[46] S. Deerwester, S. Dumais, T. Landauer, G. Furnas, and R. Harshman. Indexing by latent semantic analysis, *JASIS*, 41(6):391–407, 1990.

[47] P. Domingos and M. J. Pazzani. On the the optimality of the simple Bayesian classifier under zero-one loss, *Machine Learning*, 29(2–3), 103–130, 1997.

[48] P. Domingos. MetaCost: A general method for making classifiers cost-sensitive, *ACM KDD Conference*, pages 155–164, 1999.

[49] H. Drucker, D. Wu, and V. Vapnik. Support vector machines for spam categorization, *IEEE Transactions on Neural Networks*, 10(5):1048–1054, 1999.

[50] R. Duda, P. Hart, and W. Stork. *Pattern Classification*, Wiley Interscience, 2000.

[51] S. Dumais, J. Platt, D. Heckerman, and M. Sahami. Inductive learning algorithms and representations for text categorization, *CIKM Conference*, pages 148–155, 1998.

[52] S. Dumais and H. Chen. Hierarchical classification of web content, *ACM SIGIR Conference*, pages 256–263, 2000.

[53] C. Elkan. The foundations of cost-sensitive learning, *IJCAI Conference*, pages 973–978, 2001.

[54] R. Fisher. The use of multiple measurements in taxonomic problems, *Annals of Eugenics*, 7(2):179–188, 1936.

[55] R. El-Yaniv and O. Souroujon. Iterative double clustering for unsupervised and semi-supervised learning, *NIPS Conference*, pages 121–132, 2002.

[56] Y. Freund and R. Schapire. A decision-theoretic generalization of on-line learning and an application to boosting. In *Proceedings of Second European Conference on Computational Learning Theory*, pages 23–37, 1995.

[57] Y. Freund, R. Schapire, Y. Singer, and M. Warmuth. Using and combining predictors that specialize, *Proceedings of the 29th Annual ACM Symposium on Theory of Computing*, pages 334–343, 1997.

[58] S. Gao, W. Wu, C.-H. Lee, and T.-S. Chua. A maximal figure-of-merit learning approach to text categorization, *SIGIR Conference*, pages 190–218, 2003.

[59] R. Gilad-Bachrach, A. Navot, and N. Tishby. Margin based feature selection – theory and algorithms, *ICML Conference*, pages 43–50, 2004.

[60] S. Gopal and Y. Yang. Multilabel classification with meta-level features, *ACM SIGIR Conference*, pages 315–322, 2010.

[61] L. Guthrie and E. Walker. Document classification by machine: Theory and practice, *COLING*, pages 1059–1063, 1994.

[62] L. Hamel. *Knowledge Discovery with Support Vector Machines*, Wiley, 2009.

[63] E.-H. Han, G. Karypis, and V. Kumar. Text categorization using weighted-adjusted k-nearest neighbor classification, *PAKDD Conference*, pages 53–65, 2001.

[64] E.-H. Han and G. Karypis. Centroid-based document classification: Analysis and experimental results, *PKDD Conference*, pages 424–431, 2000.

[65] D. Hardin, I. Tsamardinos, and C. Aliferis. A theoretical characterization of linear SVM-based feature selection, *ICML Conference*, 2004.

[66] S. Haykin. *Neural Networks and Learning Machines*, Prentice Hall, 2008.

[67] T. Hofmann. Probabilistic latent semantic indexing, *ACM SIGIR Conference*, pages 50–57, 1999.

[68] P. Howland, M. Jeon, and H. Park. Structure preserving dimension reduction for clustered text data based on the generalized singular value decomposition, *SIAM Journal of Matrix Analysis and Applications*, 25(1):165–179, 2003.

[69] P. Howland and H. Park. Generalizing discriminant analysis using the generalized singular value decomposition, *IEEE Transactions on Pattern Analysis and Machine Intelligence*, 26(8):995–1006, 2004.

[70] D. Hull, J. Pedersen, and H. Schutze. Method combination for document filtering, *ACM SIGIR Conference*, pages 279–287, 1996.

[71] R. Iyer, D. Lewis, R. Schapire, Y. Singer, and A. Singhal. Boosting for document routing, *CIKM Conference*, pages 70–77, 2000.

[72] M. James. *Classification Algorithms*, Wiley Interscience, 1985.

[73] D. Jensen, J. Neville, and B. Gallagher. Why collective inference improves relational classification, *ACM KDD Conference*, page 593–598, 2004.

[74] T. Joachims. A probabilistic analysis of the rocchio algorithm with TFIDF for text categorization, *ICML Conference*, pages 143–151, 1997.

[75] T. Joachims. Text categorization with support vector machines: Learning with many relevant features, *ECML Conference*, pages 137–142, 1998.

[76] T. Joachims. Transductive inference for text classification using support vector machines, *ICML Conference*, pages 200–209, 1999.

[77] T. Joachims. A statistical learning model of text classification for support vector machines, *ACM SIGIR Conference*, pages 128–136, 2001.

[78] T. Joachims. Making large scale SVMs practical, In *Advances in Kernel Methods, Support Vector Learning*, pages 169–184, MIT Press, Cambridge, MA, 1998.

[79] T. Joachims. Training linear SVMs in linear time, *KDD*, pages 217–226, 2006.

[80] D. Johnson, F. Oles, T. Zhang and T. Goetz. A decision tree-based symbolic rule induction system for text categorization, *IBM Systems Journal*, 41(3):428–437, 2002.

[81] I. T. Jolliffee. *Principal Component Analysis*, Springer, 2002.

[82] T. Kalt and W. B. Croft. A new probabilistic model of text classification and retrieval, *Technical Report IR-78, University of Massachusetts Center for Intelligent Information Retrieval*, 1996. http://ciir.cs.umass.edu/publications/index.shtml

[83] G. Karypis and E.-H. Han. Fast supervised dimensionality reduction with applications to document categorization and retrieval, *ACM CIKM Conference*, pages 12–19, 2000.

[84] T. Kawatani. Topic difference factor extraction between two document sets and its application to text categorization, *ACM SIGIR Conference*, pages 137–144, 2002.

[85] Y.-H. Kim, S.-Y. Hahn, and B.-T. Zhang. Text filtering by boosting naive Bayes classifiers, *ACM SIGIR Conference*, pages 168–175, 2000.

[86] D. Koller and M. Sahami. Hierarchically classifying documents using very few words, *ICML Conference*, pages 170–178, 2007.

[87] S. Lam and D. Lee. Feature reduction for neural network based text categorization, *DASFAA Conference*, pages 195–202, 1999.

[88] W. Lam and C. Y. Ho. Using a generalized instance set for automatic text categorization, *ACM SIGIR Conference*, pages 81–89, 1998.

[89] W. Lam and K.-Y. Lai. A meta-learning approach for text categorization, *ACM SIGIR Conference*, pages 303–309, 2001.

[90] K. Lang. Newsweeder: Learning to filter netnews, *ICML Conference*, page 331–339, 1995.

[91] L. S. Larkey and W. B. Croft. Combining classifiers in text categorization, *ACM SIGIR Conference*, pages 289–297, 1996.

[92] D. Lewis and J. Catlett. Heterogeneous uncertainty sampling for supervised learning, *ICML Conference*, pages 148–156, 1994.

[93] D. Lewis and M. Ringuette. A comparison of two learning algorithms for text categorization, *SDAIR*, pages 83–91, 1994.

[94] D. Lewis. Naive (Bayes) at forty: The independence assumption in information retrieval, *ECML Conference*, pages 4–15, 1998.

[95] D. Lewis. An evaluation of phrasal and clustered representations on a text categorization task, *ACM SIGIR Conference*, pages 37–50, 1992.

[96] D. Lewis and W. Gale. A sequential algorithm for training text classifiers, *SIGIR Conference*, pages 3–12, 1994.

[97] D. Lewis and K. Knowles. Threading electronic mail: A preliminary study, *Information Processing and Management*, 33(2):209–217, 1997.

[98] H. Li and K. Yamanishi. Document classification using a finite mixture model, *Annual Meeting of the Association for Computational Linguistics*, pages 39–47, 1997.

[99] Y. Li and A. Jain. Classification of text documents, *The Computer Journal*, 41(8):537–546, 1998.

[100] B. Liu, W. Hsu, and Y. Ma. Integrating classification and association rule mining, *ACM KDD Conference*, pages 80–86, 1998.

[101] B. Liu and L. Zhang. A survey of opinion mining and sentiment analysis. In *Mining Text Data*, Ed. C. Aggarwal, C. Zhai, Springer, 2011.

[102] N. Littlestone. Learning quickly when irrelevant attributes abound: A new linear-threshold algorithm. *Machine Learning*, 285–318, 1988.

[103] P. Long and R. Servedio. Random classification noise defeats all convex potential boosters, *ICML Conference*, pages 287–304, 2008.

[104] Y. Lu, Q. Mei, and C. Zhai. Investigating task performance of probabilistic topic models: an empirical study of PLSA and LDA, *Information Retrieval*, 14(2):178-203.

[105] S. J. Pan and Q. Yang. A survey on transfer learning, *IEEE Transactons on Knowledge and Data Engineering*, 22(10):1345–1359, 2010.

[106] S. A. Macskassy and F. Provost. Classification in networked data: A toolkit and a univariate case study, *Journal of Machine Learning Research*, 8(May):935–983, 2007.

[107] A. McCallum. Bow: A toolkit for statistical language modeling, text retrieval, classification and clustering and `http://www.cs.cmu.edu/~mccallum/bow`, 1996.

[108] A. McCallum, K. Nigam. A comparison of event models for naive Bayes text classification, *AAAI Workshop on Learning for Text Categorization*, 1998.

[109] A. McCallum, R. Rosenfeld, and T. Mitchell, A. Ng. Improving text classification by shrinkage in a hierarchy of classes, *ICML Conference*, pages 359–367, 1998.

[110] A.K. McCallum. "MALLET: A Machine Learning for Language Toolkit," http://mallet.cs.umass.edu, 2002.

[111] T. M. Mitchell. *Machine Learning*, WCB/McGraw-Hill, 1997.

[112] T. M. Mitchell. The role of unlabeled data in supervised learning, *Proceedings of the Sixth International Colloquium on Cognitive Science*, 1999.

[113] D. Mladenic, J. Brank, M. Grobelnik, and N. Milic-Frayling. Feature selection using linear classifier weights: Interaction with classification models, *ACM SIGIR Conference*, pages 234–241, 2004.

[114] K. Myers, M. Kearns, S. Singh, and M. Walker. A boosting approach to topic spotting on subdialogues, *ICML Conference*, 2000.

[115] H. T. Ng, W. Goh, and K. Low. Feature selection, perceptron learning, and a usability case study for text categorization, *ACM SIGIR Conference*, pages 67–73, 1997.

[116] A. Y. Ng and M. I. Jordan. On discriminative vs. generative classifiers: A comparison of logistic regression and naive Bayes, *NIPS*. pages 841–848, 2001.

[117] K. Nigam, A. McCallum, S. Thrun, and T. Mitchell. Learning to classify text from labeled and unlabeled documents, *AAAI Conference*, pages 792–799, 1998.

[118] H.-J. Oh, S.-H. Myaeng, and M.-H. Lee. A practical hypertext categorization method using links and incrementally available class information, pages 264–271, *ACM SIGIR Conference*, 2000.

[119] G. Qi, C. Aggarwal, and T. Huang. Towards semantic knowledge propagation from text corpus to web images, *WWW Conference*, pages 297–306, 2011.

[120] G. Qi, C. Aggarwal, Y. Rui, Q. Tian, S. Chang, and T. Huang. Towards cross-category knowledge propagation for learning visual concepts, *CVPR Conference*, pages 897–904, 2011.

[121] X. Qi and B. Davison. Classifiers without borders: incorporating fielded text from neighboring web pages, *ACM SIGIR Conference*, pages 643–650, 2008.

[122] J. R. Quinlan, Induction of decision trees, *Machine Learning*, 1(1):81–106, 1986.

[123] H. Raghavan and J. Allan. An interactive algorithm for asking and incorporating feature feedback into support vector machines, *ACM SIGIR Conference*, pages 79–86, 2007.

[124] S. E. Robertson and K. Sparck-Jones. Relevance weighting of search terms. *Journal of the American Society for Information Science*, 27(3):129–146, 1976.

[125] J. Rocchio. Relevance feedback information retrieval. In *The Smart Retrieval System- Experiments in Automatic Document Processing*, G. Salton, Ed., Prentice Hall, Englewood Cliffs, NJ, pages 313–323, 1971.

[126] M. Ruiz and P. Srinivasan. Hierarchical neural networks for text categorization, *ACM SIGIR Conference*, pages 281–282, 1999.

[127] F. Sebastiani. Machine learning in automated text categorization, *ACM Computing Surveys*, 34(1):1–47, 2002.

[128] M. Sahami. Learning limited dependence Bayesian classifiers, *ACM KDD Conference*, pages 335–338, 1996.

[129] M. Sahami, S. Dumais, D. Heckerman, and E. Horvitz. A Bayesian approach to filtering junk e-mail, *AAAI Workshop on Learning for Text Categorization. Tech. Rep. WS-98-05, AAAI Press.* http://robotics.stanford.edu/users/sahami/papers.html

[130] T. Salles, L. Rocha, G. Pappa, G. Mourao, W. Meira Jr., and M. Goncalves. Temporally-aware algorithms for document classification, *ACM SIGIR Conference*, pages 307–314, 2010.

[131] G. Salton. *An Introduction to Modern Information Retrieval*, Mc Graw Hill, 1983.

[132] B. Scholkopf and A. J. Smola. *Learning with Kernels: Support Vector Machines, Regularization, Optimization, and Beyond*, Cambridge University Press, 2001.

[133] I. Steinwart and A. Christmann. *Support Vector Machines*, Springer, 2008.

[134] R. Schapire and Y. Singer. BOOSTEXTER: A Boosting-based system for text categorization, *Machine Learning*, 39(2/3):135–168, 2000.

[135] H. Schutze, D. Hull, and J. Pedersen. A comparison of classifiers and document representations for the routing problem, *ACM SIGIR Conference*, pages 229–237, 1995.

[136] B. Settles. *Active Learning*, Morgan and Claypool, 2012.

[137] R. Shapire, Y. Singer, and A. Singhal. Boosting and Rocchio applied to text filtering, *ACM SIGIR Conference*, pages 215–223, 1998.

[138] J. Shavlik and T. Eliassi-Rad. Intelligent agents for web-based tasks: An advice-taking approach, *AAAI-98 Workshop on Learning for Text Categorization. Tech. Rep. WS-98-05, AAAI Press*, 1998. http://www.cs.wisc.edu/~shavlik/mlrg/publications.html

[139] V. Sindhwani and S. S. Keerthi. Large scale semi-supervised linear SVMs, *ACM SIGIR Conference*, pages 477–484, 2006.

[140] N. Slonim and N. Tishby. The power of word clusters for text classification, *European Colloquium on Information Retrieval Research (ECIR)*, 2001.

[141] N. Slonim, N. Friedman, and N. Tishby. Unsupervised document classification using sequential information maximization, *ACM SIGIR Conference*, pages 129–136, 2002.

[142] J.-T. Sun, Z. Chen, H.-J. Zeng, Y. Lu, C.-Y. Shi, and W.-Y. Ma. Supervised latent semantic indexing for document categorization, *ICDM Conference*, pages 535–538, 2004.

[143] P.-N. Tan, M. Steinbach, and V. Kumar. *Introduction to Data Mining.* Pearson, 2005.

[144] V. Vapnik. *Estimations of dependencies based on statistical data*, Springer, 1982.

[145] V. Vapnik. *The Nature of Statistical Learning Theory*, Springer, New York, 1995.

[146] A. Weigand, E. Weiner, and J. Pedersen. Exploiting hierarchy in text catagorization. *Information Retrieval*, 1(3):193–216, 1999.

[147] S. M. Weiss, C. Apte, F. Damerau, D. Johnson, F. Oles, T. Goetz, and T. Hampp. Maximizing text-mining performance, *IEEE Intelligent Systems*, 14(4):63–69, 1999.

[148] S. M. Weiss and N. Indurkhya. Optimized rule induction, *IEEE Expert*, 8(6):61–69, 1993.

[149] E. Wiener, J. O. Pedersen, and A. S. Weigend. A neural network approach to topic spotting, *SDAIR*, pages 317–332, 1995.

[150] J. Xu and B. W. Croft, Improving the effectiveness of information retrieval with local context analysis, *ACM Transactions on Information Systems*, 18(1), Jan, 2000. pp. 79–112.

[151] G.-R. Xue, D. Xing, Q. Yang, Y. Yu. Deep classification in large-scale text hierarchies, *ACM SIGIR Conference*, 2008.

[152] J. Yan, N. Liu, B. Zhang, S. Yan, Z. Chen, Q. Cheng, W. Fan, W.-Y. Ma. OCFS: optimal orthogonal centroid feature selection for text categorization, *ACM SIGIR Conference*, 2005.

[153] Y. Yang, L. Liu. A re-examination of text categorization methods, *ACM SIGIR Conference*, 1999.

[154] Y. Yang, J. O. Pederson. A comparative study on feature selection in text categorization, *ACM SIGIR Conference*, 1995.

[155] Y. Yang, C.G. Chute. An example-based mapping method for text categorization and retrieval, *ACM Transactions on Information Systems*, 12(3), 1994.

[156] Y. Yang. Noise Reduction in a Statistical Approach to Text Categorization, *ACM SIGIR Conference*, 1995.

[157] Y. Yang. A Study on Thresholding Strategies for Text Categorization, *ACM SIGIR Conference*, 2001.

[158] Y. Yang, T. Ault, T. Pierce. Combining multiple learning strategies for effective cross-validation, *ICML Conference*, 2000.

[159] J. Zhang, Y. Yang. Robustness of regularized linear classification methods in text categorization, *ACM SIGIR Conference*, 2003.

[160] T. Zhang, A. Popescul, B. Dom. Linear prediction models with graph regularization for web-page categorization, *ACM KDD Conference*, 2006.

[161] C. Zhai, *Statistical Language Models for Information Retrieval (Synthesis Lectures on Human Language Technologies)*, Morgan and Claypool Publishers, 2008.

[162] S. Zhu, K. Yu, Y. Chi, Y. Gong. Combining content and link for classification using matrix factorization, *ACM SIGIR Conference*, 2007.

[163] X. Zhu and A. Goldberg. *Introduction to Semi-Supervised Learning*, Morgan and Claypool, 2009.

Chapter 12

Multimedia Classification

Shiyu Chang

University of Illinois at Urbana-Champaign
Urbana, IL
chang87@illinois.edu

Wei Han

University of Illinois at Urbana-Champaign
Urbana, IL
weihan3@illinois.edu

Xianming Liu

University of Illinois at Urbana-Champaign
Urbana, IL
xliu102@illinois.edu

Ning Xu

University of Illinois at Urbana-Champaign
Urbana, IL
xingxu2@illinois.edu

Pooya Khorrami

University of Illinois at Urbana-Champaign
Urbana, IL
pkhorra2@illinois.edu

Thomas S. Huang

University of Illinois at Urbana-Champaign
Urbana, IL
t-huang1@illinois.edu

12.1   Introduction

Many machine classification problems resemble human decision making tasks, which are multi-modal in nature. Humans have the capability to combine various types of sensory data and associate them with natural language entities. Complex decision tasks such as person identification often heavily depend on such synergy or fusion of different modalities. For that reason, much effort on machine classification methods goes into exploiting the underlying relationships among modalities and constructing an effective fusion algorithm. This is a fundamental step in the advancement of artificial intelligence, because the scope of learning algorithms is not limited to one type of sensory data.

This chapter is about multimedia classification, a decision making task involving several types of data such as audio, image, video (time sequence of images), and text. Although other modalities, e.g., haptic data, may fall into the category of multimedia, we only discuss audio, visual, and text because they represent a broad spectrum of both human decision tasks and machine learning applications.

12.1.1   Overview

Advances in modern digital media technologies has made generating and storing large amounts of multimedia data more feasible than ever before. The World Wide Web has become a huge, diverse, dynamic, and interactive medium for obtaining knowledge and sharing information. It is a rich and gigantic repository of multimedia data collecting a tremendous number of images, music, videos etc. The emerging need to understand large, complex, and information-rich multimedia contents on the Web is crucial to many data-intensive applications in the field of business, science, medical analysis, and engineering.

The Web (or a multimedia database in general) stores a large amount of data whose modality is divided into multiple categories including audio, video, image, graphics, speech, text, document, and hypertext, which contains text markups and linkages. Machine classification on the Web is an entire process of extracting and discovering knowledge from multimedia documents on a computer-based methodology. Figure 12.1 illustrates the general procedure for multimedia data learning.

Multimedia classification and predictive modeling are the most fundamental problems that resemble human decision making tasks. Unlike learning algorithms that focus within a single domain,

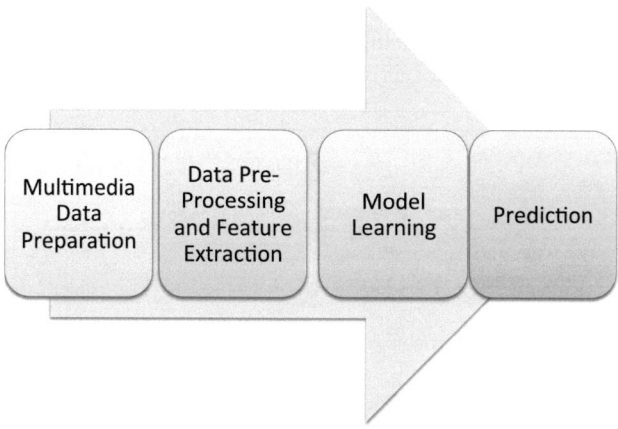

FIGURE 12.1 (See color insert.): Flowchart of general multimedia learning process.

many questions are open in information fusion. Multimedia classification studies how to fuse dependent sources, modalities, or samples efficiently to achieve a lower classification error. Most of the learning algorithms can be applied to multimedia domain in a straightforward manner. However, the specificity of an individual model of data is still very significant in most real-world applications and multimedia system designs. To shed more light on the problems of multimedia classification, we will first introduce the three fundamental types of data including audio, visual, and text.

Audio and visual are essential sensory channels for humans to acquire information about the world. In multimedia classification, the fusion of audio and visual has been extensively studied, and achieved different levels of improvement in tasks including event detection, person identification, and speech recognition, when compared to methods based on only one modality. In this chapter we survey several important related topics and various existing audio-visual fusion methods.

Text is quite different from any sensory modality. It is in the form of natural language, a special medium for human to exchange knowledge. There is an unproven hypothesis in neuroscience that the human brain associates patterns observed in different sensory data to entities that are tied to basic units in natural language [43]. This hypothesis coincides with the recent efforts on *ontology*, a graphical knowledge representation that consists of abstract entities with semantically meaningful links, and learning algorithms to build the connection among natural language, ontology, and sensory data (mostly images).

In this chapter, we mainly focus on decision making from multi-modality of data in an application-oriented manner instead of illustrating different learning algorithms. The remainder of this chapter is organized as follows. In Section 12.2, we illustrate commonly used features for the three fundamental data types. Audio-visual fusion methods and potential applications are introduced in Section 12.3. Section 12.4 explains how to use ontological structure to enhance multimedia machine classification accuracy. We then discuss geographical classification using multimedia data in Section 12.5. Finally, we conclude this chapter and discuss future research directions in the field of Multimedia.

12.2 Feature Extraction and Data Pre-Processing

Existing learning algorithms in a variety of fields frequently rely on vectorized data, because each sample input can be easily expressed mathematically as an individual point abstracted in some

Euclidean space. Although in the modern digital world, data can be naturally represented as vectors, finding discriminative features in a meaningful feature space is extremely important. We call the process of studying meaningful data representation and discovering knowledge from the data as feature extraction.

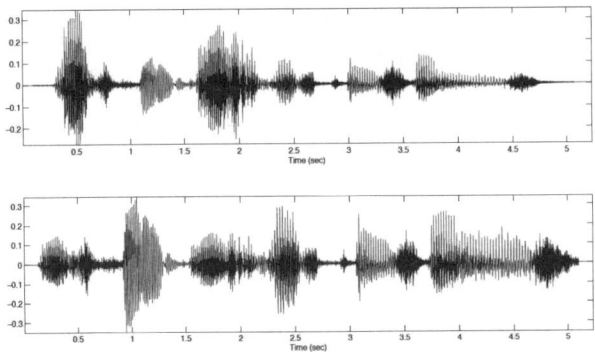

FIGURE 12.2 (See color insert.): Time-domain waveform of the same speech from different persons [80].

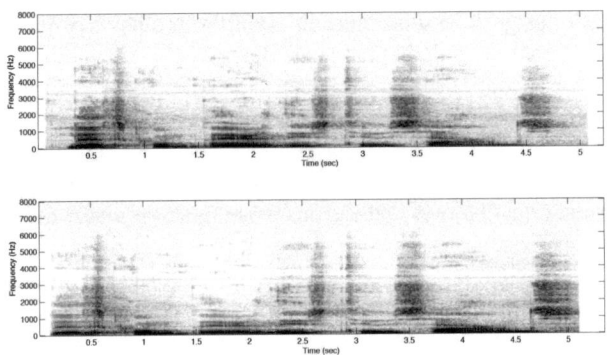

FIGURE 12.3 (See color insert.): Frequency response of the same speech in Figure 12.2 [80].

In general, good features can make the learning problem trivial. Conversely, bad representations will make the problem unsolvable in the extreme case. A simple example in the context of speech recognition is illustrated in Figure 12.2. Both waveforms are exactly the same sentence spoken by different people. One can easily observe large deviations between the two, which make it difficult to visually determine whether the two signals indicate the same speech without the explicit help of professionals in the field. However, if we transfer them into the frequency domain using a short time Fourier Transform, the harmonics shown in Figure 12.3 on different frequencies over time are almost identical to each other. This simple example suggests that raw signal data is rarely used as input for machine classification algorithms. We now briefly introduce some common features that have been widely used in the multimedia community. This simple example demonstrates how raw signal data is quite limited in its ability to model the content that is relevant for machine classification algorithms.

12.2.1 Text Features

Text data preprocessing and feature extraction are very important for text categorization, a conventional classification task that assigns pre-defined categories to text content. However, unlike images and audios, text documents do not have a natural feature representation. To alleviate the problem, the most popular and the simplest feature is a bag-of-words, which vectorizes the text documents. Bag-of-words feature extraction uses the relative frequency of occurrences of a set of keywords as a quantitative measure to convert each text document to a vector. The reason for using relative frequencies instead of absolute word counts is to reduce the effect of length difference among documents.

Furthermore, feature extraction usually involves some data preprocessing steps such as data cleaning and normalization. By removing unwanted details and "normalizing" data, learning performances can be significantly improved. For instance, in text documents, punctuation symbols and non-alphanumeric characters are usually filtered because they do not contribute to feature discrimination. Moreover, words that occur too frequently such as "the," "a," "is," and "at" are also discarded because of their frequent prevalence in documents. The most common way to remove such common English words is by applying "stop-list" at the preprocessing stage. Another common trick to apply for data preprocessing is variant unifying. Variant refers to different forms of the same word, for example, "go," "goes," "went," "gone," and "going." This can be solved by *stemming*. This corresponds to the replacement of these words by a standard one. More sophisticated methods can not only be applied to the words with the same semantic meaning, but also can improve discrimination by considering word co-occurrence. The interested reader can refer to Chapter 10 in this book.

12.2.2 Image Features

Compared to text features, image features are usually large. Consider a modern image taken from a smartphone, in which the typical resolution is five mega-pixels (2592×1936). Without any pre-processing or compression, each image will be vectorized into a $5,018,112$-dimensional Euclidean space. Typical large-scale image classification problems deal with millions or tens of millions of image samples. The storage for data of this size is nearly infeasible for most commodity desktop hardware. Moreover, such a huge dimensionality will introduce the curse of dimensionality. An increase in the number of parameters leads to instability and overfitting problems in model estimation [7]. Image data preprocessing and feature extraction are usually application-specific. Common techniques for preprocessing image data include downsampling, converting RGB data to 256-level gray scale, and normalizing data vectors to unit length. For particular applications such as face recognition or object recognition, cropping the desired region and aligning each face/object image will boost the classification performance significantly.

One unique problem in image classification compared to text is the so-called semantic ambiguity or semantic gap. Machine classification uses high level concepts or labels to categorize different data. A good semantic understanding of the feature space is critical for classification. Text is a form of natural language, and a special medium for humans to exchange knowledge. Therefore, the semantics of the feature representation is clear. Unfortunately, in the case of image data, the semantic knowledge is hard to interpret. In other words, samples with similar high level features may be further apart, while samples from different classes may cluster together [53], [83].

In the field of image processing, computer vision and pattern recognition, bridging the semantic gap between high-level concepts and low-level feature representations has been studied for a long time. The most effective way is to extract features in an application-specific way. There is no such feature extraction method that will work *universally* for all classification topics. Some example features include color, shape, texture and spatial layout information. A simple example is classifying "sky" and "sea." These two concepts are difficult to distinguish due to their resemblance in color. However their spatial layout in the image can be a useful tool for classification: sky often appears at the top of image, while sea is at the bottom.

Another characteristic difficulty for image categorization is that samples from the same class usually contain large variations. A typical example is that of images of a person's face. Humans can easily distinguish between different people without having seen the person before. However, this rather simple task is far more difficult for computer-based algorithms. The images of the same person in a database might have been taken under different conditions including viewpoint, lighting condition, image quality, occlusion, resolution, etc. Figure 12.4 illustrates huge differences for images of a previous US President.

FIGURE 12.4 (See color insert.): Example images of President George W. Bush showing huge intra-person variations.

Many sophisticated feature extraction and machine classification methods have been proposed in the image domain in recent years. In this subsection, we will mainly focus on the four fundamentals: color, texture, shape, and spatial relation.

1. **Color:** Color features are widely used in image classification to capture the spatial color distribution of each given image. Colors are defined on different color spaces. Commonly used color spaces include RGB, HSV, YCrCb, etc. These color spaces are very close to human perception, and detailed description on different color spaces can be found in [74]. Color-covariance, color histogram, color moments, and color coherence are all commonly used color features defined on these color spaces.

   One of the most widely used features in image classification is color. Images represent color data using color spaces such RGB, HSV, or YCrCb, each of which closely models human perception. A more detailed description of the different color spaces can be found in [74]. These spaces are often used when constructing color histograms or modeling color-covariance, color moments, and color coherence.

2. **Texture:** Texture features are designed for capturing a specific pattern appearing in an image. For objects containing certain textures on its surface (i.e. fruit skin, clouds, trees, bricks, etc.), the extracted features provide important information for image classification. Some important texture features include Gabor filtering [100], wavelet transform [98], and other local statistical features [90], [50]. Some easily computable textural features based on graytone spatial dependencies can be found in [31]. There applications have been shown in category identification, photo-micrographs, panchromatic aerial photographs, and multi-spectral imagery.

3. **Shape:** The shape is a well-defined concept for extracting the information of desired objects in images. Shape features include Fourier descriptors, aspect ratio, circularity, moment invariants, consecutive boundary segments, etc. [65]. Despite the importance of shape features for classifying different objects, when compared with color and texture feature, shape is less efficient due to the difficulty of obtaining an accurate segmentation.

4. **Spatial Relation:** Besides the aforementioned image features, spatial location is also useful in region classification. Consider the previous example of an object's spatial layout information for effective classification. Most of the existing work simply defines spatial location as "upper," "bottom," "top," according to where the object appears in an image [34], [68]. If an image contains more than one object, relative spatial relationship is more important than absolute spatial location in deriving semantic features. The most common structure used to represent directional relationship between objects such as "left/right," "up/down," is 2D-string [17].

In recent years, the most efficient image features have included HOG (Histogram of Oriented Gradients) [22], SIFT (Scale Invariant Feature Transform) [54], and its variations [6,10,67]. They are local features, which characterize the visual content within a local range instead of the whole images. They are widely used in object detection, recognition, and retrieval. We briefly introduce these two popular features here.

1. **HOG:** The HOG is a descriptor that characterizes the object appearance and shape by evaluating the distribution of intensity gradients or edge directions. For an image, it is first divided into small spatial regions (called "cells"). For each cell, a 1-D histogram of gradient directions or edge orientations over the pixels of the cell is calculated [22]. The descriptor is then constructed by the combinations of these histograms, which encodes the local spatial information. To further improve the accuracy, the local histogram can be contrast-normalized by accumulating a measure of the intensity (referred as "energy") over a larger spatial region (called "block") of the image. In implementation, the cells and blocks on dense grids are generated by sliding windows at different scales.

   The key idea behind HOG is to find a "template" of the average shape for a certain type of targets (e.g., pedestrian), by "viewing" and picking up all the possible examples. The HOG descriptor works well in various tasks, due to its high computational efficiency. It can capture the local shapes, and the tolerance in translation and contrast. It is widely used in human face detection and object recognition.

2. **SIFT:** Scale Invariant Feature Transform [54] is the most widely used image feature in image classification, object detection, image matching, etc.. It detects the key points (referred to as "interest points") in an image, and constructs a scale and rotation invariant descriptor for each interest point.

   SIFT corresponds to the detector and descriptor. First, in the detector, interesting points (also referred to as "key points") are obtained by searching the points at which the DoG (difference of Gaussian) values assume extrema with respect to both the spatial coordinates in the image domain and the scale level in the scale space, inspired by the scale selection in scale space [48]. At each interest point detected, a position-dependent descriptor is calculated similarly to the HOG, but in a more sophisticated way.

   To achieve the scale invariance, the size of the local neighborhood (corresponds to the "block" size in HOG) is normalized according to the scale. To achieve the rotation invariance, the principal orientation is determined by counting all the orientations of gradients within its neighborhood, which is used to orient the grid. Finally, the descriptor is calculated with respect to the principal orientation. Figure 12.5 shows an example of the SIFT detectors and the matching between two different images.

   One of the most important extensions of SIFT is the Dense version of SIFT features [10]. Instead of detecting interest points in the first step, it samples points densely (e.g., in uniform grids) from the image, and assigns a scale for each point alternatively. It has been reported that in some situations Dense SIFT outperforms the original SIFT feature.

FIGURE 12.5 (See color insert.): Samples of SIFT detector from [51]: interest points detected from two images of a similar scene. For each point, the circle indicates the scale. It also shows the matching between the similar SIFT points in the two images.

12.2.3 Audio Features

Unlike text and images, audio is a continuous media type. Audio data can be in the form of radio, speech, music, or spoken language. For the purpose of audio analysis and classification, low-level audio features of the sound signals should first be extracted. The frequency based features for audio classification can be grouped into three different categories: time-domain features, frequency domain features, and psycho-acoustic features [97].

1. **Time-domain features**:

   (a) *Short-time energy*: The energy of audio waveform within a frame.

   (b) *Energy statistics*: The energy statistics of a particular audio waveform usually includes mean, standard deviation, dynamic range, etc.

   (c) *Silence ration*: The percentage of low-energy audio frames.

   (d) *Zero crossing rate (ZCR)*: ZCR is a commonly used temporal feature. It counts the number of times audio waves across zero axis within a frame.

   (e) *Pause rate*: The rate of stopping rate of speech due to separation of sentences.

2. **Frequency-domain features**:

   (a) *Pitch*: It measures the fundamental frequency of audio signals.

   (b) *Subband energy ratio*: It is a histogram-like energy distribution over different frequencies.

   (c) *Spectral statistics*: The most important spectral statistics are frequency centroid (FC) and bandwidth (BW). FC indicates weighted average of all frequency components of an audio frame. For a particular frame, BW is the weighted average of the squared difference between each frequency and its frequency centroid.

3. **Psycho-acoustic features**:

   (a) *Four-Hz modulation energy*: Usually, speech contains a characteristic energy modulation peak around the four hertz syllabic rate. It calculates the energy of four-Hz modulation.

   (b) *Spectral roll-off point*: The spectral roll-off point is used to identify voiced and unvoiced speech by calculating the 95 percent of the power spectrum.

Audio and visual data can be coherent in a unified learning framework. Utilizing different modalities of data could potentially enhance classification results. It is also related to video classification, since video itself contains both a visual and an audio component. The detailed methods will be discussed later in this chapter.

12.2.4 Video Features

Video classification is even more complicated than images, mainly because of the multiple data modalities involved. Video is typically treated as a continuous media composed of moving images, audios, and texts (such as scripts). A heuristic approach for video classification is to treat the visual and audio parts separately, and use fusion algorithms to combine the results into a final decision. For some types of audios such as news reports, and movies, the text information is also useful (e.g., subtitles, scripts). For instance, in the sports video event detection, both the visual (the detections of items such as the goal, players, ball, lines) and audio (e.g., the cheers from audience, the whistling from the referees) features are extracted and utilized in classification.

Without loss of generality we focus on the visual features widely used in video classifications in this subsection. Most visual features of images can be applied to videos, including color, textures, et al. As the popularity of local features, the combination of Bag-of-Word and SIFT descriptors is widely used in video classification, retrieval, and event detection [84], due to its high efficiency and robustness.

Moreover, the most important video information is the temporal axis. The movement in the spatial-temporal space forms motion. On the micro, various motion features have been proposed to capture the local movements of objects, e.g., Optical Flow [33] and STIP [44]. The behaviors are decomposed into basic movements consisting of motions. At the macro-level, the trajectory is widely adopted to classify events and analyze behaviors [96]. A detailed survey on the features used in automatic video classification is presented in [13].

12.3 Audio Visual Fusion

The idea of fusing audio and visual information is not a new one. In the early seventies, research was conducted on designing speech recognition systems with visual input. Later, the idea evolved into a rich field of study. The main motivation of audio visual fusion lies behind the bimodal nature of human perception, particularly speech perception. Demonstrated in early studies [88], humans utilize both audio and visual signals for decoding the speech contents. The McGurk effect found in [62], where the subjects tend to perceive the phoneme /da/ when the actual sound is /ga/ on a video of people saying /ba/, further implied the existence of a fusion mechanism. In speech recognition, the present of visual speech proved helpful for human subjects to interpret speech signals especially in noisy environments [88]. A widely accepted explanation is that the two information sources, audio and visual signals compensate each other to decode the correct speech, since ambiguities of different phonemes exist in each individual modality. For example, people often find it difficult to distinguish between the consonants /b/ and /d/ on the phone, i.e., with acoustic signal alone in the communication. Such confusion can be easily disambiguated with the visibility of lip movement. As the acoustic ambiguity goes higher with environmental noise, the visual signal becomes more beneficial for speech recognition. Beside speech recognition, fusion is also beneficial for other tasks in which the audio and visual information are correlated, e.g., person identification, event detection, and recognition.

In this section we describe the core fusion algorithms and briefly introduce some important applications.

12.3.1 Fusion Methods

The fusion of multiple information for decision making is a general research problem in its own right. For instance, in the case of Audio Visual Speech Recognition (AVSR), two modalities provide two information streams about the speech. While the decision can be made on each single modality, the goal of fusion is to have better decisions than both single modality-based decision makers. There is a theoretical justification of information fusion from a statistical learning point of view in [55].

Existing fusion approaches can be grouped into two categories: feature-level fusion and decision-level fusion, sometimes also referred to as early fusion and late fusion, respectively. The difference lies in the stage at which fusion occurs in the overall classification procedure. Feature-level fusion occurs at an early stage. It combines the audio and visual information into a joint feature space before the classification. On the other hand, decision-level fusion relies on multiple classifiers for each modality and combines the decisions into one.

Feature-level fusion searches for a transformation from the two individual feature spaces to a joint one, usually with a lower number of dimensions. The transformation is restricted to be linear in many existing methods, including plain feature concatenation [2], addition or multiplication weighting [16], followed by a linear dimensionality reduction, e.g., PCA or LDA. Nonlinear methods are often based on kernel learning. Recently, there is a new trend of using deep learning architectures for a nonlinear feature transformation, which has advanced the state of the art on several benchmarks [72].

Decision-level fusion has the advantage that it explicitly models the reliability of each modality. For different speech content and noise levels, the discriminative power of audio and visual information can be quite different. It is important to model reliability as well. The weighting of the likelihood or scores returned from unimodal classifiers is a common approach. In [55], a probabilistic justification is provided for weighting in the log likelihood domain, with the assumption of a mismatch between training and testing data. In practice, the weights can be set as equal [30], or tuned on validation dataset. Better performance is often achieved if the weights are adaptive (i.e., dynamically adjusted to different environments).

12.3.2 Audio Visual Speech Recognition

Automatic speech recognition (ASR) has been intensively studied in the past three decades. Recently, with the emergence of deep learning methods, significant improvements have been made to the performance of ASR in benchmark datasets and real world applications. However, current state-of-the art ASR systems are still sensitive to the vast variations in speech, including the noise from different channels and environments. The robustness of ASR systems under noisy environments can be improved by additional signals that correlate with the speech. The most informative source among these signals is visual speech: the visual appearance of speech articulators, e.g., the lips. In the literature, ASR with additional visual speech input is refereed to as Audio Visual Speech Recognition (AVSR).

The structure of a typical AVSR system is depicted in Figure 12.6 as a flow chart. The raw audio/video signals from sensors first go though the acoustic and visual front end, respectively. The goal of the front end is to transform the signal into a more compact representation without losing much of the information (that is helpful for decoding the speech). At this point, information fusion is performed on the two modalities, and the combined information will be fed into the speech recognizer. Among the large number of AVSR systems in the past two decades, there are three major aspects [75] in which they differ: the visual front end, the fusion strategy, and the speech recognition method. Note that the acoustic front end is not mentioned here, because most AVSR systems adopt commonly used acoustic features in speech recognition. Similarly, the speech recognition methods are used directly from the speech recognition community, and differ only on different applications, e.g., small or large vocabulary recognition. In this section, we concentrated on large vocabulary speech recognition, because it is the more challenging task.

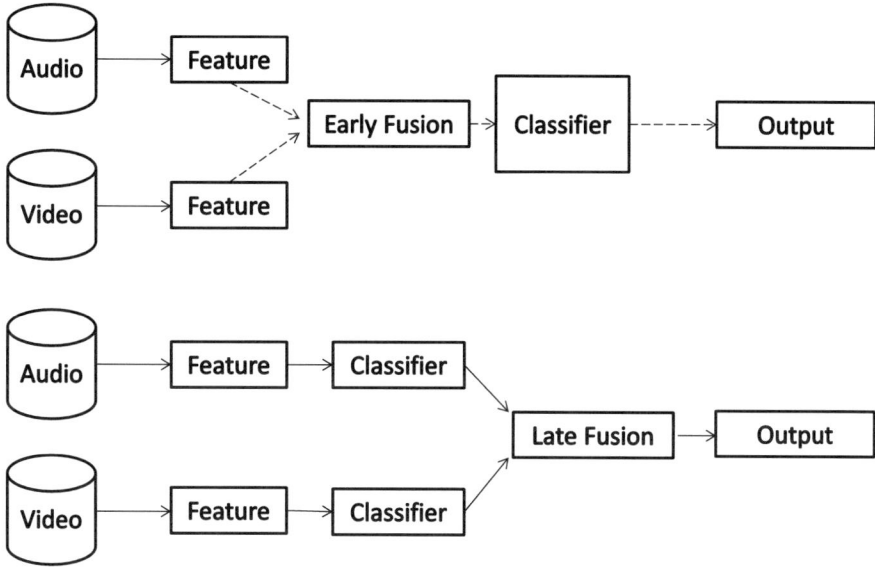

FIGURE 12.6: Flowchart of typical AVSR systems.

12.3.2.1 Visual Front End

The visual front end consists of a two-stage processing of the images. First, pose normalization is needed to extract the region of interest from facial images. Typically, the mouth region is highly correlated to speech. Then, the normalized images are further transformed into the desired feature representation. In the literature, a large variety of image processing and computer vision methods have been applied in the visual front end for AVSR. Some examples are face/mouth/lips detection and tracking, shape and appearance modeling for face/mouth, and global/local image features. The diversity of front end methods increase the difficulty in comparing different AVSR systems. So far, no single method has been proven to be strictly superior to others.

Normalization

Most pose normalization methods in AVSR extract lips from the image. The algorithms for lips localization mainly fall into two categories [76]. The first category makes use of the prior information about the whole face. Two representative methods are active shape model (ASM) and active appearance model (AAM) [19]. While the ASM represents the face as a set of landmarks, AAM further includes the textures inside the face boundary. These methods can be extended to 3D [24] to further normalize the out-of-plane head rotations. The other category only uses the local appearance around the lips [1]. The informative aspects of lips are the shape of outer and inner contours, and the image appearance inside the contours. Existing AVSR front ends can be shape-based [56], appearance-based (pixel) [61], or a combination of the two.

Robust pose normalization and part detection are, in general, difficult tasks. This is especially true in uncontrolled environments with varying illumination and background. Most of the current AVSR research simplifies the normalization stage by controlled environment and human labeling. For example, many audio visual speech corpora are recorded in limited pose variation and constant lighting settings [66]. The cropping window for mouth regions is usu-

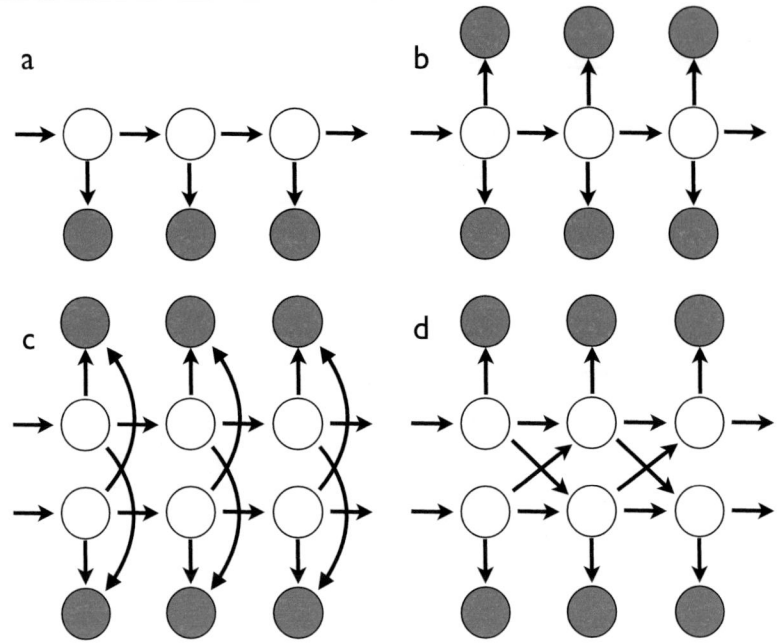

FIGURE 12.7: Graphical model illustration of audio visual models, a): HMM, b): MSHMM, c): FHMM, d): CHMM.

ally provided, while other regions of the faces can be ignored in AVSR. The simplification of facial image processing helps AVSR researchers concentrate on feature design and audio visual fusion, and enable meaningful benchmarks for different methods.

Features The raw features after normalization can be classified as shape-based and appearance-based. Frequently used shape features are the geometric property of the lip contour, and model parameters of a parametric lip model. Appearance-based features take raw pixels from the region of interest from normalized face images.

As the typical dimension of image pixels is too large for the statistical modeling in a traditional speech recognizer, feature transforms are adopted to reduce the dimension. Some popular approaches are Principal Component Analysis (PCA) [26], Discrete Cosine Transform (DCT) [71], wavelet [77], Linear Discriminative Analysis (LDA) [61], etc. Although dimensionality reduction is not necessary for the contour based approaches, PCA is often applied for the purpose of more compact features. The combination of shape and appearance features can be viewed as another multimodal fusion task: each feature carries different information of the same source. In AVSR systems, direct concatenation is the commonly used approach [16].

12.3.2.2 Decision Fusion on HMM

A special property of the fusion for AVSR is that one needs to calculate the likelihood of a sequence instead of independent data samples. Therefore, assumptions about the temporal relationship between audio and visual streams are needed. Previous studies have found a lack of synchrony in audio visual signals of speech, because speech articulators often start to move to a ready position before the actual production of the sound [12]. Taking this into account, the two-stream audio visual

Hidden Markov Model is usually loosened to allow state asynchrony. The various types of HMMs proposed for the audio visual fusion have been summarized in [70] using the notion of a graphical model, as shown in Figure 12.7.

Theoretically, audio visual HMM is a special case of Dynamical Bayesian Model (DBN) with two observation streams. Various decision fusion methods differ in the statistical dependency among observation and hidden state streams, e.g., the edges in the DBN graph. The audio visual product HMM [70] assumes one hidden state stream and two conditional independent observation streams. Therefore, the states are synchronous, and the likelihoods are multiplied together. Factorial HMM [29] uses one hidden state stream for each modality but the two streams are assumed to be independent. The coupled HMM further imposes dependencies among audio and visual hidden states. Therefore, it has the capability to incorporate audio visual asynchrony, at least in theory.

12.3.3 Other Applications

Audio visual fusion is important in the context of a number of other multimedia classification tasks. For example, in the detection and recognition of high level events in video, a major challenge is to take cues from different modalities into account. Many major datasets for video event detection/recognition, including TRECVID [85], KTH Human Action Dataset [82], Columbia consumer video (CCV) database [63], etc, have audio. A comprehensive review of video event classification can be found in [64].

Audio visual fusion studies the joint distribution of the two modalities, or the conditional distribution related to a third modality. An example is the text in AVSR. Other aspects of the joint or conditional distribution can also lead to interesting applications, including but not limited to: audio visual speech synthesis [99], speech to lips conversion [11], lip reading, and sound source localization [39]. They often share some of the core techniques, e.g., visual front end, feature fusion, and decision fusion, as discussed in this section.

12.4   Ontology-Based Classification and Inference

In the field of computer vision and multimedia processing, the recognition of object categories in images or videos is one of the most challenging problems. Existing object recognition algorithms usually treat the categories as completely separate and independent, both visually and semantically. For example, the traditional way of performing object recognition is to first extract some low level features, e.g., SIFT [54], HOG [22], etc., and then perform multi-class classification with a set of binary SVM classifiers in the one-against-rest setting [102]. There are several disadvantages in the traditional scheme:

1. It is highly inefficient when the number of categories becomes large. This is often the case in real applications.

2. The relations between categories are ignored. Such relations do exist in real applications. For example, it is unreasonable to train a classifier to distinguish *cars* from *vehicles*. Furthermore, *trucks* and *buses* are more related to one another than *trucks* and *cats*.

3. Sometimes, the classes are high level concepts. As suggested in [15], it is difficult to capture the semantics of high level concepts by using low level features.

All the aforementioned points can be addressed by incorporating ontologies. Ontologies are a formal and explicit representation of knowledge that consists of a concept lexicon, concept properties, and relations among concepts [8]. A well-defined ontology provides a suitable format and a common-shared terminology for knowledge description. Hierarchy is the simplest ontology, where each node represents one concept, and nodes are linked by directed edges representing the *Subclass* relation in a tree structure. Ontologies have already shown their usefulness in many application domains, such as Art Architecture Thesaurus (AAT) for art, and the Gene ontology [4] for health care.

Knowledge can be modeled by ontologies. For example, a *car* is a *wheeled vehicle* and a *motorcycle* is also a *wheeled vehicle*. Thus, both should incorporate a *wheel*. This means that one can determine semantic relationships between class labels that are assigned to the observed visual object instances during visual object recognition. In fact, humans use this knowledge when learning the visual appearance of the objects. For instance, when one encounters a new car model, it is not sensible to learn all the appearance details. It is enough to remember that it looks like a car, as well as the discriminative details. This can help in learning the visual appearance of new object types and speed up the recognition process. Both advantages are very desirable in object recognition.

Motivated by these ideas, some initial research has been done on utilizing different aspects of ontology for many applications. We first introduce some popular ontology applied in multimedia. Then, we provide more technical details for using the ontological relations. More specifically, the *Subclass* relation and *Co-occurrence* relation are described. We also include some algorithms that use the ontology inherently. Finally, we make some concluding remarks.

12.4.1 Popular Applied Ontology

One of the most popular semantic networks for the English language is WordNet [27]. It is a large lexical database of English. Nouns, verbs, adjectives and adverbs are grouped into sets of cognitive synonyms (synsets), each expressing a distinct concept. Synsets are interlinked by means of conceptual-semantic and lexical relations. The most frequently encoded relation among synsets is 1) the super-subordinate relation (also called hyperonymy, hyponymy, or ISA relation); 2) Meronymy, the part-whole relation is present between synsets like chair and back, backrest, seat, and leg. WordNet's structure makes it a useful tool for computational linguistics and natural language processing.

Several video ontologies have also been constructed for domain-specific videos. The most well-known one is LSCOM, developed by Naphade et al. [69]. There are about 1000 concepts in LSCOM, which are organized into seven categories: programs categories, locations, people, objects, activities/events, and graphics. LSCOM is well known for its multiple utility criteria, coverage, observability, and feasibility. Bertini et al. [8] developed a domain-specific ontology for soccer video annotation. They proposed a methodology to link the linguist ontology with visual information by clustering the video highlight clips, and regarded the cluster centers as linguistic concept instances. In [87], Snoek et al. introduced the challenge problem for video annotation, and defined 101 semantic concepts. To represent semantic concepts in medical education videos, Fan. et.al. [57] constructed a medical ontology, including both concepts and hierarchical relations among concepts.

In addition, a few other complex multimedia ontological schemes have also been explored in the literature. Bloehdorn et al. [9] combined multimedia ontology with low-level MPEG-7 visual descriptions. Tsinaraki et al. [94] presented a novel way for the interoperability of Web Ontology Language (OWL) with MPEG-7 Multimedia Description Schemes (MPEG-7 MDS). However, many promising machine learning algorithms could not make full use of these ontological definitions, due to the lack of mathematical description of the ontologies.

12.4.2 Ontological Relations

An ontology is created by linking individual concepts with their mutual relations. Therefore, ontological relations are important for understanding the dependence of concepts. Much work has been done by utilizing different ontological relations. It has been shown that such prior knowledge can help boost machine understanding, and also bridge the semantic gap between low and high level concepts to some extent.

12.4.2.1 Definition

The most applicable and frequently used relations are *Subclass* relations and *Co-occurrence* relations. The definition of the two relations are given as follows:

1. *Subclass*: also called hyperonymy, hyponymy, or "IS-A" relation. It links more general synsets such as *furniture* to increasingly specific ones such as *bed*.

2. *Co-occurrence*: It is a linguistics term that refers to the frequent occurrence of two terms from a text corpus alongside each other in a certain order. For example, *indoor* and *furniture* will typically be correlated by their occurrence in a corpus.

Ontology-based algorithms utilize one of the two relations, or both of them, according to specific applications. The ways that the relations are used are very different, and have their own pros and cons. We will then introduce these algorithms based on the type of relations they use.

12.4.2.2 Subclass Relation

Hierarchy is one of the simplest ontologies where nodes are linked by directed edges representing *Subclass* relations in a tree. Because of the hierarchical structure, the *Subclass* relation can improve the efficiency of algorithms when the number of categories is large, and possibly improve the accuracy.

Dumais [25] proposed a top-down divide-and-conquer method, which trains models only based on the samples associated with their children nodes. While testing, only test samples predicted as positive in the parent nodes are examined in the child nodes. The advantage of such a framework is that whenever the algorithm picks one edge to continue traversing the tree, it automatically prunes the rest of the edges. As a result, the search space and computational complexity are greatly reduced. The disadvantage is also obvious, in that it does not provide any mechanism to prevent error propagation from higher to lower-level nodes caused by its greedy decision process.

Deng et al. [23] introduced a new database called "ImageNet," a large scale ontology of images built upon the backbone of the WordNet structure. Similar to WordNet, synsets of images in ImageNet are interlinked by several types of relations, the "IS-A" (*Subclass*) relation being the most comprehensive and useful. They also proposed a simple object classification method that they called the "tree-max classifier." This method propagates the detection of low-level (concrete) concepts upwards to their high-level (abstract) ancestors. Their experiments showed that the performance of the tree-max classifier was consistently better than the independent classifier. This result indicates that a simple way of exploiting the *Subclass* hierarchy can already provide substantial improvement for the image classification task without additional training or model learning.

Tsai et al. [91] proposed a hierarchical image feature extraction method that extracts image features based on the location of the current node in the hierarchy to fit the images under the current node and to better distinguish its subclasses. The feature extraction method is based on Hierarchical Gaussianization (HG) and Discriminant Attribute Projection (DAP) [105]. Furthermore, they proposed to propagate the images according to the scores of the current node's children nodes instead of the current node, which is an idea similar to the "tree-max classifier."

Cai et al. [14] formulated the hierarchical classification problem in the tensor space and encoded the path from root concept to target concept by a fixed-length binary vector. The original classification task is then transformed to a multi-class classification problem. One problem with their algorithm is the scalability. The high dimensionality of the new representation in tensor space increases the complexity of the new problem, compared to the original one.

12.4.2.3   Co-Occurrence Relation

There is generally no unified ontological structure for the *Co-occurrence* relation. Therefore, the design of such algorithms is variable.

Li et al. [46] proposed a high-level image representation, called the "Object Bank," where an image is represented as a scale-invariant response map of a large number of pre-trained generic object detectors, blind to the testing dataset or visual task. They leveraged a simple linear hypothesis classifier combined with a sparse-coding scheme to ensure both feature sparsity and object sparsity in the Object Bank representation, and also to grasp the semantic relations among these Objects (concepts).

Similarly, Smith et al. [86] proposed a two-step Discriminative Model Fusion (DMF) method for video annotation, which attempted to mine the relations among concepts by generating model vectors, based on detection of scores from individual detectors. The model vectors are then fed to a contextual Support Vector Machine (SVM) to refine the detection.

Tsai et al. [92] proposed a novel representation, called *Compositional Object Pattern*. This representation characterizes object-level patterns conveying much richer semantic information than low level visual features. Such an approach can help in more effective album classification, by interpreting the rich semantics in this kind of data. They first mined frequent object patterns in the training set, and then ranked them by their discriminating power. The album feature is then set to the frequencies of these discriminative patterns. Such frequencies are called *Compositional Object Pattern Frequency (COPF)* and classified by traditional classifiers such as SVM.

12.4.2.4   Combination of the Two Relations

In real applications, the *Subclass* relation and *Co-occurrence* relation usually co-exist in the ontology. Thus, much research has been done to leverage both relations simultaneously.

Marszalek et al. [59] proposed to use the semantics of image labels to integrate prior knowledge about the *Subclass* and *Co-occurrence* relationships into visual appearance learning. To create the semantic hierarchical classifier for object detection, they first queried WordNet with the class labels and extracted the knowledge in the form of a semantic hierarchy. This hierarchy is used for reasoning, and to organize and train the binary SVM detectors that are obtained by a state-of-the-art image classification method. Then, the trained hierarchical classifier can be used to efficiently recognize a large number of object categories.

Zha et al. [101] proposed a semantic ontology scheme for video annotation. They first leveraged LSCOM to construct the concept lexicon and described concept property as the weights of different modalities that were obtained manually or by a data-driven approach. They modeled *Subclass* and *Co-occurrence* relations separately. In particular, they propagated the initial detection of concepts to co-occurring concepts based on how likely they co-occur. On the other hand, the subclass relation is exploited by a hierarchical Bayesian network. Then these two refinements are concatenated to incorporate both relations.

Gao et al. [28] proposed to partition concepts into salient objects and scenes, in which the detection of salient objects utilizes object taxonomy, such as a subclass hierarchy, and to recognize scenes. A naive Bayesian framework modeling co-occurrence relation among salient object and scene is used. They claimed that their framework for hierarchical probabilistic image concept reasoning is scalable to diverse image contents with large within-category variations.

Tsai et al. [93] proposed a novel ontological visual inference framework that utilizes subclass and co-occurrence relations in the ontology to enhance multi-label image annotation. These two kinds of relations in the ontology are learned automatically: subclass relation is obtained via graph traversal techniques on WordNet's hypernymy/hyponymy graph, and the statistical co-occurrence relation is learned from the training set. Their ontological inference framework then seeks the combination of labels that satisfies all the subclass constraints and scores the highest with co-occurrence reward or penalty.

12.4.2.5 Inherently Used Ontology

Without the notion of an ontology, some works on multi-label classification also exploit correlations among concepts.

Tsochantaridis et al. [95] proposed to generalize multi-class *Support Vector Machine learning* for interdependent and structured output spaces. Qi et al. [78] adopted a Gibbs Random Field to model the interaction between concepts, where the unary and pairwise potential function are learned simultaneously. Qi et al. [79] proposed to iteratively ask for the labels of the selected images to minimize the classification error rate under the active learning scenario.

There are clear advantages of leveraging ontology in traditional machine learning problems. This section provided details about the popularly applied ontologies as well as the very important ontological relations, such as *Subclass* and *Co-occurrence*. Many ontology-based algorithms are introduced and compared based on the relations they used. The result in their work showed the effectiveness and efficiency of ontology in diverse machine learning applications.

12.5 Geographical Classification with Multimedia Data

In recent years, there has been increasing geographically tagged multimedia on community Web sites such as Flickr,[1] and social networks (e.g., Facebook, Google+, Instagram), especially after the popularity of smart mobile devices. For instance, Flickr has reported over 100 million geotagged photos. Effectively organizing, searching, and discovering knowledge from such geotagged data are in great need. The research in geographical classification of multimedia is based on this.

Typically, geographical information has been studied in multimedia and computer vision research, within a contextual modeling framework. However, there are many possibilities for utilizing the geography as context, with different data modalities and purposes. The data types available for Geo-Information processing include GIS (Geographical Information System) databases, aerial or remote sensing images, unstructured geo-reference Web resources (e.g., images from blogs), and structured Web multimedia collections such as Flickr, Instagram et al., [58]. The purposes of these kinds of data may vary from geo-location annotation [32, 45], retrieval [38], landmark recognition [18, 104], to mining and recommendation (e.g., the tourism guide system) [35, 37].

In this section, we focus on the problem of classification/recognition of structured Web multimedia that provides organized meta-data, especially for landmark recognition. The problem of geographical classification or landmark recognition aims at determining the location of an image, leveraging collections of geo-located images in the training set. The basic approach is to directly match feature points upon the indexing structure, or learn visual appearance models. It is an important research topic in various applications.

In the remainder of this section, we will first talk about the data and information available for Web multimedia, and then discuss the challenges in the geographical classification. Two detailed applications are explored, corresponding to the geo-classification of Web images, and of videos.

12.5.1 Data Modalities

The types of information available in structured Web multimedia for geographical classification include but ate not limited to the following:

1. *GPS information:* Mobile phones and smart devices, which record location information when taking photos. The GPS coordinates allow mapping images onto the globe, if the globe were

---

[1] www.flickr.com

(a)   (b)   (c)

FIGURE 12.8 (See color insert.): The challenges in landmark image recognition. (a) and (b) are photos of Drum Tower of Xi'an and Qianmen Gate of Beijing, respectively. Though of different locations the visual appearances are similar due to historical and style reasons. (c) shows three disparate viewing angles of Lincoln Memorial in Washington, DC.

to be treated as a rectangular coordinate system, and the position is determined as latitude and longitude.

2. *Text information:* Typically, the textual meta-data such as the descriptions of Flickr images, or articles in personal blogs, contain named entities of positions (e.g., "Sydney Opera," "Eiffel tower"). This is essential in indicating the geo-location of images. An example is [35], which extracted city names from MSN blogs and photo albums.

3. *Visual features:* Most problems deal with matching images to locations (e.g., landmark names). Visual features are of high importance for landmark recognition. Popular visual features include local points (e.g., SIFT [54]) and bag-of-words representations [84].

4. *Collaborative features:* With the popularity of social networks, collaborative features are also enhanced, especially for social media. Examples of collaborative features include *Hashing Tags* on Twitter labeled by users, and user comments or reposts. The problem of utilizing the collaborative features more efficiently still remains relatively unexplored, since some of them are highly unstructured and noisy.

Geographical classification always needs to fuse different modalities of information, to accurately predict geo-tags.

12.5.2 Challenges in Geographical Classification

The challenges of geographical classification of multimedia data are mainly due to the following two aspects:

1. *Visual ambiguity:* Different geo-locations may have similar visual appearance, because of historical or style reasons. More specifically, the building from the same period, or of the same style may not be distinguishable enough in terms of their visual features. This makes the construction of the visual appearance model for each landmark or class even more challenging.

2. *Geo-location ambiguity:* Even the same geo-location or landmark may look different from different visual angles. Though the advanced scale and rotation visual features such as SIFT [54] can handle the in-plane rotation, current image descriptors cannot address the out-of-plane rotation, i.e., the viewing angle changes.

Figure 12.8 shows two examples of these challenges. To deal with the visual ambiguity challenge, it is feasible to make use of multi-modality information, to eliminate these challenges. Classical approaches include fusing GPS position and visual features [37], or textual information and visual features [35].

For geo-location ambiguity, learning the viewing-angle level models for each landmark is widely adopted. This can be further integrated into a voting or boosting framework. In the following, we introduce two related topics, corresponding to the geo-classification for Web images and videos, respectively.

12.5.3   Geo-Classification for Images

The works on landmark recognition and geographical classification usually resorted to supervised methods based on content or context similarities. The early work on geographical classification started from the *"Where am I"* contest at ICCV 2005 [89]. Early approaches mainly relied on the feature points matching, e.g., local interest points and images regions across multiple images, based on a series of geo-tagged training images. In [103], features such as SIFT [54], SURF [6], and MSER [60] are utilized to match the test image to one or more training images. Furthermore, Kennedy et al. [37] extracted popular landmark images on Flickr by constructing a graph using SIFT point matching.

Other approaches use more effective visual features instead of interest point matching. Schindler et al. [81] extended the feature matching algorithm by introducing a large scale vocabulary tree. A vocabulary tree is a structure obtained by hierarchically clustering the data points on each level, typically, the SIFT [54] feature. By introducing the vocabulary tree, geographical classification or recognition algorithms are able to deal with large scale data. In [81], they used 30,000 images and constructed a tree with 10^6 nodes.

FIGURE 12.9 (See color insert.): The same landmark will differ in appearance from different viewing angles. Typical approaches use clustering to get the visual information for each viewing angle. From Liu et al. [52]

To reduce the effect of ambiguities in landmark classification (see Figure 12.9), meta-data are introduced and the classification is performed using multiple features. In [37, 38], the images are first pre-processed according to their GPS information, to eliminate the geo-location ambiguity. Typically, the images of the same landmark will be clustered into different viewing angles to tackle the visual ambiguity. Cristani et al. [21] integrated both the visual and geographical information in a statistical graph model, by a joint inference. Ji et al. [36] proposed to build the vocabulary tree considering context (e.g., textual information), and then applied to the geo-location classification task. Crandall et al. further [20] proposed to combine multiple features like visual, textual, and temporal features to assign photos into the worldwide geographical map.

12.5.3.1 Classifiers

The k-NN (k Nearest Neighbors) [32,41] and the Support Vector Machine (SVM) are commonly used. Moreover, researchers also modeled the image collections in a graph in which the nodes are connected according to their content or context similarities. PageRank [73] or HITS-like [42] algorithms are used to propagate and classify the images' geographical information following a probabilistic manner. Cristani et al. [21] utilized the probabilistic topic model, the location-dependent pLSA (LD-pLSA), to jointly model the visual and geographical information, which aims to coherently categorize images using a dual-topic modeling. Recently, structural SVM is also utilized [47] in recognizing the landmarks.

12.5.4 Geo-Classification for Web Videos

While research on using geotagged images has gathered significant momentum, this is not quite as true for geotagged videos. Unlike Web images such as Flickr, the Web videos are expensive to label. Therefore, training data is limited, and so are the benchmarks. Anther challenge is that a video clip is usually composed of multiple scenes. This increases the difficulty of directly applying known algorithms for image geo-classification to videos directly.

Pioneer works on geo-classification of videos include [3, 5, 40], which employ the spatial and temporal properties for search and classification. The typical approach is to perform a spatial-temporal filtering to the "fields of view" relative to a specific position. Recently, Liu et al. [52] proposed to transfer the geo-information from geo-tagged images (e.g., Flickr) to Web videos, to tackle lack of training data. Due to the cross-domain difficulty, a transfer learning algorithm selected credible samples and features to build the final classifiers for Web videos. Another advantage of videos is the consistency within video shots. Thus, they integrated the frame-level classification by sliding window smoothing into video-shot level results. However, geographical classification of videos is still an open problem. Efforts on both data collection and algorithm designing are emerging.

12.6 Conclusion

In this chapter we present current advances in the field of multimedia classification, with details on feature selection, information fusion, and three important application domains. These correspond to audio-visual fusion, ontology based classification, and geographical classification. In particular, we emphasized the benefit of information fusion for multiple content forms. Brief overviews of the state-of-the-art methods were provided by this chapter, though a detailed exposition would be beyond the scope of a single chapter. Because of its great practical value, multimedia classification remains an active and fast growing research area. It also serves as a good test-bed for new methods in audio/image processing and data science, including general data classification methods in the rest of this book.

Bibliography

[1] Waleed H Abdulla, Paul WT Yu, and Paul Calverly. Lips tracking biometrics for speaker recognition. *International Journal of Biometrics*, 1(3):288–306, 2009.

[2] Ali Adjoudani and Christian Benoit. On the integration of auditory and visual parameters in an HMM-based ASR. *NATO ASI Series F Computer and Systems Sciences*, 150:461–472, 1996.

[3] Sakire Arslan Ay, Lingyan Zhang, Seon Ho Kim, Ma He, and Roger Zimmermann. GRVS: A georeferenced video search engine. In *Proceedings of the 17th ACM International Conference on Multimedia*, pages 977–978. ACM, 2009.

[4] Michael Ashburner, Catherine A Ball, Judith A Blake, David Botstein, Heather Butler, J Michael Cherry, Allan P Davis, Kara Dolinski, Selina S Dwight, Janan T Eppig, et al. Gene ontology: Tool for the unification of biology. *Nature Genetics*, 25(1):25–29, 2000.

[5] Sakire Arslan Ay, Roger Zimmermann, and Seon Ho Kim. Viewable scene modeling for geospatial video search. In *Proceedings of the 16th ACM International Conference on Multimedia*, pages 309–318. ACM, 2008.

[6] Herbert Bay, Tinne Tuytelaars, and Luc Van Gool. Surf: Speeded up robust features. In *Computer Vision–ECCV 2006*, pages 404–417. Springer, 2006.

[7] Richard Bellman. *Adaptive Control Processes: A Guided Tour*, volume 4. Princeton University Press Princeton, NJ, 1961.

[8] Marco Bertini, Alberto Del Bimbo, and Carlo Torniai. Automatic video annotation using ontologies extended with visual information. In *Proceedings of the 13th Annual ACM International Conference on Multimedia*, pages 395–398. ACM, 2005.

[9] Stephan Bloehdorn, Kosmas Petridis, Carsten Saathoff, Nikos Simou, Vassilis Tzouvaras, Yannis Avrithis, Siegfried Handschuh, Yiannis Kompatsiaris, Steffen Staab, and Michael G Strintzis. Semantic annotation of images and videos for multimedia analysis. In *The Semantic Web: Research and Applications*, pages 592–607. Springer, 2005.

[10] Anna Bosch, Andrew Zisserman, and Xavier Munoz. Scene classification via pLSA. In *Computer Vision–ECCV 2006*, pages 517–530. Springer, 2006.

[11] Christoph Bregler, Michele Covell, and Malcolm Slaney. Video rewrite: Driving visual speech with audio. In *Proceedings of the 24th Annual Conference on Computer Graphics and Interactive Techniques*, pages 353–360. ACM Press/Addison-Wesley Publishing Co., 1997.

[12] Christoph Bregler and Yochai Konig. "Eigenlips" for robust speech recognition. In *IEEE International Conference on Acoustics, Speech, and Signal Processing, 1994. ICASSP-94.*, volume 2, pages II–669. IEEE, 1994.

[13] Darin Brezeale and Diane J. Cook. Automatic video classification: A survey of the literature. *IEEE Transactions on Systems, Man, and Cybernetics, Part C*, 38(3):416–430, 2008.

[14] Lijuan Cai and Thomas Hofmann. Hierarchical document categorization with support vector machines. In *Proceedings of the Thirteenth ACM International Conference on Information and Knowledge Management*, pages 78–87. ACM, 2004.

[15] Liangliang Cao, Jiebo Luo, and Thomas S Huang. Annotating photo collections by label propagation according to multiple similarity cues. In *Proceedings of the 16th ACM International Conference on Multimedia*, pages 121–130. ACM, 2008.

[16] Michael T Chan. HMM-based audio-visual speech recognition integrating geometric and appearance-based visual features. In *2001 IEEE Fourth Workshop on Multimedia Signal Processing*, pages 9–14. IEEE, 2001.

[17] S K Chang, Q Y Shi, and C W Yan. Iconic indexing by 2-d strings. *IEEE Transactions on Pattern Analysis and Machine Intelligence*, 9(3):413–428, May 1987.

[18] David M Chen, Georges Baatz, K Koser, Sam S Tsai, Ramakrishna Vedantham, Timo Pylvanainen, Kimmo Roimela, Xin Chen, Jeff Bach, Marc Pollefeys, et al. City-scale landmark identification on mobile devices. In *IEEE Conference on Computer Vision and Pattern Recognition (CVPR), 2011*, pages 737–744. IEEE, 2011.

[19] Timothy F Cootes, Gareth J Edwards, and Christopher J. Taylor. Active appearance models. *IEEE Transactions on Pattern Analysis and Machine Intelligence*, 23(6):681–685, 2001.

[20] David J Crandall, Lars Backstrom, Daniel Huttenlocher, and Jon Kleinberg. Mapping the world's photos. In *Proceedings of the 18th International Conference on World Wide Web*, pages 761–770. ACM, 2009.

[21] Marco Cristani, Alessandro Perina, Umberto Castellani, and Vittorio Murino. Geo-located image analysis using latent representations. In *IEEE Conference on Computer Vision and Pattern Recognition, 2008. CVPR 2008*, pages 1–8. IEEE, 2008.

[22] Navneet Dalal and Bill Triggs. Histograms of oriented gradients for human detection. In *IEEE Computer Society Conference on Computer Vision and Pattern Recognition, 2005. CVPR 2005*, volume 1, pages 886–893. IEEE, 2005.

[23] Jia Deng, Wei Dong, Richard Socher, Li-Jia Li, Kai Li, and Li Fei-Fei. Imagenet: A large-scale hierarchical image database. In *IEEE Conference on Computer Vision and Pattern Recognition, 2009. CVPR 2009*, pages 248–255. IEEE, 2009.

[24] Fadi Dornaika and Jörgen Ahlberg. Fast and reliable active appearance model search for 3-D face tracking. *IEEE Transactions on Systems, Man, and Cybernetics, Part B: Cybernetics*, 34(4):1838–1853, 2004.

[25] Susan Dumais and Hao Chen. Hierarchical classification of web content. In *Proceedings of the 23rd Annual International ACM SIGIR Conference on Research and Development in Information Retrieval*, pages 256–263. ACM, 2000.

[26] Stéphane Dupont and Juergen Luettin. Audio-visual speech modeling for continuous speech recognition. *IEEE Transactions on Multimedia*, 2(3):141–151, 2000.

[27] Christiane Fellbaum (Ed.) *WordNet: An Electronic Lexical Database*, Springer, 2010.

[28] Yul Gao and Jianping Fan. Incorporating concept ontology to enable probabilistic concept reasoning for multi-level image annotation. In *Proceedings of the 8th ACM International Workshop on Multimedia Information Retrieval*, pages 79–88. ACM, 2006.

[29] Zoubin Ghahramani and Michael I Jordan. Factorial hidden Markov models. *Machine Learning*, 29(2-3):245–273, 1997.

[30] Hervé Glotin, D Vergyr, Chalapathy Neti, Gerasimos Potamianos, and Juergen Luettin. Weighting schemes for audio-visual fusion in speech recognition. In *2001 IEEE International Conference on Acoustics, Speech, and Signal Processing, 2001. Proceedings, (ICASSP'01)*, volume 1, pages 173–176. IEEE, 2001.

[31] Robert M Haralick, K Shanmugam, and Its'Hak Dinstein. Textural features for image classification. *IEEE Transactions on Systems, Man and Cybernetics*, SMC-3(6):610–621, November 1973.

[32] James Hays and Alexei A Efros. Im2gps: estimating geographic information from a single image. In *IEEE Conference on Computer Vision and Pattern Recognition, 2008. CVPR 2008*, pages 1–8. IEEE, 2008.

[33] Berthold KP Horn and Brian G Schunck. Determining optical flow. *Artificial Intelligence*, 17(1):185–203, 1981.

[34] J Jeon, V Lavrenko, and R Manmatha. Automatic image annotation and retrieval using cross-media relevance models. In *Proceedings of the 26th Annual International ACM SIGIR Conference on Research and Development in Informaion Retrieval*, SIGIR '03, pages 119–126, New York, NY, USA, 2003. ACM.

[35] Rongrong Ji, Xing Xie, Hongxun Yao, and Wei-Ying Ma. Mining city landmarks from blogs by graph modeling. In *Proceedings of the 17th ACM International Conference on Multimedia*, pages 105–114. ACM, 2009.

[36] Rongrong Ji, Hongxun Yao, Qi Tian, Pengfei Xu, Xiaoshuai Sun, and Xianming Liu. Context-aware semi-local feature detector. *ACM Transactions on Intelligent Systems and Technology (TIST)*, 3(3):44, 2012.

[37] Lyndon Kennedy, Mor Naaman, Shane Ahern, Rahul Nair, and Tye Rattenbury. How flickr helps us make sense of the world: context and content in community-contributed media collections. In *Proceedings of the 15th International Conference on Multimedia*, pages 631–640. ACM, 2007.

[38] Lyndon S Kennedy and Mor Naaman. Generating diverse and representative image search results for landmarks. In *Proceedings of the 17th International Conference on World Wide Web*, pages 297–306. ACM, 2008.

[39] Einat Kidron, Yoav Y Schechner, and Michael Elad. Pixels that sound. In *IEEE Computer Society Conference on Computer Vision and Pattern Recognition, 2005. CVPR 2005*, volume 1, pages 88–95. IEEE, 2005.

[40] Seon Ho Kim, Sakire Arslan Ay, Byunggu Yu, and Roger Zimmermann. Vector model in support of versatile georeferenced video search. In *Proceedings of the First Annual ACM SIGMM Conference on Multimedia Systems*, pages 235–246. ACM, 2010.

[41] Jim Kleban, Emily Moxley, Jiejun Xu, and BS Manjunath. Global annotation on georeferenced photographs. In *Proceedings of the ACM International Conference on Image and Video Retrieval*, page 12. ACM, 2009.

[42] Jon M Kleinberg. Authoritative sources in a hyperlinked environment. *Journal of the ACM (JACM)*, 46(5):604–632, 1999.

[43] Sydney M Lamb. *Pathways of the brain: The neurocognitive basis of language*, volume 170. John Benjamins Publishing, 1999.

[44] Ivan Laptev. On space-time interest points. *International Journal of Computer Vision*, 64(2-3):107–123, 2005.

[45] Martha Larson, Mohammad Soleymani, Pavel Serdyukov, Stevan Rudinac, Christian Wartena, Vanessa Murdock, Gerald Friedland, Roeland Ordelman, and Gareth JF Jones. Automatic tagging and geotagging in video collections and communities. In *Proceedings of the 1st ACM International Conference on Multimedia Retrieval*, page 51. ACM, 2011.

[46] Li-Jia Li, Hao Su, Li Fei-Fei, and Eric P Xing. Object bank: A high-level image representation for scene classification & semantic feature sparsification. In *Advances in Neural Information Processing Systems*, pages 1378–1386, 2010.

[47] Yunpeng Li, David J Crandall, and Daniel P Huttenlocher. Landmark classification in large-scale image collections. In *IEEE 12th International Conference on Computer Vision 2009*, pages 1957–1964. IEEE, 2009.

[48] Tony Lindeberg. Feature detection with automatic scale selection. *International Journal of Computer Vision*, 30(2):79–116, 1998.

[49] Tony Lindeberg. Scale invariant feature transform. *Scholarpedia*, 7(5):10491, 2012.

[50] Fang Liu and R W Picard. Periodicity, directionality, and randomness: Wold features for image modeling and retrieval. *IEEE Transactions on Pattern Analysis and Machine Intelligence*, 18(7):722–733, 1996.

[51] Xian-Ming Liu, Hongxun Yao, Rongrong Ji, Pengfei Xu, Xiaoshuai Sun, and Qi Tian. Learning heterogeneous data for hierarchical web video classification. In *Proceedings of the 19th ACM International Conference on Multimedia*, pages 433–442. ACM, 2011.

[52] Xian-Ming Liu, Yue Gao, Rongrong Ji, Shiyu Chang, and Thomas Huang. Localizing web videos from heterogeneous images. In *Workshops at the Twenty-Seventh AAAI Conference on Artificial Intelligence*, 2013.

[53] Ying Liu, Dengsheng Zhang, Guojun Lu, and Wei-Ying Ma. A survey of content-based image retrieval with high-level semantics. *Pattern Recognition*, 40(1):262–282, 2007.

[54] David G Lowe. Distinctive image features from scale-invariant keypoints. *International Journal of Computer Vision*, 60(2):91–110, 2004.

[55] Simon Lucey. *Audio-Visual Speech Processing*. PhD thesis, Citeseer, 2002.

[56] Juergen Luettin, Neil A Thacker, and Steve W Beet. Active shape models for visual speech feature extraction. *NATO ASI Series F Computer and Systems Sciences*, 150:383–390, 1996.

[57] Hangzai Luo and Jianping Fan. Building concept ontology for medical video annotation. In *Proceedings of the 14th Annual ACM International Conference on Multimedia*, pages 57–60. ACM, 2006.

[58] Jiebo Luo, Dhiraj Joshi, Jie Yu, and Andrew Gallagher. Geotagging in multimedia and computer vision – a survey. *Multimedia Tools and Applications*, 51(1):187–211, 2011.

[59] Marcin Marszalek and Cordelia Schmid. Semantic hierarchies for visual object recognition. In *IEEE Conference on Computer Vision and Pattern Recognition, 2007. CVPR'07*, pages 1–7. IEEE, 2007.

[60] Jiri Matas, Ondrej Chum, Martin Urban, and Tomás Pajdla. Robust wide-baseline stereo from maximally stable extremal regions. *Image and Vision Computing*, 22(10):761–767, 2004.

[61] Iain Matthews, Gerasimos Potamianos, Chalapathy Neti, and Juergen Luettin. A comparison of model and transform-based visual features for audio-visual LVCSR. In *Proceedings of the International Conference and Multimedia Expo*, pages 22–25, 2001.

[62] Harry McGurk and John MacDonald. Hearing lips and seeing voices. *Nature*, 264:746–748, 1976.

[63] Yu-Gang Jiang, Guangnan Ye, Shih-Fu Chang, Daniel Ellis, and Alexander C. Loui. Consumer video understanding: A benchmark database and an evaluation of human and machine performance. In *Proceedings of ACM International Conference on Multimedia Retrieval (ICMR), oral session*, 2011.

[64] Yu-Gang Jiang, Subhabrata Bhattacharya, Shih-Fu Chang, and Mubarak Shah. High-level event recognition in unconstrained videos. *International Journal of Multimedia Information Retrieval*, 2(2):73–101, 2012.

[65] Rajiv Mehrotra and James E. Gary. Similar-shape retrieval in shape data management. *Computer*, 28(9):57–62, 1995.

[66] Kieron Messer, Jiri Matas, Josef Kittler, Juergen Luettin, and Gilbert Maitre. XM2VTSDB: The extended M2VTS database. In *Second International Conference on Audio and Video-Based Biometric Person Authentication*, pages 72–77, 1999.

[67] Krystian Mikolajczyk and Cordelia Schmid. Scale & affine invariant interest point detectors. *International Journal of Computer Vision*, 60(1):63–86, 2004.

[68] Aleksandra Mojsilovic, Jose Gomes, and Bernice Rogowitz. Isee: Perceptual features for image library navigation, *Proceedings of SPIE 4662*, *Human Vision and Electronic Imaging VII*, pages 266–277, 2002.

[69] Milind Naphade, John R Smith, Jelena Tesic, Shih-Fu Chang, Winston Hsu, Lyndon Kennedy, Alexander Hauptmann, and Jon Curtis. Large-scale concept ontology for multimedia. *Multimedia, IEEE*, 13(3):86–91, 2006.

[70] Ara V Nefian, Luhong Liang, Xiaobo Pi, Xiaoxing Liu, and Kevin Murphy. Dynamic Bayesian networks for audio-visual speech recognition. *EURASIP Journal on Advances in Signal Processing*, 2002(11):1274–1288, 2002.

[71] Ara V Nefian, Luhong Liang, Xiaobo Pi, Liu Xiaoxiang, Crusoe Mao, and Kevin Murphy. A coupled HMM for audio-visual speech recognition. In *2002 IEEE International Conference on Acoustics, Speech, and Signal Processing (ICASSP)*, volume 2, pages II–2013. IEEE, 2002.

[72] Jiquan Ngiam, Aditya Khosla, Mingyu Kim, Juhan Nam, Honglak Lee, and Andrew Y Ng. Multimodal deep learning. In *Proceedings of the 28th International Conference on Machine Learning (ICML-11), ICML*, volume 11, 2011.

[73] Lawrence Page, Sergey Brin, Rajeev Motwani, and Terry Winograd. *The PageRank Citation Ranking: Bringing Order to the Web*. Technical Report. Stanford InfoLab, Palo Alto, CA, 1999.

[74] Konstantinos N Plataniotis and Anastasios N Venetsanopoulos. *Color Image Processing and Applications*. New York: Springer-Verlag New York, Inc., 2000.

[75] Gerasimos Potamianos, Eric Cosatto, Hans Peter Graf, and David B Roe. Speaker independent audio-visual database for bimodal ASR. In *Audio-Visual Speech Processing: Computational & Cognitive Science Approaches*, pages 65–68, 1997.

[76] Gerasimos Potamianos, Chalapathy Neti, Juergen Luettin, and Iain Matthews. Audio-visual automatic speech recognition: An overview. *Issues in Visual and Audio-Visual Speech Processing*, 22:23, MIT Press, Cambridge, MA. 2004.

[77] Gerasimos Potamianos, Hans Peter Graf, and Eric Cosatto. An image transform approach for HMM based automatic lipreading. In *Proceedings, 1998 International Conference on Image Processing, 1998. ICIP 98*, pages 173–177. IEEE, 1998.

[78] Guo-Jun Qi, Xian-Sheng Hua, Yong Rui, Jinhui Tang, Tao Mei, and Hong-Jiang Zhang. Correlative multi-label video annotation. In *Proceedings of the 15th International Conference on Multimedia*, pages 17–26. ACM, 2007.

[79] Guo-Jun Qi, Xian-Sheng Hua, Yong Rui, Jinhui Tang, and Hong-Jiang Zhang. Two-dimensional multilabel active learning with an efficient online adaptation model for image classification. *IEEE Transactions on Pattern Analysis and Machine Intelligence*, 31(10):1880–1897, 2009.

[80] Paris Smaragdis. Machine learning for signal processing, In *Course slides for CS598PS Fall 2013*.

[81] Grant Schindler, Matthew Brown, and Richard Szeliski. City-scale location recognition. In *IEEE Conference on Computer Vision and Pattern Recognition, 2007. CVPR'07*, pages 1–7. IEEE, 2007.

[82] Christian Schuldt, Ivan Laptev, and Barbara Caputo. Recognizing human actions: a local SVM approach. In *Proceedings of the 17th International Conference on Pattern Recognition, 2004. ICPR 2004*, volume 3, pages 32–36. IEEE, 2004.

[83] Ishwar K Sethi, Ioana L Coman, and Daniela Stan. Mining association rules between low-level image features and high-level concepts. In *Proceedings of SPIE 4384, Data Mining and Knowledge Discovery: Theory, Tools, and Technology III*, pages 279–290, 2001.

[84] Josef Sivic and Andrew Zisserman. Video Google: A text retrieval approach to object matching in videos. In *Proceedings of the Ninth IEEE International Conference on Computer Vision 2003*, pages 1470–1477. IEEE, 2003.

[85] Alan F Smeaton, Paul Over, and Wessel Kraaij. Evaluation campaigns and trecvid. In *MIR '06: Proceedings of the 8th ACM International Workshop on Multimedia Information Retrieval*, pages 321–330, ACM Press, New York, 2006.

[86] John R Smith, Milind Naphade, and Apostol Natsev. Multimedia semantic indexing using model vectors. In *ICME'03, Proceedings, of the 2003 International Conference on Multimedia and Expo, 2003*, volume 2, pages II–445. IEEE, 2003.

[87] Cees GM Snoek, Marcel Worring, Jan C Van Gemert, Jan-Mark Geusebroek, and Arnold WM Smeulders. The challenge problem for automated detection of 101 semantic concepts in multimedia. In *Proceedings of the 14th Annual ACM International Conference on Multimedia*, pages 421–430. ACM, 2006.

[88] William H Sumby and Irwin Pollack. Visual contribution to speech intelligibility in noise. *The Journal of the Acoustical Society of America*, 26:212, 1954.

[89] Richard Szeliski. Where am i? In *Proceedings of ICCV Computer Vision Conference*. IEEE, 2005.

[90] Hideyuki Tamura, Shunji Mori, and Takashi Yamawaki. Textural features corresponding to visual perception. *IEEE Transactions on System, Man and Cybernetic*, 8(6):460–473, 1978.

[91] Min-Hsuan Tsai, Shen-Fu Tsai, and Thomas S Huang. Hierarchical image feature extraction and classification. In *Proceedings of the International Conference on Multimedia*, pages 1007–1010. ACM, 2010.

[92] Shen-Fu Tsai, Liangliang Cao, Feng Tang, and Thomas S Huang. Compositional object pattern: a new model for album event recognition. In *Proceedings of the 19th ACM International Conference on Multimedia*, pages 1361–1364. ACM, 2011.

[93] Shen-Fu Tsai, Hao Tang, Feng Tang, and Thomas S Huang. Ontological inference framework with joint ontology construction and learning for image understanding. In *2012 IEEE International Conference on Multimedia and Expo (ICME)*, pages 426–431. IEEE, 2012.

[94] Chrisa Tsinaraki, Panagiotis Polydoros, and Stavros Christodoulakis. Interoperability support for ontology-based video retrieval applications. In *Image and Video Retrieval*, pages 582–591. Springer, 2004.

[95] Ioannis Tsochantaridis, Thomas Hofmann, Thorsten Joachims, and Yasemin Altun. Support vector machine learning for interdependent and structured output spaces. In *Proceedings of the Twenty-First International Conference on Machine Learning*, page 104. ACM, 2004.

[96] Pavan Turaga, Rama Chellappa, Venkatramana S Subrahmanian, and Octavian Udrea. Machine recognition of human activities: A survey. *IEEE Transactions on Circuits and Systems for Video Technology*, 18(11):1473–1488, 2008.

[97] Hualu Wang, Ajay Divakaran, Anthony Vetro, Shih-Fu Chang, and Huifang Sun. Survey of compressed-domain features used in audio-visual indexing and analysis. *Journal of Visual Communication and Image Representation*, 14(2):150–183, June 2003.

[98] James Z Wang, Jia Li, and Gio Wiederhold. Simplicity: Semantics-sensitive integrated matching for picture libraries. *IEEE Transactions on Pattern Analysis and Machine Intelligence*, 23:947–963, 2001.

[99] Lijuan Wang, Yao Qian, Matthew R Scott, Gang Chen, and Frank K Soong. Computer-assisted audiovisual language learning. *Computer*, 45(6):38–47, 2012.

[100] Wei Ying Ma and B S Manjunath. Netra: A toolbox for navigating large image databases. In *Multimedia Systems*, pages 568–571, 1999.

[101] Zheng-Jun Zha, Tao Mei, Zengfu Wang, and Xian-Sheng Hua. Building a comprehensive ontology to refine video concept detection. In *Proceedings of the International Workshop on Multimedia Information Retrieval*, pages 227–236. ACM, 2007.

[102] Jianguo Zhang, Marcin Marszalek, Svetlana Lazebnik, and Cordelia Schmid. Local features and kernels for classification of texture and object categories: A comprehensive study. *International Journal of Computer Vision*, 73(2):213–238, 2007.

[103] Wei Zhang and Jana Kosecka. Image based localization in urban environments. In *Third International Symposium on 3D Data Processing, Visualization, and Transmission*, pages 33–40. IEEE, 2006.

[104] Yan-Tao Zheng, Ming Zhao, Yang Song, Hartwig Adam, Ulrich Buddemeier, Alessandro Bissacco, Fernando Brucher, Tat-Seng Chua, and Hartmut Neven. Tour the world: Building a web-scale landmark recognition engine. In *IEEE Conference on Computer Vision and Pattern Recognition, 2009. CVPR 2009*, pages 1085–1092. IEEE, 2009.

[105] Xi Zhou, Na Cui, Zhen Li, Feng Liang, and Thomas S Huang. Hierarchical Gaussianization for image classification. In *2009 IEEE 12th International Conference on Computer Vision*, pages 1971–1977. IEEE, 2009.

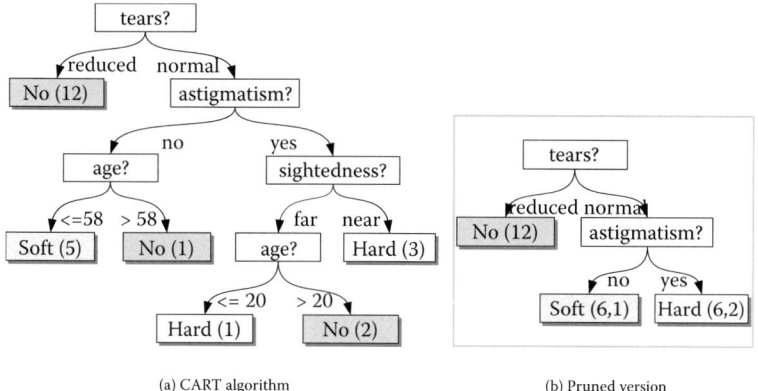

(a) CART algorithm

(b) Pruned version

FIGURE 4.1: 3-class decision trees for contact lenses recommendation.

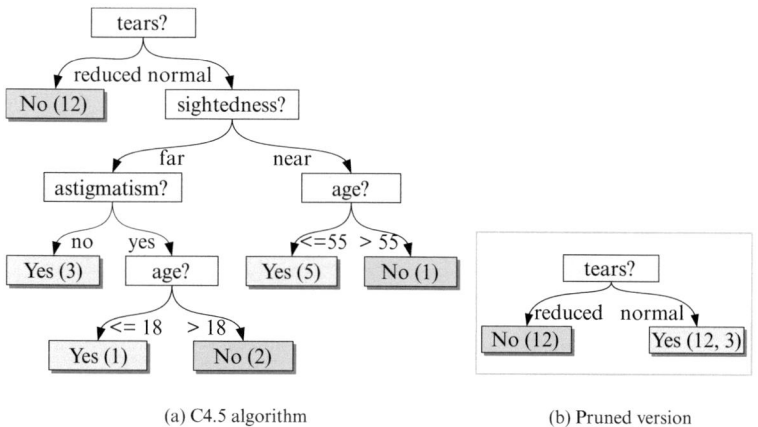

(a) C4.5 algorithm

(b) Pruned version

FIGURE 4.2: 2-class decision trees for contact lenses recommendation.

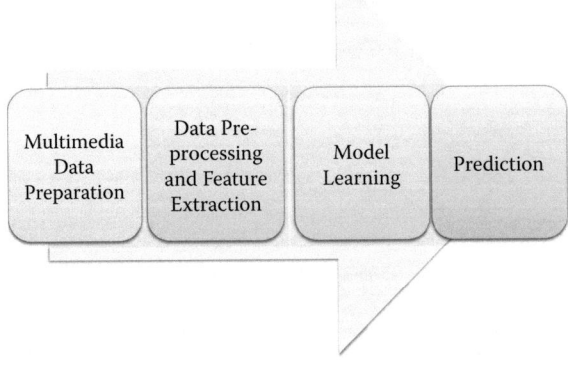

FIGURE 12.1: Flowchart of general multimedia learning process.

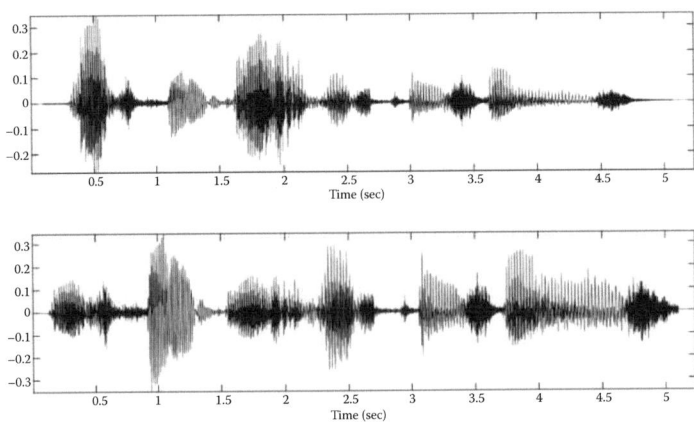

FIGURE 12.2: Time domain waveform of the same speech from different person [80].

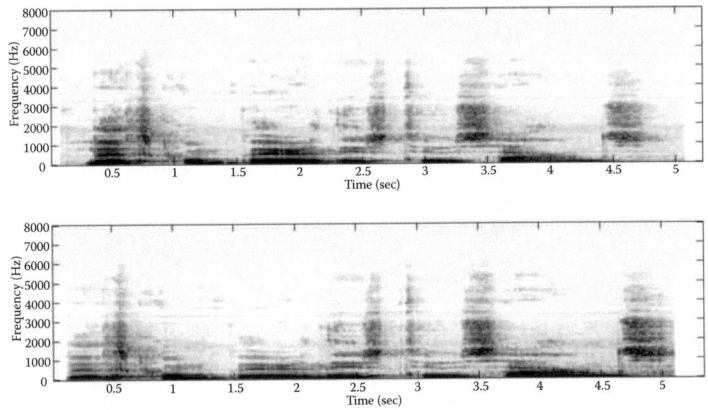

FIGURE 12.3: Frequency response of the same speech in Figure 12.2 [80].

FIGURE 12.4: Example images of President George W. Bush showing huge intra-person variations.

FIGURE 12.5: Samples of SIFT detector from [51]: interest points detected from two images of a similar scene. For each point, the circle indicates the scale. It also shows the matching between the SIFT points in the two images.

(a) The Drum Tower of Xi'an (b) Qianmen Gate of Beijing (c) Different viewing angle of "Lincoln Memorial"

FIGURE 12.8: The challenges of landmark recognition: (a) and (b) are photos of Drum Tower of Xi'an and Qianmen Gate of Beijing, respectively. Though of different locations, the visual appearances are similar due to historical and style reasons. (c) shows three disparate viewing angles of the Lincoln Memorial in Washington, DC.

FIGURE 12.9: The same landmark will differ in appearance from different viewing angles. Typical approaches use clustering to get the visual information for each viewing angle. (From Liu et al. [52].

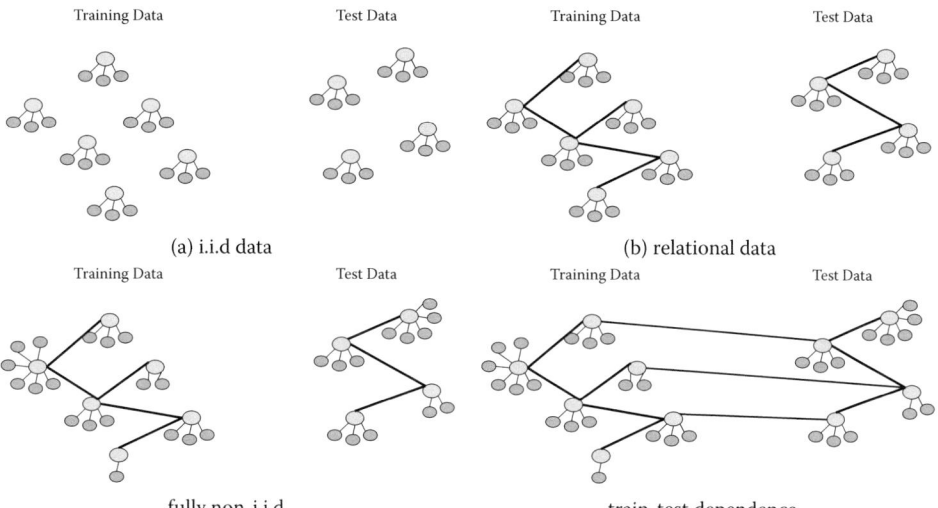

FIGURE 15.1: (a) An illustration of the common i.i.d. supervised learning setting. Here each instance is represented by a subgraph consisting of a label node (blue) and several local feature nodes (purple). (b) The same problem, cast in the relational setting, with links connecting instances in the training and training sets, respectively. The instances are no longer independent. (c) A relational learning problem in which each node has a varying number of local features and relationships, implying that the nodes are neither independent nor identically distributed. (d) The same problem, with relationships (links) between the training and test set.

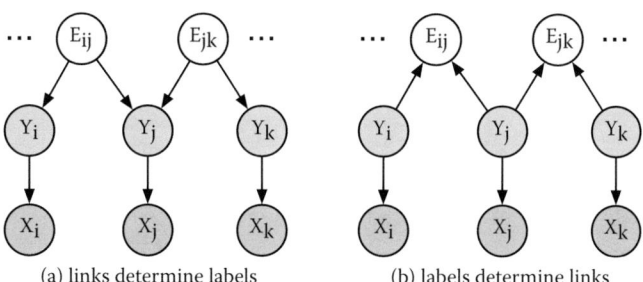

FIGURE 15.2: Example BN for collective classification. Label nodes (green) determine features (purple), which are represented by a single vector-varied variable. An edge variable (yellow) is defined for all potential edges in the data graph. In (a), labels are determined by link structure, representing contagion. In (b), links are functions of labels, representing homophily. Both structures are acyclic.

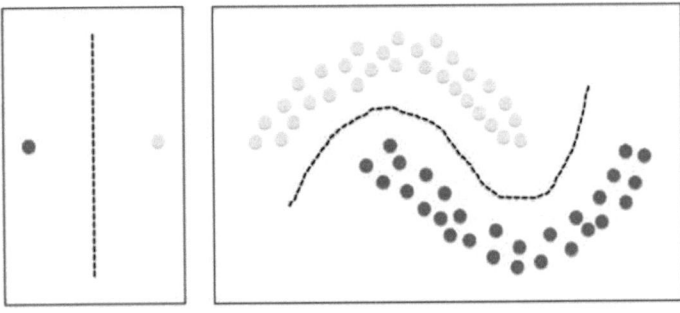

FIGURE 20.1: Unlabeled examples and prior belief.

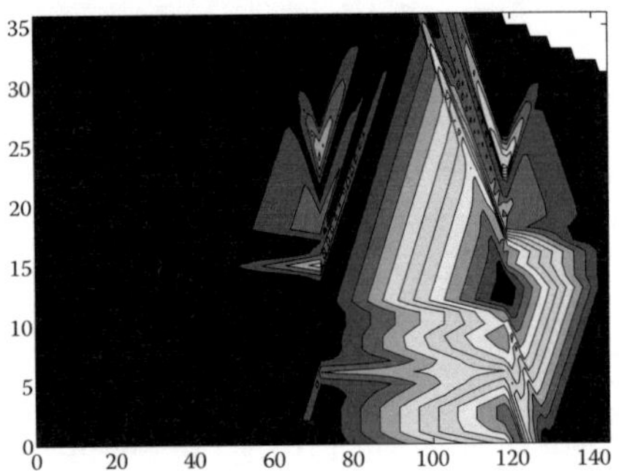

(a) WiFi RSS received by device A in T_1.

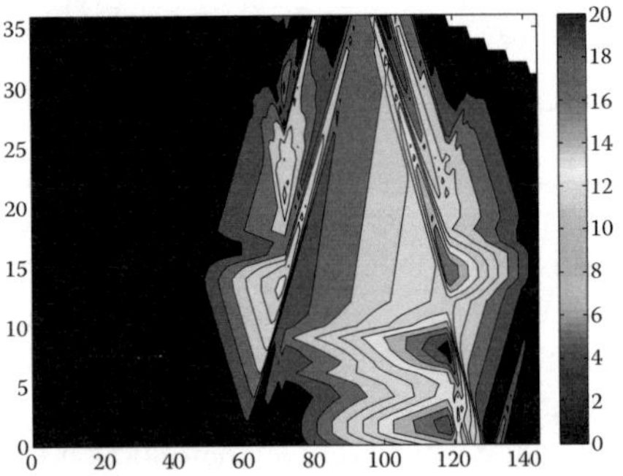

(b) WiFi RSS received by device A in T_2.

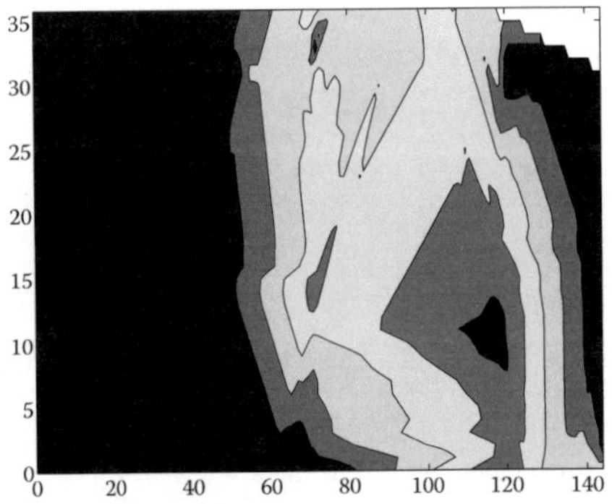

(c) WiFi RSS received by device B in T_1.

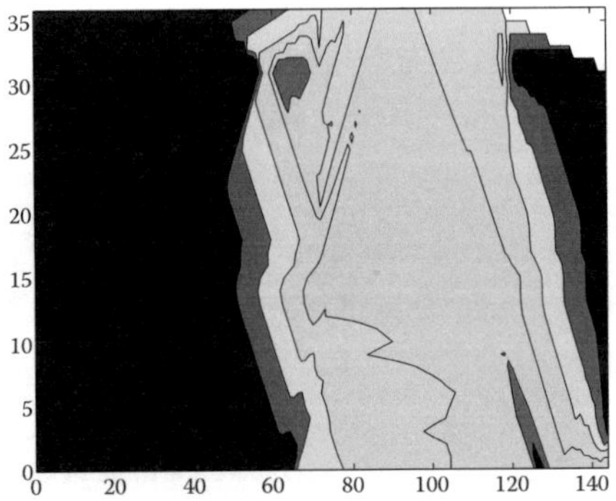

(d) WiFi RSS received by device B in T_2.

FIGURE 21.1: Contours of RSS values over a 2-dimensional environment collected from a same access point in difference time periods and received by different mobile devices. Different colors denote different values of signal strength.

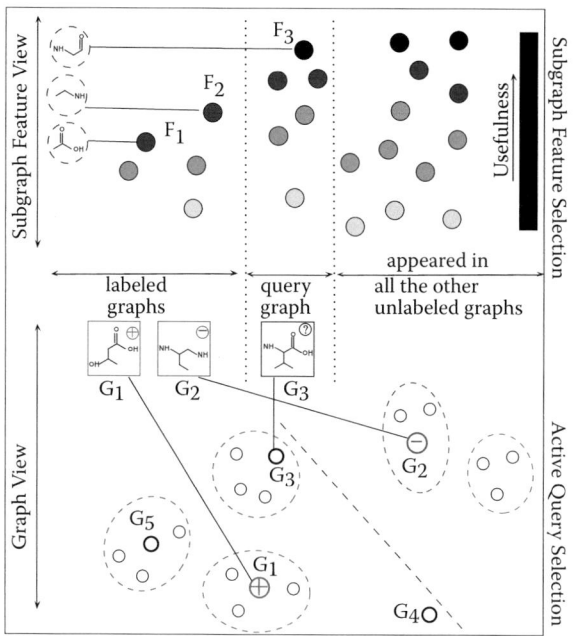

FIGURE 22.2: Motivation of active learning on small graphs [70].

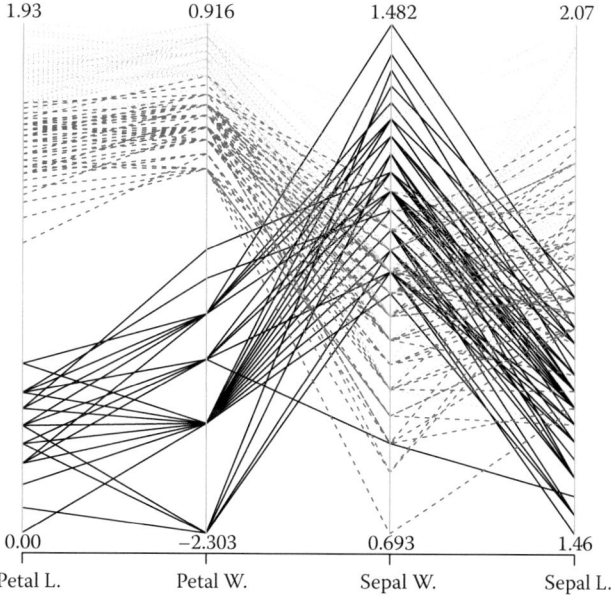

FIGURE 23.2: Parallel coordinates. In this example, each object has four dimensions and represents the characteristics of a species of iris flower (petal and sepal width and length in logarithmic scale). The three types of lines represent the three kinds of iris. With parallel coordinates, it is easy to see common patterns among flowers of the same kind; however, edge cluttering is already visible even with a small dataset.

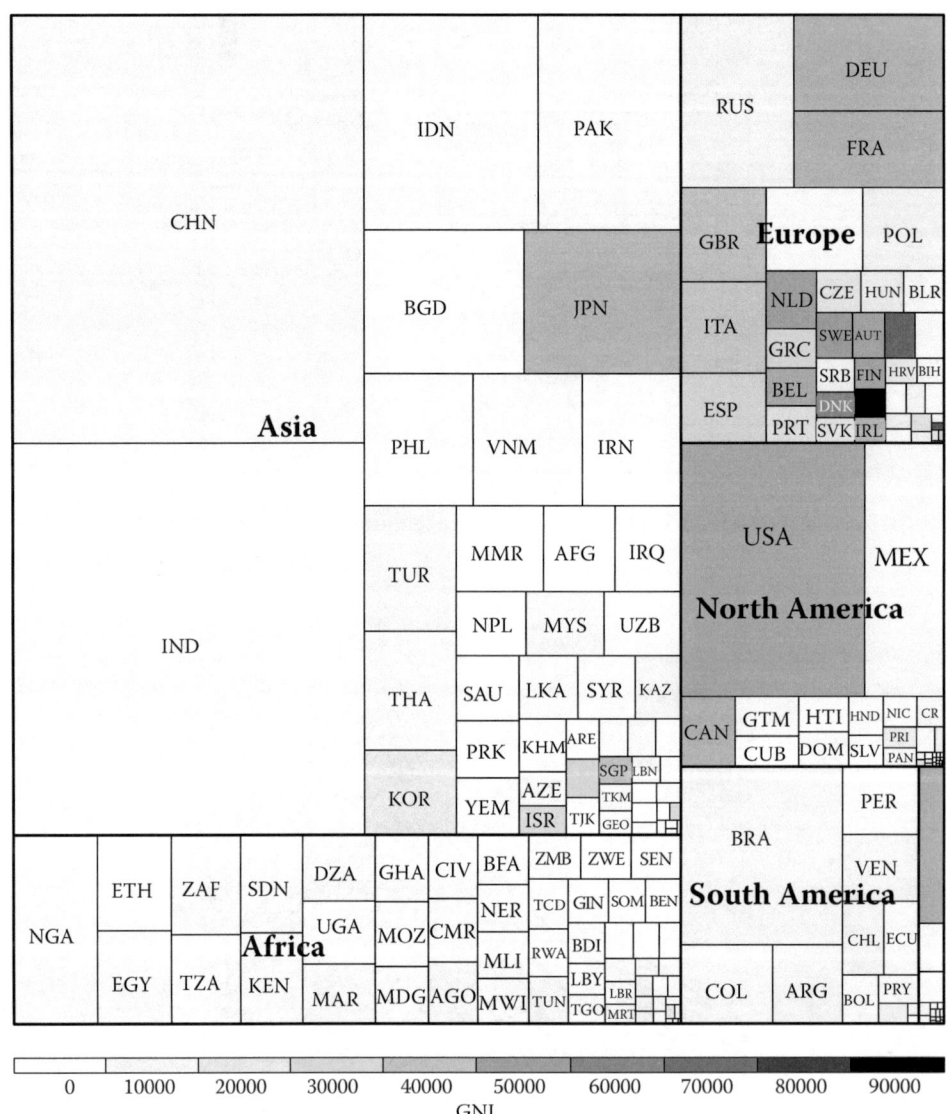

FIGURE 23.7: Treemaps. This plot represents a dataset of 2010 about population size and gross national income for each country. The size of each node of the treemap is proportional to the size of the population, while the shade of blue of each box represents the gross national income of that country. The countries of a continent are grouped together into a rectangular area.

Chapter 13

Time Series Data Classification

Dimitrios Kotsakos

Dept. of Informatics and Telecommunications
University of Athens
Athens, Greece
dimkots@di.uoa.gr

Dimitrios Gunopulos

Dept. of Informatics and Telecommunications
University of Athens
Athens, Greece
dg@di.uoa.gr

13.1 Introduction

Time-series data is one of the most common forms of data encountered in a wide variety of scenarios such as the stock markets, sensor data, fault monitoring, machine state monitoring, environmental applications, or medical data. The problem of classification finds numerous applications in the time series domain, such as the determination of predefined groups of entities that are most similar to a time series entity whose group is still unknown. Timeseries classification has numerous applications in diverse problem domains:

- Financial Markets: In financial markets, the values of the stocks represent time-series which continually vary with time. The classification of a new time series of unknown group can provide numerous insights into descisions about this specific time series.

- Medical Data: Different kinds of medical data such as EEG readings are in the form of time series. The classification of such time series, e.g., for a new patient, can provide insights into similar treatment or aid the domain experts in the decisions that have to be made, as similar behavior may indicate similar diseases.

- Machine State Monitoring: Numerous forms of machines create sensor data, which provides a continuous idea of the states of these objects. These can be used in order to provide an idea of the underlying behaviors and the groups a time series belongs to.

- Spatio-temporal Data: Trajectory data can be considered a form of multi-variate time series data, in which the X- and Y-coordinates of objects correspond to continuously varying series. The behavior in these series can be used in order to determine the class a trajectory belongs to, and then decide, e.g., if a specific trajectory belongs to a pedestrian or to a vehicle.

Time series data falls in the class of contextual data representations. Many kinds of data such as time series data, discrete sequences, and spatial data fall in this class. Contextual data contain two kinds of attributes:

- *Contextual Attribute:* For the case of time-series data, this corresponds to the time dimension. The time dimension provides the reference points at which the behavioral values are measured. The timestamps could correspond to actual time values at which the data points are measured, or they could correspond to indices at which these values are measured.

- *Behavioral Attribute:* This could correspond to any kind of behavior that is measured at the reference point. Some examples include stock ticker values, sensor measurements such as temperature, or other medical time series.

Given a time series object of unknown class, the determination of the class it belongs to is extremely challenging because of the difficulty in defining similarity across different time series, which may be scaled and translated differently both on the temporal and behavioral dimension. Therefore, the concept of similarity is very important for time series data classification. Therefore, this chapter will devote a section to the problem of time series similarity measures.

In the classification problem the goal is to separate the classes that characterize the dataset by a function that is induced from the available labeled examples. The ultimate goal is to produce a classifier that classifies unseen examples, or examples of unknown class, with high precision and recall, while being able to scale well. A significant difference between time series data classification and classification of objects in Euclidean space is that the time series to be classified to pre-defined or pre-computed classes may not be of equal length to the time series already belonging to these classes. When this is not the case, so all time series are of equal length, standard classification techniques can be applied by representing each time series as a vector and using a traditional L_p-norm distance. With such an approach, only similarity in time can be exploited, while similarity in shape and similarity in change are disregarded. In this study we split the classification process in two basic steps. The first one is the choice of the similarity measure that will be employed, while the second one concerns the classification algorithm that will be followed. Xing et al. in their sequence classification survey [42] argue that the high dimensionality of the feature space and their sequential nature make sequence classification a challenging task. These facts apply in the time series classification task as well, as time series are essentially sequences.

The majority of the studies and approaches proposed in the literature for time series classification is application dependent. In most of them the authors try to improve the performance of a specific feature. From this point of view, the more general time series classification approaches are of their own interest and importance.

13.2   Time Series Representation

In this study we employ the raw data representation, where a time series T of length n is represented as an ordered sequence of values $T = [t_1, t_2, ..., t_n]$. T is called raw representation of the time series data. Regarding the dimensionality of the data points t_i, for each time series T, each $t_i \in T$ can be n-dimensional, where $n \geq 1$. For example, when considering time series that represent trajectories of moving objects in the two-dimensional plane, $n = 2$, while for financial time series that represent stock prices, $n = 1$.

13.3   Distance Measures

When it comes to time series data mining, the distance or similarity measure that will be used to compare time series objects must be carefully selected and defined, as time series exhibit different characteristics than traditional data types. In this section we briefly describe the most commonly used distance measures that have been used in the literature to facilitate similarity search on time series data. Ding et al. performed an extensive evaluation over a variety of representations and distance/similarity measures for time series, using a nearest neighbor classifier (1NN) on labeled data, where the correct class label of each time series is known a priori [9].

13.3.1 L_p-Norms

The L_p-norm is a distance metric, since it satisfies all of the non-negativity, identity, symmetry and the triangle inequality conditions. An advantage of the L_p-norm is that it can be computed in linear time to the length of the trajectories under comparison, thus its time complexity is O(n), n being the length of the time series. In order to use the L_p-norm, the two time series under comparison must be of the same length.

The Minkowski of order p or the L_p-norm distance, being the generalization of Euclidean distance, is defined as follows:

$$L_p - \text{norm}(T_1, T_2) = D_{M,p}(T_1, T_2) = \sqrt[p]{\sum_{i=1}^{n}(T_{1i} - T_{2i})^p}.$$ (13.1)

The Euclidean Distance between two one-dimensional time series T_1 and T_2 of length n is a special case of the L_p-norm for $p = 2$ and is defined as:

$$D_E(T_1, T_2) = L_2 - \text{norm} = \sqrt{\sum_{i=1}^{n}(T_{1i} - T_{2i})^2}.$$ (13.2)

L_1-norm (p=1) is named the Manhattan distance or city block distance.

13.3.2   Dynamic Time Warping (DTW)

Dynamic Time Warping (DTW) is a well known and widely used shape-based distance measure. DTW computes the warping path $W = w_1, w_2, ..., w_K$ with $\max(m, n) \leq K \leq m + n - 1$ of minimum distance for two time series of lengths m and n.

DTW stems from the speech processing community [33] and has been very popular in the literature of time series distance measures. With use of dynamic programming, DTW between two one-dimensional time series T_1 and T_2 of length m and n, respectively, can be computed as follows:

(a) $D_{\mathrm{DTW}}(T_1, T_2) = 0$, if $m = n = 0$

(b) $D_{\mathrm{DTW}}(T_1, T_2) = \infty$, if $m = n = 0$

(c) $D_{\mathrm{DTW}}(T_1, T_2) = \mathrm{dist}(T_{11}, T_{21}) + minFactor$, otherwise

where *minFactor* is computed as:

$$minFactor = \min \begin{cases} D_{\mathrm{DTW}}(\mathrm{Rest}(T_1), \mathrm{Rest}(T_2)) \\ D_{\mathrm{DTW}}(\mathrm{Rest}(T_1), T_2) \\ D_{\mathrm{DTW}}(T_1, \mathrm{Rest}(T_2)) \end{cases}$$

where $\mathrm{dist}(T_{1,1}, T_{2,1})$ is typically the L_2-norm. The Euclidean Distance, as long as all L_p-norms, described in 13.3.1, performs a one-to-one mapping between the data points of the time series under comparison. Thus, it can be seen as a special case of the DTW distance, which performs a one-to-many mapping. DTW is a more robust distance measure than L_p-norm because it allows time shifting and thus matches similar shapes even if they have a time phase difference. An important point is that DTW does not satisfy the triangular inequality, which could be a problem while indexing time series. However, in the literature there are several lower bounds that serve as solutions for indexing DTW offering faster performance [36]. Another advantage of DTW over L_p-norm is that DTW can handle different sampling intervals in the time series. This is a very important feature especially for long time series that span many years as the data sampling strategy may change over a long time.

A variation of DTW is the Fréchet distance. Intuitively, the Fréchet distance is defined as the minimum length of a leash a dog-owner needs to have to be connected to his dog at all times for a given trajectory of the owner and the dog, over all possible monotone movements of the owner and the dog along their trajectories.

When comparing two trajectories — or curves — using the Fréchet distance, they are represented as the trajectories of the dog and the owner. Fréchet distance is used to calculate distances between curves and takes into account the location and ordering of the points along the curves [16], [2]. The Fréchet distance is essentially the L_{inf} equivalent of the Dynamic Time Warping distance. Eiter and Manilla proposed a discrete variation and a dynamic programming polynomial time algorithm for the computation of the Fréchet distance [12]. Driemel et al. give almost linear time approximation algorithms for Fréchet distance [11].

13.3.3 Edit Distance

EDIT distance (ED) comes from the field of string comparison and measures the number of insert, delete, and replace operations that are needed to make two strings of possibly different lengths identical to each other. More specifically, the EDIT distance between two strings S_1 and S_2 of length m and n, respectively, is computed as follows:

(a) $D_{ED}(S_1, S_2) = m$, if $n = 0$

(b) $D_{ED}(S_1, S_2) = n$, if $m = 0$

(c) $D_{ED}(S_1, S_2) = D_{ED}(\mathrm{Rest}(S_1), \mathrm{Rest}(S_2))$, if $S_{1_1} = S_{2_1}$

(d) $D_{ED}(S_1, S_2) = \min \begin{cases} D_{ED}(\text{Rest}(S_1), \text{Rest}(S_2)) + 1 \\ D_{ED}(\text{Rest}(S_1), S_2) + 1 \\ D_{ED}(S_1, \text{Rest}(S_2)) + 1. \end{cases}$ otherwise

Although ED for strings is proven to be a metric distance, the two ED-related time series distance measures DTW and Longest Common Subsequence (LCSS) that will be described in the next subsection are proven not to follow the triangle inequality. Lei Chen [7] proposed two extensions to EDIT distance, namely Edit distance with Real Penalty (ERP) to support local time shifting and Edit Distance on Real sequence (EDR) to handle both local time shifting and noise in time series and trajectories. Both extensions have a high computational cost, so Chen proposes various lower bounds, indexing and pruning techniques to retrieve similar time series more efficiently. Both ERP and DTW can handle local time shifting and measure the distance between two out-of-phase time series effectively. An advantage that ERP has over DTW is that the former is a metric distance function, whereas the latter is not. DTW does not obey triangle inequality, and therefore traditional index methods cannot be used to improve efficiency in DTW-based applications. On the other hand, ERP is proven to be a metric distance function [7], and therefore traditional access methods can be used. EDR, as a distance function, proves to be more robust than Euclidean distance, DTW, and ERP and more accurate than LCSS as will be described in the next subsection. EDR is not a metric, thus the author proposes three non-constraining pruning techniques (mean value Q-grams, near triangle inequality, and histograms) to improve retrieval efficiency.

13.3.4 Longest Common Subsequence (LCSS)

The Longest Common Subsequence (LCSS) distance is a variation of EDIT distance described in 13.3.3 [37]. LCSS allows time series to stretch in the time axis and does not match all elements, thus being less sensitive to outliers than L_p-norms and DTW. Specifically, the LCSS distance between two real-valued sequences S_1 and S_2 of length m and n respectively is computed as follows:

(a) $LCSS_{\delta,\varepsilon}(S_1, S_2) = 0$, if $n = 0$ or $m = 0$

(c) $LCSS_{\delta,\varepsilon}(S_1, S_2) = 1 + LCSS_{\delta,\varepsilon}(\text{HEAD}(S_1), \text{HEAD}(S_2))$
   if $|S_{1,m} - S_{2,m}| < \varepsilon$ and $|m - n| \leq \delta$

(d) $\max \begin{cases} LCSS_{\delta,\varepsilon}(\text{HEAD}(S_1), S_2) \\ LCSS_{\delta,\varepsilon}(S_1, \text{HEAD}(S_2)) \end{cases}$, otherwise

where $\text{HEAD}(S_1)$ is the subsequence $[S_{1,1}, S_{1,2}, ..., S_{1,m-1}]$, δ is an integer that controls the maximum distance in the time axis between two matched elements, and ε is a real number $0 < \varepsilon < 1$ that controls the maximum distance that two elements are allowed to have to be considered matched.

Apart from being used in time series classification, LCSS distance is often used in domains like speech recognition and text pattern mining. Its main drawback is that it often is needed to scale are transform the one sequence to the other. A detailed study of LCSS variations and algorithms is presented in [5].

13.4 *k*-NN

The well-known and popular traditional *k*-NN classifier is not different when it comes to time series classification. The most challenging part in this setting is again the choice of the distance or similarity measure that will be used to compute the distance or similarity between two time series objects. The intuition is that a time series object is assigned the class label of the majority of its k nearest neighbors. k-NN has been very popular in the classification literature, due to both its simplicity and the highly accurate results it produces.

Prekopcsák and Lemire study the problem of choosing a suitable distance measure for nearest neighbor time series classification [28]. They compare variations of the Mahalanobis distance against Dynamic Time Warping (DTW), and find that the latter is superior in most cases, while the Mahalanobis distance is one to two orders of magnitude faster. The task of learning one Mahalanobis distance per class is recommended by the authors, in order to achieve better performance. The authors experimented with a traditional 1-NN classifier and came to the conclusion that the class-specific covariance-shrinkage estimate and the class-specific diagonal Mahalanobis distances achieve the best results regarding classification error and running time. Another interesting observation is that in 1-NN classification, the DTW is at least two orders of magnituted slower than the Mahalanobis distance for all 17 datasets the authors experimented on.

Ding et al. [9] use the k-NN classifier with $k = 1$ on labelled data as a golden measure to evaluate the efficacy of a variety of distance measures in their attempt to extensively validate representation methods and similarity measures over many time series datasets. Specifically, the idea is to predict the correct class label of each time series with an 1-NN classifier as the label of the object that is most similar to the query object. With this approach, the authors evaluate directly the effectiveness of each distance measure as its choice is very important for the k-NN classifier in all settings, let alone time series classification, where the choice of a suitable distance measure is not a trivial task.

In a more recent work, similar to k-NN classification, Chen et al. propose generative latent source model for *online* binary time series classification in a social network with the goal of detecting trends as early as possible [6]. The proposed classifier uses weighted majority voting to classify the unlabeled time series, by assigning the time series to be classified to the class indicated by the labeled data. The labeled time series vote in favor of the ground truth model they belong to. The authors applied the described model for prediction of virality of news topics in the Twitter social network, while they do not take the definitions of *virality* or *trend* into account and use them only as a black box ground truth. The experimental results indicated that weighted majority voting can often detect or predict trends earlier than Twitter.

13.4.1 Speeding up the k-NN

In order to improve the time needed to classify a time series using the k-NN classifier, an index structure can be used to store the labeled objects. However, it can be proved that neither DTW nor LCSS fail to satisfy the the triangle inequality, which is a basic metric property, thus they are not metrics and a traditional index cannot be used. More specifically, in an example scenario with three time series A, B, and C where:

$$A = 1,1,5; B = 2,2,5; C = 2,2,2,2,2,5 \qquad (13.3)$$

the DTW computation yields:

$$DTW(A,B) = 2, DTW(B,C) = 0, DTW(A,C) = 5 \qquad (13.4)$$

thus $DTW(A,B) + DTW(B,C) < DTW(A,C)$. Similarly if:

$$A = 1,1,1,1; B = 2,2,2,2; C = 3,3,3,3 \qquad (13.5)$$

with $\varepsilon = 1$, the LCSS computation yields:

$$LCSS_\varepsilon(A,B) = 4, LCSS_\varepsilon(B,C) = 4, LCSS_\varepsilon(A,C) = 0. \qquad (13.6)$$

As LCSS is not a distance function but a similarity measure, we can turn it into a distance:

$$D_{LCSS}(TS_i, TS_j) = 1 - \frac{LCSS(TS_i, TS_j)}{max\{length(TS_i), length(TS_j)\}}. \tag{13.7}$$

By applying the above formula on the time series of the example we obtain:

$$D_{LCSS}(A,B) = 0, D_{LCSS}(B,C) = 0, D_{LCSS}(A,C) = 1. \tag{13.8}$$

Again, $D_{LCSS}(A,B) + D_{LCSS}(B,C) < D_{LCSS}(A,C)$.

Alternatively, instead of using a traditional index structure that requires the distance measure to be a metric, Vlachos et al. [40] have proposed to examine if the used distance measure satisfies a pseudo-metric property. In their work they show that LCSS indeed has a pseudo-metric property that allows us to bound $D_{LCSS}(TS_i, TS_j)$ if we know $D(TS_i, TS_k)$ and $D(TS_k, TS_j)$ for a third time series TS_k. This fact allows tree-based intex structures to be used and speeds up the identification of the k nearest neighbors to a query time series TS_q.

Another speeding up technique is to lower bound the used distance measure. For DTW, $LB_{improved}$ proposed in [25] is one of the most recent lower bounding techniques, while other lower bounds have been proposed in [21] and [1]. Vlachos et al. have proposed lower bounding techniques for LCSS in [38] and [39]

13.5 Support Vector Machines (SVMs)

Support Vector Machine (SVMs) [8] is a powerful machine learning approach that has been used widely in the literature for solving binary classification problems. The SVM algorithm constructs a hyper plane or set of hyper planes in high dimensional space without any assumptions regarding the distribution of the studied datasets. SVMs use a hypothesis space that consists of linear functions in a high dimensional feature space and the training phase derives from optimization theory. SVM classification achieves a global minimum solution, while involving a relatively fast training phase, which becomes of increasing importance as datasets of ever larger scale become available. SVMs have proven to perform well in tackling overfitting problems.

SVMs are reported to perform non-optimally when it comes to time series classification, because of the not-trivial task to identify intra-class similarities between the training objects, whereas distance-based classification methods like k-NN can overcome this difficulty by using DTW-distance. As described above, DTW is insensitive to shifts, length variations, and deformations and captures the similarity between time series despite of the presence of such characteristics.

More specifically, SVM operates on a training dataset $\mathcal{D}_{training}$ that contains N training time series and their corresponding class labels, such that $\mathcal{D}_{training} = (ts_0, l_0), (ts_1, l_1), ...(ts_{N-1}, l_{N-1})$. The Support Vector Machine algorithm separates the sets in a space of much higher dimensionality, since the classes may not be linary separable in the original space. In order to do this mapping, SVMs use kernel functions that represent a dot product of the input data points mapped into the higher dimensional feature space by a transformation ϕ.

Köknar and Latecki focus on the problem of classification of imbalanced time series datasets [23]. They argue that it is impossible to insert synthetic data points in *feature spaces*, while they propose to insert synthetic data points in *distance spaces*. Doing so, they then use a Support Vector Machines classifier for time series classification and they report significant improvements of classification rates of points belonging to minority classes, called rare events. The authors also report an overall improvement in the accuracy of SVM after the insertion of the synthetic data points. The main contribution of [23] was the insertion of synthetic data points and the over-sampling of minority class in imbalanced data sets. Synthetic data points can be added in all distance spaces, even in non-metric ones, so that elastic sequence matching algorithms like DTW can be used. Huerta et al. extend SVMs to classify multidimensional time series by using concepts from nonlinear dynamical systems theory [17], and more specifically the Rössler oscillator, which is a three-dimensional dynamical system. The authors experimented on both real and synthetic benchmark datasets and the proposed dynamical SVMs proved to achieve in several cases better results than many proposed kernel-based approaches for multidimensional time series classification. Sasirekha and Kumar applied SVM classification on real-life long-term ECG time series recordings and concluded that SVMs can act as a good predictor of the heart rate signal [34]. In their work they propose and compare a variety of feature extraction techniques and compare the respective accuracy values by comparing the corresponding Heart Rate Variability (HRV) features. The experimental results show that SVM classification for heartbeat time series performs well in detecting different attacks and is comparable to those reported in literature. Kampouraki et al. [20] also applied Gaussian kernel-based SVM classification with a variety of statistical and wavelet HRV features to tackle the heartbeat time series classification problem, arguing that the SVM classifier performs better in this specific application than other proposed approaches based on neural networks. In their study, SVMs proved to be more robust to cases where the signal-to-noise ratio was very low. In the experimental evaluation, the authors compared SVM classification to learning vector quantization (LVQ) neural network [22] and a Levenberg-Marquardt (LM) minimization back-propagation neural network [15], and the SVMs achieved better precision and recall values, even when the signals to be classified were noisy. Grabocka et al. propose a method that improves SVM time series classification performance by transforming the training set by developing new training objects based on the support vector instances [14]. In their experimental evaluation on many time series datasets, the enhanced method outperforms both the DTW-based k-NN and the traditional SVM classifiers. The authors propose a data-driven, localized, and selective new-instance generation method based on Virtual Support Vectors (VSV), which have been used for image classification. In a brief outline, the method uses a deformation algorithm to transform the time series, by splitting each time series instance into many regions and transforming each of them separately. The experimental results show an accuracy improvement between 5% and 11% over traditional SVM classification, while producing better mean error rates than the DTW-based k-NN classifier. However, the proposed approach of boosting SVM accuracy has the trade-off of being worse in terms of time complexity, because of the presence of more training examples.

As all works mentioned above underline, the main issue in applying SVMs to time series classification is the choice of the kernel function that leads to the minimum generalization error and at the same time keeps the classification error of the time series training set low. Various kernel approaches have been proposed in the literature, some of which have been used to classify sequences that have similar characteristics with time series (e.g. [18, 41] or for handwriting recognition [4].

13.6 Classification Trees

Classification or decision trees, being a widely used classifier, create a model that predicts the value of a target variable based on several input variables [27]. Classification trees are popular mainly because of the comprehensibility they offer regarding the classification task. Starting with the root node, a classification tree building algorithm searches among the features for the binary distinction that provides the most information about the classes. Then, the same process is repeated for each of the resulting new nodes, continuing the recursion until a stopping criterion is reached. The resulting tree is often too large, a fact that leads to overfitting problems, so a pruning process reduces the tree size in order to both improve classification accuracy and to reduce classification complexity. The most popular decision tree algorithms are ID3 (Iterative Dichotomiser 3) [29], C4.5, which is an extension of ID3 [30], CART (Classification and Regression Tree) [27], and CHAID (CHi-squared Automatic Interaction Detector) [35], which performs multi-level splits when computing classification trees. Douzal-Chouakria and Amblard describe an extension to the traditional decision tree classifier, where the metric changes from one node to another and the split criterion in each node extracts the most characteristic subsequences of the time series to be split [10]. Geurts argues that many time series classification problems can be solved by identifying and exploiting local patterns in the time series. In [13] the author performs time series segmentation to extract prototypes that can be represented by sets of numerical values so that a traditional classifier can be applied. More specifically, the proposed method extracts patterns from time series, which then are combined in decision trees to produce classification rules. The author applies the proposed approach on a variety of synthetic and real datasets and proves that pattern extraction used this way can improve classification accuracy and interpretability, as the decision trees' rules can provide a user with a broad overview of classes and datasets. With a similar goal in mind, namely interpretability of classification results, Rodríguez and Alonso present a method that uses decision trees for time series classification [32]. More specifically, two types of trees are presented: a) interval-based trees, where each decision node evaluates a function in an interval, and b) DTW-based trees, where each decision node has a reference example, which the time series to classify is compared against, e.g., $DTW(ts_i, ts_{ref}) < Threshold$. The decision trees can also be constructed by extracting global and local features (e.g., global: mean, length, maximum; local: local maximum, slope, etc.) or by extracting simple or complex patterns. A complex pattern is a sequence of simple patterns and a simple pattern is a time series whose Euclidean Distance to the examined one is small enough. Most decision tree algorithms are known to have high variance error [19]. In their work, Jović et al. examined the performance of decision tree ensembles and experimented on biomedical datasets. Decision tree ensembles included AdaBoost.M1 and Multiboost applied to the traditional decision tree algorithm C4.5. The classification accuracy improved and the authors argue that decision tree ensembles are equivalent if not better than SVM-based classifiers that have been commonly used for time series classification in the biomedical domain, which is one of the most popular ones for time series classification.

As in all cases where classification trees are chosen as a mechanism to classify data of unknown labels, in the time series domain the main issue is find appropriate split criteria that separate the input dataset and avoid overfitting. Some time series specific issues regarding the choice of a split criterion are related to the open nature of the distance between two time series objects. Time series classification trees that use the same split criterion for all nodes may fail to identify unique characteristics that distinguish time series classes in different parts of the tree. Moreover, distance-based approaches should not consider only the cases where the distance is computed on the whole time series, since there may be cases where subsequences exhibit certain separation characteristics.

However, most of the works that use decision trees for time series classification focus on comprehensibility of the classification results, rather than running time or accuracy, precision, or recall. Thus, a general conclusion that can be drawn regarding decision trees for time series classification is the following: classification trees for time series are good and suitable when interpretability of results are the main concern. When classification accuracy matters, other classifiers, like k-NN or SVMs, achieve better results.

13.7   Model-Based Classification

A method to identify the most similar time series to a given one is by using a model to describe the time series. Kotsifakos et al. [24] used *Hidden Markov Models (HMMs)* to model the underlying structure of the time series. An HMM is a doubly stochastic process that contains a finite set of states, each one emitting a value by following a different probability distribution. Moreover, an HMM is described by a set of *transition probabilities* that define how the process moves from one state to another. When a dataset consists of sets of similar time series, then HMMs can be used to perform search or classification tasks. Each group of similar time series forms a class and each class is represented by an HMM. Then, given a query time series TS_q, the classification algorithm assigns TS_q to the class that maximizes the probability of having generated it. More specifically, regarding time series representation, an HMM is a dynamic probabilistic model that describes a time series and deals with uncertainty. It models the joint probability of a collection of hidden and observed discrete random variables, so it consists of directly unobservable variables that form the hidden layer of the model, each of them representing a different timestamp, assuming the Markov property. The Markov property holds when the conditional probability distribution of a future state of the studied process depends only on the present state and not on the time series values that preceded it. HMMs incorporate as well an observable layer that consists of observable variables that are affected by the corresponding hidden ones. For a time series $TS = ts_t, t \in [0, N]$, the HMM representing TS is defined by:

- a set of transition probabilities $TP = tp_{ij}$ that represent the probability of transitioning from state i to state j

- and a set of observation probabilities $OP = op_{jk}$ that represent the probability of observing the value k at state j.

Moreover, there is a set of prior probabilities $Pr = p_j$ that represent the probability for the first state of the model. HMMs in most proposed approaches assume a Gaussian probability distribution OP_j for each state j. The most common approach for learning Hidden Markov Models is the Expectation Maximization (EM) algorithm [26], which after performing a recursive estimation converges to a local maximum of the likelihood. More specifically, the EM algorithm is used by the popular BaumWelch algorithm in order to find the unknown parameters of a HMM [31]. It relies on the assumption that the i-th hidden variable depends only on the $(i-1)$-th hidden variable and not on previous ones, as the current observation variable does. So, given a set of observed time series that get transformed into a set of observed feature vectors, the Baum-Welch algorithm finds the maximum likelihood estimate of the paramters of the corresponding HMM. More formally, if X_t is a hidden random variable at time t, the probability $P(X_t|X_{t-1})$ does note depend on the value of t. A hidden Markov chain is described by $\theta = (TP, OP, Pr)$. The BaumWelch algorithm computes $\theta = \max_\theta P(TS|\theta)$, namely the parameters that maximize the probability of the observation sequence TS.

In [24], Kotsifakos et al. evaluated and compared the accuracy achieved by search-by-model and search-by-example methods in large time series databases. Search-by-example included DTW and constrained-DTW methods. In the search-by-model approach, the authors used a two-step training phase that consisted of *(i) initialization* and *(ii) refinement*. In the *initialization* step an HMM is computed for each class in the dataset. In the *refinement* step, the Viterbi algorithm is applied on each training time series per class in order to identify the best state sequence for each of them. The transition probabilities used in this step are the ones computed at the initialization step of the algorithm. After an HMM has been trained for each class in the training set, a probability distribution is defined, which represents the probability of each time series in the testing set assuming that it belongs to the corresponding class. Given a time series TS, the probability of TS belonging to class C_i, $P(TS|TS \in C_i), i = 0, ..., |C|$ is computed with the dynamic programming Forward algorithm [31]. In the experimental evaluation on 20 time series datasets, the model-based classification method yielded more accurate results than the search-by-example technique when there were a lot of training examples per model. However, when there were few training examples per class, search-by-example proved to be more accurate.

In [3], Antonucci and De Rosa also use HMMs for time series classification, while coping with the problems imposed by the short length of time series in some applications. The EM algorithm calculates unreliable estimates when the amount of training data is small. To tackle this problem, the authors describe an imprecise version of the learning algorithm and propose an HMM-based classifier that returns multiple classes when the highest-likelihood intervals corresponding to the returned class overlap.

13.8 Distributed Time Series Classification

Zeinalipour-Yazti et al. define and solve the distributed spatio-temporal similarity search problem, which given a a query trajectory Q aims to identify the trajectories that are most similar to Q [44]. Trajectories are essentially time series, in that they consist of multidimensional points ordered in the time dimension. Trajectories are usually two-dimensional time series, when movement in the two-dimensional plane is considered. In the described setting each trajectory is segmented over a distributed network of nodes, whether they are sensors or vehicles. The authors propose two algorithms that compute local lower and upped bounds on the similarity between the distributed segments of the time series and the query time series Q. In a similar spirit, in a more recent work, Zeinalipour-Yazti et al. describe a setting where the distributed power of smartphones has been used to provide a distributed solution to a complex time series classification problem [43]. The difference to the previously mentioned setting is that in the latter each trajectory or trace is in its entirety stored in a specific node, whereas in the former each trajectory was segmented across the network. Specifically, in the proposed framework the authors compare a query trace Q against a dataset of traces that are generated from distributed smartphones. Zeinalipour-Yazti et al. proposed a top-K query processing algorithm that uses distributed trajectory similarity measures that are insensitive to noise, in order to identify the most identical time series to the query trace Q.

13.9 Conclusion

Time series data mining literature includes a lot of works devoted to time series classification tasks. As time series datasets and databases grow larger in size and as applications are evolving, performance and accuracy of time series classification algorithms become ever more important for real life applications. Although most time series classification approaches adopt and adapt techniques from general data classification studies, some essential modifications to these techniques have to be introduced in order to cope with the nature of time series distance and time series representation. The problems and the benefits of time series classification have found applications in economics, health care, and multimedia among others. In this chapter, we highlighted the major issues and the most influential corresponding techniques that deal with the time series classification task.

Acknowledgements

This work was supported by European Union and Greek National funds through the Operational Program "Education and Lifelong Learning"of the National Strategic Reference Framework (NSRF) — Research Funding Programs: Heraclitus II fellowship, THALIS — GeomComp, THALIS — DISFER, ARISTEIA — MMD, ARISTEIA — INCEPTION" and the EU FP7 project INSIGHT (www.insight-ict.eu).

Bibliography

[1] Ghazi Al-Naymat, Sanjay Chawla, and Javid Taheri. Sparsedtw: A novel approach to speed up dynamic time warping. In *Proceedings of the Eighth Australasian Data Mining Conference-Volume 101*, pages 117–127. Australian Computer Society, Inc., 2009.

[2] Helmut Alt and Michael Godau. Computing the fréchet distance between two polygonal curves. *International Journal of Computational Geometry & Applications*, 5(01–02):75–91, 1995.

[3] Alessandro Antonucci and Rocco De Rosa. Time series classification by imprecise hidden markov models. In *WIRN*, pages 195–202, 2011.

[4] Claus Bahlmann, Bernard Haasdonk, and Hans Burkhardt. Online handwriting recognition with support vector machines—a kernel approach. In *Proceedings of Eighth International Workshop on Frontiers in Handwriting Recognition, 2002*, pages 49–54. IEEE, 2002.

[5] Lasse Bergroth, Harri Hakonen, and Timo Raita. A survey of longest common subsequence algorithms. In *SPIRE 2000. Proceedings of Seventh International Symposium on String Processing and Information Retrieval, 2000*, pages 39–48. IEEE, 2000.

[6] George H Chen, Stanislav Nikolov, and Devavrat Shah. A latent source model for online time series classification. *arXiv preprint arXiv:1302.3639*, 2013.

[7] Lei Chen, M Tamer Özsu, and Vincent Oria. Robust and fast similarity search for moving object trajectories. In *Proceedings of the 2005 ACM SIGMOD International Conference on Management of Data*, pages 491–502. ACM, 2005.

[8] Corinna Cortes and Vladimir Vapnik. Support-vector networks. *Machine Learning*, 20(3):273–297, 1995.

[9] Hui Ding, Goce Trajcevski, Peter Scheuermann, Xiaoyue Wang, and Eamonn Keogh. Querying and mining of time series data: experimental comparison of representations and distance measures. *Proceedings of the VLDB Endowment*, 1(2):1542–1552, 2008.

[10] Ahlame Douzal-Chouakria and Cécile Amblard. Classification trees for time series. *Pattern Recognition*, 45(3):1076–1091, 2012.

[11] Anne Driemel, Sariel Har-Peled, and Carola Wenk. Approximating the fréchet distance for realistic curves in near linear time. *Discrete & Computational Geometry*, 48(1):94–127, 2012.

[12] Thomas Eiter and Heikki Mannila. Computing discrete fréchet distance. *Tech. Report CS-TR-2008–0010, Christian Doppler Laboratory for Expert Systems, Vienna, Austria*, 1994.

[13] Pierre Geurts. Pattern extraction for time series classification. In *Principles of Data Mining and Knowledge Discovery*, pages 115–127. Springer, 2001.

[14] Josif Grabocka, Alexandros Nanopoulos, and Lars Schmidt-Thieme. Invariant time-series classification. In *Machine Learning and Knowledge Discovery in Databases*, pages 725–740. Springer, 2012.

[15] Martin T Hagan and Mohammad B Menhaj. Training feedforward networks with the marquardt algorithm. *IEEE Transactions on Neural Networks*, 5(6):989–993, 1994.

[16] Sariel Har-Peled et al. New similarity measures between polylines with applications to morphing and polygon sweeping. *Discrete & Computational Geometry*, 28(4):535–569, 2002.

[17] Ramón Huerta, Shankar Vembu, Mehmet K Muezzinoglu, and Alexander Vergara. Dynamical SVM for time series classification. In *Pattern Recognition*, pages 216–225. Springer, 2012.

[18] Tommi Jaakkola, David Haussler, et al. Exploiting generative models in discriminative classifiers. *Advances in Neural Information Processing Systems*, pages 487–493, 1999.

[19] Alan Jović, Karla Brkić, and Nikola Bogunović. Decision tree ensembles in biomedical time-series classification. In *Pattern Recognition*, pages 408–417. Springer, 2012.

[20] Argyro Kampouraki, George Manis, and Christophoros Nikou. Heartbeat time series classification with support vector machines. *IEEE Transactions on Information Technology in Biomedicine*, 13(4):512–518, 2009.

[21] Eamonn Keogh and Chotirat Ann Ratanamahatana. Exact indexing of dynamic time warping. *Knowledge and Information Systems*, 7(3):358–386, 2005.

[22] Teuvo Kohonen. The self-organizing map. *Proceedings of the IEEE*, 78(9):1464–1480, 1990.

[23] Suzan Köknar-Tezel and Longin Jan Latecki. Improving SVM classification on imbalanced time series data sets with ghost points. *Knowledge and Information Systems*, 28(1):1–23, 2011.

[24] Alexios Kotsifakos, Vassilis Athitsos, Panagiotis Papapetrou, Jaakko Hollmén, and Dimitrios Gunopulos. Model-based search in large time series databases. In *PETRA*, page 36, 2011.

[25] Daniel Lemire. Faster retrieval with a two-pass dynamic-time-warping lower bound. *Pattern Recognition*, 42(9):2169–2180, 2009.

[26] Geoffrey McLachlan and Thriyambakam Krishnan. *The EM Algorithm and Extensions*, volume 382. John Wiley & Sons, 2007.

[27] L Breiman, JH Friedman, RA Olshen, and Charles J Stone. *Classification and Regression Trees*. Wadsworth International Group, 1984.

[28] Zoltán Prekopcsák and Daniel Lemire. Time series classification by class-specific mahalanobis distance measures. *Advances in Data Analysis and Classification*, 6(3):185–200, 2012.

[29] J. Ross Quinlan. Induction of decision trees. *Machine learning*, 1(1):81–106, 1986.

[30] John Ross Quinlan. *C4.5: Programs for Machine Learning*, volume 1. Morgan Kaufmann, 1993.

[31] Lawrence R Rabiner. A tutorial on hidden markov models and selected applications in speech recognition. *Proceedings of the IEEE*, 77(2):257–286, 1989.

[32] Juan J Rodríguez and Carlos J Alonso. Interval and dynamic time warping-based decision trees. In *Proceedings of the 2004 ACM symposium on Applied computing*, pages 548–552. ACM, 2004.

[33] Nick Roussopoulos, Stephen Kelley, and Frédéric Vincent. Nearest neighbor queries. *ACM sigmod record*, 24(2):71–79, 1995.

[34] A Sasirekha and P Ganesh Kumar. Support vector machine for classification of heartbeat time series data. *International Journal of Emerging Science and Technology*, 1(10):38–41, 2013.

[35] Merel van Diepen and Philip Hans Franses. Evaluating chi-squared automatic interaction detection. *Information Systems*, 31(8):814–831, 2006.

[36] Michail Vlachos, Dimitrios Gunopulos, and Gautam Das. Rotation invariant distance measures for trajectories. In *Proceedings of the Tenth ACM SIGKDD International Conference on Knowledge Discovery and Data Mining*, pages 707–712. ACM, 2004.

[37] Michail Vlachos, Dimitrios Gunopulos, and George Kollios. Robust similarity measures for mobile object trajectories. In *2002. Proceedings of the 13th International Workshop on Database and Expert Systems Applications*, pages 721–726. IEEE, 2002.

[38] Michail Vlachos, Marios Hadjieleftheriou, Dimitrios Gunopulos, and Eamonn Keogh. Indexing multi-dimensional time-series with support for multiple distance measures. In *Proceedings of the ninth ACM SIGKDD International Conference on Knowledge Discovery and Data Mining*, pages 216–225. ACM, 2003.

[39] Michail Vlachos, Marios Hadjieleftheriou, Dimitrios Gunopulos, and Eamonn Keogh. Indexing multidimensional time-series. *The VLDB Journal—The International Journal on Very Large Data Bases*, 15(1):1–20, 2006.

[40] Michail Vlachos, George Kollios, and Dimitrios Gunopulos. Discovering similar multidimensional trajectories. In *2002 Proceedings on the 18th International Conference on Data Engineering*, pages 673–684. IEEE, 2002.

[41] Chris Watkins. Dynamic alignment kernels. *Advances in Neural Information Processing Systems*, pages 39–50, 1999.

[42] Zhengzheng Xing, Jian Pei, and Eamonn Keogh. A brief survey on sequence classification. *SIGKDD Explor. Newsl.*, 12(1):40–48, November 2010.

[43] Demetrios Zeinalipour-Yazti, Christos Laoudias, Costandinos Costa, Michail Vlachos, M Andreou, and Dimitrios Gunopulos. Crowdsourced trace similarity with smartphones. *IEEE Transactions on Knowledge and Data Engineering*, 25(6):1240–1253, 2012.

[44] Demetrios Zeinalipour-Yazti, Song Lin, and Dimitrios Gunopulos. Distributed spatio-temporal similarity search. In *Proceedings of the 15th ACM International Conference on Information and Knowledge Management*, pages 14–23. ACM, 2006.

Chapter 14

Discrete Sequence Classification

Mohammad Al Hasan

Indiana University Purdue University
Indianapolis, IN
alhasan@cs.iupui.edu

14.1 Introduction

Sequence classification is an important research task with numerous applications in various domains, including bioinformatics, e-commerce, health informatics, computer security, and finance. In bioinformatics domain, classification of protein sequences is used to predict the structure and function of newly discovered proteins [12]. In e-commerce, query sequences from a session are used to distinguish committed shoppers from the window shoppers [42]. In the Web application domain, a sequence of queries is used to distinguish Web robots from human users [49]. In health informatics, sequence of events is used in longitudinal studies for predicting the risk factor of a patient for a given disease [17]. In the security domain, the sequence of a user's system activities on a terminal is monitored for detecting abnormal activities [27]. In the financial world, sequences of customer activities, such as credit card transactions, are monitored to identify potential identity theft, sequences of bank transactions are used for classifying accounts that are used for money laundering [33], and so on. See [36] for a survey on sequential pattern mining

Because of its importance, sequence classification has found the deserved attention of the researchers from various domains, including data mining, machine learning, and statistics. Earlier works on this task are mostly from the field of statistics that are concerned with classifying sequences of real numbers that vary with time. This gave rise to a series of works under the name of *time series classification* [15], which has found numerous applications in real life. Noteworthy examples include classification of ECG signals for classification between various states of a patient [20], and classification of stock market time series [22]. However, in recent years, particularly in data mining, the focus has shifted towards classifying discrete sequences [25, 28], where a sequence is considered as a sequence of *events*, and each event is composed of one or a set of *items* from an alphabet. Examples can be a sequence of queries in a Web session, a sequence of events in a manufacturing process, a sequence of financial transactions from an account, and so on. Unlike time series, the events in a discrete sequence are not necessarily ordered based on their temporal position. For instance, a DNA sequence is composed of four amino acids A, C, G, T, and a DNA segment, such as $ACCGTTACG$, is simply a string of amino acids without any temporal connotation attached to their order in the sequence.

In this chapter, our discussion will be focused on the classification of discrete sequences only. This is not limiting, because any continuous sequence can be easily discretized using appropriate methods [13]. Excluding a brief survey [52], we have not found any other works in the existing literature that summarize various sequence classification methods in great detail as we do in this chapter.

The rest of the chapter is organized as follows. In Section 14.2, we discuss some background materials. In Section 14.3, we introduce three different categories in which the majority of sequence classification methods can be grouped. We then discuss sequence classification methods belonging to these categories in each of the following three sections. In Section 14.7 we discuss some of the methods that overlap across multiple categories based on the grouping defined in Section 14.3. Section 14.8 provides a brief overview of a set of sequence classification methods that uses somewhat non-typical problem definition. Section 14.9 concludes this chapter.

14.2 Background

In this section, we will introduce some of the background materials that are important to understand the discussion presented in this chapter. While some of the background materials may have been covered in earlier chapters of this book, we cover them anyway to make the chapter as independent as possible.

14.2.1 Sequence

Let, $I = \{i_1, i_2, \cdots, i_m\}$ be a set of m distinct items comprising an alphabet, Σ. An *event* is a nonempty unordered collection of items. An event is denoted as $(i_1 i_2 \ldots i_k)$, where i_j is an item. Number of items in an event is its *size*. A *sequence* is an ordered list of events. A sequence α is denoted as $(\alpha_1 \rightarrow \alpha_2 \rightarrow \cdots \rightarrow \alpha_q)$, where α_i is an event. For example, $AB \rightarrow E \rightarrow ACG$ is a sequence, where AB is an event, and A is an item. The *length* of the sequence $(\alpha_1 \rightarrow \alpha_2 \rightarrow \cdots \rightarrow \alpha_q)$ is q and the width is the maximum size of any α_i, for $1 \leq i \leq q$. For some sequences, each of the events contain exactly one item, i.e., the width of the sequence is 1. In that case, the sequence is simply an ordered string of items. In this paper, we call such a sequence a *string sequence* or simply a *string*. Examples of string sequences are DNA or protein sequences, query sequences, or sequences of words in natural language text.

14.2.2 Sequence Classification

Assume we are given a sequence dataset, $\mathcal{D}$, which consists of a set of input sequences. Each sequence in $\mathcal{D}$ has a class label called *class* associated with it from a list of k classes, $c_1, \cdots c_k$. We will use $\mathcal{D}_c$ to denote the subset of input-sequences in $\mathcal{D}$ with class label c. In most of the practical applications, there are at most two different class labels that are associated with the sequences in a dataset, $\mathcal{D}$. For example, in a database of DNA segments, a segment can be labeled as *coding* or *non-coding* based on whether the segment is taken from the coding or non-coding region of the DNA. In a database of financial transaction sequences, we can label a sequence either as *normal* or *anomalous*. Nevertheless, multi-class scenarios also arise in some applications. For example, in a longitudinal (temporal) study of cancer patients, a patient can be classified into three different disease stages, such as *benign*, *in situ*, and *malignant*.

The objective of sequence classification is to build a classification model using a labeled dataset $\mathcal{D}$ so that the model can be used to predict the class label of an unseen sequence. Many different classification models have been considered for classifying sequences, and their performance varies based on the application domain, and properties of the sequence data. To compute the performance of a sequence classifier we use the standard supervised classification performance metrics, such as accuracy, AUC (Area under ROC curve), precision-recall, etc. A list and a comparison among these metrics are available in [10].

There are some works that deviate from the above definition of sequence classification. Although most of the sequence classification methods consider supervised scenario, there exist methods that consider semi-supervised classification [57] of sequences. Although uncommon, there are scenarios where different items of an input sequence can be labeled with different classes. This happens mostly for temporal sequences where a given sequence evolves to a different class as the time progresses. For example, for a sequence of patient data over a long period of time, the health condition of the patient may change, and thus the patient may belong to a different class at a different time. Streaming data is another example for evolving class of a sequence. In fact, a streaming sequence can be regarded as virtually unlimited sequence, for which we can predict a sequence of class labels. This problem is called a *strong sequence classification* task, and it is considered in some works [21]. In this paper, we will discuss some of the non-traditional sequence classification tasks in Section 14.8.

14.2.3 Frequent Sequential Patterns

Most of the popular classification methods represent a data instance as a vector of features. For sequence data, a feature is typically a subsequence (or substring) of the given sequence. In data mining literature, frequently occurring subsequences of a set of sequences are called frequent sequential patterns. Below, we define them formally as some of the sequence classification methods, and use them as features for classification.

For a sequence α, if the event α_i occurs before α_j, we denote it as $\alpha_i < \alpha_j$. We say α is a subsequence of another sequence β, denoted as $\alpha \preceq \beta$, if there exists a one-to-one order-preserving function f that maps events in α to events in β, that is, 1) $\alpha_i \subseteq f(\alpha_i) = \beta_k$, and 2) if $\alpha_i < \alpha_j$ then $f(\alpha_i) < f(\alpha_j)$. For example the sequence $(B \to AC)$ is a subsequence of $(AB \to E \to ACD)$, since $B \subseteq AB$ and $AC \subseteq ACD$, and the order of events is preserved. On the other hand the sequence $(AB \to E)$ is not a subsequence of (ABE), and vice versa.

A database $\mathcal{D}$ is a collection of input-sequences. An input-sequence C is said to *contain* another sequence α, if $\alpha \preceq C$. The support of a sequence α, denoted as $\sigma(\alpha, \mathcal{D})$, is the total number of sequences in the database $\mathcal{D}$ that contain α. Given a user-specified support called the *minimum support* (denoted as *min_sup*), we say that a sequence is *frequent* if it occurs more than *min_sup* times. Given a database $\mathcal{D}$ of input-sequences, and a *min_sup*, the problem of mining sequential patterns is to find all frequent sequences (denoted as $\mathcal{F}$) in the database. Many algorithms exist for mining frequent sequential patterns [36, 55].

14.2.4   *n*-Grams

For string sequences, sometimes substrings, instead of subsequences, are used as features for classification. These substrings are commonly known as *n*-grams. Formally, an *n*-gram is a contiguous sequence of *n* items from a given sequence. An *n*-gram of size 1 is referred to as a "unigram," size 2 as a "bigram," size 3 is a "trigram," and so on.

14.3   Sequence Classification Methods

The sequence classification methods can be broadly classified into three large groups.

- **Feature-based:** The methods in the first group represent each sequence as a d-dimensional feature vector, and use traditional supervised classifiers, such as naive Bayes, decision trees, support vector machines (SVM), or random forest for performing sequence classification. The main task of these methods is selecting suitable features that are effective for the classification task. We discuss these methods in Section 14.4.

- **Distance-based:** The set of classification methods in this group design a distance/similarity function between a pair of sequences. Using this function, such a method partitions a set of input sequences into different classes. Kernel-based methods also fall into this group, as a kernel function is nothing but a similarity function between a pair of objects. We discuss these methods in Section 14.5.

- **Model-based:** The third group consists of classification methods that are based on sequential models, such as Markov model, Hidden Markov Model (HMM), or graphical model, such as conditional random field (CRF). It also include non-sequential probabilistic models, such as Naive Bayes. We discuss the methods in this group in Section 14.6.

Some of the methods may belong to multiple groups. For example, a sequence classifier using string kernel can be regarded as both a feature-based and distance-based classifier, as to compute the kernel matrix such a method first finds the feature representation of a sequence and then computes the similarity value between a pair of sequences using a kernel matrix. We discuss a collection of such methods in Section 14.7.

14.4   Feature-Based Classification

In a traditional supervised classification method, each of the data instances in the training set is represented as a 2-tuple $\langle x, y \rangle$ consisting of a feature vector ($x \in X$), and a class label ($y \in Y$).

A learning method uses these data instances for building a classification model that can be used for predicting the class label of an unseen data instance. Since each data instance is represented as a feature vector, such a classification method is called feature-based classification. Sequence data does not have a native vector representation, so it is not straightforward to classify sequences using a feature-based method. However, researchers have proposed methods that find features that capture sequential patterns that are embedded in a sequence.

Sequential data that belong to different classes not only vary in the items that they contain, but also vary in the order of the items as they occur in the sequence. To build a good classifier, a feature in a sequence classifier must capture the item orders that are prevalent in a sequence, which can be done by considering the frequencies of short subsequences in the sequences. For string sequences, the 2-grams in a sequence consider all the ordered term-pairs, 3-grams in a sequence consider all the ordered term-triples, and so on; for a string sequence of length l we can thus consider all the n-grams that have length up to l. The frequencies of these sub-sequences can be used to model all the item-orders that exist in the given sequence. For example, if $CACAG$ is a sequence, we can consider all the n-grams of length 2, 3, 4, and 5 as its feature; CA, AC, and AG are 2-grams, CAC, ACA, and CAG are 3-grams, $CACA$, and $ACAG$ are 4-grams, and the entire sequence is a 5-gram feature. Except for CA, which has a feature value of 2, all the other features have a value of 1. In case the sequences are not strings, then term order can be modeled by mining frequent sequential patterns, and the count of those patterns in a given sequence can be used as the feature value.

Feature-based classification has several challenges. One of those is, for long sequences or for a large alphabet, there are too many features to consider and many of them are not good features for classification. Complexity of many supervised classification algorithms depends on the feature dimension. Further, for many classification methods, irrelevant features can degrade classification performance. To overcome this limitation, feature selection methods are necessary for selecting a subset of sequential features that are good candidates for classification. Another challenge is that the length of various sequences in a dataset can vary considerably, and hence the variation of n-gram or sequence pattern frequency values among different sequences could be due to their length variation, instead of their class label. To negate the influence of length on feature values, various normalizations (similar to *tf-idf* in information retrieval) can be applied to make the feature values uniform across sequences of different lengths.

In machine learning, feature selection [35] is a well-studied task. Given a collection of features, this task selects a subset of features that achieves the best performance for the classification task. For sequence classification, if we exhaustively enumerate all the n-grams or if we mine a large collection of sequential patterns using a small minimum support threshold, we should apply a feature selection method for finding a concise set of good sequential features. There are two broad categories in feature selection in the domain of machine learning: they are filter-based, and wrapper-based. In this and the following section, we will discuss some methods that are similar to filter-based approach. In Section 14.4.3, we discuss a feature selection method that is built on the wrapper paradigm.

14.4.1 Filtering Method for Sequential Feature Selection

The majority of all filter approaches work as follows. In a first step, a relevance score for each feature is calculated. Subsequently, these scores are sorted and low scoring features are removed (typically some threshold is chosen for the feature scores). Below we discuss some of the scoring mechanisms considering a two-class scenario. However, many of these scoring methods can be extended for tasks with more than two classes.

In the definitions below, X_j represents a sequential feature, x_j represents an instantiation (value) of this feature, c_i denotes the class with label i, and v denotes the number of discrete values feature X_j can have:

1. **Support and Confidence:** Support and Confidence are two measures that are used for ranking

association rules in itemset pattern mining [2]. These measures have also been adapted for finding classification association rules (CAR) [32]. A CAR, as defined in [32], is simply a rule $P \rightarrow y$, where P is an itemset and y is a class label. However, in the context of sequential feature selection, P is a sequential feature instead of an itemset, so we will refer to such a rule as CR (classification rule) to avoid confusion with CAR, as the latter is specifically defined for itemsets.

The *support* of a CR $P \rightarrow y$ (denoted as $\sigma(P \rightarrow y, \mathcal{D})$) is defined as the number of cases in $\mathcal{D}$ that contain the pattern P and are also labeled with the class y. To make the *support* size invariant, we typically normalize it with the database size, i.e., $support(P \rightarrow y) = \frac{\sigma(P \rightarrow y, \mathcal{D})}{|\mathcal{D}|}$.

The *confidence* of a CR $P \rightarrow y$ is defined as the fraction of cases in $\mathcal{D}$ that contain the pattern P and have a class label y with respect to the cases that contains the pattern P. i.e., $confidence(P \rightarrow y) = \frac{\sigma(P \rightarrow y, \mathcal{D})}{\sigma(P, \mathcal{D})}$. If the dataset has only two classes (say, y_1 and y_2), and the confidence of the CR $(P \rightarrow y_1)$ is c, then the confidence of the CR $(P \rightarrow y_2)$ is $1 - c$.

Using the above definitions, both the support and the confidence of a CR is between 0 and 1. The higher the support of a CR, the more likely that the rule will be applicable in classifying unseen sequences; the higher the confidence of a rule, the more likely that the rule will predict the class accurately. So a simple way to perform feature selection is to choose a minimum support and minimum confidence value and select a sequence C if for a class y, the CR, $C \rightarrow y$, which satisfies both the above constraints. For instance, if $minsup = 0.2$ and $minconf = 0.60$ are minimum support and minimum confidence threshold, and y is one of the class labels, then the sequence C is selected as a feature if $support(C \rightarrow y) \geq 0.2$ and $confidence(C \rightarrow y) \geq 0.6$.

Support and confidence based measure is easy to understand and is computationally inexpensive. However, there are some drawbacks. First, it is not easy to select a good value for the *minsup* and *minconf* threshold, and this choice may affect the classification quality. Another limitation is that this feature selection method does not consider the multiple occurrences of a sequence feature in an input sequence, as the support and confidence measures add a weight of 1 when a sequence feature exists in an input-sequence, ignoring the plurality of the existence in that sequence. Another drawback is that this feature selection method considers each feature in isolation, ignoring dependencies among difference features.

2. **Information gain:** Information gain is a feature selection metric. It is popularly used in a decision tree classifier for selecting the split feature that is used for partitioning the feature space at a given node of the tree. However, this metric can generally be used for ranking features based on their importance for a classification task. Consider a sequence pattern X, and a dataset of input-sequences, $\mathcal{D}$; based on whether X exists or not in a input-sequence in $\mathcal{D}$, we can partition $\mathcal{D}$ into two parts, $\mathcal{D}_{X=y}$, and $\mathcal{D}_{X=n}$. For a good feature, the partitioned datasets are more pure than the original dataset, i.e., the partitions consist of input-sequences mostly from one of the classes only. This purity is quantified using the *entropy*, which we define below.

In information theory, the entropy of a partition or region $\mathcal{D}$ is defined as $H(\mathcal{D}) = -\sum_{i=1}^{k} P(c_i | \mathcal{D}) \lg P(c_i | \mathcal{D})$, where $P(C_i | D)$ is the probability of class C_i in $\mathcal{D}$, and k is the number of classes. If a dataset is pure, i.e., it consists of input sequences from only one class, then the entropy of that dataset is 0. If the classes are mixed up, and each appear with equal probability $P(c_i | \mathcal{D}) = \frac{1}{k}$, then the entropy has the highest value, $H(\mathcal{D}) = \lg k$. If a dataset is partitioned in $\mathcal{D}_y$ and $\mathcal{D}_n$, then the resulting entropy of the partitioned dataset is: $H(\mathcal{D}_y, \mathcal{D}) = \frac{|\mathcal{D}_{X=y}|}{|\mathcal{D}|} H(\mathcal{D}_{X=y}) + \frac{|\mathcal{D}_{X=n}|}{|\mathcal{D}|} H(\mathcal{D}_{X=n})$. Now information gain of the feature X is defined as $Gain(X, \mathcal{D}) = H(\mathcal{D}) - H(\mathcal{D}_{X=Y}, \mathcal{D}_{X=N})$. The higher the gain, the better the feature for sequence classification. For selecting a subset of features, we can simply rank the features based on the information gain value, and select a desired number of top-ranked features.

3. **Odds ratio**: Odds ratio measures the odds of a sequence occurring in input-sequences labeled with some class c_i, normalized by the odd that it occurs in input-sequences labeled with some class other than c_i. If X is a sequence feature, $P(X|c_i)$ denotes the probability that the sequence X occurs in class c_i, and $P(X|\bar{c}_i)$ denotes the probability that the sequence X occurs in any class other than c_i, then the odd ratio of X can be computed by $\frac{P(X|c_i)\cdot(1-P(X|\bar{c}_i))}{P(X|\bar{c}_i)\cdot(1-P(X|c_i))}$.

If the value of odd ratio is near 1, then the sequence P is a poor feature, as it is equally likely to occur in input-sequences of class c_i and also in input-sequences of classes other than c_i. An odds ratio greater (less) than 1 indicates that P is more (less) likely to occur in input sequences of class c_i than in input sequences of other classes; in that case P is a good feature. The odds ratio must be nonnegative if it is defined. It is undefined if the denominator is equal to 0.

Note that sometimes, we take the absolute value of the logarithm of odd ratio instead of odd-ratio. Then, for a poor feature, log-odd-ratio is a small number near 0, but for a good feature, it is a large positive number. To obtain a small subset of sequential features, we can simply rank the sequential features based on the log-odd-ratio, and consider a desired number of features according to the ranking.

Besides the above, there are other measures, such as Kullback-Leibler (K-L) divergence, Euclidean distance, feature-class entropy, etc., that can also be used for feature ranking.

14.4.2 Pattern Mining Framework for Mining Sequential Features

In the previous subsection, we considered a two-step scenario where a set of sequential features are mined first, and then, using filtering methods, a small set of features are selected. One of the drawbacks of this method is that it is inefficient, because a very large number of features are typically generated in the first step. Another drawback is that such methods consider each of the features independently, whereas many of the sequential features that are mined in the first step are actually highly correlated. If many of the selected features are highly correlated the performance of the classifier sometimes degrades. An alternative to the above two-step method is to unify the mining and feature selection in a single step. Lesh et al. [28] proposed such a method, named FEATUREMINE, which we describe below.

FEATUREMINE adapts a frequent sequence mining algorithm, called SPADE [55]; it takes a labeled dataset as input and mines features that are suited for sequence classification. For selecting good features it uses the following three heuristics that are domain-and-classifier independent: (1) Features should be frequent (high support); (2) Features should be distinctive of at least one class (high confidence); (3) Feature sets should not contain redundant features. The first of the above heuristics ensures that the chosen feature is not rare, otherwise, it will only be rarely useful for classifying unseen sequence instances. The second heuristic simply prefers features that are more accurate. The third heuristic reduces the size of the feature-set by ignoring a sequence that can be subsumed by one of its subsequence. If $M(f, \mathcal{D})$ is the set of input-sequences in $\mathcal{D}$ that contain the feature f, a feature f_1 subsumes a feature f_2 with respect to predicting class c, if and only if $M(f_2, \mathcal{D}_c) \subseteq M(f_1, \mathcal{D}_c)$ and $M(f_1, \mathcal{D}_{\neg c}) \subseteq M(f_2, \mathcal{D}_{\neg c})$.

FEATUREMINE accepts a labeled dataset $\mathcal{D}$, along with parameters for minimum support (s), maximum length (l), maximum width (w), and returns feature set $\mathcal{F}$ such that for every feature f_i and for any class c_j, if $length(f_i) \leq l$ and $width(f_i) \leq w$, and $support(f_i) \geq s$, and $conf(f_i \to c_j)$ is significantly greater than $\frac{|\mathcal{D}_{c_j}|}{|\mathcal{D}|}$, then either $\mathcal{F}$ contains f_i, or $\mathcal{F}$ contains a feature that subsumes f_i with respect to class c_j in dataset $\mathcal{D}$. FEATUREMINE prunes frequent patterns as it enumerates them using the SPADE algorithm by using several pruning rules: (1) if the confidence of the pattern is 1.0, FEATUREMINE does not extend the pattern any further; (2) if an item B appears in every event where item A occurs in every sequence in $\mathcal{D}$, then any feature containing a set with both A and B will be subsumed by A only. Authors show experimental results for naive Bayes, and

Winnow classifier; both the classifiers perform significantly better with the features mined from FEATUREMINE compared to a set of baseline features that have a length of one.

14.4.3   A Wrapper-Based Method for Mining Sequential Features

Kudenko and Hirsh [25] proposed a method called FGEN, for mining sequential patterns from string sequences. FGEN is conceptually similar to a wrapper method for feature selection. However, it is different than a wrapper feature selection method because the latter uses a holdout set for evaluating a feature subset, but FGEN uses the hold-out set for evaluating a single feature at any given iteration. Each of the features that FGEN generates is a macro-feature, i.e., they represent information containing more than a single subsequence. A feature F is a Boolean feature that is written as $s_1^{n_1} \wedge s_2^{n_2} \wedge \cdots \wedge s_k^{n_k}$, where each of the s_i's is a sequence (or substring), and each of the n_i's is an integer. For a given sequence S, the feature F is true if each of the s_i occurs at least n_i times in the sequence S. For example, if $F = AC^2 \wedge CB^3$, then $F(ACBBACBCB) = true$, but $F(ACBAC) = false$. If a feature F is evaluated to true for a given sequence s, it is said that F subsumes s.

The FGEN algorithm consists of two main parts: (1) feature building; and (2) generalization. Features are built incrementally, starting with a feature that corresponds to just a single example (seed) of the target class. Initially, the set of potential seeds consists of all the input sequences of the target class. At every iteration, the seed sequence is chosen from the potential seed-list, to be the one whose feature subsumes the greatest number of sequences remaining in the seed-list. Once a feature is chosen all the input sequences that are subsumed by the feature are removed from the seed-list. Thus, every iteration of feature generation builds a feature and removes a list of input sequences from the potential seed-list; the process continues until the potential seed-list is empty.

The seed feature generated from a seed sequence denotes the minimum frequency restrictions on all subsequences (up to some maximum length) of the seed sequences. For example, if the seed sequence is ABB, then the corresponding seed feature is defined as follows: $F = A^1 \wedge B^2 \wedge AB^1 \wedge BB^1 \wedge ABB^1$, where we consider all subsequences up to length 3. At every iteration, once a seed feature is generated, it is evaluated for accuracy against a holdout set and the feature is generalized to subsume at least one extra sequence in the seed-list. The generalization continues as long as the accuracy on the hold-out set increases, in a hill climbing fashion.

At the end of the feature building step, a dedicated generalization step is used to generalize each of the features (say, F) by decrementing a frequency restriction in F. Again, the holdout set is used to perform the generalization as long as the accuracy improves. Authors show that the number of features that FGEN finally keeps is small (typically around 50), and the performance of C4.5 (decision tree classifier) and Ripper is better with FGEN feature than a comprehensive list of features using all n-grams upto a given length.

14.5   Distance-Based Methods

Distance based classification methods compute a distance score between a pair of sequences, and then use those distances for classifying an unseen sequence. Typically, a distance based classification method is used in applications where a non-standard distance metric is required for achieving good classification performance. On many occasions, such metrics encode domain knowledge within the definition of the distance. For instance, while classifying protein sequences, a standard distance metric that considers each amino acid as an independent entity may not work well. This is due to the fact that a protein sequence may be composed of various functional domains, and two protein sequences may share only a few of those domains; in that case even if the overall similarity between these two proteins is weak, if the matched functional domains are highly significant, these proteins may belong to the same class. Similarly, when classifying genome sequences, similarity metric should consider only the coding part of the DNA, and discard a significant part of the genome, such as junk DNA, and tandem repeats. In such scenarios, finding a suitable distance metric is a pre-condition for achieving good classification performance.

One of the most common distance-based classifiers is k-NN (nearest neighbors) classifier, which is lazy, i.e., it does not pre-compute a classification model. For an unseen sequence instance, u, k-NN computes u's distance to all the input-sequences in $\mathcal{D}$, and finds the k nearest neighbors of u. It then predicts the majority of the class labels of k-NN as the class label of u.

SVM (support vector machines) is another classification model, which can also be regarded as a distance based classifier. However, unlike k-NN, SVM computes the similarity instead of distance between a pair of sequences. SVM calls such a similarity function a kernel function $(K(x, y))$, which is simply a dot product between the feature representation of a pair of sequences x, and y, in a Hilbert space [44]. Given a gram matrix, which stores all the pair-wise similarities between the sequences in the training set in a matrix, SVM learning method finds the maximum-margin hyperplane to separate two classes. Once a suitable kernel function is defined, SVM method for sequence classification is the same for any other classification task. So the main challenge of using SVM for sequence classification is to define a suitable kernel function, and to speed up the computation of the gram matrix.

Below, we discuss a collection of distance metrics that can be used for measuring the distance between a pair of sequences. It is important to note that though we use the term "metric," many of the similarities measurements may not be "metrics" using their mathematical definition. Also, note that we sometimes denote the metric as a similarity metric, instead of distance metric, if the metric computes the similarity between a pair of sequences.

14.5.0.1 Alignment-Based Distance

The most popular distance metric for sequence data is Levenshtein distance or edit distance [18]. It denotes the number of edits needed to transform one string into the other, with the allowable edit operations being insertion, deletion, or substitution of a single character. For a pair of sequences, this distance can be computed by performing the global alignment between the sequences. For two sequences of length l_1 and l_2, the global alignment cost is $O(l_1 l_2)$. This is costly for sequences that are long, such as protein or DNA sequences. Another limitation of this distance metric is that it captures the optimal global alignment between two sequences, whereas for the classification of sequences from most of the domains, local similarities between two sequences play a more critical role. Also the dependency of edit distance metric on the length of the sequence makes it a poor metric if the length of the sequences in the data set varies significantly. To capture the local similarities between two sequences, Smith-Waterman local alignment score [18] can be used. Instead of aligning the entire sequence, the local alignment algorithm aligns similar regions between two sequences. The algorithm compares segments of all possible lengths and optimizes the similarity measure. Though it overcomes some of the problems of global alignment, it is as costly as the global alignment algorithm.

During the 1980s and 1990s two popular tools, known as FASTA [37], and BLAST [3], were developed to improve the scalability of similarity search from biological sequence database. Both the above tools accept a query sequence, and return a set of statistically significant similar sequences from a sequence database. The benefits of these tools over an alignment based method is that the former are highly scalable, as they adopt smart heuristics for finding similar segments after sacrificing the strict optimality. Also, the tools, specifically BLAST, have various versions (such as PSI-BLAST, PHI-BLAST, BLAST-tn) that are customized based on the kind of sequences and the kind of scoring matrix used. For a given set of sequences, FASTA and BLAST can also be used to find a set of similar sequences that are pair-wise similar. For a pair of similar sequence, BLAST also returns the bit score (also known as BLAST-score), which represents the similarity strength that can be used for sequence classification.

14.5.0.2 Keyword-Based Distance

To model the effect of local alignment explicitly, some sequence similarity metrics consider n-gram based method, where a sequence is simply considered as a bag of short segments of fixed length (say, n); thus a sequence can be represented as a vector, in which each component corresponds to the frequency of one of the n-length segments. Then the similarity between two sequences is measured using any metric that measures the similarity between two vectors, such as dot product, Euclidean distance, or Jaccard coefficient. This approach is also known as keyword-based method, as one can consider each sequence as a document and each n-gram as a keyword in the document. The biggest advantage of similarity computation using keyword-based method is that it is fast. Another advantage is that this method represents a sequence using an $\mathbb{R}^n$ vector, which can accommodate most of the classification algorithms that work only on vector-based data.

14.5.0.3 Kernel-Based Similarity

In recent years, kernel-based sequence similarity metrics also got popular [29, 30, 34, 45, 46]. In [29, 30], the authors present several families of k-gram based string kernels, such as, restricted gappy kernels, substitution kernels, and wildcard kernels, all of which are based on feature spaces indexed by k-length subsequences (k-mers) from the string alphabet. These kernels are generally known as k-spectrum kernels. Given an alphabet $\mathcal{A}$ from which the items of sequences is chosen, the simplest of the k-spectrum kernels transforms a sequence x with a function $\Phi_k(x) = (\phi_a(x))_{a \in \mathcal{A}^k}$, where $\phi_a(x)$ is the number of times a occurs in x. The kernel function is the dot product of the feature vectors $K(x,y) = \Phi_k(x) \cdot \Phi_k(y)$. [31] shows how to compute $K(x,y)$ in $O(kn)$ time using a suffix tree data structure. Sonnenburg et al. [46] propose a fast k-spectrum kernel with mismatching.

One disadvantage of a kernel based method is that they are not interpretable, which is a limitation for classification tasks in exploratory domains, such as bioinformatics. Sonnenburg et al. [45] propose a method to learn interpretable SVMs using a set of string kernels. Their idea is to use a weighted linear combination of base kernels, where each base kernel uses a distinctive set of features. The weights represent the importance of the features. After learning the SVM model, the user can have an insight into the importance of different features. Other kernels for sequential classification are polynomial-like kernels [41], Fisher's kernel [19], and diffusion kernel [40].

14.5.0.4 Model-Based Similarity

Probabilistic models, such as HMM (Hidden Markov Model), are also used for finding similarity metric. For classifying a given set of sequences into different classes, such a method trains one HMM for each of the input sequences. Then the similarity between two sequences can be obtained from the similarity (or distance) between the corresponding HMMs. In the past, few authors have proposed approaches for computing the distance between two HMMs [38]; early approaches were based on the Euclidean distance of the discrete observation probability, others on entropy, or on co-emission probability of two HMM models, or on the Bayes probability of error [6].

14.5.0.5 Time Series Distance Metrics

Many of the early distance metrics of sequences are defined for time series data; examples include Euclidean distance [22], and dynamic time warping distance [8]. Time series distance functions are, however, more appropriate for continuous data, rather than discrete data. We refer the readers to the previous chapter, which covers time series data classification in detail.

14.6 Model-Based Method

A generative model for sequence classification assumes that the sequences in each class c_i are being generated by a model M_{c_i}. M_{c_i} is defined over some alphabet Σ, and for any string $s \in \Sigma^*$, M_{c_i} specifies the probability $P^{M_{c_i}}(s|c_i)$ that the given sequence s is generated by the model M_{c_i}; in other words $P^{M_{c_i}}(s|c_i)$ is the likelihood that the given sequence belongs to class i. A classifier can then be constructed as below. First, using a training dataset, a probabilistic model M_{c_i} is trained for each class c_i, for $i = 1$ to k using the sequences belonging to class c_i. Then, to predict an unlabeled sequence (say, s) we simply use Bayes rule: $class(s) = \arg\max_k \frac{P^{M_{c_i}}(s|c_i) \cdot P(c_i)}{\sum_{j=1}^{k} P^{M_{c_j}}(s|c_j) \cdot P(c_j)}$.

Various probabilistic tools can be used to build the generative model, M_{c_i}. The simplest is a naive Bayes classifier, which assumes that the features are independent, so the likelihood probability is simply the multiplication of the likelihood of finding a feature in a sequence given that the sequence belongs to the class c_i. In [5], the authors show different variants of naive Bayes for building generative models for protein sequence classification. The first variant treats each protein sequence as if it were simply a bag of amino acids. The second variant (NB n-grams) applies the naive Bayes classifier to a bag of n-grams ($n > 1$). Thus for the second variant, if we choose $n = 1$, it becomes the first variant. Note that the second variant, NB n-grams, violates the naive Bayes assumption of independence, because the neighboring n-grams overlap along the sequence and two adjacent n-grams have $n-1$ elements in common. The third variant overcomes the above problem by constructing an undirected graphical probabilistic model for n-grams that uses a junction tree algorithm. Thus, the third variant is similar to the second variant except that the likelihood of each n-grams for some class is corrected so that the independence assumption of naive Bayes can be upheld. Their experiments show that the third variant performs better than the others while classifying protein sequences.

Another tool for building generative models for sequences is k-order Markov chains [54]. For this, a sequence is modeled as a graph in which each sequence element is represented by a node, and a direct dependency between two neighboring elements is represented by an edge in the graph. More generally, Markov models of order k capture the dependency between the current element s_{k+1} and its k preceding elements $[s_{k+1}...s_1]$ in a sequence. The higher the value of k, the more complex the model is. As shown in [53, 54], the joint probability distribution for a $(k-1)$-order Markov model (MM) follows directly from the junction tree Theorem [11] and the definition of conditional probability.

$$P(s = s_1 s_2 \cdots s_n, c_j) = \frac{\prod_{i=1}^{n} P(s = s_i \cdots s_{i+k-1}, c_j)}{\prod_{i=1}^{n} P(s = s_i \cdots s_{i+k-2}, c_j)}$$

$$= P(s = s_1 \cdots s_{k-1}, c_j) \cdot \prod_{i=k}^{n} P(s = s_i | s_{i-1} \cdots s_{i-k+1}, c_j)$$

Like any other generative model, the probability $P(s = s_i | s_{i-1} \cdots s_{i-k+1}, c_j)$ can be estimated from data using the counts of the subsequences $s_i \cdots s_{i-k+1}, c_j$ and $s_{i-1} \cdots s_{i-k+1}, c_j$. Once all the model parameters (probabilities) are known, a new sequence s can be assigned to the most likely class based on the generative models for each class.

Yakhnenko et al. [53] trains a $(k-1)$-order Markov model discriminatively instead of using a conventional generative setting so that the classification power of the Markov model can be increased. The difference between a generative model and a discriminative model is as below: a generative model, as we have discussed above, models the probability distribution of the process generating the sequence from each class; then the classification is performed by examining the likelihood of each class producing the observed features in the data, and assigning the sequence to the most likely class. On the other hand, a discriminative model directly computes the class membership probabilities (or model class boundaries) without modeling the underlying class feature densities; to achieve this, it finds the parameter values that maximize a conditional likelihood function [39]. Thus the training step uses the count of various k-grams to iteratively update the parameter values (condition likelihood probabilities) to achieve a local optimal solution using some terminating condition. [53] used gradient descent method for updating the parameters. In their work, Yakhnenko et al. show that a discriminative Markov model performs better than a generative Markov model for sequence classification.

Hidden Markov Model (HMM) is also popular for sequence clustering and classification. Predominant use of HMM in sequence modeling is to build profile HMM that can be used for building probabilistic models of a sequence family. It is particularly used in bioinformatics for aligning a protein sequence with a protein family, a task commonly known as *multiple sequence alignment*. A profile HMM has three hidden states, match, insertion, and deletion; match and insertion are emitting states, and deletion is a non-emitting state. In the context of protein sequences, the emission probability of a match state is obtained from the distribution of residues in the corresponding column, and the emission probability of an insertion state is set to the background distribution of the residues in the dataset. Transition probabilities among various states are decided based on various gap penalty models [14]. To build a profile HMM from a set of sequences, we can first align them, and then use the alignment to learn the transition and emission probabilities by counting. Profile HMMs can also be built incrementally by aligning each new example with a profile HMM, and then updating the HMM parameters from the alignment that includes the new example.

Once built, a profile HMM can be used for sequence classification [24, 48, 56], In the classification setting, a distinct profile HMM is build for each of the classes using the training examples of the respective classes. For classifying an unseen sequence, it is independently aligned with the profile-HMM of each of the classes using dynamic programming; the sequence is then classified to the class that achieves the highest alignment score. One can also estimate the log-likelihood ratio to decide which class a sequence should belong to. For this, given an HMM H_i for the class i, assume $P(s|H_i)$ denotes the probability of s under the HMM, and $P(s|H_0)$ denotes the probability of s under the null model. Then the likelihood ratio can be computed as:

$$\mathcal{L}(s) = \log \frac{P(s|H_i) \cdot P(H_i)}{P(s|H_0) \cdot P(H_0)} = \log \frac{P(x|H_i)}{P(x|H_0)} + \log \frac{P(H_i)}{P(H_0)}$$

A positive value of $\mathcal{L}(s)$ indicates that the sequence s belongs to the class i. The constant factor $\log P(H_i)/P(H_0)$ is called log prior odds, which provides an *a priori* means for biasing the decision [24]. In a recent work, Soullard and Artieres [47] use HMM together with HCRF (Hidden Conditional Random Field) for sequence classification.

14.7 Hybrid Methods

Many of the recent methods for sequence classification belong to multiple groups, which we call hybrid methods. We discuss some of these methods below.

A common hybrid is a distance based method that is equipped with model-based methodology, where the model is used for computing a distance or similarity. One of the earliest among these works is the work by Jaakkola et al. [19], which uses a hidden Markov model (HMM) for computing a kernel called Fisher kernel, which can be used along with an SVM classifier. Such usage of HMM is called discriminative setup of HMM, which is conceptually similar to the discriminative Markov model of Yakhnenko et al. [53] that we discussed in the previous section. Jaakkola et al. begin with an HMM trained from positive examples of class i to model a given protein family. Then they use this HMM to map each new protein sequence (say, s) that they want to classify into a fixed length vector (called score vector) and compute the kernel function on the basis of the Euclidean distance between the score vector of s and the score vectors of known positive and negative examples of the protein family i. The resulting discriminative function is given by $\mathcal{L}(s) = \sum_{i:s_i \in X_i} \lambda_i K(s, s_i) - \sum_{i:s_i \in X_0} \lambda_i K(s, s_i)$, where λ_i are estimated from the positive training examples (X_i) and negative training examples (X_0).

In [43], the authors proposed a method that combines distance-based and feature-based methods. It is a feature-based method that uses a distance-based method for super-structuring and abstraction. It first finds all the k-grams of a sequence, and represents the sequences as a bag of k-grams. However, instead of using each of the k-grams as a unique feature, it partitions the k-grams into different sets so that each of the sets contain a group of most *similar* features. Each such group can be considered as an abstract feature. To construct the abstract features, the method takes a sequence dataset as a bag-of-k-grams, and clusters the k-grams using agglomerative hierarchical clustering until m (user-defined) abstract features are obtained. In this clustering step, it computes a distance between each of the existing abstraction by using an information gain criterion.

Blasiak and Rangwala [9] proposed another hybrid method that combines distance-based and feature-based methods. They employ a scheme that uses an HMM variant to map a sequence to a set of fixed-length description vectors. The parameters of an HMM variant is learnt using various inference algorithms, such as Baum-Welch, Gibbs sampling, and a variational method. From the fixed-length description, any feature based traditional classification method can be used to classify a sequence into different classes.

In another work, Aggarwal [1] proposed a method that uses wavelet decomposition of a sequence to map it in the wavelet space. The objective of this transformation is to exploit the multi-resolution property of wavelet to create a scheme which considers sequence features that capture similarities between two sequences in different levels of granularity. Although the distance is computed in the wavelet space, for the classification, the method uses a rule-based classifier.

14.8 Non-Traditional Sequence Classification

We defined the sequence classification problem in Section 14.2.2. However, there are many variations to this definition. In this section we discuss some of the classification methods that consider alternative definitions of sequence classification.

14.8.1 Semi-Supervised Sequence Classification

In a traditional supervised classification task, the classification model is built using labeled training instances. However, finding labeled data is costly. So, some classification systems use both labeled and unlabeled data to build a classification model that is known as a semi-supervised classification system. Many works have shown that the presence of unlabeled examples improves the classification performance [7, 58, 59]. Semi-supervised classification is considered in many of the existing works on sequence classification; we discuss some of the representative ones below.

For semi-supervised sequence classification, Weston et al. [51] propose a simple and scalable cluster kernel technique for incorporating unlabeled data for designing string kernels for sequence classification. The basic idea of cluster kernels is that two points in the same cluster or region should have a small distance between each other. The neighborhood kernel uses averaging over a neighborhood of sequences defined by a local sequence similarity measure, and the bagged kernel uses bagged clustering of the full sequence dataset. These two kernels are used to modify a base string kernel so that unlabeled data are considered in the kernel computation. Authors show that the modified kernels greatly improve the performance of the classification of protein sequences.

Zhong et al. [57] propose an HMM-based semi-supervised classification for time series data (sequence of numbers); however, the method is general and can be easily made to work for discrete sequences as long as they can be modeled with an HMM. The method uses labeled data to train the initial parameters of a first order HMM, and then uses unlabeled data to adjust the model in an EM process. Wei et al. [50] adopt one nearest neighbor classifier for semi-supervised time series classification. The method handles a scenario where only small amounts of positively labeled instances are available.

14.8.2 Classification of Label Sequences

For many sequential data, each of the items in a sequence is given a label, instead a single label for the entire sequence. More formally, a label sequence classification problem can be formulated as follows. Let $\{(s_i, y_i)\}_{i=1}^{N}$ be a set of N training instances, where $s_i = s_{i,1} s_{i,2} \cdots s_{i,k}$, and $y = y_{i,1} y_{i,2} \cdots y_{i,k}$. For example, in part-of-speech tagging, one (s_i, y_i) pair might consist of $x_i = \langle$ do you want fries with that$\rangle$ and $y_i = \langle$verb pronoun verb noun prep pronoun$\rangle$ The goal is to construct a classifier h that can correctly predict a new label sequence $y = h(x)$ given an input sequence x. Many of the earlier works on classifying label sequences appear in natural language processing [16, 26].

Hidden Markov Model is one of the most suitable models for classification of label sequences. In an HMM, the labels are considered the hidden states, and the sequence item at a position is the emission symbol at that corresponding hidden state. Using training examples, the HMM parameters (transition and emission probabilities) can be learned. Given a test sequence, the most likely label sequence can be obtained from the learned HMM model by using Viterbi algorithm. Besides HMM, other graphical models, particularly conditional random field (CRF), which is an undirected graphical model, is also used for predicting label sequences [26].

More recent works to solve this problem adopt sophisticated techniques. For instance, Altun et al. [4] proposed a method called Hidden Markov Support Vector machine, which overcomes some of the limitations of an HMM-based method. Specifically, this method adopts a discriminative approach to HMM modeling similar to some of the other works that we discussed [19, 53] in other sections. The label sequence problem has also been solved by using recurrent neural networks [16].

14.8.3 Classification of Sequence of Vector Data

Another variant of sequence classification is such that input sequence can be thought of as $\mathbf{x} = \mathbf{x_1} \cdots \mathbf{x_t}$ where each of the $\mathbf{x_t}$ is a d-dimensional real-valued vector. Common examples of such a setup is image sequences in a video, and time series of vector-based data. Kim and Pavlovic [23] proposed a hybrid method for classifying a dataset of such sequences. Their method, called large margin Hidden Markov Model, uses a real-valued confidence function $f(c, \mathbf{x})$. The confidence function predicts the class label of a new sequence $\mathbf{x}^*$ as $c^* = \arg\max_c f(c, \mathbf{x}^*)$, effectively defining the model's class boundaries. HMM or any other probabilistic model that avails a posterior probability $P(c, \mathbf{x})$ can be turned into a classifier with a log-likelihood confidence, namely $f(c, (x)) = \log P(c, (x))$ In a large margin framework, given the true class label y of the training data, it learns the confidence function $f(\cdot)$ so that the margin between true and incorrect classes, defined as $f(y, \mathbf{x}) - f(c, \mathbf{x})$, is greater than 1 for $\forall c \neq y$, which leads to a max-margin optimization problem. They solve an approximate version of this problem with an instance of a convex program, solvable by efficient gradient search.

14.9 Conclusions

Sequence classification is a well-developed research task with many effective solutions, however, some challenges still remain. One of them is that in many domains such a classification task suffers from *class imbalance*, i.e., for a sequence classification task in those domains the prior probability of the minority class can be a few magnitudes smaller that that of the majority class. A common example of this scenario can be a task that distinguishes between normal and anomalous sequences of credit card transactions of a customer. In a typical iid training dataset, the population of the anomalous class is rare, and most of the classification methods, to some extend, suffer performance loss due to this phenomenon. More research is required to overcome this limitation. Another challenge is to obtain a scalable sequence classification system, preferably on distributed systems (such as mapreduce) so that large scale sequence classification problems that appear in Web and e-commerce domains can be solved effectively.

To conclude, in this chapter we provided a survey of discrete sequence classification. We grouped the methodologies of sequence classification into three major groups: feature-based, distance-based, and model based, and discussed various classification methods that fall in these groups. We also discussed a few methods that overlap across multiple groups. Finally, we discussed some variants of sequence classification and discussed a few methods that solve those variants.

Bibliography

[1] Charu C. Aggarwal. On effective classification of strings with wavelets. In *Proceedings of the Eighth ACM SIGKDD International Conference on Knowledge Discovery and Data Mining*, KDD '02, pages 163–172, 2002.

[2] Rakesh Agrawal and Ramakrishnan Srikant. Fast algorithms for mining association rules in large databases. In *Proceedings of the 20th International Conference on Very Large Data Bases*, VLDB '94, pages 487–499, 1994.

[3] S. Altschul, W. Gish, W. Miller, E. Myers, and D. Lipman. Basic local alignment search tool. *Journal of Molecular Biology*, 215:403–410, 1990.

[4] Yasemin Altun, Ioannis Tsochantaridis, and Thomas Hofmann. Hidden Markov support vector machines. In *Proceedings of International Conference on Machine Learning*, ICML'03, 2003.

[5] Carson Andorf, Adrian Silvescu, Drena Dobbs, and Vasant Honavar. Learning classifiers for assigning protein sequences to gene ontology functional families. In *Fifth International Conference on Knowledge Based Computer Systems (KBCS)*, page 256, 2004.

[6] C. Bahlmann and H. Burkhardt. Measuring HMM similarity with the Bayes probability of error and its application to online handwriting recognition. In *Proceedings of the Sixth International Conference on Document Analysis and Recognition*, ICDAR '01, pages 406–411, 2001.

[7] Mikhail Belkin, Partha Niyogi, and Vikas Sindhwani. Manifold regularization: A geometric framework for learning from labeled and unlabeled examples. *Journal of Machine Learning Research*, 7:2399–2434, 2006.

[8] Donald J. Berndt and James Clifford. Using dynamic time warping to find patterns in time series. In *KDD Workshop'94*, pages 359–370, 1994.

[9] Sam Blasiak and Huzefa Rangwala. A Hidden Markov Model variant for sequence classification. In *Proceedings of the Twenty-Second International Joint Conference on Artificial Intelligence - Volume Two*, IJCAI'11, pages 1192–1197. AAAI Press, 2011.

[10] Rich Caruana and Alexandru Niculescu-Mizil. Data mining in metric space: An empirical analysis of supervised learning performance criteria. In *Proceedings of the Tenth ACM SIGKDD International Conference on Knowledge Discovery and Data Mining*, KDD '04, pages 69–78, 2004.

[11] R. G. Cowell, A. P. Dawid, S. L. Lauritzen, and D. J. Spiegelhalter. *Probabilistic Networks and Expert Systems*. Springer, 1999.

[12] Mukund Deshpande and George Karypis. Evaluation of techniques for classifying biological sequences. In *Proceedings of the 6th Pacific-Asia Conference on Advances in Knowledge Discovery and Data Mining*, PAKDD '02, pages 417–431, 2002.

[13] Elena S. Dimitrova, Paola Vera-Licona, John McGee, and Reinhard C. Laubenbacher. Discretization of time series data. *Journal of Computational Biology*, 17(6):853–868, 2010.

[14] Richard Durbin, Eddy Sean R., Anders Krogh, and Graeme Mitchison. *Biological Sequence Analysis: Probabilistic Models of Proteins and Nucleic Acids*. Cambridge University Press, 1988.

[15] Pierre Geurts. Pattern extraction for time series classification. In *Proceedings of the 5th European Conference on Principles of Data Mining and Knowledge Discovery*, PKDD '01, pages 115–127, 2001.

[16] Alex Graves, Santiago Fernández, Faustino Gomez, and Jürgen Schmidhuber. Connectionist temporal classification: Labelling unsegmented sequence data with recurrent neural networks. In *Proceedings of the 23rd International Conference on Machine Learning*, ICML '06, pages 369–376, 2006.

[17] M. Pamela Griffin and J. Randall Moorman. Toward the early diagnosis of neonatal sepsis and sepsis-like illness using novel heart rate analysis. *Pediatrics*, 107(1):97–104, 2001.

[18] Dan Gusfield. *Algorithms on Strings, Trees, and Sequences: Computer Science and Computational Biology*. Cambridge University Press, 1997.

[19] Tommi Jaakkola, Mark Diekhans, and David Haussler. Using the Fisher kernel method to detect remote protein homologies. In *Proceedings of the Seventh International Conference on Intelligent Systems for Molecular Biology*, pages 149–158, 1999.

[20] Wei Jiang, S.G. Kong, and G.D. Peterson. ECG signal classification using block-based neural networks. In *Proceedings of the 2005 IEEE International Joint Conference on Neural Networks, 2005. IJCNN '05*, volume 1, pages 326–331, 2005.

[21] Mohammed Waleed Kadous and Claude Sammut. Classification of multivariate time series and structured data using constructive induction. *Machine Learning*, 58(2-3):179–216, 2005.

[22] Eamonn Keogh and Shruti Kasetty. On the need for time series data mining benchmarks: A survey and empirical demonstration. In *Proceedings of the Eighth ACM SIGKDD International Conference on Knowledge Discovery and Data Mining*, KDD '02, pages 102–111, 2002.

[23] Minyoung Kim and Vladimir Pavlovic. Sequence classification via large margin Hidden Markov Models. *Data Mining and Knowledge Discovery*, 23(2):322–344, 2011.

[24] Anders Krogh, Michael Brown, I. Saira Mian, Kimmen Sjolander, and David Haussler. Hidden Markov Models in computational biology: Applications to protein modeling. *Journal of Molecular Biology*, 235:1501–1531, 1994.

[25] Daniel Kudenko and Haym Hirsh. Feature generation for sequence categorization. In *Proceedings of the Fifteenth National/Tenth Conference on Artificial Intelligence/Innovative Applications of Artificial Intelligence*, AAAI '98/IAAI '98, pages 733–738, 1998.

[26] John D. Lafferty, Andrew McCallum, and Fernando C. N. Pereira. Conditional random fields: Probabilistic models for segmenting and labeling sequence data. In *Proceedings of the Eighteenth International Conference on Machine Learning*, ICML '01, pages 282–289, 2001.

[27] Terran Lane and Carla E. Brodley. Temporal sequence learning and data reduction for anomaly detection. *ACM Transactions on Information Systems Security*, 2(3):295–331, 1999.

[28] Neal Lesh, Mohammed J Zaki, and Mitsunori Ogihara. Mining features for sequence classification. In *Proceedings of the Fifth ACM SIGKDD International Conference on Knowledge Discovery and Data Mining*, pages 342–346. ACM, 1999.

[29] C. Leslie and R. Kuang. Fast string kernels using inexact matching for protein sequences. *Journal of Machine Learning Research*, 5:1435–1455, 2004.

[30] C. S. Leslie, E. Eskin, A. Cohen, J. Weston, and W. S. Noble. Mismatch string kernels for discriminative protein classification. *Bioinformatics*, 20(4):467–476, 2004.

[31] Christina S. Leslie, Eleazar Eskin, and William Stafford Noble. The spectrum kernel: A string kernel for SVM protein classification. In *Pacific Symposium on Biocomputing*, pages 566–575, 2002.

[32] B. Liu, W. Hsu, and Y. Ma. Integrating classification and association rule mining. In *Proceedings of the 4th International Conference on Knowledge Discovery and Data Mining (KDD'98)*, pages 80–86, 1998.

[33] Xuan Liu, Pengzhu Zhang, and Dajun Zeng. Sequence matching for suspicious activity detection in anti-money laundering. In *Proceedings of the IEEE ISI 2008 PAISI, PACCF, and SOCO International Workshops on Intelligence and Security Informatics*, PAISI, PACCF and SOCO '08, pages 50–61, 2008.

[34] Huma Lodhi, Craig Saunders, John Shawe-Taylor, Nello Cristianini, and Chris Watkins. Text classification using string kernels. *Journal of Machine Learning Research*, 2:419–444, March 2002.

[35] Luis Carlos Molina, Lluís Belanche, and Àngela Nebot. Feature selection algorithms: A survey and experimental evaluation. In *Proceedings of the 2002 IEEE International Conference on Data Mining*, ICDM '02, pages 306–313, 2002.

[36] Carl H. Mooney and John F. Roddick. Sequential pattern mining–Approaches and algorithms. *ACM Comput. Surv.*, 45(2):19:1–19:39, March 2013.

[37] W. R. Pearson and D. J. Lipman. Improved tools for biological sequence comparison. *Proceedings of the National Academy of Sciences*, 85(8):2444–2448, 1988.

[38] L. R. Rabiner. Readings in speech recognition. Chapter A tutorial on hidden Markov models and selected applications in speech recognition, pages 267–296. Morgan Kaufmann Publishers, SanFrancisco, CA, 1990.

[39] D. Rubinstein and T. Hastie. Discriminative vs informative learning. In *Proceedings of the ACM SIGKDD International Conference on Knowledge Discovery and Data Mining*, KDD '97, 1997.

[40] B. Scholkopf, K. Tsuda, and J-P. Vert. *Kernel Methods in Computational Biology*, pages 171-192. MIT Press, 2004.

[41] Rong She, Fei Chen, Ke Wang, Martin Ester, Jennifer L. Gardy, and Fiona S. L. Brinkman. Frequent-subsequence-based prediction of outer membrane proteins. In *Proceedings of the Ninth ACM SIGKDD International Conference on Knowledge Discovery and Data Mining*, KDD '03, pages 436–445, 2003.

[42] Yelong Shen, Jun Yan, Shuicheng Yan, Lei Ji, Ning Liu, and Zheng Chen. Sparse hidden-dynamics conditional random fields for user intent understanding. In *Proceedings of the 20th International Conference on World Wide Web*, WWW '11, pages 7–16, 2011.

[43] Adrian Silvescu, Cornelia Caragea, and Vasant Honavar. Combining super-structuring and abstraction on sequence classification. In *Proceedings of the 2009 Ninth IEEE International Conference on Data Mining*, ICDM '09, pages 986–991, 2009.

[44] Alex J. Smola and Bernhard Schlkopf. *Learning with Kernels: Support Vector Machines, Regularization, Optimization, and Beyond*. MIT Press, 2001.

[45] S. Sonnenburg, G. Rätsch, and C. Schäfer. Learning interpretable SVMS for biological sequence classification. In *Proceedings of the 9th Annual International Conference on Research in Computational Molecular Biology*, RECOMB'05, pages 389–407, 2005.

[46] Sören Sonnenburg, Gunnar Rätsch, and Bernhard Schölkopf. Large scale genomic sequence SVM classifiers. In *Proceedings of the 22nd International Conference on Machine Learning*, ICML '05, pages 848–855, 2005.

[47] Yann Soullard and Thierry Artieres. Iterative refinement of HMM and HCRF for sequence classification. In *Proceedings of the First IAPR TC3 Conference on Partially Supervised Learning*, PSL'11, pages 92–95, 2012.

[48] Prashant Srivastava, Dhwani Desai, Soumyadeep Nandi, and Andrew Lynn. HMM-mode - improved classification using profile Hidden Markov Models by optimising the discrimination threshold and modifying emission probabilities with negative training sequences. *BMC Bioinformatics*, 8(1):104, 2007.

[49] Pang-Ning Tan and Vipin Kumar. Discovery of web robot sessions based on their navigational patterns. *Data Mining and Knowledge Discovery*, 6(1):9–35, 2002.

[50] Li Wei and Eamonn Keogh. Semi-supervised time series classification. In *Proceedings of the 12th ACM SIGKDD International Conference on Knowledge Discovery and Data Mining*, KDD '06, pages 748–753, 2006.

[51] Jason Weston, Christina Leslie, Eugene Ie, Dengyong Zhou, Andre Elisseeff, and William Stafford Noble. Semi-supervised protein classification using cluster kernels. *Bioinformatics*, 21(15):3241–3247, 2005.

[52] Zhengzheng Xing, Jian Pei, and Eamonn Keogh. A brief survey on sequence classification. *ACM SIGKDD Explorations Newsletter*, 12(1):40–48, 2010.

[53] Oksana Yakhnenko, Adrian Silvescu, and Vasant Honavar. Discriminatively trained Markov model for sequence classification. In *Proceedings of the Fifth IEEE International Conference on Data Mining*, pages 498–505, IEEE, 2005.

[54] Zheng Yuan. Prediction of protein subcellular locations using markov chain models. *FEBS Letters*, pages 23–26, 1999.

[55] Mohammed J. Zaki. SPADE: An efficient algorithm for mining frequent sequences. *Machine Learning*, 42(1-2):31–60, 2001.

[56] Yuan Zhang and Yanni Sun. HMM-frame: Accurate protein domain classification for metagenomic sequences containing frameshift errors. *BMC Bioinformatics*, 12(1):198, 2011.

[57] Shi Zhong. Semi-supervised sequence classification with HMMs. *IJPRAI*, 19(2):165–182, 2005.

[58] Dengyong Zhou, Jiayuan Huang, and Bernhard Schölkopf. Learning from labeled and unlabeled data on a directed graph. In *Proceedings of the 22nd International Conference on Machine Learning*, ICML '05, pages 1036–1043, 2005.

[59] Xiaojin Zhu and Zoubin Ghahramani. Learning from labeled and unlabeled data with label propagation. Technical Report CMUCALD-02-107, Carnegie Mellon University, 2002.

Chapter 15

Collective Classification of Network Data

Ben London

University of Maryland
College Park, MD 20742
blondon@cs.umd.edu

Lise Getoor

University of Maryland
College Park, MD 20742
getoor@cs.umd.edu

15.1 Introduction

Network data has become ubiquitous. Communication networks, social networks, and the World Wide Web are becoming increasingly important to our day-to-day life. Moreover, networks can be defined implicitly by certain structured data sources, such as images and text. We are often interested in inferring hidden attributes (i.e., labels) about network data, such as whether a Facebook user will adopt a product, or whether a pixel in an image is part of the foreground, background, or some specific object. Intuitively, the network should help guide this process. For instance, observations and inference about someone's Facebook friends should play a role in determining their adoption probability. This type of joint reasoning about label correlations in network data is often referred to as *collective classification.*

Classic machine learning literature tends to study the supervised setting, in which a classifier is learned from a fully-labeled *training set*; classification performance is measured by some form of statistical accuracy, which is typically estimated from a held-out *test set*. It is commonly assumed that data points (i.e., feature-label pairs) are generated independently and identically from an underlying distribution over the domain, as illustrated in Figure 15.1(a). As a result, classification is performed independently on each object, without taking into account any underlying network between the objects. Classification of network data does not fit well into this setting. Domains such as Webpages, citation networks, and social networks have naturally occurring relationships between objects. Because of these connections (illustrated in Figure 15.1(b)), their features and labels are likely to be correlated. Neighboring points may be more likely to share the same label (a phenomenon sometimes referred to as *social influence* or *contagion*), or links may be more likely between instances of the same class (referred to as *homophily* or *assortativity*). Models that classify each object independently are ignoring a wealth of information, and may not perform well.

Classifying real network data is further complicated by heterogenous networks, in which nodes may not have uniform local features and degrees (as illustrated in Figure 15.1(c)). Because of this, we cannot assume that nodes are identically distributed. Also, it is likely that there is not a clean split between the training and test sets (as shown in Figure 15.1(d)), which is common in relational datasets. Without independence between training and testing, it may be difficult to isolate training accuracy from testing accuracy, so the statistical properties of the estimated model are not straightforward.

In this article, we provide an overview of existing approaches to collective classification. We begin by formally defining the problem. We then examine several approaches to collective classification: iterative wrappers for local predictors, graph-based regularization and probabilistic graphical models. To help ground these concepts in practice, we review some common feature engineering techniques for real-world problems. Finally, we conclude with some interesting applications of collective classification.

15.2 Collective Classification Problem Definition

Fix a graph $G \triangleq (\mathcal{V}, \mathcal{E})$ on n nodes $\mathcal{V} \triangleq \{1, \ldots, n\}$, with edges $\mathcal{E} \subseteq \mathcal{V} \times \mathcal{V}$. For the purposes of this chapter, assume that the structure of the graph is given or implied by an observed network topology. For each node i, we associate two random variables: a set of local features X_i and a label Y_i, whose (possibly heterogeneous) domains are X_i and $\mathcal{Y}_i$ respectively. Assume that the local features of the entire network, $\mathbf{X} \triangleq \{X_1, \ldots, X_n\}$, are observed. In some cases, a subset of the labels, $\mathbf{Y} \triangleq \{Y_1, \ldots, Y_n\}$, are observed as well; we denote the labeled and unlabeled subsets by $\mathbf{Y}^\ell \subseteq \mathbf{Y}$ and $\mathbf{Y}^u \subseteq \mathbf{Y}$, respectively, where $\mathbf{Y}^\ell \cap \mathbf{Y}^u = \emptyset$ and $\mathbf{Y}^\ell \cup \mathbf{Y}^u = \mathbf{Y}$. Given $\mathbf{X}$ and $\mathbf{Y}^\ell$, the collective classification task is to infer $\mathbf{Y}^u$.

In general, collective classification is a combinatorial optimization problem. The objective function varies, depending on the choice of model; generally, one minimizes an energy function that depends on parametric assumptions about the generating distribution. Here we describe several common approaches: iterative classification, label propagation, and graphical models.

Throughout this document, we employ the following notation. Random variables will be denoted by uppercase letters, e.g., X, while realizations of variables will be indicated by lowercase, e.g., x. Sets or vectors of random variables (or realizations) will be denoted by bold, e.g., $\mathbf{X}$ (or $\mathbf{x}$). For a

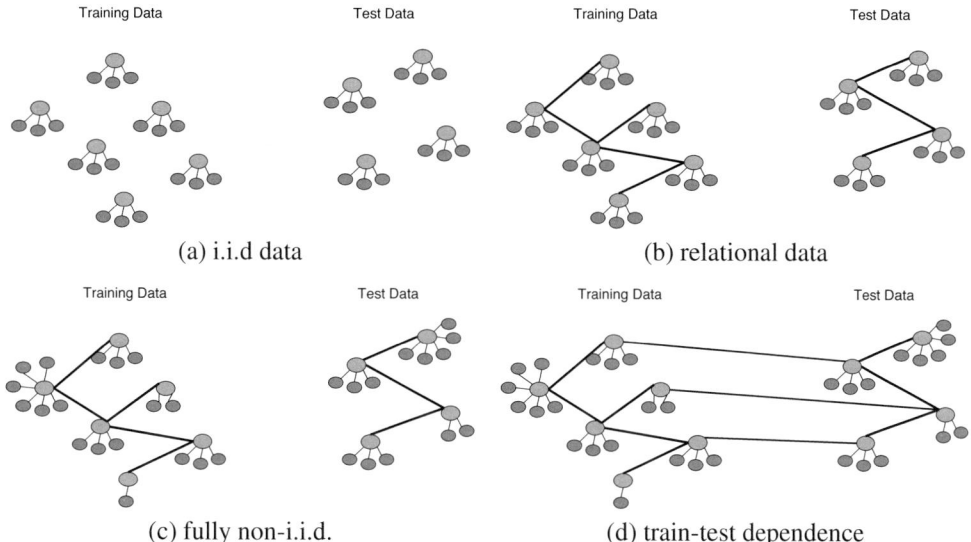

Training Data   Test Data   Training Data   Test Data

(a) i.i.d data   (b) relational data

Training Data   Test Data   Training Data   Test Data

(c) fully non-i.i.d.   (d) train-test dependence

FIGURE 15.1 (See color insert.): (a) An illustration of the common i.i.d. supervised learning setting. Here each instance is represented by a subgraph consisting of a label node (blue) and several local feature nodes (purple). (b) The same problem, cast in the relational setting, with links connecting instances in the training and testing sets, respectively. The instances are no longer independent. (c) A relational learning problem in which each node has a varying number of local features and relationships, implying that the nodes are neither independent nor identically distributed. (d) The same problem, with relationships (links) between the training and test set.

node i, let $\mathcal{N}_i$ denote the set of indices corresponding to its (open) neighborhood; that is, the set of nodes adjacent to i (but not including it).

15.2.1 Inductive vs. Transductive Learning

Learning scenarios for collective classification broadly fall into two main categories: *inductive* and *transductive*. In inductive learning, data is assumed to be drawn from a distribution over the domain; that is, a sentence, image, social network, or some other data structure is generated according to a distribution over instances of said structure. Given a number of labeled structures drawn from this distribution, the objective is to learn to predict on new draws. In the transductive setting, the problem domain is fixed, meaning the data simply exists. The distribution from which the data was drawn is therefore irrelevant, since there is no randomness over what values could occur. Instead, randomness comes from which nodes are labeled, which happens via some stochastic sampling process. Given a labeled subset of the data, the goal is to learn to correctly predict the remaining instances.

In the inductive setting, one commonly assumes that draws of test examples are independent of the training examples. However, this may not hold with relational data, since the data-generating process may inject some dependence between draws from the distribution. There may be dependencies between nodes in the train and test data, as illustrated in Figure 15.1. The same is true of the transductive setting, since the training data may just be a labeled portion of one large network. The dependencies between training and testing must be considered when computing certain metrics, such as train and test accuracy [21].

Since collective methods leverage the correlations between adjacent nodes, researchers typically assume that a small subset of labels are given during prediction. In the transductive setting, these nodes are simply the training set; in the inductive setting, this assumes that draws from the distribution over network structures are partially labeled. However, inductive collective classification is still possible even if no labels are given.

15.2.2 Active Collective Classification

An interesting subproblem in collective classification is how one acquires labels for training (or prediction). In supervised and semi-supervised learning, it is commonly assumed that annotations are given *a priori*—either adversarially, as in the online model, or agnostically, as in the *probably approximately correct* (PAC) model. In these settings, the learner has no control over which data points are labeled.

This motivates the study of *active* learning. In active learning, the learner is given access to an *oracle*, which it can query for the labels of certain examples. In active collective classification, the learning algorithm is allowed to ask for the labels of certain nodes, so as to maximize its performance using the minimal number of labels. How it decides *which* labels to query for is an open problem that is generally NP-hard; nonetheless, researchers have proposed many heuristics that work well in practice [4, 21, 27]. These typically involve some trade-off between propagation of information from an acquired label and coverage of the network.

The label acquisition problem is also relevant during inference, since the predictor might have access to a label oracle at test time. This form of *active inference* has been very successful in collective classification [3, 35]. Queries can sometimes be very sophisticated: a query might return not only a node's label, but also the labels of its neighbors; if the data graph is uncertain, a query might also return information about a node's local relationships (e.g., friends, citations, influencers, etc.). This has been referred to as *active surveying* [30, 38].

Note that an optimal label acquisition strategy for active learning may not be optimal for active inference, so separate strategies may be beneficial. However, the relative performance improvements of active learning and inference are easily conflated, making it difficult to optimize the acquisition strategies. [21] proposes a relational active learning framework to combat this problem.

15.3   Iterative Methods

Collective classification can in some sense be seen as achieving agreement amongst a set of interdependent, local predictions. Viewed as such, some approaches have sought ways to iteratively combine and revise individual node predictions so as to reach an equilibrium. The collective inference algorithm is essentially just a wrapper (or *meta-algorithm*) for a local prediction subroutine. One of the benefits of this technique is that local predictions can be made efficiently, so the complexity of collective inference is effectively the number of iterations needed for convergence. Though neither convergence nor optimality is guaranteed, in practice, this approach typically converges quickly to a good solution, depending on the graph structure and problem complexity. The methods presented in this section are representative of this iterative approach.

15.3.1   Label Propagation

A natural assumption in network classification is that adjacent nodes are likely to have the same label, (i.e., contagion). If the graph is weighted, then the edge weights $\{w_{i,j}\}_{(i,j) \in \mathcal{E}}$ can be interpreted as the strength of the associativity. Weights can sometimes be derived from observed features $\mathbf{X}$ via a similarity function. For example, a *radial basis function* can be computed between adjacent nodes (i, j) as

$$w_{i,j} \triangleq \exp\left(-\frac{\|X_i - X_j\|_2^2}{\sigma^2}\right), \tag{15.1}$$

where σ is a parameter that determines the width of the Gaussian.

Suppose the labels are binary, with $\mathcal{Y}_i \in \{\pm 1\}$ for all $i = 1, \ldots, n$. If node i is unlabeled, we could predict a score for $Y_i = 1$ (or $Y_i = -1$) as the weighted average of its neighbors' labels, i.e.,

$$Y_i \leftarrow \left(\frac{1}{\sum_{j \in \mathcal{N}_i} w_{i,j}}\right) \sum_{j \in \mathcal{N}_i} w_{i,j} Y_j.$$

One could then clamp Y_i to $\{\pm 1\}$ using its sign, $\mathrm{sgn}(Y_i)$. While we probably will not know all of the labels of $\mathcal{N}_i$, if we already had predictions for them, we could use these, then iterate until the predictions converge. This is precisely the idea behind a method known as *label propagation*. Though the algorithm was originally proposed by [48] for general transductive learning, it can easily be applied to network data by constraining the similarities according to a graph. An example of this is the *modified adsorption* algorithm [39].

Algorithm 15.1 provides pseudocode for a simple implementation of label propagation. The algorithm assumes that all labels are k-valued, meaning $|\mathcal{Y}_i| = k$ for all $i = 1, \ldots, n$. It begins by constructing a $n \times k$ label matrix $\mathbf{Y} \in \mathbb{R}^{n \times k}$, where entry i, j corresponds to the probability that $Y_i = j$. The label matrix is initialized as

$$\mathbf{Y}_{i,j} = \begin{cases} 1 & \text{if } Y_i \in \mathbf{Y}^\ell \text{ and } Y_i = j, \\ 0 & \text{if } Y_i \in \mathbf{Y}^\ell \text{ and } Y_i \neq j, \\ 1/k & \text{if } Y_i \in \mathbf{Y}^u. \end{cases} \tag{15.2}$$

It also requires an $n \times n$ *transition matrix* $\mathbf{T} \in \mathbb{R}^{n \times n}$; semantically, this captures the probability that a label propagates from node i to node j, but it is effectively just the normalized edge weight, defined as

$$\mathbf{T}_{i,j} = \begin{cases} \frac{w_{i,j}}{\sum_{j \in \mathcal{N}_i} w_{i,j}} & \text{if } j \in \mathcal{N}_i, \\ 0 & \text{if } j \notin \mathcal{N}_i. \end{cases} \tag{15.3}$$

The algorithm iteratively multiplies $\mathbf{Y}$ by $\mathbf{T}$, thereby propagating label probabilities via a weighted average. After the multiply step, the unknown rows of $\mathbf{Y}$, corresponding to the unknown labels, must be normalized, and the known rows must be clamped to their known values. This continues until the values of $\mathbf{Y}$ have stabilized (i.e., converged to within some sufficiently small ε of change), or until a maximum number of iterations has been reached.

Algorithm 15.1 Label propagation

1: Initialize $\mathbf{Y}$ per Equation (15.2)
2: Initialize $\mathbf{T}$ per Equation (15.3)
3: **repeat**
4:      $\mathbf{Y} \leftarrow \mathbf{TY}$
5:      Normalize unknown rows of $\mathbf{Y}$
6:      Clamp known rows of $\mathbf{Y}$ using known labels
7: **until** convergence or maximum iterations reached
8: Assign to Y_i the jth label, where j is the highest value in row i of $\mathbf{Y}$.

One interesting property of this formulation of label propagation is that it is guaranteed to converge to a unique solution. In fact, there is a closed-form solution, which we will describe in Section 15.4.

15.3.2 Iterative Classification Algorithms

While label propagation is surprisingly effective, its predictor is essentially just a weighted average of neighboring labels, which may sometimes fail to capture complex relational dynamics. A more sophisticated approach would be to use a richer predictor. Suppose we have a classifier h that has been trained to classify a node i, given its features X_i and the features $\mathbf{X}_{\mathcal{N}_i}$ and labels $\mathbf{Y}_{\mathcal{N}_i}$ of its neighbors. *Iterative classification* does just that, applying local classification to each node, conditioned on the current predictions (or ground truth) on its neighbors, and iterating until the local predictions converge to a global solution. Iterative classification is an "algorithmic framework," in that it is it is agnostic to the choice of predictor; this makes it a very versatile tool for collective classification.

[6] introduced this approach and reported impressive gains in classification accuracy. [31] further developed the technique, naming it "iterative classification," and studied the conditions under which it improved classification performance [14]. Researchers [24, 25, 28] have since proposed various improvements and extensions to the basic algorithm we present.

Algorithm 15.2 Iterative classification

1: **for** $Y_i \in \mathbf{Y}^u$ **do** [Bootstrapping]
2:      $Y_i \leftarrow h(X_i, \mathbf{X}_{\mathcal{N}_i}, \mathbf{Y}^\ell_{\mathcal{N}_i})$
3: **end for**
4: **repeat** [update predicted labels]
5:      $\pi \leftarrow \text{GenPerm}(n)$ # generate permutation π over $1, \ldots, n$
6:      **for** $i = 1, \ldots, n$ **do**
7:          **if** $Y_{\pi(i)} \in \mathbf{Y}^u$ **then**
8:              $Y_{\pi(i)} \leftarrow h(X_{\pi(i)}, \mathbf{X}_{\mathcal{N}_{\pi(i)}}, \mathbf{Y}_{\mathcal{N}_{\pi(i)}})$
9:          **end if**
10:      **end for**
11: **until** convergence or maximum iterations reached

Algorithm 15.2 depicts pseudo-code for a simple iterative classification algorithm. The algorithm begins by initializing all unknown labels $\mathbf{Y}^u$ using only the features $(X_i, \mathbf{X}_{\mathcal{N}_i})$ and *observed* neighbor labels $\mathbf{Y}^\ell_{\mathcal{N}_i} \subseteq \mathbf{Y}_{\mathcal{N}_i}$. (This may require a specialized initialization classifier.) This process is sometimes referred to as *bootstrapping*. It then iteratively updates these values using the current predictions as well as the observed features and labels. This process repeats until the predictions have stabilized, or until a maximum number of iterations has been reached.

Clearly, the order in which nodes are updated affects the predictive accuracy and convergence rate, though there is some evidence to suggest that iterative classification is fairly robust to a number of simple ordering strategies—such as random ordering, ascending order of neighborhood diversity and descending order of prediction confidences [11]. Another practical issue is when to incorporate the predicted labels from the previous round into the the current round of prediction. Some researchers [28, 31] have proposed a "cautious" approach, in which only predicted labels are introduced gradually. More specifically, at each iteration, only the top k most confident predicted labels are used, thus ignoring less confident, potentially noisy predictions. At the start of the algorithm, k is initialized to some small number; then, in subsequent iterations, the value of k is increased, so that in the last iteration all predicted labels are used.

One benefit of iterative classification is that it can be used with any local classifier, making it extremely flexible. Nonetheless, there are some practical challenges to incorporating certain classifiers. For instance, many classifiers are defined on a predetermined number of features, making it difficult to accommodate arbitrarily-sized neighborhoods. A common workaround is to aggregate the neighboring features and labels, such as using the proportion of neighbors with a given label, or the most frequently occurring label. For classifiers that return a vector of scores (or conditional probabilities) instead of a label, one typically uses the label that corresponds to the maximum score. Some of the classifiers used included: naïve Bayes [6, 31], logistic regression [24], decision trees, [14] and weighted-majority [25].

Iterative classification prescribes a method of inference, but it does not instruct how to train the local classifiers. Typically, this is performed using traditional, non-collective training.

15.4 Graph-Based Regularization

When viewed as a transductive learning problem, the goal of collective classification is to complete the labeling of a partially-labeled graph. Since the problem domain is fixed (that is, the data to be classified is known), there is no need to learn an inductive model;[1] simply the predictions for the unknown labels will suffice. A broad category of learning algorithms, known as *graph-based regularization* techniques, are designed for this type of *model-free* prediction. In this section, we review these methods.

For the remainder of this section, we will employ the following notation. Let $\mathbf{y} \in \mathbb{R}^n$ denote a vector of labels corresponding to the nodes of the graph. For the methods considered in this section, we assume that the labels are binary; thus, if the ith label is known, then $y_i \in \{\pm 1\}$, and otherwise, $y_i = 0$. The learning objective is to produce a vector $\mathbf{h} \in \mathbb{R}^n$ of predictions that minimizes the L2 distance to $\mathbf{y}$ for the known labels. We can formulate this as a weighted inner product using a diagonal matrix $\mathbf{C} \in \mathbb{R}^{n \times n}$, where the (i, i) entry is set to 1 if the ith label is observed and 0 otherwise.[2] The error can thus be expressed as

$$(\mathbf{h} - \mathbf{y})^\top \mathbf{C} (\mathbf{h} - \mathbf{y}).$$

Unconstrained graph-based regularization methods can be generalized using the following abstraction (due to [8]). Let $\mathbf{Q} \in \mathbb{R}^{n \times n}$ denote a symmetric matrix, whose entries are determined based on the structure of the graph $\mathcal{G}$, the local attributes $\mathbf{X}$ (if available), and the observed labels $\mathbf{Y}^\ell$. We will give several explicit definitions for $\mathbf{Q}$ shortly; for the time being, it will suffice to think of $\mathbf{Q}$ as a *regularizer* on $\mathbf{h}$. Formulated as an unconstrained optimization, the learning objective is

$$\arg\min_{\mathbf{h}} \mathbf{h}^\top \mathbf{Q} \mathbf{h} + (\mathbf{h} - \mathbf{y})^\top \mathbf{C} (\mathbf{h} - \mathbf{y}).$$

One can interpret the first term as penalizing certain label assignments, based on observed information; the second term is simply the prediction error with respect to the training labels. Using vector calculus, we obtain a closed-form solution to this optimization as

$$\mathbf{h}^\star = (\mathbf{C}^{-1} \mathbf{Q} + \mathbf{I})^{-1} \mathbf{y} = (\mathbf{Q} + \mathbf{C})^{-1} \mathbf{C} \mathbf{y}, \tag{15.4}$$

where $\mathbf{I}$ is the $n \times n$ identity matrix. This is fairly efficient to compute for moderate-sized networks; the time complexity is dominated by $O(n^3)$ operations for the matrix inversion and multiplication. For prediction, the "soft" values of $\mathbf{h}$ can be clamped to $\{\pm 1\}$ using the sign operator.

The effectiveness of this generic approach comes down to how one defines the regularizer, $\mathbf{Q}$. One of the first instances is due to [49]. In this formulation, $\mathbf{Q}$ is a *graph Laplacian*, constructed

---

[1]This is not to say that inductive models are not useful in the transductive setting. Indeed, many practitioners apply model-based approaches to transductive problems [37].

[2]One could also apply different weights to certain nodes; or, if $\mathbf{C}$ were not diagonal, one could weight errors on certain combinations of nodes differently.

as follows: for each edge $(i,j) \in \mathcal{E}$, define a weight matrix $\mathbf{W} \in \mathbb{R}^{n \times n}$, where each element $w_{i,j}$ is defined using the radial basis function in (15.1); define a diagonal matrix $\mathbf{D} \in \mathbb{R}^{n \times n}$ as

$$d_{i,i} \triangleq \sum_{j=1}^{n} w_{i,j};$$

one then computes the regularizer as

$$\mathbf{Q} \triangleq \mathbf{D} - \mathbf{W}.$$

One could alternately define the regularizer as a *normalized* Laplacian,

$$\mathbf{Q} \triangleq \mathbf{I} - \mathbf{D}^{-\frac{1}{2}} \mathbf{W} \mathbf{D}^{-\frac{1}{2}},$$

per [47]. [15] extended this method for heterogeneous networks—that is, graphs with multiple types of nodes and edges. Another variant, due to [43], sets

$$\mathbf{Q} \triangleq (\mathbf{I} - \mathbf{A})^{\top} (\mathbf{I} - \mathbf{A}),$$

where $\mathbf{A} \in \mathbb{R}^{n \times n}$ is a row-normalized matrix capturing the local pairwise similarities. All of these formulations impose a *smoothness* constraint on the predictions, that "similar" nodes—where similarity can be defined by the Gaussian in (15.1) or some other kernel—should be assigned the same label.

There is an interesting connection between graph-based regularization and label propagation. Under the various parameterizations of $\mathbf{Q}$, one can show that (15.4) provides a closed-form solution to the label propagation algorithm in Section 15.3.1 [48]. This means that one can compute certain formulations of label propagation without directly computing the iterative algorithm. Heavily optimized linear algebra solvers can be used to compute (15.4) quickly. Another appealing aspect of these methods is their strong theoretical guarantees [8].

15.5 Probabilistic Graphical Models

Graphical models are powerful tools for joint, probabilistic inference, making them ideal for collective classification. They are characterized by a graphical representation of a probability distribution P, in which random variables are nodes in a graph $\mathcal{G}$. Graphical models can be broadly categorized by whether the underlying graph is directed (e.g., Bayesian networks or collections of local classifiers) or undirected (e.g., Markov random fields). In this section, we discuss both kinds.

Collective classification in graphical models involves finding the assignment $\mathbf{y}^u$ that maximized the conditional likelihood of $\mathbf{Y}^u$, given evidence $(\mathbf{X} = \mathbf{x}, \mathbf{Y}^\ell = \mathbf{y}^\ell)$; i.e.,

$$\underset{\mathbf{y}}{\arg\max}\, P(\mathbf{Y}^u = \mathbf{y}^u \mid \mathbf{X} = \mathbf{x}, \mathbf{Y}^\ell = \mathbf{y}^\ell), \tag{15.5}$$

where $\mathbf{y} = (\mathbf{y}^\ell, \mathbf{y}^u)$. This type of inference—known alternately as *maximum a posteriori* (MAP) or *most likely explanation* (MPE)—is known to be NP-hard in general graphical models, though there are certain exceptions in which it can be computed efficiently, and many approximation algorithms that perform well in most settings. We will review selected inference algorithms where applicable.

15.5.1 Directed Models

The fundamental directed graphical model is a *Bayesian network* (also called *Bayes net*, or *BN*).

Definition 15.5.1 (Bayesian Network) *A* Bayesian network *consists of a set of random variables* $\mathbf{Z} \triangleq \{Z_1, \ldots, Z_n\}$, *a directed, acyclic graph (DAG)* $\mathcal{G} \triangleq (\mathcal{V}, \mathcal{E})$, *and a set of* conditional probability distributions *(CPDs)*, $\{P(Z_i \,|\, \mathbf{Z}_{\mathcal{P}_i})\}_{i=1}^{n}$, *where* $\mathcal{P}_i$ *denotes the indices corresponding to the causal parents of* Z_i. *When multiplied, the CPDs describe the joint distribution of* $\mathbf{Z}$; *i.e.,* $P(\mathbf{Z}) = \prod_i P(Z_i \,|\, \mathbf{Z}_{\mathcal{P}_i})$.

BNs model causal relationships, which are captured by the directionalities of the edges; an edge $(i, j) \in \mathcal{E}$ indicates that Z_i influences Z_j. For a more thorough review of BNs, see [18] or Chapter X of this book.

Though BNs are very popular in machine learning and data mining, they can only be used for models with fixed structure, making them inadequate for problems with arbitrary relational structure. Since collective classification is often applied to arbitrary data graphs—such as those found in social and citation networks—some notion of *templating* is required. In short, templating defines subgraph patterns that are *instantiated* (or, *grounded*) by the data graph; model parameters (CPDs) are thus *tied* across different instantiations. This allows directed graphical models to be used on complex relational structures.

One example of a templated model is *probabilistic relational models* (PRMs) [9, 12]. A PRM is a directed graphical model defined by a relational database schema. Given an input database, the schema is instantiated by the database records, thus creating a BN. This has been shown to work for some collective classification problems [12, 13], and has the advantage that a full joint distribution is defined.

To satisfy the requirement of acyclicity when the underlying data graph is undirected, one constructs a (templated) BN or PRM as follows. For each potential edge $\{i, j\}$ in data graph, define a binary random variable $E_{i,j}$. Assume that a node's features are determined by its label. If we further assume that its label is determined by its neighbors' labels (i.e., contagion), then we draw a directed edge from each $E_{i,j}$ to the corresponding Y_i and Y_j, as illustrated in Figure 15.2a. On the other hand, if we believe that a node's label determines who it connects to (i.e., homophily), then we draw an edge from each Y_i to all $E_{i,\cdot}$, as shown in Figure 15.2b. The direction of causality is a modeling decision, which depends on one's prior belief about the problem. Note that, in both cases, the resulting graphical model is acyclic.

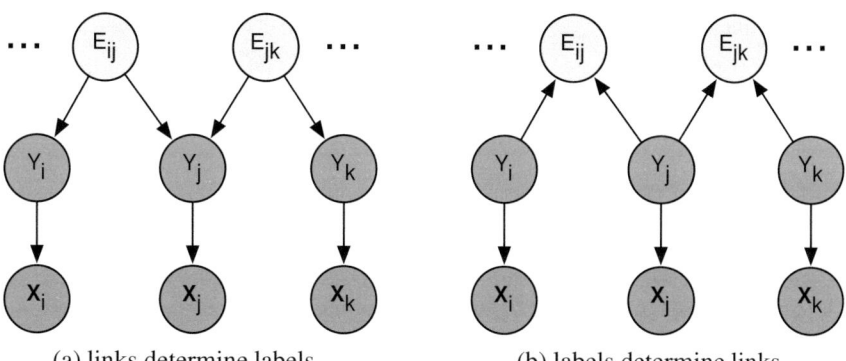

(a) links determine labels (b) labels determine links

FIGURE 15.2 (See color insert.): Example BN for collective classification. Label nodes (green) determine features (purple), which are represented by a single vector-valued variable. An edge variable (yellow) is defined for all potential edges in the data graph. In (a), labels are determined by link structure, representing contagion. In (b), links are functions of labels, representing homophily. Both structures are acyclic.

Another class of templated, directed graphical models is *relational dependency networks* (RDNs) [32]. RDNs have the advantage of supporting graph cycles, though this comes at the cost of *consistency*; that is, the product of an RDN's CPDs does not necessarily define a valid probability distribution. RDN inference is therefore only approximate, but can be very fast. Learning RDNs is also fast, because it reduces to independently learning a set of CPDs.

15.5.2 Undirected Models

While directed graphical models are useful for their representation of causality, sometimes we do not need (or want) to explicitly define causality; sometimes we only know the interactions between random variables. This is where undirected graphical models are useful. Moreover, undirected models are strictly more general than directed models; any directed model can be represented by an undirected model, but there are distributions induced by undirected models that cannot be reproduced in directed models. Specifically, graph structures involving cycles are representable in undirected models, but not in directed models.

Most undirected graphical models fall under the umbrella of Markov random fields (MRFs), sometimes called Markov networks [40].

Definition 15.5.2 (Markov random field) *A Markov random field (MRF) is defined by a set of random variables* $\mathbf{Z} \triangleq \{Z_1,\ldots,Z_n\}$, *a graph* $\mathcal{G} \triangleq (\mathcal{V}, \mathcal{E})$, *and a set of clique potentials* $\{\phi_c : \mathrm{dom}(c) \to \mathbb{R}\}_{c \in C}$, *where* C *is a set of predefined cliques and* $\mathrm{dom}(c)$ *is the domain of clique c. (To simplify notation, assume that potential* ϕ_c *only operates on the set of variables contained in clique c.) The potentials are often defined as a log-linear combination of features* f_c *and weights* w_c, *such that* $\phi_c(\mathbf{z}) \triangleq \exp(w_c \cdot f_c(\mathbf{z}))$. *An MRF defines a probability distribution P that factorizes as*

$$P(\mathbf{Z} = \mathbf{z}) = \frac{1}{\Phi} \prod_{c \in C} \phi_c(\mathbf{z}) = \frac{1}{\Phi} \exp\left(\sum_{c \in C} w_c \cdot f_c(\mathbf{z})\right),$$

where $\Phi \triangleq \sum_{\mathbf{z}'} \prod_{c \in C} \phi_c(\mathbf{z}')$ *is a normalizing constant. This model is said to be* Markovian *because any variable* Z_i *is independent of any non-adjacent variables, conditioned its neighborhood* $\mathbf{Z}_{\mathcal{N}_i}$ *(sometimes referred to as its* Markov blanket*).*

For collective classification, one can define a *conditional* MRF (sometimes called a CRF), whose conditional distribution is

$$P(\mathbf{Y}^u = \mathbf{y}^u \mid \mathbf{X} = \mathbf{x}, \mathbf{Y}^\ell = \mathbf{y}^\ell) = \frac{1}{\Phi} \prod_{c \in C} \phi_c(\mathbf{x}, \mathbf{y}) = \frac{1}{\Phi} \exp\left(\sum_{c \in C} w_c \cdot f_c(\mathbf{x}, \mathbf{y})\right).$$

For relational tasks, such as collective classification, one typically defines the cliques via templating. Similar to a PRM (see Section 15.5.1), a clique template is just a subgraph pattern—although in this case it is a fully-connected, undirected subgraph. The types of templates used directly affects model complexity, in that smaller templates correspond to a simpler model, which usually generalizes better to unseen data. Thus, MRFs are commonly defined using low-order templates, such as singletons, pairs, and sometimes triangles. Examples of templated MRFs include *relational MRFs* [40], *Markov logic networks* [36], and *hinge-loss Markov random fields* [2].

To make this concrete, we consider the class of *pairwise* MRFs. A pairwise MRF has features and weights for all singleton and pairwise cliques in the graph; thus, its distribution factorizes as

$$P(\mathbf{Z} = \mathbf{z}) = \frac{1}{\Phi} \left(\prod_{i \in \mathcal{V}} \phi_i(\mathbf{z})\right) \left(\prod_{\{i,j\} \in \mathcal{E}} \phi_{i,j}(\mathbf{z})\right)$$
$$= \frac{1}{\Phi} \exp\left[\left(\sum_{i \in \mathcal{V}} w_i \cdot f_i(\mathbf{z})\right) + \left(\sum_{\{i,j\} \in \mathcal{E}} w_{i,j} \cdot f_{i,j}(\mathbf{z})\right)\right].$$

(Since it is straightforward to derive the posterior distribution for collective classification, we omit it here.) If we assume that the domains of the variables are discrete, then it is common to define the features as basis vectors indicating the state of the assignment. For example, if $|Z_i| = k$ for all i, then $f_i(\mathbf{z})$ is a length-k binary vector, whose j^{th} entry is equal to one if z_i is in the jth state and zero otherwise; similarly, $f_{i,j}(\mathbf{z})$ has length k^2 and the only nonzero entry corresponds to the joint state of (z_i, z_j). To make this MRF templated, we simply replace all w_i with a single w_{single}, and all $w_{i,j}$ with a single w_{pair}.

It is important to note that the data graph does not necessarily correspond to the graph of the MRF; there is some freedom in how one defines the relational features $\{f_{i,j}\}_{\{i,j\}\in\mathcal{E}}$. However, when using a pairwise MRF for collective classification, it is natural to define a relational feature for each edge in the data graph. Defining $f_{i,j}$ as a function of (Y_i, Y_j) models the dependence between labels. One may alternately model (X_i, Y_i, X_j, Y_j) to capture the pairwise interactions of both features and labels.

15.5.3 Approximate Inference in Graphical Models

Exact inference in general graphical models is intractable, depending primarily on the structure of the underlying graph. Specifically, inference is exponential in the graph's *treewidth*. For structures with low treewidth—such as chains and trees—exact inference can be relatively fast. Unfortunately, these tractable settings are rare in collective classification, so inference is usually computed approximately. In this section, we review some commonly used approximate inference algorithms for directed and undirected graphical models.

15.5.3.1 Gibbs Sampling

Gibbs sampling is a general framework for approximating a distribution, when the distribution is presumed to come from a specified family, such as Gaussian, Poisson, etc. It is a *Markov chain Monte Carlo* (MCMC) algorithm, in that it iteratively samples from the current estimate of the distribution, constructing a Markov chain that converges to the target (stationary) distribution. Gibbs sampling is efficient because it samples from the conditional distributions of the individual variables, instead of the joint distribution over all variables. To make this more concrete, we examine Gibbs sampling for inference in directed graphical models, in the context of collective classification.

Pseudocode for a Gibbs sampling algorithm (based on [10, 25]) is given in Algorithm 15.3. At a high level, the algorithm works by iteratively sampling from the posterior distribution of each Y_i, i.e., random draws from $P(Y_i | \mathbf{X}, \mathbf{Y}_{\mathcal{N}_i})$. Like iterative classification, it initializes the posteriors using some function of the local and neighboring features, as well as any observed neighboring labels; this could be the (normalized) output of a local predictor. It then iteratively samples from each of the current posteriors and uses the sampled values to update the probabilities. This process is repeated until the posteriors converge, or until a maximum number of iterations is reached. Upon terminating, the samples for each node are averaged to obtain the approximate marginal label probabilities, $P(Y_i = y)$ (which can be used for prediction by choosing the label with the highest marginal probability). In practice, one typically sets aside a specified number of initial iterations as "burn-in" and averages over the remaining samples. In the limit of infinite data, the estimates should asymptotically converge to the true distribution.

Gibbs sampling is a popular method of approximate (marginal) inference in directed graphical models, such as BNs and PRMs. While each iteration of Gibbs sampling is relatively efficient, many iterations are required to obtain an accurate estimate of the distribution, which may be impractical. Thus, there is a trade-off between the running time and the accuracy of the approximate marginals.

Algorithm 15.3 Gibbs Sampling

1: **for** $i = 1, \ldots, n$ **do** [bootstrapping]
2: Initialize $P(Y_i | \mathbf{X}, \mathbf{Y}_{\mathcal{N}_i})$ using local features X_i and the features $\mathbf{X}_{\mathcal{N}_i}$ and *observed* labels $\mathbf{Y}^\ell_{\mathcal{N}_i} \subseteq \mathbf{Y}_{\mathcal{N}_i}$ of its neighbors
3: **end for**
4: **for** $i = 1, \ldots, n$ **do** [initialize samples]
5: $\mathcal{S}_i \leftarrow \emptyset$
6: **end for**
7: **repeat**[sampling]
8: $\pi \leftarrow \text{GENPERM}(n)$ # generate permutation π over $1, \ldots, n$
9: **for** $i = 1, \ldots, n$ **do**
10:   Sample $s \sim P(Y_i | \mathbf{X}, \mathbf{Y}_{\mathcal{N}_i})$
11:   $\mathcal{S}_i \leftarrow \mathcal{S}_i \cup s$
12:   Update $P(Y_i | \mathbf{X}, \mathbf{Y}_{\mathcal{N}_i})$ using $\mathcal{S}_i$
13: **end for**
14: **until** convergence or maximum iterations reached
15: **for** $i = 1, \ldots, n$ **do** [compute marginals]
16: Remove first T samples (i.e., burn-in) from $\mathcal{S}_i$
17: **for** $y \in \mathcal{Y}_i$ **do**
18:   $P(Y_i = y) \leftarrow \frac{1}{|\mathcal{S}_i|} \sum_{s \in \mathcal{S}_i} \mathbb{1}[y = s]$
19: **end for**
20: **end for**

15.5.3.2 Loopy Belief Propagation (LBP)

For certain undirected graphical models, exact inference can be computed efficiently via *message passing*, or *belief propagation* (BP), algorithms. These algorithms follow a simple iterative pattern: each variable passes its "beliefs" about its neighbors' marginal distributions, then uses the incoming messages about its own value to updates its beliefs. Convergence to the true marginals is guaranteed for tree-structured MRFs, but is not guaranteed for MRFs with cycles. That said, BP can still be used for approximate inference in general, so-called "loopy" graphs, with a minor modification: messages are discounted by a constant factor (sometimes referred to as *damping*). This algorithm is known as *loopy belief propagation* (LBP).

The LBP algorithm shown in Algorithm 15.4 [45, 46] assumes that the model is a pairwise MRF with singleton potentials defined on each (X_i, Y_i) and pairwise potentials on each adjacent (Y_i, Y_j). We denote by $m_{i \rightarrow j}(y)$ the message sent by Y_i to Y_j regarding the belief that $Y_j = y$; $\phi_i(y)$ denotes the ith local potential function evaluated for $Y_i = y$, and similarly for the pairwise potentials; $\alpha \in (0, 1]$ denotes a constant discount factor. The algorithm begins by initializing all messages to one. It then iterates over the message passing pattern, wherein Y_i passes its beliefs $m_{i \rightarrow j}(y)$ to all $j \in \mathcal{N}_i$, using the incoming messages $m_{k \rightarrow i}(y)$, for all $k \in \mathcal{N}_i \setminus j$, from the previous iteration. Similar to Gibbs sampling, the algorithm iterates until the messages stabilize, or until a maximum number of iterations is reached, after which we compute the approximate marginal probabilities.

15.6 Feature Construction

Thus far, we have used rather abstract problem representations, using an arbitrary data graph $\mathcal{G} \triangleq (\mathcal{V}, \mathcal{E})$, feature variables $\mathbf{X}$, and label variables $\mathbf{Y}$. In practice, how one maps a real-world

---

Algorithm 15.4 Loopy Belief Propagation

1: **for** $\{i,j\} \in \mathcal{E}$ **do** [initialize messages]
2:      **for** $y \in \mathcal{Y}_j$ **do**
3:          $m_{i \rightarrow j}(y) \leftarrow 1$
4:      **end for**
5: **end for**
6: **repeat**[message passing]
7:      **for** $\{i,j\} \in \mathcal{E}$ **do**
8:          **for** $y \in \mathcal{Y}_j$ **do**
9:              $m_{i \rightarrow j}(y) \leftarrow \alpha \sum_{y'} \phi_{i,j}(y',y) \phi_i(y') \prod_{k \in \mathcal{N}_i \setminus j} m_{k \rightarrow i}(y')$
10:          **end for**
11:      **end for**
12: **until** convergence or maximum iterations reached
13: **for** $i = 1, \ldots, n$ **do** [compute marginals]
14:      **for** $y \in \mathcal{Y}_i$ **do**
15:          $P(Y_i = y) \leftarrow \alpha \phi_i(y) \prod_{j \in \mathcal{N}_i} m_{j \rightarrow i}(y)$
16:      **end for**
17: **end for**

---

problem to these representations can have a greater impact than the choice of model or inference algorithm. This process, sometimes referred to as *feature construction* (or *feature engineering*), is perhaps the most challenging aspect of data mining. In this section, we explore various techniques, motivated by concrete examples.

15.6.1 Data Graph

Network structure is *explicit* in certain collective classification problems, such as categorizing users in a social network or pages in a Web site. However, there are other problems that exhibit *implicit* network structure.

An example of this is image segmentation, a common computer vision problem. Given an observed image—i.e., an $m \times n$ matrix of pixel intensities—the goal is to label each pixel as being part of a certain object, from a predetermined set of objects. The structure of the image naturally suggests a grid graph, in which each pixel (i,j) is a node, adjacent to (up to) four neighboring pixels. However, one could also define the neighborhood using the diagonally adjacent pixels, or use wider concentric rings. This decision reflects on one's prior belief of how many pixels directly interact with (i,j); that is, its Markov blanket.

A related setting is part-of-speech (POS) tagging in natural language processing. This task involves tagging each word in a sentence with a POS label. The linear structure of a sentence is naturally represented by a path graph; i.e., a tree in which each vertex has either one or two neighbors. One could also draw edges between words separated by two hops, or, more generally, n hops, depending on one's belief about sentence construction.

The data graph can also be inferred from distances. Suppose the task is to predict which individuals in a given population will contract the flu. Infection is obviously related to proximity, so it is natural to construct a network based on geographic location. Distances can be thresholded to create unweighted edges (e.g., "close" or "not close"), or they can be incorporated as weights (if the model supports it).

Structure can sometimes be inferred from other structure. An example of this is found in document classification in citation networks. In a citation network, there are *explicit* links from papers citing other papers. There are also *implicit* links between two papers that cite the same paper. These "co-citation" edges complete a triangle between three connected nodes.

Links can also be discovered as part of the inference problem [5, 26]. Indeed, collective classification and link prediction are complimentary tasks, and it has been shown that coupling these predictions can improve overall accuracy [29].

It is important to note that not all data graphs are *unimodal*—that is, they may involve multiple types of nodes and edges. In citation networks, authors write papers; authors are affiliated with institutions; papers cite other papers; and so on.

15.6.2 Relational Features

Methods based on local classifiers, such as ICA, present a unique challenge to relational feature engineering: while most local classifiers require fixed-length feature vectors, neighborhood sizes are rarely uniform in real relational data. For example, a paper can have any number of authors. If the number of neighbors is bounded by a reasonably small constant (a condition often referred to as *bounded degree*), then it is possible to represent neighborhood information using a fixed-length vector. However, in naturally occurring graphs, such as social networks, this condition is unlikely.

One solution is to aggregate neighborhood information into a fixed number of statistics. For instance, one could count the number of neighbors exhibiting a certain attribute; for a set of attributes, this amounts to a histogram. For numerical or ordinal attributes, one could also use the mean, mode, minimum or maximum. Another useful statistic is the number of triangles, which reflects the connectivity of the neighborhood. More complex, domain-specific aggregates are also possible. Within the inductive logic programming community, aggregation has been studied as a means for *propositionalizing* a relational classification problem [17, 19, 20]. In the machine learning community, [33, 34] have studied aggregation extensively.

Aggregation is also useful for defining relational features in graphical models. Suppose one uses a pairwise MRF for social network classification. For each pair of nodes, one could obviously consider the similarities of their local features; yet one could also consider the similarities of their neighborhood structures. A simple metric is the number of common neighbors. To compensate for varying neighborhood sizes, one could normalize the intersection by the union, which is known as *Jaccard similarity*,

$$J(\mathcal{N}_i, \mathcal{N}_j) \triangleq \frac{\mathcal{N}_i \cap \mathcal{N}_j}{\mathcal{N}_i \cup \mathcal{N}_j}.$$

This can be generalized to the number (or proportion) of common neighbors who exhibit a certain attribute.

15.7 Applications of Collective Classification

Collective classification is a generic problem formulation that has been successfully applied to many application domains. Over the course of this chapter, we have already discussed a few popular applications, such as document classification [41, 44], image segmentation [1] and POS tagging [22]. In this section, we briefly review some other uses.

One such problem, which is related to POS tagging, is *optical character recognition* (OCR). The goal of OCR is to automatically convert a digitized stream of handwritten characters into a text file. In this context, each node represents a scanned character; its observed features X_i are a vectorized pixel grid, and the label Y_i is chosen from the set of ASCII (or ISO) characters. While this can be predicted fairly well using a local classifier, it has been shown that considering intra-character relationships can be beneficial [42], since certain characters are more (or less) likely to occur before (or after) other characters.

Another interesting application is activity detection in video data. Given a recorded video sequence (say, from a security camera) containing multiple actors, the goal is to label each actor as performing a certain action (from a predefined set of actions). Assuming that bounding boxes and tracking (i.e., identity maintenance) are given, one can bolster local reasoning with spatiotemporal relational reasoning. For instance, it is often the case that certain actions associate with others: if an actor is crossing the street, then other actors in the proximity are likely crossing the street; similarly, if one actor is believed to be talking, then other actors in the proximity are likely either talking or listening. One could also reason about action transitions: if an actor is walking at time t, then it is very likely that they will be walking at time $t + 1$; however, there is a small probability that they may transition to a related action, such as crossing the street or waiting. Incorporating this high-level relational reasoning can be considered a form of collective classification. This approach has been used in a number of publications [7, 16] to achieve current state-of-the-art performance on benchmark datasets.

Collective classification is also used in computational biology. For example, researchers studying protein-protein interaction networks often need to annotate proteins with their biological function. Discovering protein function experimentally is expensive. Yet, protein function is sometimes correlated with interacting proteins; so, given a set of labeled proteins, one can reason about the remaining labels using collective methods [23].

The final application we consider is viral marketing, which is interesting for its relationship to *active* collective classification. Suppose a company is introducing a new product to a population. Given the social network and the individuals' (i.e., local) demographic features, the goal is to determine which customers will be interested in the product. The mapping to collective classification is straightforward. The interesting subproblem is in how one acquires labeled training data. Customer surveys can be expensive to conduct, so companies want to acquire the smallest set of user opinions that will enable them to accurately predict the remaining user opinions. This can be viewed as an active learning problem for collective classification [3].

15.8 Conclusion

Given the recent explosion of relational and network data, collective classification is quickly becoming a mainstay of machine learning and data mining. Collective techniques leverage the idea that connected (related) data objects are in some way correlated, performing joint reasoning over a high-dimensional, structured output space. Models and inference algorithms range from simple iterative frameworks to probabilistic graphical models. In this chapter, we have only discussed a few; for greater detail on these methods, and others we did not cover, we refer the reader to [25] and [37]. Many of the algorithms discussed herein have been implemented in NetKit-SRL,[3] an open-source toolkit for mining relational data.

Acknowledgements

This material is based on work supported by the National Science Foundation under Grant No. 0746930 and Grant No. IIS1218488.

---

[3]http://netkit-srl.sourceforge.net

Bibliography

[1] D. Anguelov, B. Taskar, V. Chatalbashev, D. Koller, D. Gupta, G. Heitz, and A. Ng. Discriminative learning of Markov random fields for segmentation of 3D scan data. In *International Conference on Computer Vision and Pattern Recognition*, pages 169–176, 2005.

[2] Stephen H. Bach, Bert Huang, Ben London, and Lise Getoor. Hinge-loss Markov random fields: Convex inference for structured prediction. In *Uncertainty in Artificial Intelligence*, 2013.

[3] Mustafa Bilgic and Lise Getoor. Effective label acquisition for collective classification. In *ACM SIGKDD International Conference on Knowledge Discovery and Data Mining*, pages 43–51, 2008. Winner of the KDD'08 Best Student Paper Award.

[4] Mustafa Bilgic, Lilyana Mihalkova, and Lise Getoor. Active learning for networked data. In *Proceedings of the 27th International Conference on Machine Learning (ICML-10)*, 2010.

[5] Mustafa Bilgic, Galileo Mark Namata, and Lise Getoor. Combining collective classification and link prediction. In *Workshop on Mining Graphs and Complex Structures in International Conference of Data Mining*, 2007.

[6] Soumen Chakrabarti, Byron Dom, and Piotr Indyk. Enhanced hypertext categorization using hyperlinks. In *International Conference on Management of Data*, pages 307–318, 1998.

[7] W. Choi, K. Shahid and S. Savarese. What are they doing?: Collective activity classification using spatio-temporal relationship among people. In *VS*, 2009.

[8] Corinna Cortes, Mehryar Mohri, Dmitry Pechyony, and Ashish Rastogi. Stability analysis and learning bounds for transductive regression algorithms. *CoRR*, abs/0904.0814, 2009.

[9] N. Friedman, L. Getoor, D. Koller, and A. Pfeffer. Learning probabilistic relational models. In *International Joint Conference on Artificial Intelligence*, 1999.

[10] S. Geman and D. Geman. *Stochastic relaxation, Gibbs distributions, and the Bayesian restoration of images*. Morgan Kaufmann Publishers Inc., San Francisco, CA, 1990.

[11] L. Getoor. Link-based classification. In *Advanced Methods for Knowledge Discovery from Complex Data*, Springer, 2005.

[12] Lise Getoor, Nir Friedman, Daphne Koller, and Ben Taskar. Learning probabilistic models of link structure. *Journal of Machine Learning Research*, 3:679–707, 2002.

[13] Lise Getoor, Eran Segal, Benjamin Taskar, and Daphne Koller. Probabilistic models of text and link structure for hypertext classification. In *IJCAI Workshop on Text Learning: Beyond Supervision*, 2001.

[14] D. Jensen, J. Neville, and B. Gallagher. Why collective inference improves relational classification. In *Proceedings of the 10th ACM SIGKDD International Conference on Knowledge Discovery and Data Mining*, pages 593–598, 2004.

[15] Ming Ji, Yizhou Sun, Marina Danilevsky, Jiawei Han, and Jing Gao. Graph regularized transductive classification on heterogeneous information networks. In *Proceedings of the 2010 European Conference on Machine Learning and Knowledge Discovery in Databases: Part I*, ECML PKDD'10, pages 570–586, Berlin, Heidelberg, 2010. Springer-Verlag.

[16] Sameh Khamis, Vlad I. Morariu, and Larry S. Davis. Combining per-frame and per-track cues for multi-person action recognition. In *European Conference on Computer Vision*, pages 116–129, 2012.

[17] A. Knobbe, M. deHaas, and A. Siebes. Propositionalisation and aggregates. In *Proceedings of the Fifth European Conference on Principles of Data Mining and Knowledge Discovery*, pages 277–288, 2001.

[18] D. Koller and N. Friedman. *Probabilistic Graphical Models: Principles and Techniques*. MIT Press, 2009.

[19] S. Kramer, N. Lavrac, and P. Flach. Propositionalization approaches to relational data mining. In S. Dzeroski and N. Lavrac, editors, *Relational Data Mining*. Springer-Verlag, New York, 2001.

[20] M. Krogel, S. Rawles, F. Zeezny, P. Flach, N. Lavrac, and S. Wrobel. Comparative evaluation of approaches to propositionalization. In *International Conference on Inductive Logic Programming*, pages 197–214, 2003.

[21] Ankit Kuwadekar and Jennifer Neville. Relational active learning for joint collective classification models. In *Proceedings of the 28th International Conference on Machine Learning*, 2011.

[22] John D. Lafferty, Andrew McCallum, and Fernando C. N. Pereira. Conditional random fields: Probabilistic models for segmenting and labeling sequence data. In *Proceedings of the International Conference on Machine Learning*, pages 282–289, 2001.

[23] Stanley Letovsky and Simon Kasif. Predicting protein function from protein/protein interaction data: a probabilistic approach. *Bioinformatics*, 19:197–204, 2003.

[24] Qing Lu and Lise Getoor. Link based classification. In *Proceedings of the International Conference on Machine Learning*, 2003.

[25] S. Macskassy and F. Provost. Classification in networked data: A toolkit and a univariate case study. *Journal of Machine Learning Research*, 8(May):935–983, 2007.

[26] Sofus A. Macskassy. Improving learning in networked data by combining explicit and mined links. In *Proceedings of the Twenty-Second Conference on Artificial Intelligence*, 2007.

[27] Sofus A. Macskassy. Using graph-based metrics with empirical risk minimization to speed up active learning on networked data. In *Proceedings of the 15th ACM SIGKDD International Conference on Knowledge Discovery and Data Mining*, 2009.

[28] Luke K. Mcdowell, Kalyan M. Gupta, and David W. Aha. Cautious inference in collective classification. In *Proceedings of AAAI*, 2007.

[29] Galileo Mark Namata, Stanley Kok, and Lise Getoor. Collective graph identification. In *ACM SIGKDD International Conference on Knowledge Discovery and Data Mining*, 2011.

[30] Galileo Mark Namata, Ben London, Lise Getoor, and Bert Huang. Query-driven active surveying for collective classification. In *Workshop on Mining and Learning with Graphs*, 2012.

[31] Jennifer Neville and David Jensen. Iterative classification in relational data. In *Workshop on Statistical Relational Learning, AAAI*, 2000.

[32] Jennifer Neville and David Jensen. Relational dependency networks. *Journal of Machine Learning Research*, 8:653–692, 2007.

[33] C. Perlich and F. Provost. Aggregation-based feature invention and relational concept classes. In *ACM SIGKDD International Conference on Knowledge Discovery and Data Mining*, 2003.

[34] C. Perlich and F. Provost. Distribution-based aggregation for relational learning with identifier attributes. *Machine Learning Journal*, 62(1-2):65–105, 2006.

[35] Matthew J. Rattigan, Marc Maier, David Jensen Bin Wu, Xin Pei, JianBin Tan, and Yi Wang. Exploiting network structure for active inference in collective classification. In *Proceedings of the Seventh IEEE International Conference on Data Mining Workshops*, ICDMW '07, pages 429–434. IEEE Computer Society, 2007.

[36] Matt Richardson and Pedro Domingos. Markov logic networks. *Machine Learning*, 62 (1-2):107–136, 2006.

[37] Prithviraj Sen, Galileo Mark Namata, Mustafa Bilgic, Lise Getoor, Brian Gallagher, and Tina Eliassi-Rad. Collective classification in network data. *AI Magazine*, 29(3):93–106, 2008.

[38] Hossam Sharara, Lise Getoor, and Myra Norton. Active surveying: A probabilistic approach for identifying key opinion leaders. In *The 22nd International Joint Conference on Artificial Intelligence (IJCAI '11)*, 2011.

[39] Partha Pratim Talukdar and Koby Crammer. New regularized algorithms for transductive learning. In *ECML/PKDD (2)*, pages 442–457, 2009.

[40] B. Taskar, P. Abbeel, and D. Koller. Discriminative probabilistic models for relational data. In *Proceedings of the Annual Conference on Uncertainty in Artificial Intelligence*, 2002.

[41] B. Taskar, V. Chatalbashev, and D. Koller. Learning associative Markov networks. In *Proceedings of the International Conference on Machine Learning*, 2004.

[42] B. Taskar, C. Guestrin, and D. Koller. Max-margin Markov networks. In *Neural Information Processing Systems*, 2003.

[43] Mingrui Wu and Bernhard Schölkopf. Transductive classification via local learning regularization. *Journal of Machine Learning Research - Proceedings Track*, 2:628–635, 2007.

[44] Yiming Yang, S. Slattery, and R. Ghani. A study of approaches to hypertext categorization. *Journal of Intelligent Information Systems*, 18 (2-3):219–241, 2002.

[45] J.S. Yedidia, W.T. Freeman, and Y. Weiss. Constructing free-energy approximations and generalized belief propagation algorithms. In *IEEE Transactions on Information Theory*, pages 2282–2312, 2005.

[46] J.S. Yedidia, W.T. Freeman, and Y. Weiss. Generalized belief propagation. In *Neural Information Processing Systems*, 13:689–695, 2000.

[47] Dengyong Zhou, Olivier Bousquet, Thomas Navin Lal, Jason Weston, and Bernhard Schölkopf. Learning with local and global consistency. In *NIPS*, pages 321–328, 2003.

[48] Xiaojin Zhu and Zoubin Ghahramani. Learning from labeled and unlabeled data with label propagation. Technical report, Carnegie Mellon University, 2002.

[49] Xiaojin Zhu, Zoubin Ghahramani, and John Lafferty. Semi-supervised learning using Gaussian fields and harmonic functions. In *Proceedings of the 20th International Conference on Machine Learning (ICML-03)*, pages 912–919, 2003.

Chapter 16

Uncertain Data Classification

Reynold Cheng

The University of Hong Kong
ckcheng@cs.hku.hk

Yixiang Fang

The University of Hong Kong
yxfang@cs.hku.hk

Matthias Renz

University of Munich
renz@dbs.ifi.lmu.de

16.1 Introduction

In emerging applications such as location-based services (LBS), sensor networks, and biological databases, the values stored in the databases are often *uncertain* [11, 18, 19, 30]. In an LBS, for example, the location of a person or a vehicle sensed by imperfect GPS hardware may not be exact. This *measurement error* also occurs in the temperature values obtained by a thermometer, where 24% of measurements are off by more than 0.5°C, or about 36% of the normal temperature range [15]. Sometimes, uncertainty may be injected to the data by the application, in order to provide a better protection of privacy. In demographic applications, partially aggregated data sets, rather than personal data, are available [3]. In LBS, since the exact location of a person may be sensitive, researchers have proposed to "anonymize" a location value by representing it as a region [20]. For these applications, data uncertainty needs to be carefully handled, or else wrong decisions can be made. Recently, this important issue has attracted a lot of attention from the database and data mining communities [3, 9–11, 18, 19, 30, 31, 36].

In this chapter, we investigate the issue of classifying data whose values are not certain. Similar to other data mining solutions, most classification methods (e.g., decision trees [27, 28] and naive Bayes classifiers [17]) assume exact data values. If these algorithms are applied on uncertain data (e.g., temperature values), they simply ignore them (e.g., by only using the thermometer reading). Unfortunately, this can severely affect the accuracy of mining results [3, 29]. On the other hand, the use of uncertainty information (e.g., the derivation of the thermometer reading) may provide new insight about the mining results (e.g., the probability that a class label is correct, or the chance that an association rule is valid) [3]. It thus makes sense to consider uncertainty information during the data mining process.

How to consider uncertainty in a classification algorithm, then? For ease of discussion, let us consider a database of n objects, each of which has a d-dimensional feature vector. For each feature vector, we assume that the value of each dimension (or "attribute") is a number (e.g., temperature). A common way to model uncertainty of an attribute is to treat it as a probability density function, or *pdf* in short [11]. For example, the temperature of a thermometer follows a Gaussian pdf [15]. Given a d-dimensional feature vector, a natural attempt to handle uncertainty is to replace the pdf of an attribute by its *mean*. Once all the n objects are processed in this way, each attribute value becomes a single number, and thus any classification algorithm can be used. We denote this method as AVG. The problem of this simple method is that it ignores the variance of the pdf. More specifically, AVG cannot distinguish between two pdfs with the same mean but a big difference in variances. As shown experimentally in [1, 23, 29, 32, 33], this problem impacts the effectiveness of AVG.

Instead of representing the pdf by its mean value, a better way could be to consider all its possible values during the classification process. For example, if an attribute's pdf is distributed in the range $[30, 35]$, each real value in this range has a non-zero chance to be correct. Thus, every single value in $[30, 35]$ should be considered. For each of the n feature vectors, we then consider all the possible instances — an enumeration of d attribute values based on their pdfs. Essentially, a database is expanded to a number of *possible worlds*, each of which has a single value for each attribute, and has a probability to be correct [11, 18, 19, 30]. Conceptually, a classification algorithm (e.g., decision tree) can then be applied to each possible world, so that we can obtain a classification result for each possible world. This method, which we called PW, provides information that is not readily available for AVG. For example, we can compute the probability that each label is assigned to each object; moreover, the label with the highest probability can be assigned to the object. It was also pointed out in previous works (e.g., [3, 32, 33]) that compared with AVG, PW yields a better classification result.

Unfortunately, PW is computationally very expensive. This is because each attribute has an infinitely large number of possible values, which results in an infinitely large number of possible worlds. Even if we approximate a pdf with a set of points, the number of possible worlds can still be exponentially large [11, 18, 19, 30]. Moreover, the performance of PW does not scale with the database size. In this chapter, we will investigate how to incorporate data uncertainty in several well-known algorithms, namely decision tree, rule-based, associative, density-based, nearest-neighbor-based, support vector, and naive Bayes classification. These algorithms are redesigned with the goals that (1) their effectiveness mimic that of PW, and (2) they are computationally efficient.

The rest of this chapter is as follows. In Section 16.2, we will describe uncertainty models and formulate the problem of classifying uncertain data. Section 16.3 describes the details of several uncertain data classification algorithms. We conclude in Section 16.4.

16.2 Preliminaries

We now discuss some background information that is essential to the understanding of the classification algorithms. We first describe the data uncertainty models assumed by these algorithms, in Section 16.2.1. Section 16.2.2 then explains the framework that is common to all the classification algorithms we discussed here.

16.2.1 Data Uncertainty Models

Let us first explain how to model the uncertainty of a d-dimensional feature vector of object o. Suppose that this feature vector is represented by $\{A_1, A_2, \cdots A_d\}$, where A_i $(i = 1, \ldots, d)$ is a real number or a set of values (i.e., a "categorical value"). Notice that A_i may not be exact. Specifically, we represent the feature vector of o as a d-dimensional vector $x = (x^1, x^2, \cdots x^d)$, where each x^j $(j = 1, \ldots, d)$ is a random variable. We consider two common forms of x^j, namely (1) probability density function (pdf) and (2) probability mass function (pmf):

- **pdf:** x_j is a real value in $[l, r]$, whose pdf is denoted by $f^j(x)$, with $\int_l^r f^j(x) dx = 1$. Some common pdfs are Gaussian and uniform distribution [11]. For example, a tympanic (ear) thermometer measures the temperature of the ear drum via an infrared sensor. The reading of this thermometer can be modeled by a Gaussian pdf, with mean $37°C$ (i.e., the normal human body temperature) and $[l, r] = [37 - 0.7, 37 + 0.7]$ (i.e., the range of body temperature).

- **pmf:** The domain of x_j is a set $\{v_1, v_2, \cdots v_m\}$, where v_i could be a number, or the name of some category. Here, x^j is a vector $p = \{p_1, p_2, \cdots p_m\}$ such that the probability that $x^j = v_k$ is equal to p_k, and $\sum_{k=1}^{m} p_k = 1 (1 \leq k \leq m)$. For instance, a medical dataset contains information of patients suffering from tumor. The type of tumor could be either *benign* or *malignant*; each type has a probability to be true. If we let A_j be an attribute that denotes the tumor type of a patient, then x_j is a vector $\{p_1, p_2\}$, where p_1 and p_2 are the probabilities that $A_j = benign$ and $A_j = malignant$, respectively.

These two models about the uncertainty of an attribute are commonly used by uncertain data classification algorithms. Next, let us give an overview of these algorithms.

16.2.2 Classification Framework

Let us now describe the framework of the classification algorithms that will be discussed in this chapter. Let $C = \{C_1, C_2, \cdots C_L\}$ be the set of the class labels, where L is the number of classes. A *training dataset*, $Train\_D$, contains n d-dimensional objects $x_1, x_2, \cdots x_n$. Each object x_i, with feature vector $(x_i^1, x_i^2, \cdots x_i^d)(1 \leq i \leq n)$, is associated with a label $c_i \in C$. Each dimension $x_i^j (1 \leq j \leq d)$ of object x_i can be uncertain, and is associated with either a pdf or a pmf. The goal of classification is to construct a model, or *classifier*, based on $Train\_D$, and use it to predict the class label of a set of unseen objects (called *test object*) in a *testing dataset* (or $Test\_D$).

Specifically, a classifier M has to be first constructed by some method (e.g., decision tree construction algorithm). This process is also known as *supervised learning*, because we have to provide labels for objects in $Train\_D$. After M is generated, the relationship between the feature vectors of $Train\_D$ and labels in C is established [17]. In particular, M maps the feature vector of each object in $Train\_D$ to a probability distribution P on the labels in C.

In the second step, for each test object y_j with feature vector $(y_j^1, y_j^2, \cdots y_j^d)$ in *Test_D*, $M(y_j^1, y_j^2, \cdots y_j^d)$ yields a probability distribution P_j of class labels for y_j. We then assign y_j the class label, say, c_0, whose probability is the highest, i.e., $c_0 = \arg\max_{c \in C} \{P_j(c)\}$.

The two common measures of a classifier M are:

- *Accuracy*: M should be highly accurate in its prediction of labels for *Train_D*. For example, M should have a high percentage of test tuples predicted correctly on *Train_D*, compared with the ground truth. To achieve a high accuracy, a classifier should be designed to avoid underfit or overfit problems.

- *Performance*: M should be efficient, i.e., the computational cost should be low. Its performance should also scale with the database size.

As discussed before, although AVG is efficient, it has a low accuracy. On the other hand, PW has a higher accuracy, but is extremely inefficient. We next examine several algorithms that are designed to satisfy both of these requirements.

16.3 Classification Algorithms

We now describe several important classification algorithms that are designed to handle uncertain data:

- Decision trees (Section 16.3.1);

- Rule-based classification (Section 16.3.2);

- Associative classification (Section 16.3.3);

- Density-based classification (Section 16.3.4);

- Nearest-neighbor-based classification (Section 16.3.5);

- Support vector classification (Section 16.3.6); and

- Naive Bayes classification (Section 16.3.7).

The above algorithms are derived from the classical solutions that classify exact data. For each algorithm, we will discuss its framework, and how it handles uncertain data efficiently.

16.3.1 Decision Trees

Overview: Decision tree classification is a well-studied technique [27, 28]. A decision tree is a tree-like structure that consists of internal nodes, leaf nodes, and branches. Each internal node of a decision tree represents a test on an unseen test tuple's feature (attribute), while each leaf node represents a class or a probability distribution of classes. Using a decision tree, classification rules can be extracted easily. A decision tree is easy to understand and simple to use. Moreover, its construction does not require any domain knowledge or parameter values.

In traditional decision tree classification algorithms (e.g., ID3 and C4.5) [27, 28], a feature (an attribute) of a tuple is either categorical or numerical. Its values are usually assumed to be certain. Recent works [23, 24, 32, 33] have extended this technique to support uncertain data. Particularly, the authors in [32, 33] studied the problem of constructing decision trees for binary classification on

id	class	mean	probability distribution				
			-10	-1.0	0.0	+1.0	+10
1	A	+2.0		8/11			3/11
2	A	-2.0	1/9	8/9			
3	A	+2.0		5/8		1/8	2/8
4	B	-2.0	5/19	1/19		13/19	
5	B	+2.0			1/35	30/35	4/35
6	B	-2.0	3/11			8/11	

FIGURE 16.1: An example uncertain database.

data with uncertain numerical attributes. In this section, we focus on the *Uncertain Decision tree* (or UDT) proposed in [32].

Input: A set of labeled objects whose uncertain attributes are associated with pdf models.

Algorithm: Let us first use an example to illustrate how decision trees can be used for classifying uncertain data. Suppose that there are six tuples with given labels *A* or *B* (Figure 16.1). Each tuple has a single uncertain attribute. After sampling its pdf for a few times, its probability distribution is approximated by five values with probabilities, as shown in the figure. Notice that the mean values of attributes tuples 1, 3 and 5 are the same, i.e., 2.0. However, their probability distributions are different.

We first use these tuples as a training dataset, and consider the method AVG. Essentially, we only use these tuples' mean values to build a decision tree based on traditional algorithms such as C4.5 [28] (Figure 16.2(a)). In this example, the *split point* is $x = -2$. Conceptually, this value determines the branch of the tree used for classifying a given tuple. Each leaf node contains the information about the probability of an object to be in each class. If we use this decision tree to classify tuple 1, we will reach the right leaf node by traversing the tree, since its mean value is 2.0. Let $P(A)$ and $P(B)$ be the probabilities of a test tuple for belonging to classes *A* and *B*, respectively. Since $P(A) = 2/3$, which is greater than $P(B) = 1/3$, we assign tuple 1 with label "*A*." Using a similar approach, we label tuples 1, 3, and 5 as "*A*," and name tuples 2, 4, 6 as "*B*." The number of tuples classified correctly, i.e., tuples 1, 3, 4 and 6, is 4, and the accuracy of AVG in this example is 2/3.

The form of the decision tree trained by UDT is in fact the same as the one constructed through AVG. In our previous example, the decision tree generated by UDT is shown in Figure 16.2(b). We can see that the only difference from (a) is that the split point is $x = -1$, which is different from that

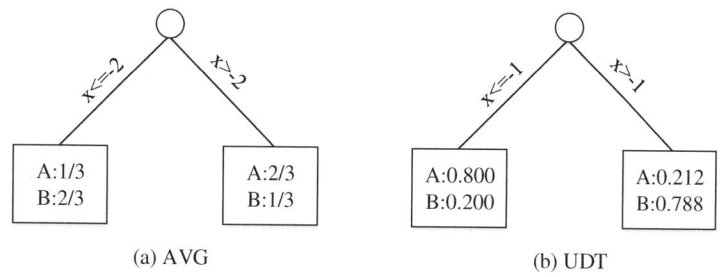

FIGURE 16.2: Decision trees constructed by (a) AVG and (b) UDT.

of AVG. Let us now explain (1) how to use UDT to train this decision tree; and (2) how to use it to classify previously unseen tuples.

1. Training Phase. This follows the framework of C4.5 [28]. It builds decision trees from the root to the leaf nodes in a top-down, recursive, and divide-and-conquer manner. Specifically, the training set is recursively split into smaller subsets according to some *splitting criterion*, in order to create subtrees. The splitting criterion is designed to determine the best way to partition the tuples into individual classes. The criterion not only tells us which attribute to be tested, but also how to use the selected attribute (by using the *split point*). We will explain this procedure in detail soon. Once the split point of a given node has been chosen, we partition the dataset into subsets according to its value. In Figure 16.2(a), for instance, we classify tuples with $x > -2$ to the right branch, and put tuples with $x \leq -2$ to the left branch. Then we create new sub-nodes with the subsets recursively. The commonly used splitting criteria are based on indexes such as information gain (entropy), gain ratio, and Gini index. Ideally, a good splitting criterion should make the resulting partitions at each branch as "pure"as possible (i.e., the number of distinct objects that belong to the branch is as few as possible).

Let us now explain how to choose an attribute to split, and how to split it. First, assume that an attribute is chosen to split, and the goal is to find out how to choose the best split point for an uncertain attribute. To do this, UDT samples points from the pdf of the attribute, and considers them as *candidate split points*. It then selects the split point with the lowest *entropy*. The computation of entropy $H(q, A_j)$ for a given split point q over an attribute A_j is done as follows: first split the dataset into different subsets according to the split point. Then for each subset, we compute its probabilities of belonging to different classes. Then its entropy $H(q, A_j)$ can be computed based on these probabilities. The optimal split point is the one that minimizes the entropy for A_j. Notice that this process can potentially generate more candidate split points than that of AVG, which is more expensive but can be more accurate [32]. In the above example, to choose the best split point, we only have to check 2 points in AVG since the attributes concerned only have two distinct values, 2 and -2. On the other hand, we have to check five distinct values in UDT.

How to choose the attribute for which a split point is then applied? This is done by first choosing the optimal split points for all the attributes respectively, and then select the attribute whose entropy based on its optimal split point is the minimum among those of other attributes. After an attribute and a split point have been chosen for a particular node, we split the set S of objects into two subsets, L and R. If the pdf of a tuple contains the split point x, we split it into two fractional tuples [28] t_L and t_R, and add them to L and R, respectively. For example, for a given tuple x_i, let the interval of its j- attribute A_j be $[l_{i,j}, r_{i,j}]$, and the split point be $l_{i,j} \leq p \leq r_{i,j}$. After splitting, we obtain two fractional tuples, whose attributes have intervals $[l_{i,j}, p]$ and $(p, r_{i,j}]$. The pdf of attribute A_j of a fractional tuple x_f is the same as the one defined for x_i, except that the pdf value is equal to zero outside the interval of the fractional tuple's attribute (e.g., $[l_{i,j}, p]$). After building the tree by the above steps, some branches of a decision tree may reflect anomalies in the training data due to noise or outliers. In this case, traditional pruning techniques, such as *pre-pruning* and *post-pruning* [28], can be adopted to prune these nodes.

2. Testing Phase. For a given test tuple with uncertainty, there may be more than one path for traversing the decision tree in a top-down manner (from the root node). This is because a node may only cover part of the tuple. When we visit an internal node, we may split the tuple into two parts at its split point, and distribute each part recursively down the child nodes accordingly. When the leaf nodes are reached, the probability distribution at each leaf node contributes to the final distribution for predicting its class label.[1]

To illustrate this process, let us consider the decision tree in Figure 16.2(b) again, where the split point is $x = -1$. If we use this decision tree to classify tuple 1 by its pdf, its value is either -1 with probability 8/11, or 10 with probability 3/11. After traversing from the root of the tree, we will reach

---

[1] In AVG, there is only one path tracked from root to leaf node.

the left leaf node with probability 8/11, and the right leaf node with probability 3/11. Since the left leaf node has a probability 0.8 of belonging to class A, and right leaf node has a probability 0.212 of belonging to class A, $P(A) = (8/11) \times 0.8 + (3/11) \times 0.212 = 0.64$. We can similarly compute $P(B) = (8/11) \times 0.2 + (3/11) \times 0.788 = 0.36$. Since $P(A) > P(B)$, tuple 1 is labelled as "A."

Efficient split point computation. By considering pdf during the decision tree process, UDT promises to build a more accurate decision tree than AVG. Its main problem is that it has to consider a lot more split points than AVG, which reduces its efficiency. In order to accelerate the speed of finding the best split point, [32] proposes several strategies to prune candidate split points. To explain, suppose that for a given set S of tuples, the interval of the j-th attribute of the i-th tuple in S is represented as $[l_{i,j}, r_{i,j}]$, associated with pdf $f_i^j(x)$. The set of *end-points* of objects in S on attribute A_j can be defined as $Q_j = \{q | (q = l_{i,j}) \vee (q = r_{i,j})\}$. Points that lie in $(l_{i,j}, r_{i,j})$ are called *interior points*. We assume that there are v end-points, $q_1, q_2 \cdots q_v$, sorted in ascending order. Our objective is to find an optimal split point in $[q_1, q_v]$.

The end-points define $v - 1$ disjoint intervals: $(q_k, q_{k+1}]$, where $1 \leq k \leq v - 1$. [32] defines three types of interval: *empty*, *interval*, and *heterogeneous*. An interval $(q_k, q_{k+1}]$ is called *empty* if the probability on this interval $\int_{q_k}^{q_{k+1}} f_i^j(x)dx = 0$ for all $x_i \in S$. An interval $(q_k, q_{k+1}]$ is *homogeneous* if there exists a class label $c \in C$ such that the probability on this interval $\int_{q_k}^{q_{k+1}} f_i^j(x)dx \neq 0 \Rightarrow C_i = c$ for all $x_i \in S$. An interval $(q_k, q_{k+1}]$ is *heterogeneous* if it is neither empty nor homogeneous. We next discuss three techniques for pruning candidate split points.

(i) Pruning empty and homogeneous intervals. [32] shows that empty and homogeneous intervals do not need to be considered. In other words, the interior points of empty and homogeneous intervals can be ignored.

(ii) Pruning by bounding. This strategy removes heterogeneous intervals through a bounding technique. The main idea is to compute the entropy $H(q, A_j)$ for all end-points $q \in Q_j$. Let H_j^* be the minimum value. For each heterogeneous interval $(a, b]$, a lower bound, L_j, of $H(z, A_j)$ over all candidate split points $z \in (q_k, q_{k+1})$ is computed. An efficient method of finding L_j without considering all the candidate split points inside the interval is detailed in [33]. If $L_j \geq H_j^*$, none of the candidate split points within the interval $(q_k, q_{k+1}]$ can give an entropy smaller than H_j^*; therefore, the whole interval can be pruned. Since the number of end-points is much smaller than the total number of candidate split points, many heterogeneous intervals are pruned in this manner, which also reduces the number of entropy computations.

(iii) End-point sampling. Since many of the entropy calculations are due to the determination of entropy at the end-points, we can sample a portion (say, 10%) of the end-points, and use their entropy values to derive a pruning threshold. This method also reduces the number of entropy calculations, and improves the computational performance.

Discussions: From Figures 16.1 and 16.2, we can see that the accuracy of UDT is higher than that of AVG, even though its time complexity is higher than that of AVG. By considering the sampled points of the pdfs, UDT can find better split points and derive a better decision tree. In fact, the accuracy of UDT depends on the number of sampled points; the more points are sampled, the accuracy of UDT is closer to that of PW. In the testing phase, we have to track one or several paths in order to determine its final label, because an uncertain attribute may be covered by several tree nodes. Further experiments on ten real datasets taken from the UCI Machine Learning Repository [6] show that UDT builds more accurate decision trees than AVG does. These results also show that the pruning techniques discussed above are very effective. In particular, the strategy that combines the above three techniques reduces the number of entropy calculations to 0.56%-28% (i.e., a pruning effectiveness ranging from 72% to 99.44%). Finally, notice that UDT has also been extended to handle pmf in [33]. This is not difficult, since UDT samples a pdf into a set of points. Hence UDT can easily be used to support pmf.

We conclude this section by briefly describing two other decision tree algorithms designed for uncertain data. In [23], the authors enhance the traditional decision tree algorithms and extend measures, including entropy and information gain, by considering the uncertain data's intervals and pdf values. It can handle both numerical and categorical attributes. It also defines a "probabilistic entropy metric" and uses it to choose the best splitting point. In [24], a decision-tree based classification system for uncertain data is presented. This tool defines new measures for constructing, pruning, and optimizing a decision tree. These new measures are computed by considering pdfs. Based on the new measures, the optimal splitting attributes and splitting values can be identified.

16.3.2 Rule-Based Classification

Overview: Rule-based classification is widely used because the extracted rules are easy to understand. It has been studied for decades [12], [14]. A rule-based classification method often creates a set of IF-THEN rules, which show the relationship between the attributes of a dataset and the class labels during the training stage, and then uses it for classification. An IF-THEN rule is an expression of the form

IF *condition* THEN *conclusion*.

The former "IF"-part of a rule is the rule antecedent, which is a conjunction of the attribute test conditions. The latter "THEN"-part is the rule consequent, namely the predicted label under the condition shown in the "IF"-part. If a tuple can satisfy the *condition* of a rule, we say that the tuple is covered by the rule. For example, a database that can predict whether a person will buy computers or not may contain attributes *age* and *profession*, and two labels *yes* and *no* of the class *buy_computer*. A typical rule may be $(age = youth) \wedge (profession = student) \Rightarrow buy\_computer = yes$, which means that for a given tuple about a customer, if its values of attributes *age* and *profession* are *youth* and *student*, respectively, we can predict that this customer will buy computers because the corresponding tuple about this customer is covered by the rule.

Most of the existing rule-based methods use the sequential covering algorithm to extract rules. The basic framework of this algorithm is as follows. For a given dataset consisting of a set of tuples, we extract the set of rules from one class at a time, which is called the procedure of Learn_One_Rule. Once the rules extracted from a class are learned, the tuples covered by the rules are removed from the dataset. This process repeats on the remaining tuples and stops when there is not any tuple. The classical rule-based method RIPPER [12] has been extended for uncertain data classification in [25], [26] which is called uRule. We will introduce uRule in this section.

Input: A set of labeled objects whose uncertain attributes are associated with pdf models or pmf models.

Algorithm: The algorithm uRule extracts the rules from one class at a time by following the basic framework of the sequential covering approach from the training data. That means we should first extract rules from class C_1 by Learn_One_Rule, and then repeat this procedure for all the other classes $C_2, C_3 \cdots C_L$ in order to obtain all the rules. We denote the set of training instances with class label C_k as D_k, $1 \leq k \leq L$. Notice that these generated rules may perform well on the training dataset, but may not be effective on other data due to overfitting. A commonly used solution for this problem is to use a validation set to prune the generated rules and thus improve the rules' quality. Hence, the Learn_One_Rule procedure includes two phases: growing and pruning. It splits the input dataset into two subsets: *grow_Dataset* and *prune_Dataset* before generating rules. The *grow_Dataset* is used to generate rules, and the *prune_Dataset* is used to prune some conjuncts, and boolean expressions between attributes and values, from these rules.

For a given sub-training dataset $D_k(1 \leq k \leq L)$, the Learn_One_Rule procedure performs the following steps: (1) Split D_k into *grow_Dataset* and *prune_Dataset*; (2) Generate a rule from *grow_Dataset*; (3) Prune this rule based on *prune_Dataset*; and (4) Remove objects covered by the pruned rule from D_k and add the pruned rule to the final rule set. The above four steps will be repeated until there is not any tuple in the D_k. The output is a set of rules extracted from D_k.

1. Rules Growing Phase. uRule starts with an initial rule: $\{\} \Rightarrow C_k$, where the left side is an empty set and the right side contains the target class label C_k. New conjuncts will subsequently be added to the left set. uRule uses the probabilistic information gain as a measure to identify the best conjunct to be added into the rule antecedent (We will next discuss this measure in more detail). It selects the conjunct that has the highest probabilistic information gain, and adds it as an antecedent of the rule to the left set. Then tuples covered by this rule will be removed from the training dataset D_k. This process will be repeated until the training data is empty or the conjuncts of the rules contain all the attributes. Before introducing probability information gain, uRule defines the probability cardinalities for attributes with pdf model as well as pmf model.

(a) pdf model. For a given attribute associated with pdf model, its values are represented by a pdf with an interval, so a conjunct related to an attribute associated with pdf model covers an interval. For example, a conjunct of a rule for the attribute *income* may be $1000 \leq income < 2000$. The maximum and the minimum values of the interval are called end-points. Suppose that there are N end-points for an attribute after eliminating duplication; then this attribute can be divided into N+1 partitions. Since the leftmost and rightmost partitions do not contain data instances at all, we only need to consider the remaining N-1 partitions. We let each candidate conjunct involve a partition when choosing the best conjunct for an attribute, so there are N-1 candidate conjuncts for this attribute.

For a given rule R extracted from D_k, we call the instances in D_k that are covered and classified correctly by R positive instances, and call the instances in D_k that are covered, but are not classified correctly by R, negative instances. We denote the set of positive instances of R w.r.t. D_k as P_k, and denote the set of negative instances of R w.r.t. D_k as N_k.

For a given attribute A_j with pdf model, the probability cardinality of all the positive instances over one of its intervals $[a,b)$ can be computed as $\sum_{l=1}^{|P_k|} P(x_l^j \in [a,b) \wedge c_l = C_k)$, and the probability cardinality of all the negative instances over the interval $[a,b)$ can be computed as $\sum_{l=1}^{|N_k|} P(x_l^j \in [a,b) \wedge c_l = C_k)$. We denote the former as $PC^+([a,b))$ and the latter as $PC^-([a,b))$.

(b) pmf model. For a given attribute associated with pmf model, its values are represented by a set of discrete values with probabilities. Each value is called as a candidate split point. A conjunct related to an attribute associated with pmf model only covers one of its values. For example, a conjunct of a rule for the attribute *profession* may be *profession=student*. To choose the best split point for an attribute, we have to consider all its candidate split points. Similar to the case of pdf model, we have the following definitions.

For a given attribute A_j with pmf model, the probability cardinality of all the positive instances over one of its values v can be computed as $\sum_{l=1}^{|P_k|} P(x_l^j = v \wedge c_l = C_k)$, and the probability cardinality of all the negative instances over value v can be computed as $\sum_{l=1}^{|N_k|} P(x_l^j = v \wedge c_l = C_k)$. We denote the former as $PC^+(v)$ and the latter as $PC^-(v)$.

Now let us discuss how to choose the best conjunct for a rule R in the case of a pdf model (the method for pmf model is similar). We consider all the candidate conjuncts related to each attribute. For each candidate conjunct, we do the following two steps:

- (1) We compute the probability cardinalities of the positive and negative instances of R over the covered interval of this conjunct, using the method shown in (a). Let us denote the probability cardinalities on R's positive and negative instances as $PC^+(R)$ and $PC^-(R)$.
- (2) We form a new rule R' by adding this conjunct to R. Then we compute the probability cardinalities of R''s positive and negative instances over the covered interval of this conjunct respectively. Let us denote the former as $PC^+(R')$ and the latter as $PC^-(R')$. Then we use *Definition* 1 to compute the probabilistic information gain between R and R'.

Among all the candidate conjuncts of all the attributes, we choose the conjunct with the highest probabilistic information gain, which is defined as *Definition* 1, as the best conjunct of R, and then add it to R.

Definition 1. Let R' be a rule by adding a conjunct to R. The probabilistic information gain between R and R' over the interval or split point related to this conjunct is

$$ProbInfo(R,R') = PC^+(R') \cdot \left[log_2 \frac{PC^+(R')}{PC^+(R') + PC^-(R')} - log_2 \frac{PC^+(R)}{PC^+(R) + PC^-(R)} \right]. \quad (16.1)$$

From *Definition* 1, we can see that the probabilistic information gain is proportional to $PC^+(R')$ and $\frac{PC^+(R')}{PC^+(R')+PC^-(R')}$, so it prefers rules that have high accuracies.

2. Rule Pruning Phase. After extracting a rule from *grow_Dataset*, we have to conduct pruning on it because it may perform well on *grow_Dataset*, but it may not perform well in the test phase. So we should prune some conjuncts of it based on its performance on the validation set *prune_Dataset*. For a given rule R in the pruning phase, uRule starts with its most recently added conjuncts when considering pruning. For each conjunct of R's condition, we consider a new rule R' by removing it from R. Then we use R' and R to classify the *prune_Dataset*, and get the set of positive instances and the set of negative instances of them. If the proportion of positive instances of R' is larger than that of R, it means that the accuracy can be improved by removing this conjunct, so we remove this conjunct from R and repeat the same pruning steps on other conjuncts of R, otherwise we keep it and stop the pruning.

3. Classification Phase. Once the rules are learnt from the training dataset, we can use them to classify unlabeled instances. In traditional rule-based classifications such as RIPPER [12], they often use the first rule that covers the instance to predict the label. However, an uncertain data object may be covered by several rules, and so we have to consider the weight of an object covered by different rules (conceptually, the weight here is the probability that an instance is covered by a rule; for details, please read [25]), and then assign the class label of the rule with the highest weight to the instance.

Discussions: The algorithm uRule follows the basic framework of the traditional rule-based classification algorithm RIPPER [12]. It extracts rules from one class at a time by the Learn_One_Rule procedure. In the Learn_One_Rule procedure, it splits the training dataset into *grow_Dataset*, and *prune_Dataset*, generates rules from *grow_Dataset* and prunes rules based on *prune_Dataset*. To choose the best conjunct when generating rules, uRule proposes the concept of probability cardinality, and uses a measure called probabilistic information gain, which is based on probability cardinality, to choose the best conjunct for a rule. In the test phase, since a test object may be covered by one or several rules, uRule has considered the weight of coverage for final classification. Their experimental results show that this method can achieve relative stable accuracy even though the extent of uncertainty reaches 20%. So it is very robust in cases which contain large ranges of uncertainty.

16.3.3 Associative Classification

Overview: Associative classification is based on frequent pattern mining. The basic idea is to search for strong associations between frequent patterns (conjunctions of a set of attribute-value pairs) and class labels [17]. A strong association is generally expressed as a rule in the form $p_1 \wedge p_2 \wedge \cdots \wedge p_s \Rightarrow c_k$, called association rule, where the rule consequent $c_k \in C$ is a class label. Typical associative classification algorithms are, for example, CBA [21], CMAR [22], and HARMONY [35]. Strong associations are usually detected by identifying patterns covering a set of attribute-value pairs and a class label associated with this set of attribute-value pairs that frequently appear in a given database.

The basic algorithm for the detection of such rules works as follows: Given a collection of attribute-value pairs, in the first step the algorithm identifies the complete set of frequent patterns from the training dataset, given the user-specified minimum support threshold and/or discriminative measurements like the minimum confidence threshold. Then, in the second step a set of rules are selected based on several covering paradigms and discrimination heuristics. Finally, the rules for classification extracted in the first step can be used to classify novel database entries or for training other classifiers. In many studies, associative classification has been found to be more accurate than some traditional classification algorithms, such as decision tree, which only considers one attribute at a time, because it explores highly confident associations among multiple attributes.

In the case of uncertain data, specialized associative classification solutions are required. The work called HARMONY proposed in [35] has been extended for uncertain data classification in [16], which is called uHARMONY. It efficiently identifies discriminative patterns directly from uncertain data and uses the resulting classification features/rules to help train either SVM or rule-based classifiers for the classification procedure.

Input: A set of labeled objects whose uncertain attributes are associated with pmf models.

Algorithm: uHARMONY follows the basic framework of algorithm HARMONY [35]. The first step is to search for frequent patterns. Traditionally an itemset y is said to be supported by a tuple x_i if $y \subseteq x_i$. For a given certain database D consisting of n tuples $x_1, x_2, \cdots, x_n$. The label of x_i is denoted by c_i. $|\{x_i | x_i \subseteq D \wedge y \subseteq x_i\}|$ is called the support of itemset y with respect to D, denoted by sup_y. The support value of itemset y under class c is defined as $|\{x_i | c_i = c \wedge x_i \subseteq D \wedge y \subseteq x_i\}|$, denoted by sup_x^c. y is said to be frequent if $sup_x \geq sup_{min}$, where sup_{min} is a user specified minimum support threshold. The confidence of y under class c can be defined as $conf_y^c = sup_y^c / sup_y$.

However, for a given uncertain database, there exists a probability of $y \subseteq x_i$ when y contains at least one item of uncertain attribute, and the support of y is no longer a single value but a probability distribution function instead. uHARMONY defines pattern frequentness by means of the expected support and uses the expected confidence to represent the discrimination of found patterns.

1. Expected Confidence. For a given itemset y on the training dataset, its expected support $E(sup_y)$ is defined as $E(sup_y) = \sum_{i=1}^{n} P(y \subseteq x_i)$. Its expected support on class c is defined as $E(sup_y^c) = \sum_{i=1}^{n} (P(y \subseteq x_i) \wedge c_i = c)$. Its expected confidence can be defined as *Definition* 1.

Definition 1. Given a set of objects and the set of possible worlds W with respect to it, the expected confidence of an itemset y on class c is

$$E(conf_y^c) = \sum_{w_i \in W} conf_{y,w_i}^c \times P(w_i) = \sum_{w_i \in W} \frac{sup_{y,w_i}^c}{sup_{y,w_i}} \times P(w_i) \tag{16.2}$$

where $P(w_i)$ stands for the probability of world w_i; $conf_{y,w_i}^c$ is the confidence of y on class c in world w_i, while $sup_{y,w_i}(sup_{y,w_i}^c)$ is the support of x (on class c) in world w_i.

For a given possible world w_i, $conf_{y,w_i}^c$, sup_{y,w_i}^c, and sup_{y,w_i} can be computed using the traditional method as introduced in the previous paragraph. Although we have $conf_y^c = sup_y^c / sup_y$ on a certain database, the expected confidence $E(conf_y^c)=E(sup_y^c / sup_y)$ of itemset y on class c is not equal to $E(sup_y^c)/E(sup_y)$ on an uncertain database. For example, Figure 16.3 shows an uncertain database about computer purchase evaluation with a certain attribute on looking and an uncertain attribute on quality. The column named *Evaluation* represents the class labels. For the example of tuple 1, the probabilities of *Bad*, *Medium*, and *Good* qualities are 0.8, 0.1 and 0.1 respectively.

Let us consider the class $c=Unacceptable$, and the itemset $y=\{Looking = -, Quality = -\}$, which means the values of the attributes *Looking* and *Quality* are *Bad* in the itemset y. In order to compute the expected confidence $E(conf_y^c)$, we have to consider two cases that contain y (notice that we will not consider cases that do not contain y since their expected support and expected confidence is 0.). Let *case*1 denote the first case in which the first two tuples contain y and the

ID	Looking	Quality	Evaluation
1	-	{-:0.8,/:0.1,+:0.1}	Unacceptable
2	-	{-:0.1,/:0.8,+:0.1}	Acceptable
3	+	{-:0.1,/:0.8,+:0.1}	Good
4	+	{-:0.1,/:0.1,+:0.8}	Very Good

FIGURE 16.3: Example of an uncertain database (+:Good, /:Medium, -:Bad).

other two tuples do not contain y, so the probability of *case*1 is $P(case1) = 0.8 \times 0.1$. Since only one tuple in w_1, i.e., tuple 1, has the label $c = Unacceptable$, $sup_{y,case1}{}^c = 1$ and $sup_{y,case1} = 2$. Let *case*2 denote the second case in which only the first one tuple contains y and the other three tuples do not contain y, so the probability of *case*2 is $P(case2) = 0.8 \times 0.9$ $sup_{y,case2}{}^c = 1$ and $sup_{y,case2} = 1$. So we can calculate $E(conf_y{}^c) = 1.0 \times (0.8 \times 0.9) + 0.5 \times (0.8 \times 0.1) = 0.76$, while $E(sup_y{}^c)/E(sup_y) = 0.8/(0.8 + 0.1) = 0.89$. Therefore, the calculation of expected confidence is non-trivial and requires careful computation.

Unlike probabilistic cardinalities like probabilistic information gain discussed in Section 16.3.2 which may be not precise and lack theoretical explanations and statistical meanings, the expected confidence is statistically meaningful while providing a relatively accurate measure of discrimination. However, according to the *Definition* 1, if we want to calculate the expected confidence of a given itemset y, the number of possible worlds $|W|$ is extremely large and it is actually exponential. The work [16] proposes some theorems to compute the expected confidence efficiently.

Lemma 1. Since $0 \leq sup_y{}^c \leq sup_y \leq n$, we have: $E(conf_y{}^c) = \sum_{w_i \in W} conf_{y,w_i}{}^c \times P(w_i) =$

$$\sum_{i=0}^{n} \sum_{j=0}^{i} \frac{j}{i} \times P(sup_y = i \wedge sup_y{}^c = j) = \sum_{i=0}^{n} \frac{E_i(sup_y{}^c)}{i} = \sum_{i=0}^{n} E_i(conf_y{}^c), \text{ where } E_i(sup_y{}^c) \text{ and } E_i(conf_y{}^c)$$

denote the part of expected support and confidence of itemset y on class c when $sup_y = i$.

Given $0 \leq j \leq n$, let $E_j(sup_y{}^c) = \sum_{i=0}^{n} E_{i,j}(sup_y{}^c)$ be the expected support of y on class c on the first j objects of *Train_D*, and $E_{i,j}(sup_y{}^c)$ be the part of expected support of y on class c with support of i on the first j objects of *Train_D*.

Theorem 1. Let p_i denote $P(y \subseteq x_i)$ for each object $x_i \in Train_D$, and $P_{i,j}$ be the probability of y having support of i on the first j objects of *Train_D*. We have $E_{i,j}(sup_y{}^c) = p_j \times E_{i-1,j-1}(sup_y{}^c) + (1 - p_j) \times E_{i,j-1}(sup_y{}^c)$ when $c_j \neq c$, and $E_{i,j}(sup_y{}^c) = p_j \times E_{i-1,j-1}(sup_y{}^c + 1) + (1 - p_j) \times E_{i,j-1}(sup_y{}^c)$ when $c_j = c$, where $0 \leq i \leq j \leq n$. $E_{i,j}(sup_y{}^c) = 0$ for $\forall j$ where $i = 0$, or where $j < i$. (For the details of the proof, please refer to [16].)

Notice that $P_{i,j}$ can be computed in a similar way as shown in *Therorem* 1 (for details, please refer to [16]). Based on the above theory, we can compute the expected confidence with $O(n^2)$ time complexity and $O(n)$ space complexity. In addition, the work [16] also proposes some strategies to compute the upper bound of expected confidence, which makes the computation of expected confidence more efficient.

2. Rules Mining. The steps of mining frequent itemsets of uHARMONY are very similar with HARMONY [35]. Before running the itemset mining algorithm, it sorts the attributes to place all the certain attributes before uncertain attributes. The uncertain attributes will be handled at last when extending items by traversing the attributes. During the mining process, we start with all the candidate itemsets, each of which contains only one attribute-value pair, namely a conjunct. For each candidate itemset y, we first compute its expected confidence $E(conf_y{}^c)$ under different classes, and then we assign the class label c^* with the maximum confidence as its class label. We denote this candidate rule as $y \Rightarrow c^*$ and then count the number of covered instances by this rule. If the number of covered instances by this itemset is not zero, we output a final rule $y \Rightarrow c^*$ and remove y from the set of candidate itemsets. Based on this mined rule, we compute a set of new candidate itemsets by adding all the possible conjuncts related to other attributes to y. For each new candidate itemset, we repeat the above process to generate all the rules.

After obtaining a set of itemsets, we can use them as rules for classification directly. For each test instance, by summing up the product of the confidence of each itemset on each class and the probability of the instance containing the itemset, we can assign the class label with the largest value to it. In addition, we can convert the mined patterns to training features of other classifiers such as SVM for classification [16].

Discussions: The algorithm uHARMONY follows the basic framework of HARMONY to mine discriminative patterns directly and effectively from uncertain data. It proposes to use expected confidence of a discovered itemset to measure the discrimination. If we compute the expected confidence according to its original definition, which considers all the simple combinations of those training objects, i.e., the possible worlds, the time complexity is exponential, which is impractical in practice. So uHARMONY proposes some theorems to compute expected confidence efficiently. The overall time complexity and space complexity of computing the expected confidence of a given itemset y are at most $O(n^2)$ and $O(n)$, respectively. Thus, it can be practical in real applications. These mined patterns can be used as classification features/rules to classify new unlabeled objects or help to train other classifiers (i.e., SVM). Their experimental results show that uHARMONY can outperform other uncertain data classification algorithms such as decision tree classification [23] and rule-based classification [25] with 4% to 10% improvement on average in accuracy on 30 categorical datasets.

16.3.4 Density-Based Classification

Overview: The basis of density-based classification are density estimation methods that provide an effective intermediate representation of potentially erroneous data capturing the information about the noise in the underlying data, and are in general useful for uncertain data mining [1]. Data mining in consideration of density in the underlying data is usually associated with clustering [17], such as density-based clustering DBSCAN [13] or hierarchical density-based clustering OPTICS [5]. These concepts form clusters by dense regions of objects in the data space separated by region of low density (usually interpreted as noise) yielding clusters of arbitrary shapes that are robust to noise. The idea of using object density information for mutual assignment of objects can also be used for classification. In the work [1], a new method is proposed for handling error-prone and missing data with the use of density estimation.

Input: A set of labeled objects whose uncertain attributes are associated with pdf models.

Algorithm: The presence of uncertainty or errors in the data may change the performance of classification process greatly. This can be illustrated by the two. dimensional binary classification example as shown in Figure 16.4. Suppose the two uncertain training samples y and z, whose uncertainties are illustrated by the oval regions, respectively, are used to classify the uncertain object x. A nearest-neighbor classifier would pick the class label of data point y. However, the data point z may have a much higher probability of being the nearest neighbor to x than the data point y, because x lies within the error boundary of z and outside the error boundary of y. Therefore, it is important to design a classification method that is able to incorporate potential uncertainty (errors) in the data in order to improve the accuracy of the classification process.

1. Error-Based Density Estimation. The work proposed by Aggarwal [1] introduces a general and scalable approach to uncertain data mining based on multi-variate density estimation. The kernel density estimation $\bar{f}(x)$ provides a continuous estimate of the density of the data at a given point x. An introduction of kernel density estimation is given in Section 16.3.7. Specifically, the value of the density at point x is the sum of smoothed values of kernel functions $K'_h(\cdot)$ associated with each

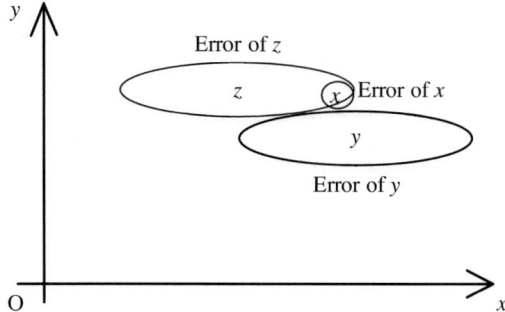

FIGURE 16.4: Effect of errors on classification.

point in the data set. Each kernel function is associated with a kernel width h, which determines the level of smoothing created by the function. The kernel estimation is defined as follows:

$$\overline{f}(x) = \frac{1}{n} \sum_{i=1}^{n} K_h'(x - x_i) \tag{16.3}$$

A Gaussian kernel with width h can be defined as Equation(16.4).

$$K_h'(x - x_i) = (1/\sqrt{2\pi} \cdot h) \cdot e^{\frac{-(x-\widehat{x_i})^2}{2h^2}} \tag{16.4}$$

The overall effect of kernel density estimation is to replace each discrete point x_i by a continuous function $K_h'(x - x_i)$, which peaks at $\widehat{x}_i$ and has a variance that is determined by the smoothing parameter h.

The presence of attribute-specific errors may change the density estimates. In order to take these errors into account, we need to define error-based kernels. The error-based kernel can be defined as:

$$Q_h'(x - \widehat{x}_i, \varphi(x_i)) = \left(1/\sqrt{2\pi} \cdot (h + \varphi(x_i))\right) \cdot e^{\frac{-(x-\widehat{x_i})^2}{2 \cdot (h^2 + \varphi(x_i)^2)}}, \tag{16.5}$$

where $\varphi(x_i)$ denotes the estimated error associated with data point x_i.

As in the previous case, the error-based density at a given data point is defined as the sum of the error-based kernels over different data points. Consequently, the error-based density $\overline{f^Q}(x, \varphi(x_i))$ at point x_i is defined as:

$$\overline{f^Q}(x, \varphi(x_i)) = (1/N) \cdot \sum_{i=1}^{N} Q_h'(x - x_i, \varphi(x_i)). \tag{16.6}$$

This method can be generalized for very large data sets and data streams by condensing the data into micro-clusters; specifically, here we need to adapt the concept of micro-cluster by incorporating the error information. In the following we will define an error-based micro-cluster as follows:

Definition 1: An error-based micro-cluster for a set of d-dimensional points $MC=\{x_1, x_2, \cdots x_m\}$ with order $1, 2, \cdots, m$ is defined as a $(3 \cdot d + 1)$ tuple $(\overline{CF2^x}(MC), \overline{EF2^x}(MC), \overline{CF1^x}(MC), n(MC))$, wherein $\overline{CF2^x}(MC)$, $\overline{EF2^x}(MC)$, and $\overline{CF1^x}(MC)$ each correspond to a vector of d entries. The definition of each of these entries is as follows.

- For each dimension, the sum of the squares of the data values is maintained in $\overline{CF2^x}(MC)$. Thus, $\overline{CF2^x}(MC)$ contains d values. The p-th entry of $\overline{CF2^x}(MC)$ is equal to $\sum_{i=1}^{m} (x_i^p)^2$.

- For each dimension, the sum of the square of the errors in data values is maintained in $\overline{EF2^x}(MC)$. Thus, $\overline{EF2^x}(MC)$ contains d values. The p-th entry of $\overline{EF2^x}(MC)$ is equal to $\sum_{i=1}^{m} (\varphi_p(x_i))^2$.
- For each dimension, the sum of the data values is maintained in $\overline{CF1^x}(MC)$. Thus, $\overline{CF1^x}(MC)$ contains d values. The p-entry of $\overline{CF1^x}(MC)$ is equal to $\sum_{i=1}^{m} x_i^p$.
- The number of points in the data is maintained in $n(MC)$.

In order to compress large data sets into micro-clusters, we can create and maintain the clusters using a single pass of data as follows. Supposing the number of expected micro-clusters for a given dataset is q, we first randomly choose q centroids for all the micro-clusters that are empty in the initial stage. For each incoming data point, we assign it to its closest micro-cluster centroid using the nearest neighbor algorithm. Note that we never create a new micro-cluster, which is different from [2]. Similarly, these micro-clusters will never be discarded in this process, since the aim of this micro-clustering process is to compress the data points so that the resulting statistic can be held in a main memory for density estimation. Therefore, the number of micro-clusters q is defined by the amount of main memory available.

In [1] it is shown that for a given micro-cluster $MC=\{x_1, x_2, \cdots x_m\}$, its true error $\Delta(MC)$ can be computed efficiently. Its value along the j-th dimension can be defined as follows:

$$\Delta_j(MC) = \frac{\overline{CF2_j^x}(MC)}{m} - \frac{\overline{CF1_j^x}(MC)^2}{m} + \frac{\overline{EF2_j^x}(MC)}{m}, \quad (16.7)$$

where $\overline{CF2_j^x}(MC)$, $\overline{CF1_j^x}(MC)^2$ and $\overline{EF2_j^x}(MC)$ are the j-th entry of the statistics of the corresponding error-based micro-cluster MC as defined above.

The true error of a micro-cluster can be used to compute the error-based kernel function for a micro-cluster. We denote the centroid of the cluster MC by $c(MC)$. Then, we can define the kernel function for the micro-cluster MC in an analogous way to the error-based definition of a data point.

$$Q'_h(x - c(MC), \Delta(MC)) = \left(1/\sqrt{2\pi} \cdot (h + \Delta(MC))\right) \cdot e^{\frac{-(x - \hat{x}_i)^2}{2 \cdot (h^2 + \Delta(MC)^2)}}. \quad (16.8)$$

If the training dataset contains the micro-clusters $MC_1, MC_2, \cdots MC_{m'}$ where m' is the number of micro-clusters, then we can define the density estimate at x as follows:

$$\overline{f^Q}(x, \Delta(MC)) = (1/N) \cdot \sum_{i=1}^{m'} n(MC_i) Q'_h(x - c(MC_i), \Delta(x_i)). \quad (16.9)$$

This estimate can be used for a variety of data mining purposes. In the following, it is shown how these error-based densities can be exploited for classification by introducing an error-based generalization of a multi-variate classifier.

2. Density-Based Classification. The density-based classification designs a density-based adaption of rule-based classifiers. For the classification of a test instance, we need to find relevant classification rules for it. The challenge is to use the density-based approach in order to find the particular subsets of dimensions that are most discriminatory for the test instance. So we need to find those subsets of dimensions, namely subspaces, in which the instance-specific local density of dimensions for a particular class is significantly higher than its density in the overall data. Let D_i denote the objects that belong to class i. For a particular test instance x, a given subspace S, and a given training dataset D, let $g(x, S, D)$ denote the density of S and D. The process of computing the density over a given subspace S is equal to the computation of the density of the full-dimensional space while incorporating only the dimensions in S.

In the test phase, we compute the micro-clusters together with the corresponding statistic of each data set $D_1, \ldots, D_L$ where L is the number of classes in the preprocessing step. Then, the compressed micro-clusters for different classes can be used to generate the accuracy density over different subspaces by using Equations (16.8) and (16.9). To construct the final set of rules, we can use an iterative approach to find the most relevant subspaces for classification. For the relevance of a particular subspace S we define the density-based local accuracy $A(x, S, C_i)$ as follows:

$$A(x, S, C_i) = \frac{|D_i| \cdot g(x, S, D_i)}{|Train\_D| \cdot g(x, S, Train\_D)}. \tag{16.10}$$

The dominant class $dom(x, S)$ for point x in dataset S can be defined as the class with the highest accuracy. That is,

$$dom(x, S) = \arg\max_{1 \le i \le L} A(x, S, C_i). \tag{16.11}$$

To determine all the highly relevant subspaces, we follow a bottom-up methodology to find those subspaces S that yield a high accuracy of the dominant class $dom(x, S)$ for a given test instance x. Let $T_j (1 \le j \le d)$ denote the set of all j-th dimensional subspaces. We will compute the relevance between x and each subspace in T_j ($1 \le j \le d$) in an iterative way, starting with T_1 and then proceeding with $T_2, T_3, \cdots T_d$. The subspaces with high relevance can be used as rules for final classification, by assigning to a test instance x the label of the class that is the most frequent dominant class of all highly relevant subspaces of x. Notice that to make this process efficient, we can impose some constraints. In order to decide whether we should consider a subspace in the set of $(j + 1)$-th dimension spaces, at least one subset of it needs to satisfy the accuracy threshold requirements. Let $L_j (L_j \subseteq T_j)$ denote the subspaces that have sufficient discriminatory power for that instance, namely the accuracy is above threshold α. In order to find such subspaces, we can use Equations (16.10) and (16.11). We iterate over increasing values of j, and join the candidate set L_j with the set L_1 in order to determine T_{j+1}. Then, we can construct L_{j+1} by choosing subspaces from T_{j+1} whose accuracy is above α. This process can be repeated until the set T_{j+1} is empty. Finally, the sets of dimensions $L_{all} = \cup_j L_j$ can be used to predict the class label of the test instance x.

To summarize the density-based classification procedure, the steps of classifying a test instance are as follows. (1) compute the micro-clusters for each data set $D_1, \ldots D_L$ only once in the preprocessing step; (2) compute the highly relevant subspaces for this test instance following the bottom-up methodology and obtain a set of subspaces L_{all}; (3) assign the label of the majority class in L_{all} to the test instance.

Discussions: Unlike other uncertain data classification algorithms [8, 29, 32], the density-based classification introduced in this section not only considers the distribution of data, but also the errors along different dimensions when estimating the kernel density. It proposes a method for error-based density estimation, and it can be generalized to large data sets cases. In the test phase, it designs a density-based adaption of the rule-based classifiers. To classify a test instance, it finds all the highly relevant subspaces and then assign the label of the majority class of these subspaces to the test instance. In addition, the error-based density estimation discussed in this algorithm can also be used to construct accurate solutions for other cases, and thus it can be applied for a variety of data management and mining applications when the data distribution is very uncertain or contains a lot of errors. In an experimental evaluation it is demonstrated that the error-based classification method not only performs fast, but also can achieve higher accuracy compared to methods that do not consider error at all, such as the nearest-neighbor-based classification method that will be introduced next.

16.3.5 Nearest Neighbor-Based Classification

Overview: The nearest neighbor-based classification method has been studied since the 1960s. Although it is a very simple algorithm, it has been applied to a variety of applications due to its high accuracy and easy implementation. The normal procedure of nearest neighbor classification method

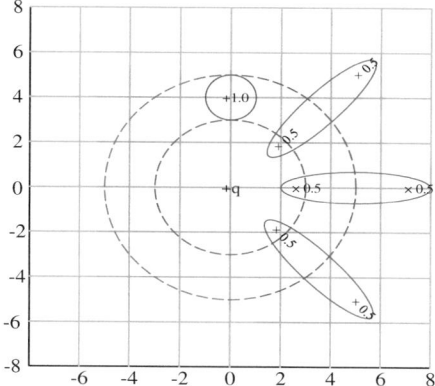

FIGURE 16.5: Example of a comparison between the nearest neighbor and class.

is as follows: Given a set of training instances, to classify an unlabeled instance q, we have to search for the nearest-neighbor $NN_{Train\_D}(q)$ of q in the training set, i.e., the training instance having the smallest distance to q, and assign the class label of $NN_{Train\_D}(q)$ to q. The nearest neighbor-based classification can be generalized easily to take into account the k nearest-neighbor easily, which is called k nearest neighbor classification (KNN, for short). In this case, the classification rule assigns to the unclassified instance q the class containing most of its k nearest neighbors. The KNN classification is more robust against noise compared to the simple NN classification variant. The work [4] proposes a novel nearest neighbor classification approach that is able to handle uncertain data called Uncertain Nearest Neighbor (UNN, for short) classification, which extends the traditional nearest neighbor classification to the case of data with uncertainty. It proposes a concept of nearest-neighbor class probability, which is proved to be a much more powerful uncertain object classification method than the straight-forward consideration of nearest-neighbor object probabilities.

Input: A set of labeled objects whose uncertain attributes are associated with pdf models.

Algorithm: To extend the nearest neighbor method for uncertain data classification, it is straight-forward to follow the basic framework of the traditional nearest neighbor method and replace the distance metric method with distance between means (i.e., the expected location of the uncertain instances) or the expected distance between uncertain instances. However, there is no guarantee on the quality of the classification yielded by this naive approach. The example as shown in Figure 16.5 demonstrates that this naive approach is defective because it may misclassify. Suppose we have one certain unclassified object q, three uncertain training objects whose support is delimited by ellipsis belonging to class="red" and another uncertain training object whose support is delimited by a circle belonging to another class="blue". Suppose the class="blue" object consists of one normally distributed uncertain object whose central point is (0,4), while the class="red" objects are bimodal uncertain objects. Following the naive approach, one can easily see that q will be assigned to the "blue" class. However, the probability that a "red"-class object is closer to q than a "blue"-class one is $1 - 0.5^3 = 0.875$. This also indicates that the probability that the nearest neighbor of object q is a "red"-class object is 87.5%. Thus, the performance of the naive approach is quite poor as distribution information (such as the variances) of the uncertain objects is ignored.

1. Nearest Neighbor Rule for Uncertain Data. To avoid the flaws of the naive approach following the above observations, the approach in [4] proposes the concept of most probable class. It simultaneously considers the distance distributions between q and all uncertain training objects in $Train\_D$. The probability $Pr(NN_{Train\_D}(q) = c)$ that the object q will be assigned to class c by means of the nearest neighbor rule can be computed as follows:

$$Pr(NN_{Train\_D}(q) = c) = \int_{D'} Pr(I_D) \cdot I_c(NN_{I_D}(q))dI_D, \qquad (16.12)$$

where function $I_c(\cdot)$ outputs 1 if its argument is c, and 0 otherwise, $NN_{I_D}(q)$ represents the label of the nearest neighbor of q in set I_D, $Pr(I_D)$ represents the occurrence probability of set I_D, and D' represents all the possible combinations of training instances in $Train\_D$. It is easy to see that the probability $Pr(NN_{Train\_D}(q) = c)$ is the summation of the occurrence probabilities of all the outcomes I_D of the training set $Train\_D$ for which the nearest neighbor object of q in I_D has class c. Note that if the domain involved in the training dataset doesn't concide with the real number field, we can transfer the integration as summations and get Equation(16.13).

$$Pr(NN_{Train\_D}(q) = c) = \sum_{I_D \in D'} Pr(I_D) \cdot I_c(NN_{I_D}(q))dI_D. \qquad (16.13)$$

Therefore, if we want to classify q, we should compute the probabilities $Pr(NN_{Train\_D}(q) = c)$ for all the classes first, and then assign it the class label c^*, which is called the *most probable class* of q and can be computed as follows:

$$c^* = \arg\max_c Pr(NN_{Train\_D}(q) = c). \qquad (16.14)$$

2. Efficient Strategies. To classify a test object, the computational cost will be very large if we compute the probabilities using the above equations directly, so [4] introduces some efficient strategies such as the uncertain nearest neighbor (UNN) rule. Let D_c denote the subset of $Train\_D$ composed of the objects having class label c, $D_k(q,c)$ denote the distance between object q and class c (k is the number of nearest neighbors), and $p_i(R) = Pr(d(q,x_i) \leq R)$ denote the cumulative density function representing the relative likelihood for the distance between q and training set object x_i to assume value less than or equal to R. Then we have

$$p_i(R) = Pr(d(q,x_i) \leq R) = \int_{B_R(q)} f^{x_i}(v)dv \qquad (16.15)$$

where $f^{x_i}(v)$ denotes the probability for x_i to assume value v, $B_R(q)$ denotes the set of values $\{v \in Train\_D : d(v,q) \leq R\}$.

Then, the cumulative density function associated with $D_k(q,c)$ can be computed as Equation (16.16), which is one minus the probability that no object of the class c lies within distance R from q.

$$Pr(D_k(q,c) \leq R) = 1 - \left(\prod_{x_i \in D_c} (1 - p_i(R)) \right) \qquad (16.16)$$

Let us consider the case of a binary classification problem for uncertain data. Suppose that there are two classes, i.e., c and c', and a test object q. We define the *nearest neighbor class* of the object q as c if

$$Pr(D_k(q,c) \leq D_k(q,c')) \geq 0.5 \qquad (16.17)$$

holds, and c' otherwise. This means that class c is the nearest neighbor of q if the probability that the nearest neighbor of q comes from class c is greater than the probability that it comes from class c'. In particular, this probability can be computed by means of the following one-dimensional integral.

$$Pr(D_k(q,c) \leq D_k(q,c')) = \int_0^{+\infty} Pr(D_k(q,c) = R) \cdot Pr(D_k(q,c) > R)dR. \qquad (16.18)$$

The *Uncertain Nearest Neighbor Rule* (UNN): Given a training dataset, it assigns the label of its nearest neighbor class to the test object q. In the UNN rule, the distribution of the closet class tends to overshadow noisy objects and so it will be more robust than traditional nearest neighbor classification. The UNN rule can be generalized for handling more complex cases such as multiclass classification problems, etc. Besides, [4] proposes a theorem that the uncertain nearest neighbor rule outputs the most probable class of the test object (for proof, please refer to [4]). This indicates that the UNN rule performs equivalent with the most probable class.

In the test phase, [4] also proposes some efficient strategies. Suppose each uncertain object x is associated with a finite region $SUP(x)$, containing the support of x, namely the region such that $Pr(x \notin SUP(x)) = 0$ holds. Let the region $SUP(x)$ be a hypersphere, then $SUP(x)$ can be identified by means of its center $c(x)$ and its radius $r(x)$ where $c(x)$ is an instance and $r(x)$ is a positive real number. Then, the *minimum distance* $mindist(x_i, x_j)$ and the *maximum distance* $maxdist(x_i, x_j)$ between x_i and x_j can be defined as follows:

$$mindist(x_i,x_j) = \min\{d(v,w) : v \in SUP(x_i) \wedge w \in SUP(x_j)\} = \max\{0, d(c(x_i),c(x_j)) - r(x_i) - r(x_j)\}.$$
(16.19)

$$maxdist(x_i,x_j) = \max\{d(v,w) : v \in SUP(x_i) \wedge w \in SUP(x_j)\} = d(c(x_i),c(x_j)) + r(x_i) + r(x_j).$$
(16.20)

Let $inner_{D_c}(q,R)$ denote the set $\{x \in D_c : maxdist(q,x) \leq R\}$, that is the subset of D_c composed of all the object x whose maximum distance from q is not greater than R. Let R_c^q denote the positive real number $R_c^q = \min\{R \geq 0 : |inner_{D_c}(q,R)| \geq k\}$ representing the smallest radius R for which there exist at least k objects of the class c having maximum distance from q not greater than R. Moreover, we can define R_{max}^q and R_{min}^q as follows:

$$R_{max}^q = \min\{R_c^q, R_{c'}^q\} R_{min}^q = \min_{x \in D} mindist(q,x).$$
(16.21)

To compute the probability reported in Equation (16.18) for binary classification, a specific finite domain can be considered instead of an infinite domain such that the probability can be computed efficiently as follows:

$$Pr(D_k(q,c) < D_k(q,c')) = \int_{R_{min}^q}^{R_{max}^q} Pr(D_k(q,c) = R) \cdot Pr(D_k(q,c') > R) dR.$$
(16.22)

In addition, some other efficient strategies for computing the above integral have also been proposed, such as the Histogram technique (for details, please refer to [4]).

3. Algorithm Steps. The overall steps of the uncertain nearest neighbor classification algorithm can be summarized as follows, for a given dataset $Train_D$, with two classes c and c' and integer $k > 0$. To classify a certain test object q, we process it according to the following steps. (1) Determine the value $R_{max}^q = \min\{R_c^q, R_{c'}^q\}$; (2) determine the set D^q composed of the training set objects x_i such that $mindist(q,x_i) \leq R_{max}^q$; (3) if in D^q there are less than k objects of the class c' (c, resp.), then return c (c', resp.) with associated probability 1 and exit; (4) determine the value $R_{min}^q = \min_i(mindist(q,x_i))$ by considering only the object x_i belonging to D^q; (5) determine the nearest neighbor probability p of class c w.r.t. class c'. If $p \geq 0.5$, return c with associated probability, otherwise return c' with associated probability 1-p.

Discussions: As we noticed from the simple example shown in Figure 16.5, the naive approach that follows the basic framework of traditional nearest neighbor classification and is easy to implement can not guarantee the quality of classification in the uncertain data cases. The approach proposed in [4] extends the nearest neighbor-based classification for uncertain data. The main contribution of this algorithm is to introduce a novel classification rule for the uncertain setting, i.e., the Uncertain Nearest Neighbor (UNN) rule. UNN relies on the concept of nearest neighbor class, rather than on that of neatest neighbor object. For a given test object, UNN will assign the label of its nearest neighbor class to it. This probabilistic class assignment approach is much more robust and achieves better results than the straight-forward approach assigning to q the class of the most probable nearest neighbor of a test object q. To classify a test object efficiently, some novel strategies are proposed. Further experiments on real datasets demonstrate that the accuracy of nearest neighbor-based methods can outperform some other methods such as density-based method [1] and decision tree [32], among others.

16.3.6 Support Vector Classification

Overview: Support vector machine (SVM) has been proved to be very effective for the classification of linear and nonlinear data [17]. Compared with other classification methods, the SVM classification method often has higher accuracy due to its ability to model complex nonlinear decision boundaries, enabling it to cope with linear and non-linear separable data. The problem is that the training process is quite slow compared to the other classification methods. However, it can perform well even if the number of training samples is small, due to its anti-overfitting characteristic. The SVM classification method has been widely applied for many real applications such as text classification, speaker identification, etc.

In a nutshell, the binary classification procedure based on SVM works as follows [17]: Given a collection of training data points, if these data points are linearly separable, we can search for the maximum marginal hyperplane (MMH), which is a hyperplane that gives the largest separation between classes from the training datasets. Once we have found the MMH, we can use it for the classification of unlabeled instances. If the training data points are non-linearly separable, we can use a nonlinear mapping to transform these data points into a higher dimensional space where the data is linearly separable and then searches for the MHH in this higher dimensional space. This method has also been generalized for multi-class classification. In addition, SVM classification has been extended to handle data with uncertainty in [8].

Input: A set of labeled objects whose uncertain attributes are associated with pdf models.

Algorithm: Before giving the details of support vector classification for uncertain data, we need the following statistical model for prediction problems as introduced in [8].

1. A Statistical Model. Let x_i be a data point with noise information whose label is c_i, and let its original uncorrupted data be x_i'. By examining the process of generating uncertainty, we can assume that the data x_i' is generated according to a particular distribution $p(x_i', c_i | \theta)$ where θ is an unknown parameter. Given x_i', we can assume that x_i is generated from x_i', but independent of c_i according to a distribution $p(x_i | \theta', \sigma_i, x_i')$ where θ' is another possibly unknown parameter and σ_i is a known parameter, which is our estimate of the uncertainty for x_i. So the joint probability of (x_i', x_i, c_i) can be obtained easily and then the joint probability of (x_i, c_i) can be computed by integrating out the unobserved quantity x_i':

$$p(x_i, c_i) = \int p(x_i', c_i | \theta) p(x_i | \theta, \sigma_i, x_i') dx_i' : \tag{16.23}$$

Let us assume that we model uncertain data by means of the above equation. Then, if we have observed some data points, we can estimate the two parameters θ and θ' by using maximum-likelihood estimates as follows:

$$\max_{\theta, \theta'} \sum_i \ln p(x_i, c_i | \theta, \theta') = \max_{\theta, \theta'} \int p(x_i', c_i | \theta) p(x_i | \theta, \sigma_i, x_i') dx_i'. \tag{16.24}$$

The integration over the unknown true input x_i' might become quite difficult. But we can use a more tractable and easier approach to solve the maximum-likelihood estimate. Equation (16.25) is an approximation that is often used in practical applications.

$$\max_{\theta, \theta'} \sum_i \ln \sup_{x_i'} [p(x_i', c_i | \theta) p(x_i | \theta, \sigma_i, x_i')] \tag{16.25}$$

For the prediction problem, there are two types of statistical models, generative models and discriminative models (conditional models). In [8] the discriminative model is used assuming that $p(x_i', c_i|\theta) = p(x_i')p(c_i|\theta, x_i')$. Now, let us consider regression problems with Gaussian noise as an example:

$$p(x_i', c_i|\theta) \sim p(x_i')\exp\left(-\frac{(\theta^T x_i' - c_i)^2}{2\sigma^2}\right), \ p(x_i|\theta', \sigma_i, x_i') \sim \exp\left(-\frac{\|x_i - x_i'\|^2}{2\sigma_i^2}\right) \quad (16.26)$$

so that the method in 16.25 becomes:

$$\theta = \arg\min_{\theta} \sum_i \inf_{x_i'} \left[\frac{(\theta^T x_i' - c_i)^2}{2\sigma^2} + \frac{\|x_i - x_i'\|^2}{2\sigma_i^2}\right]. \quad (16.27)$$

For binary classification, where $c_i \in \{\pm 1\}$, we consider the logistic conditional probability model for c_i, while still assuming Gaussian noise in the input data. Then we can obtain the joint probabilities and the estimate of θ.

$$p(x_i', c_i|\theta) \sim p(x_i')\exp\left(\frac{1}{1+\exp(-\theta^T x_i' c_i)}\right), \ p(x_i|\theta', \sigma_i, x_i') \sim \exp\left(-\frac{\|x_i - x_i'\|^2}{2\sigma_i^2}\right)$$

$$\theta = \arg\min_{\theta} \sum_i \inf_{x_i'} \left[\ln(1 + e^{-\theta^T x_i' c_i}) + \frac{\|x_i - x_i'\|^2}{2\sigma_i^2}\right]: \quad (16.28)$$

2. Support Vector Classification. Based on the above statistic model, [8] proposes a total support vector classification (TSVC) algorithm for classifying input data with uncertainty, which follows the basic framework of traditional SVM algorithms. It assumes that data points are affected by an additive noise, i.e., $x_i' = x_i + \Delta x_i$ where noise Δx_i satisfies a simple bounded uncertainty model $\|\Delta x_i\| \leq \delta_i$ with uniform priors.

The traditional SVM classifier construction is based on computing a separating hyperplane $\mathbf{w}^T x + b = 0$, where $\mathbf{w} = (w_1, w_2, \cdots w_n)$ is a weight vector and b is a scalar, often referred to as a bias. Like in the traditional case, when the input data is uncertain data, we also have to distinguish between linearly separable and non-linearly separable input data. Consequently, TSVC will address the following two problems separately.

(1) Separable case: To search for the maximum marginal hyperplane (MMH) we have to replace the parameter θ in Equation (16.26) and (16.28) with a weight vector $\mathbf{w}$ and a bias b. Then, the problem of searching for MMH can be formulated as follows.

$$\min_{\mathbf{w}, b, \Delta x_i, i=1, \cdots, n} \frac{1}{2}\|\mathbf{w}\|^2$$
$$\text{subject to} \quad c_i(\mathbf{w}^T(x_i + \Delta x_i) + b) \geq 1, \|\Delta x_i\| \leq \delta_i, i = 1, \cdots, n. \quad (16.29)$$

(2) Non-separable case: We replace the square loss in Equation (16.26) or the logistic loss in Equation (16.28) with the margin-based hinge-loss $\xi_i = \max\{0, 1 - c_i(\mathbf{w}^T x_i + b)\}$, which is often used in the standard SVM algorithm, and we get

$$\min_{\mathbf{w}, b, \xi_i, \Delta x_i, i=1, \cdots, n} C\sum_{i=1}^n \xi_i + \frac{1}{2}\|\mathbf{w}\|^2$$
$$\text{subject to} \quad c_i(\mathbf{w}^T(x_i + \Delta x_i) + b) \geq 1 - \xi_i, \xi_i \geq 0, \|\Delta x_i\| \leq \delta_i, i = 1, \cdots, n. \quad (16.30)$$

To solve the above problems, TSVC proposes an iterative approach based on an alternating optimization method [7] and its steps are as follows.

Initialize $\Delta x_i = 0$, repeat the following two steps until a termination criterion is met:
1. Fix $\Delta x_i, i = 1, \cdots, n$ to the current value, solve problem in Equation (16.30) for $\mathbf{w}$, b, and $\boldsymbol{\xi}$, where $\boldsymbol{\xi} = (\xi_1, \xi_2, \cdots \xi_n)$.

2. Fix **w**, b to the current value, solve problem in Equation (16.30) for $\Delta x_i, i = 1, \cdots, n$, and ξ.

The first step is very similar to standard SVM, except for the adjustments to account for input data uncertainty. Similar to how traditional SVMs are usually optimized, we can compute **w** and b [34]. The second step is to compute the Δx_i. When there are only linear functions or kernel functions needed to be considered, TSVC proposes some more efficient algorithms to handle them, respectively (for details, please refer to [8]).

Discussions: The authors developed a new algorithm (TSVC), which extends the traditional support vector classification to the input data, which is assumed to be corrupted with noise. They devise a statistical formulation where unobserved data is modelled as a hidden mixture component, so the data uncertainty can be well incorporated in this model. TSVC attempts to recover the original classifier from the corrupted training data. Their experimental results on synthetic and real datasets show that TSVC can always achieve lower error rates than the standard SVC algorithm. This demonstrates that the new proposed approach TSVC, which allows us to incorporate uncertainty of the input data, in fact obtains more accurate predictors than the standard SVM for problems where the input data is affected by noise.

16.3.7 Naive Bayes Classification

Overview: Naive Bayes classifier is a widely used classification method based on Bayes theory. It assumes that the effect of an attribute value on a given class is independent of the values of the other attributes, which is also called the assumption of conditional independence. Based on class conditional density estimation and class prior probability, the posterior class probability of a test data point can be derived and the test data will be assigned to the class with maximum posterior class probability. It has been extended for uncertain data classification in [29].

Input:[2] A training dataset $Train\_D$, a set of labels $C = \{C_1, C_2, \cdots C_L, \}$. $x_{n_k} = (x^1_{n_k}, x^2_{n_k}, \cdots x^d_{n_k})$, $n_k = 1, 2, \cdots N_k$ represent the d-dimensional training objects (tuples) of class $C_k (1 \leq k \leq L)$ where N_k is the number of objects in class C_k, and each dimension is associated with a pdf model. The total number of training objects is $n = \sum_{k=1}^{L} N_k$.

Algorithm: Before introducing the naive Bayes classification for uncertain data, let us go though the naive Bayes classification for certain data.

1. Certain Data Case. Suppose each dimension of an object is a certain value. Let $P(C_k)$ be the prior probability of the class C_k, and $P(x|C_k)$ be the conditional probability of seeing the evidence x if the hypothesis C_k is true. For a a d-dimensional unlabeled instance $x = (x^1, x^2, \cdots x^d)$, to build a classifier to predict its unknown class label based on naive Bayes theorem, we can build a classifier shown in (16.31).

$$P(C_k|x) = \frac{P(x|C_k)P(C_k)}{\sum_{k'} P(x|C_{k'})P(C_{k'})}. \tag{16.31}$$

Since the assumption of conditional independence is assumed in naive Bayes classifiers, we have:

$$P(x|C_k) = \prod_{j=1}^{d} P(x^j|C_k). \tag{16.32}$$

The kernel density estimation, a non-parametric way of estimating the probability density function of a random variable, is often used to estimate the class conditional density. For the set of

---

[2]The input format of naive Bayes classification is a little different from other algorithms. This is because we have to estimate the prior and conditional probabilities of different classes separately.

instances with class label C_k, we can estimate the probability density function of their j-th dimension by using kernel density estimation as follows:

$$\widehat{f}_{h_k^j}(x^j|C_k) = \frac{1}{N_k h_k^j} \sum_{n_k=1}^{N_k} K(\frac{x^j - x_{n_k}^j}{h_k^j}) \tag{16.33}$$

where the function K is some kernel with the bandwidth h_k^j. A common choice of K is the standard Gaussian function with zero mean and unit variance, i.e., $K(x) = \frac{1}{\sqrt{2\pi}} e^{-\frac{1}{2}x^2}$.

To classify $x = (x^1, x^2, \cdots x^d)$ using a naive Bayes model with Equation (16.32), we need to estimate the class condition density $P(x^j|C_k)$. We use $\widehat{f}_{h_k^j}(x^j)$ as the estimation. From Equations (16.32) and (16.33), we get:

$$P(x|C_k) = \prod_{j=1}^{d} \{ \frac{1}{N_k h_k^j} \sum_{n_k=1}^{N_k} K(\frac{x^j - x_{n_k}^j}{h_k^j}) \}. \tag{16.34}$$

We can compute $P(C_k|x)$ using Equation (16.31) and (16.34) and predict the label of x as $y = \arg\max_{C_k \in C} P(C_k|x)$.

2. Uncertain Data Case. In this case, each dimension $x_{n_k}^j (1 \leq j \leq d)$ of a tuple in the training dataset is uncertain, i.e., represented by a probability density function $p_{n_k}^j$. To classify an unlabeled uncertain data $x = (x^1, x^2, \cdots x^d)$, where each attribute is modeled by a pdf $p^j (j = 1, 2, \cdots d)$, [29] proposes a *Distribution-based* naive Bayes classification for uncertain data classification. It follows the basic framework of the above algorithm and extends some parts of it for handling uncertain data.

The key step of the *Distribution-based* method is the estimation of class conditional density on uncertain data. It learns the class conditional density from uncertain data objects represented by probability distributions. They also use kernel density estimation to estimate the class conditional density. However, we may not be able to estimate $P(x^j|C_k)$ using $\widehat{f}_{h_k^j}(x^j|C_k)$ without any redesign, because the value of x^j is modeled by a pdf p^j and the kernel function K is defined for scalar-valued parameters only. So, we need to extend Equation (16.33) to create a kernel-density estimation for x^j. Since x^j is a probability distribution, it is natural to replace K in Equation (16.33) using its expected value, and we can get:

$$\begin{aligned}
\widehat{f}_{h_k^j}(x^j) &= \frac{1}{N_k h_k^j} \sum_{n_k=1}^{N_k} E[K(\frac{x^j - x_{n_k}^j}{h_k^j})] \\
&= \frac{1}{N_k h_k^j} \sum_{n_k=1}^{N_k} \int \int K(\frac{x^j - x_{n_k}^j}{h_k^j}) p^j(x^j) p_{nk}^j(x_{nk}^j) dx^j dx_{nk}^j.
\end{aligned} \tag{16.35}$$

Using the above equation to estimate $P(x^j|C_k)$ in Equation (16.32) gives:

$$P(x|C_k) = \prod_{j=1}^{d} \{ \frac{1}{N_k h_k^j} \sum_{n_k=1}^{N_k} \int \int K(\frac{x^j - x_{n_k}^j}{h_k^j}) p^j(x^j) p_{nk}^j(x_{nk}^j) dx^j dx_{nk}^j \}. \tag{16.36}$$

The *Distribution-based* method presents two possible methods, i.e., formula-based method and sample-based method, to compute the double integral in Equation (16.36).

(1). Formula-Based Method. In this method, we first derive the formula for kernel estimation for uncertain data objects. With this formula, we can then compute the kernel density and run the naive Bayes method to perform the classification. Suppose x and x_{n_k} are uncertain data objects with multivariate Gaussian distributions, i.e., $x \sim N(\mu, \Sigma)$ and $X_{n_k} \sim N(\mu_{n_k}, \Sigma_{n_k})$. Here, $\mu = (\mu^1, \mu^2, \cdots \mu^d)$ and $\mu_{n_k} = (\mu_{n_k}^1, \mu_{n_k}^2, \cdots \mu_{n_k}^d)$ are the means of x and x_{n_k}, while Σ and Σ_{n_k} are their covariance matrixes, respectively. Because of the independence assumption, Σ and Σ_{n_k} are diagonal matrixes. Let σ^j

and $\sigma_{n_k}^j$ be the standard deviations of the j-th dimension for x and x_{n_k}, respectively. Then, $x^j \sim N(\mu^j, \sigma^j \cdot \sigma^j)$ and $x^j \sim N(\mu^j, \sigma^j \cdot \sigma^j)$. To classify x using naive Bayes model, we compute all the class condition densities $P(x|C_k)$ based on Equation (16.36). Since $x_{n_k}^j$ follows Gaussian distribution, we have:

$$p_{n_k}^j(x_{n_k}^j) = \frac{1}{\sigma_{n_k}^j \sqrt{2\pi}} \exp(-\frac{1}{2}(\frac{x_{n_k}^j - \mu_{n_k}^j}{\sigma_{n_k}^j})) \qquad (16.37)$$

and similarly for x^j (by omitting all subscripts in Equation (16.37)).

Based on formulae (16.36) and (16.37), we have:

$$P(x|C_k) = \prod_{j=1}^{d} \left\{ \sum_{n_k=1}^{N_k} \frac{\exp(-\frac{1}{2}(\frac{\mu^j - \mu_{n_k}^j}{v_{k,n_k}^j})^2)}{N_k v_{k,n_k}^j \sqrt{2\pi}} \right\} \qquad (16.38)$$

where $v_{k,n_k}^j = \sqrt{h_k^j \cdot h_k^j + \sigma^j \cdot \sigma^j + \sigma_{n_k}^j \cdot \sigma_{n_k}^j}$. It is easy to see the time complexity of computing $P(x|C_k)$ is $O(N_k d)$, so the time complexity of the formula-based method is $\sum_{k=1}^{K} O(N_k d) = O(nd)$.

(2). Sample-based method. In this method, every training and testing uncertain data object is represented by sample points based on their own distributions. When using kernel density estimation for a data object, every sample point contributes to the density estimation. The integral of density can be transformed into the summation of the data points' contribution with their probability as weights. Equation (16.36) is thus replaced by:

$$P(x|C_k) = \prod_{j=1}^{d} \frac{1}{N_k h_k^j} \sum_{n_k=1}^{N_k} \sum_{c=1}^{s} \sum_{d=1}^{s} K(\frac{x_c^j - x_{n_k,d}^j}{h_k^j}) P(x_c^j) P(x_{n_k,d}^j) \qquad (16.39)$$

where x_c^j represents the c-th sample point of uncertain test data object x along the j-th dimension. $x_{n_k,d}^j$ represents the d-th sample point of uncertain training data object x_{n_k} along the j-th dimension. $P(x_c^j)$ and $P(x_{n_k,d}^j)$ are probabilities according to x and x_{n_k}'s distribution, respectively. s is the number of samples used for each of x^j and $x_{k_n}^j$ along the j-th dimension, and gets the corresponding probability of each sample point. After computing the $P(x|C_k)$ for x with each class C_k, we can compute the posterior probability $P(C_k|x)$ based on Equation (16.31) and x can be assigned to the class with maximum $P(C_k|x)$. It is easy to see the time complexity of computing $P(x|C_k)$ is $O(N_k ds^2)$, so the time complexity of the formula-based method is $\sum_{k=1}^{K} O(N_k ds^2) = O(nds^2)$.

Discussions: In this section, we have discussed a naive Bayes method, i.e., *distribution-based* naive Bayes classification, for uncertain data classification. The critical step of naive Bayes classification is to estimate the class conditional density, and kernel density estimation is a common way for that. The *distribution-based* method extends the kernel density estimation method to handle uncertain data by replacing the kernel function using its expectation value. Compared with AVG, the *distribution-based* method has considered the whole distributions, rather than mean values of instances, so more accurate classifiers can be learnt, which means we can achieve better accuracies. Besides, it reduces the extended kernel density estimation to the evaluation of double-integrals. Then, two methods, i.e. formula-based method and sample-based method, are proposed to solve it. Although the *distribution-based* naive Bayes classification may achieve better accuracy, it performs slowly because it has to compute two summations over s sample points each, while AVG uses a single value in the place of this double summation. The further experiments on real data datasets in [29] show that *distribution-based* method can always achieve higher accuracy than AVG. Besides, the formula-based method generally gives higher accuracy than the sample-based method.

16.4   Conclusions

As one of the most essential tasks in data mining and machine learning, classification has been studied for decades. However, the solutions developed for classification often assume that data values are exact. This may not be adequate for emerging applications (e.g., LBS and biological databases), where the uncertainty of their databases cannot be ignored. In this chapter, we survey several recently-developed classification algorithms that consider uncertain data as a "first-class citizen." In particular, they consider the entire uncertainty information that is available, i.e., every single possible value in the pdf or pmf of the attributes involved, in order to achieve a higher effectiveness than AVG. They are also more efficient than PW, since they do not have to consider all the possible worlds.

We believe that there are plenty of interesting issues to consider in this area, including:

- A systematic comparison among the algorithms that we have studied here is lacking. It would be interesting to derive and compare their computational complexities. More experiments should also be conducted on these solutions, so that we know which algorithm performs better under which situations. Also, most of the datasets used in the experiments are synthetic (e.g., the pdf is assumed to be uniform). It would be important to test these solutions on data obtained from real applications or systems.

- Some classification algorithms apply to pmf only, while others (e.g., naive Bayes classifier) can be used for pdf with particular distributions (e.g., Gaussian). It would be interesting to examine a general and systematic method to modify a classification algorithm, so that it can use both pdf and pmf models.

- Most of the classification algorithms consider the uncertainty of attributes, represented by pdf or pmf. As a matter of fact, in the uncertain database literature, there are a few other well-studied uncertainty models, including tuple uncertainty [19], logical model [30], c-tables [18], and Bayesian models. These models differ in the complexity of representing uncertainty; e.g., some models assume entities (e.g., pdf and pmf) are independent, while others (e.g., Bayesian) handle correlation among entities. It would be interesting to consider how classification algorithms should be designed to handle these uncertainty models.

- All the algorithms studied here only work on one database snapshot. However, if these data changes (e.g., in LBS, an object is often moving, and the price of a stock in a stock market system fluctuates), it would be important to develop efficient algorithms that can incrementally update the classification results for a database whose values keep changing.

- It would be interesting to examine how the uncertainty handling techniques developed for classification can also be applied to other mining tasks, e.g., clustering, frequent pattern discovery, and association rule mining.

Bibliography

[1] C. C. Aggarwal. On density based transforms for uncertain data mining. In *ICDE Conference Proceedings*. IEEE, pages 866–875, 2007.

[2] C. C. Aggarwal, J. Han, J. Wang, and P Yu. A framework for clustering evolving data streams. In *VLDB Conference Proceedings*, 2003.

[3] C. C. Aggarwal and Philip S. Yu. A survey of uncertain data algorithms and applications. In *IEEE Transactions on Knowledge and Data Engineering*, 21(5):609–623, 2009.

[4] F. Angiulli and F. Fassetti. Nearest neighbor-based classification of uncertain data. In *ACM Transactions on Knowledge Discovery from Data*, 7(1):1, 2013.

[5] M. Ankerst, M. M. Breuning, H. P. Kriegel, and J. Sander. OPTICS: Ordering Points To Identify the Clustering Structure. In *ACM SIGMOD International Conference on Management of Data*. pages 49–60, 1999.

[6] A. Asuncion and D. Newman. UCI machine learning repository. [Online]. Available: `http://www.ics.uci.edu/~mlearn/MLRepository,html`, 2007.

[7] J. Bezdek and R. Hathaway. Convergence of alternating optimization. In *Neural, Parallel Science Computation*, 11:351-368, 2003.

[8] J. Bi and T. Zhang. Support vector classification with input data uncertainty. In *NIPS*, 2004.

[9] J. Chen and R. Cheng. Efficient evaluation of imprecise location-dependent queries. In *International Conference on Data Engineering*, pages 586–595, Istanbul, Turkey, April 2007.

[10] R. Cheng, J. Chen, and X. Xie. Cleaning uncertain data with quality guarantees. In *Proceedings of the VLDB Endowment*, 1(1):722–735, August 2008.

[11] R. Cheng, D. Kalashnikov, and S. Prabhakar. Evaluating probabilistic queries over imprecise data. In *Proceedings of the ACM Special Interest Group on Management of Data*, pages 551–562, June 2003.

[12] W. W. Cohen. Fast effective rule induction. In *Proceedings of 12th International Conference on Machine Learning*, pages 115–125, 1995.

[13] M. Ester, H. P. Kriegel, J. Sander, and X. Xu. A density-based algorithm for discovering clusters in large spatial databases with noise. In *Proceedings of the Second International Conference on Knowledge Discovery and Data Mining (KDD-96)*, pages 226–231, 1996.

[14] J. Furnkranze and G. Widmer. Incremental reduced error pruning. In *Machine Learning: Proceedings of 12th Annual Conference*, 1994.

[15] Gray L. Freed and J. Kennard Fraley. 25% "error rate" in ear temperature sensing device. *Pediatrics*, 87(3):414–415, March 2009.

[16] C. Gao and J. Wang. Direct mining of discriminative patterns for classifying uncertain data. In *Proceedings of the 16th ACM SIGKDD International Conference on Knowledge Discovery and Data Mining*, pages 861–870, 2010.

[17] J. Han and K. Micheline. Data Mining: Concepts and Techniques. Morgan Kaufman Publishers Inc., San Francisco, CA, 2005.

[18] T. Imielinski and W. Lipski Jr. Incomplete information in relational databases. In *ICDE*, 2008.

[19] B. Kanagal and A. Deshpande. Lineage processing over correlated probabilistic database. In *SIGMOD*, 2010.

[20] D. Lin. E. Bertino, R. Cheng, and S. Prabhakar. Position transformation: A location privacy protection method for moving objects. In *Transactions on Data Privacy: Foundations and Technologies (TDP)*, 2(1): 21–46, April 2009.

[21] B. Liu, W. Hsu, and Y. Ma. Integrating classification and association rule mining. In *Proceedings of the 4th ACM SIGKDD International Conference on Knowledge Discovery and Data Mining*, pages 80–86, 1998.

[22] W. Li, J. Han and J. Pei. CMAR: Accurate and efficient classification based on multiple class-association rules. In *Proceedings of the 2001 IEEE International Conference on Data Mining*, pages 369–376, 2001.

[23] B. Qin, Y. Xia, and F. Li. DTU: A decision tree for uncertain data. In *Proceedings of the 13th Pacific-Asia Conference on Knowledge Discovery and Data Mining*, pages 4–15, 2009.

[24] B. Qin, Y. Xia, R. Sathyesh, J. Ge, and S. Prabhakar. Classifying uncertain data with decision tree. In *DASFAA*, pages 454–457, 2011.

[25] B. Qin, Y. Xia, S. Prabhakar and Y. Tu. A rule-based classification algorithm for uncertain data. In *The Workshp on Management and Mining of Uncertain Data in ICDE*, pages 1633–1640, 2009.

[26] B. Qin, Y. Xia, and S. Prabhakar. Rule induction for uncertain data. In *Knowledge and Information Systems*. 29(1):103–130, 2011.

[27] J. Ross Quinlan. Induction of decision trees. *Machine Learning*,1(1):81–106, 1986.

[28] J. Ross Quinlan. *C4.5: Programs for Machine Learning*. Morgan Kaufmann, 1993.

[29] J. Ren, S. D. Lee, X. Chen, B. Kao, R. Cheng, and D. Chueng Naive Bayes classification of uncertain data. In *ICDM Conference Proceedings*, pages 944–949, 2009.

[30] A. D. Sarma, O. Benjelloun, A. Halevy, and J. Widom. Working models for uncertain data. In *ICDE*, 2008.

[31] L. Sun, R. Cheng, D. W. Cheung, and J. Cheng. Mining uncertain data with probabilistic guarantees. In the *16th ACM SIGKDD Conference on Knowledge Discovery and Data Mining*, pages 273–282, Washington D.C., USA, July 2010.

[32] S. Tsang, B. Kao, K. Y. Yip, W. S. Ho, and S. D. Lee. Decision trees for uncertain data. In *ICDE*, IEEE, pages 441–444, 2009.

[33] S. Tsang, B. Kao, K. Y. Yip, W. S. Ho, and S. D. Lee. Decision trees for uncertain data. In *IEEE Transactions on Knowledge and Data Engineering*, 23(1):64–78, 2011.

[34] V. N. Vapnik. Statistical Learning Theory. John Wiley & Sons, Inc., New York, 1988.

[35] J. Wang and G. Karypis On mining instance-centric classification rules. In *IEEE Transactions on Knowledge and Data Engineering*, 18(11):1497–1511, 2006.

[36] L. Wang, D. W. Cheung, R. Cheng, and S. D. Lee. Efficient mining of frequent itemsets on large uncertain databases. In the *IEEE Transactions on Knowledge and Data Engineering*, 24(12):2170–2183, Dec 2012.

Chapter 17

Rare Class Learning

Charu C. Aggarwal

IBM T. J. Watson Research Center
Yorktown Heights, NY 10598
`charu@us.ibm.com`

17.1 Introduction

The problem of rare class detection is closely related to outlier analysis [2]. In *unsupervised* outlier analysis, no supervision is used for the anomaly detection process. In such scenarios, many of the anomalies found correspond to noise, and may not be of any interest to an analyst. It has been observed [35, 42, 61] in diverse applications such as system anomaly detection, financial fraud, and Web robot detection that *the nature of the anomalies is often highly specific to particular kinds*

445

of abnormal activity in the underlying application. In such cases, unsupervised outlier detection methods may often discover noise, which may not be specific to that activity, and therefore may also not be of any interest to an analyst. The goal of supervised outlier detection and rare class detection is to incorporate application-specific knowledge into the outlier analysis process, so as to obtain more meaningful anomalies with the use of learning methods. Therefore, the rare class detection problem may be considered the closest connection between the problems of classification and outlier detection. In fact, while classification may be considered the supervised analogue of the clustering problem, the rare class version of the classification problem may be considered the supervised analogue of the outlier detection problem. This is not surprising, since the outliers may be considered rare "unsupervised groups" for a clustering application.

In most real data domains, some examples of normal or abnormal data may be available. This is referred to as *training data*, and can be used to create a *classification model*, which distinguishes between normal and anomalous instances. Because of the rare nature of anomalies, such data is often limited, and it is hard to create robust and generalized models on this basis. The problem of classification has been widely studied in its own right, and numerous algorithms are available in the literature [15] for creating supervised models from training data. In many cases, different kinds of abnormal instances may be available, in which case the classification model may be able to distinguish between them. For example, in an intrusion scenario, different kinds of intrusion anomalies are possible, and it may be desirable to distinguish among them.

This problem may be considered a very difficult special case (or variation) of the classification problem, depending upon the following possibilities, which may be present either in isolation or in combination.

- *Class Imbalance:* The distribution between the normal and rare class will be very skewed. From a practical perspective, this implies that the optimization of classification accuracy may not be meaningful, especially since the misclassification of positive (outlier) instances is less desirable than the misclassification of negative (normal) instances. In other words, false positives are more acceptable than false negatives. This leads to cost-sensitive variations of the classification problem, in which the objective function for classification is changed.

- *Contaminated Normal Class Examples (Positive-Unlabeled Class Problem):* In many real scenarios, the data may originally be present in unlabeled form, and manual labeling is performed for annotation purposes. In such cases, only the positive class is labeled, and the remaining "normal" data contains some abnormalities. This is natural in large scale applications such as the Web and social networks, in which the sheer volume of the underlying data makes contamination of the normal class more likely. For example, consider a social networking application, in which it is desirable to determine spam in the social network feed. A small percentage of the documents may be spam. In such cases, it may be possible to recognize and label some of the documents as spam, but many spam documents may remain in the examples of the normal class. Therefore, the "normal" class may also be considered an unlabeled class. In practice, however, the unlabeled class is predominantly the normal class, and the anomalies in it may be treated as contaminants. The classification models need to be built to account for this. Technically, this case can be considered a form of partial supervision [40], though it can also be treated as a difficult special case of full supervision, in which the normal class is more noisy and contaminated. Standard classifiers can be used on the positive-unlabeled version of the classification problem, as long as the relative frequency of contaminants is not extreme. In cases where the unlabeled class does not properly reflect the distribution in the test instances, the use of such unlabeled classes can actually *harm* classification accuracy [37].

A different flavor of incomplete supervision refers to missing training data about an *entire* class, rather than imperfect or noisy labels. This case is discussed below.

- *Partial Training Information (Semi-supervision or novel class detection):* In many applications, examples of one or more of the anomalous classes may not be available. For example, in an intrusion detection application, one may have examples of the normal class, and *some* of the intrusion classes, as new kinds of intrusions arise with time. In some cases, examples of one or more normal classes are available. A particularly commonly studied case is the *one class variation*, in which only examples of the normal class are available. The only difference between this extreme case and the unsupervised scenario is that the examples of the normal class are typically guaranteed to be free of outlier class examples. In many applications, this is a very natural consequence of the extreme rarity of the outlier class. For example, in a bioterrorist attack scenario, no examples of anomalous classes may be available, since no such event may have occurred in the first place. Correspondingly, the examples of the training class are also guaranteed to be free of outliers. This particular special case, in which the training data contains only normal classes, is much closer to the unsupervised version of the outlier detection problem. This will be evident from the subsequent discussion in the chapter.

It is evident that most of the above cases are either a special case, or a variant of the classification problem, which provides different challenges. Furthermore, it is possible for some of these conditions to be present in combination. For example, in an intrusion detection application, labeled data may be available for some of the intrusions, but no labeled information may be available for other kinds of intrusions. Thus, this scenario requires the determination of both rare classes and novel classes. In some cases, rare class scenarios can be reduced to partially supervised scenarios, when only the rare class is used for training purposes. Therefore, the boundaries between these scenarios are often blurred in real applications. Nevertheless, since the techniques for the different scenarios can usually be combined with one another, it is easier to discuss each of these challenges separately. Therefore, a section will be devoted to each of the aforementioned variations.

A particular form of supervision is *active learning*, when human experts may intervene in order to identify relevant instances. Very often, active learning may be accomplished by providing an expert with candidates for rare classes, which are followed by the expert explicitly labeling these pre-filtered examples. In such cases, *label acquisition* is combined with *model construction* in order to progressively incorporate more human knowledge into the rare class detection. Such a human-computer cooperative approach can sometimes provide more effective results than automated techniques.

Paucity of training data is a common problem, when the class distribution is imbalanced. Even in a modestly large training data set, only a small number of rare instances may be available. Typically, it may be expensive to acquire examples of the rare class. Imbalanced class distributions could easily lead to training algorithms that show differentially overfitting behavior. In other words, the algorithm may behave robustly for the normal class, but may overfit the rare class. Therefore, it is important to design the training algorithms, so that overfitting is avoided.

This chapter is organized as follows. The next section will discuss the problem of rare-class detection in the fully supervised scenario. The semi-supervised case of classification with positive and unlabeled data will be studied in Section 17.3. Section 17.4 will discuss the problem of novel class detection. This is also a form of semi-supervision, though it is of a different kind. Methods for human supervision are addressed in Section 17.5. A discussion of other work, and a summary of its relationship to the methods discussed in this chapter, is presented in Section 17.6. The conclusions and summary are presented in Section 17.7.

17.2 Rare Class Detection

The problem of *class imbalance* is a common one in the context of rare class detection. The straightforward use of evaluation metrics and classifiers that are not cognizant of this class imbalance may lead to very surprising results. For example, consider a medical application in which it is desirable to identify tumors from medical scans. In such cases, 99% of the instances may be normal, and only 1% are abnormal.

Consider the trivial classification algorithm, in which every instance is labeled as normal without even examining the feature space. Such a classifier would have a very high absolute accuracy of 99%, but would not be very useful in the context of a real application. In fact, many forms of classifiers (which are optimized for absolute accuracy) may show a degradation to the trivial classifier. For example, consider a k-nearest neighbor classifier, in which the majority class label in the neighborhood is reported as the relevant class label. Because of the inherent bias in the class distribution, the majority class may very often be normal even for abnormal test instances. Such an approach fails because it does not account for the *relative* behavior of the test instances with respect to the original class distribution. For example, if 49% of the training instances among the k-nearest neighbors of a test instance are anomalous, then that instance is *much* more likely to be anomalous *relative* to its original class distribution. By allowing changes to the classification criterion, such as reporting non-majority anomalous classes as the relevant label, it is possible to improve the classification accuracy of anomalous classes. However, the *overall* classification accuracy may degrade. Of course, the question arises whether the use of measures such as overall classification accuracy is meaningful in the first place. Therefore, the issue of *evaluation* and *model construction* are closely related in the supervised scenario. The first step is to identify how the rare class distribution relates to the objective function of a classification algorithm, and the algorithmic changes required in order to incorporate the modifications to the modeling assumptions.

There are two primary classes of algorithms that are used for handling class imbalance:

- *Cost Sensitive Learning:* The objective function of the classification algorithm is modified in order to weight the errors in classification differently for different classes. Classes with greater rarity have higher costs. Typically, this approach requires algorithm-specific changes to different classifier models in order to account for costs.

- *Adaptive Re-sampling:* The data is re-sampled so as to magnify the *relative proportion* of the rare classes. Such an approach can be considered an *indirect form* of cost-sensitive learning, since data re-sampling is equivalent to implicitly assuming higher costs for misclassification of rare classes.

Both these methodologies will be discussed in this section. For the case of the cost-sensitive problem, it will also be discussed how classification techniques can be heuristically modified in order to approximately reflect costs. A working knowledge of classification methods is assumed in order to understand the material in this section. The reader is also referred to [15] for a description of the different types of classifiers.

For the discussion in this section, it is assumed that the training data set is denoted by $\mathcal{D}$, and the labels are denoted by $L = \{1, \ldots k\}$. Without loss of generality, it can be assumed that the normal class is indexed by 1. The ith record is denoted by $\overline{X_i}$, and its label l_i is drawn from L. The number of records belonging to the ith class are denoted by N_i, and $\sum_{i=1}^{k} N_i = N$. The class imbalance assumption implies that $N_1 >> N - N_1$. While imbalances may exist between other anomalous classes too, the major imbalance occurs between the normal and the anomalous classes.

17.2.1 Cost Sensitive Learning

In cost sensitive learning, the goal is to learn a classifier, which maximizes the weighted accuracy over the different classes. The *misclassification cost* of the ith class is denoted by c_i. Some models [14] use a $O(k \times k)$ cost matrix to represent the full spectrum of misclassification behavior. In such models, the cost is dependent not only on the class identity of the misclassified instance, but is also dependent on the specific class label *to which* it is misclassified. A simpler model is introduced here, which is more relevant to the rare class detection problem. Here the cost only depends on the origin class, and not on a combination of the origin and destination class. The goal of the classifier is to learn a training model that minimizes the *weighted misclassification rate*.

The choice of c_i is picked in an application specific manner, though numerous heuristics exist to pick the costs in an automated way. The work in [70] proposes methods to learn the costs directly in a data driven manner. Other simpler heuristic rules are used often in many practical scenarios. For example, by choosing the value of c_i to be proportional to $1/N_i$, the *aggregate* impact of the instances of each class on the weighted misclassification rate is the same, in spite of the imbalance between the classes. Such methods are at best rule-of-thumb techniques for addressing imbalance, though more principled methods also exist in the literature. Many such methods will be discussed in this chapter.

17.2.1.1 MetaCost: A Relabeling Approach

A general framework known as MetaCost [14] uses a *relabeling* approach to classification. This is a *meta-algorithm*, which can be applied to any classification algorithm. In this method, the idea is to relabel some of the training instances in the data, by using the costs, so that normal training instances, which have a reasonable probability of classifying to the rare class, are relabeled to that rare class. Of course, rare classes may also be relabeled to a normal class, but the cost-based approach is *intended to* make this less likely. Subsequently, a classifier can be used on this more balanced training data set. The idea is to use the costs in order to move the decision boundaries in a cost-sensitive way, so that normal instances have a greater chance of misclassification than rare instances, and the *expected misclassification cost* is minimized.

In order to perform the relabeling, the classifier is applied to each instance of the training data and its classification prediction is combined with costs for re-labeling. Then, if a classifier predicts class label i with probability $p_i(\overline{X})$ for the data instance $\overline{X}$, then the expected misclassification cost of the prediction of $\overline{X}$, *under the hypothesis that it truly belonged to r*, is given by $\sum_{i \neq r} c_i \cdot p_i(\overline{X})$. Clearly, one would like to minimize the expected misclassification cost of the prediction. Therefore, the *MetaCost* approach tries different hypothetical classes for the training instance, and relabels it to the class that minimizes the expected misclassification cost. A key question arises as to how the probability $p_i(\overline{X})$ may be estimated from a classifier. This probability clearly depends upon the specific classifier that is being used. While some classifiers explicitly provide a probability score, not all classifiers provide such probabilities. The work in [14] proposes a bagging approach [6] in which the training data is sampled with replacement (bootstrapping), and a model is repeatedly constructed on this basis. The training instances are repeatedly classified with the use of such a bootstrap sample. The fraction of predictions (or votes) for a particular class across different training data samples are used as the classification probabilities.

The challenge of such an approach is that relabeling training data is always somewhat risky, especially if the bagged classification probabilities do not reflect intrinsic classification probabilities. In fact, each bagged classification model-based prediction is *highly correlated* to the others (since they share common training instances), and therefore the aggregate estimate is not a true probability.

In practice, the estimated probabilities are likely to be *very* skewed towards one of the classes, which is typically the normal class. For example, consider a scenario in which a rare class instance (with global class distribution of 1%) is present in a local region with 15% concentration of rare class instances. Clearly, this rare instance shows informative behavior in terms of *relative* concentration of the rare class in the locality of the instance. A vanilla 20-nearest neighbor classifier will virtually *always*[1] classify this instance to a normal class in a large bootstrapped sample. This situation is not specific to the nearest neighbor classifier, and is likely to occur in many classifiers, when the class distribution is very skewed. For example, an unmodified Bayes classifier will usually assign a lower probability to the rare class, because of its much lower *a-priori* probability, which is factored into the classification. Consider a situation where a hypothetically perfect Bayes classifier has a prior probability of 1% and a posterior probability of 30% for the correct classification of a rare class instance. Such a classifier will typically assign far fewer than 30% of the votes to the rare class in a bagged prediction, especially[2] when large bootstrap samples are used. In such cases, the normal class will win every time in the bagging because of the prior skew. This means that the bagged classification probabilities can sometimes be close to 1 for the normal class in a skewed class distribution.

This suggests that the effect of cost weighting can sometimes be overwhelmed by the erroneous skews in the probability estimation attained by bagging. In this particular example, even with a cost ratio of 100 : 1, the rare class instance will be wrongly relabeled to a normal class. This moves the classification boundaries in the opposite direction of what is desired. In fact, in cases where the unmodified classifier degrades to a trivial classifier of always classifying to the normal class, the expected misclassification cost criterion of [14] will result in relabeling all rare class instances to the normal class, rather than the intended goal of selective relabeling in the other direction. In other words, relabeling may result in a further *magnification* of the errors arising from class skew. This leads to degradation of classification accuracy, *even from a cost-weighted perspective*.

In the previous example, if the *fraction* of the 20-nearest neighbors belonging to a class are used as its probability estimate for relabeling, then much more robust results can be obtained with *Meta-Cost*. Therefore, the effectiveness of *MetaCost* depends on the quality of the probability estimate used for re-labeling. Of course, if good probability estimates are directly available from the training model in the first place, then a test instance may be directly predicted using the expected misclassification cost, rather than using the indirect approach of trying to "correct" the training data by re-labeling. This is the idea behind weighting methods, which will be discussed in the next section.

17.2.1.2 Weighting Methods

Most classification algorithms can be modified in natural ways to account for costs with some simple modifications. The primary driving force behind these modifications is to implicitly treat each training instance with a weight, where the weight of the instance corresponds to its misclassification cost. This leads to a number of simple modifications to the underlying classification algorithms. In most cases, the weight is not used explicitly, but the underlying classification model is changed to reflect such an implicit assumption. Some methods have also been proposed in the literature [69] in order to incorporate the weights explicitly into the learning process. In the following, a discussion is provided about the natural modifications to the more common classification algorithms.

17.2.1.3 Bayes Classifiers

The modification of the Bayes classifier provides the simplest case for cost-sensitive learning. In this case, changing the weight of the example only changes the a-priori probability of the class,

---

[1]The probability can be (approximately) computed from a binomial distribution to be at least equal to $\sum_{i=0}^{9} \binom{20}{i} \cdot 0.15^i \cdot 0.85^{20-i}$ and is greater than 0.999.

[2]The original idea of bagging was not designed to yield class probabilities [6]. Rather, it was designed to perform robust prediction for instances where either class is an almost equally good fit. In cases where one of the classes has a "reasonably" higher (absolute) probability of prediction, the bagging approach will simply boost that probability to almost 1, when counted in terms of the number of votes. In the rare class scenario, it is expected for unmodified classifiers to misclassify rare classes to normal classes with "reasonably" higher probability.

and all other terms within the Bayes estimation remain the same. Therefore, this is equivalent to multiplying the Bayes probability in the unweighted case with the cost, and picking the largest one. Note that this is the same criterion that is used in *MetaCost*, though the latter uses this criterion for *relabeling* training instances, rather than predicting test instances. When good probability estimates are available from the Bayes classifier, the test instance can be directly predicted in a cost-sensitive way.

17.2.1.4 Proximity-Based Classifiers

In nearest neighbor classifiers, the classification label of a test instance is defined to be the majority class from its k nearest neighbors. In the context of *cost-sensitive* classification, the *weighted* majority label is reported as the relevant one, where the weight of an instance from class i is denoted by c_i. Thus, fewer examples of the rare class need to be present in a neighborhood of a test instance, in order for it to be reported as the relevant one. In a practical implementation, the number of k-nearest neighbors for each class can be multiplied with the corresponding cost for that class. The majority class is picked *after* the weighting process. A discussion of methods for k-nearest neighbor classification in the context of data classification may be found in [72].

17.2.1.5 Rule-Based Classifiers

In rule-based classifiers, frequent pattern mining algorithms may be adapted to determine the relevant rules at a given level of support and confidence. A rule relates a condition in the data (e.g., ranges on numeric attributes) to a class label. The support of a rule is defined as the number of training instances which are relevant to that rule. The confidence of a rule is the fractional probability that the training instance belongs to the class on the right-hand side, if it satisfies the conditions on the left-hand side. Typically, the data is first discretized, and all the relevant rules are mined from the data, at pre-specified levels of support and confidence. These rules are then prioritized based on the underlying confidence (and sometimes also the support). For a given test instances, all the relevant rules are determined, and the results from different rules can be combined in a variety of ways (e.g., majority class from relevant rules, top matching rule etc.) in order to yield the final class label.

Such an approach is not difficult to adapt to the cost-sensitive case. The main adaptation is that the weights on the different training examples need to be used during the computation of measures such as the support or the confidence. Clearly, when rare examples are weighted more heavily, the confidence of a rule will be much higher, when its right hand side corresponds to a rare class because of the weighting. This will result in the selective emphasis of rules corresponding to rare instances. Some methods for using rule-based methods in imbalanced data classification are proposed in [26], [28].

17.2.1.6 Decision Trees

In decision trees, the training data is recursively partitioned, so that the instances of different classes are successively separated out at lower levels of the tree. The partitioning is performed by using conditions on one or more features in the data. Typically, the split criterion uses the various entropy measures such as the gini-index for deciding the choice of attribute and the position of the split. For a node containing a fraction of instances of different classes denoted by $p_1 \ldots p_k$, its gini-index is denoted by $1 - \sum_{i=1}^{k} p_i^2$. Better separations of different classes leads to a lower gini-index. The split attribute and partition point is decided as one that minimizes the gini-index of the children nodes. By using costs as weights for the instances, the computation of the gini-index will be impacted so as to selectively determine regions of the data containing higher proportions of the rare class. Some examples of cost-sensitive decision trees are discussed in [64, 65].

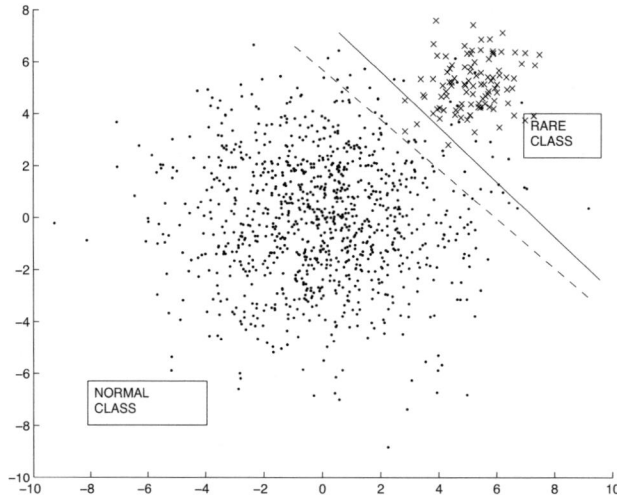

FIGURE 17.1: Optimal hyperplanes will change because of weighting of examples.

17.2.1.7 SVM Classifier

SVM classifiers work by learning hyperplanes, which optimally separate the two classes in order to minimize the expected error. Thus, SVM classifiers can be modeled as an optimization problem, where the goal is to learn the coefficients of the underlying hyperplane. For example, a two-class example has been illustrated in Figure 17.1. The optimal separator hyperplane for the two classes is illustrated in the same figure with the solid line. However, it is possible to change the optimization model by incorporating weights (or costs) into the optimization problem. This shifts the decision boundary, so as to allow erroneous classification of a larger number of normal instances, while correctly classifying more rare instances. The result would be a reduction in the overall classification accuracy, but an increase in the cost-sensitive accuracy. For example, in the case of Figure 17.1, the optimal separator hyperplane would move from the solid line to the dotted line in the figure. The issue of class-boundary re-alignment for SVMs in the context of imbalanced data sets has been explored in detail in [63,66]. While these models are not designed with the use of example re-weighting, they achieve similar goals by using class-biased penalties during the SVM model creation.

17.2.2 Adaptive Re-Sampling

In adaptive re-sampling, the different classes are differentially sampled in order to enhance the impact of the rare class on the classification model. Sampling can be performed either with or without replacement. Either the rare class can be oversampled, or the normal class can be under-sampled, or both. The classification model is learned on the re-sampled data. The sampling probabilities are typically chosen in proportion to their misclassification costs. This enhances the proportion of the rare costs in the sample used for learning. It has generally been observed [16] that under-sampling has a number of advantages over over-sampling. When under-sampling is used, the sampled training data is much smaller than the original data set. In some variations, all instances of the rare class are used in combination with a small sample of the normal class [12, 34]. This is also referred to as *one-sided selection*. Under-sampling also has the advantage of being efficient without losing too much information, because:

- The model construction phase for a smaller training data set requires much less time.

- The normal class is less important for modeling purposes, and most of the rare class is included for modeling. Therefore, the discarded instances do not take away too much from the modeling effectiveness.

17.2.2.1 Relation between Weighting and Sampling

Since cost-sensitive learning can be logically understood as methods that weigh examples differently, a question arises as to how these methods relate to one another. Adaptive re-sampling methods can be understood as methods that sample the data in proportion to their weights, and then treat all examples equally. From a practical perspective, this may often lead to similar models in the two cases, though sampling methods may throw away some of the relevant data. It should also be evident that a direct weight-based technique retains more information about the data, and is therefore likely to be more accurate. This seems to be the case from many practical experiences with real data [10]. On the other hand, adaptive re-sampling has distinct *efficiency* advantages because it works with a much smaller data set. For example, for a data set containing 1% of labeled anomalies, it is possible for a re-sampling technique to work effectively with 2% of the original data, when the data is re-sampled into an equal mixture of the normal and anomalous classes. This translates to a performance improvement of a factor of 50.

17.2.2.2 Synthetic Over-Sampling: SMOTE

Over-sampling methods are also used in the literature, though less frequently so than under-sampling. One of the problems of over-sampling the minority class is that a larger number of samples with replacement leads to repeated samples of the same record. This could lead to over-fitting, and does not necessarily help the effectiveness of the classifier. In other to address this issue, it was suggested [8] that synthetic over-sampling could be used to create the over-sampled examples in a way that provides better effectiveness. The *SMOTE* approach works as follows. For each minority instance, its k nearest neighbors are found. Then, depending upon the level of over-sampling required, a fraction of them are chosen randomly. A synthetic data example is generated on the line segment connecting that minority example to its nearest neighbor. The exact position of the example is chosen uniformly at random along the line segment. The *SMOTE* algorithm has been shown to provide more robust over-sampling than a vanilla over-sampling approach. This approach forces the decision region of the re-sampled data to become more general than one in which only members from the rare classes in the *original* training data are over-sampled.

17.2.2.3 One Class Learning with Positive Class

It is possible to take adaptive re-sampling to its logical extreme by not including any examples of the normal class. This artificially transforms the problem to the semi-supervised scenario, though the nature of the semi-supervision is quite different from naturally occurring scenarios. In most natural forms of semi-supervision, the positive class is missing, and copious examples of the normal class may be available. Here the normal class examples are removed from the data. This problem is also different from the positive-unlabeled classification problem. Such a problem may sometimes occur naturally in scenarios where the background class is too diverse or noisy to be sampled in a meaningful way.

In such cases, unsupervised models can be constructed on the subset of the data corresponding to the positive class. The major difference is that *higher fit* of the data to the positive class corresponds to greater outlier scores. This is the reverse of what is normally performed in outlier detection. The assumption is that the representative data contains only anomalies, and therefore outliers are more likely to be similar to this data. Proximity-based classifiers are very natural to construct in the one-class scenario, since the propensity of a test instance to belong to a class can be naturally modeled in terms of distances.

In the case of SVM classifiers, it is possible to create a two-class distribution by using the origin as one of the classes [59]. Typically, a kernel function is used in order to transform the data into a new space in which the dot product corresponds to the value of the kernel function. In such a case, an SVM classifier will naturally create a hyperplane that separates out the combination of features which describe the one class in the data. However, the strategy of using the origin as the second class in combination with a feature transformation is not necessarily generic and may not work well in all data domains. This differential behavior across different data sets has already been observed in the literature. In some cases, the performance of vanilla one-class SVM methods is quite poor, without careful changes to the model [55]. Other one-class methods for SVM classification are discussed in [31, 43, 55, 62].

17.2.2.4 Ensemble Techniques

A major challenge of under-sampling is the loss of the training data, which can have a detrimental effect on the quality of the classifier. A natural method to improve the quality of the prediction is to use ensemble techniques, in which the data instances are repeatedly classified with different samples, and then the majority vote is used for predictive purposes. In many of these methods, all instances from the rare class are used, but the majority class is under-sampled [12, 39]. Therefore, the advantages of selective sampling may be retained without a significant amount of information loss from the sampling process. In addition, a special kind of ensemble known as the *sequential ensemble* has also been proposed in [39]. In the sequential ensemble, the choice of the majority class instances picked in a given iteration depends upon the behavior of the classifier during previous iterations. Specifically, only majority instances that are correctly classified by the classifier in a given iteration are not included in future iterations. The idea is to reduce the redundancy in the learning process, and improve the overall robustness of the ensemble.

17.2.3 Boosting Methods

Boosting methods are commonly used in classification in order to improve the classification performance on difficult instances of the data. The well known *AdaBoost* algorithm [58] works by associating each training example with a *weight*, which is updated in each iteration, depending upon the results of the classification in the last iteration. Specifically, instances that are misclassified are given higher weights in successive iterations. The idea is to give higher weights to "difficult" instances that may lie on the decision boundaries of the classification process. The overall classification results are computed as a combination of the results from different rounds. In the tth round, the weight of the ith instance is $D_t(i)$. The algorithm starts off with equal weight of $1/N$ for each of the N instances, and updates them in each iteration. In practice, it is always assumed that the weights are normalized in order to sum to 1, though the approach will be described below in terms of (unscaled) relative weights for notational simplicity. In the event that the ith iteration is misclassified, then its (relative) weight is increased to $D_{t+1}(i) = D_t(i) \cdot e^{\alpha_t}$, whereas in the case of a correct classification, the weight is decreased to $D_{t+1}(i) = D_t(i) \cdot e^{-\alpha_t}$. Here α_t is chosen as the function $(1/2) \cdot \ln((1 - \varepsilon_t)/\varepsilon_t)$, where ε_t is the fraction of incorrectly predicted instances on a weighted basis. The final result for the classification of a test instance is a weighted prediction over the different rounds, where α_t is used as the weight for the tth iteration.

In the imbalanced and cost-sensitive scenario, the *AdaCost* method has been proposed [20], which can update the weights based on the cost of the instances. In this method, instead of updating the misclassified weights for instance i by the factor e^{α_t}, they are instead updated by $e^{\beta_-(c_i) \cdot \alpha_t}$, where c_i is the cost of the ith instance. Note that $\beta_-(c_i)$ is a function of the cost of the ith instance and serves as the "adjustment" factor, which accounts for the weights. For the case of correctly classified instances, the weights are updated by the factor $e^{-\beta_+(c_i) \cdot \alpha_t}$. Note that the adjustment factor is different depending upon whether the instance is correctly classified. This is because for the case of costly instances, it is desirable to increase weights more than less costly instances in case of misclassification. On the other hand, in cases of correct classification, it is desirable to reduce weights less for more costly instances. In either case, the adjustment is such that costly instances get relatively higher weight in later iterations. Therefore $\beta_-(c_i)$ is a non-decreasing function with cost, whereas $\beta_+(c_i)$ is a non-increasing function with cost. A different way to perform the adjustment would be to use the same exponential factor for weight updates as the original *Adaboost* algorithm, but this weight is further multiplied with the cost c_i [20], or other non-decreasing function of the cost. Such an approach would also provide higher weights to instances with larger costs. The use of boosting in weight updates has been shown to significantly improve the effectiveness of the imbalanced classification algorithms.

Boosting methods can also be combined with synthetic oversampling techniques. An example of this is the *SMOTEBoost* algorithm, which combines synthetic oversampling with a boosting approach. A number of interesting comparisons of boosting algorithms are presented in [27, 29]. In particular, an interesting observation in [29] is that the effectiveness of the boosting strategy is dependent upon the quality of the learner that it works with. When the boosting algorithm starts off with a weaker algorithm to begin with, the final (boosted) results are also not as good as those derived by boosting a stronger algorithm.

17.3 The Semi-Supervised Scenario: Positive and Unlabeled Data

In many data domains, the positive class may be easily identifiable, though examples of the negative class may be much harder to model simply because of their diversity and inexact modeling definition. Consider, for example, a scenario where it is desirable to classify or collect all documents that belong to a rare class. In many scenarios, such as the case of Web documents, the types of the documents available are too diverse, and it is hard to define a representative negative sample of documents from the Web.

This leads to numerous challenges at the *data acquisition stage*, where it is unknown what kinds of negative examples one might collect for contrast purposes. The problem is that the universe of instances in the negative class is rather large and diverse, and the collection of a representative sample may be difficult. For very large scale collections such as the Web and social networks [68], this scenario is quite common. A number of methods are possible for negative data collection, none of which are completely satisfactory in terms of being *truly representative* of what one might encounter in a real application. For example, for Web document classification, one simple option would be to simply crawl a random subset of documents off the Web. Nevertheless, such a sample would contain contaminants that do belong to the positive class, and it may be hard to create a purely negative sample, unless a significant amount of effort is invested in creating a clean sample. The amount of human effort involved in human labeling in rare class scenarios is especially high because the vast majority of examples are negative, and a manual process of filtering out the positive examples would be too slow and tedious. Therefore, a simple solution is to use the sampled background collection as the unlabeled class for training, but this may contain positive contaminants. This could lead to two different levels of challenges:

- The contaminants in the negative class can reduce the effectiveness of a classifier, though it is still better to use the contaminated training examples rather than completely discard them.

- The collected training instances for the unlabeled class may not reflect the true distribution of documents. In such cases, the classification accuracy may actually be *harmed* by using the negative class [37].

A number of methods have been proposed in the literature for this variant of the classification problem, which can address the aforementioned issues.

While some methods in the literature treat this as a new problem, which is distinct from the fully supervised classification problem [40], other methods [18] recognize this problem as a noisy variant of the classification problem, to which traditional classifiers can be applied with some modifications. An interesting and fundamental result proposed in [18] is that the accuracy of a classifier trained on this scenario differs by only a constant factor from the true conditional probabilities of being positive. The underlying assumption is that the labeled examples in the positive class are picked randomly from the positive examples in the combination of the two classes. These results provide strong support for the view that learning from positive and unlabeled examples is essentially equivalent to learning from positive and negative examples.

There are two broad classes of methods that can be used in order to address this problem. In the first class of methods, heuristics are used in order to identify training examples that are negative. Subsequently, a classifier is trained on the positive examples, together with the examples, which have already been identified to be negative. A less common approach is to assign weights to the unlabeled training examples [36, 40]. The second case is a special one of the first, in which each weight is chosen to be binary. It has been shown in the literature [41] that the second approach is superior. An SVM approach is used in order to learn the weights. The work in [71] uses the weight vector in order to provide robust classification estimates.

17.3.1 Difficult Cases and One-Class Learning

While the use of the unlabeled class provides some advantage to classification in most cases, this is not always true. In some scenarios, the unlabeled class in the training data reflects the behavior of the negative class in the test data very poorly. In such cases, it has been shown that the use of the negative class actually *degrades* the effectiveness of classifiers. In such cases, it has been shown in [37] that the use of one-class learners provides more effective results than the use of a standard classifier. Thus, in such situations, it may be better to simply discard the training class examples, which do not truly reflect the behavior of the test data. Most of the one-class SVM classifiers discussed in the previous section can be used in this scenario.

17.4 The Semi-Supervised Scenario: Novel Class Detection

The previous section discussed cases where it is difficult to obtain a clean sample of normal data, when the background data is too diverse or contaminated. A more common situation in the context of outlier detection is one in which no training data is available about one or more of the anomalous classes. Such situations can arise, when the anomalous class is so rare that it may be difficult to collect concrete examples of its occurrence, even when it is recognized as a concrete possibility. Some examples of such scenarios are as follows:

- In a bio-terrorist attack application, it may be easy to collect normal examples of environmental variables, but no explicit examples of anomalies may be available, if such an attack has never occurred.

- In an intrusion or viral attack scenario, many examples of normal data and previous intrusions or attacks may be available, but new forms of intrusion may arise over time.

This is truly a semi-supervised version of the problem, since training data is available about some portions of the data, but not others. Therefore, such scenarios are best addressed with a combination of supervised and unsupervised techniques. It is also important to distinguish this problem from one-class classification, in which instances of the *positive class* are available. In the one-class classification problem, it is desirable to determine other examples, which are as *similar* as possible to the training data, whereas in the novel class problem, it is desirable to determine examples, that are as *different* as possible from the training data.

In cases where only examples of the normal class are available, the only difference from the unsupervised scenario is that the training data is guaranteed to be free of outliers. The specification of normal portions of the data makes the determination of further outliers easier, because this data can be used in order to construct a model of what the normal data looks like. Another distinction between unsupervised outlier detection and one-class novelty detection is that novelties are often defined in a temporal context, and eventually become a normal part of the data.

17.4.1 One Class Novelty Detection

Since the novel-class detection problem is closely related to the one-class problem in cases where only the normal class is specified, it is natural to question whether it is possible to adapt some of the one-class detection algorithms to this scenario. The major difference in this case is that it is desirable to determine classes that are as *different* as possible from the specified training class. This is a more difficult problem, because a data point may be different from the training class in several ways. If the training model is not exhaustive in describing the corresponding class, it is easy for mistakes to occur.

For example, nearest neighbor models are easy to adapt to the one class scenario. In the one-class models discussed in the previous section, it is desirable to determine data points which are as *close* as possible to the training data. In this case, the opposite is desired, where it is desirable to determine data points which are as *different* as possible from the specified training data. This is of course no different from the unsupervised methods for creating proximity-based outlier detection methods. *In fact, any of the unsupervised models for outlier detection can be used in this case.* The major difference is that the training data is guaranteed to contain only the normal class, and therefore the outlier analysis methods are likely to be more robust. Strictly speaking, when only examples of the normal class are available, the problem is hard to distinguish from the unsupervised version of the problem, at least from a methodological point of view. From a *formulation* point of view, the training and test records are not distinguished from one another in the unsupervised case (any record can be normal or an anomaly), whereas the training (only normal) and test records (either normal or anomaly) are distinguished from one another in the semi-supervised case.

One class SVM methods have also been adapted to novelty detection [60]. The main difference from *positive example training-based* one-class detection is that the class of interest lies on the *opposite* side of the separator as the training data. Some of the one-class methods such as SVMs are unlikely to work quite as well in this case. This is because a one-class SVM may really only be able to model the class present in the training data (the normal class) well, and may not easily be able to design the best separator for the class that is most *different* from the normal class. Typically, one-class SVMs use a kernel-based transformation along with reference points such as the origin in order to determine a synthetic reference point for the other class, so that a separator can be defined. If the transformation and the reference point is not chosen properly, the one-class SVM is unlikely to provide robust results in terms of identifying the outlier. One issue with the one-class SVM is that the anomalous points (of interest) and the training data now need to lie on *opposite* sides of the separator. This is a more difficult case than one in which the anomalous points (of interest) and the training data need to lie on the same side of the separator (as was discussed in a previous section on positive-only SVMs). The key difference here is that the examples of interest are not available on the *interesting* side of the separator, which is poorly modeled.

It has been pointed out that the use of the origin as a prior for the anomalous class [7] can lead to incorrect results, since the precise nature of the anomaly is unknown a-priori. Therefore, the work in [7] attempts to determine a linear or non-linear decision surface, which wraps around the surfaces of the normal class. Points that lie outside this decision surface are anomalies. It is important to note that this model essentially uses an indirect approach such as SVM to model the dense regions in the data. Virtually all unsupervised outlier detection methods attempt to model the normal behavior of the data, and can be used for novel class detection, especially when the only class in the training data is the normal class. *Therefore the distinction between normal-class only variations of the novel class detection problem and the unsupervised version of the problem are limited and artificial, especially when other labeled anomalous classes do not form a part of the training data.* Numerous analogues of unsupervised methods have also been developed for novelty detection, such as extreme value methods [56], direct density ratio estimation [23], and kernel-based PCA methods [24]. This is not surprising, given that the two problems are different only at a rather superficial level. In spite of this, the semi-supervised version of the (normal-class only) problem seems to have a distinct literature of its own. This is somewhat unnecessary, since any of the unsupervised algorithms can be applied to this case. The main difference is that the training and test data are distinguished from one another, and the outlier score is computed for a test instance with respect to the training data. Novelty detection can be better distinguished from the unsupervised case in temporal scenarios, where novelties are defined *continuously* based on the past behavior of the data. This is discussed in Chapter 9 on streaming classification in this book.

17.4.2 Combining Novel Class Detection with Rare Class Detection

A more challenging scenario arises, when labeled rare classes are present in the training data, but novel classes may also need to be detected. Such scenarios can arise quite often in many applications such as intrusion detection, where partial knowledge is available about *some* of the anomalies, but others may need to be modeled in an unsupervised way. Furthermore, it is important to distinguish different kinds of anomalies from one another, whether they are found in a supervised or unsupervised way. The labeled rare classes already provide important information about *some* of the outliers in the data. This can be used to determine different kinds of outliers in the underlying data, and distinguish them from one another. This is important in applications, where it is not only desirable to determine outliers, but also to obtain an understanding of the kind of outlier that is discovered. The main challenge in these methods is to seamlessly combine unsupervised outlier detection methods with fully supervised rare class detection methods. For a given test data point two decisions need to be made, in the following order:

1. Is the test point a natural fit for a model of the training data? This model also includes the currently occurring rare classes. A variety of unsupervised models such as clustering can be used for thus purpose. If not, it is immediately flagged as an outlier, or a novelty.

2. If the test point is a fit for the training data, then a classifier model is used to determine whether it belongs to one of the rare classes. Any cost-sensitive model (or an ensemble of them) can be used for this purpose.

Thus, this model requires a combination of unsupervised and supervised methods in order to determine the outliers in the data. This situation arises more commonly in online and streaming scenarios, which will be discussed in the next section.

17.4.3 Online Novelty Detection

The most common scenario for novel class detection occurs in the context of *online* scenarios in concept drifting data streams. In fact, novel class detection usually has an implicit assumption of *temporal* data, since classes can be defined as novel only in terms of what has already been seen in the *past*. In many of the batch-algorithms discussed above, this temporal aspect is not fully explored, since a single snapshot of training data is assumed. Many applications such as intrusion detection are naturally focussed on a streaming scenario. In such cases, novel classes may appear at any point in the data stream, and it may be desirable to distinguish different kinds of novel classes from one another [5, 46–48]. Furthermore, when new classes are discovered, these kinds of anomalies may recur over time, albeit quite rarely. In such cases, the effectiveness of the model can be improved by keeping a memory of the *rarely recurring classes*. This case is particularly challenging because aside from the temporal aspects of modeling, it is desirable to perform the training and testing in an online manner, in which only one pass is allowed over the incoming data stream. This scenario is a true amalgamation of supervised and unsupervised methods for anomaly detection, and is discussed in detail in Chapter 9 on streaming classification.

In the streaming scenario containing only unlabeled data, unsupervised clustering methods [3,4] can be used in order to identify significant novelties in the stream. In these methods, *novelties occur as emerging clusters in the data, which eventually become a part of the normal clustering structure of the data*. Both the methods in [3, 4] have statistical tests to identify, when a newly incoming instance in the stream should be considered a novelty. Thus, the output of these methods provides an understanding of the natural complementary relationship between the clusters (normal unsupervised models) and novelties (temporal abnormalities) in the underlying data.

17.5   Human Supervision

The issue of human supervision arises in the problem of combining rare class detection with unsupervised outlier detection. This is a case where the label acquisition process is performed by the human in conjunction with rare class detection. Therefore, this problem lies on the interface of unsupervised outlier analysis and rare class detection. In this case, a human expert may intervene in the outlier detection process in order to further improve the effectiveness of the underlying algorithms. One of the major challenges in outlier detection is that the anomalies found by an algorithm that is either purely unsupervised or only partially supervised may not be very useful. The incorporation of human supervision can augment the limited knowledge of outlier analysis algorithms. An unsupervised or supervised outlier detection algorithm may present pre-filtered results to a user, and the user can provide their feedback on this small number of pre-filtered examples. This process would not be possible to perform manually on the original data set, which may be large, and in which the vast majority of examples are normal [52]. This is related to the concept of active learning, which is discussed in detail in a later chapter of this book. Active learning is particularly helpful in the context of rare class learning analysis, because of the high cost of finding rare examples.

An interesting procedure for active learning from unlabeled data is proposed in [52]. An iterative procedure is used in order to label some of the examples in each iteration. In each iteration, a number of interesting instances are identified, for which the addition of labels would be helpful for further classification. These are considered the "important" instances. The human expert provides labels for these examples. These are then used in order to classify the data set with the augmented labels. The first iteration is special, in which a purely unsupervised approach is used for learning. These procedures are performed iteratively until the addition of further examples is no longer deemed helpful for further classification. The overall procedure is illustrated in Figure 17.2. It should be noted that this approach can also be used in scenarios in which a small number of positive examples are available to begin with.

A key question arises as to *which* examples should be presented to the user for the purposes of labeling. It is clear that examples that are very obviously positive or negative (based on current models) are not particularly useful to present to the user. Rather, it is the examples with the greatest

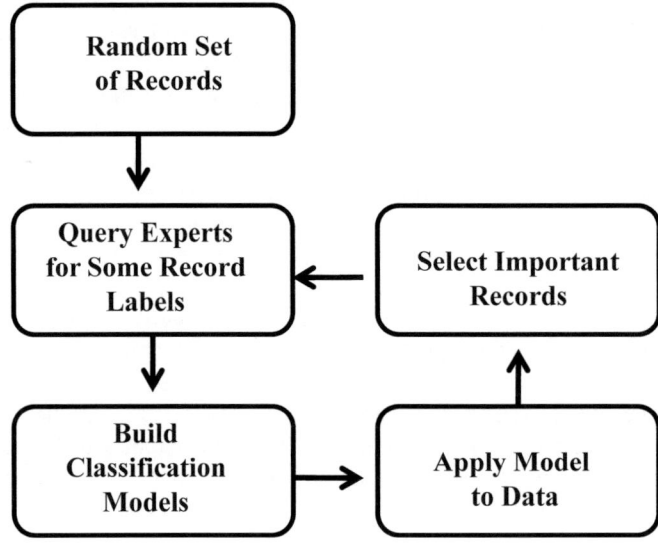

FIGURE 17.2: The overall procedure for active learning.

uncertainty or *ambiguity*, that should be presented to the user in order to gain the greatest knowledge about the decision boundaries between the different classes. It is expected that the selected examples should lie on the decision boundaries, in order to maximize the learning of the contours separating different classes, with the use of least amount of expert supervision, which can be expensive in many scenarios.

A common approach to achieving this goal in active learning is the principle of *query by committee* [57]. . In these methods, an ensemble of classifiers is learned, and the greatest *disagreement* among them is used to select data points that lie on the decision boundary. A variety of such criteria based on ensemble learning are discussed in [49]. It is also possible to use the model characteristics directly in order to select such points. For example, two primary criteria that can be used for selection are as follows [52]:

- *Low Likelihood:* These are data points that have low fit to the model describing the data. For example, if an EM algorithm is used for modeling, then these are points that have low fit to the underlying model.

- *High Uncertainty:* These are points that have the greatest uncertainty in terms of the component of the model to which they belong. In other words, in an EM model, such a data point would show relatively similar soft probabilities for different components of the mixture.

All data points are ranked on the basis of the two aforementioned criteria. The lists are merged by alternating between them, and adding the next point in the list, which has not already been added to the merged list. Details of other relevant methods such as *interleaving* are discussed in [52].

17.6 Other Work

Supervision can be incorporated in a variety of ways, starting from partial supervision to complete supervision. In the case of complete supervision, the main challenges arise in the context of class imbalance and cost-sensitive learning [10, 11, 19]. The issue of evaluation is critical in cost-sensitive learning because of the inability to model the effectiveness with measures such as the absolute accuracy. Methods for interpreting ROC curves and classification accuracy in the presence of costs and class imbalance are discussed in [17, 21, 30, 53, 54]. The impact of class imbalance is relevant even for feature selection [50, 73], because it is more desirable to select features that are more indicative of the rare class.

A variety of general methods have been proposed for cost-sensitive learning, such as *Meta-Cost* [14], weighting [69], and sampling [10, 12, 16, 34, 69]. Weighting methods are generally quite effective, but may sometimes be unnecessarily inefficient, when most of the training data corresponds to the background distribution. In this context, sampling methods can significantly improve the efficiency. Numerous cost-sensitive variations of different classifiers have been proposed along the lines of weighting, and include the Bayes classifier [69], nearest neighbor classifier [72], decision trees [64, 65], rule-based classifiers [26, 28], and SVM classifiers [63, 66].

Ensemble methods for improving the robustness of sampling are proposed in [12, 39]. Since the under-sampling process reduces the number of negative examples, it is natural to use an ensemble of classifiers that combine the results of classifiers trained on different samples. This provides more robust results, and ameliorates the instability that arises from under-sampling. The major problem in over-sampling of the minority class is the over-fitting obtained by re-sampling duplicate instances. Therefore, a method known as *SMOTE* creates synthetic data instances in the neighborhood of the rare instances [8].

The earliest work on boosting rare classes was proposed in [33]. This technique is designed for imbalanced data sets, and the intuition is to boost the positive training instances (rare classes) faster than the negatives. Thus, it increases the weight of false negatives more than that of the false positives. However, it is not cost-sensitive, and it also decreases the weight of true positives more than that of true negatives, which is not desirable. The *AdaCost* algorithm proposed in this chapter was proposed in [20]. Boosting techniques can also be combined with sampling methods, as in the case of the *SMOTEBoost* algorithm [9]. An evaluation of boosting algorithms for rare class detection is provided in [27]. Two new algorithms for boosting are also proposed in the same paper. The effect of the base learner on the final results of the boosting algorithm are investigated in [29]. It has been shown that the final result from the boosted algorithm is highly dependent on the quality of the base learner.

A particular case that is commonly encountered is one in which the instances of the positive class are specified, whereas the other class is unlabeled [18, 36, 37, 40, 41, 68, 71]. Since the unlabeled class is predominantly a negative class with contaminants, it is essentially equivalent to a fully supervised problem, with some loss in accuracy, which can be quantified [18]. In some cases, when the collection mechanisms for the negative class are not reflective of what would be encountered in test instances, the use of such instances may harm the performance of the classifier. In such cases, it may be desirable to discard the negative class entirely and treat the problem as a one-class problem [37]. However, as long as the training and test distributions are not too different, it is generally desirable to also use the instances from the negative class.

The one-class version of the problem is an extreme variation in which only positive instances of the class are used for training purpose. SVM methods are particularly popular for one-class classification [31, 43, 55, 59, 62]. Methods for one-class SVM methods for scene classification are proposed in [67]. It has been shown that the SVM method is particularly sensitive to the data set used [55].

An important class of semi-supervised algorithms is known as *novelty detection*, in which no training data is available about some of the anomalous classes. This is common in many scenarios such as intrusion detection, in which the patterns in the data may change over time, and may therefore lead to novel anomalies (or intrusions). These problems are a combination of the supervised and unsupervised case, and numerous methods have been designed for the streaming scenario [5,47,48]. The special case, where only the normal class is available, is not very different from the unsupervised scenario, other than the fact that it may have an underlying temporal component. Numerous methods have been designed for this case, such as single-class SVMs [7,60], minimax probability machines [32], kernel-based PCA methods [56], direct density ratio estimation [23], and extreme value analysis [24]. Single class novelty detection has also been studied extensively in the context of the first story detection in text streams [76]. The methods for the text streaming scenario are mostly highly unsupervised, and use standard clustering or nearest-neighbor models. In fact, a variety of stream clustering methods [3, 4] discover newly forming clusters (or emerging novelties) as part of their output of the overall clustering process. A detailed survey of novelty detection methods may be found in [44, 45].

Human supervision is a natural goal in anomaly detection, since most of the anomalous instances are not interesting, and it is only by incorporating user feedback that the interesting examples can be separated from noisy anomalies. Methods for augmenting user-specified examples with automated methods are discussed in [74, 75]. These methods also add artificially generated examples to the training data, in order to increase the number of positive examples for the learning process. Other methods are designed for *selectively* presenting examples to a user, so that only the relevant ones are labeled [52]. A nearest-neighbor method for active learning is proposed in [22]. The effectiveness of active learning methods for selecting good examples to present to the user is critical in ensuring minimal human effort. Such points should lie on the decision boundaries separating two classes [13]. Methods that use query by committee to select such points with ensembles are discussed in [49,57]. A selective sampling method that uses active learning in the context of outlier detection is proposed in [1]. A method has also been proposed in [38] as to how unsupervised outlier detection algorithms can be leveraged in conjunction with limited human effort in order to create a labeled training data set.

17.7 Conclusions and Summary

This chapter discusses the problem of rare class learning. In many real scenarios, training data is available, which can be used in order to greatly enhance the effectiveness of the outlier detection process. Many of the standard classification algorithms in the literature can be adapted to this problem, especially when full supervision is available. The major challenge of using the standard classification algorithms is that they may not work very well in scenarios where the distribution of classes is imbalanced. In order to address this issue, sampling and re-weighting can be used quite effectively.

The partially supervised variations of the problem are diverse. Some of these methods do not provide any labels on the normal class. This corresponds to the fact that the normal class may be contaminated with an unknown number of outlier examples. Furthermore, in some cases, the distribution of the normal class may be very different in the training and test data. One-class methods can sometimes be effective in addressing such issues.

Another form of partial supervision is the identification of novel classes in the training data. Novel classes correspond to scenarios in which the labels for some of the classes are completely missing from the training data. In such cases, a combination of unsupervised and supervised methods need to be used for the detection process. In cases where examples of a single normal class are available, the scenario becomes almost equivalent to the unsupervised version of the problem.

Supervised methods are closely related to active learning in which human experts may intervene in order to add more knowledge to the outlier detection process. Such combinations of automated filtering with human interaction can provide insightful results. The use of human intervention sometimes provides the more insightful results, because the human is involved in the entire process of label acquisition and final anomaly detection.

Bibliography

[1] N. Abe, B. Zadrozny, and J. Langford. Outlier detection by active learning, *ACM KDD Conference*, pages 504–509, 2006.

[2] C. C. Aggarwal, *Outlier Analysis*, Springer, 2013.

[3] C. C. Aggarwal, J. Han. J. Wang, and P. Yu. A framework for clustering evolving data streams, *VLDB Conference*, pages 81–92, 2003.

[4] C. C. Aggarwal and P. Yu. On Clustering massive text and categorical data streams, *Knowledge and Information Systems*, 24(2):171–196, 2010.

[5] T. Al-Khateeb, M. Masud, L. Khan, C. Aggarwal, J. Han, and B. Thuraisingham. Recurring and novel class detection using class-based ensemble, *ICDM Conference*, 2012.

[6] L. Brieman. Bagging predictors, *Machine Learning*, 24:123–140, 1996.

[7] C. Campbell and K. P. Bennett. A linear-programming approach to novel class detection, *NIPS Conference*, 2000.

[8] N. V. Chawla, K. W. Bower, L. O. Hall, and W. P. Kegelmeyer. SMOTE: synthetic minority over-sampling technique, *Journal of Artificial Intelligence Research (JAIR)*, 16:321–356, 2002.

[9] N. Chawla, A. Lazarevic, L. Hall, and K. Bowyer. SMOTEBoost: Improving prediction of the minority class in boosting, *PKDD*, pages 107–119, 2003.

[10] N. V. Chawla, N. Japkowicz, and A. Kotcz. Editorial: Special issue on learning from imbalanced data sets, *ACM SIGKDD Explorations Newsletter*, 6(1):1–6, 2004.

[11] N. V. Chawla, D. A. Cieslak, L. O. Hall, and A. Joshi. Automatically countering imbalance and its empirical relationship to cost. *Data Mining and Knowledge Discovery*, 17(2):225–252, 2008.

[12] P. K. Chan and S. J. Stolfo. Toward scalable learning with non-uniform class and cost distributions: A case study in credit card fraud detection. *KDD Conference*, pages 164–168, 1998.

[13] D. Cohn, R. Atlas, and N. Ladner. Improving generalization with active learning, *Machine Learning*, 15:201–221.

[14] P. Domingos. MetaCost: A general framework for making classifiers cost-sensitive, *ACM KDD Conference*, pages 165–174, 1999.

[15] R. Duda, P. Hart, and D. Stork, *Pattern Classification*, Wiley, 2001.

[16] C. Drummond and R. Holte. C4.5, class imbalance, and cost sensitivity: Why undersampling beats oversampling. *ICML Workshop on Learning from Imbalanced Data Sets*, pages 1–8, 2003.

[17] C. Drummond and R. Holte. Explicitly representing expected cost: An alternative to ROC representation. *ACM KDD Conference*, pages 198–207, 2001.

[18] C. Elkan and K. Noto. Learning classifiers from only positive and unlabeled data, *ACM KDD Conference*, pages 213–220, 2008.

[19] C. Elkan. The foundations of cost-sensitive learning, *IJCAI*, pages 973–978, 2001.

[20] W. Fan, S. Stolfo, J. Zhang, and P. Chan. AdaCost: Misclassification cost sensitive boosting, *ICML Conference*, pages 97–105, 1999.

[21] T. Fawcett. ROC Graphs: Notes and Practical Considerations for Researchers, Technical Report HPL-2003-4, Palo Alto, CA: HP Laboratories, 2003.

[22] J. He and J. Carbonell. Nearest-Neighbor-Based Active Learning for Rare Category Detection. CMU Computer Science Department, Paper 281, 2007. `http://repository.cmu.edu/compsci/281`

[23] S. Hido, Y. Tsuboi, H. Kashima, M. Sugiyama, and T. Kanamori. Statistical outlier detection using direct density ratio estimation. *Knowledge and Information Systems*, 26(2):309–336, 2011.

[24] H. Hoffmann. Kernel PCA for novelty detection, *Pattern Recognition*, 40(3):863–874, 2007.

[25] G. Lanckriet, L. Ghaoui, and M. Jordan. Robust novelty detection with single class MPM, *NIPS*, 2002.

[26] M. Joshi, R. Agarwal, and V. Kumar. Mining needles in a haystack: Classifying rare classes via two-phase rule induction, *ACM SIGMOD Conference*, pages 91–102, 2001.

[27] M. Joshi, V. Kumar, and R. Agarwal. Evaluating boosting algorithms to classify rare classes: Comparison and improvements. *ICDM Conference*, pages 257–264, 2001.

[28] M. Joshi and R. Agarwal. PNRule: A new framework for learning classifier models in data mining (A case study in network intrusion detection), *SDM Conference*, 2001.

[29] M. Joshi, R. Agarwal, and V. Kumar. Predicting rare classes: Can boosting make any weak learner strong? *ACM KDD Conference*, pages 297–306, 2002.

[30] M. Joshi. On evaluating performance of classifiers for rare classes, *ICDM Conference*, pages 641–644, 2003.

[31] P. Juszczak and R. P. W. Duin. Uncertainty sampling methods for one-class classifiers. *ICML Workshop on Learning from Imbalanced Data Sets*, 6(1), 2003.

[32] G. Lanckriet, L. Ghaoui, and M. Jordan. Robust novelty detection with single class MPM, *NIPS*, 2002.

[33] G. Karakoulas and J. Shawe-Taylor. Optimising classifiers for imbalanced training sets, *NIPS*, pages 253–259, 1998.

[34] M. Kubat and S. Matwin. Addressing the curse of imbalanced training sets: One sided selection. *ICML Conference*, pages 179–186, 1997.

[35] T. Lane and C. Brodley. Temporal sequence learning and data reduction for anomaly detection, *ACM Transactions on Information and Security*, 2(3):295–331, 1999.

[36] W. Lee and B. Liu. Learning with positive and unlabeled examples using weighted logistic regression. *ICML Conference*, 2003.

[37] X. Li, B. Liu, and S. Ng. Negative training data can be harmful to text classification, *EMNLP*, pages 218–228, 2010.

[38] L. Liu, and X. Fern. Constructing training sets for outlier detection, *SDM Conference*, 2012.

[39] X.-Y. Liu, J. Wu, and Z.-H. Zhou. Exploratory undersampling for class-imbalance learning. *IEEE Transactions on Systems, Man and Cybernetics – Part B, Cybernetics*, 39(2):539–550, April 2009.

[40] B. Liu, W. S. Lee, P. Yu, and X. Li. Partially supervised classification of text, *ICML Conference*, pages 387–394, 2002.

[41] B. Liu, Y. Dai, X. Li, W. S. Lee, and P. Yu. Building text classifiers using positive and unlabeled examples. *ICDM Conference*, pages 179–186, 2003.

[42] X. Liu, P. Zhang, and D. Zeng. Sequence matching for suspicious activity detection in anti-money laundering. *Lecture Notes in Computer Science*, 5075:50–61, 2008.

[43] L. M. Manevitz and M. Yousef. One-class SVMs for document classification, *Journal of Machine Learning Research*, 2:139–154, 2001.

[44] M. Markou and S. Singh. Novelty detection: A review, Part 1: Statistical approaches, *Signal Processing*, 83(12):2481–2497, 2003.

[45] M. Markou and S. Singh. Novelty detection: A review, Part 2: Neural network-based approaches, *Signal Processing*, 83(12):2481–2497, 2003.

[46] M. Masud, Q. Chen, L. Khan, C. Aggarwal, J. Gao, J, Han, and B. Thuraisingham. Addressing concept-evolution in concept-drifting data streams. *ICDM Conference*, 2010.

[47] M. Masud, T. Al-Khateeb, L. Khan, C. Aggarwal, J. Gao, J. Han, and B. Thuraisingham. Detecting recurring and novel classes in concept-drifting data streams. *ICDM Conference*, 2011.

[48] M. Masud, Q. Chen, L. Khan, C. Aggarwal, J. Gao, J. Han, A. Srivastava, and N. Oza. Classification and adaptive novel class detection of feature-evolving data streams, *IEEE Transactions on Knowledge and Data Engineering*, 25(7):1484–1497, 2013.

[49] P. Melville and R. Mooney. Diverse ensembles for active learning, *ICML Conference*, pages 584–591, 2004.

[50] D. Mladenic and M. Grobelnik. Feature selection for unbalanced class distribution and naive bayes. *ICML Conference*, pages 258-267, 1999.

[51] S. Papadimitriou, H. Kitagawa, P. Gibbons, and C. Faloutsos. LOCI: Fast outlier detection using the local correlation integral. *ICDE Conference*, pages 315-326, 2003.

[52] D. Pelleg and A. Moore. Active learning for anomaly and rare category detection, *NIPS Conference*, 2004.

[53] F. Provost and T. Fawcett. Analysis and visualization of classifier performance: Comparison under imprecise class and cost distributions, *ACM KDD Conference*, pages 43–48, 1997.

[54] F. Provost, T. Fawcett, and R. Kohavi. The case against accuracy estimation for comparing induction algorithms, *ICML Conference*, pages 445-453, 1998.

[55] B. Raskutti and A. Kowalczyk. Extreme rebalancing for SVMS: A case study. *SIGKDD Explorations*, 6(1):60–69, 2004.

[56] S. Roberts. Novelty detection using extreme value statistics, *IEEE Proceedings on Vision, Image and Signal Processing*, 146(3):124–129, 1999.

[57] H. Seung, M. Opper, and H. Sompolinsky. Query by committee. *ACM Workshop on Computational Learning Theory*, pages 287–294, 1992.

[58] R. Schapire and Y. Singer. Improved boosting algorithms using confidence-rated predictions. *Annual Conference on Computational Learning Theory*, 37(3):297–336, 1998.

[59] B. Scholkopf, J. C. Platt, J. Shawe-Taylor, A. J. Smola, and R. C. Williamson. Estimating the support of a high-dimensional distribution. *Neural Computation*, 13(7):1443–1472, 2001.

[60] B. Scholkopf, R. C. Williamson, A. J. Smola, J. Shawe-Taylor, and J. C. Platt. Support-vector method for novelty detection, *NIPS Conference*, 2000.

[61] P.-N. Tan and V. Kumar. Discovery of web robot sessions based on their navigational patterns, *Data Mining and Knowledge Discovery*, 6(1):9–35, 2002.

[62] D. Tax. One Class Classification: Concept-learning in the Absence of Counter-examples, *Doctoral Dissertation, University of Delft*, Netherlands, 2001.

[63] Y. Tang, Y.-Q. Zhang, N. V. Chawla, and S. Krasser. SVMs Modeling for highly imbalanced classification, *IEEE Transactions on Systems, Man and Cybernetics — Part B: Cybernetics*, 39(1):281–288, 2009.

[64] K. M. Ting. An instance-weighting method to induce cost-sensitive trees. *IEEE Transactions on Knowledge and Data Engineering*, 14(3):659–665, 2002.

[65] G. Weiss and F. Provost. Learning when training data are costly: The effect of class distribution on tree induction, *Journal of Artificial Intelligence Research*, 19(1):315–354, 2003.

[66] G. Wu and E. Y. Chang. Class-boundary alignment for imbalanced dataset learning. *Proceedings of the ICML Workshop on Learning from Imbalanced Data Sets*, pages 49–56, 2003.

[67] R. Yan, Y. Liu, R. Jin, and A. Hauptmann. On predicting rare classes with SVM ensembles in scene classification. *IEEE International Conference on Acoustics, Speech and Signal Processing*, 3:21–24, 2003.

[68] H. Yu, J. Han, and K. C.-C. Chang. PEBL: Web page classification without negative examples. *IEEE Transactions on Knowledge and Data Engineering*, 16(1):70–81, 2004.

[69] B. Zadrozny, J. Langford, and N. Abe. Cost-sensitive learning by cost-proportionate example weighting, *ICDM Conference*, pages 435–442, 2003.

[70] B. Zadrozny and C. Elkan. Learning and making decisions when costs and probabilities are unknown, *KDD Conference*, pages 204–213, 2001.

[71] D. Zhang and W. S. Lee. A simple probabilistic approach to learning from positive and unlabeled examples. *Annual UK Workshop on Computational Intelligence*, pages 83–87, 2005.

[72] J. Zhang and I. Mani. KNN Approach to unbalanced data distributions: A case study involving information extraction. *Proceedings of the ICML Workshop on Learning from Imbalanced Datasets*, 2003.

[73] Z. Zheng, X. Wu, and R. Srihari. Feature selection for text categorization on imbalanced data. *SIGKDD Explorations*, 6(1):80–89, 2004.

[74] C. Zhu, H. Kitagawa, S. Papadimitriou, and C. Faloutsos. OBE: Outlier by example, *PAKDD Conference*, 36(2):217–247, 2004.

[75] C. Zhu, H. Kitagawa, and C. Faloutsos. Example-based robust outlier detection in high dimensional datasets, *ICDM Conference*, 2005.

[76] http://www.itl.nist.gov/iad/mig/tests/tdt/tasks/fsd.html

Chapter 18

Distance Metric Learning for Data Classification

Fei Wang

IBM T. J. Watson Research Center
Yorktown Heights, NY
fwang@us.ibm.com

18.1 Introduction

Distance metric learning is a fundamental problem in data mining and knowledge discovery, and it is of key importance for many real world applications. For example, information retrieval utilizes the learned distance metric to measure the relevance between the candidate data and the query; clinical decision support uses the learned distance metric to measure pairwise patient similarity [19, 23, 24]; pattern recognition can use the learned distance metric to match most similar patterns. In a broader sense, distance metric learning lies in the heart of many data classification problems. As long as a proper distance metric is learned, we can always adopt *k-Nearest Neighbor* (*kNN*) classifier [4] to classify the data. In recent years, many studies have demonstrated [12, 27, 29], either theoretically or empirically, that learning a good distance metric can greatly improve the performance of data classification tasks.

TABLE 18.1: The Meanings of Various Symbols That Will be Used Throughout This Chapter

Symbol	Meaning
n	number of data
d	data dimensionality
$\mathbf{x}_i$	the i-th data vector
$\mathbf{X}$	data matrix
$\mathbf{M}$	precision matrix of the generalized Mahalanobis distance
$\mathbf{w}_i$	the i-th projection vector
$\mathbf{W}$	projection matrix
$\mathcal{N}_i$	the neighborhood of $\mathbf{x}_i$
$\phi(\cdot)$	nonlinear mapping used in kernel methods
$\mathbf{K}$	kernel matrix
$\mathbf{L}$	Laplacian matrix

In this survey, we will give an overview of the existing supervised distance metric learning approaches and point out their strengths and limitations, as well as present challenges and future research directions. We focus on supervised algorithms because they are under the same setting as data classification. We will categorize those algorithms from the aspect of linear/nonlinear, local/global, transductive/inductive, and also the computational technology involved.

In the rest of this chapter, we will first introduce the definition of distance metric in Section 18.2. Then we will overview existing supervised metric learning algorithms in Section 18.3, followed by discussions and conclusions in Section 18.5. Table 18.1 summarizes the notations and symbols that will be used throughout the paper.

18.2 The Definition of Distance Metric Learning

Before describing different types of distance metric learning algorithms, we first define necessary notations and concepts on distance metric learning.

Throughout the paper, we use X to represent a set of data points. If $\mathbf{x}, \mathbf{y}, \mathbf{z} \in X$ are data vectors with the same dimensionality, we call $\mathcal{D} : X \times X \to \mathbb{R}$ a **Distance Metric** if it satisfies the following four properties:[1]

- Nonnegativity: $\mathcal{D}(\mathbf{x}, \mathbf{y}) \geqslant 0$

- Coincidence: $\mathcal{D}(\mathbf{x}, \mathbf{y}) = 0$ if and only if $\mathbf{x} = \mathbf{y}$

- Symmetry: $\mathcal{D}(\mathbf{x}, \mathbf{y}) = \mathcal{D}(\mathbf{y}, \mathbf{x})$

- Subadditivity: $\mathcal{D}(\mathbf{x}, \mathbf{y}) + \mathcal{D}(\mathbf{y}, \mathbf{z}) \geqslant \mathcal{D}(\mathbf{x}, \mathbf{z})$

If we relax the coincidence condition to *if* $\mathbf{x} = \mathbf{y} \Rightarrow \mathcal{D}(\mathbf{x}, \mathbf{y}) = 0$, then $\mathcal{D}$ is called a **Pseudo Metric**. There are many well-known distance metrics, such as the Euclidean and Mahalanobis distance below:

---

[1]http://en.wikipedia.org/wiki/Metric_(mathematics)

- *Euclidean distance*, which measures the distance between **x** and **y** by

$$\mathcal{D}(\mathbf{x}, \mathbf{y}) = \sqrt{(\mathbf{x} - \mathbf{y})^\top (\mathbf{x} - \mathbf{y})}. \tag{18.1}$$

- *Mahalanobis distance*,[2] which measures the distance between **x** and **y** by

$$\mathcal{D}(\mathbf{x}, \mathbf{y}) = \sqrt{(\mathbf{x} - \mathbf{y})^\top \mathbf{S}(\mathbf{x} - \mathbf{y})} \tag{18.2}$$

where **S** is the inverse of the data covariance matrix (also referred to as the precision matrix).[3]

Most of the recent distance metric learning algorithms can be viewed as learning a *generalized Mahalanobis distance* defined as below:

Generalized Mahalanobis Distance (GMD). *A GMD measures the distance between data vectors* **x** *and* **y** *by*

$$\mathcal{D}(\mathbf{x}, \mathbf{y}) = \sqrt{(\mathbf{x} - \mathbf{y})^\top \mathbf{M}(\mathbf{x} - \mathbf{y})} \tag{18.3}$$

where **M** *is some arbitrary* Symmetric Positive Semi-Definite *(SPSD) matrix.*

The major goal of learning a GMD is to learn a proper **M**. As **M** is SPSD, we can decompose **M** as $\mathbf{M} = \mathbf{U}\Lambda\mathbf{U}^\top$ with eigenvalue decomposition, where **U** is a matrix collecting all eigenvectors of **M**, and Λ is a diagonal matrix with all eigenvalues of **M** on its diagonal line. Let $\mathbf{W} = \mathbf{U}\Lambda^{1/2}$, then we have

$$\begin{aligned}
\mathcal{D}(\mathbf{x}, \mathbf{y}) &= \sqrt{(\mathbf{x} - \mathbf{y})^\top \mathbf{W}\mathbf{W}^\top(\mathbf{x} - \mathbf{y})} = \sqrt{(\mathbf{W}^\top(\mathbf{x} - \mathbf{y}))^\top(\mathbf{W}^\top(\mathbf{x} - \mathbf{y}))} \\
&= \sqrt{(\widetilde{\mathbf{x}} - \widetilde{\mathbf{y}})^\top(\widetilde{\mathbf{x}} - \widetilde{\mathbf{y}})}
\end{aligned} \tag{18.4}$$

where $\widetilde{\mathbf{x}} = \mathbf{W}^\top \mathbf{x}$. Therefore GMD is equivalent to the Euclidean distance in some projected space. Based on this observation, we define distance metric learning as follows:

Distance Metric Learning. *The problem of learning a distance function* $\mathcal{D}$ *for a pair of data points* **x** *and* **y** *is to learn a mapping function* f, *such that* $f(\mathbf{x})$ *and* $f(\mathbf{y})$ *will be in the Euclidean space and* $\mathcal{D}(\mathbf{x}, \mathbf{y}) = \|f(\mathbf{x}) - f(\mathbf{y})\|$, *where* $\|\cdot\|$ *is the* ℓ_2 *norm.*

With this definition, we can also categorize whether a distance metric learning algorithm is **linear** or **nonlinear** based on whether the projection is linear or nonlinear.

18.3 Supervised Distance Metric Learning Algorithms

In this section we will review the existing supervised distance metric learning algorithms, which learn distance metrics on both data points and their supervision information, which is usually in the form of (1) *Data labels*, which indicate the class information each training data point belongs to. The assumption is that distance between data points with the same label should be closer to distance between data points from different labels. (2) *Pairwise constraints* indicate whether a pair of data points should belong to the same class (*must-links*) or not (*cannot-links*). Before going into the details of those algorithms, we first categorize those supervised methodologies into different types according to their characteristics, which is shown in Table 18.2.

---

[2]http://en.wikipedia.org/wiki/Mahalanobis_distance
[3]http://en.wikipedia.org/wiki/Covariance_matrix

TABLE 18.2: Supervised Distance Metric Learning Algorithms

	Linear	Nonlinear	Local	Global	ED	QP	GD	Other Optimization
NCA [7]	$\checkmark$		$\checkmark$				$\checkmark$	
ANMM [25]	$\checkmark$		$\checkmark$		$\checkmark$			
KANMM [25]		$\checkmark$	$\checkmark$		$\checkmark$			
LMNN [26]	$\checkmark$		$\checkmark$			$\checkmark$		
KLMNN [26]		$\checkmark$	$\checkmark$			$\checkmark$		
LDA [6]	$\checkmark$			$\checkmark$	$\checkmark$			
KLDA [14]		$\checkmark$		$\checkmark$	$\checkmark$			
LSI [28]	$\checkmark$			$\checkmark$			$\checkmark$	
ITML [3]	$\checkmark$			$\checkmark$				$\checkmark$
MMDA [11]	$\checkmark$			$\checkmark$		$\checkmark$		
KMMDA [11]		$\checkmark$		$\checkmark$		$\checkmark$		
RCA [18]	$\checkmark$			$\checkmark$	$\checkmark$			
KRCA [20]		$\checkmark$		$\checkmark$	$\checkmark$			

Note: ED stands for Eigenvalue Decomposition, QP stands for Quadratic Programming, GD stands for Gradient Descent, and all other abbreviations can be found in the main text.

18.3.1   Linear Discriminant Analysis (LDA)

Linear Discriminant Analysis (LDA) [6] is one of the most popular supervised linear embedding methods. It seeks for the projection directions under which the data from different classes are well separated. More concretely, supposing that the data set belongs to C different classes, LDA defines the *compactness matrix* and *scatterness matrix* as

$$\Sigma_C = \frac{1}{C}\sum_c \frac{1}{n_c}\sum_{\mathbf{x}_i \in c}(\mathbf{x}_i - \bar{\mathbf{x}}_c)(\mathbf{x}_i - \bar{\mathbf{x}}_c)^\top \tag{18.5}$$

$$\Sigma_S = \frac{1}{C}\sum_c (\bar{\mathbf{x}}_c - \bar{\mathbf{x}})(\bar{\mathbf{x}}_c - \bar{\mathbf{x}})^\top. \tag{18.6}$$

The goal of LDA is to find a $\mathbf{W}$ that maximizes

$$\min_{\mathbf{W}^\top \mathbf{W}=\mathbf{I}} \frac{tr(\mathbf{W}^\top \Sigma_C \mathbf{W})}{tr(\mathbf{W}^\top \Sigma_S \mathbf{W})}. \tag{18.7}$$

By expanding the numerator and denominator of the above expression, we can observe that the numerator corresponds to the sum of distances between each data point to its class center after projection, and the denominator represents the sum of distances between every class center to the entire data center after projection. Therefore, minimizing the objective will maximize the between-class scatterness while minimizing the within-class scatterness after projection. Solving problem (18.7) directly is hard, some researchers [8, 10] have done research on this topic. LDA is a linear and global method. The learned distance between $\mathbf{x}_i$ and $\mathbf{x}_j$ is the Euclidean distance between $\mathbf{W}^\top \mathbf{x}_i$ and $\mathbf{W}^\top \mathbf{x}_j$.

Kernelization of LDA: Similar to the case of PCA, we can extend LDA to the nonlinear case via the kernel trick, which is called *Kernel Discriminant Analysis (KDA)) [14]*. After mapping the data into the feature space using ϕ, we can compute the compactness and scatterness matrices as

$$\Sigma_C^\phi = \frac{1}{C}\sum_c \frac{1}{n_c}\sum_{\mathbf{x}_i \in c}(\phi(\mathbf{x}_i) - \bar{\phi}_c)(\phi(\mathbf{x}_i) - \bar{\phi}_c)^\top \tag{18.8}$$

$$\Sigma_S = \frac{1}{C}\sum_c (\bar{\phi}_c - \bar{\phi})(\bar{\phi}_c - \bar{\phi})^\top. \tag{18.9}$$

Suppose the projection matrix we want to get is $\mathbf{W}^\phi$ in the feature space, then with the representer theorem

$$\mathbf{W}^\phi = \Phi\alpha \tag{18.10}$$

where $\Phi = [\phi(\mathbf{x}_1), \phi(\mathbf{x}_2), \cdots, \phi(\mathbf{x}_n)]$ and α is the coefficient vector over all $\phi(\mathbf{x}_i)$ for $1 \le i \le n$. We define $\mathbf{K} = \Phi^\top\Phi$ as the kernel matrix.

Then

$$
\begin{aligned}
(\mathbf{W}^\phi)^\top \Sigma_C^\phi \mathbf{W}^\phi &= \alpha^\top \left[\frac{1}{C} \sum_{c=1}^C \frac{1}{n_c} \sum_{\mathbf{x}_i \in c} \Phi^\top (\phi(\mathbf{x}_i) - \bar{\phi}_c)(\phi(\mathbf{x}_i) - \bar{\phi}_c)^\top \Phi \right] \alpha \\
&= \alpha^\top \left[\frac{1}{C} \sum_{c=1}^C \frac{1}{n_c} \sum_{\mathbf{x}_i \in c} (\mathbf{K}_{\cdot i} - \bar{\mathbf{K}}_{\cdot c})(\mathbf{K}_{\cdot i} - \bar{\mathbf{K}}_{\cdot c})^\top \right] \alpha \\
&= \alpha^\top \mathbf{M}_C \alpha
\end{aligned}
\tag{18.11}
$$

where $\mathbf{K}_{\cdot i} =$ and $\bar{\mathbf{K}}_{\cdot c} = \frac{1}{n_c} \sum_{\mathbf{x}_i \in c} \mathbf{K}_{\cdot i}$.

$$
\begin{aligned}
(\mathbf{W}^\phi)^\top \Sigma_S^\phi \mathbf{W}^\phi &= \alpha^\top \left[\frac{1}{C} \sum_c (\bar{\mathbf{K}}_{\cdot c} - \bar{\mathbf{K}}_{\cdot *})(\bar{\mathbf{K}}_{\cdot c} - \bar{\mathbf{K}}_{\cdot *})^\top \right] \alpha \\
&= \alpha^\top \mathbf{M}_S \alpha
\end{aligned}
\tag{18.12}
$$

where $\mathbf{K}_{\cdot *} = \frac{1}{n} \sum_{i=1}^n \mathbf{K}_{\cdot i}$. Therefore we can get α by solving

$$\min_{\alpha^\top\alpha = \mathbf{I}} \frac{tr\left(\alpha^\top \mathbf{M}_C \alpha\right)}{tr\left(\alpha^\top \mathbf{M}_S \alpha\right)}. \tag{18.13}$$

18.3.2 Margin Maximizing Discriminant Analysis (MMDA)

MMDA can be viewed as an extension of the *Support Vector Machine* (SVM) algorithm [21]. Supposing data points $\mathbf{X} = [\mathbf{x}_1, \mathbf{x}_2, \cdots, \mathbf{x}_n]$ come from two classes, the label of $\mathbf{x}_i$ is $l_i \in \{-1, +1\}$. As a recap, the goal of SVM is to find the maximum-margin hyperplane that divides the points with $l_i = 1$ from those with $l_i = -1$ (thus it is a supervised method). Any hyperplane can be written as a point $\mathbf{x}$ satisfying

$$\mathbf{w}^\top\mathbf{x} - b = 0 \tag{18.14}$$

where $b/\|\mathbf{w}\|$ corresponds to the distance of the hyperplane from the origin. SVM aims to choose the $\mathbf{w}$ and b to maximize the distance between the parallel hyperplanes that are as far apart as possible while still separating the data, which is usually referred to as the margin of the two classes. These hyperplanes can be described by the equations

$$\mathbf{w}^\top\mathbf{x} - b = 1 \quad \text{or} \quad \mathbf{w}^\top\mathbf{x} - b = -1. \tag{18.15}$$

The distance between the two parallel hyperplanes is $2/\|\mathbf{w}\|$. Then if the data from two classes are clearly separated, the goal of SVM is to solve the following optimization problem to find the hyperplane that maximizes the margin between two classes

$$\min_{\mathbf{w},b} \quad \frac{1}{2}\|\mathbf{w}\|^2 \tag{18.16}$$

$$s.t. \quad l_i(\mathbf{w}^\top\mathbf{x}_i - b) \ge 1 \quad (\forall i = 1, 2, \cdots, n).$$

However in reality the two classes may not be perfectly separable, i.e., there might be some overlapping between them. Then we need *soft margin* SVM, which aims at solving

$$\min_{\mathbf{w},b,\xi} \quad \frac{1}{2}\|\mathbf{w}\|^2 + C\sum_{i=1}^{n}\xi_i \tag{18.17}$$

$$s.t. \quad l_i(\mathbf{w}^\top \mathbf{x}_i - b) \geqslant 1 - \xi_i \quad (\forall i = 1,2,\cdots,n)$$

where $\{\xi_i\} \geqslant 0$ are slack variables used to penalize the margin on the overlapping region.

MMDA aims to solve more than one projection directions, which aims to solve the following optimization problem

$$\min_{\mathbf{W},b,\xi_r \geq 0} \quad \frac{1}{2}\sum_{r=1}^{d}\|\mathbf{w}_r\|^2 + \frac{C}{n}\sum_{r=1}^{d}\sum_{i=1}^{n}\xi_{ri} \tag{18.18}$$

$$s.t. \quad \forall i = 1,\ldots,n, \quad r = 1,\ldots,d$$

$$l_i\left((\mathbf{w}^r)^T \mathbf{x}_i + b\right) \geq 1 - \xi_{ri},$$

$$\mathbf{W}^T\mathbf{W} = \mathbf{I}.$$

Therefore MMDA is a global and linear approach. One can also apply the kernel trick to make it nonlinear; the details can be found in [11]. The learned distance between $\mathbf{x}_i$ and $\mathbf{x}_j$ is just the Euclidean distance between $\mathbf{W}^\top\mathbf{x}_i$ and $\mathbf{W}^\top\mathbf{x}_j$.

18.3.3 Learning with Side Information (LSI)

Both LDA and MMDA use data labels as the supervision information. As we introduced at the beginning of Section 18.3, another type of supervision information we considered is pairwise constraints. The data label information is more strict in the sense that we can convert data labels into pairwise constraints, but not vice versa.

One of the earliest researches that makes use of pairwise constraints for learning a proper distance metric is the *Learning with Side-Information (LSI)* approach [28]. We denote the set of *must-link* constraints as $\mathcal{M}$ and the set of *cannot-link* constraints as $\mathcal{C}$; then the goal of LSI is to solve the following optimization problem

$$\max_{\mathbf{M}} \sum_{(\mathbf{x}_i,\mathbf{x}_j)\in\mathcal{C}} (\mathbf{x}_i - \mathbf{x}_j)^\top \mathbf{M}(\mathbf{x}_i - \mathbf{x}_j) \tag{18.19}$$

$$s.t. \quad \sum_{(\mathbf{x}_u,\mathbf{x}_v)\in\mathcal{M}} (\mathbf{x}_u - \mathbf{x}_v)^\top \mathbf{M}(\mathbf{x}_u - \mathbf{x}_v) \leqslant 1$$

$$\mathbf{M} \succeq 0.$$

This is a quadratic optimization problem and [28] proposed an iterative projected gradient ascent method to solve it. As $\mathbf{M}$ is positive semi-definite, we can always factorize it as $\mathbf{M} = \mathbf{W}\mathbf{W}^\top$. Thus LSI is a global and linear approach. The learned distance formulation is exactly the general Mahalanobis distance with precision matrix $\mathbf{M}$.

18.3.4 Relevant Component Analysis (RCA)

Relevant Component Analysis (RCA) [18] is another representative distance metric learning algorithm utilizing pairwise data constraints. The goal of RCA is to find a transformation that amplifies *relevant variability* and suppresses *irrelevant variability*. We consider that data variability is correlated with a specific task if the removal of this variability from the data deteriorates (on average) the results of clustering or retrieval. Variability is irrelevant if it is maintained in the data but not correlated with the specific task [18]. We also define small clusters called *chunklets*, which are connected components derived by all the must-links. The specific steps involved in RCA include:

- Construct *chunklets* according to equivalence (must-link) constraints, such that the data in each chunklet are connected by must-link constraints pairwisely.

- Assume a total of p points in k chunklets, where chunklet j consists of points $\{\mathbf{x}_{ji}\}_{i=1}^{n_j}$ and its mean is $\bar{\mathbf{m}}_j$. RCA computes the following weighted within-chunklet covariance matrix:

$$\mathbf{C} = \frac{1}{p}\sum_{j=1}^{k}\sum_{i=1}^{n_j}(\mathbf{x}_{ji}-\bar{\mathbf{m}}_j)(\mathbf{x}_{ji}-\bar{\mathbf{m}}_j)^{\top}. \tag{18.20}$$

- Compute the whitening transformation $\mathbf{W}=C^{1/2}$, and apply it to the original data points: $\tilde{\mathbf{x}} = \mathbf{W}\mathbf{x}$. Alternatively, use the inverse of $\mathbf{C}$ as the precision matrix of a generalized Mahalanobis distance.

Therefore, RCA is a global, linear approach.

18.3.5 Information Theoretic Metric Learning (ITML)

Information theoretic objective is one mechanism to develop a supervised distance metric. *Information Theoretic Metric Learning (ITML)* [3] is one such representative algorithm. Suppose we have an initial generalized Mahalanobis distance parameterized by precision matrix $\mathbf{M}_0$, a set $\mathcal{M}$ of must-link constraints, and a set $\mathcal{C}$ of cannot-link constraints. ITML solves the following optimization problem

$$\min_{\mathbf{M}\succeq 0} \quad d_{logdet}(\mathbf{M},\mathbf{M}_0) \tag{18.21}$$
$$s.t. \quad (\mathbf{x}_i-\mathbf{x}_j)^{\top}\mathbf{M}(\mathbf{x}_i-\mathbf{x}_j) \geqslant l, \ (\mathbf{x}_i,\mathbf{x}_j) \in \mathcal{C}$$
$$(\mathbf{x}_u-\mathbf{x}_v)^{\top}\mathbf{M}(\mathbf{x}_u-\mathbf{x}_v) \leqslant u, \ (\mathbf{x}_u,\mathbf{x}_v) \in \mathcal{M}$$

where

$$d_{logdet}(\mathbf{M},\mathbf{M}_0) = tr(\mathbf{M}\mathbf{M}_0^{-1}) - \log det(\mathbf{M}\mathbf{M}_0^{-1}) - n \tag{18.22}$$

where d_{logdet} is the LogDet divergence, which is also known as Stein's loss. It can be shown that Stein's loss is the unique scale invariant loss-function for which the uniform minimum variance unbiased estimator is also a minimum risk equivariant estimator [3]. The authors in [3] also proposed an efficient *Bregman projection* approach to solve problem (18.21). ITML is a global and linear approach. The learned distance metric is the Mahalanobis distance with precision matrix $\mathbf{M}$.

18.3.6 Neighborhood Component Analysis (NCA)

All the supervised approaches we introduced above are global methods. Next we will also introduce several representative local supervised metric learning algorithms. First we will overview *Neighborhood Component Analysis (NCA)* [7]. Similar as in SNE, described in unsupervised metric learning, each point $\mathbf{x}_i$ selects another point $\mathbf{x}_j$ as its neighbor with some probability p_{ij}, and inherits its class label from the point it selects. NCA defines the probability that point i selects point j as a neighbor:

$$p_{ij} = \frac{\exp\left(-\|\mathbf{W}^{\top}\mathbf{x}_i - \mathbf{W}^{\top}\mathbf{x}_j\|^2\right)}{\sum_{k\neq i}\exp\left(-\|\mathbf{W}^{\top}\mathbf{x}_i - \mathbf{W}^{\top}\mathbf{x}_k\|^2\right)}. \tag{18.23}$$

Under this stochastic selection rule, NCA computes the probability that point i will be correctly classified

$$p_i = \sum_{j\in L_i} p_{ij} \tag{18.24}$$

where $\mathcal{L}_i = \{j | l_i = l_j\})$ that is the set of points in the same class as point i.

The objective NCA maximizes is the expected number of points correctly classified under this scheme:

$$\mathcal{J}(\mathbf{W}) = \sum_i p_i = \sum_i \sum_{j \in \mathcal{L}_i} p_{ij} \qquad (18.25)$$

[7] proposed a truncated gradient descent approach to minimize $\mathcal{J}(\mathbf{W})$. NCA is a local and linear approach. The learned distance between $\mathbf{x}_i$ and $\mathbf{x}_j$ is the Euclidean distance between $\mathbf{W}^\top \mathbf{x}_i$ and $\mathbf{W}^\top \mathbf{x}_j$.

18.3.7   Average Neighborhood Margin Maximization (ANMM)

Average Neighborhood Margin Maximization (ANMM) [25] is another local supervised metric learning approach, which aims to find projection directions where the local class discriminability is maximized. To define local discriminability, [25] first defines the following two types of neighborhoods:

Definition 1(*Homogeneous Neighborhoods*). For a data point $\mathbf{x}_i$, its ξ *nearest homogeneous neighborhood* $\mathcal{N}_i^o$ is the set of ξ most similar[4] data, which are in the same class with $\mathbf{x}_i$.

Definition 2(*Heterogeneous Neighborhoods*). For a data point $\mathbf{x}_i$, its ζ *nearest heterogeneous neighborhood* $\mathcal{N}_i^e$ is the set of ζ most similar data, which are not in the same class with $\mathbf{x}_i$.

Then the *average neighborhood margin* γ_i for $\mathbf{x}_i$ is defined as

$$\gamma_i = \sum_{k: \mathbf{x}_k \in \mathcal{N}_i^e} \frac{\|\mathbf{y}_i - \mathbf{y}_k\|^2}{|\mathcal{N}_i^e|} - \sum_{j: \mathbf{x}_j \in \mathcal{N}_i^o} \frac{\|\mathbf{y}_i - \mathbf{y}_j\|^2}{|\mathcal{N}_i^o|},$$

where $|\cdot|$ represents the cardinality of a set. This margin measures the difference between the average distance from $\mathbf{x}_i$ to the data points in its heterogeneous neighborhood and the average distance from it to the data points in its homogeneous neighborhood. The maximization of such a margin can push the data points whose labels are different from $\mathbf{x}_i$ away from $\mathbf{x}_i$ while pulling the data points having the same class label with $\mathbf{x}_i$ towards $\mathbf{x}_i$.

Therefore, the total *average neighborhood margin* can be defined as

$$\gamma = \sum_i \gamma_i = \sum_i \left(\sum_{k: \mathbf{x}_k \in \mathcal{N}_i^e} \frac{\|\mathbf{y}_i - \mathbf{y}_k\|^2}{|\mathcal{N}_i^e|} - \sum_{j: \mathbf{x}_j \in \mathcal{N}_i^o} \frac{\|\mathbf{y}_i - \mathbf{y}_j\|^2}{|\mathcal{N}_i^o|} \right) \qquad (18.26)$$

and the *ANMM criterion* is to maximize γ. By replacing $\mathbf{y}_i = \mathbf{W}^\top \mathbf{x}_i$, [25] obtains the optimal $\mathbf{W}$ by performing eigenvalue decomposition of some discriminability matrix. Thus ANMM is a local and linear approach. The learned distance between $\mathbf{x}_i$ and $\mathbf{x}_j$ is the Euclidean distance between $\mathbf{W}^\top \mathbf{x}_i$ and $\mathbf{W}^\top \mathbf{x}_j$. The authors in [25] also proposed a kernelized version of ANMM to handle nonlinear data called KANMM, thus KANMM is a local and nonlinear approach.

18.3.8   Large Margin Nearest Neighbor Classifier (LMNN)

The last local supervised metric learning approach we want to introduce is the *Large Margin Nearest Neighbor Classifier (LMNN)* [26]. The goal of LMNN is similar to that of ANMM, i.e., to pull the data with the same labels closer while pushing data with different labels far apart. LMNN

---

[4]In this paper two data vectors are considered to be similar if the Euclidean distance between them is small; two data tensors are considered to be similar if the Frobenius norm of their difference tensor is small.

deploys a different margin formulation. Specifically, LMNN defines the pull energy term as

$$\varepsilon_{pull} = \sum_{\mathbf{x}_j \in \mathcal{N}_i^{lo}} \left\| \mathbf{W}^\top (\mathbf{x}_i - \mathbf{x}_j) \right\|^2 \tag{18.27}$$

which is the sum of pairwise distances between a data point $\mathbf{x}_i$ and the data point in $\mathbf{x}_i$'s homogeneous neighborhood after projection. LMNN defines the push energy as

$$\varepsilon_{push} = \sum_i \sum_{\mathbf{x}_j \in \mathcal{N}_i^{lo}} \sum_l (1 - \delta_{il}) \left[1 + \left\| \mathbf{W}^\top (\mathbf{x}_i - \mathbf{x}_j) \right\|^2 - \left\| \mathbf{W}^\top (\mathbf{x}_i - \mathbf{x}_l) \right\|^2 \right]_+ \tag{18.28}$$

where $\delta_{il} = 1$ is the labels of $\mathbf{x}_i$ and $\mathbf{x}_l$ are the same, and $\delta_{il} = 0$ otherwise. The intuition is to require that the data points from different classes should be at least separated from it by the distance 1. This formulation is very similar to the margin formulation in multiclass SVM [2] LMNN also pushes the data with different labels to at least distance 1 from its homogeneous neighborhood. The goal of LMNN is to minimize

$$\varepsilon = \mu \varepsilon_{pull} + (1 - \mu) \varepsilon_{push}. \tag{18.29}$$

The authors in [26] proposed a semi-definite programming technique to solve for $\mathbf{M} = \mathbf{W}\mathbf{W}^\top$. Thus LMNN is a local and linear approach. The learned distance between $\mathbf{x}_i$ and $\mathbf{x}_j$ is the Euclidean distance between $\mathbf{W}^\top \mathbf{x}_i$ and $\mathbf{W}^\top \mathbf{x}_j$.

18.4 Advanced Topics

Although supervised metric learning has successfully been applied in many applications, it is human labor intensive and time consuming to get the supervision information. Also most of the above mentioned approaches require eigenvalue decomposition or quadratic optimization. In this section we will briefly review two advanced topics in metric learning: semi-supervised approaches and online learning.

18.4.1 Semi-Supervised Metric Learning

Semi-supervised approaches aim to learn a proper distance metric from the data where the supervision information is only available on a small portion of the data. Those algorithms utilize both data with and without supervision information in the learning process. Therefore one straightforward way one can think of to construct a semi-supervised algorithm is to deploy an objective in the form of $\lambda_1 \mathcal{L}(\mathcal{D}) + \lambda_2 \mathcal{U}(\mathcal{D})$, where $\mathcal{U}(\mathcal{D})$ is constructed on the entire data, and $\mathcal{L}(\mathcal{D})$ is constructed on the data with supervision information only. Finally, we put some constraints on the learned distance metric to balance both parts.

18.4.1.1 Laplacian Regularized Metric Learning (LRML)

Laplacian Regularized Metric Learning (*LMRL*) [9] is one semi-supervised distance metric learning approach. LRML adopts data smoothness term as the unsupervised term $\mathcal{U}(\mathcal{D})$; in terms of the supervised term, LRML chooses an ANMM type of objective as $\mathcal{L}(\mathcal{D})$. The optimization problem that LRML aims to solve is

$$\min_{\mathbf{M}} \underbrace{t}_{\mathcal{U}(\mathcal{D})} + \underbrace{\gamma_1 t_2 - \gamma_2 t_3}_{\mathcal{L}(\mathcal{D})} \qquad (18.30)$$

$$s.t. \quad t_1 \leqslant t$$
$$\mathbf{M} \succeq 0$$

where the smoothness term is

$$t_1 = \sum_{i,j} \|\mathbf{W}^\top \mathbf{x}_i - \mathbf{W}^\top \mathbf{x}_j\|^2 \omega_{ij} = tr(\mathbf{W}^\top \mathbf{X} \mathbf{L} \mathbf{X}^\top \mathbf{W}) = tr(\mathbf{X} \mathbf{L} \mathbf{X}^\top \mathbf{M}) \qquad (18.31)$$

where $\mathbf{M} = \mathbf{W} \mathbf{W}^\top$. The supervised terms consisting of compactness and scatterness are

$$t_2 = \sum_{(\mathbf{x}_i, \mathbf{x}_j) \in \mathcal{M}} \|\mathbf{W}^\top \mathbf{x}_i - \mathbf{W}^\top \mathbf{x}_j\|^2 = tr\left[\mathbf{M} \sum_{(\mathbf{x}_i, \mathbf{x}_j) \in \mathcal{M}} (\mathbf{x}_i - \mathbf{x}_j)(\mathbf{x}_i - \mathbf{x}_j)^\top\right] \qquad (18.32)$$

$$t_3 = \sum_{(\mathbf{x}_i, \mathbf{x}_j) \in \mathcal{C}} \|\mathbf{W}^\top \mathbf{x}_i - \mathbf{W}^\top \mathbf{x}_j\|^2 = tr\left[\mathbf{M} \sum_{(\mathbf{x}_i, \mathbf{x}_j) \in \mathcal{C}} (\mathbf{x}_i - \mathbf{x}_j)(\mathbf{x}_i - \mathbf{x}_j)^\top\right] \qquad (18.33)$$

where $\mathcal{M}$ and $\mathcal{C}$ are the sets of must-links and cannot-links, respectively.

[9] proposed a semi-definite programming approach for solving problem (18.30). LRML is a mixture of local (its unsupervised part) and global (its supervised part) approach, and it is linear. The learned distance is the Mahalanobis distance with precision matrix $\mathbf{M}$.

18.4.1.2 Constraint Margin Maximization (CMM)

Similarly, *Constraint Margin Maximization (CMM)* [22] selects the same supervised term as LRML, but a different PCA-type unsupervised term in its objective. Specifically, the optimization problem CMM aims to solve is

$$\max_{\mathbf{W}} \underbrace{t_4}_{\mathcal{U}(\mathcal{D})} - \underbrace{\gamma_1 t_2 - \gamma_2 t_3}_{\mathcal{L}(\mathcal{D})} \qquad (18.34)$$

$$s.t. \quad \mathbf{W}^\top \mathbf{W} = \mathbf{I}$$

where the unsupervised term is $t_4 = tr(\mathbf{W}^\top \mathbf{X} \Sigma \mathbf{X}^\top \mathbf{W})$ is the PCA-like term. Note that before we apply CMM, all data points need to be centered, i.e., their mean should be subtracted from the data matrix. The intuition of CMM is to maximally unfold the data points in the projection space while at the same time satisfying those pairwise constraints. The authors in [22] showed that the optimal $\mathbf{W}$ can be obtained by eigenvalue decomposition to some matrix. Therefore CMM is a global and linear approach. Wang et. al. [22] also showed how to derive its kernelized version for handling nonlinear data. The learned distance between $\mathbf{x}_i$ and $\mathbf{x}_j$ is the Euclidean distance between $\mathbf{W}^\top \mathbf{x}_i$ and $\mathbf{W}^\top \mathbf{x}_j$.

18.4.2 Online Learning

Most of the distance metric learning approaches involve expensive optimization procedures such as eigen-decomposition and semi-definite programming. One way to make those algorithms more efficient is the *online learning* [16] strategy, which incorporates the data points into the learning process in a sequential manner. More concretely, online learning updates the learned distance metric iteratively. At each iteration, only one or a small batch of data are involved. Another scenario where the online learning strategy can be naturally applied is to learn distance metrics for streaming data, where the data are coming in a streaming fashion so that the distance metric needs to be updated iteratively. Next we present two examples of online distance metric learning approaches.

18.4.2.1 Pseudo-Metric Online Learning Algorithm (POLA)

Pseudo-Metric Online Learning (*POLA*) [17] falls into the category of supervised metric learning with pairwise constraints. More specifically, there is a must-link constraint set $\mathcal{M}$ and a cannot-link constraint set $\mathcal{C}$. POLA assigns a label l_{ij} for each pair $(\mathbf{x}_i, \mathbf{x}_j)$ in $\mathcal{M}$ or $\mathcal{C}$, such that if $(\mathbf{x}_i, \mathbf{x}_j) \in \mathcal{M}$, $l_{ij} = 1$; if $(\mathbf{x}_i, \mathbf{x}_j) \in \mathcal{C}$, $l_{ij} = -1$. Then it introduces a threshold b and constructs the following constraints

$$\forall (\mathbf{x}_i, \mathbf{x}_j) \in \mathcal{M}, \, l_{ij} = 1 \quad \Rightarrow \quad (\mathbf{x}_i - \mathbf{x}_j)^\top \mathbf{M}(\mathbf{x}_i - \mathbf{x}_j) \leqslant b - 1 \tag{18.35}$$

$$\forall (\mathbf{x}_i, \mathbf{x}_j) \in \mathcal{C}, \, l_{ij} = -1 \quad \Rightarrow \quad (\mathbf{x}_i - \mathbf{x}_j)^\top \mathbf{M}(\mathbf{x}_i - \mathbf{x}_j) \geqslant b + 1 \tag{18.36}$$

which can be unified as

$$l_{ij}[b - (\mathbf{x}_i - \mathbf{x}_j)^\top \mathbf{M}(\mathbf{x}_i - \mathbf{x}_j)] \geqslant 1 \tag{18.37}$$

Note that this formulation is similar to the constraint in standard SVM. Then the objective of POLA is the follow hinge loss

$$\mathcal{I}_{ij}(\mathbf{M}, b) = \max_{\mathcal{C}^1_{ij} \& \mathcal{C}^2} (0, l_{ij}[(\mathbf{x}_i - \mathbf{x}_j)^\top \mathbf{M}(\mathbf{x}_i - \mathbf{x}_j) - b] + 1). \tag{18.38}$$

The two constraint sets are defined as

$$\mathcal{C}^1_{ij} = \left\{ (\mathbf{M}, b) \in \mathbb{R}^{n^2 \times 1} : \mathcal{I}_{ij}(\mathbf{M}, b) = 0 \right\} \tag{18.39}$$

$$\mathcal{C}^2 = \left\{ (\mathbf{M}, b) \in \mathbb{R}^{n^2 \times 1} : \mathbf{M} \succeq 0, \, b \geqslant 1 \right\} \tag{18.40}$$

POLA operates in an iterative way: First POLA initializes $\mathbf{M}$ as a zero matrix, then at each step, it randomly picks one data pair from the constraint set (either $\mathcal{M}$ or $\mathcal{C}$), and then does projections on $\mathcal{C}^1_{ij}$ and $\mathcal{C}^2$ alternatively. By projecting $\mathbf{M}$ and b onto $\mathcal{C}^1_{ij}$, POLA gets the updating rules for $\mathbf{M}$ and b as

$$\widehat{\mathbf{M}} = \mathbf{M} - l_{ij} \alpha_{ij} \mathbf{u}_{ij} \mathbf{u}_{ij}^\top \tag{18.41}$$

$$\widehat{b} = b + \alpha_{ij} l_{ij} \tag{18.42}$$

where

$$\mathbf{u}_{ij} = \mathbf{x}_i - \mathbf{x}_j \tag{18.43}$$

$$\alpha_{ij} = \frac{\mathcal{I}_{ij}(\mathbf{M}, b)}{\|\mathbf{u}_{ij}\|_2^4 + 1}. \tag{18.44}$$

By projecting $(\mathbf{M}, b)$ onto $\mathcal{C}^2$, POLA updates $\mathbf{M}$ as

$$\widehat{\mathbf{M}} = \mathbf{M} - \lambda \mu \mu^\top \tag{18.45}$$

where $\lambda = \min\{\tilde{\lambda}, 0\}$ and $(\tilde{\lambda}, \mu)$ are the smallest eigenvalue-eigenvector pair of $\widehat{\mathbf{M}}$. Therefore, POLA incorporates the data in constraint sets in a sequential manner.

18.4.2.2 Online Information Theoretic Metric Learning (OITML)

Another technique we want to review here is the *Online Information Theoretic Metric Learning* (*OITML*) approach [3]. This method also falls into the category of supervised metric learning. It is the online version of the ITML approach we introduced in Section 18.3.5. Suppose at time $t + 1$, we need to randomly pick a pair of data from the constraint set, and minimize the following objective

$$\mathbf{M}_{t+1} = arg \min_{\mathbf{M}} \mathcal{R}(\mathbf{M}, \mathbf{M}_t) + \eta_t \ell(\mathcal{D}_t, \widehat{\mathcal{D}}_t) \tag{18.46}$$

where $\widehat{\mathcal{D}}_t = (\mathbf{x}_i - \mathbf{x}_j)^\top \mathbf{M}(\mathbf{x}_i - \mathbf{x}_j)$ and $\mathcal{R}(\mathbf{M}, \mathbf{M}_t) = d_{logdet}(\mathbf{M}, \mathbf{M}_t)$ is the logdet divergence. [3] showed that $\mathbf{M}_{t+1}$ can be updated with the following rule

$$\mathbf{M}_{t+1} \leftarrow \mathbf{M}_t - \frac{2\eta_t(\mathcal{D}_t - \widehat{\mathcal{D}}_t)\mathbf{M}_t(\mathbf{x}_i - \mathbf{x}_j)(\mathbf{x}_i - \mathbf{x}_j)^\top \mathbf{M}_t}{1 + 2\eta_t(\mathcal{D}_t - \widehat{\mathcal{D}}_t)(\mathbf{x}_i - \mathbf{x}_j)^\top \mathbf{M}_t(\mathbf{x}_i - \mathbf{x}_j)} \tag{18.47}$$

where

$$\eta_t = \left\{ \begin{array}{l} \eta_0, \ \mathcal{D}_t - \widehat{\mathcal{D}}_t \leqslant 0 \\ \min \left\{ \eta_0, \frac{1}{2(\mathcal{D}_t - \widehat{\mathcal{D}}_t)} \left(\frac{1}{(\mathbf{x}_i - \mathbf{x}_j)^\top (\mathbf{I} + (\mathbf{M}_t^{-1} - \mathbf{I})^{-1})(\mathbf{x}_i - \mathbf{x}_j)} \right) \right\} , \ \text{otherwise.} \end{array} \right. \tag{18.48}$$

POLA and OITML are only two examples of online distance metric learning.

18.5 Conclusions and Discussions

Till now we have introduced the state-of-the-art supervised distance metric learning algorithms of recent years. All of those supervised metric learning approaches we listed above require some pre-specified free parameters, and all of them involve some expensive computational procedures such as eigenvalue decomposition or semi-definite programming. One exception is the ITML approach, as it deploys a Bregman projection strategy that may make the solution relatively efficient. However, ITML is sensitive to the initial choice of $\mathbf{M}_0$, which makes it difficult to apply in the case when we do not have enough prior knowledge. In practice, according to the type of supervision information provided, we can select a proper supervised metric learning approach that can handle that supervision information. However, in many real world applications, it may be expensive and time consuming to get that supervision information. Next we survey semi-supervised approaches that can leverage both labeled and unlabeled information. For the future of distance metric learning research, we believe the following directions are promising.

- *Large scale distance metric learning*. Most of the existing distance metric learning approaches involve computationally expensive procedures. How can we make distance metric learning efficient and practical on large-scale data? Promising solutions include online learning or distributed learning. We have introduced the most recent works on online distance metric learning in Section 4.1. For parallel/distributed distance metric algorithms, as the major computational techniques involved are eigenvalue decomposition and quadratic programming, we can adopt parallel matrix computation/optimization algorithms [1,15] to make distance metric learning more scalable and efficient.

- *Empirical evaluations*. Although a lot of distance metric learning algorithms have been proposed, there is still a lack of systematic comparison and proof points on the utility of many distance metric learning algorithms in real world applications. Such empirical discussion will be helpful to showcase the practical value of distance metric learning algorithms. Some recent works have started developing and applying distance metric learning on healthcare for measuring similarity among patients [19, 23, 24].

- *General formulation.* As can be seen from this survey, most of the existing distance metric learning algorithms suppose the learned distance metric is *Euclidean* in the projected space. Such an assumption may not be sufficient for real world applications as there is no guarantee that Euclidean distance is most appropriate to describe the pairwise data relationships. There are already some initial effort towards this direction [5, 13], and this direction is definitely worth exploring.

Bibliography

[1] Yair Censor. *Parallel Optimization: Theory, algorithms, and applications.* Oxford University Press, 1997.

[2] Koby Crammer and Yoram Singer. On the algorithmic implementation of multiclass kernel-based vector machines. *Journal of Machine Learning Research*, 2:265–292, 2001.

[3] Jason V. Davis, Brian Kulis, Prateek Jain, Suvrit Sra, and Inderjit S. Dhillon. Information-theoretic metric learning. In *Proceedings of International Conference on Machine Learning*, pages 209–216, 2007.

[4] Richard O. Duda, Peter E. Hart, and David G. Stork. *Pattern Classification (2nd Edition).* Wiley-Interscience, 2nd edition, 2001.

[5] Charles Elkan. Bilinear models of affinity. Personal note, 2011.

[6] Keinosuke Fukunaga. *Introduction to Statistical Pattern Recognition, Second Edition (Computer Science and Scientific Computing Series).* Academic Press, 1990.

[7] Jacob Goldberger, Sam Roweis, Geoff Hinton, and Ruslan Salakhutdinov. Neighborhood component analysis. In *Advances in Neural Information Processing Systems*, 17:513–520, 2004.

[8] Y. Guo, S. Li, J. Yang, T. Shu, and L. Wu. A generalized foley-sammon transform based on generalized fisher discriminant criterion and its application to face recognition. *Pattern Recognition Letters*, 24(1-3):147–158, 2003.

[9] Steven C. H. Hoi, Wei Liu, and Shih-Fu Chang. Semi-supervised distance metric learning for collaborative image retrieval. In *Proceedings of IEEE Computer Society Conference on Computer Vision and Pattern Recognition*, pages 1–7, 2008.

[10] Yangqing Jia, Feiping Nie, and Changshui Zhang. Trace ratio problem revisited. *IEEE Transactions on Neural Networks*, 20(4):729–735, 2009.

[11] András Kocsor, Kornél Kovács, and Csaba Szepesvári. Margin maximizing discriminant analysis. In *Proceedings of European Conference on Machine Learning*, volume 3201 of *Lecture Notes in Computer Science*, pages 227–238. Springer, 2004.

[12] Brian Kulis. Metric learning. *Tutorial at International Conference on Machine Learning*, 2010.

[13] Zhen Li, Liangliang Cao, Shiyu Chang, John R. Smith, and Thomas S. Huang. Beyond mahalanobis distance: Learning second-order discriminant function for people verification. In *Prcoeedings of Computer Vision and Pattern Recognition Workshops (CVPRW), 2012 IEEE Computer Society Conference on*, pages 45–50, 2012.

[14] S. Mika, G. Ratsch, J. Weston, B. Schölkopf, and K. R. Müllers. Fisher discriminant analysis with kernels. *Neural Networks for Signal Processing IX, 1999. Proceedings of the 1999 IEEE Signal Processing Society Workshop*, pages 41–48, 1999.

[15] Jagdish J. Modi. Parallel algorithms and matrix computation. Oxford University Press, 1989.

[16] Shai Shalev-Shwartz. Online Learning: Theory, Algorithms, and Applications. The Hebrew University of Jerusalem. Phd thesis, July 2007.

[17] Shai Shalev-Shwartz, Yoram Singer, and Andrew Y. Ng. Online and batch learning of pseudo-metrics. In *Proceedings of International Conference on Machine Learning*, pages 94–101, 2004.

[18] N. Shental, T. Hertz, D. Weinshall, and M. Pavel. Adjustment learning and relevant component analysis. In *Proceedings of European Conference on Computer Vision*, pages 776–790, 2002.

[19] Jimeng Sun, Daby Sow, Jianying Hu, and Shahram Ebadollahi. Localized supervised metric learning on temporal physiological data. In *ICPR*, pages 4149–4152, 2010.

[20] Ivor W. Tsang, Pak ming Cheung, and James T. Kwok. Kernel relevant component analysis for distance metric learning. In *IEEE International Joint Conference on Neural Networks (IJCNN*, pages 954–959, 2005.

[21] Vladimir N. Vapnik. *The Nature of Statistical Learning Theory*. Springer New York Inc., New York, NY, 1995.

[22] Fei Wang, Shouchun Chen, Changshui Zhang, and Tao Li. Semi-supervised metric learning by maximizing constraint margin, *CIKM Conference*, pages 1457–1458, 2008.

[23] Fei Wang, Jimeng Sun, and Shahram Ebadollahi. Integrating distance metrics learned from multiple experts and its application in patient similarity assessment. In *SDM*, 2011.

[24] Fei Wang, Jimeng Sun, Jianying Hu, and Shahram Ebadollahi. imet: Interactive metric learning in healthcare applications. In *SDM*, 2011.

[25] Fei Wang and Changshui Zhang. Feature extraction by maximizing the average neighborhood margin. In *Proceedings of IEEE Computer Society Conference on Computer Vision and Pattern Recognition*, 2007.

[26] Kilian Q. Weinberger, John Blitzer, and Lawrence K. Saul. Distance metric learning for large margin nearest neighbor classification. In *Advances in Neural Information Processing Systems*, 2005.

[27] Michael Werman, Ofir Pele, and Brian Kulis. Distance functions and metric learning. *Tutorial at European Conference on Computer Vision*, 2010.

[28] Eric P. Xing, Andrew Y. Ng, Michael I. Jordan, and Stuart Russell. Distance metric learning, with application to clustering with side-information. In *Advances in Neural Information Processing Systems 15*, 15:505–512, 2002.

[29] Liu Yang and Rong Jin. Distance metric learning: A comprehensive survey. Technical report, Department of Computer Science and Engineering, Michigan State University, 2006.

Chapter 19

Ensemble Learning

Yaliang Li

State University of New York at Buffalo
Buffalo, NY
`yaliangl@buffalo.edu`

Jing Gao

State University of New York at Buffalo
Buffalo, NY
`jing@buffalo.edu`

Qi Li

State University of New York at Buffalo
Buffalo, NY
`qli22@buffalo.edu`

Wei Fan

Huawei Noah's Ark Lab
Hong Kong
`david.weifan@huawei.com`

19.1   Introduction

People have known the benefits of combining information from multiple sources since a long time ago. From the old story of "blind men and an elephant," we can learn the importance of capturing a big picture instead of focusing on one single perspective. In the story, a group of blind men touch different parts of an elephant to learn what it is like. Each of them makes a judgment about the elephant based on his own observation. For example, the man who touched the ear said it was like a fan while the man who grabbed the tail said it was like a rope. Clearly, each of them just got a partial view and did not arrive at an accurate description of the elephant. However, as they captured different aspects about the elephant, we can learn the big picture by integrating the knowledge about all the parts together.

Ensemble learning can be regarded as applying this "crowd wisdom" to the task of classification. Classification, or supervised learning, tries to infer a function that maps feature values into class labels from training data, and apply the function to data with unknown class labels. The function is called a model or classifier. We can regard a collection of some possible models as a hypothesis space $\mathcal{H}$, and each single point $h \in \mathcal{H}$ in this space corresponds to a specific model. A classification algorithm usually makes certain assumptions about the hypothesis space and thus defines a hypothesis space to search for the correct model. The algorithm also defines a certain criterion to measure the quality of the model so that the model that has the best measure in the hypothesis space will be returned. A variety of classification algorithms have been developed [8], including Support Vector Machines [20,22], logistic regression [47], Naive Bayes, decision trees [15,64,65], k-nearest neighbor algorithm [21], and Neural Networks [67,88]. They differ in the hypothesis space, model quality criteria and search strategies.

In general, the goal of classification is to find the model that achieves good performance when predicting the labels of future unseen data. To improve the generalization ability of a classification model, it should not overfit the training data; instead, it should be general enough to cover unseen cases. Ensemble approaches can be regarded as a family of classification algorithms, which are developed to improve the generalization abilities of classifiers. It is hard to get a single classification model with good generalization ability, which is called a strong classifier, but ensemble learning can transform a set of weak classifiers into a strong one by their combination. Formally, we learn T classifiers from a training set D: $\{h_1(\mathbf{x}), \dots, h_T(\mathbf{x})\}$, each of which maps feature values $\mathbf{x}$ into a class label y. We then combine them into an ensemble classifier $H(\mathbf{x})$ with the hope that it achieves better performance.

There are two major factors that contribute to the success of ensemble learning. First, theoretical analysis and real practice have shown that the expected error of an ensemble model is smaller than that of a single model. Intuitively, if we know in advance that $h_1(\mathbf{x})$ has the best prediction performance on future data, then without any doubt, we should discard the other classifiers and choose $h_1(\mathbf{x})$ for future predictions. However, we do not know the true labels of future data, and thus we are unable to know in advance which classifier performs the best. Therefore, our best bet should be the prediction obtained by combining multiple models. This simulates what we do in real life—When making investments, we will seldom rely on one single option, but rather distribute the money across multiple stocks, plans, and accounts. This will usually lead to lower risk and higher gain. In many other scenarios, we will combine independent and diverse opinions to make better decisions.

A simple example can be used to further illustrate how the ensemble model achieves better performance [74]. Consider five completely independent classifiers and suppose each classifier has a prediction accuracy of 70% on future data. If we build an ensemble classifier by combining these five classifiers using majority voting, i.e., predict the label as the one receiving the highest votes among classifiers, then we can compute the probability of making accurate classification as: $10 \times 0.7^3 \times$

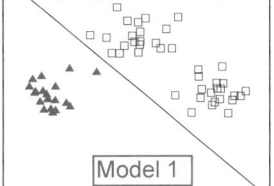 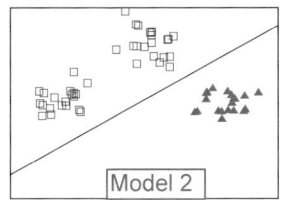 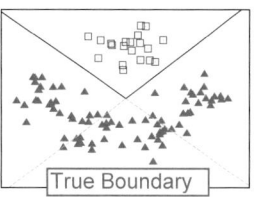

FIGURE 19.1: Ensemble approach overcomes the limitation of hypothesis space.

$0.3^2 + 5 \times 0.7^4 \times 0.3 + 0.7^5 = 83.7\%$ (the sum of the probability of having 3, 4, and 5 classifiers predicting correctly). If we now have 101 such classifiers, following the same principle, we can derive that the majority voting approach on 101 classifiers can reach an accuracy of 99.9%. Clearly, the ensemble approach can successfully cancel out independent errors made by individual classifiers and thus improve the classification accuracy.

Another advantage of ensemble learning is its ability to overcome the limitation of hypothesis space assumption made by single models. As discussed, a single-model classifier usually searches for the best model within a hypothesis space that is assumed by the specific learning algorithm. It is very likely that the true model does not reside in the hypothesis space, and then it is impossible to obtain the true model by the learning algorithm. Figure 19.1 shows a simple example of binary classification. The rightmost plot shows that the true decision boundary is V-shaped, but if we search for a classifier within a hypothesis space consisting of all the linear classifiers, for example, models 1 and 2, we are unable to recover the true boundary. However, by combining multiple classifiers, ensemble approaches can successfully simulate the true boundary. The reason is that ensemble learning methods combine different hypotheses and the final hypothesis is not necessarily contained in the original hypothesis space. Therefore, ensemble methods have more flexibility in the hypothesis they could represent.

Due to these advantages, many ensemble approaches [4, 23, 52, 63, 74, 75, 90] have been developed to combine complementary predictive power of multiple models. Ensemble learning is demonstrated as useful in many data mining competitions (e.g., Netflix contest,[1] KDD cup,[2] ICDM contest[3]) and real-world applications. There are two critical components in ensemble learning: Training base models and learning their combinations, which are discussed as follows.

From the majority voting example we mentioned, it is obvious that the base classifiers should be accurate and independent to obtain a good ensemble. In general, we do not require the base models to be highly accurate—as long as we have a good amount of base classifiers, the weak classifiers can be boosted to a strong classifier by combination. However, in the extreme case where each classifier is terribly wrong, the combination of these classifiers will give even worse results. Therefore, at least the base classifiers should be better than random guessing. Independence among classifiers is another important property we want to see in the collection of base classifiers. If base classifiers are highly correlated and make very similar predictions, their combination will not improve anymore. In contrast, when base classifiers are independent and make diverse predictions, the independent errors have better chances to be canceled out. Typically, the following techniques have been used to generate a good set of base classifiers:

- Obtain different bootstrap samples from the training set and train a classifier on each bootstrap sample;

---

[1]http://www.netflixprize.com/
[2]http://www.kddcup-orange.com/
[3]http://www.cs.uu.nl/groups/ADA/icdm08cup/

- Extract different subsets of examples or subsets of features and train a classifier on each subset;

- Apply different learning algorithms on the training set;

- Incorporate randomness into the process of a particular learning algorithm or use different parametrization to obtain different prediction results.

To further improve the accuracy and diversity of base classifiers, people have explored various ways to prune and select base classifiers. More discussions about this can be found in Section 19.6.

Once the base classifiers are obtained, the important question is how to combine them. The combination strategies used by ensemble learning methods roughly fall into two categories: Un-weighted and weighted. Majority voting is a typical un-weighted combination strategy, in which we count the number of votes for each predicted label among the base classifiers and choose the one with the highest votes as the final predicted label. This approach treats each base classifier as equally accurate and thus does not differentiate them in the combination. On the other hand, weighted combination usually assigns a weight to each classifier with the hope that higher weights are given to more accurate classifiers so that the final prediction can bias towards the more accurate classifiers. The weights can be inferred from the performance of base classifiers or the combined classifier on the training set.

In this book chapter, we will provide an overview of the classical ensemble learning methods discussing their base classifier generation and combination strategies. We will start with Bayes optimal classifier, which considers all the possible hypotheses in the whole hypothesis space and combines them. As it cannot be practically implemented, two approximation methods, i.e., Bayesian model averaging and Bayesian model combination, have been developed (Section 19.2). In Section 19.3, we discuss the general idea of bagging, which combines classifiers trained on bootstrap samples using majority voting. Random forest is then introduced as a variant of bagging. In Section 19.4, we discuss the boosting method, which gradually adjusts the weights of training examples so that weak classifiers can be boosted to learn accurately on difficult examples. AdaBoost will be introduced as a representative of the boosting method. Stacking is a successful technique to combine multiple classifiers, and its usage in top performers of the Netflix competition has attracted much attention to this approach. We discuss its basic idea in Section 19.5. After introducing these classical approaches, we will give a brief overview of recent advances in ensemble learning, including new ensemble learning techniques, ensemble pruning and selection, and ensemble learning in various challenging learning scenarios. Finally, we conclude the book chapter by discussing possible future directions in ensemble learning. The notations used throughout the book chapter are summarized in Table 19.1.

TABLE 19.1: Notation Summary

notation	meaning
$\mathbb{R}^n$	feature space
$\mathcal{Y}$	class label set
m	number of training examples
$\mathcal{D} = \{\mathbf{x}_i, y_i\}_{i=1}^{m}$ ($\mathbf{x}_i \in \mathbb{R}^n$, $y_i \in \mathcal{Y}$)	training data set
$\mathcal{H}$	hypothesis space
h	base classifier
H	ensemble classifier

19.2 Bayesian Methods

In this section, we introduce Bayes optimal classification and its implementation approaches including Bayesian model averaging and Bayesian model combination.

19.2.1 Bayes Optimal Classifier

Ensemble methods try to combine different models into one model in which each base model is assigned a weight based on its contribution to the classification task. As we discussed in the Introduction, two important questions in ensemble learning are: Which models should be considered and how to choose them from the hypothesis space? How to assign the weight to each model? One natural solution is to consider all the possible hypotheses in the hypothesis space and infer model weights from training data. This is the basic principle adopted by the Bayes optimal classifier.

Specifically, given a set of training data $\mathcal{D} = \{\mathbf{x}_i, y_i\}_{i=1}^m$, where $\mathbf{x}_i \in \mathbb{R}^n$ and $y_i \in \mathcal{Y}$, our goal is to obtain an ensemble classifier that assigns a label to unseen data $\mathbf{x}$ in the following way:

$$y = \arg\max_{y \in \mathcal{Y}} \sum_{h \in \mathcal{H}} p(y|\mathbf{x}, h) p(h|\mathcal{D}). \tag{19.1}$$

In this equation, y is the predicted label of $\mathbf{x}$, $\mathcal{Y}$ is the set of all possible labels, $\mathcal{H}$ is the hypothesis space that contains all possible hypothesis h, and $p()$ denotes probability functions. We can see that the Bayes optimal classifier combines all possible base classifiers, i.e., the hypotheses, by summing up their weighted votes. The posterior probability $p(h|\mathcal{D})$ is adopted to be their corresponding weights.

We assume that training examples are drawn independently. By Bayes' theorem, the posterior probability $p(h|\mathcal{D})$ is given by:

$$P(h|\mathcal{D}) = \frac{p(h)p(\mathcal{D}|h)}{p(\mathcal{D})} = \frac{p(h)\Pi_{i=1}^m p(\mathbf{x}_i|h)}{p(\mathcal{D})}. \tag{19.2}$$

In this equation, $p(h)$ is the prior probability reflecting the degree of our belief that h is the "correct" model prior to seeing any data. $p(\mathcal{D}|h) = \Pi_{i=1}^m p(\mathbf{x}_i|h)$ is the likelihood, i.e., how likely the given training set is generated under the model h. The data prior $p(\mathcal{D})$ is the same for all hypotheses, so it could be ignored when making the label predictions.

To sum up, Bayes optimal classifier makes a prediction for $\mathbf{x}$ as follows:

$$y = \arg\max_{y \in \mathcal{Y}} \sum_{h \in \mathcal{H}} p(y|\mathbf{x}, h) p(\mathcal{D}|h) p(h). \tag{19.3}$$

By this approach, all the models are combined by considering both prior knowledge and data likelihood. In other words, the Bayes optimal classifier combines all the hypotheses in the hypothesis space, and each hypothesis is given a weight reflecting the probability that the training data would be sampled under the hypothesis if that hypothesis were true. In [61], it is pointed out that Bayes optimal classifier can reach optimal classification results under Bayes theorem, and on average, no other ensemble method can outperform it.

Although Bayes optimal classifier is the ideal ensemble method, unfortunately, it cannot be practically implemented. The prior knowledge about each model $p(h)$ is usually unknown. For most tasks, the hypothesis space is too large to iterate over and many hypotheses only output a predicted label rather than a probability. Even though in some classification algorithms, it is possible to estimate the probability from training data, computing an unbiased estimate of the probability of the training data given a hypothesis, i.e., $p(\mathcal{D}|h)$, is difficult. The reason is that when we calculate

$p(\mathbf{x}_i|h)$, we usually have some assumptions that could be biased. Therefore, in practice, some algorithms have been developed to approximate Bayes optimal classifier. Next, we will discuss two popular algorithms, i.e., Bayesian model averaging and Bayesian model combination, which effectively implement the basic principles of Bayes optimal classifier.

19.2.2 Bayesian Model Averaging

Bayesian model averaging [46] approximates Bayes optimal classifier by combining a set of sampled models from the hypothesis space. Models are typically sampled using a Monte Carlo sampling technique. Or a simpler way is to use the model trained on a random subset of training data as a sampled model.

From Equation 19.3, we notice that there are two probabilities to compute: One is the prior probability $p(h)$, and the other is the likelihood $p(\mathcal{D}|h)$. When no prior knowledge is available, we simply use uniform distribution without normalization, i.e., $p(h) = 1$ for all hypotheses. As for the likelihood, we can infer it based on the hypothesis's performance on training data. Assume that the error rate of hypothesis h is $\varepsilon(h)$, then the following equation is used to compute each training example's probability given h:

$$p(\mathbf{x}_i|h) = \exp\{\varepsilon(h)\log(\varepsilon(h)) + (1 - \varepsilon(h))\log(1 - \varepsilon(h))\}.$$

Suppose the size of the random subset of training data is m', then we can get

$$p(\mathcal{D}_t|h) = \exp\{m' \cdot (\varepsilon(h)\log(\varepsilon(h)) + (1 - \varepsilon(h))\log(1 - \varepsilon(h)))\}.$$

Algorithm 19.1 Bayesian Model Averaging

Input: Training data $\mathcal{D} = \{\mathbf{x}_i, y_i\}_{i=1}^m$ ($\mathbf{x}_i \in \mathbb{R}^n$, $y_i \in \mathcal{Y}$)
Output: An ensemble classifier H

1: **for** $t \leftarrow 1$ to T **do**
2:     Construct a subset of training data $\mathcal{D}_t$ with size m' by randomly sampling in $\mathcal{D}$
3:     Learn a base classifier h_t based on $\mathcal{D}_t$
4:     Set the prior probability $p(h_t) = 1$
5:     Calculate h_t's error rate for predicting on $\mathcal{D}_t$, which is denoted as $\varepsilon(h_t)$
6:     Calculate the likelihood $p(\mathcal{D}_t|h) = \exp\{m' \cdot (\varepsilon(h)\log(\varepsilon(h)) + (1 - \varepsilon(h))\log(1 - \varepsilon(h)))\}$
7:     Weight of h_t is set as $weight(h_t) = p(\mathcal{D}_t|h)p(h)$
8: **end for**
9: Normalize all the weights to sum to 1
10: **return** $H(\mathbf{x}) = \arg\max_{y \in \mathcal{Y}} \sum_{t=1}^T p(y|\mathbf{x}, h) \cdot weight(h_t)$

Bayesian model averaging algorithm is shown in Algorithm 19.1 [1]. In [42], it was proved that when the representative hypotheses are drawn and averaged using the Bayes theorem, Bayesian model averaging has an expected error that is bounded to be at most twice the expected error of the Bayes optimal classifier.

Despite its theoretical correctness, Bayesian model averaging may encounter over-fitting problems. Bayesian model averaging prefers the hypothesis that by chance has the lowest error on training data rather than the hypothesis that truly has the lowest error [26]. In fact, Bayesian model averaging conducts a selection of classifiers instead of combining them. As the uniform distribution is used to set the prior probability, the weight of each base classifier is equivalent to its likelihood on training data. Typically, none of these base classifiers can fully characterize the training data, so most of them receive a small value for the likelihood term. This may not be a problem if base classifiers are exhaustively sampled from the hypothesis space, but this is rarely possible when we have limited training examples. Due to the normalization, the base classifier that captures data distribution the best will receive the largest weight and has the highest impact on the ensemble classifier. In most cases, this classifier will dominate the output and thus Bayesian model averaging conducts model selection in this sense.

Example. We demonstrate Bayesian model averaging step by step using the following example. Suppose we have a training set shown in Table 19.2, which contains 10 examples with 1-dimensional feature values and corresponding class labels ($+1$ or -1).

TABLE 19.2: Training Data Set

x	1	2	3	4	5	6	7	8	9	10
y	+1	+1	+1	-1	-1	-1	-1	+1	+1	+1

We use a set of simple classifiers that only use a threshold to make class predictions: If **x** is above a threshold, it is classified into one class; if it is on or below the threshold, it is classified into the other class. By learning from the training data, the classifier will decide the optimal θ (the decision boundary) and the class labels to be assigned to the two sides so that the error rate will be minimized.

We run the algorithm for five rounds using $m' = 5$. Altogether we construct five training sets from the original training set and each of them contains five randomly selected examples. We first learn a base classifier h_t based on each sampled training set, and then calculate its training error rate $\varepsilon(h_t)$, the likelihood $p(\mathcal{D}_t|h)$ and $weight(h_t)$ using the formula described in Algorithm 19.1. Table 19.3 shows the training data for each round and the predicted labels on the selected examples made by the base classifiers.

TABLE 19.3: Base Model Output of Bayesian Model Averaging

Round t	Weak Classifier h_t						$\varepsilon(h_t)$	$weight(h_t)$
1	x	2	3	6	7	8	0.2	0.013
	y	+1	+1	-1	-1	-1		
2	x	1	2	3	4	5	0	0.480
	y	+1	+1	+1	-1	-1		
3	x	1	3	7	9	10	0.2	0.013
	y	+1	+1	+1	+1	+1		
4	x	2	5	6	8	9	0.2	0.013
	y	-1	-1	-1	+1	+1		
5	x	1	3	4	6	7	0	0.480
	y	+1	+1	-1	-1	-1		

Based on the model output shown in Table 19.3, an ensemble classifier can be constructed by following Bayesian model averaging and its prediction results are shown in Table 19.4.

TABLE 19.4: Ensemble Classifier Output by Bayesian Model Averaging

x	1	2	3	4	5	6	7	8	9	10
y	+1	+1	+1	-1	-1	-1	-1	-1	-1	-1

The highlighted points are incorrectly classified. Thus, the ensemble classifier has an error rate of 0.3. It is in fact the same classifier we obtained from Rounds 2 and 5. The weights of these two base classifiers are so high that the influence of the others is ignored in the final combination. Some techniques have been proposed to solve this dominating weight problem, for example, the Bayesian model combination method that will be discussed next.

Algorithm 19.2 Bayesian Model Combination

Input: Training data $\mathcal{D} = \{\mathbf{x}_i, y_i\}_{i=1}^m$ ($\mathbf{x}_i \in \mathbb{R}^n$, $y_i \in \mathcal{Y}$)
Output: An ensemble classifier H

1: **for** $t \leftarrow 1$ to T **do**
2:     Construct a subset of training data $\mathcal{D}_t$ with size m' by randomly sampling in $\mathcal{D}$
3:     Learn a base classifier h_t based on $\mathcal{D}_t$
4:     Set the initial weight $weight(h_t) = 0$
5: **end for**
6: $SumWeight = 0$
7: $z = -\infty$ (used to maintain numerical stability)
8: Set the iteration number for computing weights: $iteration$
9: **for** $iter \leftarrow 1$ to $iteration$ **do**
10:     For each weak classifier, draw a temp weight
    $TempWeight(h_t) = -\log(RandUniform(0,1))$
11:     Normalize $TempWeight$ to sum to 1
12:     Combine the base classifiers as $H' = \sum_{t=1}^T h_t \cdot TempWeight(h_t)$
13:     Calculate the error rate of H' for predicting the data in $\mathcal{D}$
14:     Calculate the likelihood
    $p(\mathcal{D}|H') = \exp\{m \cdot (\varepsilon(H')\log(\varepsilon(H')) + (1 - \varepsilon(H'))\log(1 - \varepsilon(H')))\}$
15:     **if** $p(\mathcal{D}|H') > z$ **then**
16:         For each base classifier, $weight(h_t) = weight(h_t)\exp\{z - p(\mathcal{D}|H')\}$
17:         $z = p(\mathcal{D}|H')$
18:     **end if**
19:     $w = \exp\{p(\mathcal{D}|H') - z\}$
20:     For each base classifier $weight(h_t) = weight(h_t)\frac{SumWeight}{SumWeight+w} + w \cdot TempWeight(h_t)$
21:     $SumWeight = SumWeight + w$
22: **end for**
23: Normalize all the weights to sum to 1
24: **return** $H(\mathbf{x}) = \arg\max_{y \in \mathcal{Y}} \sum_{t=1}^T p(y|\mathbf{x},h) \cdot weight(h_t)$

19.2.3 Bayesian Model Combination

To overcome the over-fitting problem of Bayesian model averaging, Bayesian model combination directly samples from the space of possible ensemble hypotheses instead of sampling each hypothesis individually. By applying this strategy, Bayesian model combination finds the combination of base classifiers that simulates the underlying distribution of training data the best. This is different from Bayesian model averaging, which selects one base classifier with the best performance on training data.

The procedure of Bayesian model combination is similar to that of Bayesian model averaging. The only difference is that Bayesian model combination regards all the base classifiers as one model and iteratively calculates their weights simultaneously. The details of this algorithm can be found in Algorithm 19.2 [1]. Although Bayesian model combination is more computationally costly compared with Bayesian model averaging, it gives better results in general [62].

Example. We demonstrate how Bayesian Model Combination works using the same example in Table 19.2. In order to compare with Bayesian model averaging, we use the same random samples and the same base classifiers. We set $iteration = 3$. At each iteration, we draw a temp weight and normalize it (Table 19.5). Then, we combine the base classifiers by their temp weights and calculate the likelihood (shown in Table 19.6). We update $weight(h_t)$ as described in Algorithm 19.2 (Table 19.7).

TABLE 19.5: *TempWeight* for Three Iterations

Classifier	$TempWeight_1$	$TempWeight_2$	$TempWeight_3$
1	0.09	0.36	0.16
2	0.04	0.00	0.10
3	0.58	0.07	0.23
5	0.24	0.11	0.37
5	0.05	0.45	0.14

TABLE 19.6: H' for Three Iterations

iter		Classifier H'										$P(\mathcal{D}|H')$
1	x	1	2	3	4	5	6	7	8	9	10	
	y	+1	+1	+1	+1	+1	+1	+1	+1	+1	+1	0.0078
2	x	1	2	3	4	5	6	7	8	9	10	
	y	+1	+1	+1	-1	-1	-1	-1	-1	-1	-1	0.0122
3	x	1	2	3	4	5	6	7	8	9	10	
	y	+1	+1	+1	-1	-1	-1	-1	+1	+1	+1	1

It is clear that Bayesian model combination incurs more computation compared with Bayesian model averaging. Instead of sampling each hypothesis individually, Bayesian model combination samples from the space of possible ensemble hypotheses. It needs to compute an ensemble classifier and update several sets of parameters during one iteration. *Weight$_{final}$* shows the normalized *Weight$_3$* that we will use to calculate the final classifier. It is indeed a weighted combination of all sets of *TempWeight*. Since the last *TempWeight* set gives the best results with the highest likelihood, the final weight *Weight$_{final}$* is closer to it. The prediction results made by the ensemble classifier of Bayesian model averaging are shown in Table 19.8. This classifier classifies all the points correctly. Comparing with Bayesian model averaging, Bayesian model combination gives better results.

19.3   Bagging

In this section, we introduce a very popular ensemble approach, i.e., Bagging, and then discuss Random Forest, which adapts the idea of Bagging to build an ensemble of decision trees.

19.3.1   General Idea

Bootstrap aggregation (i.e., Bagging) [11] is an ensemble method that adopts Bootstrap sampling technique [28] in constructing base models. It generates new data sets by sampling from the original data set with replacement, and trains base classifiers on the sampled data sets. To get an ensemble classifier, it simply combines all the base classifiers by majority voting. The general procedure of Bagging is illustrated in Algorithm 19.3. Although it is a simple approach, it has been shown to be a powerful method empirically and theoretically. When Bagging was introduced, [11] presented an explanation about why Bagging is effective with unstable weak classifiers. In [35], the authors studied Bagging through a decomposition of statistical predictors.

Now let's look at the details of this approach. The bootstrap sampling procedure is as follows. Given a set of training data $\mathcal{D} = \{\mathbf{x}_i, y_i\}_{i=1}^m$ ($\mathbf{x}_i \in \mathbb{R}^n$, $y_i \in \mathcal{Y}$ and m is the size of the training set), we sample data from $\mathcal{D}$ with replacement to form a new data set $\mathcal{D}'$. The size of $\mathcal{D}'$ will be kept

TABLE 19.7: *Weight* for Three Iterations

Classifier	$Weight_1$	$Weight_2$	$Weight_3$	$Weight_{final}$
1	0.09	0.41	0.27	0.19
2	0.04	0.03	0.10	0.08
3	0.58	0.36	0.32	0.23
5	0.24	0.23	0.43	0.31
5	0.05	0.48	0.26	0.19

TABLE 19.8: Final Results of Bayesian Model Combination

x	1	2	3	4	5	6	7	8	9	10
y	+1	+1	+1	-1	-1	-1	-1	+1	+1	+1

the same as that of $\mathcal{D}$. Some of the examples appear more than once in $\mathcal{D}'$ while some examples in $\mathcal{D}$ do not appear in $\mathcal{D}'$. For a particular example $\mathbf{x}_i$, the probability that it appears k times in $\mathcal{D}'$ follows a Poisson distribution with $\lambda = 1$ [11]. By setting $k = 0$ and $\lambda = 1$, we can get that $\mathbf{x}_i$ does not appear in $\mathcal{D}'$ with a probability of $\frac{1}{e}$, so $\mathbf{x}_i$ appears in $\mathcal{D}'$ with a probability of $1 - \frac{1}{e} \approx 0.632$. $\mathcal{D}'$ is expected to have 63.2% unique data of $\mathcal{D}$ while the rest are duplicates. After sampling T data sets using bootstrap sampling, we train a classifier on each of the sampled data sets $\mathcal{D}'$ and combine their output by majority voting. For each example $\mathbf{x}_i$, its final prediction by Bagging is the class label with the highest number of predictions made by base classifiers.

We need to be careful in selecting learning algorithms to train base classifiers in Bagging. As $\mathcal{D}'$ has 63.2% overlap with the original data set $\mathcal{D}$, if the learning algorithm is insensitive to the change on training data, all the base classifiers will output similar predictions. Then the combination of these base classifiers cannot improve the performance of ensemble. To ensure high diversity of base classifiers, Bagging prefers unstable learning algorithms such as Neural Networks, Decision Trees rather than stable learning algorithms such as K-Nearest Neighbor.

Typically, Bagging adopts majority voting to combine base classifiers. However, if the base classifiers output prediction confidence, weighted voting or other combination methods are also possible. If the size of data set is relatively small, it is not easy to get base classifiers with high diversity because base classifiers' diversity mainly comes from data sample manipulation. In such cases, we could consider to introduce more randomness, such as using different learning algorithms in base models.

Example. We demonstrate Bagging step by step using the toy example shown in Table 19.2. In this experiment, we set $T = 5$ and $|\mathcal{D}'| = |\mathcal{D}| = 10$. Therefore, we construct five training sets and each one of them contains ten random examples. Since we draw samples with replacement, it is possible to have some examples repeating in the training set. We then learn base classifiers from these training sets. Table 19.9 shows the training data for each round and the prediction result of the base classifier applied on the corresponding training set.

Based on the base model output shown in Table 19.9, we can calculate the final result by majority voting. For example, for $x = 1$, there are three classifiers that predict its label to be +1 and two classifiers that predict its label as -1, so its final predicted label is +1. The label predictions made by Bagging on this example dataset are shown in Table 19.10.

We can see that only one point ($x = 4$) is incorrectly classified (highlighted in the table). The ensemble classifier has an error rate of only 0.1, so Bagging achieves better performance than the base models.

Algorithm 19.3 Bagging Algorithm

Input: Training data $\mathcal{D} = \{\mathbf{x}_i, y_i\}_{i=1}^m$ ($\mathbf{x}_i \in \mathbb{R}^n$, $y_i \in \mathcal{Y}$)
Output: An ensemble classifier H

1: **for** $t \leftarrow 1$ to T **do**
2:     Construct a sample data set $\mathcal{D}_t$ by randomly sampling with replacement in $\mathcal{D}$
3:     Learn a base classifier h_t based on $\mathcal{D}_t$
4: **end for**
5: **return** $H(\mathbf{x}) = \arg\max_{y \in \mathcal{Y}} \sum_{t=1}^T \mathbf{1}(h_t(\mathbf{x}) = y)$

TABLE 19.9: Training Sets and Base Model Predictions

1	x	1	1	2	2	2	3	5	5	7	8
	y	+1	+1	+1	+1	+1	+1	-1	-1	-1	-1
2	x	2	4	6	7	7	7	8	8	8	9
	y	-1	-1	-1	-1	-1	-1	+1	+1	+1	+1
3	x	2	2	3	3	6	6	7	8	9	10
	y	+1	+1	+1	+1	+1	+1	+1	+1	+1	+1
4	x	4	4	5	6	6	7	8	9	10	10
	y	-1	-1	-1	-1	-1	-1	+1	+1	+1	+1
5	x	1	1	2	3	5	6	6	7	8	9
	y	+1	+1	+1	+1	-1	-1	-1	-1	-1	-1

19.3.2 Random Forest

Random Forest [13, 45] can be regarded as a variant of Bagging approach. It follows the major steps of Bagging and uses decision tree algorithm to build base classifiers. Besides Bootstrap sampling and majority voting used in Bagging, Random Forest further incorporates random feature space selection into training set construction to promote base classifiers' diversity.

Specifically, the following describes the general procedure of Random Forest algorithm:

- Given $\mathcal{D} = \{\mathbf{x}_i, y_i\}_{i=1}^m$, we construct a Bootstrap data set $\mathcal{D}'$ by random sampling with replacement.

- Build a decision tree on $\mathcal{D}'$ by applying *LearnDecisionTree* function and passing the following parameters *LearnDecisionTree*($data = \mathcal{D}', iteration = 0, ParentNode = root$). *LearnDecisionTree* is a recursive function that takes the dataset, iteration step, and parent node index as input and returns a partition of the current dataset.

Specifically, *LearnDecisionTree* function conducts the following steps at each node:

- Check whether the stopping criterion is satisfied. Usually we can choose one or multiple choices of the following criteria: a) All the examples at the current node have the same class label; b) the impurity of current node is below a certain threshold (impurity can be represented by entropy, Gini index, misclassification rate or other measures; more details will be discussed later); c) there are no more features available to split the data at the current node to improve the impurity; or d) the height of the tree is greater than a pre-defined number. If the stopping criterion is satisfied, then we stop growing the tree, otherwise, continue as follows.

- Randomly sample a subset of n' features from the whole feature space $\mathbb{R}^n$ so that each example becomes $\{\mathbf{x}_i', y_i\}$ where $\mathbf{x}_i' \in \mathbb{R}^{n'}$ and $n' \leq n$. We denote the dataset in the subspace as $\hat{\mathcal{D}}_{current}$.

- Find the best feature q^* to split the current dataset that achieves the biggest gain in impurity. Specifically, suppose $i(node)$ denotes the impurity of a node, and *LeftChildCode*

TABLE 19.10: Ensemble Output by Bagging

x	1	2	3	4	5	6	7	8	9	10
y	+1	+1	+1	+1	-1	-1	-1	+1	+1	+1

and *RightChildNode* are the child nodes of the current node. We are trying to find a feature q^* to maximize $i(node) - P_L \cdot i(LeftChildNode) - P_R \cdot i(RightChildNode)$, where P_L and P_R are the fraction of the data that go to the corresponding child node if the split feature q^* is applied. We can use one of the following impurity measures: a) Entropy: $i(node) = -\sum_{y_i \in \mathcal{Y}} p(y_i) \log p(y_i)$; b) Gini index: $i(node) = 1 - \sum_{y_i \in \mathcal{Y}} p(y_i)^2$; c) misclassification rate: $i(node) = 1 - \max_{y_i \in \mathcal{Y}} p(y_i)$.

- After we select the splitting feature, we will split the data at the current node v into two parts and assign them to its child nodes *LeftChildNode* and *RightChildNode*. We denote these two datasets as $\mathcal{D}_L$ and $\mathcal{D}_R$. Label v is the parent node of *LeftChildNode* and *RightChildNode* under the split feature q^*.

- Call the function $LearnDecisionTree(\mathcal{D}_L, iteration = iteration + 1, ParentNode = v)$ and $LearnDecisionTree(\mathcal{D}_R, iteration = iteration + 1, ParentNode = v)$ to continue the tree construction.

Random Forest algorithm builds multiple decision trees following the above procedure and takes majority voting of the prediction results of these trees. The algorithm is summarized in Algorithm 19.4. To incorporate diversity into the trees, Random Forest approach differs from the traditional decision tree algorithm in the following aspects when building each tree: First, it infers the tree from a Bootstrap sample of the original training set. Second, when selecting the best feature at each node, Random Forest only considers a subset of the feature space. These two modifications of decision tree introduce randomness into the tree learning process, and thus increase the diversity of base classifiers. In practice, the dimensionality of the selected feature subspace n' controls the randomness. If we set $n' = n$ where n is the original dimensionality, then the constructed decision tree is the same as the traditional deterministic one. If we set $n' = 1$, at each node, only one feature will be randomly selected to split the data, which leads to a completely random decision tree. In [13], it was suggested that n' could be set as the logarithm of the dimensionality of original feature space, i.e., $n' = \log(n)$.

Note that although the subset of feature space is randomly selected, the choice of the feature used to split a node is still deterministic. In [56], the authors introduce the VR-tree ensemble method, which further randomizes the selection of features to split nodes. At each node, a Boolean indicator variable is adopted: If the indicator variable is true, the node will be constructed in the deterministic way, i.e., among all possible features, we select the one that achieves the biggest impurity gain; if the indicator variable is false, a feature will be randomly selected to split the node. If the feature space is large, this will benefit the learning process by reducing the computation cost and increasing the diversity among all decision trees.

In practice, Random Forest is simple yet powerful. The diversity among the trees ensures good performance of Random Forest. By tuning the subspace dimensionality n', different tradeoffs between computation resources and diversity degrees can be achieved. The decision tree building process can be implemented in parallel to reduce running time. There are other variations of Random Forest approaches that have been demonstrated to be effective in density estimation [31, 55] and anomaly detection [57].

Algorithm 19.4 Random Forest

Input: Training data $\mathcal{D} = \{\mathbf{x}_i, y_i\}_{i=1}^{m}$ ($\mathbf{x}_i \in \mathbb{R}^n$, $y_i \in \mathcal{Y}$)
Output: An ensemble classifier H

1: **for** $t \leftarrow 1$ to T **do**
2:     Construct a sample data set $\mathcal{D}_t$ by randomly sampling with replacement in $\mathcal{D}$
3:     Learn a decision tree h_t by applying
    $LearnDecisionTree(\mathcal{D}_t, iteration = 0, ParentNode = root)$:
4:         If stop criterion is satisfied, return
5:         Randomly sample features in the whole feature space $\mathbb{R}^n$ to get a new data set
    $\hat{\mathcal{D}}_{current} = RandomSubset(\mathcal{D}_{current})$
6:         Find the best feature q^* according to impurity gain
7:         Split data $(\mathcal{D}_L, \mathcal{D}_R) = split(\mathcal{D}_{current}, q^*)$
8:         Label the new parent node $v = parent.newchild(q^*)$
9:         Conduct $LearnDecisionTree(\mathcal{D}_L, iteration = iteration + 1, ParentNode = v)$ and
    $LearnDecisionTree(\mathcal{D}_R, iteration = iteration + 1, ParentNode = v)$
10: **end for**
11: **return** $H(\mathbf{x}) = \arg\max_{y \in \mathcal{Y}} \sum_{t=1}^{T} \mathbf{1}(h_t(\mathbf{x}) = y)$

19.4 Boosting

Boosting is a widely used ensemble approach, which can effectively boost a set of weak classifiers to a strong classifier by iteratively adjusting the importance of examples in the training set and learning base classifiers based on the weight distribution. In this section, we review the general procedure of Boosting and its representative approach: AdaBoost.

19.4.1 General Boosting Procedure

We wish to learn a strong classifier that has good generalization performance, but this is a difficult task. In contrast, weak classifiers, which have performance only comparable to random guessing, are much easier to obtain. Therefore, people wonder if it is possible to use a set of weak classifiers to create a single strong classifier [48]. More formally speaking, are weakly learnable and strongly learnable equivalent? In [69], it was proved that the answer to this question is yes, and this motivates the development of Boosting.

Boosting is a family of algorithms that could convert weak classifiers to a strong one [70]. The basic idea of Boosting is to correct the mistakes made in weak classifiers gradually. Let's consider a binary classification task, in which the training set only has three examples, $\mathbf{x}_1$, $\mathbf{x}_2$, and $\mathbf{x}_3$. Boosting starts with a base classifier, say h_1. Suppose h_1 makes correct predictions on $\mathbf{x}_1$ and $\mathbf{x}_2$, but predicts wrongly on $\mathbf{x}_3$. (Note that in this binary classification task, we can always find base classifiers with $\frac{1}{3}$ error rate. If the error rate of some weak classifier is $\frac{2}{3}$, by simply flipping, we can get base classifiers with $\frac{1}{3}$ error rate.) From h_1, we can derive a weight distribution for training examples to make the wrongly-classified examples more important. In this example, as h_1 makes an error with $\mathbf{x}_3$, the weight distribution will assign a higher value to $\mathbf{x}_3$. Based on this weight distribution, we train another classifier, say h_2. Now suppose h_2 misclassifies $\mathbf{x}_2$ and the weight distribution will be adjusted again: The weight for $\mathbf{x}_2$ is increased while $\mathbf{x}_3$'s weight is decreased and $\mathbf{x}_1$ has the lowest weight. According to the updated weight distribution, another classifier h_3 is obtained. By combing h_1, h_2 and h_3, we can get a strong classifier that makes correct predictions for all the three training examples.

Algorithm 19.5 Boosting Algorithm

Input: Training data $\mathcal{D} = \{\mathbf{x}_i, y_i\}_{i=1}^m$ ($\mathbf{x}_i \in \mathbb{R}^n$, $y_i \in \{+1, -1\}$)
Output: An ensemble classifier H

1: Initialize the weight distribution W_1
2: **for** $t \leftarrow 1$ to T **do**
3:     Learn weak classifier h_t based on $\mathcal{D}$ and W_t
4:     Evaluate weak classifier $\varepsilon(h_t)$
5:     Update weight distribution W_{t+1} based on $\varepsilon(h_t)$
6: **end for**
7: **return** $H = Combination(\{h_1, \ldots, h_T\})$

As summarized in Algorithm 19.5, Boosting approaches learn different weak classifiers iteratively by adjusting the weight of each training example. During each round, the weight of misclassified data will be increased and base classifiers will focus on those misclassified ones more and more. Under this general framework, many Boosting approaches have been developed, including AdaBoost [34], LPBoost [6], BrownBoost [32] and LogitBoost [71]. In the following, we will discuss the most widely used Boosting approach, Adaboost, in more detail.

19.4.2 AdaBoost

We will first describe AdaBoost in the context of binary classification. Given a training set $\mathcal{D} = \{\mathbf{x}_i, y_i\}_{i=1}^m$ ($\mathbf{x}_i \in \mathbb{R}^n$, $y_i \in \{+1, -1\}$), our goal is to learn a classifier that could classify unseen data with high accuracy. AdaBoost [34, 66] derives an ensemble model H by combining different weak classifiers. During the training process, the weight of each training example is adjusted based on the learning performance in the previous round, and then the adjusted weight distribution will be fed into the next round. This is equivalent to inferring classifiers from training data that are sampled from the original data set based on the weight distribution.

The detailed procedure of AdaBoost is discussed as follows. At first, the weight distribution W is initialized as $W_1(i) = \frac{1}{m}$, i.e., we initialize W_1 by uniform distribution when no prior knowledge is provided. We learn a base classifier h_t at each iteration t, given the training data $\mathcal{D}$ and weight distribution W_t. Among all possible $h \in \mathcal{H}$, we choose the best one that has the lowest classification error. In binary classification ($\mathcal{Y} = \{+1, -1\}$), if a classifier h has worse performance than random guessing (error rate $\varepsilon(h) \geq 0.5$), by simply flipping the output, we can turn h into a good classifier with a training error $1 - \varepsilon(h)$. In this scenario, choosing the best classifier is equivalent to picking either the best or worst one by considering all $h \in \mathcal{H}$. Without loss of generality, we assume that all the base classifiers have better performance compared with random guessing, i.e., $\varepsilon(h) \leq 0.5$. At each round, we pick the best model that minimizes the training error.

AdaBoost algorithm adopts weighted voting strategy to combine base classifiers. We derive a base classifier h_t at iteration t and calculate its weight in combination as

$$\alpha_t = \frac{1}{2} \ln \frac{1 - \varepsilon(h_t)}{\varepsilon(h_t)},$$

which is computed according its training error. This weight assignment has the following properties: 1) if $\varepsilon(h_1) \leq \varepsilon(h_2)$, then we have $\alpha_{h_1} \geq \alpha_{h_2}$, i.e., a higher weight will be assigned to the classifier with smaller training error; and 2) as $\varepsilon(h_t) \leq 0.5$, α_t is always positive.

An important step in each round of AdaBoost is to update the weight distribution of training examples based on the current base classifier according to the update equation

$$W_{t+1}(i) = \frac{W_t(i) \exp\{-\alpha_t \cdot h_t(\mathbf{x}_i) y_i\}}{Z_t},$$

where $Z_t = \sum_{i'=1}^{m} W_t(i') \exp\{-\alpha_t \cdot h_t(\mathbf{x}'_i)y'_i\}$ is a normalization term to ensure that the sum of $W_{t+1}(i)$ is 1. From this equation, we can see that if a training example is misclassified, its weight will be increased and then in the next iteration the classifier will pay more attention to this example. Instead, if an example is correctly classified, its corresponding weight will decrease. Specifically, if $\mathbf{x}_i$ is wrongly classified, $h_t(\mathbf{x}_i) \cdot y_i$ is -1 and $\alpha_t \geq 0$ so that $-\alpha_t \cdot h_t(\mathbf{x}_i)y_i$ is positive. As $\exp\{-\alpha_t \cdot h_t(\mathbf{x}_i)y_i\} > 1$, the new weight $W_{t+1}(i) > W_t(i)$. Similarly, we can see that if $\mathbf{x}_i$ is correctly classified, $W_{t+1}(i) < W_t(i)$, i.e., the new weight decreases.

The above procedure constitutes one iteration in AdaBoost. We repeat these steps until some stopping criterion is satisfied. We can set the number of iterations T as the stop criterion, or simply stop when we cannot find a base classifier that is better than random guessing. The whole algorithm is summarized in Algorithm 19.6.

Algorithm 19.6 AdaBoost in Binary Classification

Input: Training data $\mathcal{D} = \{\mathbf{x}_i, y_i\}_{i=1}^{m}$ ($\mathbf{x}_i \in \mathbb{R}^n$, $y_i \in \{+1, -1\}$)
Output: An ensemble classifier H
1: Initialize the weight distribution $W_1(i) = \frac{1}{m}$
2: **for** $t \leftarrow 1$ to T **do**
3:     Learn a base classifier $h_t = \arg\min_h \varepsilon(h)$ where $\varepsilon(h) = \sum_{i=1}^{m} W_t(i) \cdot \mathbf{1}(h(\mathbf{x}_i) \neq y_i)$
4:     Calculate the weight of h_t: $\alpha_t = \frac{1}{2} \ln \frac{1-\varepsilon(h_t)}{\varepsilon(h_t)}$
5:     Update the weight distribution of training examples: $W_{t+1}(i) = \frac{W_t(i) \exp\{-\alpha_t \cdot h_t(\mathbf{x}_i)y_i\}}{\sum_{i'=1}^{m} W_t(i') \exp\{-\alpha_t \cdot h_t(\mathbf{x}'_i)y'_i\}}$
6: **end for**
7: **return** $H = \sum_{t=1}^{T} \alpha_t \cdot h_t$;

As misclassified data receive more attention during the learning process, AdaBoost can be sensitive to noise and outliers. When applying to noisy data, the performance of AdaBoost may not be satisfactory. In practice, we may alleviate such a problem by stopping early (set T as a small number), or reducing the weight increase on misclassified data. Several follow-up approaches have been developed to address this issue. For example, MadaBoost [24] improves AdaBoost by depressing large weights, and FilterBoost [10] adopts log loss functions instead of exponential loss functions.

Now let us see how to extend AdaBoost to multi-class classification in which $\mathcal{Y} = \{1, 2, \ldots, K\}$. We can generalize the above algorithm [33] by changing the way of updating weight distribution: If $h_t(\mathbf{x}_i) = y_i$, then $W_{t+1}(i) = \frac{W_t(i) \cdot \exp\{-\alpha_t\}}{Z_t}$; else, $W_{t+1}(i) = \frac{W_t(i) \cdot \exp\{\alpha_t\}}{Z_t}$. With this modification, we can directly apply the other steps in AdaBoost on binary classification to the multi-class scenario if the base classifiers can satisfy $\varepsilon(h) < 0.5$. However, when there are multiple classes, it may not be easy to get weak classifiers with $\varepsilon(h) < 0.5$. To overcome this difficulty, an alternative way is to convert multi-class problem with K classes into K binary classification problems [72], each of which determines whether $\mathbf{x}$ belongs to the k-th class or not.

Note that the algorithm requires that the learning algorithm that is used to infer base classifiers should be able to learn on training examples with weights. If the learning algorithm is unable to learn from weighted data, an alternative way is re-sampling, which samples training data according to the weight distribution and then applies the learning algorithm. Empirical results show that there is no clear performance difference between learning with weight distribution and re-sampling training data.

There are many theoretical explanations to AdaBoost and Boosting. In [71] a margin-based explanation to AdaBoost was introduced, which has nice geometric intuition. Meanwhile, it was shown that AdaBoost algorithm can be interpreted as a stagewise estimation procedure for fitting an additive logistic regression model [71]. As for Boosting, population theory was proposed [14] and it was considered as the Gauss-Southwell minimization of a loss function [7].

Example. We demonstrate AdaBoost algorithm using the data set in Table 19.2. In this example, we still set $T = 3$. At each round, we learn a base classifier on the original data, and then calculate α_t based on weighted error rate. Next, we update weight distribution $W_{t+1}(i)$ as described in Algorithm 19.6. We repeat this procedure for T times. Table 19.11 shows the training data for each round and the results of the base classifiers.

TABLE 19.11: AdaBoost Training Data and Prediction Results

t		Weak Classifier h_t										α_t
	x	1	2	3	4	5	6	7	8	9	10	
1	y	+1	+1	+1	-1	-1	-1	-1	-1	-1	-1	0.42
	W_t	.07	.07	.07	.07	.07	.07	.07	.17	.17	.17	
	x	1	2	3	4	5	6	7	8	9	10	
2	y	-1	-1	-1	-1	-1	-1	-1	+1	+1	+1	0.65
	W_t	.17	.17	.17	.04	.04	.04	.04	.11	.11	.11	
	x	1	2	3	4	5	6	7	8	9	10	
3	y	+1	+1	+1	+1	+1	+1	+1	+1	+1	+1	0.75
	W_t	.10	.10	.10	.12	.12	.12	.12	.07	.07	.07	

At the first round, the base classifier makes errors on points $x = 8, 9, 10$, so these examples' weights increase accordingly. Then at Round 2, the base classifier pays more attention to these points and classifies them correctly. Since points $x = 4, 5, 6, 7$ are correctly predicted by both classifiers, they are considered "easy" and have lower weights, i.e., lower penalty if they are misclassified. From Table 19.11, we can construct an ensemble classifier as shown in Table 19.12, which makes correct predictions on all the examples.

TABLE 19.12: Ensemble Predictions Made by AdaBoost

x	1	2	3	4	5	6	7	8	9	10
y	+1	+1	+1	-1	-1	-1	-1	+1	+1	+1

19.5 Stacking

In this section, we introduce Stacked Generalization (Stacking), which learns an ensemble classifier based on the output of multiple base classifiers.

19.5.1 General Stacking Procedure

We discussed Bagging and Boosting in the previous sections. Bagging adopts Bootstrap sampling to learn independent base learners and takes the majority as final prediction. Boosting updates weight distribution each round, learns base classifiers accordingly, and then combines them according to their corresponding accuracy. Different from these two approaches, Stacking [12, 68, 89] learns a high-level classifier on top of the base classifiers. It can be regarded as a meta learning approach in which the base classifiers are called first-level classifiers and a second-level classifier is learnt to combine the first-level classifiers.

The general procedure of Stacking is illustrated in Algorithm 19.7 [90], which has the following three major steps:

- Step 1: Learn first-level classifiers based on the original training data set. We have several

Algorithm 19.7 Stacking

Input: Training data $\mathcal{D} = \{\mathbf{x}_i, y_i\}_{i=1}^m$ ($\mathbf{x}_i \in \mathbb{R}^n$, $y_i \in \mathcal{Y}$)
Output: An ensemble classifier H

1: Step 1: Learn first-level classifiers
2: **for** $t \leftarrow 1$ to T **do**
3:     Learn a base classifier h_t based on $\mathcal{D}$
4: **end for**
5: Step 2: Construct new data sets from $\mathcal{D}$
6: **for** $i \leftarrow 1$ to m **do**
7:     Construct a new data set that contains $\{\mathbf{x}_i', y_i\}$, where $\mathbf{x}_i' = \{h_1(\mathbf{x}_i), h_2(\mathbf{x}_i), \ldots, h_T(\mathbf{x}_i)\}$
8: **end for**
9: Step 3: Learn a second-level classifier
10: Learn a new classifier h' based on the newly constructed data set
11: **return** $H(\mathbf{x}) = h'(h_1(\mathbf{x}), h_2(\mathbf{x}), \ldots, h_T(\mathbf{x}))$

choices to learn base classifiers: 1) We can apply Bootstrap sampling technique to learn independent classifiers; 2) we can adopt the strategy used in Boosting, i.e., adaptively learn base classifiers based on data with a weight distribution; 3) we can tune parameters in a learning algorithm to generate diverse base classifiers (homogeneous classifiers); 4) we can apply different classification methods and/or sampling methods to generate base classifiers (heterogeneous classifiers).

- Step 2: Construct a new data set based on the output of base classifiers. Here, the output predicted labels of the first-level classifiers are regarded as new features, and the original class labels are kept as the labels in the new data set. Assume that each example in $\mathcal{D}$ is $\{\mathbf{x}_i, y_i\}$. We construct a corresponding example $\{\mathbf{x}_i', y_i\}$ in the new data set, where $\mathbf{x}_i' = \{h_1(\mathbf{x}_i), h_2(\mathbf{x}_i), \ldots, h_T(\mathbf{x}_i)\}$.

- Step 3: Learn a second-level classifier based on the newly constructed data set. Any learning method could be applied to learn the second-level classifier.

Once the second-level classifier is generated, it can be used to combine the first-level classifiers. For an unseen example $\mathbf{x}$, its predicted class label of stacking is $h'(h_1(\mathbf{x}), h_2(\mathbf{x}), \ldots, h_T(\mathbf{x}))$, where $\{h_1, h_2, \ldots, h_T\}$ are first-level classifiers and h' is the second-level classifier.

We can see that Stacking is a general framework. We can plug in different learning approaches or even ensemble approaches to generate first or second level classifiers. Compared with Bagging and Boosting, Stacking "learns" how to combine the base classifiers instead of voting.

Example. We show the basic procedure of Stacking using the data set in Table 19.2. We set $T = 2$ at the first step of Algorithm 19.7 and show the results of the two base classifiers trained on original data in Table 19.13.

TABLE 19.13: Stacking Step 1 on the Toy Example

t		Weak Classifier h_t									
1	x	1	2	3	4	5	6	7	8	9	10
	y	+1	+1	+1	-1	-1	-1	-1	-1	-1	-1
2	x	1	2	3	4	5	6	7	8	9	10
	y	-1	-1	-1	-1	-1	-1	-1	+1	+1	+1

Then we can construct a new data set based on the output of base classifiers. Since there are two base classifiers, our new x_i' has two dimensions: $x_i' = (x_i^1, x_i^2)$, where x_i^1 is x_i's predicted label from the first classifier, and x_i^2 is the predicted label from the second classifier. The new data set is shown in Table 19.14.

TABLE 19.14: Stacking Step 2 on the Toy Example

x	(1,-1)	(1,-1)	(1,-1)	(-1,-1)	(-1,-1)
y	+1	+1	+1	-1	-1
x	(-1,-1)	(-1,-1)	(-1,1)	(-1,1)	(-1,1)
y	-1	-1	+1	+1	+1

Note that in the illustrations of various ensemble algorithms on this toy example shown in the previous sections, we always use one-dimensional data on which we simply apply a threshold-based classifier. In this Stacking example, we can easily extend the threshold-based classifier to two dimensions:

$$h(x) = \begin{cases} -1, & \text{if } x^1 = -1 \text{ and } x^2 = -1 \\ +1, & \text{otherwise.} \end{cases} \tag{19.4}$$

The final results on the toy example obtained by this second-level classifier are shown in Table 19.15. We can see that the ensemble classifier classifies all the points correctly.

TABLE 19.15: Stacking Final Result on the Toy Example

x	1	2	3	4	5	6	7	8	9	10
y	+1	+1	+1	-1	-1	-1	-1	+1	+1	+1

19.5.2 Stacking and Cross-Validation

In Algorithm 19.7, we use the same data set $\mathcal{D}$ to train first-level classifiers and prepare training data for second-level classifiers, which may lead to over-fitting. To solve this problem, we can incorporate the idea of cross validation [49] in stacking. K-fold cross validation is the most commonly used technique to evaluate classification performance. To evaluate the prediction ability of a learning algorithm, we conduct the following K-fold cross validation procedure: We partition training data into K disjoint subsets and run the learning algorithm K times. Each time we learn a classifier from $K - 1$ subsets and use the learnt model to predict on the remaining one subset and obtain the learning accuracy. The final prediction accuracy is obtained by averaging among the K runs.

As shown in Algorithm 19.8, the idea of cross validation can be combined with Stacking to avoid using the same training set for building both first and second level classifiers. Instead of using all the training examples to get first-level classifiers, we now partition data in $\mathcal{D}$ into K subsets and get K classifiers, each of which is trained only on $K - 1$ subsets. Each classifier is applied to the remaining one subset and the output of all the first-level classifiers constitute the input feature space for the second-level classifier. Note that over-fitting is avoided because the data to train first-level classifiers and the data to receive predicted labels from first-level classifiers are different. After we build the second-level classifier based on the first-level classifiers' predicted labels, we can re-train first-level classifiers on the whole training set $\mathcal{D}$ so that all the training examples are used. Applying the second-level classifier on the updated first-level classifier output will give us the final ensemble output.

Algorithm 19.8 Stacking with K-fold Cross Validation

Input: Training data $\mathcal{D} = \{\mathbf{x}_i, y_i\}_{i=1}^m$ ($\mathbf{x}_i \in \mathbb{R}^n$, $y_i \in \mathcal{Y}$)
Output: An ensemble classifier H

1: Step 1: Adopt cross validation approach in preparing a training set for second-level classifier
2: Randomly split $\mathcal{D}$ into K equal-size subsets: $\mathcal{D} = \{\mathcal{D}_1, \mathcal{D}_2, \ldots, \mathcal{D}_K\}$
3: **for** $k \leftarrow 1$ to K **do**
4:     Step 1.1: Learn first-level classifiers
5:     **for** $t \leftarrow 1$ to T **do**
6:         Learn a classifier h_{kt} from $\mathcal{D} \setminus \mathcal{D}_k$
7:     **end for**
8:     Step 1.2: Construct a training set for second-level classifier
9:     **for** $\mathbf{x}_i \in \mathcal{D}_k$ **do**
10:         Get a record $\{\mathbf{x}_i', y_i\}$, where $\mathbf{x}_i' = \{h_{k1}(\mathbf{x}_i), h_{k2}(\mathbf{x}_i), \ldots, h_{kT}(\mathbf{x}_i)\}$
11:     **end for**
12: **end for**
13: Step 2: Learn a second-level classifier
14: Learn a new classifier h' from the collection of $\{\mathbf{x}_i', y_i\}$
15: Step 3: Re-learn first-level classifiers
16: **for** $t \leftarrow 1$ to T **do**
17:     Learn a classifier h_t based on $\mathcal{D}$
18: **end for**
19: **return** $H(\mathbf{x}) = h'(h_1(\mathbf{x}), h_2(\mathbf{x}), \ldots, h_T(\mathbf{x}))$

19.5.3 Discussions

Empirical results [19] show that Stacking has robust performance and often out-performs Bayesian model averaging. Since its introduction [89], Stacking has been successfully applied to a wide variety of problems, such as regression [12], density estimation [79] and spam filtering [68]. Recently, Stacking has shown great success in the Netflix competition, which was an open competition on using user history ratings to predict new ratings of films. Many of the top teams employ ensemble techniques and Stacking is one of the most popular approaches to combine classifiers among teams. In particular, the winning team [5] adopted Stacking (i.e., blending) to combine hundreds of different models, which achieved the best performance.

There are several important practical issues in Stacking. It is important to consider what types of feature to create for the second-level classifier's training set, and what type of learning methods to use in building the second-level classifier [89]. Besides using predicted class labels of first-level classifiers, we can consider using class probabilities as features [82]. The advantage of using conditional probabilities as features is that the training set of second-level classifier will include not only predictions but also prediction confidence of first-level classifiers. In [82], the authors further suggest using multi-response linear regression, a variant of least-square linear regression, as the second-level learning algorithm. There are other choices of second-level classification algorithms. For example, Feature-Weighted Linear Stacking [76] combines classifiers using a linear combination of meta features.

19.6 Recent Advances in Ensemble Learning

Ensemble methods have emerged as a powerful technique for improving the robustness as well as the accuracy of base models. In the past decade, there have been numerous studies on the problem of combining competing models into a committee, and the success of ensemble techniques has been observed in multiple disciplines and applications. In this section, we discuss several advanced topics and recent development in the field of ensemble learning.

Ensemble Learning Techniques. We have discussed the variants and follow-up work of bagging, boosting, Bayesian model averaging, and stacking in the previous sections. Besides these approaches, some other techniques have been developed to generate an ensemble model. In [36], an importance sampling learning ensemble (ISLE) framework was proposed to provide a unified view of existing ensemble methods. Under this framework, an ensemble model is a linear model in a very high dimensional space in which each point in the space represents a base model and the coefficients of the linear model represent the model weights. The ISLE framework provides a way to sample various "points" in the hypothesis space to generate the base models, and then we can learn the model coefficients using regularized linear regression. Based on this ISLE framework, a rule ensemble approach [37] has been proposed to combine a set of simple rules. Each rule consists of a conjunction of statements based on attribute values, and thus the advantage of rule ensemble technique is that it provides a much more interpretable model compared with other ensemble approaches. In [29, 84], the authors analyze ensemble of kernel machines such as SVM. Especially in [84], the authors suggest two ways to develop SVM-based ensemble approaches from the error decomposition point of view: One is to employ low-bias SVMs as base learners in a bagged ensemble, and the other is to apply bias-variance analysis to construct a heterogeneous, diverse set of accurate and low-bias classifiers. In recent years, there have been many tutorials, surveys, and books on ensemble learning [4, 23, 43, 52, 63, 74, 75, 90], that summarize methodologies, analysis, and applications of various ensemble learning approaches.

Performance Analysis. There have been many efforts on analyzing the performance of ensemble approaches. The main theory developed to explain the success of ensemble learning is based on bias-variance error decomposition [27]. Specifically, the prediction error of a learning algorithm can be expressed as the sum of three non-negative errors: Intrinsic noise, bias, and variance. Intrinsic noise is unavoidable error of the task, bias is the difference between the expected output and the true value, and variance measures the fluctuations in the predictions using different training sets of the same size. Under this error decomposition framework, bagging is an approach that greatly reduces variance especially when its base classifiers are unstable with large variance, and boosting reduces both bias and variance on weak classifiers. There is analysis of bias-variance decomposition on other ensemble methods, such as neural networks [83] and SVM [84]. The effectiveness of ensemble approaches can also be explained from the perspective of learning margins, i.e., ensemble approaches can enlarge the margin and thus improve the generalization abilities [59].

Ensemble Diversity and Pruning. As the diversity of base models plays an important role in building a good ensemble model, many approaches have been proposed to create or select a set of diverse base models for ensemble learning. In [60], a set of diverse models are constructed using artificially constructed training examples. We may want to remove the models that are inaccurate or redundant. In [54, 81], several diversity measures are studied and their relationships with classification accuracy are explored. A special issue of the journal *Information Fusion* was dedicated to the issue of diversity in ensemble learning [53]. After generating a set of base classifiers, we can still remove inaccurate, irrelevant, or redundant ones to ensure better ensemble quality. Moreover, reducing the size of the ensemble can help alleviate the usage of storage and computational resources. Therefore, another line of research is to select a subset of classifiers from the whole collection so that the combination of the selected classifiers can achieve comparable or even better performance compared with the ensemble using all the classifiers. These approaches are usually referred to as ensemble pruning, which can be roughly categorized into the following groups [90]: 1) Ordering-based pruning, which orders the models according to a certain criterion and then selects the top models, 2) clustering-based pruning, which clusters similar models and prunes within each cluster to maintain a set of diverse models, and 3) optimization-based pruning, which tries to find a subset of classifiers by optimizing a certain objective function with regard to the generalization ability of the ensemble built on selected classifiers.

Ensemble Learning in Different Scenarios. Ensemble learning has also been studied in the context of challenging learning scenarios. In a stream environment where data continuously arrive, some ensemble approaches [38, 39, 50, 73, 86] have been developed to handle concept drifts, i.e., the fact that data distributions change over time. Many classification tasks encounter the class imbalance problem in which examples from one class dominate the training set. Classifiers learnt from this training set will perform well on the majority class but poorly on the other classes [87]. One effective technique to handle class imbalance is to create a new training set by over-sampling minority examples or under-sampling majority examples [3, 16]. Naturally such sampling techniques can be combined with ensemble approaches to improve the performance of the classifier on minority examples [17, 39, 44]. Ensemble methods have also been adapted to handle cost-sensitive learning [25, 30, 85], in which misclassification on different classes incurs different costs and the goal is to minimize the combined cost. Moreover, ensemble learning has been shown to be useful in not only classification, but also many other learning tasks and applications. One very important lesson learnt from the widely known Netflix competition is the effectiveness of ensemble techniques in collaborative filtering [5, 51]. The top teams all somewhat involve blending the predictions of a variety of models that provide complementary predictive powers. In unsupervised learning tasks such as clustering, many methods have been developed to integrate various clustering solutions into one clustering solution by reaching consensus among base models [41, 80]. Recently, people began to explore ensemble learning techniques for combining both supervised and unsupervised models [2, 40, 58]. The objective of these approaches is to reach consensus among base models considering both predictions made by supervised models and clustering constraints given by unsupervised models. Another relevant topic is multi-view learning [9, 18, 77, 78], which assumes different views share similar target functions and learns classifiers from multiple views of the same objects by minimizing classification error as well as inconsistency among classifiers.

19.7 Conclusions

In this chapter, we gave an overview of ensemble learning techniques with a focus on the most popular methods including Bayesian model averaging, bagging, boosting, and stacking. We discussed various base model generation and model combination strategies used in these methods. We also discussed advanced topics in ensemble learning including alternative techniques, performance analysis, ensemble pruning and diversity, and ensemble used in other learning tasks.

Although the topic of ensemble learning has been studied for decades, it still enjoys great attention in many different fields and applications due to its many advantages in various learning tasks. Nowadays, rapidly growing information and emerging applications continuously pose new challenges for ensemble learning. In particular, the following directions are worth investigating in the future:

- Big data brings big challenges to data analytics. Facing the daunting scale of big data, it is important to adapt ensemble techniques to fit the needs of large-scale data processing.

For example, ensemble models developed on parallel processing and streaming platforms are needed.

- Another important issue in ensemble learning is to enhance the comprehensibility of the model. As data mining is used to solve problems in different fields, it is essential to have the results interpretable and accessible to users with limited background knowledge in data mining.

- Although numerous efforts on theoretical analysis of ensemble learning have been made, there is still a largely unknown space to explore to fully understand the mechanisms of ensemble models. Especially the analysis so far has been focusing on traditional approaches such as bagging and boosting, but it is also important to explain ensemble approaches from theoretical perspectives in emerging problems and applications.

New methodologies, principles, and technologies will be developed to address these challenges in the future. We believe that ensemble learning will continue to benefit real-world applications in providing an effective tool to extract highly accurate and robust models from gigantic, noisy, dynamically evolving, and heterogeneous data.

Bibliography

[1] Wikipedia. http://en.wikipedia.org/wiki/Ensemble_learning.

[2] A. Acharya, E. R. Hruschka, J. Ghosh, and S. Acharyya. C 3e: A framework for combining ensembles of classifiers and clusterers. In *Proceedings of the International Workshop on Multiple Classifier Systems (MCS'11)*, pages 269–278, 2011.

[3] G. Batista, R. Prati, and M. C. Monard. A study of the behavior of several methods for balancing machine learning training data. *SIGKDD Explorations Newsletter*, 6:20–29, 2004.

[4] E. Bauer and R. Kohavi. An empirical comparison of voting classification algorithms: Bagging, boosting, and variants. *Machine Learning*, 36(1-2):105–139, 2004.

[5] R. M. Bell and Y. Koren. Lessons from the Netflix prize challenge. *SIGKDD Explorations Newsletter*, 9(2):75–79, 2007.

[6] K. P. Bennett. Linear programming boosting via column generation. In *Machine Learning*, pages 225–254, 2002.

[7] P. J. Bickel, Y. Ritov, and A. Zakai. Some theory for generalized boosting algorithms. *The Journal of Machine Learning Research*, 7:705–732, 2006.

[8] C. M. Bishop. *Pattern recognition and Machine Learning*. Springer, 2006.

[9] A. Blum and T. Mitchell. Combining labeled and unlabeled data with co-training. In *Proceedings of the Annual Conference on Computational Learning Theory (COLT'98)*, pages 92–100, 1998.

[10] J. K. Bradley and R. Schapire. Filterboost: Regression and classification on large datasets. *Advances in Neural Information Processing Systems*, 20:185–192, 2008.

[11] L. Breiman. Bagging predictors. *Machine Learning*, 24(2):123–140, 1996.

[12] L. Breiman. Stacked regressions. *Machine Learning*, 24(1):49–64, 1996.

[13] L. Breiman. Random forests. *Machine Learning*, 45(1):5–32, 2001.

[14] L. Breiman. Population theory for boosting ensembles. *The Annals of Statistics*, 32(1):1–11, 2004.

[15] L. Breiman, J. H. Friedman, C. J. Stone, and R. A. Olshen. *Classification and Regression Trees*. Chapman and Hall/CRC, 1984.

[16] N. V. Chawla, K. W. Bowyer, L. O. Hall, and W. P. Kegelmeyer. SMOTE: Synthetic minority over-sampling technique. *Journal of Artificial Intelligence Research*, 16:321–357, 2002.

[17] N. V. Chawla, A. Lazarevic, L. O. Hall, and K. W. Bowyer. Smoteboost: Improving prediction of the minority class in boosting. In *Proceedings of the European Conference on Principles and Practice of Knowledge Discovery in Databases (PKDD'03)*, pages 107–119, 2003.

[18] C. M. Christoudias, R. Urtasun, and T. Darrell. Multi-view learning in the presence of view disagreement. *CoRR*, abs/1206.3242, 2012.

[19] B. Clarke. Comparing Bayes model averaging and stacking when model approximation error cannot be ignored. *The Journal of Machine Learning Research*, 4:683–712, 2003.

[20] C. Cortes and V. Vapnik. Support-vector networks. *Machine Learning*, 20(3):273–297, 1995.

[21] T. Cover and P. Hart. Nearest neighbor pattern classification. *IEEE Transactions on Information Theory*, 13:21–27, 1967.

[22] N. Cristianini and J. Shawe-Taylor. *An Introduction to Support Vector Machines: And Other Kernel-Based Learning Methods*. Cambridge University Press, 2000.

[23] T. Dietterich. Ensemble methods in machine learning. In *Proceedings of the International Workshop on Multiple Classifier Systems (MCS'00)*, pages 1–15, 2000.

[24] C. Domingo and O. Watanabe. Madaboost: A modification of adaboost. In *Proceedings of the Thirteenth Annual Conference on Computational Learning Theory (COLT'00)*, pages 180–189, 2000.

[25] P. Domingos. Metacost: A general method for making classifiers cost-sensitive. In *Proceedings of the ACM SIGKDD International Conference on Knowledge Discovery and Data Mining (KDD'99)*, pages 155–164, 1999.

[26] P. Domingos. Bayesian averaging of classifiers and the overfitting problem. In *Proceedings of the International Conference on Machine Learning (ICML'00)*, pages 223–230, 2000.

[27] P. Domingos. A unified bias-variance decomposition and its applications. In *Proceedings of the International Conference on Machine Learning (ICML'00)*, pages 231–238, 2000.

[28] B. Efron and R. Tibshirani. *An Introduction to the Bootstrap*. Chapman and Hall, London, 1993.

[29] T. Evgeniou, M. Pontil, and A. Elisseeff. Leave one out error, stability, and generalization of voting combinations of classifiers. *Machine Learning*, 55(1):71–97, 2004.

[30] W. Fan, S. J. Stolfo, J. Zhang, and P. K. Chan. Adacost: Misclassification cost-sensitive boosting. In *Proceedings of the International Conference on Machine Learning (ICML'99)*, pages 97–105, 1999.

[31] W. Fan, H. Wang, P. S. Yu, and S. Ma. Is random model better? on its accuracy and efficiency. In *Proc. of the IEEE International Conference on Data Mining (ICDM'03)*, pages 51–58, 2003.

[32] Y. Freund. An adaptive version of the boost by majority algorithm. *Machine Learning*, 43(3):293–318, 2001.

[33] Y. Freund and R. E. Schapire. Experiments with a new boosting algorithm. In *Proceedings of the International Conference on Machine Learning (ICML'96)*, pages 148–156, 1996.

[34] Y. Freund and R. E. Schapire. A decision-theoretic generalization of on-line learning and an application to boosting. *Journal of Computer and System Sciences*, 55(1):119–139, 1997.

[35] J. H. Friedman and P. Hall. On bagging and nonlinear estimation. *Journal of Statistical Planning and Inference*, 137(3):669–683, 2007.

[36] J. H. Friedman and B. E. Popescu. Importance sampled learning ensembles. Technical report, Statistics, Stanford University, 2003.

[37] J. H. Friedman and B. E. Popescu. Predictive learning via rule ensembles. *Annals of Applied Statistics*, 3(2):916–954, 2008.

[38] J. Gao, W. Fan, and J. Han. On appropriate assumptions to mine data streams: Analysis and practice. In *Proceedings of the IEEE International Conference on Data Mining (ICDM'07)*, pages 143–152, 2007.

[39] J. Gao, W. Fan, J. Han, and P. S. Yu. A general framework for mining concept-drifting data streams with skewed distributions. In *Proceedings of the SIAM International Conference on Data Mining (SDM'07)*, pages 3–14, 2007.

[40] J. Gao, F. Liang, W. Fan, Y. Sun, and J. Han. Graph-based consensus maximization among multiple supervised and unsupervised models. In *Advances in Neural Information Processing Systems*, pages 585–593, 2009.

[41] J. Ghosh and A. Acharya. Cluster ensembles. *WIREs Data Mining and Knowledge Discovery*, 1:305–315, 2011.

[42] D. Haussler, M. Kearns, and R. Schapire. Bounds on the sample complexity of Bayesian learning using information theory and the VC dimension. In *Machine Learning*, pages 61–74, 1992.

[43] D. Hernández-Lobato, G. M.-M. Noz, and I. Partalas. Advanced topics in ensemble learning ecml/pkdd 2012 tutorial. `https://sites.google.com/site/ecml2012ensemble/`.

[44] S. Hido, H. Kashima, and Y. Takahashi. Roughly balanced bagging for imbalanced data. *Statistical Analysis and Data Mining*, 2(5-6):412–426, 2009.

[45] T. K. Ho. Random decision forests. In *Proceedings of the International Conference on Document Analysis and Recognition*, pages 278–282, 1995.

[46] J. Hoeting, D. Madigan, A. Raftery, and C. Volinsky. Bayesian model averaging: a tutorial. *Statistical Science*, 14(4):382–417, 1999.

[47] D. W. Hosmer and S. Lemeshow. *Applied Logistic Regression*. Wiley-Interscience Publication, 2000.

[48] M. Kearns. Thoughts on hypothesis boosting, 1988. `http://www.cis.upenn.edu/~mkearns/papers/boostnote.pdf`.

[49] R. Kohavi. A study of cross-validation and bootstrap for accuracy estimation and model selection. In *Proceedings of International Joint Conference on Artificial Intelligence (IJCAI'95)*, pages 1137–1145, 1995.

[50] J. Kolter and M. Maloof. Using additive expert ensembles to cope with concept drift. In *Proceedings of the International Conference on Machine Learning (ICML'05)*, pages 449–456, 2005.

[51] Y. Koren, R. Bell, and C. Volinsky. Matrix factorization techniques for recommender systems. *Computer*, 42(8):30–37, 2009.

[52] L. I. Kuncheva. *Combining Pattern Classifiers: Methods and Algorithms*. John Wiley & Sons, 2004.

[53] L. I. Kuncheva. Diversity in multiple classifier systems. *Information Fusion*, 6(1):3–4, 2005.

[54] L. I. Kuncheva and C. J. Whitaker. Measures of diversity in classifier ensembles and their relationship with the ensemble accuracy. *Machine Learning*, 51(2):181–207, 2003.

[55] F. T. Liu, K. M. Ting, and W. Fan. Maximizing tree diversity by building complete-random decision trees. In *Proceedings of the 9th Pacific-Asia conference on Advances in Knowledge Discovery and Data Mining (PAKDD'05)*, pages 605–610, 2005.

[56] F. T. Liu, K. M. Ting, Y. Yu, and Z.-H. Zhou. Spectrum of variable-random trees. *Journal of Artificial Intelligence Research*, 32:355–384, 2008.

[57] F. T. Liu, K. M. Ting, and Z.-H. Zhou. Isolation forest. In *Proceedings of the IEEE International Conference on Data Mining (ICDM'08)*, pages 413–422, 2008.

[58] X. Ma, P. Luo, F. Zhuang, Q. He, Z. Shi, and Z. Shen. Combining supervised and unsupervised models via unconstrained probabilistic embedding. In *Proceedings of the International Joint Conference on Artifical Intelligence (IJCAI'11)*, pages 1396–1401, 2011.

[59] L. Mason, P. L. Bartlett, and J. Baxter. Improved generalization through explicit optimization of margins. *Machine Learning*, 38(3):243–255, 2000.

[60] P. Melville and R. J. Mooney. Creating diversity in ensembles using artificial data. *Information Fusion*, 6:99–111, 2006.

[61] T. M. Mitchell. *Machine Learning*. McGraw-Hill, 1997.

[62] K. Monteith, J. L. Carroll, K. D. Seppi, and T. R. Martinez. Turning Bayesian model averaging into Bayesian model combination. In *Proceedings of the International Joint Conference on Neural Networks*, pages 2657–2663, 2011.

[63] R. Polikar. Ensemble based systems in decision making. *IEEE Circuits and Systems Magazine*, 6(3):21–45, 2006.

[64] J. R. Quinlan. Induction of decision trees. *Machine Learning*, 1(1):81–106, 1986.

[65] J. R. Quinlan. *C4.5: Programs for Machine Learning*. Morgan Kaufmann, 1993.

[66] R. Rojas. Adaboost and the super bowl of classifiers: A tutorial introduction to adaptive boosting, 2009. `http://www.inf.fu-berlin.de/inst/ag-ki/adaboost4.pdf`.

[67] D. E. Rumelhart, J. L. McClelland, and the. PDP Research Group, editors. *Parallel Distributed Processing: Explorations in the Microstructure of Cognition, Vol. 1: Foundations.* MIT Press, 1986.

[68] G. Sakkis, I. Androutsopoulos, G. Paliouras, V. Karkaletsis, C. D. Spyropoulos, and P. Stamatopoulos. Stacking classifiers for anti-spam filtering of e-mail. *arXiv preprint cs/0106040*, 2001.

[69] R. E. Schapire. The strength of weak learnability. *Machine Learning*, 5(2):197–227, 1990.

[70] R. E. Schapire. The boosting approach to machine learning: An overview. In D. D. Denison, M. H. Hansen, C. Holmes, B. Mallick, and B. Yu, editors, *Nonlinear Estimation and Classification.* Springer, 2003.

[71] R. E. Schapire, Y. Freund, P. Bartlett, and W. S. Lee. Boosting the margin: A new explanation for the effectiveness of voting methods. *The Annals of Statistics*, 26(5):1651–1686, 1998.

[72] R. E. Schapire and Y. Singer. Improved boosting algorithms using confidence-rated predictions. In *Machine Learning*, pages 80–91, 1999.

[73] M. Scholz and R. Klinkenberg. An ensemble classifier for drifting concepts. In *Proceedings of the ECML/PKDD Workshop on Knowledge Discovery in Data Streams*, pages 53–64, 2005.

[74] G. Seni and J. Elder. *Ensemble methods in data mining: Improving accuracy through combining predictions.* Morgan & Claypool, 2010.

[75] M. Sewell. Ensemble learning research note. Technical report, University College London, 2011.

[76] J. Sill, G. Takács, L. Mackey, and D. Lin. Feature-weighted linear stacking. *arXiv:0911.0460*, 2009.

[77] V. Sindhwani, P. Niyogi, and M. Belkin. A co-regularization approach to semi-supervised learning with multiple views. In *Proc. of the ICML'05 workshop on Learning with Multiple Views*, 2005.

[78] V. Sindhwani and D. S. Rosenberg. An RKHS for multi-view learning and manifold co-regularization. In *Proceedings of the International Conference on Machine Learning (ICML'08)*, pages 976–983, 2008.

[79] P. Smyth and D. Wolpert. Linearly combining density estimators via stacking. *Machine Learning*, 36(1-2):59–83, 1999.

[80] A. Strehl and J. Ghosh. Cluster ensembles — a knowledge reuse framework for combining multiple partitions. *Journal of Machine Learning Research*, 3:583–617, 2003.

[81] E. K. Tang, P. N. Suganthan, and X. Yao. An analysis of diversity measures. *Machine Learning*, 65(1):247–271, 2006.

[82] K. M. Ting and I. H. Witten. Issues in stacked generalization. *arXiv:1105.5466*, 2011.

[83] K. Tumer and J. Ghosh. Analysis of decision boundaries in linearly combined neural classifiers. *Pattern Recognition*, 29(2):341–348, 1996.

[84] G. Valentini and T. G. Dietterich. Bias-variance analysis of support vector machines for the development of svm-based ensemble methods. *Journal of Machine Learning Research*, 5:725–775, 2004.

[85] P. Viola and M. Jones. Fast and robust classification using asymmetric adaboost and a detector cascade. In *Advances in Neural Information Processing System 14*, pages 1311–1318, 2001.

[86] H. Wang, W. Fan, P. Yu, and J. Han. Mining concept-drifting data streams using ensemble classifiers. In *Proceedings of the ACM SIGKDD International Conference on Knowledge Discovery and Data Mining (KDD'03)*, pages 226–235, 2003.

[87] G. Weiss and F. Provost. The effect of class distribution on classifier learning. Technical Report ML-TR-43, Rutgers University, 2001.

[88] P. J. Werbos. Beyond regression: new tools for prediction and analysis in the behavioral sciences. PhD thesis, Harvard University, 1974.

[89] D. H. Wolpert. Stacked generalization. *Neural networks*, 5(2):241–259, 1992.

[90] Z.-H. Zhou. *Ensemble Methods: Foundations and Algorithms*. Chapman & Hall/CRC Machine Learning & Pattern Recognition Series, 2012.

Chapter 20

Semi-Supervised Learning

Kaushik Sinha

Wichita State University
Wichita, KS
kaushik.sinha@wichita.edu

20.1 Introduction

Consider an input space X and an output space Y, where one would like to see an example from input space X and automatically predict its output. In traditional supervised learning, a learning algorithm is typically given a training set of the form $\{(x_i, y_i)\}_{i=1}^{l}$, where each pair $(x_i, y_i) \in X \times Y$ is drawn independently at random according to an unknown joint probability distribution $P_{X \times Y}$. In the case of a supervised *classification* problem, Y is a finite set of class labels and the goal of the

511

learning algorithm is to construct a function $g : X \rightarrow Y$ that predicts the label y given x. For example, consider the problem of predicting if an email is a spam email. This is a binary classification problem where $Y = \{+1, -1\}$ and the label "$+1$" corresponds to a spam email and the label "-1" corresponds to a non-spam email. Given a training set of email-label pairs, the learning algorithm needs to learn a function that given a new email, would predict if it is a spam email.

The criterion for choosing this function g is the low probability of error, i.e., the algorithm must choose g in such a way that when an unseen pair $(x, y) \in X \times Y$ is chosen according to $P_{X \times Y}$ and only x is presented to the algorithm, the probability that $g(x) \neq y$ is minimized over a class of functions $g \in G$. In this case, the best function that one can hope for is based on the conditional distribution $P(Y|X)$ and is given by $\eta(x) = \text{sign}[\mathbb{E}(Y|X = x)]$, which is known as the so called *Bayes optimal* classifier. Note that, when a learning algorithm constructs a function g based on a training set of size l, the function g is a random quantity and it depends on the training set size l. So it is a better idea to represent the function g as g_l to emphasize its dependence on training set size l. A natural question, then, is what properties should one expect from g_l as the training set size l increases? A learning algorithm is called *consistent* if g_l converges to η (in appropriate convergence mode) as the training set size l tends to infinity. This is the best one can hope for, as it guarantees that as the training set size l increases, g_l converges to the "right" function. Of course, "convergence rate" is also important, which specifies how fast g_l converges to the right function. Thus, one would always prefer a consistent learning algorithm. It is clear that performance of such an algorithm improves as the training set size increases, or in other words, as we have capacity to label more and more examples.

Unfortunately, in many real world classification tasks, it is often a significant challenge to acquire enough labeled examples due to the prohibitive cost of labeling every single example in a given data set. This is due to the fact that labeling often requires domain specific knowledge and in many cases, a domain expert has to actually look at the data to manually label it! In many real life situations, this can be extremely expensive and time consuming. On the other hand, in many practical situations, a large pool of unlabeled examples are quite easy to obtain at no extra cost as no labeling overhead is involved. Consider the email classification problem again. To figure out if an email is a spam email, someone has to read its content to make the judgment, which may be time consuming (and expensive if one needs to be hired to do this job), while unlabeled emails are quite easy to obtain: just take all the emails from one's email account.

Traditional supervised classification algorithms only make use of the labeled data, and prove insufficient in the situations where the number of labeled examples are limited, since as pointed out before, performance of a consistent supervised algorithm improves as the number of labeled examples (training set size) increases. Therefore, if only a limited amount of labeled examples is available, performance of supervised classification algorithms may be quite unsatisfactory. As an alternative, semi-supervised learning, i.e., learning from both labeled and unlabeled, data has received considerable attention in recent years due to its potential in reducing the need for expensive labeled data. The hope is that somehow the power of unlabeled examples drawn i.i.d according the marginal distribution P_X may be exploited to complement the unavailability of labeled examples and design better classification algorithms. This can be justified in many situations, for example, as shown in the following two scenarios.

Consider the first scenario. Suppose in a binary classification problem, only two labeled examples are given, one from each class, as shown in the left panel of Figure 20.1. A natural choice to induce a classifier on the basis of this would be the linear separator as shown by a dotted line in the left panel of Figure 20.1. As pointed out in [8], a variety of theoretical formalisms including Bayesian paradigms, regularization, minimum description length, or structural risk minimization principles, and the like, have been constructed to rationalize such a choice (based on prior notion of simplicity) as to why the linear separator is the simplest structure consistent with the data. Now, consider the situation where, in addition to the two labeled examples, one is given additional unlabeled examples as shown in the right panel of Figure 20.1. It is quite evident that in light of this new set of unlabeled examples, one must re-evaluate one's prior notion of simplicity as linear separator is clearly not an ideal choice anymore. The particular geometric structure of the marginal distribution in this case suggests that the most natural classifier is a non-linear one and hence when labeled examples are limited, unlabeled examples can indeed provide useful guidance towards learning a better classifier.

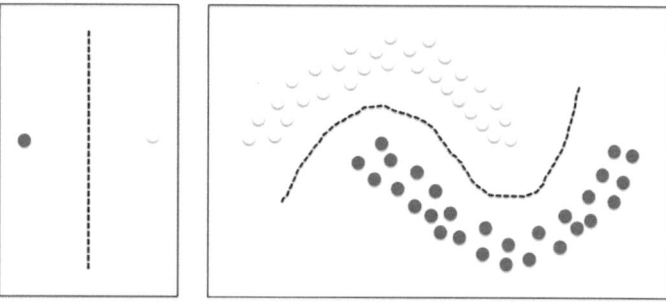

FIGURE 20.1 (See color insert.): Unlabeled examples and prior belief.

Now, consider the second scenario where the focus is again on binary classification. Here suppose the examples are generated according to two Gaussian distributions, one per class, as shown in red and green, respectively, in Figure 20.2. The corresponding Bayes-optimal decision boundary, which classifies examples into the two classes and provides minimum possible error, is shown by the dotted line. The Bayes optimal decision boundary can be obtained from the Bayes rule, if the Gaussian mixture distribution parameters (i.e., the mean and variance of each Gaussian, and the mixing parameter between them) are known. It is well known that unlabeled examples alone, when generated from a mixture of two Gaussians, are sufficient to recover the original mixture components ([41,42]). However, unlabeled examples alone cannot assign examples to classes. Labeled examples are needed for this purpose. In particular, when the decision regions are already known, only a few labeled examples would be enough to assign examples to classes. Therefore, when an infinite amount of unlabeled examples along with a few labeled examples are available, learning proceeds in two steps: (a) identify the mixture parameters and hence the decision regions from unlabeled examples, (b) use the labeled examples to assign class labels to the learned decision regions in step (a). In practice, a sufficiently large set of unlabeled examples often suffices to estimate the mixture parameters to reasonable accuracy.

The ubiquity and easy availability of unlabeled examples, together with the increased computational power of modern computers, make the paradigm of semi-supervised learning quite attractive in various applications, while connections to natural learning make it also conceptually intriguing. Vigorous research over the last decade in this domain has resulted in several models, numerous associated semi-supervised learning algorithms, and theoretical justifications. To some extent, these efforts have been documented in various dedicated books and survey papers on semi-supervised learning (see e.g., [19,52,72,74,75]).

What is missing in those documents, however, is a comprehensive treatment of various semi-supervised models, associated algorithms, and their theoretical justifications, in an unified manner. This vast gap is quite conceivable since the field of semi-supervised learning is relatively new and most of the theoretical results in this field are known only recently. The goal of this survey chapter is to close this gap as much as possible. While our understanding of semi-supervised learning is not yet complete, various recent theoretical results provide significant insight towards this direction and it is worthwhile to study these results along with the models and algorithms in an unified manner. To this end, this survey chapter aims, by no means, to cover every single semi-supervised learning algorithm developed in the past. Instead, the modest goal is to cover various popular semi-supervised

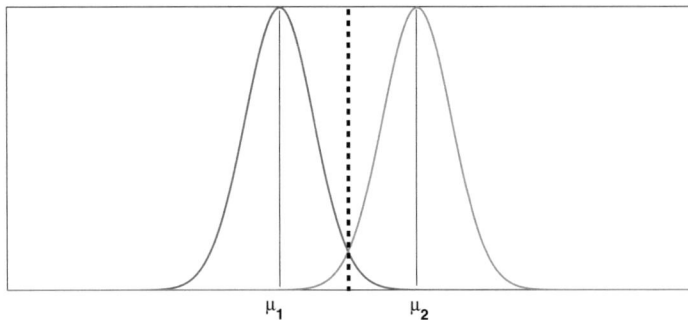

μ_1 μ_2

FIGURE 20.2: In Gaussian mixture setting, the mixture components can be fully recovered using only unlabeled examples, while labeled examples are used to assign labels to the individual components.

learning models, some representative algorithms within each of these models, and various theoretical justifications for using these models.

20.1.1 Transductive vs. Inductive Semi-Supervised Learning

Almost all semi-supervised learning algorithms can broadly be categorized into two families, namely, *transductive* an *inductive* semi-supervised learning algorithms. In a transductive setting, labels of unlabeled examples are estimated by learning a function that is defined only over a given set of labeled and unlabeled examples. In this case, no explicit classification model is developed and as a result, the algorithm cannot infer labels of any unseen example that is not part of this given set of unlabeled examples. That is, there is no out-of-sample extension in this case. In an inductive setting, on the other hand, both labeled and unlabeled examples are use to train a classification model that can infer the label of any unseen unlabeled example outside the training set. As a result, inductive semi-supervised learning algorithms are sometimes referred to as truly semi-supervised learning algorithms.

Because of the ability to handle unseen examples outside the training set, inductive semi-supervised learning algorithms are more preferable. Unfortunately, the majority of the semi-supervised learning algorithms are transductive in nature. The inductive semi-supervised learning algorithms include, e.g., manifold regularization ([8,55]), transductive support vector machine ([33]) etc.

20.1.2 Semi-Supervised Learning Framework and Assumptions

In standard semi-supervised learning, labeled examples are generated according to the joint probability distribution $P_{X \times Y}$, while the unlabeled examples are generated according to the marginal probability distribution P_X. The hope here is that the knowledge of the marginal distribution P_X in the form of unlabeled examples can be exploited for better classifier learning at the cost of a fewer labeled examples. Of course, if there is no identifiable relationship between the marginal distribution P_X and the conditional distribution $P(y|x)$, the knowledge of P_X is unlikely to be of any use and as a result, use of unlabeled examples in learning is questionable. Therefore, in order to make use of unlabeled examples meaningfully, certain assumptions need to be made.

In semi-supervised learning there are broadly two main assumptions.

1. **Manifold Assumption** ([8])

According to this assumption, the marginal distribution P_X is supported on a low dimensional manifold embedded in a high dimensional ambient space. The conditional probability distribution $P(y|x)$ varies smoothly along the geodesics in the intrinsic geometry of P_X. In other words, if two points $x_1, x_2 \in X$ are closed in the intrinsic geometry of P_X, then the conditional distributions $P(y|x_1)$ and $P(y|x_2)$ are similar and the corresponding labels are also similar.

2. **Cluster Assumption** ([17])

According to this assumption, the marginal density P_X has a cluster (multi-modal) structure, where each cluster has a homogeneous or "pure" label in the sense that marginal density is piecewise constant on each of these clusters. In particular, the cluster assumption can be interpreted as saying that two points are likely to have the same class labels if they can be connected by a path passing through high density regions only. In other words, two high density regions with different class labels must be separated by a low density region.

Both of the above assumptions are based on strong intuition and empirical evidence obtained from real life high dimensional data and have led to development of broad families of semi-supervised learning algorithms, by introducing various data dependent regularizations. For example, various graph-based algorithms, in some way or the other, try to capture the manifold structure by constructing a neighborhood graph whose nodes are labeled and unlabeled examples.

While the majority of of the semi-supervised learning algorithms are based on either the manifold assumption or the cluster assumption, there are two other principles that give rise to various other semi-supervised learning algorithms as well. These are :

1. **Co-Training**

In the co-training model ([10]), the instance space has two different yet redundant views, where each view in itself is sufficient for correct classification. In addition to that, if the underlying distribution is "compatible" in the sense that it assigns zero probability mass to the examples that differ in prediction according to the two consistent classifiers in those two views, then one can hope that unlabeled examples might help in learning.

2. **Generative Model**

In the generative model, labeled examples are generated in two steps: first, by randomly selecting a class label, and then in the second step, by selecting an example from this class. The class conditional distributions belong to a parametric family. Given unlabeled examples from the marginal (mixture) distribution, one can hope to learn the parameters by using maximum likelihood estimate as long as the mixture is identifiable, then by using maximum posteriori estimate to infer the label of an unlabeled example.

In the subsequent sections, we discuss in detail various semi-supervised learning algorithms based on manifold/cluster assumption, co-training, and generative models and also provide theoretical justification and relevant results for each of these models.

20.2 Generative Models

Generative models are perhaps the oldest semi-supervised learning methods. In a generative model, labeled instances (x, y) pairs are believed to be generated by a two step process :

- a class of labels y is chosen from a finite number of class labels (in case of binary classification $y \in \{+1, -1\}$) according to probability $p(y)$.

- once a class label is chosen, an instance from this class is generated according to the class conditional probability $p(x|y)$.

Generative models assume that marginal distribution $p(x) = \sum_y p(y)p(x|y)$ is an identifiable mixture model and the class conditional distributions $p(x|y)$ are parametric. In particular, the class conditional distributions $p(x|y, \theta)$ are parameterized by parameter vector θ. The parameterized joint distribution then can be written as $p(x, y|\theta) = p(y|\theta)p(x|y, \theta)$. A simple application of Bayes rule suggests that $p(y|x, \theta) = \frac{p(y|\theta)p(x|y, \theta)}{p(x|\theta)}$. For any θ, the label of any instance x is given by the maximum a posteriori (MAP) estimate $\arg\max_y p(y|x, \theta)$. Since θ is not known in advance, the goal of the generative semi-supervised learning is to make an estimate $\hat{\theta}$ of the parameter vector θ from data and then use

$$\hat{y} = \arg\max_y p(y|x, \hat{\theta}).$$

20.2.1 Algorithms

A practical semi-supervised learning algorithm under the generative model is proposed in [19, 45] for text classification, where each individual mixture component $p(x|y, \theta)$ is a multinomial distribution over the words in a vocabulary. In order to learn a classifier in this framework, the first step is to estimate the parametric model from data. This is done by maximizing the observable likelihood incorporating l labeled and u unlabeled examples $\arg\max_\theta p(\theta)(\theta)p(x, y|\theta)$, which is equivalent to maximizing the log likelihood:

$$l(\theta) = \log(p(\theta)) + \sum_{i=1}^{l} \log\left(p(y_i|\theta) \cdot p(x_i|y_i, \theta)\right) + \sum_{i=l+1}^{l+u} \log\left(\sum_y \left(p(y|\theta) \cdot p(x_i|y, \theta)\right)\right), \quad (20.1)$$

where the prior distribution is formed with a product of Dirichlet distributions, one for each of the class multinomial distributions and one for the overall class probabilities. Notice that the equation above contains a log of sums for the unlabeled data, which makes maximization by partial derivatives computationally intractable. As an alternative, Expectation-Maximization (EM) algorithm ([28]) is used to maximize Equation 20.1. EM is an iterative hill climbing procedure that finds the local maxima of Equation 20.1 by performing alternative E and M steps until the log likelihood converges. The procedure starts with an initial guess of the parameter vector θ. At any iteration, in the E step, the algorithm estimates the expectation of the missing values (unlabeled class information) given the model parameter estimates in the previous iteration. The M step, on the other hand, maximizes the likelihood of the model parameters using the expectation of the missing values found in the E step. The same algorithm was also used by Baluja ([5]) on a face orientation discrimination task yielding good performance.

If data are indeed generated from an identifiable mixture model and if there is a one-to-one correspondence between mixture components and classes, maximizing observable likelihood seems to be a reasonable approach to finding model parameters, and the resulting classifier using this method performs better than those trained only from labeled data. However, when these assumptions do not hold, the effect of unlabeled data is less clear and in many cases it has been reported that unlabeled data do not help or even degrade performance ([24]).

20.2.2 Description of a Representative Algorithm

This section describes the algorithmic details of a semi-supervised learning algorithm using EM under the generative model assumption proposed in [45], which suggests how unlabeled data can be used to improve a text classifier. In this setting, every document is generated according to a probability distribution defined by a set of parameters, denoted by θ. The probability distribution consists of a mixture of components $c_j \in C = \{c_1, \ldots, c_{|C|}\}$. Each component is parameterized by a disjoint subset of θ. A document, d_i, is created by first selecting a mixture component according to the mixture weights (or class prior probabilities), $P(c_j|\theta)$, and then generating a document according to the selected component's own parameters, with distribution $P(d_i|c_j;\theta)$. The parameters of an individual mixture component are a multinomial distribution over words, i.e., the collection of word probabilities, each written $\theta_{w_t|c_j}$, such that $\theta_{w_t|c_j} = P(w_t|c_j;\theta)$, where $t = \{1, \ldots, |V|\}$, V is the size of vocabulary, and $\sum_t P(w_t|c_j;\theta) = 1$. The only other parameters of the model are the mixture weights (class prior probabilities), written as θ_{c_j}, which indicate the probabilities of selecting the different mixture components. Thus, the complete collection of model parameters, θ, is a set of multinomials and prior probabilities over those multinomials: $\theta = \{\theta_{w_t|c_j} : w_t \in V, c_j \in C; \theta_{c_j} : c_j \in C\}$.

The classifier used in this approach is a naive Bayes classifier that makes the standard assumption that the words of a document are generated independently of context, that is, independently of the other words in the same document given the class label. Learning a naive Bayes text classifier consists of estimating the parameters of the generative model by using a set of labeled training data, $\mathcal{D} = \{d_1, \ldots, d_{|\mathcal{D}|}\}$. The estimate of θ is written as $\hat{\theta}$. Naive Bayes uses the maximum a posteriori estimate, thus finding $\arg\max_\theta P(\theta|\mathcal{D})$. This is the value of θ that is most probable given the evidence of the training data and a prior. Using Bayes rule, this is equivalent to finding $\arg\max_\theta P(\theta)P(\mathcal{D}|\theta)$.

In semi-supervised setting, only some subset of the documents $d_i \in \mathcal{D}^l \subset \mathcal{D}$ come with class labels $y_i \in C$, and for the rest of the documents, in subset $\mathcal{D}^u = \mathcal{D} \setminus \mathcal{D}^l$, the class labels are unknown. Learning a classifier is approached as calculating a maximum a posteriori estimate of θ, i.e. $\arg\max_\theta P(\theta)P(\mathcal{D}|\theta)$. To estimate the missing labels (of unlabeled documents), EM algorithm is applied in naive Bayes settings. Applying EM to naive Bayes is quite straightforward. First, the naive Bayes parameters, $\hat{\theta}$, are estimated from just the labeled documents. Then, the classifier is used to assign probabilistically weighted class labels to each unlabeled document by calculating expectations of the missing class labels, $P(c_j|d_i;\hat{\theta})$. This is the E-step. Next, new classifier parameters, $\hat{\theta}$, are estimated using all the document, both the original and newly labeled documents. This is the M-step. These last two steps are iterated until $\hat{\theta}$ does not change. The details are given in Algorithm 20.1.

Algorithm 20.1 EM under Generative Model assumption

Input : Collection $\mathcal{D}^l$ of labeled documents and $\mathcal{D}^u$ of unlabeled documents.
Output : A classifier, $\hat{\theta}$, that takes an unlabeled document and predicts a class label.

1. Build an initial Naive Bayes classifier, $\hat{\theta}$, from the labeled documents, $\mathcal{D}^l$, only.
   Use maximum a posteriori parameter estimation to find $\hat{\theta} = \arg\max_\theta P(\mathcal{D}|\theta)P(\theta)$.

2. Loop while classifier parameters improve, as measured by the change in $l_c(\theta|\mathcal{D};z)$ (the complete log probability of the labeled and unlabeled data, and the prior).

   (a) **(E-step)** Use the current classifier, $\hat{\theta}$, to estimate component membership of each unlabeled documents, i.e., the probability that each mixture component (and class) generated each document, $P(c_j|d_i;\hat{\theta})$.

   (b) **(M-step)** Re-estimate the classifier, $\hat{\theta}$, given the estimated component membership of each document. Use maximum a posteriori parameter estimation to find $\hat{\theta} = \arg\max_\theta P(\mathcal{D}|\theta)P(\theta)$.

20.2.3 Theoretical Justification and Relevant Results

When the examples are indeed generated from an identifiable mixture model, the hope is that the mixture components can be identified from unlabeled examples alone. Ideally, one then would only need one labeled example per component (class) to fully determine the mixture distribution (that is, to figure out which component corresponds to which class and identify the decision regions). In fact, this is the exact motivation to understand and quantify the effect of unlabeled examples in classification problems under generative model assumption in many research endeavors ([14–16, 49,50]).

Castelli and Cover ([15,16]) investigate the usefulness of unlabeled examples in the asymptotic sense, with the assumption that a number of unlabeled examples go to infinity at a rate faster than that of the labeled examples. Under the identifiable mixture model assumption and various mild assumptions, they argue that marginal distribution can be estimated from unlabeled examples alone and labeled examples are merely needed for identifying the decision regions. Under this setting, they prove that classification error decreases exponentially fast towards the Bayes optimal solution with the number of labeled examples, and only polynomially with the number of unlabeled examples, thereby, justifying the usefulness of unlabeled examples. In a similar but independent effort, Ratsaby and Venkatesh ([49], [50]) prove similar results when individual class conditional densities are Gaussian distributions.

However, both the above results inherently assume that estimators can replicate the *correct* underlying distribution that generates the data. However, in many real life learning problems, this may be a difficult assumption to satisfy. Consider the situation where true underlying marginal distribution is f, and since f is unknown, it is assumed (maybe for computational convenience) that marginal distribution belongs to the class of probability distribution $\mathcal{G}$. Now, if unlabeled examples are used to estimate the marginal distribution, as a number of unlabeled examples are added one can get better and better estimates and eventually the best possible estimate g^* within class $\mathcal{G}$, which, however, may be quite different from true underlying distribution f. The fact that g^* is the best possible estimator doesn't mean that g^* leads to the best possible classification boundary. This fact is nicely demonstrated by a toy example in [19,24], where the true underlying marginal distribution is a mixture of beta distributions but they are modeled using a mixture of Gaussian distributions, which leads to incorrect classification boundary and degraded classification performance. Cozman and Cohen ([24]) use this informal reasoning to justify the observation that in certain cases unlabeled data actually causes a degraded classification performance, as reported in the literature [5,45,54].

The informal reasoning of Cozmen and Cohen ([24]) is, to some extent, formalized in a recent work by Belkin and Sinha ([57]). Belkin and Sinha ([57]) investigate the situation when the data come from a probability distribution, which can be modeled, but not perfectly, by an identifiable mixture distribution. This seems applicable to many practical situations, when, for example, a mixture of Gaussians is used to model the data without knowing what the underlying true distribution is. The main finding of this work is that when the underlying model is different from the assumed model and unlabeled examples are used to estimate the marginal distribution under the assumed model, putting one's effort into finding the best estimator under the assumed model is not a good idea since it is different from the true underlying model anyway, and hence does not guarantee identifying the best decision boundary. Instead, a "reasonable" estimate, where the estimation error is on the order of the difference between assumed and true model is good enough, because beyond that point unlabeled examples do not provide much useful information. Depending on the size of perturbation or difference between the true underlying model and the assumed model, this work shows that the data space can be partitioned into different regimes where labeled and unlabeled examples play different roles in reducing the classification error and labeled examples are not always *exponentially* more valuable as compared to the unlabeled examples as mentioned in the literature ([14–16,49,50]).

In a recent work, Dillon et al. ([29]) study the asymptotic analysis of generative semi-supervised learning. This study allows the number of labeled examples l and number of unlabeled examples u to grow simultaneously but at a constant ratio $\lambda = l/(l+u)$. The authors consider a stochastic setting for maximizing the joint likelihood of data under the generative model, where the estimate

θ_n for the model parameter θ using $n = l + u$ examples is given by

$$\hat{\theta}_n = \text{argmax}_\theta \sum_{i=1}^{l} Z_i \log p(x_i, y_i | \theta) + \sum_{i=l+1}^{l+u} (1 - Z_i) \log p(x_i | \theta) \qquad (20.2)$$

where $Z_i \sim \text{Ber}(\lambda)$ are Bernoulli random variables with parameter λ. Under mild assumptions, the authors prove that the empirical estimate $\hat{\theta}_n$ converges to the true model parameter θ_0 with probability 1, as $n \to \infty$. In addition, they prove that $\hat{\theta}_n$ is asymptotically normal, i.e., $\sqrt{n}(\hat{\theta}_n - \theta_0)$ converges in distribution to $\mathcal{N}(0, \Sigma^{-1})$, where $\Sigma = \lambda \text{Var}_{\theta_0} (\nabla_\theta \log p(x, y | \theta_0)) + (1 - \lambda) \text{Var}_{\theta_0} (\nabla_\theta \log p(x | \theta_0))$ and that it may be used to characterize the the accuracy of $\hat{\theta}_n$ as a function of n, θ_0, λ. This result suggests that asymptotic variance is a good proxy for finite sample measures of error rates and empirical mean squared error, which the authors also verify empirically.

20.3   Co-Training

Blum and Mitchell introduce co-training ([10]) as a general term for rule based bootstrapping, in which examples are believed to have two distinct yet sufficient feature sets (views) and each rule is based entirely on either the first view or the second view. This method enjoys reasonable success in scenarios where examples can naturally be thought of as having two views. One prototypical example for application of the co-training method is the task of Web page classification. Web pages contain text that describes contents of the Web page (the first view) and have hyperlinks pointing to them (the second view). The existance of two different and somewhat redundant sources of information (two views) and the availability of unlabeled data suggest the following learning strategy ([10]) for classifying, say, faculty Web pages. First, using small set of labeled examples, find weak predictors based on each kind of information, e.g., the phrase "research interest" on a Web page may be a weak indicator that the page is a faculty home page and the phrase "my adviosr" on a link may be an indicator that the page being pointed to is a faculty Web page. In the second stage attempt is made to bootstrap from these weak predictors using unlabeled data.

In co-training model, the instance space is $X = X_1 \times X_2$, where X_1 and X_2 correspond to two different views of an example, in other words, each example x is written as a pair (x_1, x_2), where $x_1 \in X_1$ and $x_2 \in X_2$. The label space is $\mathcal{Y}$. There are distinct concept classes C_1 and C_2 defined over X_1 and X_2, respectively, i.e., C_1 consists of functions predicting label $\mathcal{Y}$ from X_1 and C_2 consists of functions predicting labels $\mathcal{Y}$ from X_2. The central assumption here is that each view in itself is sufficient for correct classification. This boils down to the existence of $c_1 \in C_1$ and $c_2 \in C_2$ such that $c_1(x) = c_2(x) = y$ for any example $x = (x_1, x_2) \in X_1 \times X_2$ observed with label $y \in \mathcal{Y}$. This means that the probability distribution P_X over X assigns zero mass to any eample (x_1, x_2) such that $c_1(x_1) \neq c_2(x_2)$. Co-training makes a crucial assumption that the two views x_1 and x_2 are conditionally independent given the label y, i.e., $P(x_1 | y, x_2) = P(x_1 | y)$ and $P(x_2 | y, x_1) = P(x_2 | y)$.

The reason why one can expect that unlabeled data might help in this context is as follows. For a given distribution P_X over X the target function $c = (c_1, c_2)$ is *compatible* with P_X if it assigns zero mass to any eample (x_1, x_2) such that $c_1(x_1) \neq c_2(x_2)$. As a result, even though C_1 and C_2 may be large concept classes with high complexity, say in terms of high VC dimension measure, for a given distribution P_X, the set of compatible target concepts may be much simpler and smaller. Thus, one might hope to be able to use unlabeled examples to get a better understanding of which functions in $C = C_1 \times C_2$ are compatible, yielding useful information that might reduce the number of labeled examples needed by a learning algorithm.

20.3.1 Algorithms

Co-training is an iterative procedure. Given a set l of labeled examples and u unlabeled examples, the algorithm first creates a smaller pool of $u', (u' < u)$ unlabeled examples and then it iterates the following procedure. First l labeled examples are used to train two distinct classifiers $h_1 : X_1 \rightarrow Y$ and $h_2 : X_2 \rightarrow Y$, where h_1 is trained based on the x_1 view of the instances and h_2 is trained based on the x_2 view of the instances. Typically, naive Bayes classifier is used for the choice of h_1 and h_2. After the two classifiers are trained, each of these classifiers h_1 and h_2 are individually allowed to choose, from the pool of u' unlabeled examples, some positive and and negative examples that they are confident about. Typically, in the implementation in [10], the number of positive examples is set to one and the number of negative examples is set to three. Each example selected in this way along with their labels predicted by h_1 and h_2, respectively, are added to the set of labeled examples and then h_1 and h_2 are retrained again based on this larger set of labeled examples. The set of unlabeled examples u' is replenished from u and the procedure iterates for a certain number of iterations.

If the conditional independence assumption holds, then co-training works reasonably well as is reported by Nigam et al. ([46]), who perform extensive empirical experiments to compare co-training and generative mixture models. Collin and Singer ([23]) suggest a refinement of the co-training algorithm in which one explicitly optimizes an objective function that measures the degree of agreement between the predictions based on x_1 view and those based on x_2 view. The objective function is further boosted by methods suggested in [23]. Unlike the co-training procedure suggested in [10], Goldman and Zhou ([31]) make no assumption about the existence of two redundant views but insist that their co-training strategy places the following requirement on each supervised learning algorithm. The hypothesis of supervised learning algorithm partitions the example space into a set of equivalence classes. In particular, they use two learners of different types that require the whole feature set (as opposed to individual views) for training and essentially use one learner's high confidence examples, identified by a set of statistical tests, from the unlabeled set to teach the other learner and vice versa. Balcan and Blum ([4]) demonstrate that co-training can be quite effective, in the sense that in some extreme cases, only one labeled example is all that is needed to learn the classifier. Over the years, many other modifications of co-training algorithms have been suggested by various researchers. See for example [20,35,36,66,68,69] for many of these methods and various empirical studies.

20.3.2 Description of a Representative Algorithm

The co-training algorithm proposed by Blum and Mitchell ([10]) is shown in Algorithm 20.2. Given a set L of labeled examples and a set U of unlabeled examples, the algorithm first creates a smaller pool U' of u unlabeled examples. It then iterates the following steps k times : Using training set L, the classifier h_1 is trained using Naive Bayes Algorithm based on the x_1 portion (view) of the data. Similarly, using training set L again, the classifier h_2 is trained using Naive Bayes Algorithm, but this time, based on the x_2 portion (view) of the data. Then, each of these two classifiers h_1 and h_2 is allowed to examine U' and select p examples it most confidently labels as positive and n examples it most confidently labels as negative. These $2p + 2n$ examples along with their predicted labels (by h_1 and h_2) are added to L and finally $2p + 2n$ examples are randomly chosen from U to replenish U'. In the implementation of [10], the algorithm parameters are set as $p = 1, n = 3, k = 30$ and $u = 75$.

20.3.3 Theoretical Justification and Relevant Results

In order to work, co-training effectively requires two distinct properties of the underlying data distribution. The first is that, in principle, there should exist low error classifiers c_1 and c_2 on each of the two views. The second is that these two views should not be *too correlated*. In particular,

Algorithm 20.2 Co-training

Input : A set L of labeled examples, a set U of unlabeled examples.

1. Create a pool U' of examples by choosing u examples at random from U.

2. Loop for k iterations.

    (a) Use L to train a classifier h_1 that considers only the x_1 portion of x.

    (b) Use L to train a classifier h_2 that considers only the x_2 portion of x.

    (c) Allow h_1 to label p positive and n negative examples from U'.

    (d) Allow h_2 to label p positive and n negative examples from U'.

    (e) Add these self labeled examples to L.

    (f) Randomly choose $2p + 2n$ examples from U to replenish U'.

for co-training to actually do anything useful, there should be at least some examples for which the hypothesis based on the first view should be confident about their labels but not the other hypothesis based on the second view, and vice versa.

Unfortunately, in order to provide theoretical justification for co-training, often, strong assumptions about the second type are made. The original work of Blum and Mitchell ([10]) makes the conditional independence assumption and provides intuitive explanation regarding why co-training works in terms of maximizing agreement on unlabeled examples between classifiers based on different views of the data. However, the conditional independence assumption in that work is quite strong and no generalization error bound for co-training or justification for the intuitive account in terms of classifier agreement on unlabeled examples is provided. Abney ([1]) shows that the strong independence assumption is often violated in data and in fact a weaker assumption in the form of "weak rule dependence" actually suffices.

Even though Collins and Singer ([23]) introduce the degree of agreement between the classifiers based on two views in the objective function, no formal justification was provided until the work of Dasgupta et. al ([26]). Dasgupta et. al ([26]) provide a generalization bound for co-training under partial classification rules. A partial classification rule either outputs a class label or outputs a special symbol $\perp$ indicating no opinion. The error of a partial rule is the probability that the rule is incorrect given that it has an opinion. In [26], a bound on the generalization error of any two partial classification rules h_1 and h_2 on two views, respectively, is given, in terms of the empirical agreement rate between h_1 and h_2. This bound formally justifies both the use of agreement in the objective function (as was done in [23]) and the use of partial rules. The bound shows the potential power of unlabeled data, in the sense that low generalization error can be achieved for complex rules with a sufficient number of unlabeled examples.

All the theoretical justification mentioned above in [1, 10, 26] requires either conditional independence given the labels assumption or the assumption of weak rule dependence. Balcan et. al ([3]) substantially relax the strength of the conditional independence assumption to just a form of "expansion" of the underlying distribution (a natural analog of the graph theoretic notion of expansion and conductance) and show that in some sense, this is a necessary condition for co-training to succeed. However, the price that needs to be paid for this relaxation is a fairly strong assumption on the learning algorithm, in the form that the hypotheses the learning algorithm produces are never "confident but wrong," i.e., they are correct whenever they are confident, which formally translates to the condition that the algorithm is able to learn from positive examples only. However, only a heuristic analysis is provided in [3] when this condition does not hold.

20.4 Graph-Based Methods

Graph-based approaches to semi-supervised learning are perhaps the most popular and widely used semi-supervised learning techniques due to their good performance and ease of implementation. Graph-based semi-supervised learning treats both labeled and unlabeled examples as vertices (nodes) of a graph and builds pairwise edges between these vertices, which are weighed by the affinities (similarities) between the corresponding example pairs. The majority of these methods usually are non-parametric, discriminative, and transductive in nature; however, some graph-based methods are inductive as well. These methods usually assume some sort of label smoothness over the graph.

Many of these graph-based methods can be viewed as estimating a function $f : V \rightarrow \mathbb{R}$ on the graph, where V is the set of nodes in a graph, satisfying the following two requirements: (a) the value of f at any node, for which the label is known, should be close to the given label for that node, and (b) f should be smooth on the whole graph; in particular this means, that if two nodes on the graph are "close" then the corresponding labels predicted by f should be similar. These two requirements can be formulated in a regularization framework involving two terms, where the first term is loss function that measures how well the function f predicts the labels of the nodes for which labels are already known (labeled examples) and the second term is a regularizer that assigns higher penalty for the function being non-smooth. Many of the graph-based methods fall under this regularization framework and differ only in the choice of loss function and/or regularizer. However, in most of the cases the regularizer has a common format. Let f_i denote the value of function f at the ith node (i.e., i^{th} example) and let W be a $(l+u) \times (l+u)$ weight matrix such that its $(i,j)^{th}$ entry w_{ij} is the weight (similarity measure) between the ith and j^{th} node (example x_i and x_j, respectively). Then the following is a common form of regularizer

$$f^T L f = f^T (D - W) f = \frac{1}{2} \sum_{i,j=1}^{l+u} w_{ij} (f_i - f_j)^2 \qquad (20.3)$$

where D is a diagonal matrix whose i^{th} entry is $d_i = \sum_{j=1}^{l+u} w_{ij}$ and $L = D - W$ is the graph Laplacian.

20.4.1 Algorithms

Various graph-based techniques fall under four broad categories: graph cut-based method, graph-based random walk methods, graph transduction methods, and manifold regularization methods. Sometimes the boundary between these distinctions is a little vague. Nevertheless, some representative algorithms from each of these subgroups are given below. In recent years, large scale graph-based learning has also emerged as an important area ([40]). Some representative work along this line is also provided below.

20.4.1.1 Graph Cut

Various graph-based semi-supervised learning methods are based on graph cuts ([11,12,34,37]). Blum and Chawla ([11]) formulate the semi-supervised problem as a graph mincut problem, where in the binary case positive labels act as sources and negative labels act as sinks. The objective here is to find a minimum set of edges that blocks the flow from sources to sinks. Subsequently, the nodes connected to sources are labeled as positive examples and the the nodes connected to the sinks are labeled as negative examples. As observed in [32], graph mincut problem is equivalent to solving for the lowest energy configuration in the Markov Random Field, where the energy is expressed as

$$E(f) = \frac{1}{2} \sum_{i,j} w_{ij} |f_i - f_j| = \frac{1}{4} \sum_{i,j} w_{ij} (f_i - f_j)^2$$

where $f_i \in \{+1, -1\}$ are binary labels. Solving the lowest energy configuration in this Markov random field produces a partion of the entire (labeled and unlabeled) dataset that maximally optimizes self consistency, subject to the constraint that configuration must agree with the labeled examples. This technique is extended in [12] by adding randomness to the graph structure to address some of the shortcomings of [11] as well as to provide better theoretical justification from both the Markov random field perspective and from sample complexity considerations.

Kveton et al. ([37]) propose a different way of performing semi-supervised learning based on graph cut by first computing a harmonic function solution on the data adjacency graph (similar to [73]), then by learning a maximum margin discriminator conditioned on the labels induced by this solution.

20.4.1.2 Graph Transduction

In graph transduction method a graph is constructed whose nodes are both labeled and unlabeled examples. Graph transduction refers to predicting the labels of the unlabeled examples (the nodes of the already constructed graph) by exploiting underlying graph structure and thus lacking the ability to predict labels of unseen examples that are not part of this graph.

The Gaussian random fields and harmonic function methods of Zhu et al. ([73]) is a continuous relaxation of the discrete Markov random fields of [11]. The optimization problem here has the form

$$\underset{f \in \mathbb{R}^{l+u}}{\arg \min} \sum_{i,j=1}^{l+u} w_{ij}(f_i - f_j)^2 \text{ such that } f_i = y_i \text{ for } i = 1, \dots, l. \tag{20.4}$$

Solution of the above optimization problem has a harmonic property, meaning that the value of f at each unlabeled example is the average of f at the neighboring examples :

$$f_j = \frac{1}{d_j} \sum_{i \sim j} w_{ij} f_i \text{ for } j = l+1, \dots, l+u$$

where the notation $i \sim j$ is used to represent the fact that node i and node j are neighbors.

Belkin et al. ([6]) propose the following method using Tikhonov regularization, where the regularization parameter λ controls the effect of data fitting term and smoothness term.

$$\underset{f \in \mathbb{R}^{l+u}}{\arg \min} \sum_{i=1}^{l} (f_i - y_i)^2 + \lambda \sum_{i,j=1}^{l+u} w_{ij}(f_i - f_j)^2 \tag{20.5}$$

Zhou et. al ([65]) propose an iterative method that essentially solves the following optimization problem.

$$\underset{f \in \mathbb{R}^{l+u}}{\arg \min} \sum_{i=1}^{l+u} (f_i - y_i)^2 + \lambda \sum_{i,j=1}^{l+u} w_{ij} \left(\frac{f_i}{\sqrt{d_i}} - \frac{f_j}{\sqrt{d_j}} \right)^2$$

where $y_i = 0$ for the unlabeled examples i.e., $i = l+1, \dots, l+u$. Here, the smoothness term incorporates the local changes of function between nearby points. The local variation is measured on each edge. However, instead of simply defining the local variation on an edge by the difference of function values on the two ends of the edge, the smoothness term essentially splits the function value at each point among the edges attached to it before computing the local changes where the value assigned to each edge is proportional to its weight.

Belkin et. al ([7]) use eigenvectors of the graph Laplacian L to form basis vectors and the classification function is obtained as a linear combination of these basis vectors where the coefficients are obtained by using labeled examples. With the hope that data lie on the low dimensional manifold, only the first k (for reasonable choice of k) basis vectors are used. In particular, suppose

$v_1, \ldots, v_k \in \mathbb{R}^{l+u}$ are the first k eigenvectors of the graph Laplacian. Then, in [7], essentially the following least squared problem is solved :

$$\hat{\alpha} = \arg\min_{\alpha} \sum_{i=1}^{l} \left(y_i - \sum_{j=1}^{k} \alpha_j v_j(x_i) \right)^2 \tag{20.6}$$

where the notation $v_j(x_i)$ means the value of the jth eigenvector on example x_i.

Note that in all the above cases, it is straightforward to get a closed form solution of the above optimization problems.

Zhou et al. ([70]) propose an algorithm where instead of the widely used regularizer $f^T L f$, an iterated higher order graph Laplacian regularizer in the form of $f^T L^m f$, which corresponds to Sobolev semi-norm of order m, is used. The optimization problem in this framework takes the form

$$\arg\min_{f \in \mathbb{R}^{l+u}} \sum_{i=1}^{l} (f_i - y_i)^2 + \lambda f^T L^m f$$

where again, λ controls the effect of data fitting term and smoothness term.

20.4.1.3 Manifold Regularization

Belkin et al. ([8]) propose a family of learning algorithms based on a form of regularization that allows us to exploit the geometry of the marginal distribution when the support of the marginal distribution is a compact manifold $\mathcal{M}$. This work explicitly makes the specific assumption about the connection between the marginal and the conditional distribution in the form that if two points $x_1, x_2 \in X$ are closed in intrinsic geometry of the marginal distribution P_X, then the conditional distributions $P(y|x_1)$ and $P(y|x_2)$ are similar. In other words, the conditional probability distribution $P(y|x)$ varies smoothly along the geodesics in the intrinsic geometry of P_X. Note that this is nothing but the manifold assumption. To incorporate this fact, the regularizer in their formulation consists of an additional appropriate penalty term for the intrinsic structure of P_X. In fact when P_X is unknown, the optimization problem becomes :

$$f^* = \arg\min_{f \in \mathcal{H}_K} \frac{1}{l} \sum_{i=1}^{l} V(x_i, y_i, f) + \gamma_A \|f\|_K^2 + \gamma_I \int_{x \in \mathcal{M}} \|\nabla_{\mathcal{M}} f\|^2 dP_X(x)$$

where $\mathcal{H}_K$ is the reproducing kernel Hilbert space (RKHS), $\|f\|_K$ is the RKHS norm of f, and $\nabla_{\mathcal{M}} f$ is the gradient of f along the manifold $\mathcal{M}$ where the integral is taken over the marginal distribution. This additional term is smoothness penalty corresponding to the probability distribution and may be approximated on the basis of labeled and unlabeled examples using a graph Laplacian associated to the data. The regularization parameter γ_A controls the complexity of the function in the ambient space while γ_I controls the complexity of the function in the intrinsic geometry of P_X. Given finite labeled and unlabeled examples, the above problem takes the form

$$
\begin{aligned}
f^* &= \arg\min_{f \in \mathcal{H}_K} \frac{1}{l} \sum_{i=1}^{l} V(x_i, y_i, f) + \gamma_A \|f\|_K^2 + \frac{\gamma_I}{(u+l)^2} \sum_{i,j=1}^{l+u} (f(x_i) - f(x_j))^2 w_{ij} \\
&= \arg\min_{f \in \mathcal{H}_K} \frac{1}{l} \sum_{i=1}^{l} V(x_i, y_i, f) + \gamma_A \|f\|_K^2 + \frac{\gamma_I}{(u+l)^2} f^T L f.
\end{aligned}
\tag{20.7}
$$

Belkin et al. ([8]) show that the minimizer of the optimization problem (20.7) admits an expansion of the form

$$f^*(x) = \sum_{i=1}^{l+u} \alpha_i K(x_i, x)$$

where $K(\cdot, \cdot)$ is a kernel function, and as a result it is an inductive procedure and has a natural out-of-sample extension unlike many other semi-supervised learning algorithms.

In [55], Sindhwani et al. describe a technique to turn transductive and standard supervised learning algorithms into inductive semi-supervised learning algorithms. They give a data dependent kernel that adapts to the geometry of data distribution. Starting with a base kernel K defined over the whole input space, the proposed method warps the RKHS specified by K by keeping the same function space but altering the norm by adding a data dependent "point-cloud-norm." This results in a new RKHS space with a corresponding new kernel that deforms the original space along a finite dimensional subspace defined by data. Using the new kernel, standard supervised kernel algorithms trained on labeled examples only can perform inductive semi-supervised learning.

20.4.1.4 Random Walk

In general, Markov random walk representation exploits any low dimensional structure present in the data in a robust and probabilistic manner. A Markov random walk is based on a locally appropriate metric that is the basis for a neighborhood graph, associated weights on the edges of this graph, and consequently the transition probabilities for the random walk. For example, if w_{ij} is the weight between i^{th} and j^{th} node on a graph, then one-step transition probabilities p_{ik} from i to k are obtained directly from these weights: $p_{ik} = w_{ik} / (\sum_j w_{ij})$.

Szummer et al. ([59]) propose a semi-sepervised learning algorithm by assuming that the starting point for the Markov random walk is chosen uniformly at random. With this assumption it is easy to evaluate the probability $P_{0|t}(i|k)$, that is, Markov process started from node i given that it is ended in node k after t time steps. These conditional probabilities define a new representation for the examples. Note that this representation is crucially affected by the time scale parameter t. When $t \to \infty$, all points become indistinguishable if the original neighborhood graph is connected. On the other hand, small values of t merge points in small clusters. In this representation t controls the resolution at which we look at the data points. Given labeled and unlabeled examples, to classify the unlabeled examples in this model, it is assumed that each example has a label or a distribution $P(y|i)$ over the class labels. These distributions are typically unknown and need to be estimated. For example the kth node, which may be labeled or unlabeled, is interpreted as a sample from a t step Markov random walk started at some other node on the graph. Since labels are associated with original starting points, posterior probability of the label for node k is given by

$$P_{\text{post}}(y|k) = \sum_i P(y|i) P_{0|t}(i|k).$$

Finally, class label for node k is chosen as the one that maximizes the posterior, i.e., $\arg\max_c P_{\text{post}}(y = c|k)$. The unknown parameters $P(y|i)$ are estimated either by maximum likelihood method with EM or maximum margin methods subjected to constraints.

Azron ([2]) propose a semi-supervised learning method in which each example is associated with a particle that moves between the examples (nodes a of a graph) according to the transition probability matrix. Labeled examples are set to be absorbing states of the Markov random walk and the probability of each particle to be absorbed by the different labeled examples, as the number of steps increases, is used to derive a distribution over associated missing labels. This algorithm is in the spirit of [59] but there is a considerable difference. In [59], the random walk carries the given labels and propagates them on the graph amongst the unlabeled examples, which results in the need to find a good number of steps t for the walk to take. In [2], however, this parameter is set to infinity.

Note that the algorithm in [65] can also be interpreted from the random walk point of view, as was done in [67].

20.4.1.5   Large Scale Learning

In recent years, graph-based semi-supervised learning algorithms have been applied to extremely large data sizes, at the level of hundreds of millions of data records or more ([40]), where the major bottleneck is the time for graph construction, which is $O(dn^2)$, where n examples (both labeled and unlabeled) lie in $\mathbb{R}^d$. For Web scale applications, the quadratic dependence is not desirable. In recent years many solutions have been proposed based on sub sampling, sparsity constraints, Nystrom approximation, approximation of eigenfunctions of 1-D graph Laplacian, landmark and anchor graphs, and so on ([27, 30, 39, 61, 62, 64]). A detailed survey and recent progresses on large scale semi-supervised learning can be found in [40].

20.4.2   Description of a Representative Algorithm

This section describes the algorithmic details of a popular graph-based algorithm: the manifold regularization algorithm ([8]) mentioned in Section 20.4.1.3. Manifold regularized framework incorporates an intrinsic smoothness penalty term in the standard regularized optimization framework and solves both the Laplacian regularized least square problem (LapRLS) where the optimization problem is of the form

$$\underset{f \in \mathcal{H}_K}{\arg \min} \frac{1}{l} \sum_{i=1}^{l} (y_i - f(x_i))^2 + \gamma_A \|f\|_K^2 + \frac{\gamma_I}{(u+l)^2} f^T L f$$

and the Laplacian Support Vector Machine problem (LapSVM) where the optimization problem is of the form

$$\underset{f \in \mathcal{H}_K}{\arg \min} \frac{1}{l} \sum_{i=1}^{l} (1 - y_i f(x_i))_+ + \gamma_A \|f\|_K^2 + \frac{\gamma_I}{(u+l)^2} f^T L f.$$

The only difference in the two expressions above is due to the choice of different loss functions. In the first case it is square loss function and in the second case it is hinge loss function where $(1 - y_i f(x_i)_+$ is interpreted as $\max(0, 1 - y_i f(x_i))$. It turns out that Representer Theorem can be used to show that the solution in both the cases is an expansion of kernel functions over both the labeled and the unlabeled examples and takes the form $f(x) = \sum_{i=1}^{l+u} \alpha_i^* K(x, x_i)$, where the coefficients $\alpha_i s$ are obtained differently for LapRLS and LapSVM.

In the case of LapRLS, the solutions $\alpha^* = [\alpha_1^*, \ldots, \alpha_{l+u}^*]$ is given by the following equation:

$$\alpha^* = \left(JK + \gamma_A lI + \frac{\gamma_I l}{(u+l)^2} LK \right)^{-1} Y \tag{20.8}$$

where Y is an $(l+u)$ dimensional label vector given by $Y = [y_1, \ldots, y_l, 0, \ldots, 0]$ and J is an $(l+u) \times (l+u)$ diagonal matrix given by $J = diag(1, \ldots, 1, 0, \ldots, 0)$, with the first l diagonal entries as 1 and the rest 0.

In case of LapSVM, the solution is given by the following equation:

$$\alpha^* = \left(2\gamma_A I + 2\frac{\gamma_I}{(u+l)^2} LK \right)^{-1} J^T Y \beta^* \tag{20.9}$$

where $J = [I \ 0]$ is an $l \times (l+u)$ matrix with I as the $l \times l$ identity matrix, $Y = diag(y_1, \ldots, y_l)$, and β^* is obtained by solving the following SVM dual problem :

$$\max_{\beta \in \mathbb{R}^l} \sum_{i=1}^{l} \beta_i - \frac{1}{2} \beta^T Q \beta \tag{20.10}$$

$$\text{subject to}: \sum_{i=1}^{l} l\beta_i y_i = 0$$

$$0 \leq \beta_i \leq \frac{1}{l} \text{ for } i = 1, \ldots, l.$$

Details of the algorithm are given in Algorithm 20.3.

Algorithm 20.3 Manifold Regularization

Input : l labeled examples $\{(x_i, y_i)\}_{i=1}^{l}$, u unlabeled examples $\{x_j\}_{j=l+1}^{l+u}$.
Output : Estimated function $f : X \rightarrow \mathbb{R}$.

1. Construct data adjacency graph with $l + u$ nodes using, for example, k-nearest neighbors. Choose edge weights W_{ij}, for example, binary weights or heat kernel weights $W_{ij} = e^{-\|x_i - x_j\|^2/t}$.

2. Choose kernel function $K(x, y)$. Compute the Gram matrix $K_{ij} = K(x_i, x_j)$.

3. Compute the Graph Laplacian matrix: $L = D - W$ where D is a diagonal matrix given by $D_{ii} = \sum_{j=1}^{l+u} W_{ij}$.

4. Choose γ_A and γ_I.

5. Compute α^* using Equation 20.8 for squared loss (Laplacian RLS) or using Equation 20.9 and 20.10 for hinge loss (Laplacian SVM).

6. Output $f^*(x) = \sum_{i=1}^{l+u} \alpha_i^* K(x_i, x)$.

20.4.3 Theoretical Justification and Relevant Results

Even though various graph-based semi-supervised learning methods are quite popular and successful in practice, their theoretical justification is not quite well understood. Note that various semi-supervised learning algorithms involving graph Laplacian have a common structure as in equation 20.5, involving graph Laplacian (according to Equation 20.3). Although this formulation intuitively makes sense and these methods have enjoyed reasonable empirical success, Nadler et al. ([43]) pointed out that with a proper scaling of $\lambda \rightarrow 0$, the solution of 20.4 degenerates to constants on unlabeled examples and "spikes" at labeled examples, in the limit, when for a fixed number of labeled examples the number of unlabeled examples increases to infinity. This shows that there is no generalization in the limit and the resulting solution is unstable in the finite labeled example case. Zhou et al. ([70]) resolve this problem by using higher order Sobolev semi-norm $\|f\|_m$ of order m such that $2m > d$, where d is the intrinsic dimension of the submanifold. The Soblolev embedding theorem guarantees that this method gives continuous solutions of a certain order in the limit of infinite unlabeled examples while keeping the number of labeled examples fixed. Zhou et al. ([70]) clarify that the manifold regularization method ([8]), however, is guaranteed to provide continuous solutions due to a different formulation of the optimization problem.

In another direction, Zhang et al. ([63]) propose and study the theoretical properties of a transductive formulation of kernel learning on graphs, which is equivalent to supervised kernel learning and includes some of the previous graph-based semi-supervised learning methods as special cases. The general form of this formulation is

$$\hat{f} = \arg\min_{f \in \mathbb{R}^{l+u}} \frac{1}{l} \sum_{i=1}^{l} V(f_i, y_i) + \lambda f^T K^{-1} f \tag{20.11}$$

where $V(\cdot, \cdot)$ is a convex loss function and $K \in \mathbb{R}^{(l+u) \times (l+u)}$ is the kernel gram matrix whose $(i, j)^{th}$ entry is the kernel function $k(x_i, x_j)$. Some of the graph Laplacian-based algorithms fall under this category when K^{-1} is replaced by graph Laplacian L. Note that kernel gram matrix is always positive semi-definite but it may be singular. In that case the correct interpretation of $f^T K^{-1} f$ is $\lim_{\mu \rightarrow 0^+} f^T (K + \mu I_{(l+u) \times (l+u)})^{-1} f$ where $I_{(l+u) \times (l+u)}$ is $(l + u) \times (l + u)$ identity matrix. Zhang et al. ([63]) show that the estimator $\hat{f}$ in Equation 20.11 converges to its limit almost surely when the number of unlabeled examples $u \rightarrow 0$. However, it is unclear what this particular limit estimator is and in case it is quite different from the true underlying function then the analysis may be uninformative. Zhang et al. ([63]) also study the generalization behavior of Equation 20.11 in the following form and provide the average predictive performance when a set of l examples $x_{i_1}, \ldots, x_{i_l}$ are randomly labeled from $l + u$ examples $x_1, \ldots, x_{(l+u)}$:

$$\mathbb{E}_{x_{i_1},\ldots,x_{i_l}} \left(\frac{1}{u} \sum_{x_i \notin \{x_{i_1},\ldots,x_{i_l}\}} V(\hat{f}_i(x_{i_1},\ldots,x_{i_l}), y_i) \right)$$

$$\leq \inf_{f \in \mathbb{R}^{l+u}} \left(\frac{1}{l+u} \sum_{i=1}^{l+u} V(f_i, y_i) + \lambda f^T K^{-1} f + \frac{\gamma^2 \mathrm{trace}(K)}{2\lambda l(l+u)} \right)$$

$$\leq \inf_{f \in \mathbb{R}^{l+u}} \left(\frac{1}{l+u} \sum_{i=1}^{l+u} V(f_i, y_i) + \frac{\gamma}{\sqrt{2l}} \sqrt{\mathrm{trace}(\frac{K}{(l+u)}) f^T K^{-1} f} \right)$$

where $\left| \frac{\partial}{\partial p} V(p, y) \leq \gamma \right|$ and the last inequality is due to optimal value of λ. The generalization bound suggests that design of transductive kernel in some sense controls the predictive performance and various kernels used in [63] are closely related to the graph Laplacian (see [63] for detailed discussion).

Zhou et. al ([71]) study the error rate and sample complexity of Laplacian eigenmap ([7]) as shown in Equation 20.6 at the limit of infinite unlabeled examples. The analysis studied here gives guidance to the choice of number of eigenvectors k in Equation 20.6. In particular, it shows that when data lies on a d-dimensional domain, the optimal choice of k is $O\left(\left(\frac{l}{\log(l)} \right)^{\frac{d}{d+2}} \right)$ yielding asymptotic error rate $O\left(\left(\frac{l}{\log(l)} \right)^{-\frac{2}{d+2}} \right)$. This result is based on integrated mean squared error. By replacing integrated mean square error with conditional mean squared error (MSE) on the labeled examples, the optimal choice of k becomes $O\left(l^{\frac{d}{d+2}} \right)$ and the corresponding mean squared error $O\left(l^{-\frac{2}{d+2}} \right)$. Note that this is the optimal error rate of nonparametric local polynomial regression on the d-dimensional unknown manifold ([9]) and it suggests that unless specific assumptions are made, unlabeled examples do not help.

20.5 Semi-Supervised Learning Methods Based on Cluster Assumption

The cluster assumption can be interpreted as saying that two points are likely to have the same class labels if they can be connected by a path passing only through a high density region. In other words, two high density regions with different class labels must be separated by a low density region. That is, the decision boundary should not cross high density regions, but instead lie in low density regions.

20.5.1 Algorithms

Various algorithms have been developed based on cluster assumption. Chapelle et. al ([17]) propose a framework for constructing kernels that implements the cluster assumption. This is achieved by first starting with an RBF kernel and then modifying the eigenspectrum of the kernel matrix by introducing different transfer functions. The induced distance depends on whether two points are in the same cluster or not.

Transductive support vector machine or TSVM ([33]) builds the connection between marginal density and the discriminative decision boundary by not putting the decision boundary in high density regions. This is in agreement with the cluster assumption. However, cluster assumption was not explicitly used in [33]. Chapelle et. al ([18]) propose a gradient descent on the primal formulation of TSVM that directly optimizes the slightly modified TSVM objective according to the cluster assumption: to find a decision boundary that avoids the high density regions. They also propose a graph-based distance that reflects cluster assumption by shrinking distances between the examples from the same clusters. Used with an SVM, this new distance performs better than standard Euclidean distance.

Szummer et al. ([60]) use information theory to explicitly constrain the conditional density $p(y|x)$ on the basis of marginal density $p(x)$ in a regularization framework. The regularizer is a function of both $p(y|x)$ and $p(x)$ and penalizes any changes in $p(y|x)$ more in the regions with high $p(x)$.

Bousquet et al. ([13]) propose a regularization-based framework that penalizes variations of function more in high density regions and less in low density regions.

Sinha et. al ([58]) propose a semi-supervised learning algorithm that exploits the cluster assumption. They show that when data is clustered, i.e., the high density regions are sufficiently separated by low density valleys, each high density area corresponds to a unique representative eigenvector of a kernel gram matrix. Linear combination of such eigenvectors (or, more precisely, of their Nystrom extensions) provide good candidates for good classification functions when the cluster assumption holds. First choosing an appropriate basis of these eigenvectors from unlabeled data and then using labeled data with Lasso to select a classifier in the span of these eigenvectors yields a classifier, that has a very sparse representation in this basis. Importantly, the sparsity corresponds naturally to the cluster assumption.

20.5.2 Description of a Representative Algorithm

According to the cluster assumption, the decision boundary should preferably not cut clusters. A way to enforce this for similarity-based classifiers is to assign low similarities to pairs of points that lie in different clusters. In this section we describe the algorithm of [17] that intends to achieve this by designing a "cluster kernel" that changes the representation of the input points such that points in the same cluster are grouped together in the new representation. Tools of spectral clustering are used for this purpose and details are given in Algorithn 20.4. This cluster kernel is used in a standard

---

Algorithm 20.4 Cluster Kernel

Input : l labeled examples $\{(x_i, y_i)\}_{i=1}^{l}$, u unlabeled examples $\{x_j\}_{j=l+1}^{l+u}$.
Output : Cluster Kernel.

1. Compute the Gram matrix K from both labeled and unlabeled examples, where, $K_{ij} = K(x_i, x_j) = e^{-\|x_i - x_j\|^2 / 2\sigma^2}$.

2. Compute $L = D^{-1/2} K D^{1/2}$, where D is a diagonal matrix and $D_{ii} = \sum_{j=1}^{l+u} K(x_i, x_j)$. Compute the eigen-decompoition $L = U\Lambda U^T$.

3. Given a transfer function ϕ, let $\tilde{\lambda}_i = \phi(\lambda_i)$, where the λ_i are the eigenvalues of L, and construct $\tilde{L} = U\tilde{\Lambda}U^T$.

4. Let $\tilde{D}$ be a diagonal matrix with $\tilde{D}_{ii} = 1/\tilde{L}_i i$ and compute cluster kernel $\tilde{K} = \tilde{D}^{1/2}\tilde{L}\tilde{D}^{1/2}$.

---

way for kernel classifiers involving labeled and unlabeled examples.

20.5.3 Theoretical Justification and Relevant Results

Unfortunately, no theoretical result is known for any particular algorithm based on cluster assumption. However, the general framework of semi-supervised learning under cluster assumption is studied by Rogollet ([51]). One of the conclusions in that work is that consideration of the whole excess risk is too ambitious as cluster assumption specifies that clusters or high density regions are supposed to have pure class labels, but does not provide any specification as to what happens outside the clusters. As a result, the effect of unlabeled examples on the rate of convergence cannot be observed on the whole excess risk, and one needs to consider cluster excess risk, which corresponds to a part of the excess risk that is interesting for this problem (within clusters). This is exactly what is done in [51] under the strong cluster assumption, which informally specifies clusters as level sets under mild conditions (for example, clusters are separated, etc.). The estimation procedure studied in this setting ([51]) proceeds in three steps,

1. Use unlabeled examples to get an estimator of the clusters (level sets).

2. Identify homogeneous regions from the clusters found in step 1, that are well connected and have certain minimum Lebesgue measure.

3. Assign a single label to each estimated homogeneous region by a majority vote on labeled examples.

Under mild conditions, the excess cluster risk is shown to be

$$\tilde{O}\left(\frac{u^{-\alpha}}{1-\theta}\right) + e^{-l(\theta\delta)^2/2}$$

where, $\alpha > 0, \delta > 0$ and $\theta \in (0,1)$. The result shows that unlabeled examples are helpful given cluster assumption. Note that the results are very similar to the ones obtained under parametric settings in the generative model case (see for examples, [14–16, 49, 50]).

In a separate work, Singh et al. ([56]) study the performance gains that are possible with semi-supervised learning under cluster assumption. The theoretical characterization presented in this work explains why in certain situations unlabeled examples can help in learning, while in other situations they may not. Informally, the idea is that if the margin γ is fixed, then given enough labeled examples, a supervised learner can achieve optimal performance and thus there is no improvement due to unlabeled examples in terms of the error rate convergence for a fixed collection of distributions. However, if the component density sets are discernible from a finite sample size u of unlabeled examples but not from a finite sample of l labeled examples ($l < u$), then semi-supervised learning can provide better performance than purely supervised learning. In fact, there are certain plausible situations in which semi-supervised learning yields convergence rates that cannot be achieved by any supervised learner. In fact, under mild conditions, Singh et al. ([56]) show that there exists a semi-supervised learner $\hat{f}_{l,u}$ (obtained using l labeled and u unlabeled examples) whose expected excess risk $\mathbb{E}(\mathcal{E}(\hat{f}_{l,u}))$ (risk of this learner minus infimum risk over all possible learners) is bounded from above as follows:

$$\mathbb{E}(\mathcal{E}(\hat{f}_{l,u})) \leq \varepsilon_2(l) + O\left(\frac{1}{u} + l\left(\frac{u}{(\log u)^2}\right)^{-1/d}\right)$$

provided $|\gamma| > C_0(m/(\log m)^2)^{-1/2}$. The quantity $\varepsilon_2(l)$ in the above expression is the finite sample upper bound of expected excess risk for any supervised learner using l labeled examples. The above result suggests that if the decision sets are discernible using unbalanced examples (i.e., the margin is large enough compared to average spacing between unlabeled examples), then there exists a semi-supervised learner that can perform as well a a supervised learner with complete knowledge of the decision sets provided $u \gg i$, so that $(i/\varepsilon_2(l))^d = O\left(u/(\log u)^2\right)$, implying that the additional term in the above bound is negligible compared to $\varepsilon_2(l)$. Comparing supervised learner lower bound and semi-supervised learner upper bound, Singh et al. ([56]) provide different margin ranges where semi-supervised learning can potentially be beneficial using unlabeled examples.

20.6   Related Areas

The focus of this survey chapter is on semi-supervised learning that aims to reduce the labeling burden when labeled training examples are hard to obtain. There are other methods in a slightly different but related context, which also aims to reduce the labeling cost when there is a direct paucity of training examples. These methods include, for example, active learning and transfer learning and are discussed in different chapters of this book. However, for completeness, we describe basic ideas of active learning and transfer learning below.

Active Learning: The standard classification setting is passive in the sense that the data collection phase is decoupled from the modeling phase and generally is not addressed in the context of classification model construction. Since data collection (labeling) is costly, and since all labeled examples are not equally informative in efficient classification model building, an alternative approach is to integrate the data collection phase and the classification model building phase. This technique is called active learning. In active learning, typically, the classification is performed in an interactive way with the learner providing well chosen examples to the user, for which the user may then provide labels. The examples are typically chosen for which the learner has the greatest level of uncertainty based on the current training knowledge and labels. This choice evidently provides the greatest additional information to the learner in cases where the greatest uncertainty exists about the current label. By choosing the appropriate examples to label, it is possible to greatly reduce the effort involved in the classification process. There exists a variety of active learning algorithms based on the criteria used for querying which example is to be labeled next. Most of these algorithms try to either reduce the uncertainty in classification or reduce the error associated with the classification process. The commonly used querying criteria are uncertainty sampling ([38]), query by committee ([53]), greatest model change ([21]), greatest error reduction ([22]), greatest variance reduction ([22]), etc.

Transfer Learning: The general principle of transfer learning is to transfer the knowledge learned in one domain (in a classification setting) to enhance the classification performance in a related but different domain where the number of training examples in the second domain is limited. Such a situation arises quite often in a cross lingual domain where the goal is to classify documents into different class labels for a particular language, say German, where the number of training examples is limited but similar documents in another language, say English, are available that contain enough training examples. This general idea of transfer learning is applied in situations where knowledge from one classification category is used to enhance learning of another category [48], or the knowledge from one data domain (e.g. text) is used to enhance the learning of another data domain (e.g., images) [25, 48, 76]. A nice survey on transfer learning methods may be found in [47].

20.7   Concluding Remarks

This chapter tries to provide a glimpse of various existing semi-supervised learning algorithms. The list is by no means exhaustive. However, it clearly gives an indication that over the last couple of years, this field has really emerged as one of the practical and important areas for research in machine learning and statistical data analysis. The main focus of this chapter is from the classification point of view. However, there are various very closely related subfields within semi-supervised learning, for example, semi-supervised clustering, semi-supervised regression, and semi-supervised structure prediction. Each of these subfields is quite mature and one can find various surveys for each of them.

Even though semi-supervised learning techniques are widely used, our theoretical understanding and analysis of various semi-supervised learning methods are yet far from complete. On the one hand, we need asymptotic results of various methods saying that in the limit the estimator under consideration converges to the true underlying classifier, which will ensure that the procedure does the "right" thing given enough labeled and unlabeled examples. On the other hand, we need to have finite sample bounds that will specifically tell us how fast the estimator converges to the true underlying classifier as a function of number of labeled and unlabeled examples. This type of results are completely missing. For example, various graph-based algorithms are quite popular and widely used but yet we hardly have any theoretical finite sample result! Another area of concern is that there are so many different semi-supervised learning algorithms that can be very different from each other under the same assumption or very similar algorithms with only a minor difference. In both the cases, performance of the algorithms can be totally different. This suggests that analysis of individual semi-supervised learning algorithms that perform well in practice, are also very important. This kind of analysis will not only help us understand the behavior of the existing semi-supervised learning algorithms but also help us design better semi-supervised algorithms in the future. In the coming years, researchers from the semi-supervised learning community need to put significant effort towards resolving these issues.

Bibliography

[1] S. Abney. Bootstrapping. In *Fortieth Annual Meeting of the Association for Computational Linguistics*, pages 360–367, 2002.

[2] A. Azran. The rendezvous algorithm: Multi-class semi-supervised learning with Markov random walks. In *Twenty-fourth International Conference on Machine Learning (ICML)*, pages 47–56 2007.

[3] M.-F. Balcan, A. Blum and K. Yang. Co-training and expansion: Towards bridging theory and practice In *Nineteenth Annual Conference on Neural Information Processing Systems (NIPS)*, 2005.

[4] M.-F. Balcan and A. Blum. An augmented PAC model for learning from labeled and unlabeled data. In O. Chapelle, B. Schölkopf and A. Zien, editors. *Semi-Supervised Learning*, MIT Press, Cambridge, MA, 2006.

[5] S. Baluja. Probabilistic modeling for face orientation discrimination: Learning from labeled and unlabeled data In *Twelfth Annual Conference on Neural Information Processing Systems (NIPS)*, pages 854–860, 1998.

[6] M. Belkin, I. Matveeva and P. Niyogi. Regularization and semi-supervised learning on large graphs. In *Seventeenth Annual Conference on Learning Theory(COLT)*, pages 624–638, 2004.

[7] M. Belkin and P. Niyogi. Semi-supervised learning on Reimannian manifold. *Machine Learning*, 56:209–239, 2004.

[8] M. Belkin, P. Niyogi and V. Sindhwani. Manifold regularization: A geometric framework for learning from labeled and unlabeled examples. *Journal of Machine Learning Research*, 7: 2399–2434, 2006.

[9] P. J. Bickel and B. Li. Local polynomial regression on unknown manifolds. *Complex Datasets and Inverse Problems: Tomography, Networks and Beyond, IMS Lecture Notes Monographs Series*, 54: 177–186, 2007.

[10] A. Blum and T. Mitchell. Combining labeled and unlabeled data with co-training. In *Eleventh Annual Conference on Learning Theory (COLT)*, pages 92–100, 1998.

[11] A. Blum and S. Chawla. Learning from labeled and unlabeled data using graph mincuts. In *Eighteenth International Conference on Machine Learning (ICML)*, pages 19–26, 2001.

[12] A. Blum, J. D. Lafferty, M. R. Rwebangira and R. Reddy. Semi-supervised learning using randomized mincuts. In *Twenty-first International Conference on Machine Learning (ICML)*, 2004.

[13] O. Bousquet, O Chapelle and M. Hein. Measure based regularization. In *Seventeenth Annual Conference on Neural Information Processing Systems (NIPS)*, 2003.

[14] V. Castelli. The relative value of labeled and unlabeled samples in pattern recognition. PhD Thesis, Stanford University, 1994.

[15] V. Castelli and T. M. Cover. On the exponential value of labeled samples. *Pattern Recognition Letters*, 16: 10–111, 1995.

[16] V. Castelli and T. M. Cover. The relative value of labeled and unlabeled samples in pattern recognition with an unknown mixing parameters. *IEEE Transactions on Information Theory*, 42(6):2102–2117, 1996.

[17] O. Chapelle, J. Weston and B. Scholkopf. Cluster kernels for semi-supervised learning. In *Sixteenth Annual Conference on Neural Information Processing Systems (NIPS)*, 2002.

[18] O. Chapelle and A. Zien. Semi-supervised classification by low density separation. In *Tenth International Workshop on Artificial Intelligence and Statistics (AISTATS)*, 2005.

[19] O. Chapelle, B. Schölkopf and A. Zien, editors. *Semi-Supervised Learning*, MIT Press, Cambridge, MA, 2006. http://www.kyb.tuebingen.mpg.de/ssl-book

[20] N. V. Chawla and G. Karakoulas. Learning from labeled and unlabeled data: An empirical study across techniques and domains. *Journal of Artifical Intelligence Research*, 23:331–366, 2005.

[21] D. Cohn, L. Atlas and R. Ladner. Improving generalization with active learning. *Machine Learning*, 5(2):201–221, 1994.

[22] D. Cohn, Z. Ghahramani and M. Jordan. Active learning with statistical models. *Journal of Artificial Intelligence Research*, 4:129–145, 1996.

[23] M. Collins and Y. Singer. Unsupervised models for named entity classification. In *In Joint SIGDAT Conference on Empirical Methods in Natural Language Processing and Very Large Corpora*, pages 100–110, 1999.

[24] F. B. Cozman and I. Cohen. Unlabeled data can degrade classification performance of generative classifiers. In *Fifteenth International Florida Artificial Intelligence Society Conference*, pages 327–331, 2002.

[25] W. dai, Y. Chen, G.-R. Xue, Q. Yang and Y. Yu. Translated learning: Transfer learning across different feature spaces. In *Twenty first Annual Conference on Neural Information Processing Systems (NIPS)*, 2007.

[26] S. Dasgupta, M. L. Littman and D. McAllester. PAC generalization bounds for co-training. In *Fifteenth Annual Conference on Neural Information Processing Systems (NIPS)*, pages 375–382, 2001.

[27] O. Delalleu, Y. Bengio and N. L. Roux. Efficient non-parametric function induction in semi-supervised learning. In *Tenth International Workshop on Artificial Intelligence and Statistics (AISTATS)*, pages 96–103, 2005.

[28] A. P. Dempster, N. M. Liard and D. B. Rubin. Maximum likelihood from incomplete data via the EM algorithm. *Journal of the Royal Statistical Society, Series B*, 39(1):1–39, 1977.

[29] J. V. Dillon, K. Balasubramanian and G. Lebanon. Asymptotic analysis of generative semi-supervised learning. In *Twenty-seventh International Conference on Machine Learning (ICML)*, pages 295–302, 2010.

[30] R. Fergus, Y. Weiss and A. Torralba. Semi-supervised learning in gigantic image collections. In *Twenty-fourth Annual Conference on Neural Information Processing Systems (NIPS)*, 2009.

[31] S. Goldman and Y. Zhou. Enhancing supervised learning with unlabeled data. In *Seventeenth International Conference on Machine Learning (ICML)*, pages 327–334, 2000.

[32] D. Greig, B. Porteous and A. Seheult. Exact maximum a posteriori estimation for binary images. *Journal of the Royal Statistical Society, Series B*, 51:271–279, 1989.

[33] T. Joachims. Transductive inference for text classification using support vector machines. In *Sixteenth International Conference on Machine Learning (ICML)*, pages 200–209, 1999.

[34] T. Joachims. Transductive learning via spectral graph partitioning. In *Twentieth International Conference on Machine Learning (ICML)*, pages 290–293, 2003.

[35] R. Johnson and T. Zhang. Two-view feature generation model for semi-supervised learning. In *Twenty-fourth International Conference on Machine Learning (ICML)*, pages 25–32, 2007.

[36] R. Jones. Learning to extract entities from labeled and unlabeled text. In PhD Thesis, Carnegie Mellon University, 2005.

[37] B. Kveton, M. Valko, A. Rahimi and L. Huang. Semi-supervised learning with max-margin graph cuts. In *Thirteenth International Conference on Artificial Intelligence and Statistics (AISTATS)*, pages 421–428, 2010.

[38] D. Lewis and J. Catlett. Heterogeneous uncertainty sampling for supervised learning. In *Eleventh International Conference on Machine Learning (ICML)*, pages 148–156, 1994.

[39] W. Liu, J. He and S.-F. Chang. Large graph construction for scalable semi-supervised learning. In *Twenty-seventh International Conference on Machine Learning (ICML)*, 2010.

[40] W. Liu, J. Wang and S.-F. Chang. Robust and scalable graph-based semi-supervised learning. *Proceedings of the IEEE*, 100(9):2624–2638, 2012.

[41] G. J. McLachlan and K. Basford. *Mixture Models*. Marcel Dekker, New York, 1988.

[42] G. J. McLachlan and T. Krishnan. *The EM Algorithms and Extensions*. John Wiley & Sons, New York, 1997.

[43] B. Nadler, N. Srebro and X. Zhou. Semi-supervised learning with the graph Laplacian: The limit of infinite unlabeled data. In *Twenty-third Annual Conference on Neural Information Processing Systems (NIPS)*, 2009.

[44] A. Ng and M. Jordan. On discriminative vs. generative classifiers: a comparison of logistic regression and naive Bayes. In *Fifteenth Annual Conference on Neural Information Processing Systems (NIPS)*, 2001.

[45] K. Nigam, A. McCullum, S. Thurn and T. Mitchell. Text classification from labeled and unlabeled documents using EM. *Machine Learning*, 39(2-3):103–134, 2000.

[46] K. Nigam and R. Ghani. Analyzing the effectiveness and applicability of co-training. *Ninth International Conference on Information and Knowledge Management*, pages 86–93, 2000.

[47] S. J. Pan and Q Yang. A survey on transfer learning. *IEEE Transactions on Knowledge Discovery and Data Engineering*, 22(10):1345–1359, 2010.

[48] G. Qi, C. Aggarwal and T. Huang Towards cross-category knowledge propagation for learning visual concepts. *Twenty-fourth International Conference on Computer Vision and Pattern Recognition*, pages 897–904, 2011.

[49] J. Ratsaby. Complexity of learning from a mixture of labeled and unlabeled examples. PhD Thesis, University of Pennsylvania, 1994.

[50] J. Ratsaby and S. S. Venkatesh. Learning from a mixture of labeled and unlabeled examples with parametric side information. In *Eighth Annual Conference on Learning Theory*, pages 412–417, 1995.

[51] P. Rigollet. Generalization error bounds in semi-supervised classification under the cluster assumption. *Journal of Machine Learning Research*, 8: 1369–1392, 2007.

[52] M. Seeger, Learning with labeled and unlabeled data, Tech. Report, 2000. http://infoscience.epfl.ch/record/161327/files/review.pdf

[53] H. Seung, M. Opper and H. Sompolinsky. Query by committee. In *Fifth Annual Workshop on Learning Theory (COLT)*, pages 287–294, 1992.

[54] B. M. Shahshahani and D. A. Landgrebe. The effect of unlabeled samples in reducing the sample size problem and mitigating Hughes phenomenon. *IEEE Transactions on Geoscience and Remote Sensing*, 32(5):1087–1095, 1994.

[55] V. Sindhwani, P. Niyogi and M. Belkin. Beyond the point cloud: from transductive to semi-supervised learning. In *Twenty-second International Conference on Machine Learning (ICML)*, pages 824–831, 2005.

[56] A. Singh, R. Nowak and X. Zhu. Unlabeled data: Now it helps, now it doesn't. In *Twenty-second Annual Conference on Neural Information Processing Systems (NIPS)*, 2008.

[57] K. Sinha and M. Belkin. The value of labeled and unlabeled examples when the model is imperfect. In *Twenty-first Annual Conference on Neural Information Processing Systems (NIPS)*, pages 1361–1368, 2007.

[58] K. Sinha and M. Belkin. Semi-supervised learning using sparse eigenfunction bases. In *Twenty third Annual Conference on Neural Information Processing Systems (NIPS)*, pages 1687–1695, 2009.

[59] M. Szummer and T. Jaakkola. Partially labeled classification with Markov random walks. In *Sixteenth Annual Conference on Neural Information Processing Systems (NIPS)*, pages 945–952, 2001.

[60] M. Szummer and T. Jaakkola. Information regularization with partially labeled data. In *Seventeenth Annual Conference on Neural Information Processing Systems (NIPS)*, pages 1025–1032, 2002.

[61] I. W. Tsang and J. T. Kwok. Large scale sparsified manifold regularization. In *Twenty-first Annual Conference on Neural Information Processing Systems (NIPS)*, pages 1401–1408, 2006.

[62] J. Wang. Semi-supervised learning for scalable and robust visual search. PhD Thesis, Columbia University, 2011.

[63] T. Zhang and R. K. Ando. Analysis of spectral kernel design based semi-supervised learning. In *Nineteenth Annual Conference on Neural Information Processing Systems (NIPS)*, 2005.

[64] K. Zhang, J. T. Kwok and B. Parvin. Prototype vector machine for large scale semi-supervised learning. In *Twenty-sixth International Conference on Machine Learning (ICML)*, pages 1233–1240, 2009.

[65] D. Zhou, O. Bousquet, T. N. Lal, J. Weston and B. Scholkopf. Learning with local and global consistency. In *Eighteenth Annual Conference on Neural Information Processing Systems (NIPS)*, pages 321–328, 2003.

[66] Y. Zhou and S. Goldman. Democratic co-learning. In *Sixteenth IEEE International Conference on Tools with Artificial Intelligence*, pages 594–602, 2004.

[67] D. Zhou and B. Schlkopf. Learning from labeled and unlabeled data using random walks. In *Twenty-sixth DAGM Symposium*, pages 237–244, 2004.

[68] Z. H. Zhou and M. Li. Tri-training: exploiting unlabeled data using three classifiers. *IEEE Transactions on Knwledge and Data Engineering*, 17:1529–1541, 2005.

[69] Z. H. Zhou, D. C. Zhan and Q. Yang. Semi-supervised learning with very few labeled examples. In *Tewnty-second AAAI Conference on Artificial Intelligence*, pages 675–680, 2007.

[70] X. Zhou and M. Belkin. Semi-supervised learning by higher order regularization. In *Fourteenth International Conference on Artificial Intelligence and Statistics (AISTATS)*, pages 892–900, 2011.

[71] X. Zhou and N. Srebro. Error analysis of Laplacian eigenmaps for semi-supervised learning. In *Fourteenth International Conference on Artificial Intelligence and Statistics (AISTATS)*, pages 901–908, 2011.

[72] X. Zhou, A. Saha and V. Sindhwani. Semi-supervised learning: Some recent advances. In *Cost-Sensitive Machine Learning,* Editors: B. Krishnapuram, S. Yu, B. Rao, Chapman & Hall/CRC Machine Learning & Pattern Recognition 2011.

[73] X. Zhu, Z. Ghahramani and J. Lafferty. Semi-supervised learning using Gaussian fields and harmonic functions. In *Twentieth International Conference on Machine Learning (ICML)*, pages 912–919, 2003.

[74] X. Zhu. Semi-Supervised Learning Literature Survey, Tech. Report, University of Wisconsin Madison, 2005.

[75] X. Zhu and A. B. Goldberg. *Introduction to Semi-Supervised Learning* (Synthesis Lectures on Artificial Intelligence and Machine Learning), Morgan and Claypool Publishers, 2009.

[76] Y. Zhu, S. J. Pan, Y. Chen. G.-R. Xue, Q. Yang and Y. Yu. Heterogeneous transfer learning for image classification. In *Special Track on AI and the Web, associated with Twenty-fourth AAAI Conference on Artificial Intelligence*, 2010.

Chapter 21

Transfer Learning

Sinno Jialin Pan

Institute for Infocomm Research
Sinpapore 138632
jspan@i2r.a-star.edu.sg

21.1 Introduction

Supervised machine learning techniques have already been widely studied and applied to various real-world applications. However, most existing supervised algorithms work well only under a common assumption: the training and test data are represented by the same features and drawn from the same distribution. Furthermore, the performance of these algorithms heavily rely on collecting high quality and sufficient labeled training data to train a statistical or computational model to make predictions on the future data [57, 86, 132]. However, in many real-world scenarios, labeled training data are in short supply or can only be obtained with expensive cost. This problem has become a major bottleneck of making machine learning methods more applicable in practice.

In the last decade, semi-supervised learning [20, 27, 63, 89, 167] techniques have been proposed to address the labeled data sparsity problem by making use of a large amount of unlabeled data to discover an intrinsic data structure to effectively propagate label information. Nevertheless, most semi-supervised methods require that the training data, including labeled and unlabeled data, and the test data are both from the same domain of interest, which implicitly assumes the training and test data are still represented in the same feature space and drawn from the same data distribution.

Instead of exploring unlabeled data to train a precise model, active learning, which is another branch in machine learning for reducing annotation effort of supervised learning, tries to design an active learner to pose queries, usually in the form of unlabeled data instances to be labeled by an oracle (e.g., a human annotator). The key idea behind active learning is that a machine learning algorithm can achieve greater accuracy with fewer training labels if it is allowed to choose the data from which it learns [71, 121]. However, most active learning methods assume that there is a budget for the active learner to pose queries in the domain of interest. In some real-world applications, the budget may be quite limited, which means that the labeled data queried by active learning may not be sufficient to learn an accurate classifier in the domain of interest.

Transfer learning, in contrast, allows the domains, tasks, and distributions used in training and testing to be different. The main idea behind transfer learning is to borrow labeled data or extract knowledge from some related domains to help a machine learning algorithm to achieve greater performance in the domain of interest [97, 130]. Thus, transfer learning can be referred to as a different strategy for learning models with minimal human supervision, compared to semi-supervised and active learning. In the real world, we can observe many examples of transfer learning. For example, we may find that learning to recognize apples might help us to recognize pears. Similarly, learning to play the electronic organ may help facilitate learning the piano. Furthermore, in many engineering applications, it is expensive or impossible to collect sufficient training data to train models for use in each domain of interest. It would be more practical if one could reuse the training data that have been collected in some related domains/tasks or the knowledge that is already extracted from some related domains/tasks to learn a precise model for use in the domain of interest. In such cases, *knowledge transfer* or *transfer learning* between tasks or domains become more desirable and crucial.

Many diverse examples in knowledge engineering can be found where transfer learning can truly be beneficial. One example is sentiment classification, where our goal is to automatically classify reviews on a product, such as a brand of camera, into polarity categories (e.g., positive, negative or neural). In literature, supervised learning methods [100] have proven to be promising and widely used in sentiment classification. However, these methods are domain dependent, which means that a model built on one domain (e.g., reviews on a specific product with annotated polarity categories) by using these methods may perform poorly on another domain (e.g., reviews on another specific product without polarity categories). The reason is that one may use different domain-specific words to express opinions in different domains. Table 21.1 shows several review sentences of two domains: *Electronics* and *Video Games*. In the *Electronics* domain, one may use words like "compact" and

TABLE 21.1: Cross-Domain Sentiment Classification Examples: Reviews of *Electronics* and *Video Games*.

	Electronics	Video Games
+	**Compact**; easy to operate; very good picture quality; looks **sharp**!	A very good game! It is action packed and full of excitement. I am very much **hooked** on this game.
+	I purchased this unit from Circuit City and I was very excited about the quality of the picture. It is really nice and **sharp**.	Very **realistic** shooting action and good plots. We played this and were **hooked**.
-	It is also quite **blurry** in very dark settings. I will never buy HP again.	The game is so **boring**. I am extremely unhappy and will probably never buy UbiSoft again.

Note: Boldfaces are domain-specific words that occur much more frequently in one domain than in the other one. "**+**" and "**-**" denote positive and negative sentiment respectively.

"sharp" to express positive sentiment and use "blurry" to express negative sentiment, while in the *Video Game* domain, the words like "hooked" and "realistic" indicate positive opinions and the word "boring" indicates negative opinion. Due to the mismatch of domain-specific words between domains, a sentiment classifier trained on one domain may not work well when directly applied to other domains. Therefore, cross-domain sentiment classification algorithms are highly desirable to reduce domain dependency and manually labeling cost by transferring knowledge from related domains to the domain of interest [18, 51, 92].

The need for transfer learning may also arise in applications of wireless sensor networks, where wireless data can be easily outdated over time or very differently received by different devices. In these cases, the labeled data obtained in one time period or on one device may not follow the same distribution in a later time period or on another device. For example, in indoor WiFi-based localization, which aims to detect a mobile's current location based on previously collected WiFi data, it is very expensive to calibrate WiFi data for building a localization model in a large-scale environment because a user needs to label a large collection of WiFi signal data at each location. However, the values of WiFi signal strength may be a function of time, device, or other dynamic factors. As shown in Figure 21.1, values of received signal strength (RSS) may differ across different time periods and mobile devices. As a result, a model trained in one time period or on one device may estimate locations poorly in another time period or on another device. To reduce the re-calibration effort, we might wish to adapt the localization model trained in one time period (the source domain) for a new time period (the target domain), or to adapt the localization model trained on a mobile device (the source domain) for a new mobile device (the target domain) with little or without additional calibration [91, 98, 152, 165].

As a third example, transfer learning has been shown to be promising for defect prediction in the area of software engineering, where the goal is to build a prediction model from data sets mined from software repositories, and the model is used to identify software defects. In the past few years, numerous effective software defect prediction approaches based on supervised machine learning techniques have been proposed and received a tremendous amount of attention [66, 83]. In practice, cross-project defect prediction is necessary. New projects often do not have enough defect data to build a prediction model. This cold-start is a well-known problem for recommender systems [116] and can be addressed by using cross-project defect prediction to build a prediction model using data from other projects. The model is then applied to new projects. However, as reported by some researchers, cross-project defect prediction often yields poor performance [170]. One of the main reasons for the poor cross-project prediction performance is the difference between the data distributions of the source and target projects. To improve the cross-project prediction performance

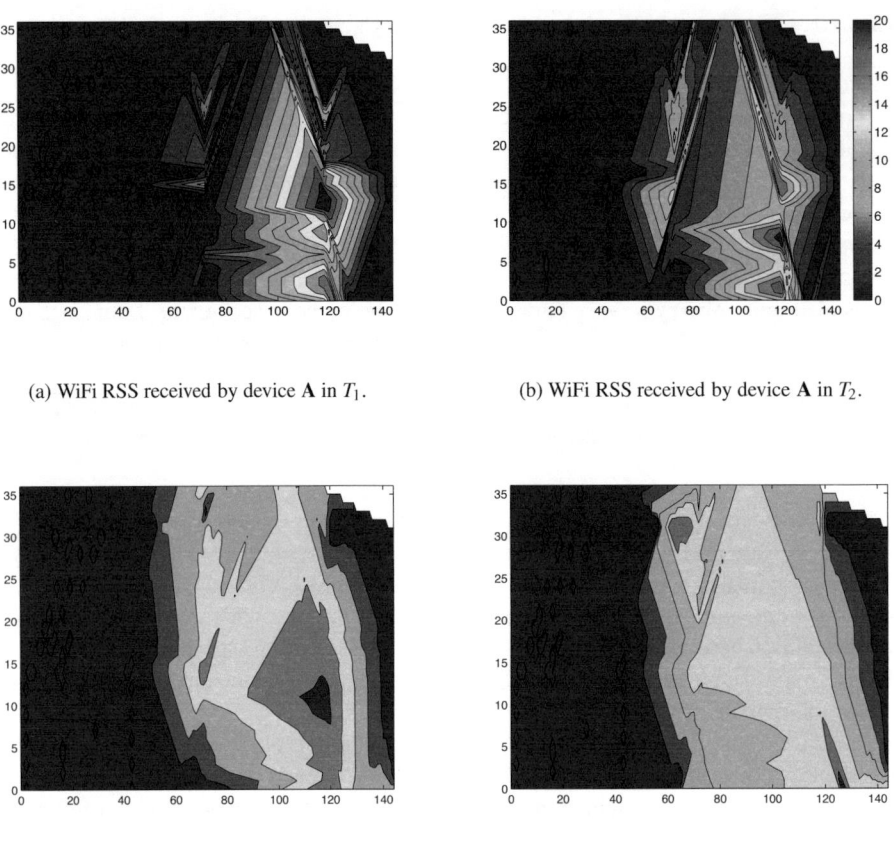

(a) WiFi RSS received by device **A** in T_1.

(b) WiFi RSS received by device **A** in T_2.

(c) WiFi RSS received by device **B** in T_1.

(d) WiFi RSS received by device **B** in T_2.

FIGURE 21.1 (See color insert.): Contours of RSS values over a 2-dimensional environment collected from a same access point in different time periods and received by different mobile devices. Different colors denote different values of signal strength.

with little additional human supervision, transfer learning techniques are again desirable, and have proven to be promising [87].

Generally speaking, transfer learning can be classified into two different fields: 1) transfer learning for classification, regression, and clustering problems [97], and 2) transfer learning for reinforcement learning tasks [128]. In this chapter, we focus on transfer learning in data classification and its real-world applications. Furthermore, as first introduced in a survey article [97], there are three main research issues in transfer learning: 1) What to transfer, 2) How to transfer, and 3) When to transfer. Specifically, "What to transfer" asks which part of knowledge can be extracted and transferred across domains or tasks. Some knowledge is domain- or task-specific, which may not be observed in other domains or tasks, while some knowledge is commonly shared by different domains or tasks, which can be treated as a bridge for knowledge transfer across domains or tasks. After discovering which knowledge can be transferred, learning algorithms need to be developed to transfer the knowledge, which corresponds to the "how to transfer" issue. Different knowledge-transfer strategies lead to specific transfer learning approaches. "When to transfer" asks in which situations it is appropriate to use transfer learning. Likewise, we are interested in knowing in which situations knowledge should not be transferred. In some situations, when the source domain and target domain are not related to each other, brute-force transfer may be unsuccessful. In the worst case, it may even hurt the performance of learning in the target domain, a situation that is often referred to as *negative transfer*. Therefore, the goal of "When to transfer" is to avoid *negative transfer* and then ensure *positive transfer*.

The rest of this chapter is organized as follows. In Section 21.2, we start by giving an overview of transfer learning including its brief history, definitions, and different learning settings. In Sections 21.3–21.4, we summarize approaches into different categories based on two transfer learning settings, namely homogenous transfer learning and heterogeneous transfer learning. In Sections 21.5–21.6, we discuss the negative transfer issue and other research issues of transfer learning. After that, we show diverse real-world applications of transfer learning in Section 21.7. Finally, we give concluding remarks in Section 21.8.

21.2 Transfer Learning Overview

In this section, we will provide an overview of transfer learning in terms of background and motivation. The notations and definitions will also be introduced.

21.2.1 Background

The study of transfer learning is motivated by the fact that people can intelligently apply knowledge learned previously to solve new problems faster or with better solutions [47]. For example, if one is good at coding in C++ programming language, he/she may learn Java programming language fast. This is because both C++ and Java are object-oriented programming (OOP) languages, and share similar programming motivations. As another example, if one is good at playing table tennis, he/she may learn playing tennis fast because the skill sets of these two sports are overlapping. Formally, from a psychological point of view, the definition of transfer learning or learning of transfer is the study of the dependency of human conduct, learning, or performance on prior experience. More than 100 years ago, researchers had already explored how individuals would transfer in one context to another context that share similar characteristics [129].

The fundamental motivation for transfer learning in the field of machine learning is the need for lifelong machine learning methods that retain and reuse previously learned knowledge such that intelligent agencies can adapt to new environments or novel tasks effectively and efficiently with little human supervision. Informally, the definition of transfer learning in the field of machine learning is the ability of a system to recognize and apply knowledge and skills learned in previous domains or tasks to new domains or novel domains, which share some commonality.

21.2.2 Notations and Definitions

In this section, we follow the notations introduced in [97] to describe the problem statement of transfer learning. A *domain* $\mathcal{D}$ consists of two components: a feature space $\mathcal{X}$ and a marginal probability distribution $P(x)$, where $x \in \mathcal{X}$. In general, if two domains are different, then they may have different feature spaces or different marginal probability distributions. Given a specific domain, $\mathcal{D} = \{\mathcal{X}, P(x)\}$, a *task* $\mathcal{T}$ consists of two components: a label space $\mathcal{Y}$ and a predictive function $f(\cdot)$ (denoted by $\mathcal{T} = \{\mathcal{Y}, f(\cdot)\}$). The function $f(\cdot)$ is a predictive function that can be used to make predictions on unseen instances $\{x^*\}$'s. From a probabilistic viewpoint, $f(x)$ can be written as

$P(y|x)$. In classification, labels can be binary, i.e., $\mathcal{Y} = \{-1, +1\}$, or discrete values, i.e., multiple classes.

For simplicity, we only consider the case where there is one source domain $\mathcal{D}_S$, and one target domain $\mathcal{D}_T$, as this is by far the most popular of the research works in the literature. The issue of knowledge transfer from multiple source domains will be discussed in Section 21.6. More specifically, we denote $\mathbf{D}_S = \{(x_{S_i}, y_{S_i})\}_{i=1}^{n_S}$ the *source domain data*, where $x_{S_i} \in X_S$ is the data instance, and $y_{S_i} \in \mathcal{Y}_S$ is the corresponding class label. Similarly, we denote $\mathbf{D}_T = \{(x_{T_i}, y_{T_i})\}_{i=1}^{n_T}$ the target domain data, where the input x_{T_i} is in X_T and $y_{T_i} \in \mathcal{Y}_T$ is the corresponding output. In most cases, $0 \le n_T \ll n_S$. Based on the notations defined above, the definition of transfer learning can be defined as follows [97]:

Definition 1 *Given a source domain $\mathcal{D}_S$ and learning task $\mathcal{T}_S$, a target domain $\mathcal{D}_T$ and learning task $\mathcal{T}_T$, transfer learning aims to help improve the learning of the target predictive function $f_T(\cdot)$ in $\mathcal{D}_T$ using the knowledge in $\mathcal{D}_S$ and $\mathcal{T}_S$, where $\mathcal{D}_S \neq \mathcal{D}_T$, or $\mathcal{T}_S \neq \mathcal{T}_T$.*

In the above definition, a domain is a pair $\mathcal{D} = \{X, P(x)\}$. Thus the condition $\mathcal{D}_S \neq \mathcal{D}_T$ implies that either $X_S \neq X_T$ or $P(x_S) \neq P(x_T)$. Similarly, a task is defined as a pair $\mathcal{T} = \{\mathcal{Y}, P(y|x)\}$. Thus the condition $\mathcal{T}_S \neq \mathcal{T}_T$ implies that either $\mathcal{Y}_S \neq \mathcal{Y}_T$ or $P(y_S|x_S) \neq P(y_T|x_T)$. When the target and source domains are the same, i.e., $\mathcal{D}_S = \mathcal{D}_T$, and their learning tasks are the same, i.e, $\mathcal{T}_S = \mathcal{T}_T$, the learning problem becomes a traditional machine learning problem. Based on whether the feature spaces or label spaces are identical or not, we can further categorize transfer learning into two settings: 1) homogenous transfer learning, and 2) heterogenous transfer learning. In the following two sections, we give the definitions of these two settings and review their representative methods, respectively.

21.3 Homogenous Transfer Learning

In this section, we first give a definition of homogenous transfer learning as follows:

Definition 2 *Given a source domain $\mathcal{D}_S$ and learning task $\mathcal{T}_S$, a target domain $\mathcal{D}_T$ and learning task $\mathcal{T}_T$, homogenous transfer learning aims to help improve the learning of the target predictive function $f_T(\cdot)$ in $\mathcal{D}_T$ using the knowledge in $\mathcal{D}_S$ and $\mathcal{T}_S$, where $X_S \cap X_T \neq \emptyset$ and $\mathcal{Y}_S = \mathcal{Y}_T$, but $P(x_S) \neq P(x_T)$ or $P(y_S|x_S) \neq P(y_T|x_T)$.*

Based on the above definition, in homogenous transfer learning, the feature spaces between domains are overlapping, and the label spaces between tasks are identical. The difference between domains or tasks is caused by the marginal distributions or predictive distributions. Approaches to homogenous transfer learning can be summarized into four categories: 1) instance-based approach, 2) feature-representation-based approach, 3) model-parameter-based approach, and 4) relational-information-based approach. In the following sections, we describe the motivations of these approaches and introduce some representative methods of each approach.

21.3.1 Instance-Based Approach

A motivation of the instance-based approach is that although the source domain labeled data cannot be reused directly, part of them can be reused for the target domain after re-weighting or re-sampling. An assumption behind the instance-based approach is that the source and target domains have a lot of overlapping features, which means that the domains share the same or similar support. Based on whether labeled data are required or not in the target domain, the instance-based approach can be further categorized into two contexts: 1) no target labeled data are available, and 2) a few target labeled data are available.

21.3.1.1 Case I: No Target Labeled Data

In the first context, no labeled data are required but a lot of unlabeled data are assumed to be available in the target domain. In this context, most instance-based methods are deployed based on an assumption that $P_S(y|x) = P_T(y|x)$, and motivated by importance sampling. To explain why importance sampling is crucial for this context of transfer learning, we first review the learning framework of empirical risk minimization (ERM) [132]. Given a task of interest, i.e., the target task, the goal of ERM is to learn an optimal parameter θ^* by minimizing the expected risk as follows:

$$\theta^* = \arg\min_{\theta \in \Theta} \mathbb{E}_{(x,y) \in P_T}[l(x,y,\theta)], \tag{21.1}$$

where $l(x,y,\theta)$ is a loss function that depends on the parameter θ. Since no labeled data are assumed to be available in the target domain, it is impossible to optimize (21.1) over target domain labeled data. It can be proved that the optimization problem (21.1) can be rewritten as follows:

$$\theta^* = \arg\min_{\theta \in \Theta} \mathbb{E}_{(x,y) \sim P_S}\left[\frac{P_T(x,y)}{P_S(x,y)} l(x,y,\theta)\right], \tag{21.2}$$

which aims to learn the optimal parameter θ^* by minimizing the weighted expected risk over source domain labeled data. Because it is assumed that $P_S(y|x) = P_T(y|x)$, by decomposing the joint distribution $P(x,y) = P(y|x)P(x)$, we obtain $\frac{P_T(x,y)}{P_S(x,y)} = \frac{P_T(x)}{P_S(x)}$. Hence, (21.2) can be further rewritten as

$$\theta^* = \arg\min_{\theta \in \Theta} \mathbb{E}_{(x,y) \sim P_S}\left[\frac{P_T(x)}{P_S(x)} l(x,y,\theta)\right], \tag{21.3}$$

where a weight of a source domain instance x is the ratio of the target and source domain marginal distributions at the data point x. Given a sample of source domain labeled data $\{(x_{S_i}, y_{S_i})\}_{i=1}^{n_S}$, by denoting $\beta(x) = \frac{P_T(x)}{P_S(x)}$, a regularized empirical objective of (21.3) can be formulated as

$$\theta^* = \arg\min_{\theta \in \Theta} \sum_{i=1}^{n_S} \beta(x_{S_i}) l(x_{S_i}, y_{S_i}, \theta) + \lambda \Omega(\theta), \tag{21.4}$$

where $\Omega(\theta)$ is a regularization term to avoid overfitting on the training sample. Therefore, a research issue on applying the ERM framework to transfer learning is how to estimate the weights $\{\beta(x)\}$'s. Intuitively, a simple solution is to first estimate $P_T(x)$ and $P_S(x)$, respectively, and thus calculate the ratio $\frac{P_T(x)}{P_S(x)}$ for each source domain instance x_{S_i}. However, density estimations on $P_T(x)$ and $P_S(x)$ are difficult, especially when data are high-dimensional and the data size is small. An alterative solution is to estimate $\frac{P_T(x)}{P_S(x)}$ directly.

In the literature, there exist various ways to estimate $\frac{P_T(x)}{P_S(x)}$ directly. Here we introduce three representative methods. For more information on this context, readers may refer to [104]. Zadrozny [158] assumed that the difference in data distributions is caused by the data generation process. Specifically, the source domain data are assumed to be sampled from the target domain data following a rejection sampling process. Let $s \in \{0,1\}$ be a selector variable to denote whether an instance in the target domain is selected to generate the source domain data or not, i.e., $s = 1$ denotes the instance is selected, otherwise unselected. In this way, the distribution of the selector variable maps the target distribution onto the source distribution as follows:

$$P_S(x) \propto P_T(x)P(s = 1|x).$$

Therefore, the weight $\beta(x)$ is propositional to $\frac{1}{P(s=1|x)}$. To estimate $\frac{1}{P(s=1|x)}$, Zadrozny proposed to consider all source domain data with labels 1's and all target domain data with labels 0's, and train a probabilistic classification model on this pseudo classification task to estimate $P(s = 1|x)$.

Huang et al. [59] proposed a different algorithm known as kernel-mean matching (KMM) to learn $\frac{P_S(x)}{P_T(x)}$ directly by matching the means between the source and target domain data in a reproducing-kernel Hilbert space (RKHS) [117]. Specifically, KMM makes use of Maximum Mean Discrepancy (MMD) introduced by Gretton et al. [52] as a distance measure between distributions. Given two samples, based on MMD, the distance between two sample distributions is simply the distance between the two mean elements in an RKHS. Therefore, the objective of KMM can be written as

$$\underset{\beta}{\arg\min} \quad \left\| \frac{1}{n_S} \sum_{i=1}^{n_S} \beta(x_{S_i})\Phi(x_{S_i}) - \frac{1}{n_T} \sum_{j=1}^{n_T} \Phi(x_{T_j}) \right\|_{\mathcal{H}}, \tag{21.5}$$

$$s.t \quad \beta(x_{S_i}) \in [0, B] \text{ and } \left| \frac{1}{n_S} \sum_{i=1}^{n_S} \beta(x_{S_i}) - 1 \right| \leq \varepsilon,$$

where B is the parameter to limit the discrepancy between $P_S(x)$ and $P_T(x)$, and ε is the nonnegative parameter to ensure the reweighted $P_S(x)$ is close to a probability distribution. It can be shown that the optimization problem (21.5) can be transformed into a quadratic programming (QP) problem, and the optimal solutions $\{\beta(x_{S_i})\}$'s of (21.5) are equivalent to the ratio values $\left\{ \frac{P_S(x_{S_i})}{P_T(x_{S_i})} \right\}$'s of (21.3) to be estimated.

As a third method, Sugiyama et al. [127] assumed that the ratio $\beta(x)$ can be estimated by the following linear model,

$$\widetilde{\beta}(x) = \sum_{\ell=1}^{b} \alpha_\ell \psi_\ell(x),$$

where $\{\psi_\ell(x)\}_{\ell=1}^{b}$ are the basic functions that are predefined, and the coefficients $\{\alpha_\ell\}_{\ell=1}^{b}$ are the parameters to be estimated. In this way, the problem of estimating $\beta(x)$ is transformed into the problem of estimating the parameters $\{\alpha_\ell\}_{\ell=1}^{b}$. By denoting $\widetilde{P}_T(x) = \widetilde{\beta}(x)P_S(x)$, the parameters can be learned by solving the following optimization problem,

$$\underset{\{\alpha_\ell\}_{\ell=1}^{b}}{\arg\min} l(P_T(x), \widetilde{P}_T(x)),$$

where $l(\cdot)$ is a loss function of the estimated target distribution $\widetilde{P}_T(x)$ to the ground truth target distribution $P_T(x)$. Different loss functions lead to various specific algorithms. For instance, Sugiyama et al. [127] proposed to use the Kullback-Leibler divergence as the loss function, while Kanamori et al. [65] proposed to use the least-squared loss as the loss function. Note that the ground truth of $P_S(x)$ and $P_T(x)$ are unknown. However, as shown in [65, 127], $P_S(x)$ and $P_T(x)$ can be eliminated when optimizing the parameters $\{\alpha_\ell\}_{\ell=1}^{b}$.

21.3.1.2 Case II: A Few Target Labeled Data

In the second context of the instance-based approach, a few target labeled data are assumed to be available. Different from the approaches in the first context, in this context, most approaches are proposed to weight the source domain data based on their contributions to the classification accuracy for the target domain.

Wu and Dietterich [142] integrated the source domain labeled data together with a few target domain labeled data into the standard Support Vector Machine (SVM) framework for improving the classification performance for the target domain as follows:

$$
\underset{w,\xi_S,\xi_T}{\arg\min} \quad \frac{1}{2}\|w\|_2^2 + \lambda_T \sum_{i=1}^{n_{T_l}} \xi_{T_i} + \lambda_S \sum_{i=1}^{n_S} \gamma_i \xi_{S_i}, \tag{21.6}
$$
$$
s.t. \quad y_{S_i} w^\top x_{S_i} \geq 1 - \xi_{S_i}, \; \xi_{S_i} \geq 0, \; i = 1,...,n_S,
$$
$$
y_{T_i} w^\top x_{T_i} \geq 1 - \xi_{T_i}, \; \xi_{T_i} \geq 0, \; i = 1,...,n_{T_l},
$$

where n_{T_l} is the number of target domain labeled data, w is the model parameter, ξ_S and ξ_T are the slack variables to absorb errors on the source and target domain data, respectively, λ_S and λ_T are the tradeoff parameters to balance the impact of different terms in the objective, and γ_i is the weight on the source domain instance x_{S_i}. There are various ways to set the values of $\{\gamma_i\}$'s. In [142], Wu and Dietterich proposed to simply set $\gamma_i = 1$ for each data point in the source domain. Jiang and Zhai [62] proposed a heuristic method to remove the "misleading" instances from the source domain, which is equivalent to setting $\gamma_i = 0$ for all "misleading" source domain instances and $\gamma_i = 1$ for the remaining instances. Note that the basic classifier used in [62] is a probabilistic model instead of SVM, but the idea is similar.

Dai et al. [38] proposed a boosting algorithm, known as TrAdaBoost, for transfer learning. TrAdaBoost is an extension of the AdaBoost algorithm [49]. The basic idea of TrAdaBoost attempts to iteratively re-weight the source domain data to reduce the effect of the "bad" source data while encouraging the "good" source data to contribute more to the target domain. Specifically, for each round of boosting, TrAdaBoost uses the same strategy as AdaBoost to update weights of the target domain labeled data, while proposing a new mechanism to decrease the weights of misclassified source domain data.

21.3.2 Feature-Representation-Based Approach

As described in the previous section, for the instance-based approach, a common assumption is that the source and target domains have a lot of overlapping features. However, in many real-world applications, the source and target domains may only have some overlapping features, which means that many features may only have support in either the source or target domain. In this case, most instance-based methods may not work well. The feature-representation-based approach to transfer learning is promising to address this issue. An intuitive idea behind the feature-representation-based approach is to learn a "good" feature representation for the source and target domains such that based on the new representation, source domain labeled data can be reused for the target domain. In this sense, the knowledge to be transferred across domains is encoded into the learned feature representation. Specifically, the feature-representation-based approach aims to learn a mapping $\varphi(\cdot)$ such that the difference between the source and target domain data after transformation, $\{\varphi(x_{S_i})\}$'s and $\{\varphi(x_{T_i})\}$'s, can be reduced. In general, there are two ways to learn such a mapping $\varphi(\cdot)$ for transfer learning. One is to encode specific domain or application knowledge into learning the mapping, the other is to propose a general method to learn the mapping without taking any domain or application knowledge into consideration.

21.3.2.1 Encoding Specific Knowledge for Feature Learning

In this section, we use sentiment classification as an example to present how to encode domain knowledge into feature learning. In sentiment classification, a domain denotes a class of objects or events in the world. For example, different types of products, such as *books*, *dvds*, and *furniture*, can be regarded as different domains. Sentiment data are the text segments containing user opinions about objects, events, and their properties of the domain. User sentiment may exist in the form of a sentence, paragraph, or article, which is denoted by x_j. Alternatively, it corresponds with a sequence of words $v_1 v_2 ... v_{x_j}$, where w_i is a word from a vocabulary V. Here, we represent user sentiment data by a bag of words with $c(v_i, x_j)$ to denote the frequency of word v_i in x_j. Without loss of generality, we use a unified vocabulary W for all domains, and assume $|W| = m$.

For each sentiment data x_j, there is a corresponding label y_j, where $y_j = +1$ if the overall sentiment expressed in x_j is positive, and $y_j = -1$ if the overall sentiment expressed in x_j is negative. A pair of sentiment text and its corresponding sentiment polarity $\{x_j, y_j\}$ is called the *labeled sentiment data*. If x_j has no polarity assigned, it is *unlabeled sentiment data*. Note that besides positive and negative sentiment, there are also neutral and mixed sentiment data in practical applications. *Mixed polarity* means user sentiment is positive in some aspects but negative in other ones. *Neutral polarity* means that there is no sentiment expressed by users. In this chapter, we only focus on positive and negative sentiment data.

For simplicity, we assume that a sentiment classifier f is a linear function as

$$y^* = f(x) = \mathbf{sgn}(w^\top x),$$

where $x \in \mathbb{R}^{m \times 1}$, $\mathbf{sgn}(w^\top x) = +1$ if $w^\top x \geq 0$, otherwise, $\mathbf{sgn}(w^\top x) = -1$, and w is the weight vector of the classifier, which can be learned from a set of training data (i.e., pairs of sentiment data and their corresponding polarity labels).

Consider the example shown in Table 21.1 as an motivating example. We use the standard bag-of-words representation to represent sentiment data of the *Electronics* and *Video Games* domains. From Table 21.2, we observe that the difference between domains is caused by the frequency of the domain-specific words. On one hand, the domain-specific words in the *Electronics* domain, such as *compact*, *sharp* and *blurry*, cannot be observed in the *Video Games* domain. On the other hand, the domain-specific words in the *Video Games* domain, such as *hooked*, *realistic* and *boring*, cannot be observed in the *Electronics* domain. Suppose that the *Electronics* domain is the source domain and the *Video Games* domain is the target domain. Apparently, based on the three training sentences in the *Electronics* domain, the weights of the features *compact* and *sharp* are positive, the weight of the feature *blurry* are negative, and the weights of the features *hooked*, *realistic* and *boring* can be arbitrary or zeros if an ℓ_1-norm regularization term is performed on w for model training. However, an ideal weight vector for the *Video Games* domain are supposed to have positive weights on the features *hooked*, *realistic* and a negative weight on the feature *boring*. Therefore, a classifier learned from the *Electronics* domain may predict poorly or randomly on the *Video Games* domain data.

TABLE 21.2: Bag-of-Words Representations of *Electronics* and *Video Games* Reviews.

		compact	sharp	blurry	hooked	realistic	boring
	+1	1	1	0	0	0	0
electronics	**+1**	0	1	0	0	0	0
	-1	0	0	1	0	0	0
	+1	0	0	0	1	0	0
video games	**+1**	0	0	0	1	1	0
	-1	0	0	0	0	0	1

Note: Only domain-specific features are considered.

Generally speaking, in sentiment classification, features can be classified into three types: 1) source domain (i.e., the *Electronics* domain) specific features, such as *compact*, *sharp*, and *blurry*, 2) target domain (i.e., the Video Game domain) specific features, such as *hooked*, *realistic*, and *boring*, and 3) domain independent features or pivot features, such as *good*, *excited*, *nice*, and *never_buy*. Based on these observations, an intuitive idea of feature learning is to align the source and target domain specific features to generate cluster- or group- based features by using the domain independent features as a bridge such that the difference between the source and target domain data based on the new feature representation can be reduced. For instance, if the domain specific features shown in Table 21.2 can be aligned in the way presented in Table 21.3, where the feature alignments are used as new features to represent the data, then apparently, a linear model learned from the source domain (i.e., the *Electronics* domain) can be used to make precise predictions on the target domain data (i.e., the *Video Game* domain).

TABLE 21.3: Using Feature Alignments as New Features to Represent Cross-Domain Data.

		sharp_hooked	compact_realistic	blurry_boring
	+1	1	1	0
electronics	**+1**	1	0	0
	-1	0	0	1
	+1	1	0	0
video games	**+1**	1	1	0
	-1	0	0	1

Therefore, there are two research issues to be addressed. A first issue is how to identify domain independent or pivot features. A second issue is how to utilize the domain independent features and domain knowledge to align domain specific features from the source and target domains to generate new features. Here the domain knowledge is that if two sentiment words co-occur frequently in one sentence or document, then their sentiment polarities tend to be the same with a high probability.

For identifying domain independent or pivot features, some researchers have proposed several heuristic approaches [18,92]. For instance, Blitzer et al. [18] proposed to select pivot features based on the term frequency in both the source and target domains and the mutual dependence between the features and labels in the source domain. The idea is that a pivot feature should be discriminative to the source domain data and appear frequently in both the source and target domains. Pan et al. [92] proposed to select domain independent features based on the mutual dependence between features and domains. Specifically, by considering all instances in the source domain with labels 1's and all instances in the target domain with labels 0's, the mutual information can be used to measure the dependence between the features and the constructed *domain labels*. The motivation is that if a feature has high mutual dependence to the domains, then it is domain specific. Otherwise, it is domain independent.

For aligning domain specific features from the source and target domains to generate cross-domain features, Blitzer et al. [19] proposed the structural correspondence learning (SCL) method. SCL is motivated by a multi-task learning algorithm, alternating structure optimization (ASO) [4], which aims to learn common features underlying multiple tasks. Specifically, SCL first identifies a set of *pivot* features of size m, and then treats each *pivot* feature as a new output vector to construct a pseudo task with non-pivot features as inputs. After that, SCL learns m linear classifiers to model the relationships between the non-pivot features and the constructed output vectors as follows,

$$y_j = \mathbf{sgn}(w_j^\top x_{np}), \ j = 1, \ldots, m,$$

where y_j is an output vector constructed from a corresponding pivot feature, and x_{np} is a vector of non-pivot features. Finally, SCL performs the singular value decomposition (SVD) on the weight matrix $W = [w_1 \ w_2 \ \ldots \ w_m] \in \mathbb{R}^{q \times m}$, where q is the number of non-pivot features, such that $W = UDV^\top$, where $U_{q \times r}$ and $V_{r \times m}$ are the matrices of the left and right singular vectors. The matrix $D_{r \times r}$ is a diagonal matrix consisting of non-negative singular values, which are ranked in non-increasing order. The matrix $U_{[1:h,:]}^\top$, where h is the number of features to be learned, is then used as a transformation to align domain-specific features to generate new features.

Pan et al. [92] proposed the Spectral Feature Alignment (SFA) method for aligning domain specific features, which shares a similar high-level motivation with SCL. Instead of constructing pseudo tasks to use model parameters to capture the correlations between domain-specific and domain-independent features, SFA aims to model the feature correlations using a bipartite graph. Specifically, in the bipartite graph, a set of nodes correspond to the domain independent features, and the other set of nodes correspond to domain specific features in either the source or target domain. There exists an edge connecting a domain specific feature and a domain independent feature, if they co-occur in the same document or within a predefined window. A number associated on an edge is the total number of the co-occurrence of the corresponding domain specific and domain independent features in the source and target domains. The motivation of using a bipartite graph to model the feature correlations is that if two domain specific features have connections to more common domain independent features in the graph, they tend to be aligned or clustered together with a higher probability. Meanwhile, if two domain independent features have connections to more common domain specific features in the graph, they tend to be aligned together with a higher probability. After the bipartite graph is constructed, the spectral clustering algorithm [88] is applied on the graph to cluster domain specific features. In this way, the clusters can be treated as new features to represent cross-domain data.

21.3.2.2 Learning Features by Minimizing Distance between Distributions

In the previous section, we have shown how to encode domain knowledge into feature learning for transfer learning. However, in many real-work scenarios, domain knowledge is not available as input. In this case, general approaches to feature learning for transfer learning are required. In this section, we first introduce a feature learning approach to transfer learning based on distribution minimization in a latent space.

Note that in many real-world applications, the observed data are controlled by only a few latent factors. If the two domains are related to each other, they may share some latent factors (or components). Some of these common latent factors may cause the data distributions between domains to be different, while others may not. Meanwhile, some of these factors may capture the intrinsic structure or discriminative information underlying the original data, while others may not. If one can recover those common latent factors that do not cause much difference between data distributions and do preserve the properties of the original data, then one can treat the subspace spanned by these latent factors as a bridge to make knowledge transfer possible. Based on this motivation, Pan et al. [90] proposed a dimensionality reduction algorithm for transfer learning, whose high-level idea can be formulated as follows:

$$\min_{\varphi} \quad \text{Dist}(\varphi(\mathbf{X}_S), \varphi(\mathbf{X}_T)) + \lambda \Omega(\varphi) \tag{21.7}$$

$$\text{s.t.} \quad \text{constraints on } \varphi(\mathbf{X}_S) \text{ and } \varphi(\mathbf{X}_T),$$

where φ is the mapping to be learned, which maps the original data to a low-dimensional space. The first term in the objective of (21.7) aims to minimize the distance in distributions between the source and target domain data, $\Omega(\varphi)$ is a regularization term on the mapping φ, and the constraints are to ensure original data properties are preserved.

Note that, in general, the optimization problem (21.7) is computationally intractable. To make it computationally solvable, Pan et al. [90] proposed to transform the optimization problem (21.7) to a kernel matrix learning problem, resulting in solving a semidefinite program (SDP). The proposed method is known as Maximum Mean Discrepancy Embedding (MMDE), which is based on the non-parametric measure MMD as introduced in Section 21.3.1.1. MMDE has proven to be effective in learning features for transfer learning. However, it has two major limitations: 1) Since it requires to solve a SDP, its computational cost is very expensive; 2) Since it formulates the kernel matrix learning problem in a transductive learning setting, it cannot generalize to out-of-sample instances. To address the limitations of MMDE, Pan et al. [95, 96] further relaxed the feature learning problem of MMDE to a generalized eigen-decomposition problem, which is very efficient and easily generalized to out-of-sample instances. Similarly, motivated by the idea of MMDE, Si et al. [125] proposed to use the Bregman divergence as the distance measure between sample distributions to minimize the distance between the source and target domain data in a latent space.

21.3.2.3 Learning Features Inspired by Multi-Task Learning

Besides learning features by minimizing distance in distributions, another important branch of approaches to learning features for transfer learning is motivated by multi-task learning [24]. In multi-task learning, given multiple tasks with a few labeled training data for each task, the goal is to jointly learn individual classifiers for different tasks by exploring latent common features shared by the tasks. Without loss of generality, for each task, we assume that the corresponding classifier is linear, and can be written as

$$f(x) = \langle \theta, (U^\top x) \rangle = \theta^\top (U^\top x),$$

where $\theta \in \mathbb{R}^{k \times 1}$ is the individual model parameter to be learned, and $U \in \mathbb{R}^{m \times k}$ is the transformation shared by all task data, which maps original data to a k-dimensional feature space, and needs to be learned as well. Note that the setting of multi-task learning is different from that of transfer learning, where a lot of labeled training data are assumed to be available in a source domain, and the focus is to learn a more precise model for the target domain. However, the idea of common feature learning under different tasks can still be borrowed for learning features for transfer learning by assuming that a few labeled training data in the target domain are available. The high-level objective of feature learning based on multi-task learning can be formulated as follows:

$$\min_{U, \theta_S, \theta_T} \sum_{t \in \{S, T\}} \sum_{i=1}^{n_t} l(U^\top x_{t_i}, y_{t_i}, \theta_t) + \lambda \Omega(\Theta, U)$$

$$\text{s.t.} \quad \text{constraints on } U, \tag{21.8}$$

where $\Theta = [\theta_S \ \theta_T] \in \mathbb{R}^{k \times 2}$ and $\Omega(\Theta, U)$ is a regularization term on Θ and U. Based on different forms of $\Omega(\Theta, U)$ and different constraints on U, approaches to learning features based on multi-task learning can be generally classified into two categories. In a first category of approaches, U is assumed to be full rank, which means that $m = k$, and Θ is sparse. A motivation behind this is that the full-rank U is only to transform the data from original space to another space of the same dimensionality, where a few *good* features underlying different tasks can be found potentially, and the sparsity assumption on Θ is to select such *good* features and ignore those that are not helpful for the source and target tasks. One of the representative approaches in this category was proposed by Argyriou et al. [6], where the $\| \cdot \|_{2,1}$ norm is proposed to regularize the matrix form of the model parameters Θ,[1] and U is assumed to be orthogonal, which means that $U^\top U = U U^\top = I$. As shown in [6], the optimization problem can be transformed to a convex optimization formulation and solved efficiently. In a follow-up work, Argyriou et al. [8] proposed a new spectral function on Θ for multi-task feature learning.

In a second category of approaches, U is assumed to be row rank, which means that $k < m$, or $k \ll m$ in practice, and there are no sparsity assumptions on Θ. In this way, U transforms the original data to *good* common feature representations directly. Representative approaches in this category include the Alternating Structure Optimization (ASO) method, which has been mentioned in Section 21.3.2.1. As described, in ASO, the SVD is performed on the matrix of the source and target model-parameters to recover a low-dimensional predictive space as a common feature space. The ASO method has been applied successfully to several applications [5, 18]. However, the proposed optimization problem is non-convex and thus a global optimum is not guaranteed to be achieved. Chen et al. [30] presented an improved formulation, called iASO, by proposing a novel regularization term on U and Θ. Furthermore, in order to convert the new formulation into a convex formulation, in [30], Chen et al. proposed a convex alternating structure optimization (cASO) algorithm to solve the optimization problem.

---

[1] The $\| \cdot \|_{2,1}$-norm of Θ is defined as $\|\Theta\|_{2,1} = \sum_{i=1}^{m} \|\Theta^i\|_2^1$, where Θ^i is the ith row of Θ.

21.3.2.4 Learning Features Inspired by Self-Taught Learning

Besides borrowing ideas from multi-task learning, a third branch of feature learning approaches to transfer learning is inspired by self-taught learning [105]. In self-taught learning, a huge number of unlabeled data are assumed to be available, whose labels can be different from those of the task of interest. The goal is to learn a set of *higher-level* features such that based on these higher-level features; a classifier trained on a few labeled training data can perform well on the task of interest. In this branch of approaches, a common assumption is that large-scale unlabeled or labeled source domain data, which can come from a single source or multiple sources, are available as inputs, and a few labeled data in the target domain are available as well. Most methods consist of three steps: 1) to learn higher-level features from the large-scale source domain data with or without their label information; 2) to represent the target domain data based on the higher-level features; 3) to train a classifier on the new representations of the target domain data with corresponding labels. A key research issue in these approaches is how to learn higher-level features. Raina et al. [105] proposed to apply sparse coding [70], which is an unsupervised feature construction method, to learn the higher-level features for transfer learning. Glorot et al. [51] proposed to apply deep learning to learn the higher-level features for transfer learning. Note that the goal of deep learning is to generate hierarchical features from lower-level input features, where the features generated in higher layers are assumed to be more higher level.

21.3.2.5 Other Feature Learning Approaches

In addition to the above three branches of feature learning methods for transfer learning, Daumé III [39] proposed a simple feature augmentation method for transfer learning in the field of Natural Language Processing (NLP). The proposed method aims to augment each of the feature vectors of different domains to a high dimensional feature vector as follows,

$$
\begin{aligned}
\widetilde{x}_S &= [x_S \, x_S \, \mathbf{0}], \\
\widetilde{x}_T &= [x_T \, \mathbf{0} \, x_T],
\end{aligned}
$$

where x_S and x_T are original feature vectors of the source and target domains, respectively, and $\mathbf{0}$ is a vector of zeros, whose length is equivalent to that of the original feature vector. The idea is to reduce the difference between domains while ensuring the similarity between data within domains is larger than that across different domains. In a follow-up work, Daumé III [40] extends the feature augmentation method in a semi-supervised learning manner. Dai et al. [36] proposed a co-clustering based algorithm to discover common feature clusters, such that label information can be propagated across different domains by using the common clusters as a bridge. Xue et al. [147] proposed a cross-domain text classification algorithm that extends the traditional probabilistic latent semantic analysis (PLSA) [58] algorithm to extract common topics underlying the source and target domain text data for transfer learning.

21.3.3 Model-Parameter-Based Approach

The first two categories of approaches to transfer learning are in the data level, where the instance-based approach tries to reuse the source domain data after re-sampling or re-weighting, while the feature-representation-based approach aims to find a good feature representation for both the source and target domains such that based on the new feature representation, source domain data can be reused. Different from these two categories of approaches, a third category of approaches to transfer learning can be referred to as the model-parameter-based approach, which assumes that the source and target tasks share some parameters or prior distributions of the hyper-parameters of the models. A motivation of the model-parameter-based approach is that a well-trained source model has captured a lot of structure, which can be transferred to learn a more precise target model. In this way, the transferred knowledge is encoded into the model parameters. In the rest of this section, we first introduce a simple method to show how to transfer knowledge across tasks or domains through model parameters, and then describe a general framework of the model-parameter-based approach.

Without loss of generality, we assume that the classifier to be learned is linear and can be written as follows,

$$f(x) = \langle \theta, x \rangle = \theta^\top x = \sum_{i=1}^{m} \theta_i x_i.$$

Given a lot of labeled training data in the source domain and a few labeled training data in the target domain, we further assume that the source model parameter θ_S is well-trained, and our goal to exploit the structure captured by θ_S to learn a more precise model parameter θ_T from the target domain training data.

Evgeniou and Pontil [48] proposed that the model parameter can be decomposed into two parts; one is referred to as a task specific parameter, and the other is referred to as a common parameter. In this way, the source and target model parameters θ_S and θ_T can be decomposed as

$$\theta_S = \theta_0 + v_S,$$
$$\theta_T = \theta_0 + v_T,$$

where θ_0 is the common parameter shared by the source and target classifiers, v_S and v_T are the specific parameters of the source and target classifiers, respectively. Evgenious and Pontil further proposed to learn the common and specific parameters by solving the following optimization problem,

$$\arg\min_{\theta_S, \theta_T} \sum_{t \in \{S,T\}} \sum_{i=1}^{n_t} l(x_{t_i}, y_{t_i}, \theta_t) + \lambda \Omega(\theta_0, v_S, v_T), \tag{21.9}$$

where $\Omega(\theta_0, v_S, v_T)$ is the regularization term on θ_0, v_S and v_T, and $\lambda > 0$ is the corresponding trade-off parameter. The simple idea presented in (21.9) can be generalized to a framework of the model-parameter-based approach as follows,

$$\arg\min_{\Theta} \sum_{t \in \{S,T\}} \sum_{i=1}^{n_t} l(x_{t_i}, y_{t_i}, \theta_t) + \lambda_1 \mathrm{tr}(\Theta^\top \Theta) + \lambda_2 f(\Theta) \tag{21.10}$$

where $\Theta = [\theta_S \ \theta_T]$, $\mathrm{tr}(\Theta^\top \Theta)$ is a regularization on θ_S and θ_T to avoid overfitting, and $f(\Theta)$ is to model the correlations between θ_S and θ_T, which is used for knowledge transfer. Different forms of $f(\Theta)$ lead to various specific methods. It can be shown that in (21.9), $f(\Theta)$ can be defined by the following form,

$$f(\Theta) = \sum_{t \in \{S,T\}} \left\| \theta_t - \frac{1}{2} \sum_{s \in \{S,T\}} \theta_s \right\|_2^2. \tag{21.11}$$

Besides using (21.11), Zhang and Yeung [161] proposed the following form to model the correlations between the source and target parameters,

$$f(\Theta) = \mathrm{tr}(\Theta^\top \Omega^{-1} \Theta), \tag{21.12}$$

where Ω is the covariance matrix to model the relationships between the source and target domains, which is unknown and needs to be learned with the constraints $\Omega \succeq 0$ and $\mathrm{tr}(\Omega) = 1$. Agarwal et al. [1] proposed to use a manifold of parameters to regularize the source and target parameters as follows:

$$f(\Theta) = \sum_{t \in \{S,T\}} \left\| \theta_t - \widetilde{\theta}_t^{\mathcal{M}} \right\|^2, \tag{21.13}$$

where $\widetilde{\theta}_S^{\mathcal{M}}$ and $\widetilde{\theta}_T^{\mathcal{M}}$ are the projections of the source parameter θ_S and target parameter θ_T on the manifold of parameters, respectively.

Besides the framework introduced in (21.10), there are a number of methods that are based on non-parametric Bayesian modeling. For instance, Lawrence and Platt [69] proposed an efficient algorithm for transfer learning based on Gaussian Processes (GP) [108]. The proposed model tries to discover common parameters over different tasks, and an informative vector machine was introduced to solve large-scale problems. Bonilla et al. [21] also investigated multi-task learning in the context of GP. Bonilla et al. proposed to use a free-form covariance matrix over tasks to model inter-task dependencies, where a GP prior is used to induce the correlations between tasks. Schwaighofer et al. [118] proposed to use a hierarchical Bayesian framework (HB) together with GP for transfer learning.

21.3.4 Relational-Information-Based Approaches

A fourth category of approaches is referred to as the relational-information-based approach. Different from the other three categories, the relational-information-based approach assumes that some relationships between objects (i.e., instances) are similar across domains or tasks; if these common relationships can be extracted, then they can be used for knowledge transfer. Note that in this category of approaches, data in the source and target domains are not required to be independent and identically distributed (i.i.d.).

Mihalkova et al. [84] proposed an algorithm known as TAMAR to transfer relational knowledge with Markov Logic Networks (MLNs) [110] across the source and target domains. MLNs is a statistical relational learning framework, which combines the compact expressiveness of first order logic with flexibility of probability. In MLNs, entities in a relational domain are represented by predicates and their relationships are represented in first-order logic. TAMAR is motivated by the fact that if two domains are related to each other, there may exist mappings to connect entities and their relationships from the source domain to the target domain. For example, a professor can be considered as playing a similar role in an academic domain to a manager in an industrial management domain. In addition, the relationship between a professor and his or her students is similar to that between a manager and his or her workers. Thus, there may exist a mapping from professor to manager and a mapping from the professor-student relationship to the manager-worker relationship. In this vein, TAMAR tries to use an MLN learned for the source domain to aid in the learning of an MLN for the target domain. In a follow-up work, Mihalkova et al. [85] extended TAMAR in a single-entity-centered manner, where only one entity in the target domain is required in training.

Instead of mapping first-order predicates across domains, Davis et al. [41] proposed a method based on second-order Markov logic to transfer relational knowledge. In second-order Markov Logic, predicates themselves can be variables. The motivation of the method is that though lower-level knowledge such as propositional logic or first-order logic is domain or task specific, higher-level knowledge such as second-order logic is general for different domains or tasks. Therefore, this method aims to generate a set of second-order logic formulas through second-order MLNs from the source domain, and use them as higher-level templates to instantiate first-order logic formulas in the target domain.

More recently, Li et al. [74] proposed a relation-information-based method for sentiment analysis across domains. In this method, syntactic relationships between topic and sentiment words are exploited to propagate label information across the source and target domains. The motivation behind this method is that though the sentiment and topic words used in the source and target domains may be different, the syntactic relationships between them may be similar or the same across domains. Based on a few sentiment and topic seeds in the target domain, together with the syntactic relationships extracted from the source domain by using NLP techniques, lexicons of topic and sentiment words can be expanded iteratively in the target domain.

Note that the relational-information-based approach introduced in this section aims to explore and exploit relationships between instances instead of the instances themselves for knowledge transfer. Therefore, the relational-information-based approach can also be applied to heterogeneous transfer learning problems that will be introduced in the following section, where the source and target feature or label spaces are different.

21.4 Heterogeneous Transfer Learning

In the previous section, we have introduced four categories of approaches to homogeneous transfer learning. Recently, some researchers have already started to consider transfer learning across heterogeneous feature spaces or non-identical label spaces. In this section, we start by giving a definition of heterogeneous transfer learning as follows,

Definition 3 *Given a source domain $\mathcal{D}_S$ and learning task $\mathcal{T}_S$, a target domain $\mathcal{D}_T$ and learning task $\mathcal{T}_T$, heterogeneous transfer learning aims to help improve the learning of the target predictive function $f_T(\cdot)$ in $\mathcal{D}_T$ using the knowledge in $\mathcal{D}_S$ and $\mathcal{T}_S$, where $X_S \cap X_T = \emptyset$ or $\mathcal{Y}_S \neq \mathcal{Y}_T$.*

Based on the definition, heterogeneous transfer learning can be further categorized into two contexts: 1) approaches to transferring knowledge across heterogeneous feature spaces, and 2) approaches to transferring knowledge across different label spaces.

21.4.1 Heterogeneous Feature Spaces

How to transfer knowledge successfully across different feature spaces is an interesting issue. It is related to multi-view learning [20], which assumes that the features for each instance can be divided into several views, each with its own distinct feature space. Though multi-view learning techniques can be applied to model multi-modality data, it requires that each instance in one view must have its correspondence in other views. In contrast, transfer learning across different feature spaces aims to solve the problem where the source and target domain data belong to two different feature spaces such as image vs. text, without correspondences across feature spaces. Recently, some heterogeneous transfer learning methods have been developed and applied to various applications, such as cross-language text classification [77, 101, 135], image classification [168], [33], [64], and object recognition [67, 115].

Transfer learning methods across heterogeneous feature spaces can be further classified into two categories. A first context of approaches is to learn a pair of feature mappings to transform the source and target domain heterogeneous data to a common latent space. Shi et al. [124] proposed a Heterogenous Spectral Mapping (HeMap) method to learn the pair of feature mappings based on spectral embedding, where label information is discarded in learning. Wang and Mahadevan [135] proposed a manifold alignment method to align heterogenous features in a latent space based on a manifold regularization term, which is denoted by DAMA in the sequel. In DAMA, label information is exploited to construct similarity matrix for manifold alignment. However, DADA only works on the data that have strong manifold structures, which limits its transferability on those data where the manifold assumption does not hold. More recently, Duan et al. [44] proposed a Heterogenous Feature Augmentation (HFA) method to augment homogeneous common features learned by a SVM-style approach with heterogeneous features of the source and target domains for transfer learning. The proposed formulation results in a semidefinite program (SDP), whose computational cost is very expensive.

Another context is to learn a feature mapping to transform heterogenous data from one domain to another domain directly. In [34, 101], the feature mappings are obtained based on some *translators* to construct corresponding features across domains. However, in general, such translators for corresponding features are not available and difficult to construct in many real-world applications. Kulis [67] proposed an Asymmetric Regularized Cross-domain transformation (ARC-t) method to learn asymmetric transformation across domains based on metric learning. Similar to DAMA, ARC-t also utilizes the label information to construct similarity and dissimilarity constraints between instances from the source and target domains, respectively. The formulated metric learning problem can be solved by an alternating optimization algorithm.

21.4.2 Different Label Spaces

In some real-world scenarios, the label spaces or categories of the source and target domain may not be the same. In this case, it is crucial to develop transfer learning methods to propagate knowledge across labels or categories. A common idea behind most existing approaches in this setting is to explore and exploit the relationships between the source and target categories such that label information can be propagated across domains. Shi et al. [123] proposed a risk-sensitive spectral partition (RSP) method to align the source and target categories based on spectral partitioning. Dai et al. [35] proposed an EigenTransfer framework to use a three-layer bipartite graph to model the relationships between instances, features, and categories. Through the three-layer bipartite graph, label information can be propagated across different categories. Quadrianto et al. [103] proposed to maximize mutual information between labels across domains to identify their correspondences. Qi et al. [102] proposed an optimization algorithm to learn a parametric matrix to model the correlations between labels across domains based on the similarities between cross-domain input data. Xiang et al. [144] proposed a novel framework named source-selection-free transfer learning (SSFTL) to achieve knowledge transfer from a Web-scale auxiliary resource, e.g., Wikipedia, for universal text classification. The idea of SSFTL is to first pre-train a huge number of source classifiers from the auxiliary resource offline, then when a target task is given, whose labels may not be observed in the auxiliary resource, SSFTL makes use of social tagging data, e.g., Flickr to bridge the auxiliary labels and the target labels, and finally select relevant source classifiers to solve the target task automatically.

21.5 Transfer Bounds and Negative Transfer

For theoretical study of transfer learning, an important issue is to recognize the limit of the power of transfer learning. So far, most theoretical studies are focused on homogeneous transfer learning. There are some research works analyzing the generalization bound in a special setting of homogeneous transfer learning where only the marginal distributions, $P_S(x)$ and $P_T(x)$, of the source and target domain data are assumed to be different [12–14, 17]. Though the generalization bounds proved in various works differ slightly, there is a common conclusion that the generalization bound of a learning model in this setting consists of two terms, one being the error bound of the learning model on the source domain labeled data, the other being the bound on the distance between the source and target domains, more specifically the distance between marginal probability distributions between domains.

In a more general setting, homogeneous transfer learning where the predictive distributions, $P_S(y|x)$ and $P_T(y|x)$, of the source and target domain data can be different, theoretical studies are more focused on the issue of transferability. That is to ask, how to avoid negative transfer and then ensure a "safe transfer" of knowledge. Negative transfer happens when the source domain/task data contribute to the reduced performance of learning in the target domain/task. Although how to avoid negative transfer is a very important issue, few research works were proposed on this issue in the past. Rosenstein et al. [114] empirically showed that if two tasks are very dissimilar, then brute-force transfer may hurt the performance of the target task.

Recently, some research works have been explored to analyze relatedness among tasks using task clustering techniques, such as [11, 15], which may help provide guidance on how to avoid negative transfer automatically. Bakker and Heskes [11] adopted a Bayesian approach in which some of the model parameters are shared for all tasks and others are more loosely connected through a joint prior distribution that can be learned from the data. Thus, the data are clustered based on the task parameters, where tasks in the same cluster are supposed to be related to each other. Hassan Mahmud and Ray [80] analyzed the case of transfer learning using Kolmogorov complexity, where some theoretical bounds are proved. In particular, they used conditional Kolmogorov complexity to measure relatedness between tasks and transfer the "right" amount of information in a sequential transfer learning task under a Bayesian framework. Eaton et al. [46] proposed a novel graph-based method for knowledge transfer, where the relationships between source tasks are modeled by a graph using transferability as the metric. To transfer knowledge to a new task, one needs to map the target task to the graph and learn a target model on the graph by automatically determining the parameters to transfer to the new learning task.

More recently, Argyriou et al. [7] considered situations in which the learning tasks can be divided into groups. Tasks within each group are related by sharing a low-dimensional representation, which differs among different groups. As a result, tasks within a group can find it easier to transfer useful knowledge. Jacob et al. [60] presented a convex approach to cluster multi-task learning by designing a new spectral norm to penalize over a set of weights, each of which is associated to a task. Bonilla et al. [21] proposed a multi-task learning method based on Gaussian Process (GP), which provides a global approach to model and learn task relatedness in the form of a task covariance matrix. However, the optimization procedure introduced in [21] is non-convex and its results may be sensitive to parameter initialization. Motivated by [21], Zhang and Yeung [161] proposed an improved regularization framework to model the negative and positive correlation between tasks, where the resultant optimization procedure is convex.

The above works [7, 11, 15, 21, 60, 161] on modeling task correlations are from the context of multi-task learning. However, in transfer learning, one may be particularly interested in transferring knowledge from one or more source tasks to a target task rather than learning these tasks simultaneously. The main concern of transfer learning is the learning performance in the target task only. Thus, we need to give an answer to the question that given a target task and a source task, whether transfer learning techniques should be applied or not. Cao et al. [23] proposed an Adaptive Transfer learning algorithm based on GP (AT-GP), which aims to adapt transfer learning schemes by automatically estimating the similarity between the source and target tasks. In AT-GP, a new semi-parametric kernel is designed to model correlations between tasks, and the learning procedure targets improving performance of the target task only. Seah et al. [120] empirically studied the negative transfer problem by proposing a predictive distribution matching classifier based on SVMs to identify the regions of relevant source domain data where the predictive distributions maximally align with that of the target domain data, and thus avoid negative transfer.

21.6 Other Research Issues

Besides the negative transfer issue, in recent years there have been several other research issues of transfer learning that have attracted more and more attention from the machine learning community, which are summarized in the following sections.

21.6.1 Binary Classification vs. Multi-Class Classification

Most existing transfer learning methods are proposed for binary classification. For multi-class classification problems, one has to first reduce the multi-class classification task into multiple binary classification tasks using the *one-vs.-rest* or *one-vs.-one* strategy, and then train multiple binary classifiers to solve them independently. Finally, predictions are made according to the outputs of all binary classifiers. In this way, the relationships between classes, which indeed can be used to further boost the performance in terms of classification accuracy, may not be fully explored and exploited. Recently, Pan et al. [94] proposed a Transfer Joint Embedding (TJE) method to map both the features and labels from the source and target domains to a common latent space such as the relationships between labels, which can be fully exploited for transfer learning in multi-class classification problems.

21.6.2 Knowledge Transfer from Multiple Source Domains

In Sections 21.3–21.4, the transfer learning methods described are focused on one-to-one transfer, which means that there is only one source domain and one target domain. However, in some real-world scenarios, we may have multiple sources at hand. Developing algorithms to make use of multiple sources for learning models in the target domain is useful in practice. Yang et al. [150] and Duan et al. [42] proposed algorithms to learn a new SVM for the target domains by adapting SVMs learned from multiple source domains. Luo et al. [79] proposed to train a classifier for use in the target domain by maximizing the consensus of predictions from multiple sources. Mansour et al. [81] proposed a distribution weighted linear combination framework for learning from multiple sources. The main idea is to estimate the data distribution of each source to reweight the data of different source domains. Yao and Doretto [155] extended TrAdaBoost in a manner of multiple source domains. Theoretical studies on transfer learning from multiple source domains have also been presented in [81], [82], [12].

21.6.3 Transfer Learning Meets Active Learning

As mentioned at the beginning of this chapter, both active learning and transfer learning aim to learn a precise model with minimal human supervision for a target task. Several researchers have proposed to combine these two techniques together in order to learn a more precise model with even less supervision. Liao et al. [76] proposed a new active learning method to select the unlabeled data in a target domain to be labeled with the help of the source domain data. Shi et al. [122] applied an active learning algorithm to select important instances for transfer learning with TrAdaBoost [38] and standard SVM. In [26], Chan and NG proposed to adapt existing Word Sense Disambiguation (WSD) systems to a target domain by using domain adaptation techniques and employing an active learning strategy [71] to actively select examples from the target domain to be annotated. Harpale and Yang [56] proposed an active learning framework for the multi-task adaptive filtering [112] problem. They first applied a multi-task learning method to adaptive filtering, and then explored various active learning approaches to the adaptive filters to improve performance. Li et al. [75] proposed a novel multi-domain active learning framework to jointly actively query data instances from all domains to be labeled to build individual classifiers for each domain. Zhao et al. [163], proposed a framework to construct entity correspondences with limited budget by using active learning to facilitate knowledge transfer across different recommender systems.

21.7 Applications of Transfer Learning

Recently, transfer learning has been applied successfully to many classification problems in various application areas, such as Natural Language Processing (NLP), Information Retrieval (IR), recommendation systems, computer vision, image analysis, multimedia data mining, bioinformatics, activity recognition, and wireless sensor networks.

21.7.1 NLP Applications

In the field of NLP, transfer learning, which is known as domain adaptation, has been widely studied for solving various tasks, such as name entity recognition [9, 39, 53, 94, 111, 140, 141], part-of-speech tagging [4, 19, 39, 62], sentiment classification [18, 51, 92], sentiment lexicon construction [74], word sense disambiguation [2, 26], coreference resolution [149], and relation extraction [61].

21.7.2 Web-Based Applications

Information Retrieval (IR) is another application area where transfer learning techniques have been widely studied and applied. Typical Web applications of transfer learning include text classification [16, 29, 36, 37, 54, 106, 137, 143, 145, 147, 153], advertising [32, 33], learn to rank [10, 28, 50, 134], and recommender systems [22, 72, 73, 99, 160].

21.7.3 Sensor-Based Applications

Transfer learning has also been explored to solve WiFi-based localization and sensor-based activity recognition problems [151]. For example, transfer learning techniques have been proposed to transfer WiFi-based localization models across time periods [91, 156, 166], space [93, 136] and mobile devices [165], respectively. Rashidi and Cook [107] and Zheng et al. [164] proposed to apply transfer learning techniques for solving indoor sensor-based activity recognition problems.

21.7.4 Applications to Computer Vision

In the past decade, transfer learning techniques have also attracted more and more attention in the fields of computer vision, image and multimedia analysis. Applications of transfer learning in these fields include image classification [68, 113, 126, 131, 142, 157], image retrieval [31, 55, 78], face verification from images [138], age estimation from facial images [162], image semantic segmentation [133], video retrieval [55, 148], video concept detection [43, 150], event recognition from videos [45], and object recognition [67, 115].

21.7.5 Applications to Bioinformatics

In Bioinformatics, motivated by the fact that different biological entities, such as organisms, genes, etc, may be related to each other from a biological point of view, some research works have been proposed to apply transfer learning techniques to solve various computational biological problems, such as identifying molecular association of phenotypic responses [159], splice site recognition of eukaryotic genomes [139], mRNA splicing [119], protein subcellular location prediction [146] and genetic association analysis [154].

21.7.6 Other Applications

Besides the above applications, Zhuo et al. [169] studied how to transfer domain knowledge to learn relational action models across domains in automated planning. Chai et al. [25] studied how to apply a GP based transfer learning method to solve the inverse dynamics problem for a robotic manipulator [25]. Alamgir et al. [3] applied transfer learning techniques to solve brain-computer interface problems. In [109], Raykar et al. proposed to jointly learn multiple different but conceptually related classifiers for computer aided design (CAD) using transfer learning. Nam et al. [87] adapted a transfer learning approach to cross-project defect prediction in the field of software engineering.

21.8   Concluding Remarks

In this chapter, we have reviewed a number of approaches to homogeneous and heterogeneous transfer learning based on different categories. Specifically, based on "what to transfer," approaches to homogeneous transfer learning can be classified into four categories, namely the instance-based approach, the feature-representation-based approach, the model-parameter-based approach, and the relational-information-based approach. Based on whether the feature spaces or label spaces between the source and target domains are different or not, heterogeneous transfer learning can be further classified into two contexts: namely transfer learning across heterogeneous feature spaces, and transfer learning across different label spaces. Furthermore, we have also discussed current theoretical studies on transfer learning and some research issues of transfer learning. Finally, we have summarized classification applications of transfer learning in diverse knowledge engineering areas.

Transfer learning is still at an early but promising stage. As described in Sections 21.5–21.6, there exist many research issues needing to be addressed. Specially, though there are some theoretical studies on homogeneous transfer learning, theoretical studies on heterogeneous transfer learning are still missing. Furthermore, most existing works on combining transfer learning and active learning consist of two steps, one for transfer learning and the other for active learning. How to integrate them into a unified framework is still an open issue. Finally, in the future, we expect to see more applications of transfer learning in novel areas.

Bibliography

[1] Arvind Agarwal, Hal Daumé III, and Samuel Gerber. Learning multiple tasks using manifold regularization. In *Advances in Neural Information Processing Systems 23*, pages 46–54. 2010.

[2] Eneko Agirre and Oier Lopez de Lacalle. On robustness and domain adaptation using svd for word sense disambiguation. In *Proceedings of the 22nd International Conference on Computational Linguistics*, pages 17–24. ACL, June 2008.

[3] Morteza Alamgir, Moritz Grosse-Wentrup, and Yasemin Altun. Multitask learning for brain-computer interfaces. In *Proceedings of the 13th International Conference on Artificial Intelligence and Statistics*, volume 9, pages 17–24. JMLR W&CP, May 2010.

[4] Rie K. Ando and Tong Zhang. A framework for learning predictive structures from multiple tasks and unlabeled data. *Journal of Machine Learning Research*, 6:1817–1853, 2005.

[5] Rie Kubota Ando and Tong Zhang. A high-performance semi-supervised learning method for text chunking. In *Proceedings of the 43rd Annual Meeting on Association for Computational Linguistics*, pages 1–9. ACL, June 2005.

[6] Andreas Argyriou, Theodoros Evgeniou, and Massimiliano Pontil. Multi-task feature learning. In *Advances in Neural Information Processing Systems 19*, pages 41–48. MIT Press, 2007.

[7] Andreas Argyriou, Andreas Maurer, and Massimiliano Pontil. An algorithm for transfer learning in a heterogeneous environment. In *Proceedings of the 2008 European Conference on Machine Learning and Knowledge Discovery in Databases*, Lecture Notes in Computer Science, pages 71–85. Springer, September 2008.

[8] Andreas Argyriou, Charles A. Micchelli, Massimiliano Pontil, and Yiming Ying. A spectral regularization framework for multi-task structure learning. In *Advances in Neural Information Processing Systems 20*, pages 25–32. MIT Press, 2008.

[9] Andrew Arnold, Ramesh Nallapati, and William W. Cohen. A comparative study of methods for transductive transfer learning. In *Workshops conducted in association with the 7th IEEE International Conference on Data Mining*, pages 77–82. IEEE Computer Society, 2007.

[10] Jing Bai, Ke Zhou, Guirong Xue, Hongyuan Zha, Gordon Sun, Belle Tseng, Zhaohui Zheng, and Yi Chang. Multi-task learning for learning to rank in web search. In *Proceeding of the 18th ACM Conference on Information and Knowledge Management*, pages 1549–1552. ACM, 2009.

[11] Bart Bakker and Tom Heskes. Task clustering and gating for Bayesian multitask learning. *Journal of Machine Learning Reserch*, 4:83–99, 2003.

[12] Shai Ben-David, John Blitzer, Koby Crammer, Alex Kulesza, Fernando Pereira, and Jenn Wortman. A theory of learning from different domains. *Machine Learning*, 79(1-2):151–175, 2010.

[13] Shai Ben-David, John Blitzer, Koby Crammer, and Fernando Pereira. Analysis of representations for domain adaptation. In *Advances in Neural Information Processing Systems 19*, pages 137–144. MIT Press, 2007.

[14] Shai Ben-David, Tyler Lu, Teresa Luu, and David Pal. Impossibility theorems for domain adaptation. In *Proceedings of the 13th International Conference on Artificial Intelligence and Statistics*, volume 9, pages 129–136. JMLR W&CP, May 2010.

[15] Shai Ben-David and Reba Schuller. Exploiting task relatedness for multiple task learning. In *Proceedings of the 16th Annual Conference on Learning Theory*, pages 825–830. Morgan Kaufmann Publishers Inc., August 2003.

[16] Steffen Bickel and Tobias Scheffer. Dirichlet-enhanced spam filtering based on biased samples. In *Advances in Neural Information Processing Systems 19*, pages 161–168. MIT Press, 2006.

[17] John Blitzer, Koby Crammer, Alex Kulesza, Fernando Pereira, and Jenn Wortman. Learning bounds for domain adaptation. In *Annual in Neural Information Processing Systems 20*, pages 129–136. MIT Press, 2008.

[18] John Blitzer, Mark Dredze, and Fernando Pereira. Biographies, Bollywood, boom-boxes and blenders: Domain adaptation for sentiment classification. In *Proceedings of the 45th Annual Meeting of the Association of Computational Linguistics*, pages 432–439. ACL, June 2007.

[19] John Blitzer, Ryan McDonald, and Fernando Pereira. Domain adaptation with structural correspondence learning. In *Proceedings of the 2006 Conference on Empirical Methods in Natural Language*, pages 120–128. ACL, July 2006.

[20] Avrim Blum and Tom Mitchell. Combining labeled and unlabeled data with co-training. In *Proceedings of the 11th Annual Conference on Learning Theory*, pages 92–100, July 1998.

[21] Edwin Bonilla, Kian Ming Chai, and Chris Williams. Multi-task Gaussian process prediction. In *Advances in Neural Information Processing Systems 20*, pages 153–160. MIT Press, 2008.

[22] Bin Cao, Nathan N. Liu, and Qiang Yang. Transfer learning for collective link prediction in multiple heterogenous domains. In *Proceedings of the 27th International Conference on Machine Learning*, pages 159–166. Omnipress, June 2010.

[23] Bin Cao, Sinno Jialin Pan, Yu Zhang, Dit-Yan Yeung, and Qiang Yang. Adaptive transfer learning. In *Proceedings of the 24th AAAI Conference on Artificial Intelligence*, pages 407–412, AAAI Press, July 2010.

[24] Rich Caruana. Multitask learning. *Machine Learning*, 28(1):41–75, 1997.

[25] Kian Ming A. Chai, Christopher K. I. Williams, Stefan Klanke, and Sethu Vijayakumar. Multi-task gaussian process learning of robot inverse dynamics. In *Advances in Neural Information Processing Systems 21*, pages 265–272. 2009.

[26] Yee Seng Chan and Hwee Tou Ng. Domain adaptation with active learning for word sense disambiguation. In *Proceedings of the 45th Annual Meeting of the Association of Computational Linguistics*, pages 49–56. ACL, June 2007.

[27] Olivier Chapelle, Bernhard Schölkopf, and Alexander Zien. *Semi-Supervised Learning*. MIT Press, Cambridge, MA, 2006.

[28] Olivier Chapelle, Pannagadatta Shivaswamy, Srinivas Vadrevu, Kilian Weinberger, Ya Zhang, and Belle Tseng. Multi-task learning for boosting with application to web search ranking. In *Proceedings of the 16th ACM SIGKDD International Conference on Knowledge Discovery and Data Mining*, pages 1189–1198. ACM, July 2010.

[29] Bo Chen, Wai Lam, Ivor W. Tsang, and Tak-Lam Wong. Extracting discriminative concepts for domain adaptation in text mining. In *Proceedings of the 15th ACM SIGKDD International Conference on Knowledge Discovery and Data Mining*, pages 179–188. ACM, June 2009.

[30] Jianhui Chen, Lei Tang, Jun Liu, and Jieping Ye. A convex formulation for learning shared structures from multiple tasks. In *Proceedings of the 26th Annual International Conference on Machine Learning*, pages 137–144. ACM, June 2009.

[31] Lin Chen, Dong Xu, Ivor W. Tsang, and Jiebo Luo. Tag-based web photo retrieval improved by batch mode re-tagging. In *Proceedings of the 23rd IEEE Conference on Computer Vision and Pattern Recognition*, pages 3440–3446, IEEE, June 2010.

[32] Tianqi Chen, Jun Yan, Gui-Rong Xue, and Zheng Chen. Transfer learning for behavioral targeting. In *Proceedings of the 19th International Conference on World Wide Web*, pages 1077–1078. ACM, April 2010.

[33] Yuqiang Chen, Ou Jin, Gui-Rong Xue, Jia Chen, and Qiang Yang. Visual contextual advertising: Bringing textual advertisements to images. In *Proceedings of the 24th AAAI Conference on Artificial Intelligence*. AAAI Press, July 2010.

[34] Wenyuan Dai, Yuqiang Chen, Gui-Rong Xue, Qiang Yang, and Yong Yu. Translated learning: Transfer learning across different feature spaces. In *Advances in Neural Information Processing Systems 21*, pages 353–360. 2009.

[35] Wenyuan Dai, Ou Jin, Gui-Rong Xue, Qiang Yang, and Yong Yu. Eigentransfer: a unified framework for transfer learning. In *Proceedings of the 26th Annual International Conference on Machine Learning*, pages 25–31. ACM, June 2009.

[36] Wenyuan Dai, Guirong Xue, Qiang Yang, and Yong Yu. Co-clustering based classification for out-of-domain documents. In *Proceedings of the 13th ACM International Conference on Knowledge Discovery and Data Mining*, pages 210–219. ACM, August 2007.

[37] Wenyuan Dai, Guirong Xue, Qiang Yang, and Yong Yu. Transferring naive Bayes classifiers for text classification. In *Proceedings of the 22nd AAAI Conference on Artificial Intelligence*, pages 540–545. AAAI Press, July 2007.

[38] Wenyuan Dai, Qiang Yang, Guirong Xue, and Yong Yu. Boosting for transfer learning. In *Proceedings of the 24th International Conference on Machine Learning*, pages 193–200. ACM, June 2007.

[39] Hal Daumé III. Frustratingly easy domain adaptation. In *Proceedings of the 45th Annual Meeting of the Association of Computational Linguistics*, pages 256–263. ACL, June 2007.

[40] Hal Daumé III, Abhishek Kumar, and Avishek Saha. Co-regularization based semi-supervised domain adaptation. In *Advances in Neural Information Processing Systems 23*, pages 478–486. 2010.

[41] Jesse Davis and Pedro Domingos. Deep transfer via second-order markov logic. In *Proceedings of the 26th Annual International Conference on Machine Learning*, pages 217–224. ACM, June 2009.

[42] Lixin Duan, Ivor W. Tsang, Dong Xu, and Tat-Seng Chua. Domain adaptation from multiple sources via auxiliary classifiers. In *Proceedings of the 26th Annual International Conference on Machine Learning*, pages 289–296. ACM, June 2009.

[43] Lixin Duan, Ivor W. Tsang, Dong Xu, and Stephen J. Maybank. Domain transfer SVM for video concept detection. In *Proceedings of the 22nd IEEE Computer Society Conference on Computer Vision and Pattern Recognition*, pages 1375–1381. IEEE, June 2009.

[44] Lixin Duan, Dong Xu, and Ivor W. Tsang. Learning with augmented features for heterogeneous domain adaptation. In *Proceedings of the 29th International Conference on Machine Learning*. icml.cc/Omnipress, June 2012.

[45] Lixin Duan, Dong Xu, Ivor W. Tsang, and Jiebo Luo. Visual event recognition in videos by learning from web data. In *Proceedings of the 23rd IEEE Conference on Computer Vision and Pattern Recognition*. IEEE, June 2010.

[46] Eric Eaton, Marie desJardins, and Terran Lane. Modeling transfer relationships between learning tasks for improved inductive transfer. In *Proceedings of the 2008 European Conference on Machine Learning and Knowledge Discovery in Databases*, Lecture Notes in Computer Science, pages 317–332. Springer, September 2008.

[47] Henry C. Ellis. *The Transfer of Learning*. The Macmillan Company, New York, 1965.

[48] Theodoros Evgeniou and Massimiliano Pontil. Regularized multi-task learning. In *Proceedings of the 10th ACM SIGKDD International Conference on Knowledge Discovery and Data Mining*, pages 109–117. ACM, August 2004.

[49] Yoav Freund and Robert E. Schapire. A decision-theoretic generalization of on-line learning and an application to boosting. In *Proceedings of the 2nd European Conference on Computational Learning Theory*, pages 23–37. Springer-Verlag, 1995.

[50] Wei Gao, Peng Cai, Kam-Fai Wong, and Aoying Zhou. Learning to rank only using training data from related domain. In *Proceeding of the 33rd International ACM SIGIR Conference on Research and Development in Information Retrieval*, pages 162–169. ACM, July 2010.

[51] Xavier Glorot, Antoine Bordes, and Yoshua Bengio. Domain adaptation for large-scale sentiment classification: A deep learning approach. In *Proceedings of the 28th International Conference on Machine Learning*, pages 513–520. Omnipress, 2011.

[52] A. Gretton, K. Borgwardt, M. Rasch, B. Schölkopf, and A. Smola. A kernel method for the two-sample problem. In *Advances in Neural Information Processing Systems 19*, pages 513–520. MIT Press, 2007.

[53] Hong Lei Guo, Li Zhang, and Zhong Su. Empirical study on the performance stability of named entity recognition model across domains. In *Proceedings of the 2006 Conference on Empirical Methods in Natural Language Processing*, pages 509–516. ACL, July 2006.

[54] Rakesh Gupta and Lev Ratinov. Text categorization with knowledge transfer from heterogeneous data sources. In *Proceedings of the 23rd National Conference on Artificial Intelligence*, pages 842–847. AAAI Press, July 2008.

[55] Sunil Kumar Gupta, Dinh Phung, Brett Adams, Truyen Tran, and Svetha Venkatesh. Nonnegative shared subspace learning and its application to social media retrieval. In *Proceedings of the 16th ACM SIGKDD International Conference on Knowledge Discovery and Data Mining*, pages 1169–1178. ACM, July 2010.

[56] Abhay Harpale and Yiming Yang. Active learning for multi-task adaptive filtering. In *Proceedings of the 27th International Conference on Machine Learning*, pages 431–438. ACM, June 2010.

[57] Trevor Hastie, Robert Tibshirani, and Jerome Friedman. *The Elements of Statistical Learning: Data Mining, Inference and Prediction*. 2nd edition, Springer, 2009.

[58] Thomas Hofmann. Probabilistic latent semantic indexing. In *Proceedings of the 22nd Annual International ACM SIGIR Conference on Research and Development in Information Retrieval*, pages 50–57. ACM, 1999.

[59] Jiayuan Huang, Alex Smola, Arthur Gretton, Karsten M. Borgwardt, and Bernhard Schölkopf. Correcting sample selection bias by unlabeled data. In *Advances in Neural Information Processing Systems 19*, pages 601–608. MIT Press, 2007.

[60] Laurent Jacob, Francis Bach, and Jean-Philippe Vert. Clustered multi-task learning: A convex formulation. In *Advances in Neural Information Processing Systems 21*, pages 745–752. 2009.

[61] Jing Jiang. Multi-task transfer learning for weakly-supervised relation extraction. In *Proceedings of the 47th Annual Meeting of the Association for Computational Linguistics and the 4th International Joint Conference on Natural Language Processing of the AFNLP*, pages 1012–1020. ACL, August 2009.

[62] Jing Jiang and ChengXiang Zhai. Instance weighting for domain adaptation in NLP. In *Proceedings of the 45th Annual Meeting of the Association of Computational Linguistics*, pages 264–271. ACL, June 2007.

[63] Thorsten Joachims. Transductive inference for text classification using support vector machines. In *Proceedings of the 16th International Conference on Machine Learning*, pages 200–209. Morgan Kaufmann Publishers Inc., June 1999.

[64] Guo Jun Qi, Charu C. Aggarwal, and Thomas S. Huang. Towards semantic knowledge propagation from text corpus to web images. In *Proceedings of the 20th International Conference on World Wide Web*, pages 297–306. ACM, March 2011.

[65] Takafumi Kanamori, Shohei Hido, and Masashi Sugiyama. A least-squares approach to direct importance estimation. *Journal of Machine Learning Research*, 10:1391–1445, 2009.

[66] Sunghun Kim, Jr., E. James Whitehead, and Yi Zhang. Classifying software changes: Clean or buggy? *IEEE Transactions on Software Engineering*, 34:181–196, March 2008.

[67] Brian Kulis, Kate Saenko, and Trevor Darrell. What you saw is not what you get: Domain adaptation using asymmetric kernel transforms. In *Proceedings of the 24th IEEE Conference on Computer Vision and Pattern Recognition*, pages 1785–1792. IEEE, June 2011.

[68] Christoph H. Lampert and Oliver Krömer. Weakly-paired maximum covariance analysis for multimodal dimensionality reduction and transfer learning. In *Proceedings of the 11th European Conference on Computer Vision*, pages 566–579, September 2010.

[69] Neil D. Lawrence and John C. Platt. Learning to learn with the informative vector machine. In *Proceedings of the 21st International Conference on Machine Learning*. ACM, July 2004.

[70] Honglak Lee, Alexis Battle, Rajat Raina, and Andrew Y. Ng. Efficient sparse coding algorithms. In *Advances in Neural Information Processing Systems 19*, pages 801–808. MIT Press, 2007.

[71] David D. Lewis and William A. Gale. A sequential algorithm for training text classifiers. In *Proceedings of the 17th Annual International ACM SIGIR Conference on Research and Development in Information Retrieval*, pages 3–12. ACM/Springer, July 1994.

[72] Bin Li, Qiang Yang, and Xiangyang Xue. Can movies and books collaborate?: Cross-domain collaborative filtering for sparsity reduction. In *Proceedings of the 21st International Jont Conference on Artificial Intelligence*, pages 2052–2057. Morgan Kaufmann Publishers Inc., July 2009.

[73] Bin Li, Qiang Yang, and Xiangyang Xue. Transfer learning for collaborative filtering via a rating-matrix generative model. In *Proceedings of the 26th Annual International Conference on Machine Learning*, pages 617–624. ACM, June 2009.

[74] Fangtao Li, Sinno Jialin Pan, Ou Jin, Qiang Yang, and Xiaoyan Zhu. Cross-domain co-extraction of sentiment and topic lexicons. In *Proceedings of the 50th Annual Meeting of the Association for Computational Linguistics*, pages 410–419. ACL, July 2012.

[75] Lianghao Li, Xiaoming Jin, Sinno Jialin Pan, and Jian-Tao Sun. Multi-domain active learning for text classification. In *Proceedings of the 18th ACM SIGKDD International Conference on Knowledge Discovery and Data Mining*, pages 1086–1094. ACM, August 2012.

[76] Xuejun Liao, Ya Xue, and Lawrence Carin. Logistic regression with an auxiliary data source. In *Proceedings of the 22nd International Conference on Machine Learning*, pages 505–512. ACM, August 2005.

[77] Xiao Ling, Gui-Rong Xue, Wenyuan Dai, Yun Jiang, Qiang Yang, and Yong Yu. Can Chinese web pages be classified with English data source? In *Proceedings of the 17th International Conference on World Wide Web*, pages 969–978. ACM, April 2008.

[78] Yiming Liu, Dong Xu, Ivor W. Tsang, and Jiebo Luo. Using large-scale web data to facilitate textual query based retrieval of consumer photos. In *Proceedings of the 17th ACM International Conference on Multimedia*, pages 55–64. ACM, October 2009.

[79] Ping Luo, Fuzhen Zhuang, Hui Xiong, Yuhong Xiong, and Qing He. Transfer learning from multiple source domains via consensus regularization. In *Proceedings of the 17th ACM Conference on Information and Knowledge Management*, pages 103–112. ACM, October 2008.

[80] M. M. Hassan Mahmud and Sylvian R. Ray. Transfer learning using Kolmogorov complexity: Basic theory and empirical evaluations. In *Advances in Neural Information Processing Systems 20*, pages 985–992. MIT Press, 2008.

[81] Yishay Mansour, Mehryar Mohri, and Afshin Rostamizadeh. Domain adaptation with multiple sources. In *Advances in Neural Information Processing Systems 21*, pages 1041–1048. 2009.

[82] Yishay Mansour, Mehryar Mohri, and Afshin Rostamizadeh. Multiple source adaptation and the Rényi divergence. In *Proceedings of the 25th Conference on Uncertainty in Artificial Intelligence*, pages 367–374. AUAI Press, June 2009.

[83] Tim Menzies, Jeremy Greenwald, and Art Frank. Data mining static code attributes to learn defect predictors. *IEEE Transactions on Software Engineering*, 33:2–13, January 2007.

[84] Lilyana Mihalkova, Tuyen Huynh, and Raymond J. Mooney. Mapping and revising Markov logic networks for transfer learning. In *Proceedings of the 22nd AAAI Conference on Artificial Intelligence*, pages 608–614. AAAI Press, July 2007.

[85] Lilyana Mihalkova and Raymond J. Mooney. Transfer learning by mapping with minimal target data. In *Proceedings of the AAAI-2008 Workshop on Transfer Learning for Complex Tasks*, July 2008.

[86] Tom M. Mitchell. *Machine Learning*. McGraw-Hill, New York, 1997.

[87] Jaechang Nam, Sinno Jialin Pan, and Sunghun Kim. Transfer defect learning. In *Proceedings of the 35th International Conference on Software Engineering*, pages 382–391. IEEE/ACM, May 2013.

[88] Andrew Y. Ng, Michael I. Jordan, and Yair Weiss. On spectral clustering: Analysis and an algorithm. In *Advances in Neural Information Processing Systems 14*, pages 849–856. MIT Press, 2001.

[89] Kamal Nigam, Andrew Kachites McCallum, Sebastian Thrun, and Tom Mitchell. Text classification from labeled and unlabeled documents using EM. *Machine Learning*, 39(2-3):103–134, 2000.

[90] Sinno Jialin Pan, James T. Kwok, and Qiang Yang. Transfer learning via dimensionality reduction. In *Proceedings of the 23rd AAAI Conference on Artificial Intelligence*, pages 677–682. AAAI Press, July 2008.

[91] Sinno Jialin Pan, James T. Kwok, Qiang Yang, and Jeffrey J. Pan. Adaptive localization in a dynamic WiFi environment through multi-view learning. In *Proceedings of the 22nd AAAI Conference on Artificial Intelligence*, pages 1108–1113. AAAI Press, July 2007.

[92] Sinno Jialin Pan, Xiaochuan Ni, Jian-Tao Sun, Qiang Yang, and Chen Zheng. Cross-domain sentiment classification via spectral feature alignment. In *Proceedings of the 19th International Conference on World Wide Web*, pages 751–760. ACM, April 2010.

[93] Sinno Jialin Pan, Dou Shen, Qiang Yang, and James T. Kwok. Transferring localization models across space. In *Proceedings of the 23rd AAAI Conference on Artificial Intelligence*, pages 1383–1388. AAAI Press, July 2008.

[94] Sinno Jialin Pan, Zhiqiang Toh, and Jian Su. Transfer joint embedding for cross-domain named entity recognition. *ACM Transactions on Information Systems*, 31(2):7:1–7:27, May 2013.

[95] Sinno Jialin Pan, Ivor W. Tsang, James T. Kwok, and Qiang Yang. Domain adaptation via transfer component analysis. In *Proceedings of the 21st International Joint Conference on Artificial Intelligence*, pages 1187–1192, July 2009.

[96] Sinno Jialin Pan, Ivor W. Tsang, James T. Kwok, and Qiang Yang. Domain adaptation via transfer component analysis. *IEEE Transactions on Neural Networks*, 22(2):199–210, 2011.

[97] Sinno Jialin Pan and Qiang Yang. A survey on transfer learning. *IEEE Transactions on Knowledge and Data Engineering*, 22(10):1345–1359, 2010.

[98] Sinno Jialin Pan, Vincent W. Zheng, Qiang Yang, and Derek H. Hu. Transfer learning for WiFi-based indoor localization. In *Proceedings of the Workshop on Transfer Learning for Complex Tasks of the 23rd AAAI Conference on Artificial Intelligence*, July 2008.

[99] Weike Pan, Evan W. Xiang, Nathan N. Liu, and Qiang Yang. Transfer learning in collaborative filtering for sparsity reduction. In *Proceedings of the 24th AAAI Conference on Artificial Intelligence*. AAAI Press, July 2010.

[100] Bo Pang, Lillian Lee, and Shivakumar Vaithyanathan. Thumbs up? Sentiment classification using machine learning techniques. In *Proceedings of the ACL-02 Conference on Empirical Methods in Natural Language Processing*, pages 79–86. ACL, July 2002.

[101] Peter Prettenhofer and Benno Stein. Cross-language text classification using structural correspondence learning. In *Proceedings of the 48th Annual Meeting of the Association for Computational Linguistics*, pages 1118–1127. ACL, July 2010.

[102] Guo-Jun Qi, Charu C. Aggarwal, Yong Rui, Qi Tian, Shiyu Chang, and Thomas S. Huang. Towards cross-category knowledge propagation for learning visual concepts. In *Proceedings of the 24th IEEE Conference on Computer Vision and Pattern Recognition*, pages 897–904. IEEE, June 2011.

[103] Novi Quadrianto, Alex J. Smola, Tiberio S. Caetano, S.V.N. Vishwanathan, and James Petterson. Multitask learning without label correspondences. In *Advances in Neural Information Processing Systems 23*, pages 1957–1965. Curran Associates, Inc., December 2010.

[104] Joaquin Quionero-Candela, Masashi Sugiyama, Anton Schwaighofer, and Neil D. Lawrence. *Dataset Shift in Machine Learning*. MIT Press, 2009.

[105] Rajat Raina, Alexis Battle, Honglak Lee, Benjamin Packer, and Andrew Y. Ng. Self-taught learning: Transfer learning from unlabeled data. In *Proceedings of the 24th International Conference on Machine Learning*, pages 759–766. ACM, June 2007.

[106] Rajat Raina, Andrew Y. Ng, and Daphne Koller. Constructing informative priors using transfer learning. In *Proceedings of the 23rd International Conference on Machine Learning*, pages 713–720. ACM, June 2006.

[107] Parisa Rashidi and Diane J. Cook. Activity recognition based on home to home transfer learning. In *Proceedings of the Workshop on Plan, Activity, and Intent Recognition of the 24th AAAI Conference on Artificial Intelligence*. AAAI Press, July 2010.

[108] Carl Edward Rasmussen and Christopher K. I. Williams. *Gaussian Processes for Machine Learning*. The MIT Press, 2005.

[109] Vikas C. Raykar, Balaji Krishnapuram, Jinbo Bi, Murat Dundar, and R. Bharat Rao. Bayesian multiple instance learning: Automatic feature selection and inductive transfer. In *Proceedings of the 25th International Conference on Machine Learning*, pages 808–815. ACM, July 2008.

[110] Matthew Richardson and Pedro Domingos. Markov logic networks. *Machine Learning Journal*, 62(1-2):107–136, 2006.

[111] Alexander E. Richman and Patrick Schone. Mining Wiki resources for multilingual named entity recognition. In *Proceedings of 46th Annual Meeting of the Association of Computational Linguistics*, pages 1–9. ACL, June 2008.

[112] Stephen Robertson and Ian Soboroff. The trec 2002 filtering track report. In *Text REtrieval Conference*, 2001.

[113] Marcus Rohrbach, Michael Stark, György Szarvas, Iryna Gurevych, and Bernt Schiele. What helps where—and why? Semantic relatedness for knowledge transfer. In *Proceedings of the 23rd IEEE Conference on Computer Vision and Pattern Recognition*, pages 910–917. IEEE, June 2010.

[114] Michael T. Rosenstein, Zvika Marx, and Leslie Pack Kaelbling. To transfer or not to transfer. In *NIPS-05 Workshop on Inductive Transfer: 10 Years Later*, December 2005.

[115] Kate Saenko, Brian Kulis, Mario Fritz, and Trevor Darrell. Adapting visual category models to new domains. In *Proceedings of the 11th European Conference on Computer Vision*, pages 213–226. Springer, September 2010.

[116] Andrew I. Schein, Alexandrin Popescul, Lyle H. Ungar, and David M. Pennock. Methods and metrics for cold-start recommendations. In *Proceedings of the 25th Annual International ACM SIGIR Conference on Research and Development in Information Retrieval*, pages 253–260. ACM, August 2002.

[117] Bernhard Scholkopf and Alexander J. Smola. *Learning with Kernels: Support Vector Machines, Regularization, Optimization, and Beyond*. MIT Press, 2001.

[118] Anton Schwaighofer, Volker Tresp, and Kai Yu. Learning Gaussian process kernels via hierarchical Bayes. In *Advances in Neural Information Processing Systems 17*, pages 1209–1216. MIT Press, 2005.

[119] Gabriele Schweikert, Christian Widmer, Bernhard Schölkopf, and Gunnar Rätsch. An empirical analysis of domain adaptation algorithms for genomic sequence analysis. In *Advances in Neural Information Processing Systems 21*, pages 1433–1440. 2009.

[120] Chun-Wei Seah, Yew-Soon Ong, Ivor W. Tsang, and Kee-Khoon Lee. Predictive distribution matching SVM for multi-domain learning. In *Proceedings of 2010 European Conference on Machine Learning and Knowledge Discovery in Databases*, Lecture Notes in Computer Science, pages 231–247. Springer, September 2010.

[121] Burr Settles. Active learning literature survey. Computer Sciences Technical Report 1648, University of Wisconsin–Madison, 2009.

[122] Xiaoxiao Shi, Wei Fan, and Jiangtao Ren. Actively transfer domain knowledge. In *Proceedings of the 2008 European Conference on Machine Learning and Knowledge Discovery in Databases*, Lecture Notes in Computer Science, pages 342–357. Springer, September 2008.

[123] Xiaoxiao Shi, Wei Fan, Qiang Yang, and Jiangtao Ren. Relaxed transfer of different classes via spectral partition. In *Preceedings of the 2009 European Conference on Machine Learning and Knowledge Discovery in Databases*, Lecture Notes in Computer Science, pages 366–381. Springer, September 2009.

[124] Xiaoxiao Shi, Qi Liu, Wei Fan, Philip S. Yu, and Ruixin Zhu. Transfer learning on heterogenous feature spaces via spectral transformation. In *Proceedings of the 10th IEEE International Conference on Data Mining*, pages 1049–1054. IEEE Computer Society, December 2010.

[125] Si Si, Dacheng Tao, and Bo Geng. Bregman divergence-based regularization for transfer subspace learning. *IEEE Transactions an Knowledge Data Engineering*, 22(7):929–942, 2010.

[126] Michael Stark, Michael Goesele, and Bernt Schiele. A shape-based object class model for knowledge transfer. In *Proceedings of 12th IEEE International Conference on Computer Vision*, pages 373–380. IEEE, September 2009.

[127] Masashi Sugiyama, Shinichi Nakajima, Hisashi Kashima, Paul Von Buenau, and Motoaki Kawanabe. Direct importance estimation with model selection and its application to covariate shift adaptation. In *Advances in Neural Information Processing Systems 20*, pages 1433–1440. MIT Press, 2008.

[128] Matthew E. Taylor and Peter Stone. Transfer learning for reinforcement learning domains: A survey. *Journal of Machine Learning Research*, 10(1):1633–1685, 2009.

[129] Edward Lee Thorndike and Robert Sessions Woodworth. The influence of improvement in one mental function upon the efficiency of the other functions. *Psychological Review*, 8:247–261, 1901.

[130] Sebastian Thrun and Lorien Pratt, editors. *Learning to learn*. Kluwer Academic Publishers, Norwell, MA, 1998.

[131] Tatiana Tommasi, Francesco Orabona, and Barbara Caputo. Safety in numbers: Learning categories from few examples with multi model knowledge transfer. In *Proceedings of the 23rd IEEE Conference on Computer Vision and Pattern Recognition*, pages 3081–3088. IEEE, June 2010.

[132] Vladimir N. Vapnik. *Statistical Learning Theory*. Wiley-Interscience, New York, 1998.

[133] Alexander Vezhnevets and Joachim Buhmann. Towards weakly supervised semantic segmentation by means of multiple instance and multitask learning. In *Proceedings of the 23rd IEEE Conference on Computer Vision and Pattern Recognition*, pages 3249–3256. IEEE, June 2010.

[134] Bo Wang, Jie Tang, Wei Fan, Songcan Chen, Zi Yang, and Yanzhu Liu. Heterogeneous cross domain ranking in latent space. In *Proceedings of the 18th ACM Conference on Information and Knowledge Management*, pages 987–996. ACM, November 2009.

[135] Chang Wang and Sridhar Mahadevan. Heterogeneous domain adaptation using manifold alignment. In *Proceedings of the 22nd International Joint Conference on Artificial Intelligence*, pages 1541–1546. IJCAI/AAAI, July 2011.

[136] Hua-Yan Wang, Vincent W. Zheng, Junhui Zhao, and Qiang Yang. Indoor localization in multi-floor environments with reduced effort. In *Proceedings of the 8th Annual IEEE International Conference on Pervasive Computing and Communications*, pages 244–252. IEEE Computer Society, March 2010.

[137] Pu Wang, Carlotta Domeniconi, and Jian Hu. Using Wikipedia for co-clustering based cross-domain text classification. In *Proceedings of the Eighth IEEE International Conference on Data Mining*, pages 1085–1090. IEEE Computer Society, 2008.

[138] Xiaogang Wang, Cha Zhang, and Zhengyou Zhang. Boosted multi-task learning for face verification with applications to web image and video search. In *Proceedings of the 22nd IEEE Computer Society Conference on Computer Vision and Pattern Recognition*, pages 142–149. IEEE, June 2009.

[139] Christian Widmer, Jose Leiva, Yasemin Altun, and Gunnar Rätsch. Leveraging sequence classification by taxonomy-based multitask learning. In *Proceedings of 14th Annual International Conference on Research in Computational Molecular Biology*, Lecture Notes in Computer Science, pages 522–534. Springer, April 2010.

[140] Tak-Lam Wong, Wai Lam, and Bo Chen. Mining employment market via text block detection and adaptive cross-domain information extraction. In *Proceedings of the 32nd International ACM SIGIR Conference on Research and Development in Information Retrieval*, pages 283–290. ACM, July 2009.

[141] Dan Wu, Wee Sun Lee, Nan Ye, and Hai Leong Chieu. Domain adaptive bootstrapping for named entity recognition. In *Proceedings of the 2009 Conference on Empirical Methods in Natural Language Processing*, pages 1523–1532. ACL, August 2009.

[142] Pengcheng Wu and Thomas G. Dietterich. Improving svm accuracy by training on auxiliary data sources. In *Proceedings of the 21st International Conference on Machine Learning*. ACM, July 2004.

[143] Evan W. Xiang, Bin Cao, Derek H. Hu, and Qiang Yang. Bridging domains using world wide knowledge for transfer learning. *IEEE Transactions on Knowledge and Data Engineering*, 22:770–783, 2010.

[144] Evan W. Xiang, Sinno Jialin Pan, Weike Pan, Jian Su, and Qiang Yang. Source-selection-free transfer learning. In *Proceedings of 22nd International Joint Conference on Artificial Intelligence*, pages 2355–2360. IJCAI/AAAI, July 2011.

[145] Sihong Xie, Wei Fan, Jing Peng, Olivier Verscheure, and Jiangtao Ren. Latent space domain transfer between high dimensional overlapping distributions. In *18th International World Wide Web Conference*, pages 91–100. ACM, April 2009.

[146] Qian Xu, Sinno Jialin Pan, Hannah Hong Xue, and Qiang Yang. Multitask learning for protein subcellular location prediction. *IEEE/ACM Transactions on Computational Biology and Bioinformatics*, 8(3):748–759, 2011.

[147] Gui-Rong Xue, Wenyuan Dai, Qiang Yang, and Yong Yu. Topic-bridged PLSA for cross-domain text classification. In *Proceedings of the 31st Annual International ACM SIGIR Conference on Research and Development in Information Retrieval*, pages 627–634. ACM, July 2008.

[148] Rong Yan and Jian Zhang. Transfer learning using task-level features with application to information retrieval. In *Proceedings of the 21st International Jont Conference on Artifical Intelligence*, pages 1315–1320. Morgan Kaufmann Publishers Inc., July 2009.

[149] Jian-Bo Yang, Qi Mao, Qiaoliang Xiang, Ivor Wai-Hung Tsang, Kian Ming Adam Chai, and Hai Leong Chieu. Domain adaptation for coreference resolution: An adaptive ensemble approach. In *Proceedings of the 2012 Joint Conference on Empirical Methods in Natural*

Language Processing and Computational Natural Language Learning, pages 744–753. ACL, July 2012.

[150] Jun Yang, Rong Yan, and Alexander G. Hauptmann. Cross-domain video concept detection using adaptive SVMS. In *Proceedings of the 15th International Conference on Multimedia*, pages 188–197. ACM, September 2007.

[151] Qiang Yang. Activity recognition: Linking low-level sensors to high-level intelligence. In *Proceedings of the 21st International Joint Conference on Artificial Intelligence*, pages 20–25. Morgan Kaufmann Publishers Inc., July 2009.

[152] Qiang Yang, Sinno Jialin Pan, and Vincent W. Zheng. Estimating location using Wi-Fi. *IEEE Intelligent Systems*, 23(1):8–13, 2008.

[153] Tianbao Yang, Rong Jin, Anil K. Jain, Yang Zhou, and Wei Tong. Unsupervised transfer classification: Application to text categorization. In *Proceedings of the 16th ACM SIGKDD International Conference on Knowledge Discovery and Data Mining*, pages 1159–1168. ACM, July 2010.

[154] Xiaolin Yang, Seyoung Kim, and Eric Xing. Heterogeneous multitask learning with joint sparsity constraints. In *Advances in Neural Information Processing Systems 22*, pages 2151–2159. 2009.

[155] Yi Yao and Gianfranco Doretto. Boosting for transfer learning with multiple sources. In *Proceedings of the 23rd IEEE Conference on Computer Vision and Pattern Recognition*, pages 1855–1862. IEEE, June 2010.

[156] Jie Yin, Qiang Yang, and L.M. Ni. Learning adaptive temporal radio maps for signal-strength-based location estimation. *IEEE Transactions on Mobile Computing*, 7(7):869–883, July 2008.

[157] Xiao-Tong Yuan and Shuicheng Yan. Visual classification with multi-task joint sparse representation. In *Proceedings of the 23rd IEEE Conference on Computer Vision and Pattern Recognition*, pages 3493–3500. IEEE, June 2010.

[158] Bianca Zadrozny. Learning and evaluating classifiers under sample selection bias. In *Proceedings of the 21st International Conference on Machine Learning*. ACM, July 2004.

[159] Kai Zhang, Joe W. Gray, and Bahram Parvin. Sparse multitask regression for identifying common mechanism of response to therapeutic targets. *Bioinformatics*, 26(12):i97–i105, 2010.

[160] Yu Zhang, Bin Cao, and Dit-Yan Yeung. Multi-domain collaborative filtering. In *Proceedings of the 26th Conference on Uncertainty in Artificial Intelligence*, pages 725–732. AUAI Press, July 2010.

[161] Yu Zhang and Dit-Yan Yeung. A convex formulation for learning task relationships in multi-task learning. In *Proceedings of the 26th Conference on Uncertainty in Artificial Intelligence*, pages 733–442. AUAI Press, July 2010.

[162] Yu Zhang and Dit-Yan Yeung. Multi-task warped gaussian process for personalized age estimation. In *Proceedings of the 23rd IEEE Computer Society Conference on Computer Vision and Pattern Recognition*, pages 2622–2629. IEEE, June 2010.

[163] Lili Zhao, Sinno Jialin Pan, Evan W. Xiang, Erheng Zhong, Zhongqi Lu, and Qiang Yang. Active transfer learning for cross-system recommendation. In *Proceedings of the 27th AAAI Conference on Artificial Intelligence*. AAAI Press, July 2013.

[164] Vincent W. Zheng, Derek H. Hu, and Qiang Yang. Cross-domain activity recognition. In *Proceedings of the 11th International Conference on Ubiquitous Computing*, pages 61–70. ACM, September 2009.

[165] Vincent W. Zheng, Sinno Jialin Pan, Qiang Yang, and Jeffrey J. Pan. Transferring multi-device localization models using latent multi-task learning. In *Proceedings of the 23rd AAAI Conference on Artificial Intelligence*, pages 1427–1432. AAAI Press, July 2008.

[166] Vincent W. Zheng, Qiang Yang, Evan W. Xiang, and Dou Shen. Transferring localization models over time. In *Proceedings of the 23rd AAAI Conference on Artificial Intelligence*, pages 1421–1426. AAAI Press, July 2008.

[167] Xiaojin Zhu. Semi-supervised learning literature survey. Technical Report 1530, Computer Sciences, University of Wisconsin-Madison, 2005.

[168] Yin Zhu, Yuqiang Chen, Zhongqi Lu, Sinno Jialin Pan, Gui-Rong Xue, Yong Yu, and Qiang Yang. Heterogeneous transfer learning for image classification. In *Proceedings of the 25th AAAI Conference on Artificial Intelligence*. AAAI Press, August 2011.

[169] Hankui Zhuo, Qiang Yang, Derek H. Hu, and Lei Li. Transferring knowledge from another domain for learning action models. In *Proceedings of 10th Pacific Rim International Conference on Artificial Intelligence*, pages 1110–1115. Springer-Verlag, December 2008.

[170] Thomas Zimmermann, Nachiappan Nagappan, Harald Gall, Emanuel Giger, and Brendan Murphy. Cross-project defect prediction: a large scale experiment on data vs. domain vs. process. In *Proceedings of the 7th joint meeting of the European software engineering conference and the ACM SIGSOFT symposium on the foundations of software engineering*, pages 91–100. ACM, August 2009.

Chapter 22

Active Learning: A Survey

Charu C. Aggarwal

IBM T. J. Watson Research Center
Yorktown Heights, NY
`charu@us.ibm.com`

Xiangnan Kong

University of Illinois at Chicago
Chicago, IL
`xkong4@uic.edu`

Quanquan Gu

University of Illinois at Urbana-Champaign
Urbana, IL
`qgu3@illinois.edu`

Jiawei Han

University of Illinois at Urbana-Champaign
Urbana, IL
`hanj@illinois.edu`

Philip S. Yu

University of Illinois at Chicago
Chicago, IL
`psyu@uic.edu`

22.1 Introduction

One of the great challenges in a wide variety of learning problems is the ability to obtain sufficient labeled data for modeling purposes. Labeled data is often expensive to obtain, and frequently requires laborious human effort. In many domains, unlabeled data is copious, though labels can be attached to such data at a specific cost in the labeling process. Some examples of such data are as follows:

- *Document Collections:* Large amounts of document data may be available on the Web, which are usually unlabeled. In such cases, it is desirable to attach labels to documents in order to create a learning model. A common approach is to manually label the documents in order to label the training data, a process that is slow, painstaking, and laborious.

- *Privacy-Constrained Data Sets:* In many scenarios, the labels on records may be sensitive information, which may be acquired at a significant query cost (e.g., obtaining permission from the relevant entity).

- *Social Networks:* In social networks, it may be desirable to identify nodes with specific properties. For example, an advertising company may desire to identify nodes in the social network that are interested in "cosmetics." However, it is rare that labeled nodes will be available in the network that have interests in a specific area. Identification of relevant nodes may only occur through either manual examination of social network posts, or through user surveys. Both processes are time-consuming and costly.

In all these cases, labels can be obtained, but only at a significant cost to the end user. An important observation is that all records are not equally important from the perspective of labeling. For example, some records may be noisy and contain no useful features that are relevant to classification. Similarly, records that cleanly belong to one class or another may be helpful, but less so than records that lie closer to the separation boundaries between the different classes.

An additional advantage of active learning methods is that they can often help in the removal of noisy instances from the data, which can be beneficial from an accuracy perspective. In fact, some studies [104] have shown that a carefully designed active learning method can sometimes provide better accuracy than is available from the base data.

Clearly, given the differential value of different records, an important question that arises in active learning is as follows:

How do we select instances from the underlying data to label, so as to achieve the most effective training for a given level of effort?

Different performance criteria may be used to quantify and fine-tune the tradeoffs between accuracy and cost, but the broader goal of all the criteria is to maximize the "bang for the buck" in spending the minimum effort in selecting examples, that maximize the accuracy as much as possible. An excellent survey on active learning may be found in [105].

Every active learning system has two primary components, one of which is already given:

- *Oracle:* This provides the responses to the underlying query. The oracle may be a human labeler, a cost driven data acquisition system, or any other methodology. It is important to note that the oracle algorithm is part of the input, though the user may play a role in its design. For example, in a multimedia application, the user may look at an image and provide a label, but this comes at an expense of human labor [115]. However, for most of the active learning algorithms, the oracle is really treated as a black box that is used directly.

- *Query System:* The job of the query system is to pose queries to the oracle for labels of specific records. It is here that most of the challenges of active learning systems are found.

Numerous strategies are possible for different active learning scenarios. At the highest level, this corresponds to the broader framework of *how* the queries are posed to the learner.

- *Membership Query Synthesis:* In this case, the learner actively synthesizes instances from the entire space, and does not necessarily sample from some underlying distribution [3]. The key here is that the learner many actually *construct* instances from the underlying space, which may not be a part of any actual pre-existing data. However, this may lead to challenges in the sense that the constructed examples may not be meaningful. For example, a synthesized image from a group of pixels will very rarely be meaningful. On the other hand, arbitrarily chosen spatial coordinates in a sea surface temperature prediction system will almost always be meaningful. Therefore, the usability of the approach clearly depends upon the underlying scenario.

- *Selective or Sequential Sampling:* In this case, the samples are drawn from the underlying data distribution, and the learner decides whether or not they should be labeled [21]. In this case, the query comes from an actual underlying data distribution, and is therefore guaranteed to make sense. In this case, the queries are sampled one by one, and a decision is made whether or not they should be queried. Such an approach is synonymous with the streaming scenario, since the decisions about querying an instance need to be made in real time in such cases. This terminology is however overloaded, since many works such as those in [104] use the term "selective sampling" to refer to another strategy described below.

- *Pool-based Sampling:* As indicated by its name, it suggests the availability of a base "pool" of instances from which to query the records of labels [74]. The task of the learner is to therefore determine instances from this pool (typically one by one), which are as informative as possible for the active learning process. This situation is encountered very commonly in practical scenarios, and also allows for relatively clean models for active learning.

The vast majority of the strategies in the literature use pool-based sampling, and in fact some works such as [104] refer to pool-based sampling as selective sampling. Therefore, this chapter will mostly focus on pool-based strategies, since these form the core of most active learning methods. Beyond these strategies, a number of other basic scenarios are possible. For example, in *batch learning*, an entire set of examples need to be labeled at a given time for the active learning process. An example of this is the *Amazon Mechanical Turk*, in which an entire set of examples is made available for labeling at a given time. This is different from the methods common in the literature, in which samples are labeled one by one, so that the learner has a chance to adjust the model, before selecting the next example. In such cases, it is usually desirable to incorporate *diversity* in the batch of labeled instances, in order to ensure that there is not too much redundancy within a particular batch of labeled instances [16, 55, 57, 119].

Active learning has numerous challenges, in that it does not always improve the accuracy of classification. While some of these issues may be related to algorithmic aspects such as sample selection bias [12], other cases are inherent to the nature of the underlying data. However, in many special cases, it has been shown [30] the number of labels needed to learn actively can be logarithmic in the usual sample complexity of passive learning.

This chapter will discuss the different methods that are used for active learning. First, we will provide a motivational example of how the selective sampling approach can describe the contours of the different class boundaries with far fewer examples. Then, we will provide a discussion of the different strategies that are commonly used for active learning. We will see that number of different scenarios are possible for the active learning process in terms of how the samples are selected.

This chapter is organized as follows. Section 22.2 provides an example of how active learning provides advantages for the learning process. We will also discuss its relationship to other methods in the literature such as semi-supervised learning. Section 22.3 discusses query strategy frameworks for active learning. Section 22.4 studies models for theoretical active learning. Section 22.5 discusses the methodologies for handling complex data types such as sequences and graphs. Section 22.6 discusses advanced topics for active learning, such as streaming data, feature learning, and class-based querying. Section 22.7 discusses the conclusions and summary.

22.2 Motivation and Comparisons to Other Strategies

The primary motivation of active learning is the paucity of training data available for learning algorithms. While it is possible to query the data randomly for labels, such an approach may not result in the best model, when each query is costly, and therefore, few labels will eventually become available. For example, consider the two class example of Figure 22.1. Here, we have a very simple division of the data into two classes, which is shown by a vertical dotted line, as illustrated in Figure 22.1(a). The two classes here are labeled A and B. Consider the case where it is possible to query only 7 examples for the two different classes. In this case, it is quite possible that the small number of allowed samples may result in a training data that is unrepresentative of the true separation between the two classes. Consider the case when an SVM classifier is used in order to construct a model. In Figure 22.1(b), we have shown a total of 7 samples randomly chosen from the underlying data. Because of the inherent noisiness in the process of picking a small number

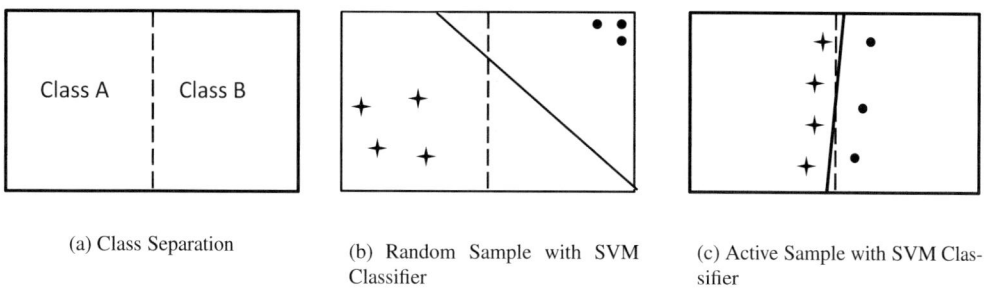

(a) Class Separation

(b) Random Sample with SVM Classifier

(c) Active Sample with SVM Classifier

FIGURE 22.1: Motivation of active learning.

of samples, an SVM classifier will be unable to accurately divide the data space. This is shown in Figure 22.1(b), where a portion of the data space is incorrectly classified, because of the error of modeling the SVM classifier. In Figure 22.1(c), we have shown an example of a well chosen set of seven instances along the decision boundary of the two classes. In this case, the SVM classifier is able to accurately model the decision regions between the two classes. This is because of the careful choice of the instances chosen by the active learning process. An important point to note is that it is particularly useful to sample instances that provide a distinct view of how the different classes are separated in the data. As will be discussed later, this general principle is used quite frequently in a variety of active learning methods, where regions of greater uncertainty are often sampled in order to obtain the relevant decision boundaries [75].

Since active learning is an approach that often uses human feedback in order to account for the lack of obtaining training examples, it is related to a number of other methods that either use human feedback or augment it with other kinds of training information. These two classes of methods are as follows:

- *Human Feedback:* Human feedback is often used to improve the accuracy of the learning process. For example, decision trees and other classifiers may be built with the active intervention of the user [4, 113]. In this case, the model itself is constructed with user-intervention rather than the choice of data examples.

- *Semi-supervised and Transfer Learning:* In this case, other *kinds* of data (e.g., unlabeled data or labeled data from other domains) are used in order to overcome the lack of training examples [4, 9, 29, 94, 97, 134].

Since both of these forms of supervision share some common principles with active learning, they will be discussed in some detail below.

22.2.1 Comparison with Other Forms of Human Feedback

Human feedback is a common approach to improving the effectiveness of classification problems. Active learning is one approach for improving the effectiveness of classification models, which is highly label-centered, and focusses *only on acquisition of labels*. Other human-centered methods used human feedback directly in order to improve the model for a given set of labels. For example, a splitting choice in a decision tree can be greatly involved with the use of human intervention. As in the case of active learning, the use of human feedback is greatly helpful in improving the quality of the model constructed with a fewer number of examples. However, the focus of the approach is completely different in terms of using human intervention at the model level rather than at the level of labels [4, 113]. A detailed discussion on the use of human feedback is provided in the chapter on visual classification in this book.

22.2.2 Comparisons with Semi-Supervised and Transfer Learning

These forms of learning are also related to the difficulty in acquisition of labels, though the approach used to improve the learning process is very different. In this case, different kinds of data are used to improve the learning process and *augment* the sparse labeled data that is already available. However, the instances that are used to augment the sparse labeled data are either unlabeled, or they are drawn from another domain. The motivations of using such methods, which are quite distinct from transfer learning, are as follows:

- *Semi-supervised Learning:* In this case, unlabeled data is used in order to learn the base distribution of the data in the underlying space [9, 94]. Once the base distribution of the data has been learned, it is combined with the labeled data in order to learn the class contours more effectively. The idea here is that the data is often either aligned along low dimensional manifolds [9], or is clustered in specific regions [94], and this can be learned effectively from the training data. This additional information helps in learning, even when the amount of labeled data available is small. For example, when the data is clustered [94], each of the dense regions typically belongs to a particular class, and a very small number of labels can map the different dense regions to the different classes.

- *Transfer Learning:* Transfer learning also uses additional labeled data from a different source, and may sometimes even be drawn from a different domain. For example, consider the case where it is desirable to classify Chinese documents with a small training collection. While it may be harder to obtain labeled Chinese documents, it is much easier to obtain labeled English documents. At the same time, data that provides correspondence between Chinese and English documents may be available. These different kinds of data may be combined together in order to provide a more effective training model for Chinese documents. Thus, the core idea here is to "transfer" knowledge from one domain to the other. However, transfer learning does not actively acquire labels in order to augment the sparse training data.

Thus, human feedback methods, transfer learning methods, and semi-supervised learning methods are all designed to handle the problem of paucity of training data. This is their shared characteristic with active learning methods. However, at the detailed level, the strategies are quite different. Both semi-supervised learning and transfer learning have been discussed in detail in different chapters of this book.

22.3 Querying Strategies

The key question in active learning algorithms is to design the precise strategies that are used for querying. At any given point, which sample should be selected so as to maximize the accuracy of the classification process? As is evident from the discussion in the previous section, it is advantageous to use strategies, so that the contours of separation between the different classes are mapped out with the use of a small number of examples. Since the boundary regions are often those in which instances of multiple classes are present, they can be characterized by class label uncertainty or disagreements between different learners. However, this may not always be the case, because instances with greater uncertainty are not representative of the data, and may sometimes lead to the selection of unrepresentative outliers. This situation is especially likely to occur in data sets that are very noisy. In order to address such issues, some models focus directly on the error itself, or try to find samples that are representative of the underlying data. Therefore, we broadly classify the querying strategies into one of three categories:

- *Heterogeneity-based models:* These models attempt to sample from regions of the space that are either more heterogeneous, or dissimilar to what has already been seen so far. Examples of such models include *uncertainty sampling*, *query-by-committee*, and *expected model change*. All these methods are based on sampling either uncertain regions of the data, or those that are dissimilar to what has been queried so far. These models only look at the heterogeneity behavior of the queried instance, rather than the effect of its addition on the performance of a classifier on the remaining unlabeled instances.

- *Performance-based models:* These models attempt to directly optimize the performance of the classifier in terms of measures such as error or variance reduction. One characteristic of these methods is that they look at the effect of adding the queried instance on the performance of the classifier on the remaining unlabeled instances.

- *Representativeness-based models:* These models attempt to create data that is as representative as possible of the underlying population of training instances. For example, density-based models are an example of such scenarios. In these cases, a product of a heterogeneity criterion and a representativeness criterion is used in order to model the desirability of querying a particular instance. Thus, these methods try to balance the representativeness criteria with the uncertainty properties of the queried instance.

Clearly, there is significant diversity in the strategies that one may use in the active learning process. These different strategies have different tradeoffs and work differently, depending upon the underlying application, analyst goal, and data distribution. This section will provide an overview of these different strategies for active learning. Throughout the following discussion, it will be assumed that there are a total of k classes, though some cases will also be analyzed in the binary scenario when k is set to 2.

22.3.1 Heterogeneity-Based Models

In these models, the idea is to learn the regions that show the greatest heterogeneity, either in terms of uncertainty of classification, dissimilarity with the current model, or disagreement between a committee of classifiers. These different techniques will be studied in this section.

22.3.1.1 Uncertainty Sampling

In uncertainty sampling, the learner attempts to label those instances for which it is least certain how to label. In a binary classification problem, the simplest possible uncertainty labeling scheme would be to use a Bayes classifier on an instance, and query for its label, if the predicted probability of the most probable label is as close to 0.5 as possible [74, 75]. The probabilities predicted by the classifier should be normalized so that they sum to 1. This is important, since many classifiers such as the unnormalized naive Bayes classifier often predict probabilities that do not sum to 1. Therefore, the following entropy-centered objective function $En(\overline{X})$ needs to be minimized for the binary class problem:

$$En(\overline{X}) = \sum_{i=1}^{k} ||p_i - 0.5||.$$

A second criterion is the difference in the predicted probabilities between the two classes. This is, however, equivalent to the first criterion, if the two probabilities have been normalized. A second pair of criteria that are especially relevant for k-ary classification is the entropy measure or the gini-index. If the predicted probabilities of the k classes are $p_1 \ldots p_k$, respectively, based on the current set of labeled instances, then the entropy measure $En(\overline{X})$ is defined as follows:

$$En(\overline{X}) = - \sum_{i=1}^{k} p_i \cdot \log(p_i).$$

Larger values of the entropy indicate greater uncertainty. Therefore, this objective function needs to be maximized. Note that an equal proportion of labels across the different classes results in the highest possible entropy. A second measure is the gini-index G.

$$G(\overline{X}) = 1 - \sum_{i=1}^{k} p_i^2.$$

As in the case of entropy, higher values of the gini-index indicate greater uncertainty. It should be pointed out that some of these measures may not work in the case of imbalanced data, where the classes are not evenly distributed. In such cases, the classes may often be associated with costs, where the cost of misclassification of i is denoted by w_i. Each probability p_i is replaced by a value proportional to $p_i \cdot w_i$, with the constant of the proportionality being determined by the probability values summing to 1.

Numerous active sampling techniques have been developed in the literature on the basis of these principles, and extensive comparisons have also been performed between these different techniques. The interested reader is referred to [27, 60, 74, 75, 102, 106] for the different techniques, and to [68, 103, 106] for the comparison of these measures. It should also be pointed out that it is not necessary to use a Bayes model that explicitly predicts probabilities. In practice, it is sufficient to use any model that provides a prediction confidence for each class label. This can be converted into a pseudo-probability for each class, and used heuristically for the instance-selection process.

22.3.1.2 Query-by-Committee

This approach [109] uses a committee of different classifiers, which are trained on the current set of labeled instances. These classifiers are then used to predict the class label of each unlabeled instance. The instance for which the classifiers disagree the most is selected as the relevant one in this scenario. At an intuitive level, the query-by-committee method achieves similar heterogeneity goals as the uncertainty sampling method, except that it does so by measuring the differences in the predictions of different classifiers, rather than the uncertainty of labeling a particular instance. Note that an instance that is classified to different classes with almost equal probability (as in uncertainty sampling) is more likely to be predicted in different classes by different classifiers. Thus, there is significant similarity between these methods at an intuitive level, though they are generally treated as very different methods in the literature. Interestingly, the method for measuring the disagreement is also quite similar between the two classes of methods. For example, by replacing the prediction probability p_i of each class i with the fraction of votes received for each class i, it is possible to obtain similar measures for the entropy and the gini-index. In addition, other probabilistic measures such as the KL-divergence have been proposed in [84] for this purpose.

The construction of the committee can be achieved by either varying the model parameters of a particular classifier (through sampling) [28, 84], or by using a bag of different classifiers [1]. It has been shown that the use of a small number of classifiers is generally sufficient [109], and the use of diverse classifiers in the committee is generally beneficial [89].

22.3.1.3 Expected Model Change

A decision theoretic-approach is to select the instance that results in the greatest change from the current model. Specifically, the instance that results in the greatest change in gradient of the objective function with respect to the model parameters is used. The intuition of such an approach is to use an instance that is most different from the current model that is already known. Thus, this is also a heterogeneity-based approach, as is the case with uncertainty sampling, and query-by-committee. Such an approach is only applicable to models where gradient-based training is used, such as discriminative probabilistic models. Let $\delta g_i(\overline{X})$ be the change in the gradient with respect to the model parameters, if the training label of the candidate instance $\overline{X}$ (with unknown label) is i. Let p_i be the posterior probability of the instance i with respect to the current label set in the training data. Then, the expected model change $C(\overline{X})$ with respect to the instance $\overline{X}$ is defined as follows:

$$C(\overline{X}) = \sum_{i=1}^{k} p_i \cdot \delta g_i(\overline{X}).$$

The instance $\overline{X}$ with the largest value of $C(\overline{X})$ is queried for the label. Numerous techniques for querying, that have been proposed using this approach may be found in [25, 47, 106, 107].

22.3.2 Performance-Based Models

The primary criticism of heterogeneity-based models is that the goal of trying to identify the most unknown regions of the space (based on the current labeling), may sometimes lead to the identification of noisy and unrepresentative regions of the data. The precise impact of using such an approach is of course highly data-dependent. There are two classes of techniques that are based on the performance of a classifier on the *remaining unlabeled instances*.

22.3.2.1 Expected Error Reduction

For the purpose of discussion, the remaining set of instances that have not yet been labeled are denoted by V. This set is used as the validation set on which the expected error reduction is computed. This approach is related to uncertainty sampling in a complementary way. Whereas uncertainty sampling *maximizes* the label uncertainty of the *queried* instance, the expected error reduction *minimizes* the expected label uncertainty of the *remaining* instances V, when the queried instance is added to the data. Thus, in the case of a binary-classification problem, we would like the labels of the instances in V to be as far away from 0.5 as possible. The idea here is that it is good to query for instances, which results in greater certainty of class label for the remaining error rate. Thus, error reduction models can also be considered as *greatest certainty* models, except that the certainty criterion is applied to the instances in V (rather than the query instance itself) after addition of the queried instance to the model. The assumption is that greater certainty of class labels *of the remaining unlabeled instances* corresponds to a lower error rate. Let $p_i(\overline{X})$ denote the posterior probability of the label i for the instance $\overline{X}$, before the queried instance is added. Let $P_j^{(\overline{X},i)}(\overline{Z})$ be the posterior probability of class label j, when the instance-label combination $(\overline{X}, i)$ is added to the model. Then, the error objective function $E(\overline{X}, V)$ for the binary class problem (i.e., $k = 2$) is defined as follows:

$$E(\overline{X}, V) = \sum_{i=1}^{k} p_i(\overline{X}) \cdot \left(\sum_{j=1}^{k} \sum_{\overline{Z} \in V} ||P_j^{(\overline{X},i)}(\overline{Z}) - 0.5|| \right). \tag{22.1}$$

The value of $E(\overline{X}, V)$ is maximized rather than minimized (as in the case of uncertainty-based models). Furthermore, the error objective is a function of both the queried instance and the set of unlabeled instances V. This result can easily be extended to the case of k-way models by using the entropy criterion, as was discussed in the case of uncertainty-based models. In that case, the expression above is modified to replace $||P_j^{(\overline{X},i)}(\overline{Z}) - 0.5||$ with the class-specific entropy term $-P_j^{(\overline{X},i)}(\overline{Z}) \cdot \log(P_j^{(\overline{X},i)}(\overline{Z}))$. Furthermore, this criterion needs to be minimized. In this context, the minimization of the expression can be viewed as the minimization of the expected loss function. This general framework has been used in a variety of different contexts in [53, 93, 100, 135].

22.3.2.2 Expected Variance Reduction

One observation about the afore-mentioned error reduction method of Equation 22.1 is that it needs to be computed in terms of the entire set of unlabeled instances in V, and a new model needs to be trained incrementally, in order to test the effect of adding a new instance. The model is therefore expensive to compute. It should be pointed out that when the error of an instance set reduces, the corresponding variance also typically reduces. The overall generalization error can be expressed as a sum of the true label noise, model bias, and variance [45]. Of these, only the last term is highly dependent on the choice of instances selected. Therefore, it is possible to reduce the variance instead of the error, and the main advantage of doing so is the reduction in computational requirements.

The main advantage of these techniques is the ability to express the variance in *closed form*, and therefore achieve greater computational efficiency. It has been shown [22, 23, 85] that the variance can be expressed in closed form for a wide variety of models such as neural networks, mixture models, or linear regression. In particular, it has been shown [85], that the output variance can be expressed in terms of the gradient with respect to the model parameters, and the Fisher Information Matrix. Interested readers are referred to [22, 23, 85, 103, 129] for details of the different variance-based methods.

22.3.3 Representativeness-Based Models

The main advantage of error-based models over uncertainty-based models is that they intend to improve the error behavior on the *aggregate*, rather than looking at the uncertainty behavior of the *queried* instance, as in heterogeneity-based models. Therefore, unrepresentative or outlier-like queries are avoided. However, representativeness can also be achieved by querying the data in such a way that the acquired instances tend to resemble the overall distribution better. This is achieved by weighting dense regions of the input space to a higher degree during the querying process. Examples of such methods include density-based models [106]. Therefore, these methods *combine* the heterogeneity behavior of the queried instance with a representativeness function from the unlabeled set V in order to decide on the queried instance. Therefore, in general, the objective function $O(\overline{X}, V)$ of such a model is expressed as the product of a heterogeneity component $H(\overline{X})$ and a representativeness component $R(\overline{X}, V)$:

$$O(\overline{X}, V) = H(\overline{X}) \cdot R(\overline{X}, V).$$

The value of $H(\overline{X})$ (assumed to be a maximization function) can be any of the heterogeneity criteria (transformed appropriately for maximization) such as the entropy criterion $En(\overline{X})$ from uncertainty sampling, or the expected model change criterion $C(\overline{X})$. The representativeness criterion $R(\overline{X}, V)$ is simply a measure of the density of $\overline{X}$ with respect to the instances in V. A simple version of this density is the average similarity of $\overline{X}$ to the instances in V [106], though it is possible to use more sophisticated methods such as kernel-density estimation to achieve the same goal. Note that such an approach is likely to ensure that the instance $\overline{X}$ is in a dense region, and is therefore not likely to be an outlier. Numerous variations of this approach have been proposed, such as those in [42, 84, 96, 106].

22.3.4 Hybrid Models

Conventional approaches for active learning usually select either informative or representative unlabeled instances. There are also some works combining multiple criteria for query selection in active learning [31, 58, 59, 120], such that the queried instances will have the following properties:

1) Informative, the queried instance will be close to the decision boundary of the learning model in terms of criteria like uncertainty, or the queried instance should be far away from existing labeled instances in order to bring new knowledge about the feature space. 2) Representative, the queried instance should be less likely to be outlier data and should be representative to a group of other unlabeled data. For example, in the work [59], the query selection is based upon both informativeness and representativeness of the unlabeled instances. A min-max framework of active learning is used to measure scores for both criteria.

22.4 Active Learning with Theoretical Guarantees

Let us recall the typical setting of active learning as follows: given a pool of unlabeled examples, an active learner is allowed to interactively query the label of any particular examples from the pool. The goal of active learning is to learn a classifier that accurately predicts the label of new examples, while requesting as few labels as possible.

It is important to contrast active learning to traditional passive learning, where labeled examples are chosen randomly. In the passive learning literature, there are well-known bounds on the number of training examples that is necessary and sufficient to learn a near-optimal classifier with a high probability. This quantity is called *sample complexity*, which depends largely on the VC dimension [14] of the hypothesis space being learned.

A natural idea is to define a similar quantity for active learning. It is called *label complexity*, i.e., the number of labeling requests that is necessary and sufficient to learn a near-optimal model with a high probability. However, not every active learning algorithm can be analyzed in terms of the label complexity. In previous sections, we introduce many active learning algorithms that are very effective empirically. However, most of these algorithms do not have any theoretical guarantee on the consistency or the label complexity [54]. In this section, we are particularly interested in active learning algorithms whose behaviors can be rigorously analyzed, i.e., that converge to an optimal hypothesis in a given hypothesis class with substantially lower label complexity than passive learning.

22.4.1 A Simple Example

Before going into the details of active learning with provable guarantee, we first present a simple example [6], which demonstrates the potential of active learning in the noise-free case when there is a perfect hypothesis with zero error rate. This is probably the simplest active learning algorithm with provable label complexity.

Consider the active learning algorithm that searches for the optimal threshold on an interval using binary search. We assume that there is a perfect threshold separating the classes, i.e., the realizable case.[1] Binary search needs $O(\log(\frac{1}{\varepsilon}))$ labeled examples to learn a threshold with error less than ε, while passive learning requires $O(\frac{1}{\varepsilon})$ labels. A fundamental drawback of this algorithm is that a small amount of adversarial noise can force the algorithm to behave badly. Thus, it is crucial to develop active learning algorithms that can work in the non-realizable case.[2]

---

[1] The target concept function is in the hypothesis class we considered.
[2] The target concept function is not in the hypothesis class we considered.

22.4.2 Existing Works

An early landmark result is the selective sampling scheme proposed in [22]. This simple active learning algorithm, designed for the realizable case, has triggered a lot of subsequent works. The seminal work of [44] analyzed an algorithm called query-by-committee, which uses an elegant sampling strategy to decide when to query examples. The core primitive required by the algorithm is the ability to sample randomly from the posterior over the hypothesis space. In some cases, such as the hypothesis class of linear separators in $\mathbb{R}^d$ and where the data is distributed uniformly over the surface of the unit sphere in $\mathbb{R}^d$, this can be achieved efficiently [46]. In particular, the authors showed that the number of labels required to achieve a generalization error ε is just $O(d \log(\frac{1}{\varepsilon}))$, which is exponentially lower than the sample complexity of a typical supervised learning algorithm $O(\frac{d}{\varepsilon})$. Later, [30] showed that a simple variant of Perceptron algorithm also achieves $O(d \log(\frac{1}{\varepsilon}))$ label complexity provided by a spherically uniform unlabeled data distribution. All the works mentioned above address active learning in the realizable case. A natural way to extend active learning to the non-realizable case (i.e., the agnostic case) is to ask the learner to return a hypothesis with error at most $L^* + \varepsilon$, where L^* is the error of the best hypothesis in the specified hypothesis class. The first rigorous analysis of agnostic active learning algorithm was developed by [6], namely A^2. later, [54] derived the label complexity of A^2 algorithm in terms of a parameter, called disagreement coefficient.

Another line of research uses importance weights to avoid the bias due to the active learning. For example, [5] considers linear representations and places some assumptions on the data generation process. However, the analysis is asymptotic rather than being a finite label complexity. To overcome this drawback, [12] proposed a new algorithm, called IWAL, satisfying PAC-style label complexity guarantees.

Due to the space limit, we do not discuss all the above methods in detail. In the remaining part of this section, we choose to introduce the IWAL algorithm [12] in detail.

22.4.3 Preliminaries

Let X be the instance space and $Y = \{\pm 1\}$ be the set of possible labels. Let H be the hypothesis class, i.e., a set of mapping functions from X to Y. We assume that there is a distribution D over all instances in X. For simplicity, we assume that H is finite ($|H| < \infty$), but does not completely agree on any single $x \in X$, i.e., $\forall x \in X, \exists h_1, h_2 \in H, h_1(x) \neq h_2(x)$. Note that $|H|$ can be replaced by VC dimension for an infinite hypothesis space. The algorithm is evaluated with respect to a given loss function $\ell : Y \times Y \to [0, \infty)$. The most common loss is $0 - 1$ loss, i.e., $\ell(z,y) = 1(y \neq z)$. The other losses include squared loss $\ell(z,y) = (y-z)^2$, hinge loss $\ell(z,y) = (1-yz)_+$ and logistic loss $\ell(z,y) = \log(1 + \exp(-yz))$. The loss of a hypothesis $h \in H$ with respect to a distribution P over $X \times Y$ is defined as

$$L(h) = \mathbb{E}_{(x,y) \sim P} \ell(h(x), y). \tag{22.2}$$

Let $h^* = \arg\min_{h \in H} L(h)$ be a hypothesis of the minimum error in H and $L^* = L(h^*)$. We have $L^* = 0$ in the realizable case, and $L^* > 0$ in the non-realizable case. The goal of active learning is to return a hypothesis $h \in H$ with an error $L(h)$ that is not much more than $L(h^*)$, using as few label queries as possible.

22.4.4 Importance Weighted Active Learning

In this section, we will discuss the problem of importance-weighted active learning.

22.4.4.1 Algorithm

In the importance weighted active learning (IWAL) framework [12], an active learner looks at the unlabeled data $x_1, x_2, \ldots, x_t$ one by one. After each new point x_t, the learner determines a probability $p_t \in [0, 1]$. Then a coin with the bias p_t is tossed, and the label y_t is queried if and only if the coin comes up heads. The query probability p_t depends on all previous unlabeled examples $x_{1:t-1}$, any previously queries labels, and the current unlabeled example x_t.

The algorithm maintains a set of labeled examples seen so far, where each example is assigned with an importance value p_t. The key of IWAL is a subroutine, which returns the probability p_t of requesting y_t, given x_t and the previous history $x_i, y_i : 1 \leq i \leq t-1$. Specifically, let $Q_t \in \{0, 1\}$ be a random variable that determines whether to query y_t or not. That is, $Q_t = 1$ indicates that the label y_t is queried, and otherwise $Q_t = 0$. Then $Q_t \in \{0, 1\}$ is conditionally independent of the current label y_t, i.e.,

$$Q_t \perp Y_t | X_{1:t}, Y_{1:t-1}, Q_{1:t-1} \tag{22.3}$$

and with conditional expectation

$$\mathbb{E}[Q_t | X_{1:t}, Y_{1:t-1}] = p_t. \tag{22.4}$$

If y_t is queried, IWAL adds $(x_t, y_t, 1/p_t)$ to the set, where $1/p_t$ is the importance of predicting y_t on x_t. The key of IWAL is how to specify a rejection threshold function to determine the query probability p_t. [12] and [13] discussed different rejection threshold functions.

The importance weighted empirical loss of a hypothesis h is defined as

$$L_T(h) = \frac{1}{T} \sum_{t=1}^{T} \frac{Q_t}{p_t} \ell(h(x_t), y_t) \tag{22.5}$$

A basic property of the above estimator is unbiasedness. It is easy to verify that $\mathbb{E}[L_T(h)] = L(h)$, where the expectation is taken over all the random variables involved.

To summarize, we show the IWAL algorithm in Algorithm 22.1.

Algorithm 22.1 Importance Weighted Active Learning (IWAL)

Input: $S_0 = \emptyset$,
for $t = 1$ to T **do**
    Receive x_t
    Set $p_t = rejection - threshold(x_t, \{x_i, y_i : 1 \leq i \leq t-1\})$
    Toss a coin $Q_t \in 0, 1$ with $\mathbb{E}[Q_t] = p_t$
    if $Q_t = 1$ **then**
        request y_t and set $S_t = S_{t-1} \cup \{(x_t, y_t, p_{min}/p_t)\}$
    else
        $S_t = S_{t-1}$
    end if
    Let $h_t = \arg\min_{h \in H} \sum_t \frac{1}{p_t} \ell(h(x_t), y_t)$
end for

22.4.4.2 Consistency

A desirable property of any learning algorithm is the consistency. The following theorem shows that IWAL is consistent, as long as p_t is bounded away from 0. Furthermore, its sample complexity is within a constant factor of supervised learning in the worst case.

Theorem 3 *[12] For all distributions D, for all finite hypothesis classes H, if there is a constant*

$p_{min} > 0$ such that $p_t > p_{min}$ for all $1 \leq t \leq T$, then with a probability that is at least $1 - \delta$, we have

$$L_T(h) \leq L(h) + \frac{\sqrt{2}}{p_{min}} \sqrt{\frac{\log|H| + log(\frac{1}{\delta})}{T}}. \tag{22.6}$$

Recall that a typical supervised learning algorithm has the following bound for sample complexity

Theorem 4 *For all distributions D, for all finite hypothesis classes H, with a probability that is at least $1 - \delta$, we have*

$$L_T(h) \leq L(h) + \sqrt{\frac{\log|H| + \log(\frac{1}{\delta})}{T}}. \tag{22.7}$$

By comparing with the above results, we can see that the sample complexity of IWAL is at most $\frac{2}{p_{min}^2}$ times the sample complexity of a typical supervised learning algorithm.

In fact, a fairly strong and large deviation bound can be given for each h_t output by IWAL as follows:

Theorem 5 *[12] For all distributions D, for all finite hypothesis classes H, with a probability that is at least $1 - \delta$, the hypothesis output by IWAL satisfies*

$$L(h_T) \leq L(h) + 2\sqrt{\frac{8}{t} \log \frac{t(t+1)|H_t|^2}{\delta}}. \tag{22.8}$$

22.4.4.3 Label Complexity

Suppose we have a stream of T examples, some of whose labels are queried. The consistency analysis tells us that at the end of the IWAL algorithm, the classifier output by IWAL is comparable to the classifier that would have been chosen by a passive supervised learning algorithm that saw all T labels. The remaining problem is how many labels of those T examples are queried by IWAL. In other words, what is the label complexity of IWAL? This subsection is devoted to proving that IWAL can yield substantial improvement on label complexity over passive learning.

The bound for this algorithm depends critically on a particular quantity, called the disagreement coefficient [54], which depends upon the hypothesis space and the example distribution. Note that the original disagreement coefficient is defined for the $0 - 1$ loss function. Here we need a general disagreement coefficient [12] for general loss functions $\ell(z, y)$. The definition of disagreement coefficient needs a few new concepts.

Definition 22.4.1 *[12] The pseudo-metric on H induced by D is defined as*

$$\rho(h_1, h_2) = \mathbb{E}_{x \sim D} \max_y |\ell(h_1(x), y) - \ell(h_2(x), y)| \tag{22.9}$$

for $h_1, h_2 \in H$. Let $B(h, r) = \{h' \in H : \rho(h, h') \leq r\}$ be the ball centered around h of radius r.

Definition 22.4.2 *[12](disagreement coefficient) The disagreement coefficient is the infimum value of θ such that for all r*

$$\sup_{h \in B(h^*, r)} \rho(h^*, h) \leq \theta r. \tag{22.10}$$

Definition 22.4.3 *[12] The slope asymmetry of a loss function ℓ is*

$$C_l = \sup_{z, z'} |\frac{\max_y \ell(z, y) - \ell(z', y)}{\min_y \ell(z, y) - \ell(z', y)}|. \tag{22.11}$$

Theorem 6 *[12] For all distributions D, for all finite hypothesis classes H, if the loss function has a slope asymmetry C_l, and the learning problem has a disagreement coefficient θ, the expected number of labels queried by IWAL after seeing T examples is at most*

$$4\theta C_l \left(L^* T + O\left(\sqrt{T \log \frac{|H|T}{\delta}} \right) \right). \tag{22.12}$$

22.5 Dependency-Oriented Data Types for Active Learning

Active learning problems can arise in the context of different kinds of complex data types such as sequences and graphs. Since these data types often have implicit or explicit dependencies between instances, this can impact the active learning process directly. This is because the dependencies between the instance translates to correlations in the node labels among different instances.

The two primary dependency-oriented data types are sequential data and graph-based data. Graph-based learning poses different kinds of challenges because of the wide variety of scenarios in which it can arise, such as the classification of many small graphs, or the classification of nodes in a single large graph. This section will discuss sequences and graphs, the two data types that are frequently explored in the context of dependency-oriented active learning.

22.5.1 Active Learning in Sequences

Sequences are defined as a succession of discrete values (tokens) $X_1 \ldots X_N$, each of which is possibly associated with a label. The problem of sequence labeling forms the core of all information extraction tasks in natural language processing, which attempt to annotate the individual words (tokens) with labels. For example, consider the following sentence:

"The bomb blast in Cairo $< place >$ left Mubarak $< person >$ wounded."

In this case, the token "Cairo" is associated with the label *place*, whereas the token "Mubarak" is associated with the label *person*. Features may also be associated with tokens of the instance, which provide useful information for classification purposes. The labels associated with the individual tokens in a sequence are not independent of one another, since the adjacent tokens (and labels) in a given sequence directly impact the label of any given token. A common approach for sequence classification is to use Hidden Markov Models (HMM), because they are naturally designed to address the dependencies between the different data instances. Another class of methods that are frequently used for active learning in sequences are Conditional Random Fields (CRF). A detailed discussion of these methods is beyond the scope of this chapter. The reader is referred to the work in [8, 28, 60, 102, 106], which provide a number of different probabilistic algorithms for active learning from sequences. The work in [106] is particularly notable, because it presents an experimental comparison of a large number of algorithms.

22.5.2 Active Learning in Graphs

Graphs represent an important scenario for active learning, since many real world networks such as the Web and social networks can be represented as graphs. In many applications involving such networks, labeled data is not readily available, and therefore active learning becomes an important approach to improve the classification process. Graph-based learning is studied in different scenar-

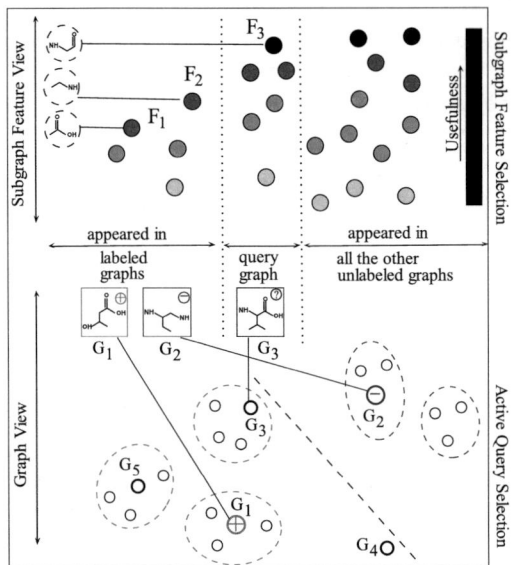

FIGURE 22.2 (See color insert.): Motivation of active learning on small graphs [70].

ios, such as the classification of many small graphs (e.g., chemical compounds) [70], or the classification of nodes in a single large graph (e.g., a social or information network) [10, 18, 49–52, 61]. Most of the specialized active learning methods in the literature are designed for the latter scenario, because the former scenario can often be addressed with straightforward adaptations of standard active learning methods for multidimensional data. However, in some cases, specific aspects of feature selection in graphs can be combined with active learning methods. Such methods will be discussed below.

22.5.2.1 Classification of Many Small Graphs

The first scenario addresses the problem of classifying graph objects within a graph database. Graphs are ubiquitous and have become increasingly important in modeling diverse kinds of objects. For many real-world classification problems, the data samples (i.e., instances) are not directly represented as feature vectors, but graph objects with complex structures. Each graph object in these applications can be associated with a label based upon certain properties. For example, in drug discovery, researchers want to discover new drugs for certain diseases by screening the chemical database. Each drug candidate, i.e., a chemical compound, is naturally represented as a graph, where the nodes correspond to the atoms and the links correspond to the chemical bonds. The chemical compound can be associated with a label corresponding to the drug activities, such as anticancer efficacy, toxicology properties, and Kinase inhibition. Label acquisition can often be expensive and time-consuming because the properties of compounds may often need to be derived through a variety of chemical processing techniques. For example, in molecular drug discovery, it requires time, efforts, and excessive resources to test drugs' anti-cancer efficacies by pre-clinical studies and clinical trials, while there are often copious amounts of unlabeled drugs or molecules available from various sources. In software engineering, human experts have to examine a program flow in order to find software bugs. All these processes are rather expensive. This is how active learning comes into play. Active learning methods could be used to select the most important drug candidates for the human experts to test the label in the lab, and could potentially save a significant amount of money and time in the drug discovery process.

Conventional active learning methods usually assume that the features of the instances are given beforehand. One issue with graph data is that the features are not readily available. Many classification approaches for small graphs require a feature extraction step to extract a set of subgraphs that are discriminative for the classification task [63,69,114,121,123]. The number of possible subgraph features that can be extracted is rather large, and the features extracted are highly dependent on the labeled instances. In the active learning settings, we can only afford to query a small number of graphs and obtain their labels. The performance of the feature extraction process depends heavily on the quality of the queried graphs in the active learning process. Meanwhile, in the active learning, we need to evaluate the importance of each instance. The performance of active learning also depends heavily on the quality of the feature extraction process. Thus the active learning and subgraph feature extraction steps are mutually beneficial.

For example, in Figure 22.2, we are given two labeled graphs (G_1 and G_2) within which we have only a small number of the useful subgraph features (F_1 and F_2). If we query the graph G_3 for the label, we are not only improving the classification performances due to the fact that G_3 is both representative and informative among the unlabeled graphs, but we are also likely to improve the performances of feature extraction, because G_3 contains new features like F_3.

The process of query selection can assist the process of feature selection in finding useful subgraph features. In other words, the better the graph object we select, the more effectively we can discover the useful subgraph features. Therefore, the work in [70] couples the active sample problem and subgraph feature selection problem, where the two processes are considered simultaneously. The idea is that the two processes influence each other since sample selection should affect feature selection and vice-versa. A method called gActive was proposed to maximize the dependency between subgraph features and graph labels using an active learning framework. A branch-and-bound algorithm is used to search for the optimal query graph and optimal features simultaneously.

22.5.2.2 Node Classification in a Single Large Graph

A second scenario addresses the problem of node classification in a single large graph. This is a very common problem in social networks, where it may be desirable to know specific properties of nodes for the purpose of influence analysis. For example, a manufacturer of a particular sporting item may wish to know nodes that correspond to interest in that sport. However, this is often difficult to know a-priori, since the interests of nodes may often need to be acquired through some querying or manual examination mechanism, which is expensive. The work in [10, 18, 49–52, 61] actively queries the labels of the nodes in order to maximize the effectiveness of classification. A particular class of active learning methods, which are popular in the graph scenario are *non-adaptive active learning methods*, in that the previous labels are not used to select the examples, and therefore the selection can be performed in batch in an unsupervised way. The downside is that the learning process is not quite as effective, because it does not use labels for the selection process. Many active learning methods for graphs [49, 51, 52, 61] belong to this category, whereas a few methods [10, 11, 18, 50] also belong to the adaptive learning category. The non-adaptive methods are generally unsupervised and attempt to use representativeness-based models such as clustering in order to select the nodes. Many of the non-adaptive methods are motivated by experimental design approaches [61, 126].

An important aspect of non-adaptive active learning of node labels in large graphs is that the structural properties of the graph play an important role in the selection of nodes for the active learning process. This is because labels are not available, and therefore only the structure of the graph may be used for the selection process. For example, the work in [15] showed that the prediction error is small when the graph cut size is large. This suggests that it is better to label nodes, such that the size of the graph cut on the labeled graph is large. The work in [51, 52] proposes a heuristic label selection method that maximizes the graph cut. Furthermore, a generalized error bound was proposed by replacing the graph cut with an arbitrary symmetric modular function. This was then leveraged to provide an improved algorithm for the use of submodular function maximization techniques. The active learning method first clusters the graph and then randomly chooses a node in each cluster. The work in [61] proposes a variance minimization approach to active learning in graphs. Specifically, the work in [61] proposes to select the most informative nodes, by proposing to minimize the prediction variance of the Gaussian field and harmonic function, which was used in [133] for semi-supervised learning in graphs. The work in [49] approaches the problem by considering the generalization error of a specific classifier on graphs. The Learning with Local and Global Consistency (LLGC) method proposed in [135] was chosen because of its greater effectiveness. A data-dependent generalization error bound for LLGC was proposed in [49] using the tool of transductive Rademacher Complexity [35]. This tool measures the richness of a class of real-valued functions with respect to a probability distribution. It was shown in [49] that the empirical transductive Rademacher complexity is a good surrogate for active learning on graphs. The work therefore selects the nodes by minimizing the empirical transductive Rademacher complexity of LLGC on a graph. The resulting active learning method is a combinatorial optimization problem, which is optimized using a sequential optimization algorithm.

Another line of research [10, 11, 18, 50] has considered adaptive active learning, where the labels for the nodes of a graph are queried and predicted in an iterative way with the use of a trained classifier from the previous labels. The work in [11] made the observation that label propagation in graphs is prone to "accumulating" errors owing to correlations among the node labels. In other words, once an error is made in the classification, this error propagates throughout the network to the other nodes. Therefore, an acquisition method is proposed, which learns where a given collective classification method makes mistakes, and then suggests acquisitions in order to correct those mistakes. It should be pointed out that such a strategy is unique to active learning in graphs, because of the edge-based relationships between nodes, which result in homophily-based correlations among node labels. The work in [11] also proposes two other methods, one of which greedily improves the objective function value, and the other adapts a viral marketing model to the label acquisition process. It was shown in [11] that the corrective model to label acquisition generally provides the best results.

The work in [10] works in scenarios where both content and structure are available with the nodes. It is assumed that the labels are acquired sequentially in batch sizes of k. The algorithm uses two learners called *CO* (content-only) and *CC* (collective classifier), and combines their predictions in order to make decisions about labeling nodes. In particular, nodes for which these two classifiers differ in their prediction are considered good candidates for labeling, because the labels of such nodes provide the most informative labels for knowing the relative importance of the different aspects (content and structure) over different parts of the data. The proposed algorithm *ALFNET* proceeds by first clustering the nodes using the network structure of the data. Then, it selects the k clusters that satisfy the following two properties:

- The two learners *CC* and *CO* differ the most in their predictions.

- The predictions of the classifiers have a distribution that does not match the distribution of the observed labels in the cluster.

An overall score is computed on the basis of these two criteria. For a batch size of k, the top-k clusters are selected. One node is selected from each of these clusters for labeling purposes.

The work in [18] proposes an active learning method that is based on the results of [51, 52] in order to design optimal query placement methods. The work in [18] proposes a method for active learning in the special case of trees, and shows that the optimal number of mistakes on the non-queries nodes is made by a simple mincut classifier. A simple modification of this algorithm also achieves optimality (within a constant factor) on the trade-off between the number of mistakes, and the number of non-queried nodes. By using spanning trees, the method can also be generalized to arbitrary graphs, though the problem of finding an optimal solution in this general case remains open.

The work in [50] proposes selective sampling for graphs, which combines the ideas of online learning and active learning. In this case, the selective sampling algorithm observes the examples in a sequential manner, and predicts its label after each observation. The algorithm can, however, choose to decide whether to receive feedback indicating whether the label is correct or not. This is because, if the label can already be predicted with high confidence, a significant amount of effort can be saved by not receiving feedback for such examples. The work in [50] uses the LLGC framework [135] for the prediction process.

22.6 Advanced Methods

Active learning can arise in the context of different scenarios, such as feature-based learning, streaming data, or other kinds of query variants. In feature-based learning, it is desirable to learn techniques for acquiring *features* rather than instance labels for classification. These different scenarios will be discussed in detail in this section.

22.6.1 Active Learning of Features

Most of the discussion in the previous sections focussed on cases where new instances are acquired in order to maximize the effectiveness of a given learner. However, there are also numerous scenarios where the labels of the instances are known, but many feature values are unknown, and require a cost of acquisition. In such scenarios, when we query the features for some instances, the "oracle" can provide their corresponding feature values. It is assumed that the values for some features of an instance are rather expensive to obtain. In some very complex scenarios, knowledge of the feature values *and* the labels may be incomplete, and either may be acquired at a given cost. Furthermore, in feature-based active learning, the acquisition of more features may be done either for the training data or for a test instance, or for both, depending upon the underlying application. In some applications, the training data may be incompletely specified in a non-homogeneous way over different instances, and the problem may be to determine the subset of features to query for a given test instance in order to maximize accuracy. In other application, the same feature may be collected for both the training data and the test data at a particular cost. An example of this case would be the installation of a sensor at a particular location, which is helpful for augmenting both the training data and the test data. Some examples of feature-based active learning scenarios are as follows:

- *Medical Applications:* Consider a medical application, where it is desirable to collect information from different medical tests on the patient. The results of each medical test may be considered a feature. Clearly, the acquisition of more features in the training data helps in better modeling, and the acquisition of more features in the test data helps in better diagnosis. However, acquisition of such features is typically expensive in most applications.

- *Sensor Applications:* In sensor applications, the collection and transmission of data is often associated with costs. In many cases, it is desirable to keep certain subsets of sensors passive, if their data is redundant with respect to other sensors and do not contribute significantly to the inference process.

- *Commercial Data Acquisition:* In many commercial applications (*e.g.*, marketing, and credit card companies), different features about customers (or customer behavior) can be acquired at a cost. The acquisition of this information can greatly improve the learning process. Therefore, feature-based active learning can help evaluate the trade-offs between acquiring more features, and performing effective classification.

Feature-based active learning is similar in spirit to instance-based active learning, but the methodologies used for querying are quite different.

One observation about missing feature values is that many of them can often be partially imputed from the correlations among the features in the existing data. Therefore, it is not useful to determine those features that can be imputed. Based on this principle, a technique was proposed in [132] in which missing feature values are first imputed. The ones among them that have the greatest uncertainty are then queried. A second approach proposed in this work uses a classifier to determine those instances that are misclassified. The feature values of the misclassified instances are then queried for labels. In incremental feature acquisition, a small set of misclassified instances are used in order to acquire their labels [88], or by using a utility function that needs to be maximized [87, 101].

Active learning for feature values is also relevant in the context of scenarios where the feature values are acquired for test instances rather than training instances. Many of the methodologies used for training instances cannot be used in this scenario, since class labels are not available in order to supervise the training process. This particular scenario was proposed in [48]. Typically standard classifiers such as naive Bayes and decision trees are modified in order to improve the classification [19, 37, 79, 110]. Other more sophisticated methods model the approach as a sequence of decisions. For such sequential decision making, a natural approach is to use Hidden Markov Models [62].

Another scenario for active learning of features is that we can actively select instances for feature selection tasks. Conventional active learning methods are mainly designed for classification tasks. The works in [80–82] studied the problem of how to selectively sample instances for feature selection tasks. The idea is that by partitioning the instances in the feature space using a KD-tree, it is more efficient to selectively sample the instances with very different features.

22.6.2 Active Learning of Kernels

Active kernel learning is loosely related to the active learning of features in the previous subsection. Most of the discussion in the previous sections focuses on designing selective sampling methods that can maximize the effectiveness of a given supervised learning method, such as SVM. Kernel-based methods are very important and effective in supervised learning, where the kernel function/matrix, instead of the feature values, plays an important role in the learning process. Since the choice of kernel functions or matrices is often critical to the performance of kernel-based methods, it becomes a more and more important research problem to automatically learn a kernel function/matrix for a given dataset. There are many research works on kernel learning, i.e., the problem of learning a kernel function or matrix from *side information* (e.g., either labeled instances or pairwise constraints). In previous kernel learning methods, such *side information* is assumed to be given beforehand. However, in some scenarios, the side information can be very expensive to acquire, and it is possible to actively select some side information to query the oracle for the ground-truth. This is how active kernel learning comes into play.

Specifically, active kernel learning corresponds to the problem of exploiting active learning to learn an optimal kernel function/matrix for a dataset with the least labeling cost. The work in [56] studied the problem of selectively sampling instance-pairs, instead of sampling instances, for a given dataset, in order to learn a proper kernel function/matrix for kernel-based methods. Given a pair of instances (x_i, x_j), the corresponding kernel value is $K_{i,j} \in \mathbb{R}$, indicating the similarity between x_i and x_j. If the two instances are in the same class, $K_{i,j}$ is usually a large positive number, while if the two instances are in different classes, $K_{i,j}$ is more likely to be a large negative number. One simple solution for active kernel learning is to select the kernel similarities that are most close to zero, i.e., $(i^*, j^*) = \arg\min_{(i,j)} |K_{i,j}|$. This solution is simply extended from the uncertainty-based active learning methods. The underlining assumption is that such instance pairs are the most informative in the kernel learning process. However, in real-world kernel-based classification methods, the instance pairs with the smallest absolute kernel values $|K_{i,j}|$ are more likely to be *must-link pairs*, i.e., both instances are in the same class. It is also crucial to query instances pairs that are more likely to be *cannot-link pairs*, where the two instances are in different classes.

The work in [56] proposed an approach based upon *min-max* principle, i.e., the assumption is that the most informative instance pairs are those resulting in a large classification margin regardless of their assigned class labels.

22.6.3 Active Learning of Classes

In most active learning scenarios, it is assumed that the instances are queried for labels. In such querying processes, we can control which instances we want to query for the labels, but cannot control which classes we will get for the queried instances. It is explicitly or implicitly assumed that: 1) The cost of getting the class label for an instance is quite expensive. 2) The cost of obtaining the instances is inexpensive or free. However, in some scenarios, such assumptions may not hold, and it may be possible to query with a *class label*, so that the oracle will feedback with some instances in the class. Thus, this is a kind of reverse query, which is relevant to certain kinds of applications [83]. For example, the work in [83] designs method for training an artificial nose which can distinguish between different kinds of vapors (the class labels). However, the key distinction here is that the instances here are *generated* for training a particular application. For example, in the "artificial nose" task, the chemical (the instances) are synthesized. Different kinds of instances are synthetically generated representing the different kinds of vapors. In traditional active learning, the instances are already *existing*. A key observation in [83] is that methods that optimize the class stability were more effective than either random sampling or methods that optimized the class accuracy.

22.6.4 Streaming Active Learning

The data stream scenario is particularly suited to active learning, since large volumes of data are often available, and the real-time stream generation process usually does not create labels of interest for specific applications. Unlike static data sets, which have often been extensively processed in order to generate labels, data stream instances are often "newly generated," and therefore there has never been an opportunity to label them. For example, in a network system, the incoming records may correspond to real time intrusions, but this cannot be known without spending (possibly manual) effort in labeling the records in order to determine whether or not they correspond to intrusions. Numerous simple methods are possible for adapting data streams to the active learning scenario. In local uncertainty sampling, active learning methods can be applied to a current segment of the data stream, with the use of existing algorithms, without considering any other segments. The weakness of this is that it fails to use any of the relevant training data from the remaining data stream. One way to reduce this is to build historical classifiers on different segments, and combine the results from the current segment with the results from the historical classifier. The work in [136] goes one step further, and combines an ensemble technique with active learning in order to perform the classification. The minimal variance principle is used for selecting the relevant sample.

The work in [26] proposes unbiased methods for active sampling in data streams. The approach designs optimal instrumental distributions to minimize the variance in the sampling process. Bayesian linear classifiers with weighted maximum likelihood are optimized online on order to estimate parameters. The approach was applied on a data stream of user-generated comments on a commercial news portal. Offline evaluation was used in order to compare various sampling strategies, and it was shown that the unbiased approach is more effective than the other methods.

The work in [41] introduces the concept of demand-driven active mining of data streams. In this model, it is assumed that a periodic refresh of the labels of the stream is performed in batch. The key question here is the choice of the timing of the model for refreshing. If the model is refreshed too slowly, it may result in a more erroneous model, which is generally undesirable. If the model is refreshed too frequently, then it will result in unnecessary computational effort without significant benefit. Therefore, the approach uses a mathematical model in order to determine when the stream patterns have changed significantly enough for an update to be reasonably considered. This is done by estimating the error of the model on the newly incoming data stream, without knowing the true labels. Thus, this approach is not, strictly speaking, a traditional active learning method that queries for the labels of individual instances in the stream.

The traditional problem of active learning from data streams, where a decision is made on each instance to be queried, is also referred to as *selective sampling*, which was discussed in the introduction section of this chapter. The term "selective sampling" is generally used by the machine learning community, whereas the term "data stream active learning" is often used by data mining researchers outside the core machine learning community. Most of the standard techniques for sample selection that are dependent on the properties of the unlabeled instance (e.g., query-by-committee), can be easily extended to the streaming scenario [43,44]. Methods that are dependent on the aggregate performance of the classifier are somewhat inefficient in this context. This problem has been studied in several different contexts such as part-of-speech tagging [28], sensor scheduling [71], information retrieval [127], and word-sense disambiguation [42].

22.6.5 Multi-Instance Active Learning

The idea in multiple instance active learning [33] is that the instances are naturally organized into "bags," each of which is assigned a label. When a label is assigned to a bag, it is implied that all instances in that bag have that label. This problem was first formalized in [107]. This problem arises quite naturally in some data domains such as protein family modeling, drug activity prediction, stock market prediction, and image retrieval. In these data domains, scenarios arise in which it is inexpensive to obtain labels for bags, but rather expensive to obtain labels at the individual instance level. The real issue here is often one of granularity. For example, in a text document, image, or a biological sequence, consider the case where individual instances correspond to portions of these objects. In such cases, it may often be easier to label the whole object, rather than portions of it, and all segments of the object inherit this label. In many cases, the labeling may simply be *available* at the higher level of granularity. It should be pointed out that multiple instance labeling naturally leads to errors in labeling of the fine-grained instances. Therefore, a major challenge in multiple-instance learning is that the learner must determine which instances of a particular label in a bag do actually belong to that label. On the flip side, these coarse labelings are relatively inexpensive, and may be obtained at a low cost.

Several scenarios have been discussed in [107] for the multi-class problem. For example, the learner may either query for bag labels, for specific instance labels, or for all instance labels in a specific bag. The appropriateness of a particular scenario of course depends upon the application setting. The work in [107] focuses on the second scenario, where specific instance labels are queried. In addition, a binary classification problem is assumed, and it is assumed that only positively labeled bags are of interest. Therefore, only instances in positive bags are queried. The approach uses multiple-instance logistic regression for classification purposes. In addition, uncertainty sampling is used, where the instances for which the labels are the least certain are sampled. It should be pointed out that the bag labels directly impact both the logistic regression modeling process, and the computation of the uncertainty values from the labels. Details of the approach are discussed in [107].

A second scenario is the case where each query may potentially correspond to multiple instances. For example, in the case of *Amazon Mechanical Turk*, the user provides a batch of instances to the learner, which are then labeled by experts at a cost. In many cases, this batch is provided in the context of a *cold start*, where no previous labeled instances are available. The cold start scenario is closer to *unsupervised* or *non-adaptive* active learning, where no labeled data is available in order to evaluate the performance of the methods. Some examples of this scenario have already been discussed in the context of active learning for node classification in networks, where such methods are very popular. Even in the case where labeled data is available, it is generally not advisable to pick the k best instances. This is because the k best instances that are picked may all be quite similar, and may not provide information that is proportional to the underlying batch size. Therefore, much of the focus of the work in the literature [16,55,57,119] is to incorporate diversity among the instances in the selection process.

22.6.6 Multi-Label Active Learning

Conventional active learning approaches mainly focus on single-label classification problems, i.e., binary classification and multi-class classification. It is usually assumed that each instance can only be assigned to *one* class among all candidate categories. Thus the different classes are mutually exclusive in the instance level. However, in some applications, such assumptions may not hold. Each instance can be associated with multiple labels simultaneously, where the classification task is to assign *a set* of labels for each instance. For example, in text classification, each text document may be assigned with multiple categories, e.g., news, politics, economy. In such classification tasks, when we query the "oracle" with an instance, the "oracle" will annotate all the labels for the queried instance. In the active learning process, we need to estimate the importance or uncertainty of an instance corresponding to all the labels.

One solution is to measure the informativeness of each unlabeled instance across all the candidate labels [17, 38, 78, 112]. For example, Yang et. al. [124] proposed an approach that selects the unlabeled data based upon the largest reduction of the expected model loss for multi-label data by summing up losses on all labels. In multi-label learning, it is also important to consider the relationship among different labels instead of treating each label independently. A second type of approaches considers the label structure of a multi-label learning problem, such as the correlations among the labels [98], the cardinality of label sets [77], and hierarchy of labels [77], to measure the informativeness of the unlabeled instances. The work in [76] proposed two methods based upon max-margin prediction uncertainty and a label cardinality inconsistency, respectively, and integrated them into an adaptive method of multi-label active learning.

22.6.7 Multi-Task Active Learning

Multi-task learning is closely related to multi-label learning scenarios. Conventional active learning mainly focuses on the scenarios of single task with single learner. However, in some applications, one instance can be involved in multiple tasks and be used by multiple learners [99]. For example, in NLP (Natural Language Processing) tasks, each instance can be involved in multiple levels of linguistic processing tasks, such as POS tagging, named entity recognition, statistical parsing. It is usually easier for the "oracle" to be queried with one instance, and feedback with labels for all the tasks, instead of being queried multiple times for different tasks. In this scenario, the key problem is how to assess the informativeness or representativeness of an unlabeled instance w.r.t. all the learning tasks. Some examples of this scenario have already been discussed in the context of active learning for NLP applications [99], which usually involve multiple tasks and the labeling costs are very high.

In the work [99], two approaches have been proposed. In the first approach, i.e., alternating selection, the parser task and named entity recognition task take turns to select important instances to query for the label iteratively. While the second approach, i.e., rank combination, combines the ranking of importance from both tasks and select important instances that are highly ranked in both tasks.

Another scenario for multi-task active learning is that the outputs of multiple tasks are coupled together and should satisfy certain constraints. For example, in relational learning, the entities classification and link prediction are two prediction tasks, where their outputs can often be related.

If an entity is predicted as the mayor of a city, the class label of the entity should be "politician" or "person", instead of other labels like "water." In such scenarios, the outputs of multiple tasks are coupled with certain constraints. These constraints could provide additional knowledge for the active learning process. The work in [128] studied the problem of active learning for multi-task learning problem with output constraints. The idea is that we should exploit not only the uncertainty of prediction in each task but also the inconsistency among the outputs on multiple tasks. If the class labels on two different tasks are mutually exclusive while the model predicts positive labels in both tasks, it indicates that the model makes an error on one of the two tasks. Such information is provided by the output constraints and can potentially help active learners to evaluate the importance of each unlabeled instance more accurately.

22.6.8 Multi-View Active Learning

Most of the discussion in the previous sections focuses on single view settings, where each instance only has one view of a set of features. However, there are also numerous scenarios where each instance is described with several different disjoint sets of features. Each set of features corresponds to one *view*, which is sufficient for learning the label concept. For example, in the Web page classification problems, we can have two different views: 1) One set of features corresponds to the text appearing on a Web page. 2) Another set of features corresponds to the anchor text in the hyper-links pointing to the Web page. Multi-view active learning is closely related to semi-supervised learning and active learning [90, 116]. The multi-view active learning problem was first studied in [90–92], where a model called co-testing was proposed. The idea is related to Query-by-Committee, which uses multiple learners and selects the set of unlabeled instances where the predictions on different learners disagree with each other. However, Query-by-Committee approaches build multiple learners on one single view, while the co-testing method focuses on multiple views and trains one learner on each view. The motivation of such model design is as follows: Suppose learners in different views predict different labels on an unlabeled instance. These kinds of unlabeled instances are called *contention points*. At least one learner could make a mistake on that instance prediction. If we query the oracle with that instance (i.e., a contention point), it is guaranteed that at least one of the learners will benefit from the label information. In this sense, the active sampling is based upon learning from mistakes (at least in one view). So the co-testing method can potentially converge to the target concept more quickly than alternative active learning methods.

The work in [116] provided a theoretical analysis on the sample complexity of multi-view active learning based upon the α-expansion assumption [7]. The sample complexity of multi-view active learning can be exponentially improved to $\widetilde{O}(\log \frac{1}{\varepsilon})$, approximately. The work in [117] extended the theoretical analysis to *Non-Realizable* cases, where the data are no longer assumed to be perfectly separable by any hypothesis in the hypothesis class because of the noise.

22.6.9 Multi-Oracle Active Learning

In conventional active learning settings, it is usually assumed that there is only one "oracle" for the system to query the labels, and the "oracle" knows ground-truth about classification tasks and can annotate all the queried samples without any errors. These assumptions may not hold in some real-world applications, where it is very hard or impossible to find such an "oracle" that knows everything about the ground-truth. It is more likely that each "oracle" (i.e., annotator) only knows or is familiar with a small part of the problem, while we could have access to multiple "oracles" with varying expertise. For example, in *Crowdsourcing* systems, such as Amazon Mechanical Turk,[3] each user or annotator may only know a small part of the task to be queried while we could query multiple users, each with different queries. In many other applications, many "oracles" may disagree

---

[3]http://www.mturk.com

with each other on the same instance. The answer from an individual "oracle" may not be accurate enough. Combining answers from multiple "oracles" would be necessary. For example, in medical imaging, the diagnosis of different doctors may be different on one patient's image, while it is often impossible to perform a biopsy on the patient in order to figure out the ground-truth.

In multi-oracle active learning, we could not only select which instance to query for the label, but also select which oracle(s) or annotator(s) to be queried with the instance. In other words, multi-oracle active learning is an instance-oracle-pair selection problem, instead of an instance-selection problem. This paradigm brings up new challenges to the active learning scenario. In order to benefit the learning model the most, we need to estimate both the importance of each unlabeled instance and the expertise of each annotator.

The work in [122] designed an approach for multi-oracle active learning. The model first estimated the importance of each instance, which is similar to conventional active learning models. The model then estimated which "oracle" is most confident about its label annotation on the instance. These two steps are performed iteratively. The high level idea of the approach is that we first decide which instance we need to query the most, then decide who has the most confident answer for the query.

The work in [111] studied both the uncertainty in the model and noises in the oracle. The idea is that the costs of labelers with low qualities are usually low in crowdsourcing applications, while we can repeatedly query different labelers for one instance to improve th labeling quality with additional costs. Thus the work [111] studied the problem of selective repeated labeling, which can potentially reduce the cost in labeling while improving the overall labeling quality, although this work focused on the cases that all oracles are equally and consistently noisy, which may not fit in many real-world cases. The work [32] extended the problem setting by allowing labelers to have different noise levels, though the noise levels are still consistent over time. It shows that the qualities of individual oracles can be properly estimated, then we can perform active learning more effectively by querying only the most reliable oracles in selective labeling process. Another work extended the setting of multi-oracle active learning in that the skill sets of multiple oracles are not assumed to be static, but are changeable by teaching each other actively. The work in [39] explored this problem setting, and proposed a self-taught active learning method from crowdsourcing. In addition to selecting the most important instance-oracle-pair, we also need to figure out which pair of oracles could be the best pair for self-teaching, i.e., one oracle has the expertise in a skill where the other oracle needs to improve the most. In order to find the most effective oracle pair, we not only need to estimate how well one oracle is on a type of skill, but also which oracle needs the most help on improving the type of skill. In this way, the active learning method can help improve the overall skill sets of the oracles and serve the labeling task with better performance.

22.6.10 Multi-Objective Active Learning

Most previous works on active learning usually involved supervised learning models with single objective optimization. The selective sampling process is mainly performed in the level of data instances, i.e., it is assumed that the labeling of the instances is very expensive. However, there are some scenarios where the learning problem involve multiple objectives in the optimization process. Each model will have multiple scores with reference to multiple objective functions. For example, in classification problems, the performance of the model can be evaluated from multiple objective/loss functions, such as empirical loss, model complexity, precision, recall, and AUC (for ROC curve). In hardware design, each model design can have different scores on multiple criteria, such as energy consumption, throughput, and chip area. In all these scenarios, the models are not evaluated according to one single objective function, but a set of multiple objective functions.

Usually there is no single model that is the best model in all objectives. Different "optimal" solutions/models can have different trade-offs on the multiple objective functions. For example, model M_1 may have better performance on "energy consumption" than another model M_2, while

M_2 may have better performance on "throughput" than model M_1. In this case, we call a model a non-dominated solution or "optimal" solution, if there is no other solution that can achieve better performance in all objectives than the model. Thus the task of multi-objective optimization is to obtain a set of multiple "optimal" models/solutions, such that all the models in the set are "optimal" and cannot be dominated by any other solution or dominate each other. Such a model set is called a "Pareto-optimal set," where there is a potentially infinite number of "optimal" solutions in the model space. It is usually infeasible to get a set of Pareto-optimal models by exhaustive search.

Moreover, in some real-world applications, the evaluation of the performance of a model/solution with reference to multiple criteria is not free, but can be rather expensive. One example is the hardware design problem. If we would like to test the performance of a model design in hardware, synthesis of only one design can take hours or even days. The major cost of the learning process is on the model evaluation step, instead of the label acquiring step. This is how active learning comes into play. We could perform active learning in the model/design space, instead of the data instance space. The work in [138] studied the problem of selectively sampling the model/design space to predict the Pareto-optimal set, and proposed a method called Pareto Active Learning (PAL). The performance on multiple objectives is modeled as a Gaussian process, and the designs are estimated to iteratively select the models that can maximize the model searching process.

22.6.11 Variable Labeling Costs

In the previous discussion, we considered cases where the costs are fixed for each training instance. This may, however, often not be the case, where some instances may be harder to label than others. For example, consider the case where the labels correspond to information extracted from text documents. In practice, the documents may correspond to varying lengths and complexity, and this may reflect the variable labeling cost associated with these documents. This interesting scenario has been discussed in [66,67,86,108]. The difference in these many scenarios is in how the labeling cost is defined, and whether it is known prior to the learning process. The work in [108] provides a detailed experimental study of the annotation times over different instances in different domains, and shows that these times can be significantly different, and can also depend on the underlying data domain.

22.6.12 Active Transfer Learning

Active learning and transfer learning are two different ways to reduce labeling costs in classification problems. Most of the previous works discussed these two problem settings separately. However, in some scenarios, it can be more effective to combine the two settings via active transfer learning. Transfer learning focuses on exploiting the labeled data from a related domain (called source domain/dataset) to facilitate the learning process in a target domain/dataset. The basic idea is that if the source dataset is very similar to the target dataset, we can save a lot of labeling costs in the target domain by transferring the existing knowledge in the source domain to the target domain. On the other hand, active learning focuses on selective sampling of unlabeled instances in the target domain.

The work in [118] first studied the problem of combining active learning with transfer learning. The idea is that both transfer learning and active learning have their own strengths and weaknesses. It is possible to combine them to form a potentially better framework. In active transfer learning, the transfer learner can have two different choices during the domain knowledge transferring. One choice is to directly transfer an instance from the source domain to the target domain following conventional transfer learning methods. A second choice is to query the oracle with an instance for the label instead of directly transfering the labels.

The work in [137] studied a second scenario where the unlabeled data in the target domain is not available. Only the source domain has unlabeled instances. The task is to actively select some unlabeled instances from the source domain and query the oracle with the instance. The queried instances should be able to benefit the target domain in the transfer learning process. The work in [20] integrated the active learning and transfer learning into one unique framework using a single convex optimization. The work in [125] provided a theoretical analysis on active transfer learning.

The work in [40] extended the active transfer learning problem with the multi-oracle settings, where there can be multiple oracles in the system. Each oracle can have different expertise, which can be hard to estimate because the quantity of instances labeled by all the participating oracles is rather small. The work in [131] extended the active transfer learning framework into the recommendation problems, where the knowledge can be actively transferred from one recommendation system to another recommendation system.

22.6.13   Active Reinforcement Learning

Active learning has also been widely used beyond supervised learning. In reinforcement learning, there are many scenarios where active learning could help improve the performance while reducing human efforts/costs during the learning process. The work in [36] studied the problem of active reinforcement learning, where transition probabilities in a Markov Decision Process (MDP) can be difficult/expensive for experts to specify. The idea is to combine offline planning and online exploration in one framework that allows oracles to specify possibly inaccurate models of the world offline.

The work in [65] studied the active imitation learning that can be reduced to I.I.D active learning. Imitation learning usually learns a target policy by observing full execution trajectories, which can be hard/expensive to generate in practice. The idea is that we could actively query the expert about the desired action at individual states, based upon the oracle's answers to previous queries.

22.7   Conclusions

This chapter discusses the methodology of active learning, which is relevant in the context of cases where it is expensive to acquire labels. While the approach has a similar motivation to many other methods, such as semi-supervised learning and transfer learning, the approach used for dealing with the paucity of labels is quite different. In active learning, instances from the training data are modeled with the use of a careful sampling approach, which maximizes the accuracy of classification, while reducing the cost of label acquisition. Active learning has also been generalized to structured data with dependencies such as sequences and graphs. The problem is studied in the context of a wide variety of different scenarios such as feature-based acquisition and data streams. With the increasing amounts of data available in the big-data world, and the tools now available to us through crowd sourcing methodologies such as Amazon Mechanical Turk, batch processing of such data sets has become an imperative. While some methodologies still exist for these scenarios, a significant scope exists for the further development of technologies for large scale active learning.

Bibliography

[1] N. Abe and H. Mamitsuka. Query learning strategies using boosting and bagging. *ICML Conference*, pages 1–9, 1998.

[2] C. Aggarwal. *Outlier Analysis*, Springer, 2013.

[3] D. Angluin. Queries and concept learning. *Machine Learning*, 2:319–342, 1988.

[4] M. Ankerst, M. Ester, and H.-P. Kriegel. Towards an effective cooperation of the user and the computer for classification, *ACM KDD Conference*, 2000.

[5] F. R. Bach. Active learning for misspecified generalized linear models. In *NIPS*, pages 65–72, 2006.

[6] M. F. Balcan, A. Beygelzimer, and J. Langford. Agnostic active learning. *Journal of Computer and System Sciences*, 75(1):78–89, 2009.

[7] M. F. Balcan, A. Blum, and K. Yang. Co-training and expansion: Towards bridging theory and practice. *NIPS*, pages 89–96, 2005.

[8] J. Baldridge and M. Osborne. Active learning and the total cost of annotation. *Proceedings of the Conference on Empirical Methods in Natural Language Processing*, pages 9–16, 2004.

[9] M. Belkin and P. Niyogi. Semi-supervised Learning on Riemannian Manifolds. *Machine Learning*, 56:209–239, 2004.

[10] M. Bilgic, L. Mihalkova, and L. Getoor. Active learning for networked data. *ICML Conference*, pp. 79–86, 2010.

[11] M. Bilgic and L.Getoor. Effective Label Acquisition for Collective Classification. *ACM KDD Conference*, pages 43–51, 2008.

[12] A. Beygelzimer, S. Dasgupta, and J. Langford. Importance weighted active learning. *ICML Conference*, pages 49–56, 2009.

[13] A. Beygelzimer, D. Hsu, J. Langford, and T. Zhang. Agnostic active learning without constraints. In *NIPS*, pp. 199–207, 2010.

[14] O. Bousquet, S. Boucheron, and G. Lugosi. Introduction to statistical learning theory. In *Advanced Lectures on Machine Learning*, pages 169–207, 2003.

[15] A. Blum and S. Chawla. Learning from labeled and unlabeled data using graph mincuts. *ICML Conference*, pages 19–26, 2001.

[16] K. Brinker. Incorporating diversity in active learning with support vector machines. *ICML Conference*, pages 59–66, 2003.

[17] K. Brinker. On active learning in multi-label classification. *Studies in Classication, Data Analysis, and Knowledge Organization*, Springer, 2006.

[18] N. Cesa-Bianchi, C. Gentile, F. Vitale, and G. Zappella. Active learning on trees and graphs. *COLT*, pages 320–332, 2010.

[19] X. Chai, L. Deng, Q. Yang, and C.X. Ling. Test-cost sentistive naive Bayes classification. *IEEE Conference on Data Mining (ICDM)*, pages 51–60, 2004.

[20] R. Chattopadhyay, W. Fan, I. Davidson, S. Panchanathan, and J. Ye. Joint transfer and batch-mode active learning. *ICML Conference*, pages 253–261, 2013

[21] D. Cohn, L. Atlas, R. Ladner, M. El-Sharkawi, R. Marks II, M. Aggoune, and D. Park. Training connectionist networks with queries and selective sampling. *Advances in Neural Information Processing Systems (NIPS)*, pages 566–573, 1990.

[22] D. Cohn, L. Atlas, and R. Ladner. Improving generalization with active learning. *Machine Learning*, 5(2):201–221, 1994.

[23] D. Cohn, Z. Ghahramani, and M. Jordan. Active learning with statistical models. *Journal of Artificial Intelligence Research*, 4:129–145, 1996.

[24] O. Chapelle, B. Scholkopf, and A. Zien. Semi-Supervised Learning. Vol. 2, *MIT Press*, Cambridge, MA, 2006.

[25] S.F. Chen and R. Rosenfeld. A survey of smoothing techniques for ME models. *IEEE Transactions on Speech and Audio Processing*, 8(1):37–50, 2000.

[26] W. Chu, M. Zinkevich, L. Li, A. Thomas, and B. Tseng. Unbiased online active learning in data streams. *ACM KDD Conference*, pages 195–203, 2011.

[27] A. Culotta and A. McCallum. Reducing labeling effort for structured prediction tasks, *AAAI Conference*, pages 746–751, 2005.

[28] I. Dagan and S. Engelson. Committee-based sampling for training probabilistic classifiers. *ICML Conference*, pages 150–157, 1995.

[29] W. Dai, Y. Chen, G.-R. Xue, Q. Yang, and Y. Yu. Translated learning: Transfer learning across different feature spaces. *Proceedings of Advances in Neural Information Processing Systems*, pages 353–360, 2008.

[30] S. Dasgupta, A. Kalai, and C. Monteleoni. Analysis of perceptron-based active learning. *Learning Theory*, pages 249–263, 2005.

[31] P. Donmez, J. G. Carbonell, and P. N. Bennett. Dual strategy active learning. *ECML conference*, pages 116–127, 2007.

[32] P. Donmez, J. Carbonell, and J. Schneider. Efficiently learning the accuracy of labeling sources for selective sampling. *KDD Conference*, pages 259–268, 2009.

[33] T. Dietterich, R. Lathrop, and T. Lozano-Perez. Solving the multiple-instance problem with axis-parallel rectangles. *Artificial Intelligence*, 89:31–71, 1997.

[34] G. Druck, B. Settles, and A. McCallum. Active learning by labeling features. *EMNLP*, pages 81–90, 2009.

[35] R. El-Yaniv and D. Pechyony. Transductive rademacher complexity and its applications. *Journal of Artificial Intelligence Research*, 35:193–234, 2009.

[36] A. Epshteyn, A. Vogel, G. DeJong. Active reinforcement learning. *ICML Conference*, pages 296–303, 2008.

[37] S. Esmeir and S. Markovitch. Anytime induction of cost-sensitive trees. *Advances in Neural Information Processing Systems (NIPS)*, pages 425–432, 2008.

[38] A. Esuli and F. Sebastiani. Active learning strategies for multi-label text classification. *ECIR Conference*, pages 102–113, 2009.

[39] M. Fang, X. Zhu, B. Li, W. Ding, and X. Wu, Self-taught active learning from crowds. *ICDM Conference*, pages 273–288, 2012.

[40] M. Fang, J. Yin, and X. Zhu. Knowledge transfer for multi-labeler active learning *ECMLPKDD Conference*, 2013

[41] W. Fan, Y. A. Huang, H. Wang, and P. S. Yu. Active mining of data streams. *SIAM Conference on Data Mining*, 2004.

[42] A. Fujii, T. Tokunaga, K. Inui, and H. Tanaka. Selective sampling for example-based word sense disambiguation. *Computational Linguistics*, 24(4):573–597, 1998.

[43] Y. Freund and R.E. Schapire. A decision-theoretic generalization of on-line learning and an application to boosting. *Journal of Computer and System Sciences*, 55(1):119–139, 1997.

[44] Y. Freund, H.S. Seung, E. Shamir, and N. Tishby. Selective samping using the query by committee algorithm. *Machine Learning*, 28:133–168, 1997.

[45] S. Geman, E. Bienenstock, and R. Doursat. Neural networks and the bias/variance dilemma. *Neural Computation*, 4:1–58, 1992.

[46] R. Gilad-Bachrach, A. Navot, and N. Tishby. Query by committee made real. In *NIPS*, 2005.

[47] J. Goodman. Exponential priors for maximum entropy models. *Human Language Technology and the North American Association for Computational Linguistics*, pages 305–312, 2004.

[48] R. Greiner, A. Grove, and D. Roth. Learning cost-sensitive active classifiers. *Artificial Intelligence*, 139, pages 137–174, 2002.

[49] Q. Gu and J. Han. Towards active learning on graphs: An error bound minimization approach. *IEEE International Conference on Data Mining*, pages 882–887, 2012.

[50] Q. Gu, C. Aggarwal, J. Liu, and J. Han. Selective sampling on graphs for classification. *ACM KDD Conference*, pages 131–139, 2013.

[51] A. Guillory and J. Bilmes. Active semi-supervised learning using submodular functions. *UAI*, 274–282, 2011.

[52] A. Guillory and J. A. Bilmes. Label selection on graphs. *NIPS*, pages 691–699, 2009.

[53] Y. Guo and R. Greiner. Optimistic active learning using mutual information. *International Joint Conference on Artificial Intelligence*, pages 823–829, 2007.

[54] S. Hanneke. A bound on the label complexity of agnostic active learning. In *ICML*, pages 353–360, 2007.

[55] S. C. H. Hoi, R. Jin, and M.R. Lyu. Large-scale text categorization by batch mode active learning. *World Wide Web Conference*, pages 633–642, 2006.

[56] S. C. H. Hoi and R. Jin. Active kernel learning. *ICML*, 2008.

[57] S. C. H. Hoi, R. Jin, J. Zhu, and M.R. Lyu. Batch mode active learning and its application to medical image classification. *International Conference on Machine Learning (ICML)*, pages 417–424, 2006.

[58] S. C. H. Hoi, R. Jin, J. Zhu, and M.R. Lyu. Semi-supervised SVM batch mode active learning for image retrieval. *IEEE Conference on CVPR*, pages 1–3, 2008.

[59] S. Huang, R. Jin, and Z. Zhou. Active learning by querying informative and representative examples. *NIPS Conference*, pages 892–200, 2011.

[60] R. Hwa. Sample selection for statistical parsing. *Computational Linguistics*, 30:73–77, 2004.

[61] M. Ji and J. Han. A variance minimization criterion to active learning on graphs. *AISTATS*, pages 556–554, 2012.

[62] S. Ji and L. Carin. Cost-sensitive feature acquisition and classification. *Pattern Recognition*, 40:1474–1485, 2007.

[63] N. Jin, C. Young, and W. Wang. GAIA: graph classification using evolutionary computation. In *SIGMOD*, pages 879–890, 2010.

[64] A. Joshi, F. Porikli, and N. Papanikolopoulos. Multi-class active learning for image classification In *CVPR Conference*, 2009.

[65] K. Judah, A. Fern, and T. Dietterich. Active imitation learning via reduction to I.I.D. active learning. In *UAI Conference*, 2012.

[66] A. Kapoor, E. Horvitz, and S. Basu. Selective supervision: Guiding supervised learning with decision theoretic active learning. *Proceedings of International Joint Conference on Artificial Intelligence (IJCAI)*, pages 877–882, 2007.

[67] R.D. King, K.E. Whelan, F.M. Jones, P.G. Reiser, C.H. Bryant, S.H. Muggleton, D.B. Kell, and S.G. Oliver. Functional genomic hypothesis generation and experimentation by a robot scientist. *Nature*, 427 (6971):247–252, 2004.

[68] C. Korner and S. Wrobel. Multi-class ensemble-based active learning. *European Conference on Machine Learning*, pp. 687–694, pages 687–694, 2006.

[69] X. Kong and P. S. Yu. Semi-supervised feature selection for graph classification. In *KDD*, pages 793–802, 2010.

[70] X. Kong, W. Fan, and P. Yu. Dual Active feature and sample selection for graph classification. *ACM KDD Conference*, pages 654–662, 2011.

[71] V. Krishnamurthy. Algorithms for optimal scheduling and management of hidden Markov model sensors. *IEEE Transactions on Signal Processing*, 50(6):1382–1397, 2002.

[72] B. Krishnapuram, D. Williams, Y. Xue, L. Carin, M. Figueiredo, and A. Hartemink. Active learning of features and labels. *ICML Conference*, 2005.

[73] A. Kuwadekar and J. Neville, Relational active learning for joint collective classification models. *ICML Conference*, pages 385–392, 2011.

[74] D. Lewis and W. Gale. A sequential algorithm for training text classifiers. *ACM SIGIR Conference*, pages 3–12, 1994.

[75] D. Lewis and J. Catlett. Heterogeneous uncertainty sampling for supervised learning. *ICML Conference*, pages 148–156, 1994.

[76] X. Li and Y. Guo. Active learning with multi-label SVM classification. *IJCAI Conference*, pages 14–25, 2013.

[77] X. Li, D. Kuang, and C.X. Ling. Active learning for hierarchical text classication. *PAKDD Conference*, pages 14–25, 2012.

[78] X. Li, D. Kuang and C.X. Ling. Multilabel SVM active learning for image classification. *ICIP Conference*, pages 2207–2210, 2004.

[79] C. X. Ling, Q. Yang, J. Wang, and S. Zhang. Decision trees with minimal costs. *ICML Conference*, pages 483–486, 2004.

[80] H. Liu, H. Motoda, and L. Yu. A selective sampling approach to active feature selection. *Artificial Intelligence*, 159(1-2):49–74, 2004

[81] H. Liu, H. Motoda, and L. Yu. Feature selection with selective sampling. *ICML Conference*, pages 395–402, 2002

[82] H. Liu, L. Yu, M. Dash, and H. Motoda. Active feature selection using classes. *PAKDD conference*, page 474–485, 2003

[83] R. Lomasky, C. Brodley, M. Aernecke, D. Walt, and M. Friedl. Active class selection. *Proceedings of the European Conference on Machine Learning*, pages 640–647, 2007.

[84] A. McCallum and K. Nigam. Employing EM in pool-based active learning for text classification. *ICML Conference*, pages 359–367, 1998.

[85] D. MacKay. Information-based objective functions for active data selection. *Neural Computation*, 4(4):590–604, 1992.

[86] D. Margineantu. Active cost-sensitive learning. *International Joint Conference on Artificial Intelligence (IJCAI)*, pages 1622–1623, 2005.

[87] P. Melville, M. Saar-Tsechansky, F. Provost, and R. Mooney. An expected utility approach to active feature-value acquisition. *IEEE ICDM Conference*, pages 2005.

[88] P. Melville, M. Saar-Tsechansky, F. Provost, and R. Mooney. Active feature-value acquisition for classifier induction. *IEEE International Conference on Data Mining (ICDM)*, pages 483–486, 2004.

[89] P. Melville and R. Mooney. Diverse ensembles for active learning. *ICML Conference*, pages 584–591, 2004.

[90] I. Muslea, S. Minton, and C. Knoblock. Active + semi-supervised learning = robust multi-view learning. *ICML Conference*, pages 435-442, 2002.

[91] I. Muslea, S. Minton, and C. Knoblock. Active learning with multiple views. *Journal of Artificial Intelligence Research*, 27(1):203–233, 2006.

[92] I. Muslea, S. Minton, and C. Knoblock. Selective sampling with redundant views. *AAAI Conference*, pages 621–626, 2000.

[93] R. Moskovitch, N. Nissim, D. Stopel, C. Feher, R. Englert, and Y. Elovici. Improving the detection of unknown computer worms activity using active learning. *Proceedings of the German Conference on AI*, pages 489–493, 2007.

[94] K. Nigam, A. McCallum, S. Thrun, and T. Mitchell. Text classification from labeled and unlabeled documents using EM. *Machine Learning*, 39(2–3):103–134, 2000.

[95] A. McCallum and K. Nigam. Employing EM pool-based active learning for text classification. *ICML Conference*, pages 350–358, 1998.

[96] H. T. Nguyen and A. Smeulders. Active learning using pre-clustering. *ICML Conference*, pages 79–86, 2004.

[97] S. J. Pan, Q. Yang. A survey on transfer learning. *IEEE Transactons on Knowledge and Data Engineering*, 22(10):1345–1359, 2010.

[98] G. Qi, X. Hua, Y. Rui, J. Tang, and H. Zhang. Two-dimensional active learning for image classification. *CVPR Conference*, 2008.

[99] R. Reichart, K. Tomanek, U. Hahn, and A. Rappoport. Multi-task active learning for linguistic annotations. *ACL Conference*, 2008.

[100] N. Roy and A. McCallum. Toward optimal active learning through sampling estimation of error reduction. *ICML Conference*, pages 441–448, 2001.

[101] M. Saar-Tsechansky, P. Melville, and F. Provost. Active feature-value acquisition. *Management Science*, 55(4):664–684, 2009.

[102] T. Scheffer, C. Decomain, and S. Wrobel. Active hidden Markov models for information extraction. *International Conference on Advances in Intelligent Data Analysis (CAIDA)*, pages 309–318, 2001.

[103] A. I. Schein and L.H. Ungar. Active learning for logistic regression: An evaluation. *Machine Learning*, 68(3):253–265, 2007.

[104] G. Schohn and D. Cohn. Less is more: Active learning with support vector machines. *ICML Conference*, pages 839–846, 2000.

[105] B. Settles. Active learning. *Synthesis Lectures on Artificial Intelligence and Machine Learning*, 6(1):1–114, 2012.

[106] B. Settles and M. Craven. An analysis of active learning strategies for sequence labeling tasks. *Proceedings of the Conference on Empirical Methods in Natural Language Processing (EMNLP)*, pages 1069–1078, 2008.

[107] B. Settles, M. Craven, and S. Ray. Multiple-instance active learning. *Advances in Neural Information Processing Systems (NIPS)*, pages 1289–1296, 2008.

[108] B. Settles, M. Craven, and L. Friedland. Active learning with real annotation costs. *NIPS Workshop on Cost-Sensitive Learning*, 2008.

[109] H. S. Seung, M. Opper, and H. Sompolinsky. Query by committee. *ACM Workshop on Computational Learning Theory*, pages 287–294, 1992.

[110] V. S. Sheng and C. X. Ling. Feature value acquisition in testing: A sequential batch test algorithm. *ICML Conference*, pages 809–816, 2006.

[111] V. S. Sheng, F. Provost, and P.G. Ipeirotis. Get another label? Improving data quality and data mining using multiple, noisy labelers. *ACM KDD Conference*, pages 614–622, 2008

[112] M. Singh, E. Curran, and P. Cunningham. Active learning for multi-label image annotation. Technical report, University College Dublin, 2009

[113] T. Soukop and I. Davidson. *Visual Data Mining: Techniques and Tools for Data Visualization*, Wiley, 2002.

[114] M. Thoma, H. Cheng, A. Gretton, J. Han, H. Kriegel, A. Smola, L. Song, P. Yu, X. Yan, and K. Borgwardt. Near-optimal supervised feature selection among frequent subgraphs. In *SDM*, pages 1075–1086, 2009.

[115] M. Wang and X. S. Hua. Active learning in multimedia annotation and retrieval: A survey. *ACM Transactions on Intelligent Systems and Technology (TIST)*, 2(2):10, 2011.

[116] W. Wang and Z. Zhou. On multi-view active learning and the combination with semi-supervised learning. *ICML Conference*, pages 1152–1159, 2008.

[117] W. Wang and Z. Zhou. Multi-view active learning in the non-realizable case. *NIPS Conference*, pages 2388–2396, 2010.

[118] X. Shi, W. Fan, and J. Ren. Actively transfer domain knowledge. *ECML Conference*, pages 342–357, 2008

[119] Z. Xu, R. Akella, and Y. Zhang. Incorporating diversity and density in active learning for relevance feedback. *European Conference on IR Research (ECIR)*, pages 246–257, 2007.

[120] Z. Xu, K. Yu, V. Tresp, X. Xu, and J. Wang. Representative sampling for text classification using support vector machines. *ECIR Conference*, pages 393–407, 2003.

[121] X. Yan, H. Cheng, J. Han, and P. Yu. Mining significant graph patterns by leap search. In *SIGMOD*, pages 433–444, 2008.

[122] Y. Yan, R. Rosales, G. Fung, and J. Dy. Active learning from crowds. In *ICML*, 2011.

[123] X. Yan, P. S. Yu, and J. Han. Graph indexing based on discriminative frequent struture analysis. *ACM Transactions on Database Systems*, 30(4):960–993, 2005.

[124] B. Yang, J. Sun, T. Wang, and Z. Chen. Effective multi-label active learning for text classication. *ACM KDD Conference*, pages 917–926, 2009.

[125] L. Yang, S. Hanneke, and J. Carbonell A theory of transfer learning with applications to active learning. *Machine Learning*, 90(2):161–189, 2013.

[126] K. Yu, J. Bi, and V. Tresp. Active learning via transductive experimental design. *ICML Conference*, pages 1081–1088, 2006.

[127] H. Yu. SVM selective sampling for ranking with application to data retrieval. *ACM KDD Conference*, pages 354–363, 2005.

[128] Y. Zhang. Multi-task active learning with output constraints. *AAAI*, 2010

[129] T. Zhang and F.J. Oles. A probability analysis on the value of unlabeled data for classification problems. *ICML Conference*, pages 1191–1198, 2000.

[130] Q. Zhang and S. Sun. Multiple-view multiple-learner active learning. *Pattern Recognition*, 43(9):3113–3119, 2010.

[131] L. Zhao, S. Pan, E. Xiang, E. Zhong, Z. Lu, and Q. Yang Active transfer learning for cross-system recommendation. *AAAI Conference*, 2013

[132] Z. Zheng and B. Padmanabhan. On active learning for data acquisition. *IEEE International Conference on Data Mining*, pages 562–569, 2002.

[133] D. Zhou, O. Bousquet, T. N. Lal, J. Weston, and B. Scholkopf. Learning with local and global consistency. *NIPS Conference*, pages 321–328, 2003.

[134] Y. Zhu, S. J. Pan, Y. Chen, G.-R. Xue, Q. Yang, and Y. Yu. Heterogeneous transfer learning for image classification. *Special Track on AI and the Web, associated with The Twenty-Fourth AAAI Conference on Artificial Intelligence*, 2010.

[135] X. Zhu, J. Lafferty, and Z. Ghahramani. Combining active learning and semisupervised learning using Gaussian fields and harmonic functions. *Proceedings of the ICML Workshop on the Continuum from Labeled to Unlabeled Data*, pages 58–65, 2003.

[136] X. Zhu, P. Zhang, X. Lin, and Y. Shi. Active learning from data streams. *ICDM Conference*, pages 757–762, 2007.

[137] Z. Zhu, X. Zhu, Y. Ye, Y. Guo, and X. Xue. Transfer active learning. *CIKM Conference*, pages 2169–2172, 2011.

[138] M. Zuluaga, G. Sergent, A. Krause, M. Puschel, and Y. Shi. Active Learning for Multi-Objective Optimization. *JMLR*, 28(1):462–470, 2013.

Chapter 23

Visual Classification

Giorgio Maria Di Nunzio

University of Padua
Padova, Italy
dinunzio@dei.unipd.it

23.1 Introduction

Extracting meaningful knowledge from very large datasets is a challenging task that requires the application of machine learning methods. This task is called data mining, the aim of which is to retrieve, explore, predict, and derive new information from a given dataset. Given the complexity of the task and the size of the dataset, users should be involved in this process because, by providing adequate data and knowledge visualizations, the pattern recognition capabilities of the human can be used to drive the learning algorithm [6]. This is the goal of Visual Data Mining [78, 85]: to present the data in some visual form, allowing the human to get insight into the data, draw conclusions, and directly interact with the data [18]. In [75], the authors define visual data mining as "the process of interaction and analytical reasoning with one or more visual representations of an abstract data that leads to the visual discovery or robust patterns in these data that form the information and knowledge utilised in informed decision making."

Visual data mining techniques have proven to be of high value in exploratory data analysis and they also have a high potential for exploring large databases [31]. This is particularly important in a context where an expert user could make use of domain knowledge to either confirm or correct a dubious classification result. An example of this interactive process is presented in [83], where the graphical interactive approaches to machine learning make the learning process explicit by visualizing the data and letting the user 'draw' the decision boundaries. In this work, parameters and model selection are no longer required because the user controls every step of the inductive process.

By means of visualization techniques, researchers can focus and analzse patterns of data from datasets that are too complex to be handled by automated data analysis methods. The essential idea is to help researchers examine the massive information stream at the right level of abstraction through appropriate visual representations and to take effective actions in real-time [46]. Interactive visual data mining has powerful implications in leveraging the intuitive abilities of the human for data mining problems. This may lead to solutions that can model data mining problems in a more intuitive and unrestricted way. Moreover, by using such techniques, the user also has much better understanding of the output of the system even in the case of single test instances [1, 3].

The research field of Visual Data Mining has witnessed a constant growth and interest. In 1999, in a Guest Editor's Introduction of *Computer Graphics and Application Journal* [85], Wong writes:

> All signs indicate that the field of visual data mining will continue to grow at an even faster pace in the future. In universities and research labs, visual data mining will play a major role in physical and information sciences in the study of even larger and more complex scientific data sets. It will also play an active role in nontechnical disciplines to establish knowledge domains to search for answers and truths.

More than ten years later, Keim presents new challenges and applications [44]:

> Nearly all grand challenge problems of the 21st century, such as climate change, the energy crisis, the financial crisis, the health crisis and the security crisis, require the analysis of very large and complex datasets, which can be done neither by the computer nor the human alone. Visual analytics is a young active science field that comes with a mission of empowering people to find solutions for complex problems from large complex datasets. By tightly integrating human intelligence and intuition with the storage and processing power of computers, many recently developed visual analytics solutions successfully help people in analyzing large complex datasets in different application domains.

In this chapter, we focus on one particular task of visual data mining, namely visual classification. The classification of objects based on previously classified training data is an important area within data mining and has many real-world applications (see Section 23.3). The chapter is organized as follows: in this introduction, we present the requirements for Visual Classification (Section 23.1.1), a set of challenges (Section 23.1.3), and a brief overview of some of the approaches organized by visualization metaphors (Section 23.1.2); in Section 23.2, we present the main visualization approaches for visual classification. For each approach, we introduce at least one of the seminal works and one application. In Section 23.3, we present some of the most recent visual classification systems that have been applied to real-world problems. In Section 23.4, we give our final remarks.

23.1.1 Requirements for Visual Classification

Shneidermann defines the "Visual Information Seeking Mantra" as a set of tasks that the user should perform [71]: overview first, zoom and filter, then details-on-demand. Along with this concept, the author proposes a type by task taxonomy of information visualizations. He lists seven tasks and seven data types. The tasks are: 1) to gain an overview of the entire collection; 2) zoom in on items of interest; 3) filter out uninteresting items; 4) select an item or group and get details when needed; 5) view relationships among items; 6) keep a history of actions to support undo, replay, and progressive refinement; 7) allow extraction of sub-collections and of the details when needed. The data types are: mono-dimensional, two-dimensional, three-dimensional, temporal, multi-dimensional, tree, and network.

In [6], Ankerst and others discuss the reasons for involving the user in the process of classification: (i) by providing adequate data and knowledge visualizations, the pattern recognition capabilities of the human can be used to increase the effectivity of the classifier construction; (ii) the users have a deeper understanding of the resulting classifier; (iii) the user can provide domain knowledge to focus the learning algorithm better. Therefore, the main goal is to get a better cooperation between the user and the system: on the one hand, the user specifies the task, focuses the search, evaluates the results of the algorithm, and feeds his domain knowledge directly into the learning algorithm; on the other hand, the machine learning algorithm presents patterns that satisfy the specified user constraints and creates appropriate visualizations.

In [9], a list of desired requirements for the visualization of the structure of classifiers is discussed. This list addresses specific requirements for what the users should be able to do when interacting with visual classification systems:

1. to quickly grasp the primary factors influencing the classification with very little knowledge of statistics;

2. to see the whole model and understand how it applies to records, rather than the visualization being specific to every record;

3. to compare the relative evidence contributed by every value of every attribute;

4. to see a characterization of a given class, that is, a list of attributes that differentiate that class from others;

5. to infer record counts and confidence in the shown probabilities so that the reliability of the classifier's prediction for specific values can be assessed quickly from the graphics;

6. to interact with the visualization to perform classification;

7. the system should handle many attributes without creating an incomprehensible visualization or a scene that is impractical to manipulate.

23.1.2 Visualization Metaphors

Representing objects in two- or three-dimensional spaces is probably the most 'natural' metaphor a visualization system can offer to model object relationships. This is how we perceive the world as humans: two objects that are 'close' to each other are probably more similar than two objects far away. The interactive visualization and navigation of such space becomes a means to browse and explore the dataset that matches predetermined characteristics. In this section, we present a brief overview of some of the approaches covered in this chapter, divided into two groups: approaches that represent objects using the metaphor of proximity to indicate similarity between objects in a two-dimensional or three-dimensional space, and other approaches that use more complex metaphors.

23.1.2.1 2D and 3D Spaces

DocMINER [10] is a system which visualizes fine-granular relationships between single objects and allows the application of different object analysis methods. The actual mapping and visualization step uses a Self-Organizing Map (see Section 23.2.5.1). Given a distance metric, objects are mapped into a two-dimensional space, so that the relative error of the distances in this 2D space regarding the true distances of the objects is minimized. High-dimensional data sets contain several attributes, and finding interesting projections can be a difficult and time-consuming task for the analyst, since the number of possible projections increases exponentially with the number of concurrently visualized attributes. VizRank [52] is a method based on K-Nearest Neighbor distance [16] which is able to rank visual projections of classified data by their expected usefulness. Usefulness of a projection can then be defined as a property that describes how well clusters with different class values are geometrically separated. The system Bead [14] represents objects as particles in a three-dimensional space and the relationships between objects are represented by their relative spatial positions. In Galaxies [84], clusters of documents are displayed by reducing the high dimensional representation to a three-dimensional scatterplot. The key measurement for understanding this visualization is the notion of document similarity. ThemeScapes [84] is a three dimensional plot that mimics terrain topology. The surface of the terrain is intended to convey relevant information about topics and themes found within the corpus: elevation depicts theme strength, while valleys, cliffs and other features represent relationships between documents. In [68], the authors present the use of three-dimensional surfaces for visualizing the clusters of the results of a search engine. The system lets users examine resulting three-dimensional shapes and immediately see differences and similarities in the results. Morpheus [61] is a tool for an interactive exploration of clusters of objects. It provides visualization techniques to present subspace clustering results such that users can gain both an overview of the detected patterns and the understanding of mutual relationships among clusters.

23.1.2.2 More Complex Metaphors

MineSet [9] provides several visualization tools that enable users to explore data and discover new patterns. Each analytical mining algorithm is coupled with a visualization tool that aids users in understanding the learned models. Perception-Based Classification [6] (PBC) was introduced as an interactive decision tree classifier based on a multidimensional visualization technique. The user can selected the split attributes and split points at each node and thus constructed the decision tree manually (see Section 23.2.6.1). This technique not only depicts the decision tree but it also provides explanations as to why the tree was constructed this way. The Evidence Visualizer [8] can display Bayes model decisions as pies and bar charts. In particular, the rows of pie charts represent each attribute, and each pie chart represents an interval or value of the attribute. ExplainD [66] is a framework for explaining decisions made by classifiers that use additive evidence. It has been applied to different linear model such as support vector machines, logistic regression, and Naïve Bayes. The main goal of this framework is to visually explain the decisions of machine-learned classifiers and the evidence for those decisions. The Class Radial Visualization [69] is an integrated visualization system that provides interactive mechanisms for a deep analysis of classification results and procedures. In this system, class items are displayed as squares and equally distributed around the perimeter of a circle. Objects to be classified are displayed as colored points in the circle and the distance between the point and the squares represents the uncertainty of assigning that object to the class. In [65], the author presents two interactive methods to improve the results of a classification task: the first one is an interactive decision tree construction algorithm with a help mechanism based on Support Vector Machines (SVM); the second one is a visualization method used to try to explain SVM results. In particular, it uses a histogram of the data distribution according to the distance to the boundary, and linked to it is a set of scatter-plot matrices or the parallel coordinates. This method can also be used to help the user in the parameter tuning step of the SVM algorithm, and significantly reduce the time needed for the classification.

23.1.3 Challenges in Visual Classification

In [44], Keim and others discuss the challenges of the future of visualization systems. Even though each individual application and task has its own requirements and specific problems to solve, there are some common challenges that may be connected to the task of visual classification. The challenges are six: scalability, uncertainty, hardware, interaction, evaluation, and infrastructure. Scalability is probably one of the most important future challenges with the forthcoming 'era of big data.' Visual solutions need to scale in size, dimensionality, data types, and levels of quality. The relevant data patterns and relationships need to be visualized on different levels of detail, and with appropriate levels of data and visual abstraction. Dealing with uncertainty in visual analytics is nontrivial because of the large amount of noise and missing values. The notion of data quality and the confidence of the algorithms for data analysis need to be appropriately represented. The analysts need to be aware of the uncertainty and be able to analyze quality properties at any stage of the data analysis process. Efficient computational methods and powerful hardware are needed to support near real time data processing and visualization for large data streams. In addition to high-resolution desktop displays, advanced display devices such as large-scale power walls and small portable personal assistants need to be supported. Visual analytics systems should adapt to the characteristics of the available output devices, supporting the visual analytics workflow on all levels of operation. Novel interaction techniques are needed to fully support the seamless intuitive visual communication with the system. User feedback should be taken as intelligently as possible, requiring as little user input as possible. A theoretically founded evaluation framework needs to be developed to assess the effectiveness, efficiency, and user acceptance of new visual analytics techniques, methods, and models. For a deeper analysis of these challenges, we suggest [46].

Interaction, evaluation, and infrastructure have recently been discussed in the ACM International Conference of Tabletops and Surfaces. In [53], the authors present the development of novel interaction techniques and interfaces for enhancing collocated multiuser collaboration so as to allow multiple users to explore large amounts of data. They build case studies where multiple users are going to interact with visualizations of a large data set like biology datasets, social networks datasets, and spatial data.

23.1.4 Related Works

In this section, we want to give the reader some complementary readings about surveys on visual data mining. Compared to ours, these surveys have different objectives and do not focus on the specific problem of visual classification. These surveys discuss issues that are very important but go beyond the scope of this chapter. For example: how to choose the appropriate visualization tool, advantages and disadvantages, strengths and weaknesses of each approach, and how to extend basic visualization approaches.

In [18], an overview of the techniques available under the light of different categorizations is presented. The role of interaction techniques is discussed, as well as the important question of how to select an appropriate visualization technique for a task.

The problem of identifying adequate visual representation is also discussed in [56]. The authors classify the visual techniques in two classes: technical and interactive techniques. For each approach they discuss advantages and disadvantages in visualizing data to be mined.

[11] presents how to integrate visualization and data mining techniques for knowledge discovery. In particular, this work looks at strengths and weaknesses of information visualization techniques and data mining techniques.

In [25], the authors present a model for hierarchical aggregation in information visualization for the purpose of improving overview and scalability of large scale visualization. A set of standard visualization techniques is presented and a discussion of how they can be extended with hierarchical aggregation functionality is given.

23.2 Approaches

In this section, we present an overview of many of the most important approaches used in data visualization that have been applied to visual classification. This survey is specifically designed to present only visual classification approaches. For each approach, we added a reference to at least one of the seminal works and one example of an application for the specific classification task. We did not enter into discussions on the appropriateness, advantages, and disadvantages of each technique, which can be found in other surveys presented in Section 23.1.4. We present the approaches in alphabetical order: nomograms, parallel coordinates, radial visualizations, scatter plots, topological maps, and trees. All the figures in this Section were produced with R,[1] and the code to reproduce these plots can be downloaded.[2]

23.2.1 Nomograms

A nomogram[3] is any graphical representation of a numerical relationship. Invented by French mathematician Maurice d'Ocagne in 1891 a nomogram has as its primary purpose to enable the user to graphically compute the outcome of an equation without needing a computer or calculator. Today, nomograms are often used in medicine to predict illness based on some evidence. For example, [57] shows the utility of such a tool to estimate the probability of diagnosis of acute myocardial infarction. In this case, the nomogram is designed in such a way that it can be printed on paper and easily used by physicians to obtain the probability of diagnosis without using any calculator or computer. There are a number of nomograms used in daily clinical practice for prognosis of outcomes of different treatments, especially in the field of oncology [43, 81]. In Figure 23.1, an example of a nomogram for predicting the probability of survival given factors like age and cholesterol level is shown.[4]

The main benefit of this approach is simple and clear visualization of the complete model and the quantitative information it contains. The visualization can be used for exploratory analysis and classification, as well as for comparing different probabilistic models.

---

[1]http://www.r-project.org/
[2]http://www.purl.org/visualclassification
[3]http://archive.org/details/firstcourseinnom00broduoft
[4]http://cran.r-project.org/web/packages/rms/index.html

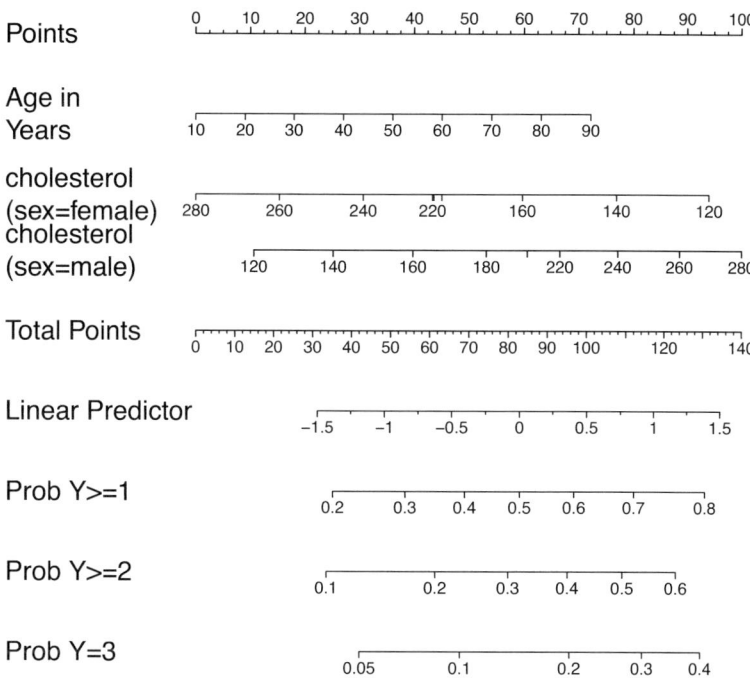

FIGURE 23.1: Nomograms. Given the age of a person and the level of cholesterol of the patient, by drawing a straight line that connects these two points on the graph, it is possible to read much information about survival probabilities ($Y >= 1$, $Y >= 2$, $Y = 3$) according to different combinations of features.

23.2.1.1 Naïve Bayes Nomogram

In [60], the authors propose the use of nomograms to visualize Naïve Bayes classifiers. This particular visualization method is appropriate for this type of classifiers since it clearly exposes the quantitative information on the effect of attribute values to class probabilities by using simple graphical objects (points, rulers, and lines). This method can also be used to reveal the structure of the Bayes classifier and the relative influences of the attribute values to the class probability and to support the prediction.

23.2.2 Parallel Coordinates

Parallel coordinates have been widely adopted for the visualization of high-dimensional and multivariate datasets [37, 38]. By using parallel axes for dimensions, the parallel coordinates technique can represent n-dimensional data in a 2-dimensional space; consequently, it can be seen as a mapping from the space R^n into the space R^2. The process to project a point of the n-dimensional space into the 2-dimensional space is the following: on a two-dimensional plane with cartesian coordinates, starting on the y-axis, n copies of the real line are placed parallel (and equidistant) to the y-axis. Each line is labeled from x_1 to x_n. A point c with coordinates $(c_1, c_2, ..., c_n)$ is represented by a polygonal line whose n vertices are at $(i-1, c_i)$ for $i = 1, ..., n$.

Since points that belong to the same class are usually close in the n-dimensional space, objects of the same class have similar polygonal lines. Therefore, one can immediately see groups of lines that correspond to points of the same class. Axes ordering, spacing, and filtering can significantly increase the effectiveness of this visualization, but these processes are complex for high dimensional datasets [86]. In [78], the authors present an approach to measure the quality of the parallel coordinates-view according to some ranking functions.

In Figure 23.2, an example of parallel coordinates to classify the Iris Dataset[5] is shown. The four-dimensional object has been projected onto four parallel coordinates. Flowers of the same kind show similar polygonal patterns; however, edge cluttering is already a problem even with this small number of objects. [6]

23.2.2.1 Edge Cluttering

Although parallel coordinates is a useful visualization tool, edge clutter prevents effective revealing of underlying patterns in large datasets [89]. The main cause of the visual clutter comes from too many polygonal lines. Clustering lines is one of the most frequently used methods to reduce the visual clutter and improve the perceptibility of the patterns in multivariate datasets. The overall visual clustering is achieved by geometrically bundling lines and forming patterns. The visualization can be enhanced by varying color and opacity according to the local line density of the curves.

Another approach to avoid edge cluttering is the angular histogram [29]. This technique considers each line-axis intersection as a vector, then both the magnitude and direction of these vectors are visualized to demonstrate the main trends of the data. Users can dynamically interact with this new plot to investigate and explore additional patterns.

23.2.3 Radial Visualizations

A radial display is a visualization paradigm in which information is laid out on a circle, ellipse, or spiral on the screen. Perhaps the earliest use of a radial display in statistical graphics was the pie chart. However, the pie chart has some limitations. In particular, when the wedges in a pie chart are almost the same size, it is difficult to determine visually which wedge is largest. A bar chart is generally better suited for this task. For example, the Evidence Visualizer [8, 13] can display Bayes model decisions as pie and bar charts. In particular, the rows of pie charts represent each attribute, and each pie chart represents an interval or value of the attribute.

Many radial techniques can be regarded as projections of a visualization from a Cartesian coordinate system into a polar coordinate system. The idea behind a radial visualization is similar to the one of parallel coordinates; however, while the space needed for Parallel Coordinates increases with the number of dimensions, the space used by a radial visualization remains fixed by the area of the circle. An example is Radviz [34, 35] where n-dimensional objects are represented by points inside a circle. The visualized attributes correspond to points equidistantly distributed along the circumference of the circle. [24] presents a survey of radial visualization, while [22] discusses advantages and drawbacks of these methods compared to classical Cartesian visualization.

---

[5]http://archive.ics.uci.edu/ml/datasets/Iris
[6]http://cran.r-project.org/web/packages/MASS/index.html

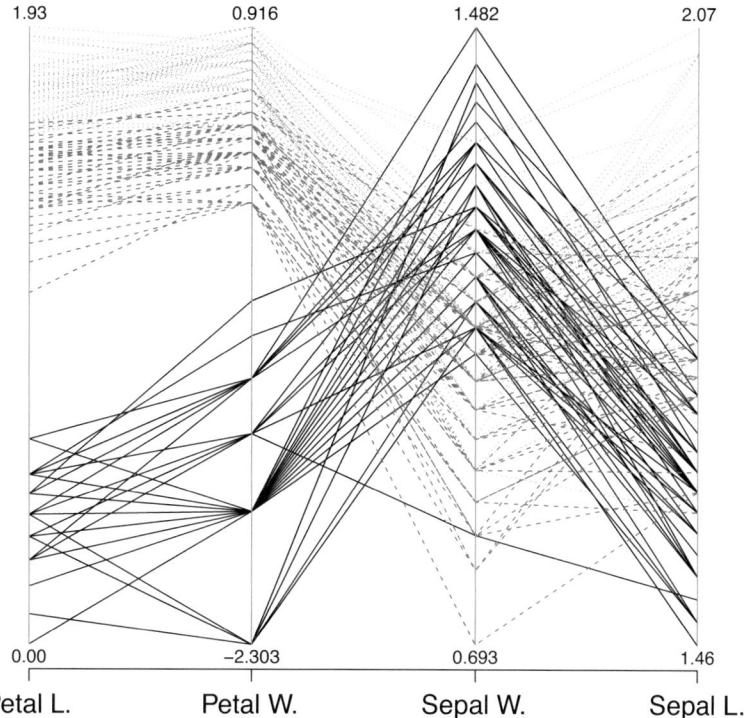

FIGURE 23.2 (See color insert.): Parallel coordinates. In this example, each object has four dimensions and represents the characteristics of a species of iris flower (petal and sepal width and length in logarithmic scale). The three types of lines represent the three kinds of iris. With parallel coordinates, it is easy to see common patterns among flowers of the same kind; however, edge cluttering is already visible even with a small dataset.

23.2.3.1 Star Coordinates

Star Coordinates represent each dimension as an axis radiating from the center of a circle to the circumference of the circle [40]. A multi-dimensional object is mapped onto one point on each axis based on its value in the corresponding dimension. StarClass [79] is a visualization tool that allows users to visualize multi-dimensional data by projecting each data object to a point on 2D display space using Star Coordinates.

In Figure 23.3, three five-dimensional objects are mapped on a star coordinate plot. Each coordinate is one of the axes radiating from the center. Objects that are similar in the originals pace have similar polygons, too.

Star coordinates have been successfully used in revealing cluster structures. In [87, 88], an approach called Hypothesis Oriented Verification and Validation by Visualization (HOV) ioffers a tunable measure mechanism to project clustered subsets and non-clustered subsets from a multidimensional space to a 2D plane. By comparing the data distributions of the subsets, users not only

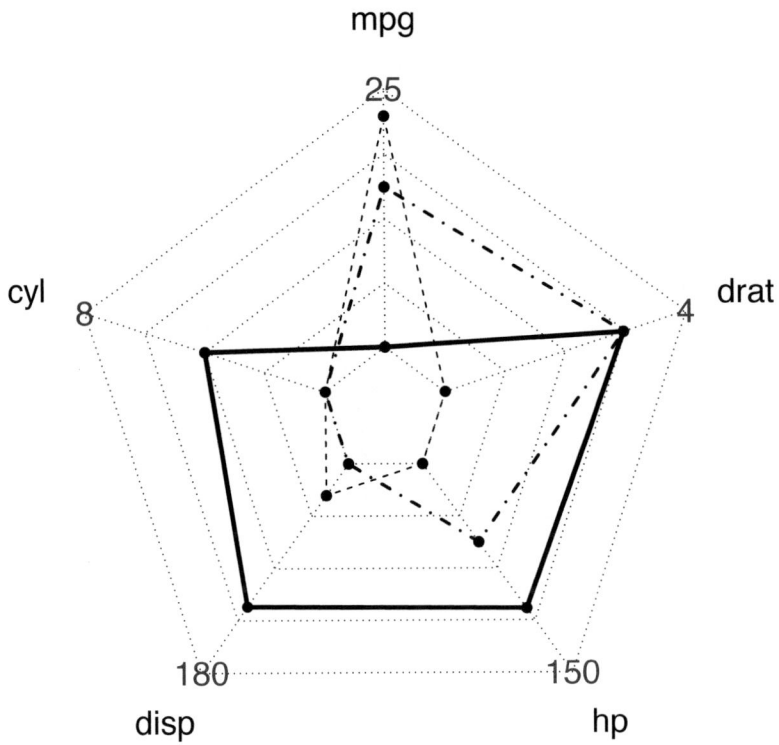

FIGURE 23.3: Star coordinates. Three five-dimensional objects are mapped on this star plot. Each coordinate is one of the axis radiating from the center. In this example, the three objects are cars described by features like: number of cylinders, horse power, and miles per gallon. The number at the end of each axis represents the maximum value for that dimension. Cars with similar features have similar polygons, too.

have an intuitive visual evaluation but also have a precise evaluation on the consistency of cluster structure by calculating geometrical information of their data distributions.

23.2.4 Scatter Plots

Scatter plots use Cartesian coordinates to display the values of two- or three-dimensional data. Since most problems in data mining involve data with large number of dimensions, dimensionality reduction is a necessary step to use this type of plot. Reduction can be performed by keeping only the most important dimensions, that is, only those that hold the most information, and by projecting some dimensions onto others. The reduction of dimensionality can lead to an increased capability of extracting knowledge from the data by means of visualization, and to new possibilities in designing efficient and possibly more effective classification schemes [82]. A survey on the methods of dimension reduction that focuses on visualizing multivariate data can be found in [26].

In Figure 23.4, a matrix of scatterplots shows all the possible combinations of features for the Iris Dataset. Even though the three species of flowers are not linearly separable, it is possible to study what pairs of features allow for a better separation. Even with this relatively few number of items, the problem of overlapping points is already visible.

In [45], the authors discuss the issue of the high degree of overlap in scatter plots in exploring large data sets. They propose a generalization of scatter plots where the analyst can control the degree of overlap, allowing the analyst to generate many different views for revealing patterns and relations from the data. In [78], an alternative solution to this problem is given by presenting a way to measure the quality of the scatter plots view according to some ranking functions. For example, a projection into a two-dimensional space may need to satisfy a certain optimality criterion that attempts to preserve distances between the class-means. In [20], a kind of projections that are similar to Fisher's linear discriminants, but faster to compute, are proposed. In [7], a type of plot that projects points on a two-dimensional plane called a similarity-dissimilarity plot is discussed. This plot provides information about the quality of features in the feature space, and classification accuracy can be predicted from the assessment of features on this plot. This approach has been studied on synthetic and real life datasets to prove the usefulness of the visualization of high dimensional data in biomedical pattern classification.

23.2.4.1 Clustering

In [19], the authors compare two approaches for projecting multidimensional data onto a two-dimensional space: Principal Component Analysis (PCA) and random projection. They investigate which of these approaches best fits nearest neighbor classification when dealing with two types of high-dimensional data: images and micro arrays. The result of this investigation is that PCA is more effective for severe dimensionality reduction, while random projection is more suitable when keeping a high number of dimensions. By using one of the two approaches, the accuracy of the classifier is greatly improved. This shows that the use of PCA and random projection may lead to more efficient and more effective nearest neighbour classification. In [70], an interactive visualization tool for high-speed power system frequency data streams is presented. A k-median approach for clustering is used to identify anomaly events in the data streams. The objective of this work is to visualize the deluge of expected data streams for global situational awareness, as well as the ability to detect disruptive events and classify them. [2] discusses a interactive approach for nearest neighbor search in order to choose projections of the data in which the patterns of the data containing the query point are well distinguished from the entire data set. The repeated feedback of the user over multiple iterations is used to determine a set of neighbors that are statistically significant and meaningful.

23.2.4.2 Naïve Bayes Classification

Naïve Bayes classifiers are one of the most used data mining approaches for classification. By using Bayes' rule, one can determine the posterior probability $Pr(c|x)$ that an object x belongs to a category c in the following way:

$$Pr(c|x) = \frac{Pr(x|c)Pr(c)}{Pr(x)} \tag{23.1}$$

where $Pr(x|c)$ is the likelihood function, $P(c)$ the prior probability of the category c, and $P(x)$ the probability of an object x.

"Likelihood projections" is an approach that uses the likelihood function $P(x|c)$ for nonlinear projections [76]. The coordinates of this "likelihood space" are the likelihood functions of the data for the various classes. In this new space, the Bayesian classifier between any two classes in the data space can be viewed as a simple linear discriminant of unit slope with respect to the axes representing the two classes. The key advantage of this space is that we are no longer restricted to considering only this linear discriminant. Classification can now be based on any suitable classifier that operates on the projected data. In [67], the likelihood space is used to classify speech audio.

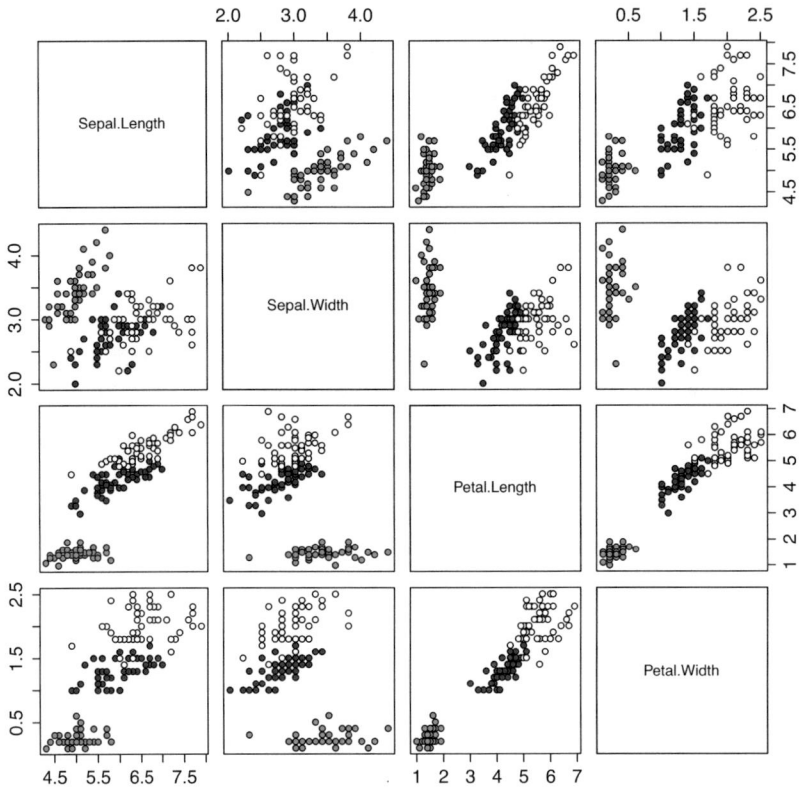

FIGURE 23.4: Scatter plots. In the Iris Dataset, flowers are represented by four-dimensional vectors. In this figure, the matrix of scatterplots presents all the possible two-dimensional combinations of features. The shade of grey of each point represents the kind of Iris. Some combinations allows for a better separation between classes; nevertheless, even with this relatively few number of items, the problem of overlapping points is already visible.

The projection of the audio data results in the transformation of diffuse, nebulous classes in high-dimensional space into compact clusters in the low-dimensional space that can be easily separated by simple clustering mechanisms. In this space, decision boundaries for optimal classification can be more easily identified using simple clustering criteria.

In [21], a similar approach is used as a visualization tool to understand the relationships between categories of textual documents, and to help users to visually audit the classifier and identify suspicious training data. When plotted on the Cartesian plane according to this formulation, the documents that belong to one category have specific shifts along the x-axis and the y-axis. This approach is very useful to compare the effect of different probabilistic models like Bernoulli, multinomial, or Poisson. The same approach can be applied to the problem of parameters optimization for probabilistic text classifiers, as discussed in [62].

23.2.5 Topological Maps

Topological maps are a means to project an n-dimensional input data into a two-dimensional data by preserving some hidden structure or relation among data [48]. The automatic systems that make this projection can automatically form two- or three-dimensional maps of features that are present in sets of input signals. If these signals are related metrically in some structured way, the same structure will be reflected in the low dimensional space. In [63], the authors show how traditional distance-based approaches fail in high-dimensional spaces and propose a framework that supports topological analysis of high dimensional document point clouds. They describe a two-stage method for topology-based projections from the original high dimensional information space to both 2D and 3D visualizations.

23.2.5.1 Self-Organizing Maps

A Self-Organizing Map (SOM) is a kind of neural network that preserves the topological properties of the input space by means of a neighborhood function [49]. It consists of units arranged as a two-dimensional or hexagonal grid where each unit represents a vector in the data space. During the training process, vectors from the dataset are presented to the map in random order and the unit with the highest response to a chosen vector and its neighborhood are adapted in such a way as to make them more responsive to similar inputs. SOMs are very useful for visualizing multidimensional data and the relationships among the data on a two-dimensional space. For example, [59] presents a typical result from the application of self-organizing maps to the problem of text classification. The grid of units represents the document collection, each unit being a class. Once the network has been trained, the grid shows how the different classes are "similar" to each other in terms of the distance on the grid.

In Figure 23.5, the result of the training of an SOM on a dataset of wines is shown. Each wine, described by a vector of 13 features, has been projected on a 5 by 5 hexagonal grid. The shape of each point (triangle, circle, cross) represents the original category of the wine; the shade of grey of each activation unit is the predicted label. [7]

Recently, SOMs have been used to study weather analysis and prediction. Since weather patterns have a geographic extent, weather stations that are geographically close to each other should reflect these patterns. In [28], data collected by weather stations in Brazil are analyzed to find weather patterns. Another important field of application of SOMs is DNA classification. In [58], the authors present an application of the hyperbolic SOM, a Self-Organizing Map that visualizes results on a hyperbolic surface. A hyperbolic SOM can perform visualization, classification, and clustering at the same time as a SOM; hyperbolic SOMs have the potential to achieve much better low-dimensional embeddings, since they offer more space due to the effect that in a hyperbolic plane the area of a circle grows asymptotically exponential with its radius. Moreover, it also incorporates links between neighboring branches which, in this particular research area, are very useful to study gene transfers in DNA.

23.2.5.2 Generative Topographic Mapping

Instead looking for spatial relations, like SOMs, one may think of correlations between the variables of the dataset. One way to capture this hidden structure is to model the distribution of the data in terms of hidden variables. An example of this approach is factor analysis, which is based on a linear transformation from data space to latent space. In [12], the authors extend this concept of a hidden variable framework into a Generative Topographic Mapping (GTM). The idea is very similar to the SOMs; however, the most significant difference between the GTM and SOM algorithms is that GTM defines an explicit probability density given by the mixture distribution of

---

[7]http://cran.r-project.org/web/packages/som/

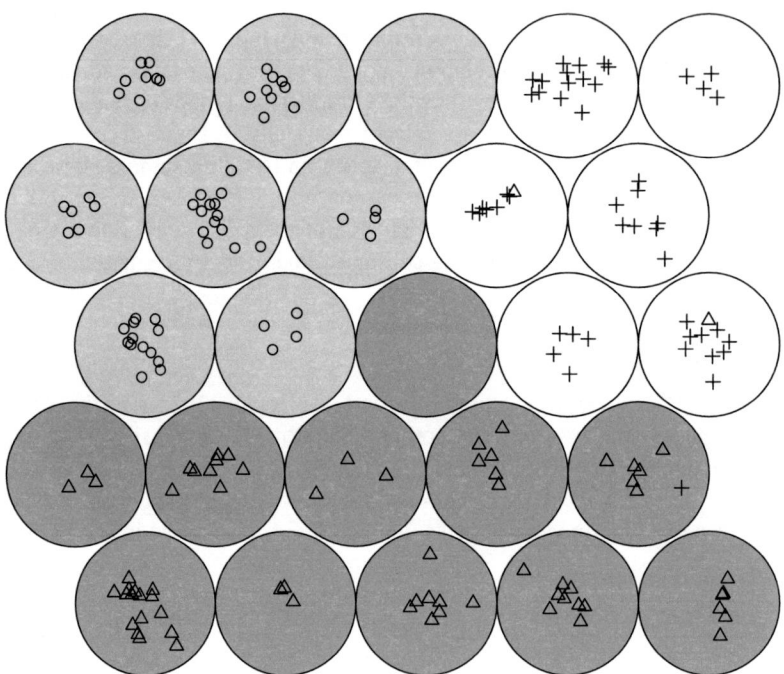

FIGURE 23.5: Self Organizing Maps. A 5×5 hexagonal SOM has been trained on a dataset of wines. Each point (triangle, circle, or cross) represents a wine that originally is described by a 13-dimensional vector. The shape of the point represents the category of the wine; the shade of grey of each activation unit (the big circles of the grid) is the predicted category. Wines that are similar in the 13-dimensional space are close to each other on this grid.

variables. As a consequence, there is a well-defined objective function given by the log likelihood, and convergence to a local maximum of the objective function is guaranteed by the use of the Expectation Maximization algorithm.

In [4], GTM is to cluster motor unit action potentials for the analysis of the behavior of the neuromuscular system. The aim of the analysis is to reveal how many motor units are active during a muscle contraction. This work compares the strength and weaknesses of GTM and principal component analysis (PCA), an alternative multidimensional projection technique. The advantage of PCA is that the method allows the visualization of objects in a Euclidian space where the perception of distance is easy to understand. On the other hand, the main advantage of the GTM is that each unit may be considered as an individual cluster, and the access to these micro-clusters may be very useful for elimination or selection of wanted or unwanted information.

23.2.6 Trees

During the 1980s, the appeal of graphical user interfaces encouraged many developers to create node-link diagrams. By the early 1990s, several research groups developed innovative methods of tree browsing that offered different overview and browsing strategies. For a history of the development of visualization tools based on trees, refer to [73]. In this section, we present four variants of visualization of trees: decision trees, tree maps, hyperbolic trees, and phylogenetic trees.

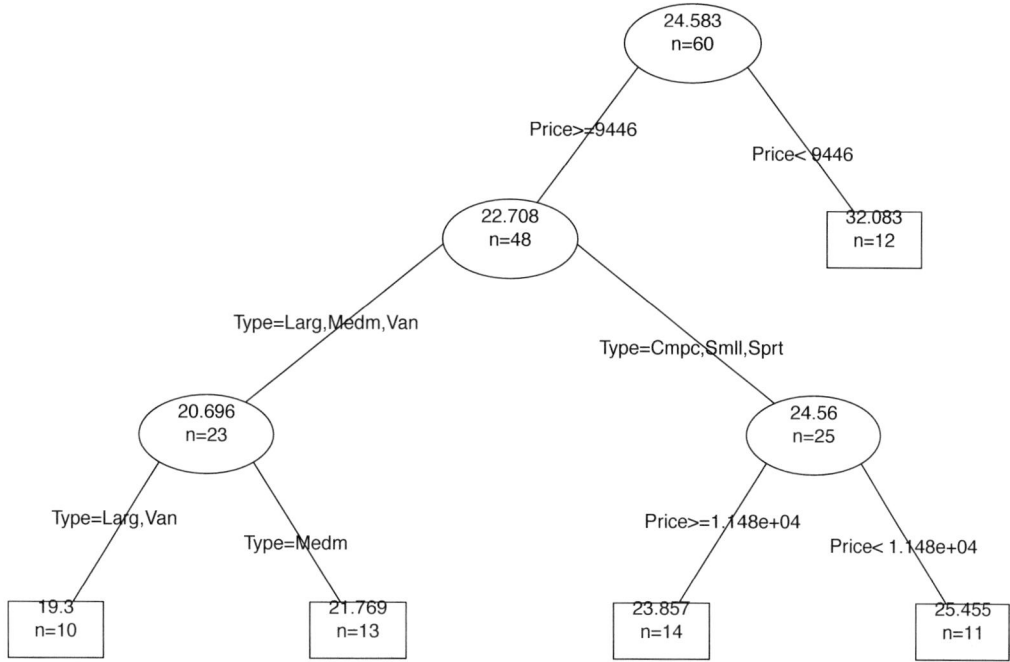

FIGURE 23.6: Decision trees. Each node of the tree predicts the average car mileage given the price, the country, the reliability, and the car type according to the data from the April 1990 issue of *Consumer Reports*. In this example, given the price and the type of the car, we are able to classify the car in different categories of gas consumption.

23.2.6.1 Decision Trees

A decision tree, also known as a classification tree or regression tree, is a technique for partitioning data into homogeneous groups. It is constructed by iteratively partitioning the data into disjoint subsets, and one class is assigned to each leaf of the tree. One of the first methods for building decision trees was CHAID [42]. This method partitions the data into mutually exclusive, and exhaustive, subsets that best describe the dependent variables.

In Figure 23.6, an example of a decision tree is shown. The dataset contains information about cars taken from the April 1990 issue of *Consumer Reports*.[8] Each node of the tree predicts the average car mileage given the price, the country, the reliability, and the car type.[9] In this example, given the price and the type of the car, we are able to classify the car in different categories of gas consumption by following a path from the root to a leaf.

Decision tree visualization and exploration is important for two reasons: (i) it is crucial to be able to navigate through the decision tree to find nodes that need to be further partitioned; (ii) exploration of the decision tree aids the understanding of the tree and the data being classified. In [5], the authors present an approach to support interactive decision tree construction. They show a method for visualizing multi-dimensional data with a class label such that the degree of impurity of each node with respect to class membership can be easily perceived by users. In [55], a conceptual model of the visualization support to the data mining process is proposed, together with a novel visualization of the decision tree classification process, with the aim of exploring human pattern recognition ability and domain knowledge to facilitate the knowledge discovery process. PaintingClass is a dif-

---

[8]http://stat.ethz.ch/R-manual/R-devel/library/rpart/html/cu.summary.html
[9]http://cran.r-project.org/web/packages/rpart/

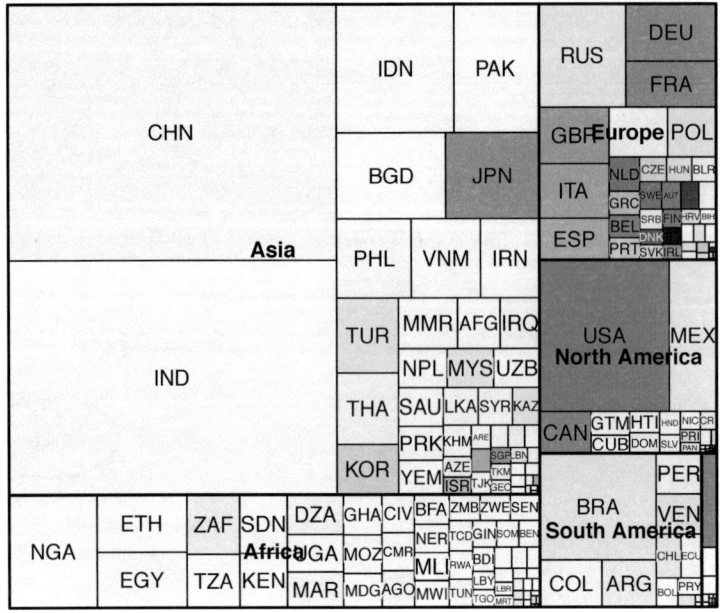

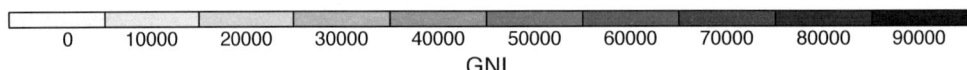

FIGURE 23.7 (See color insert.): Treemaps. This plot represents a dataset of 2010 about population size and gross national income for each country. The size of each node of the treemap is proportional to the size of the population, while the shade of blue of each box represents the gross national income of that country. The countries of a continent are grouped together into a rectangular area.

ferent interactive approach where the user interactively edits projections of multi-dimensional data and "paints" regions to build a decision tree [80]. The visual interaction of this systems combines Parallel Coordinates and Star Coordinates by showing this "dual" projection of the data.

23.2.6.2 Treemap

The Treemap visualization technique [72] makes use of the area available on the display, mapping hierarchies onto a rectangular region in a space-filling manner. This efficient use of space allows large hierarchies to be displayed and facilitates the presentation of semantic information. Each node of a tree map has a weight that is used to determine the size of a nodes-bounding box. The weight may represent a single domain property, or a combination of domain properties. A node's weight determines its display size and can be thought of as a measure of importance or degree of interest [39].

In Figure 23.7, a tree map shows the gross national income per country. Each box (the node of the tree) represents a country, and the size of the box is proportional to the size of the population of that country. The shade of grey of the box reflects the gross national income of the year 2010.[10]

---

[10]http://cran.r-project.org/web/packages/treemap/index.html

Treemaps can also be displayed in 3D [30]. For example, patent classification systems intel-
lectually organize the huge number of patents into pre-defined technology classes. To visualize the
distribution of one or more patent portfolios, an interactive 3D treemap can be generated, in which
the third dimension represents the number of patents associated with a category.

23.2.6.3 Hyperbolic Tree

Hyperbolic geometry provides an elegant solution to the problem of providing a focus and context display for large hierarchies [51]. The hyperbolic plane has the room to lay out large hierarchies,
with a context that includes as many nodes as are included by 3D approaches and with modest computational requirements. The root node is in the center with first-level nodes arranged around it in
a circle or oval. Further levels are placed in larger concentric circles or ovals, thus preserving a
two-dimensional planar approach. To ensure that the entire tree is visible, outer levels are shrunk
according to a hyperbolic formula. In [36], hyperbolic trees are used for spam classification. The
authors propose a Factors Hyperbolic Tree-based algorithm that, unlike the classical word and lexical matching algorithms, handles spam filtering in a dynamic environment by considering various
relevant factors.

23.2.6.4 Phylogenetic Trees

Phylogenetic trees are an alternative approach for the construction of object maps targeted at
reflecting similarity relationships [17]. By means of a distance matrix, the aim is inferring ancestors for a group of objects and reconstructing the evolutionary history of each object. The main
advantages of the approach are improved exploration and more clear visualization of similarity
relationships, since it is possible to build ancestry relationships from higher to lower content correlation. In [64], the authors present a phylogenetic tree to support image and text classification. They
discuss some challenges and advantages for using this type of visualization. A set of visualization
tools for visual mining of images and text is made possible by the properties offered by these trees,
complemented by the possibilities offered by multidimensional projections.

23.3 Systems

One of the most important characteristics of a visual classification system is that users should
gain insights about the data [15]. For example, how much the data within each class varies, which
classes are close to or distinct from each other, seeing which features in the data play an important
role to discriminate one class from another, and so on. In addition, the analysis of misclassified data
should provide a better understanding of which type of classes are difficult to classify. Such insight
can then be fed back to the classification process in both the training and the test phases.

In this section, we present a short but meaningful list of visual classification systems that have
been published in the last five years and that fulfill most of the previous characteristics. The aim
of this list is to address how visual classification systems support automated classification for real-
world problems.

23.3.1 EnsembleMatrix and ManiMatrix

EnsembleMatrix and ManiMatrix are two interactive visualization systems that allow users to
browse and learn properties of classifiers by comparison and contrast and build ensemble classification systems. These systems are specifically designed for Human and Computer Interaction

researchers who could benefit greatly from the ability to express user preferences about how a classifier should work.

EnsembleMatrix [77] allows users to create an ensemble classification system by discovering appropriate combination strategies. This system supplies a visual summary that spans multiple classifiers and helps users understand the models' various complimentary properties. EsnembleMatrix provides two basic mechanisms to explore combinations of classifiers: (i) partitioning, which divides the class space into multiple partitions; (ii) arbitrary linear combinations of the classifiers for each of these partitions.

The ManiMatrix (Manipulable Matrix) system is an interactive system that enables researchers to intuitively refine the behavior of classification systems [41]. ManiMatrix focuses on the manual refinement on sets of thresholds that are used to translate the probabilistic output of classifiers into classification decisions. By appropriately setting such parameters as the costs of misclassification of items, it is possible to modify the behavior of the algorithm such that it is best aligned with the desired performance of the system. ManiMatrix enables its users to directly interact with a confusion matrix and to view the implications of incremental changes to the matrix via a realtime interactive cycle of reclassification and visualization.

23.3.2 Systematic Mapping

Systematic mapping provides mechanisms to identify and aggregate research evidence and knowledge about when, how, and in what context technologies, processes, methods or tools are more appropriate for software engineering practices. [27] proposes an approach, named Systematic Mapping, based on Visual Text Mining (SM-VTM), that applies VTM to support the categorization and classification in the systematic mapping.

The authors present two different views for systematic mapping: cluster view and chronological view. Users can explore these views and interact with them, getting information to build other visual representations of a systematic map. A case study shows that there is a significant reduction of effort and time in order to conduct text categorization and classification activities in systematic mapping if compared with manual conduction. With this approach, it is possible to achieve similar results to a completely manual approach without the need of reading the documents of the collection.

23.3.3 iVisClassifier

The iVisClassifier system [15] allows users to explore and classify data based on Linear Discriminant Analysis (LDA), a supervised reduction method. Given a high-dimensional dataset with cluster labels, LDA projects the points onto a reduced dimensional representation. This low dimensional space provides a visual overview of the cluster's structure. LDA enables users to understand each of the reduced dimensions and how they influence the data by reconstructing the basis vector into the original data domain.

In particular, iVisClassifier interacts with all the reduced dimensions obtained by LDA through parallel coordinates and a scatter plot. By using heat maps, iVisClassifier gives an overview about clusters' relationships both in the original space and in the reduced dimensional space. A case study of facial recognition shows that iVisClassifier facilitates the interpretability of the computational model. The experiments showed that iVisClassifier can efficiently support a user-driven classification process by reducing human search space, e.g., recomputing LDA with a user-selected subset of data and mutual filtering in parallel coordinates and the scatter plot.

23.3.4 ParallelTopics

When analyzing large text corpora, questions pertaining to the relationships between topics and documents are difficult to answer with existing visualization tools. For example, what are the characteristics of the documents based on their topical distribution? and what documents contain multiple topics at once? ParallelTopics [23] is a visual analytics system that integrates interactive visualization with probabilistic topic models for the analysis of document collections.

ParallelTopics makes use of the Parallel Coordinate metaphor to present the probabilistic distribution of a document across topics. This representation can show how many topics a document is related to and also the importance of each topic to the document of interest. ParallelTopics also supports other tasks, which are also essential to understanding a document collection, such as summarizing the document collection into major topics, and presenting how the topics evolve over time.

23.3.5 VisBricks

The VisBricks visualization approach provides a new visual representation in the form of a highly configurable framework that is able to incorporate any existing visualization as a building block [54]. This method carries forward the idea of breaking up the inhomogeneous data into groups to form more homogeneous subsets, which can be visualized independently and thus differently.

The visualization technique embedded in each block can be tailored to different analysis tasks. This flexible representation supports many explorative and comparative tasks. In VisBricks, there are two level of analysis: the total impression of all VisBricks together gives a comprehensive high-level overview of the different groups of data, while each VisBrick independently shows the details of the group of data it represents.

23.3.6 WHIDE

The Web-based Hyperbolic Image Data Explorer (WHIDE) system is a Web visual data mining tool for the analysis of multivariate bioimages [50]. This kind of analysis spans the analysis of the space of the molecule (i.e., sample morphology) and molecular colocation or interaction. WHIDE utilizes hierarchical hyperbolic self-organizing maps (H2SOM), a variant of the SOM, in combination with Web browser technology.

WHIDE has been applied to a set of bio-images recorded to show field of view in tissue sections from a colon cancer study, to compare tissue from normal colon with tissue classified as tumor. The result of the use of WHIDE in this particular context has shown that this system efficiently reduces the complexity of the data by mapping each of the pixels to a cluster, and provides a structural basis for a sophisticated multimodal visualization, which combines topology preserving pseudo-coloring with information visualization.

23.3.7 Text Document Retrieval

In [32], the authors describe a system for the interactive creation of binary classifiers to separate a dataset of text document into relevant and non-relevant documents for improving information retrieval tasks. The problem they present is twofold: on the one hand, supervised machine learning algorithms rely on labeled data, which can be provided by domain experts; on the other hand, the optimization of the algorithms can be done by researchers. However, it is hard to find experts both in the domain of interest and in machine learning algorithms.

Therefore, the authors compare three approaches for interactive classifier training. These approaches incorporate active learning to various degrees in order to reduce the labeling effort as well as to increase effectiveness. Interactive visualization is then used for letting users explore the status of the classifier in the context of the labeled documents, as well as for judging the quality of the classifier in iterative feedback loops.

23.4 Summary and Conclusions

The exploration of large data sets is an important problem that has many complications. By means of visualization techniques, researchers can focus and analyze patterns of data from datasets that are too complex to be handled by automated data analysis methods. Interactive visual classification has powerful implications in leveraging the intuitive abilities of the human for this kind of data mining task. This may lead to solutions that can model classification problems in a more intuitive and unrestricted way.

The "Big Data Era" poses new challenges for visual classification since visual solutions will need to scale in size, dimensionality, data types, and levels of quality. The relevant data patterns and relationships will need to be visualized on different levels of detail, and with appropriate levels of data and visual abstraction.

The integration of visualization techniques with machine learning techniques is one of the many possible research paths in the future. This is confirmed by a recent workshop named "Information Visualization, Visual Data Mining and Machine Learning" [47], the aim of which was to tighten the links between the two communities in order to explore how each field can benefit from the other and how to go beyond current hybridization successes.

Bibliography

[1] Charu C. Aggarwal. Towards effective and interpretable data mining by visual interaction. *SIGKDD Explor. Newsl.*, 3(2):11–22, January 2002.

[2] Charu C. Aggarwal. On the use of human-computer interaction for projected nearest neighbor search. *Data Min. Knowl. Discov.*, 13(1):89–117, July 2006.

[3] Charu C. Aggarwal. Toward exploratory test-instance-centered diagnosis in high-dimensional classification. *IEEE Trans. on Knowl. and Data Eng.*, 19(8):1001–1015, August 2007.

[4] Adriano O. Andrade, Slawomir Nasuto, Peter Kyberd, and Catherine M. Sweeney-Reed. Generative topographic mapping applied to clustering and visualization of motor unit action potentials. *Biosystems*, 82(3):273–284, 2005.

[5] Mihael Ankerst, Christian Elsen, Martin Ester, and Hans-Peter Kriegel. Visual classification: an interactive approach to decision tree construction. In *Proceedings of the Fifth ACM SIGKDD International Conference on Knowledge Discovery and Data Mining*, KDD '99, ACM, pages 392–396, New York, NY, USA, 1999.

[6] Mihael Ankerst, Martin Ester, and Hans-Peter Kriegel. Towards an effective cooperation of the user and the computer for classification. In *KDD'00*, ACM, pages 179–188, Boston, MA, USA, 2000.

[7] Muhammad Arif. Similarity-dissimilarity plot for visualization of high dimensional data in biomedical pattern classification. *J. Med. Syst.*, 36(3):1173–1181, June 2012.

[8] Barry Becker, Ron Kohavi, and Dan Sommerfield. Visualizing the simple Baysian classifier. In Usama Fayyad, Georges G. Grinstein, and Andreas Wierse, editors, *Information Visualization in Data Mining and Knowledge Discovery*, pages 237–249. Morgan Kaufmann Publishers Inc., San Francisco, 2002.

[9] Barry G. Becker. Using mineset for knowledge discovery. *IEEE Computer Graphics and Applications*, 17(4):75–78, 1997.

[10] Andreas Becks, Stefan Sklorz, and Matthias Jarke. Exploring the semantic structure of technical document collections: A cooperative systems approach. In Opher Etzion and Peter Scheuermann, editors, *CoopIS*, volume 1901 of *Lecture Notes in Computer Science*, pages 120–125. Springer, 2000.

[11] Enrico Bertini and Denis Lalanne. Surveying the complementary role of automatic data analysis and visualization in knowledge discovery. In *Proceedings of the ACM SIGKDD Workshop on Visual Analytics and Knowledge Discovery: Integrating Automated Analysis with Interactive Exploration*, VAKD '09, ACM, pages 12–20, New York, NY, USA, 2009.

[12] Christopher M. Bishop, Markus Svensén, and Christopher K. I. Williams. GTM: The generative topographic mapping. *Neural Computation*, 10(1):215–234, 1998.

[13] Clifford Brunk, James Kelly, and Ron Kohavi. Mineset: An integrated system for data mining. In Daryl Pregibon, David Heckerman, Heikki Mannila, editors, *KDD-97*, AAAI Press, pages 135–138, Newport Beach, CA, August 14-17 1997.

[14] Matthew Chalmers and Paul Chitson. BEAD: Explorations in information visualization. In Nicholas J. Belkin, Peter Ingwersen, and Annelise Mark Pejtersen, editors, *SIGIR*, pages 330–337. ACM, 1992.

[15] Jaegul Choo, Hanseung Lee, Jaeyeon Kihm, and Haesun Park. iVisClassifier: An interactive visual analytics system for classification based on supervised dimension reduction. In *2010 IEEE Symposium on Visual Analytics Science and Technology (VAST)*, pages 27–34, 2010.

[16] T. Cover and P. Hart. Nearest neighbor pattern classification. *IEEE Transactions on Information Theory*, 13(1):21–27, 1967.

[17] A.M. Cuadros, F.V. Paulovich, R. Minghim, and G.P. Telles. Point placement by phylogenetic trees and its application to visual analysis of document collections. In *IEEE Symposium on Visual Analytics Science and Technology, 2007, VAST 2007*, pages 99–106, 2007.

[18] Maria Cristina Ferreira de Oliveira and Haim Levkowitz. From visual data exploration to visual data mining: A survey. *IEEE Trans. Vis. Comput. Graph.*, 9(3):378–394, 2003.

[19] S. Deegalla and H. Bostrom. Reducing high-dimensional data by principal component analysis vs. random projection for nearest neighbor classification. In *5th International Conference on Machine Learning and Applications, 2006, ICMLA '06*, pages 245–250, 2006.

[20] Inderjit S. Dhillon, Dharmendra S. Modha, and W.Scott Spangler. Class visualization of high-dimensional data with applications. *Computational Statistics & Data Analysis*, 41(1):59–90, 2002.

[21] Giorgio Maria Di Nunzio. Using scatterplots to understand and improve probabilistic models for text categorization and retrieval. *Int. J. Approx. Reasoning*, 50(7):945–956, 2009.

[22] Stephan Diehl, Fabian Beck, and Michael Burch. Uncovering strengths and weaknesses of radial visualizations—an empirical approach. *IEEE Transactions on Visualization and Computer Graphics*, 16(6):935–942, November 2010.

[23] Wenwen Dou, Xiaoyu Wang, R. Chang, and W. Ribarsky. Paralleltopics: A probabilistic approach to exploring document collections. In *2011 IEEE Conference on Visual Analytics Science and Technology (VAST)*, pages 231–240, 2011.

[24] G. Draper, Y. Livnat, and R.F. Riesenfeld. A survey of radial methods for information visualization. *IEEE Transactions on Visualization and Computer Graphics*, 15(5):759–776, 2009.

[25] N. Elmqvist and J. Fekete. Hierarchical aggregation for information visualization: Overview, techniques, and design guidelines. *IEEE Transactions on Visualization and Computer Graphics*, 16(3):439–454, 2010.

[26] Daniel Engel, Lars Hüttenberger, and Bernd Hamann. A survey of dimension reduction methods for high-dimensional data analysis and visualization. In Christoph Garth, Ariane Middel, and Hans Hagen, editors, *VLUDS*, volume 27 of *OASICS*, pages 135–149. Schloss Dagstuhl - Leibniz-Zentrum fuer Informatik, Germany, 2011.

[27] Katia Romero Felizardo, Elisa Yumi Nakagawa, Daniel Feitosa, Rosane Minghim, and José Carlos Maldonado. An approach based on visual text mining to support categorization and classification in the systematic mapping. In *Proceedings of the 14th International Conference on Evaluation and Assessment in Software Engineering*, EASE'10, British Computer Society, Swinton, UK, pages 34–43, 2010.

[28] José Roberto M. Garcia, Antônio Miguel V. Monteiro, and Rafael D. C. Santos. Visual data mining for identification of patterns and outliers in weather stations' data. In *Proceedings of the 13th International Conference on Intelligent Data Engineering and Automated Learning*, IDEAL'12, pages 245–252, Springer-Verlag Berlin, Heidelberg, 2012.

[29] Zhao Geng, ZhenMin Peng, R.S. Laramee, J.C. Roberts, and R. Walker. Angular histograms: Frequency-based visualizations for large, high dimensional data. *IEEE Transactions on Visualization and Computer Graphics*, 17(12):2572–2580, 2011.

[30] M. Giereth, H. Bosch, and T. Ertl. A 3d treemap approach for analyzing the classificatory distribution in patent portfolios. In *IEEE Symposium on Visual Analytics Science and Technology, VAST '08*, pages 189–190, 2008.

[31] Charles D. Hansen and Chris R. Johnson. *Visualization Handbook*. Academic Press, 1st edition, December 2004.

[32] F. Heimerl, S. Koch, H. Bosch, and T. Ertl. Visual classifier training for text document retrieval. *IEEE Transactions on Visualization and Computer Graphics*, 18(12):2839–2848, 2012.

[33] William R. Hersh, Jamie Callan, Yoelle Maarek, and Mark Sanderson, editors. *The 35th International ACM SIGIR Conference on Research and Development in Information Retrieval, SIGIR '12, Portland, OR, USA, ACM, August 12-16*, 2012.

[34] Patrick Hoffman, Georges Grinstein, Kenneth Marx, Ivo Grosse, and Eugene Stanley. DNA visual and analytic data mining. In *Proceedings of the 8th Conference on Visualization '97*, VIS '97, pages 437–ff., IEEE Computer Society Press Los Alamitos, CA, 1997.

[35] Patrick Hoffman, Georges Grinstein, and David Pinkney. Dimensional anchors: A graphic primitive for multidimensional multivariate information visualizations. In *Proceedings of the 1999 Workshop on New Paradigms in Information Visualization and Manipulation* in conjunction with the *Eighth ACM International Conference on Information and Knowledge Management*, NPIVM '99, pages 9–16, ACM, New York, NY, 1999.

[36] Hailong Hou, Yan Chen, R. Beyah, and Yan-Qing Zhang. Filtering spam by using factors hyperbolic tree. In *Global Telecommunications Conference, 2008. IEEE GLOBECOM 2008. IEEE*, pages 1–5, 2008.

[37] A. Inselberg and Bernard Dimsdale. Parallel coordinates: A tool for visualizing multi-dimensional geometry. In *Visualization '90, Proceedings of the First IEEE Conference on Visualization*, pages 361–378, 1990.

[38] Alfred Inselberg. The plane with parallel coordinates. *The Visual Computer*, 1(2):69–91, 1985.

[39] Brian Johnson. Treeviz: treemap visualization of hierarchically structured information. In *Proceedings of the SIGCHI Conference on Human Factors in Computing Systems*, CHI '92, pages 369–370, ACM, New York, NY, 1992.

[40] Eser Kandogan. Visualizing multi-dimensional clusters, trends, and outliers using star coordinates. In Doheon Lee, Mario Schkolnick, Foster J. Provost, and Ramakrishnan Srikant, editors, *KDD*, pages 107–116. ACM, 2001.

[41] Ashish Kapoor, Bongshin Lee, Desney Tan, and Eric Horvitz. Interactive optimization for steering machine classification. In *Proceedings of the SIGCHI Conference on Human Factors in Computing Systems*, CHI '10, pages 1343–1352, ACM, New York, NY, 2010.

[42] G. V. Kass. An exploratory technique for investigating large quantities of categorical data. *Journal of the Royal Statistical Society. Series C (Applied Statistics)*, 29(2):119–127, 1980.

[43] M. W. Kattan, J. A. Eastham, A. M. Stapleton, T. M. Wheeler, and P. T. Scardino. A preoperative nomogram for disease recurrence following radical prostatectomy for prostate cancer. *J Natl Cancer Inst*, 90(10):766–71, 1998.

[44] Daniel Keim and Leishi Zhang. Solving problems with visual analytics: challenges and applications. In *Proceedings of the 11th International Conference on Knowledge Management and Knowledge Technologies*, i-KNOW '11, pages 1:1–1:4, ACM, New York, NY, 2011.

[45] Daniel A. Keim, Ming C. Hao, Umeshwar Dayal, Halldor Janetzko, and Peter Bak. Generalized scatter plots. *Information Visualization*, 9(4):301–311, December 2010.

[46] Daniel A. Keim, Joern Kohlhammer, Geoffrey Ellis, and Florian Mansmann, editors. *Mastering The Information Age—Solving Problems with Visual Analytics*. Eurographics, November 2010.

[47] Daniel A. Keim, Fabrice Rossi, Thomas Seidl, Michel Verleysen, and Stefan Wrobel. Information visualization, visual data mining and machine learning (Dagstuhl Seminar 12081). *Dagstuhl Reports*, 2(2):58–83, 2012.

[48] Teuvo Kohonen. Self-organized formation of topologically correct feature maps. In James A. Anderson and Edward Rosenfeld, editors, *Neurocomputing: Foundations of research*, pages 511–521. MIT Press, Cambridge, MA, 1982.

[49] Teuvo Kohonen. *Self-Organizing Maps*. Springer Series in Information Retrieval. Springer, second edition, March 1995.

[50] Jan Kölling, Daniel Langenkämper, Sylvie Abouna, Michael Khan, and Tim W. Nattkemper. Whide—a web tool for visual data mining colocation patterns in multivariate bioimages. *Bioinformatics*, 28(8):1143–1150, April 2012.

[51] John Lamping, Ramana Rao, and Peter Pirolli. A focus+context technique based on hyperbolic geometry for visualizing large hierarchies. In *Proceedings of the SIGCHI Conference on Human Factors in Computing Systems*, CHI '95, pages 401–408, ACM Press/Addison-Wesley Publishing Co. New York, NY, 1995.

[52] Gregor Leban, Blaz Zupan, Gaj Vidmar, and Ivan Bratko. Vizrank: Data visualization guided by machine learning. *Data Min. Knowl. Discov.*, 13(2):119–136, 2006.

[53] Ioannis Leftheriotis. Scalable interaction design for collaborative visual exploration of big data. In *Proceedings of the 2012 ACM International Conference on Interactive Tabletops and Surfaces*, ITS '12, pages 271–276, ACM, New York, NY, USA, 2012.

[54] A. Lex, H. Schulz, M. Streit, C. Partl, and D. Schmalstieg. Visbricks: Multiform visualization of large, inhomogeneous data. *IEEE Transactions on Visualization and Computer Graphics*, 17(12):2291–2300, 2011.

[55] Yan Liu and Gavriel Salvendy. Design and evaluation of visualization support to facilitate decision trees classification. *International Journal of Human-Computer Studies*, 65(2):95–110, 2007.

[56] H. Ltifi, M. Ben Ayed, A.M. Alimi, and S. Lepreux. Survey of information visualization techniques for exploitation in KDD. In *IEEE/ACS International Conference on Computer Systems and Applications, 2009, AICCSA 2009*, pages 218–225, 2009.

[57] J. Lubsen, J. Pool, and E. van der Does. A practical device for the application of a diagnostic or prognostic function. *Methods of Information in Medicine*, 17(2):127–129, April 1978.

[58] Christian Martin, Naryttza N. Diaz, Jörg Ontrup, and Tim W. Nattkemper. Hyperbolic SOM-based clustering of DNA fragment features for taxonomic visualization and classification. *Bioinformatics*, 24(14):1568–1574, July 2008.

[59] Dieter Merkl. Text classification with self-organizing maps: Some lessons learned. *Neurocomputing*, 21(1–3):61–77, 1998.

[60] Martin Mozina, Janez Demsar, Michael W. Kattan, and Blaz Zupan. Nomograms for visualization of naive Bayesian classifier. In Jean-François Boulicaut, Floriana Esposito, Fosca Giannotti, and Dino Pedreschi, editors, *PKDD*, volume 3202 of *Lecture Notes in Computer Science*, pages 337–348. Springer, 2004.

[61] Emmanuel Müller, Ira Assent, Ralph Krieger, Timm Jansen, and Thomas Seidl. Morpheus: Interactive exploration of subspace clustering. In *Proceedings of the 14th ACM SIGKDD International Conference on Knowledge Discovery and Data Mining*, KDD '08, pages 1089–1092, ACM, New York, NY, USA, 2008.

[62] Giorgio Maria Di Nunzio and Alessandro Sordoni. A visual tool for Bayesian data analysis: The impact of smoothing on naive bayes text classifiers. In William R. Hersh, Jamie Callan, Yoelle Maarek, and Mark Sanderson, editors. *The 35th International ACM SIGIR Conference on Research and Development in Information Retrieval, SIGIR '12, Portland, OR, USA, August 12-16, 2012.* ACM, 2012.

[63] P. Oesterling, G. Scheuermann, S. Teresniak, G. Heyer, S. Koch, T. Ertl, and G.H. Weber. Two-stage framework for a topology-based projection and visualization of classified document collections. In *2010 IEEE Symposium on Visual Analytics Science and Technology (VAST)*, pages 91–98, 2010.

[64] J.G. Paiva, L. Florian, H. Pedrini, G.P. Telles, and R. Minghim. Improved similarity trees and their application to visual data classification. *IEEE Transactions on Visualization and Computer Graphics*, 17(12):2459–2468, 2011.

[65] François Poulet. Towards effective visual data mining with cooperative approaches. In Simeon J. Simoff et al. (ed.) *Visual Data Mining—Theory, Techniques and Tools for Visual Analytics*, pages 389–406, 2008.

[66] Brett Poulin, Roman Eisner, Duane Szafron, Paul Lu, Russell Greiner, David S. Wishart, Alona Fyshe, Brandon Pearcy, Cam Macdonell, and John Anvik. Visual explanation of evidence with additive classifiers. In *AAAI*, pages 1822–1829. AAAI Press, 2006.

[67] Bhiksha Raj and Rita Singh. Classifier-based non-linear projection for adaptive endpointing of continuous speech. *Computer Speech & Language*, 17(1):5–26, 2003.

[68] Randall M. Rohrer, John L. Sibert, and David S. Ebert. A shape-based visual interface for text retrieval. *IEEE Computer Graphics and Applications*, 19(5):40–46, 1999.

[69] Christin Seifert and Elisabeth Lex. A novel visualization approach for data-mining-related classification. In Ebad Banissi, Liz J. Stuart, Theodor G. Wyeld, Mikael Jern, Gennady L. Andrienko, Nasrullah Memon, Reda Alhajj, Remo Aslak Burkhard, Georges G. Grinstein, Dennis P. Groth, Anna Ursyn, Jimmy Johansson, Camilla Forsell, Urska Cvek, Marjan Trutschl, Francis T. Marchese, Carsten Maple, Andrew J. Cowell, and Andrew Vande Moere, editors, *Information Visualization Conference*, pages 490–495. IEEE Computer Society, 2009.

[70] B. Shneiderman. Direct manipulation: A step beyond programming languages. *Computer*, 16(8):57–69, 1983.

[71] B. Shneiderman. The eyes have it: A task by data type taxonomy for information visualizations. In *1996. Proceedings of IEEE Symposium on Visual Languages*, pages 336–343, 1996.

[72] Ben Shneiderman. Tree visualization with tree-maps: 2-d space-filling approach. *ACM Trans. Graph.*, 11(1):92–99, January 1992.

[73] Ben Shneiderman, Cody Dunne, Puneet Sharma, and Ping Wang. Innovation trajectories for information visualizations: Comparing treemaps, cone trees, and hyperbolic trees. *Information Visualization*, 11(2):87–105, 2012.

[74] Simeon J. Simoff, Michael H. Böhlen, and Arturas Mazeika, editors. *Visual Data Mining—Theory, Techniques and Tools for Visual Analytics*, volume 4404 of *Lecture Notes in Computer Science*. Springer, 2008.

[75] Simeon J. Simoff, Michael H. Böhlen, and Arturas Mazeika, editors. *Visual Data Mining—Theory, Techniques and Tools for Visual Analytics*, volume 4404 of *Lecture Notes in Computer Science*. Springer, 2008.

[76] Rita Singh and Bhiksha Raj. Classification in likelihood spaces. *Technometrics*, 46(3):318–329, 2004.

[77] Justin Talbot, Bongshin Lee, Ashish Kapoor, and Desney S. Tan. Ensemblematrix: Interactive visualization to support machine learning with multiple classifiers. In *Proceedings of the SIGCHI Conference on Human Factors in Computing Systems*, CHI '09, pages 1283–1292, ACM, New York, NY, 2009.

[78] A. Tatu, G. Albuquerque, M. Eisemann, P. Bak, H. Theisel, M. Magnor, and D. Keim. Automated analytical methods to support visual exploration of high-dimensional data. *IEEE Transactions on Visualization and Computer Graphics*, 17(5):584–597, 2011.

[79] Soon T. Teoh and Kwan-liu Ma. StarClass: Interactive visual classification using star coordinates, *Proceedings of the 3rd SIAM International Conference on Data Mining*, pages 178–185, 2003.

[80] Soon Tee Teoh and Kwan-Liu Ma. Paintingclass: Interactive construction, visualization and exploration of decision trees. In *Proceedings of the Ninth ACM SIGKDD International Conference on Knowledge Discovery and Data Mining*, KDD '03, pages 667–672, ACM, New York, NY, 2003.

[81] Clark T.G., Stewart M.E., Altman D.G., Gabra H., and Smyth J.F. A prognostic model for ovarian cancer. *British Journal of Cancer*, 85(7):944–952, October 2001.

[82] Michail Vlachos, Carlotta Domeniconi, Dimitrios Gunopulos, George Kollios, and Nick Koudas. Non-linear dimensionality reduction techniques for classification and visualization. In *Proceedings of the Eighth ACM SIGKDD International Conference on Knowledge Discovery and Data Mining*, pages 645–651, ACM, New York, NY, 2002.

[83] Malcolm Ware, Eibe Frank, Geoffrey Holmes, Mark Hall, and Ian H. Witten. Interactive machine learning: Letting users build classifiers. *Int. J. Hum. Comput. Stud.*, 56(3):281–292, March 2002.

[84] James A. Wise, James J. Thomas, Kelly Pennock, D. Lantrip, M. Pottier, Anne Schur, and V. Crow. Visualizing the non-visual: Spatial analysis and interaction with information from text documents. In Nahum D. Gershon and Stephen G. Eick, editors, *INFOVIS*, pages 51–58. IEEE Computer Society, 1995.

[85] Pak Chung Wong. Guest editor's introduction: Visual data mining. *IEEE Computer Graphics and Applications*, 19(5):20–21, 1999.

[86] Jing Yang, Wei Peng, Matthew O. Ward, and Elke A. Rundensteiner. Interactive hierarchical dimension ordering, spacing and filtering for exploration of high dimensional datasets. In *Proceedings of the Ninth annual IEEE Conference on Information Visualization*, INFOVIS'03, pages 105–112, IEEE Computer Society, Washington, DC, 2003.

[87] Ke-Bing Zhang, M.A. Orgun, R. Shankaran, and Du Zhang. Interactive visual classification of multivariate data. In *11th International Conference on Machine Learning and Applications (ICMLA), 2012*, volume 2, pages 246–251, 2012.

[88] Ke-Bing Zhang, M.A. Orgun, and Kang Zhang. A visual approach for external cluster validation. In *CIDM 2007. IEEE Symposium on Computational Intelligence and Data Mining, 2007*, pages 576–582, 2007.

[89] Hong Zhou, Xiaoru Yuan, Huamin Qu, Weiwei Cui, and Baoquan Chen. Visual clustering in parallel coordinates. *Computer Graphics Forum*, 27(3):1047–1054, 2008.

Chapter 24

Evaluation of Classification Methods

Nele Verbiest
Ghent University
Belgium
`nele.verbiest@ugent.be`

Karel Vermeulen
Ghent University
Belgium
`karelb.vermeulen@ugent.be`

Ankur Teredesai
University of Washington, Tacoma
Tacoma, WA
`ankurt@uw.edu`

24.1 Introduction

The evaluation of the quality of the classification model significantly depends on the eventual utility of the classifier. In this chapter we provide a comprehensive treatment of various classification evaluation techniques, and more importantly, provide recommendations on when to use a particular technique to maximize the utility of the classification model.

First we discuss some generical validation schemes for evaluating classification models. Given a dataset and a classification model, we describe how to set up the data to use it effectively for training and then validation or testing. Different schemes are studied, among which are cross validation schemes and bootstrap models. Our aim is to elucidate how choice of a scheme should take into account the bias, variance, and time complexity of the model.

Once the train and test datasets are available, the classifier model can be constructed and the test instances can be classified to solve the underlying problem. In Section 24.3 we list evaluation measures to evaluate this process. The most important evaluation measures are related to accuracy; these measures evaluate how well the classifier is able to recover the true class of the instances. We distinguish between discrete classifiers, where the classifier returns a class, and probabilistic classifiers, where the probability is that the instance belongs to the class. In the first case, typical evaluation measures like accuracy, recall, precision, and others are discussed, and guidelines on which measure to use in which cases are given. For probabilistic classifiers we focus on ROC curve analysis and its extension for multi-class problems. We also look into evaluation measures that are not related to accuracy, such as time complexity, storage requirements, and some special cases.

We conclude this chapter in Section 24.4 with statistical tests for comparing classifiers. After the evaluation measures are calculated for each classifier and each dataset, the question arises which classifiers are better. We emphasize the importance of using statistical tests to compare classifiers, as too often authors posit that their new classifiers are better than others based on average performances only. However, it might happen that one classifier has a better average performance but that it only outperforms the other classifiers for a few datasets. We discuss parametric statistical tests briefly, as mostly the assumptions for parametric tests are not satisfied. Next, we rigorously describe non-parametric statistical tests and give recommendations for using them correctly.

24.2   Validation Schemes

In this section we discuss how we can assess the performance of a classifier for a given data set. A classifier typically needs data (called train data) to build its model on. For instance, the nodes and thresholds in decision trees are determined using train data, or support vector machines use the train data to find support vectors. In order to evaluate the classifier built on the train data, other data, called test data, is needed. The classifier labels each example in the test data and this classification can then be evaluated using evaluation measures.

The oldest validation methods used the entire data set both for training and testing. Obviously the resulting performance will be too optimistic [35], as the classes of the instances are known in the model.

A more reasonable way to evaluate a classification method on a data set is hold-out evaluation [6], which splits the data into two parts. The train part is usually bigger than the test part; typically the training part doubles the size of the testing part. Although this design is simple and easy to use, it has some disadvantages, the main one being that the data is not fully explored, that is, the evaluation is only carried out on a small fraction of the data. Moreover, it can happen that the instances included in the testing part are too easy or too difficult to classify, resulting in a misleading high or low performance. Another problem that might occur is that instances essential to building the model are not included in the training set. This problem can be partially alleviated by running the hold-out evaluation several times, but still it might happen that some essential instances are never included in the training part or that instances difficult to classify are never evaluated in the testing part.

In order to deal with this problem, a more systematic approach, to repeat the hold-out evaluation, was developed. The widely used evaluation design called K-fold Cross Validation (K-CV, [51]), splits the data into K equal parts, and each of these parts is classified by a model built on the remaining $K - 1$ parts. The main advantage of this technique is that each data point is evaluated exactly once.

The choice of K is a trade-off between bias and variance [30]. For low values of K, the sizes of the training sets in the K-CV procedure are smaller, and the classifications are more biased depending on how the performance of the classifier changes with the instances included in the train set and with sample size. For instance, when $K = 2$, the two training sets are completely different and only cover one half of the original data, so the quality of the predictions can differ drastically for the two train sets. When $K = 5$, all train sets have 60 percent of the data in common, so the bias will be lower. For high values of K, the procedure becomes more variable due to the stronger dependence on the training data, as all training sets are very similar to one another.

Typical good values for K are 5 or 10. Another widely used related procedure is 5×2 CV [8], in which the 2-CV procedure is repeated five times. When K equals the data size, K-CV is referred to as Leave-One-Out-CV (LOOCV, [51]). In this case, each instance is classified by building a model on all remaining instances and applying the resulting model on the instance. Generalized-CV (GCV) is a simplified version of LOOCV that is less computationally expensive.

Other than splitting the data in K folds, one can also consider all subsets of a given size P, which is referred to as leave-P-out-CV or also delete-P-CV [3]. The advantage of this method is that many more combinations of training instances are used, but this method comes with a high computational cost.

Another type of methods that use more combinations of training instances are repeated learning-testing methods, also called Monte-Carlo-CV (MCCV, [54]) methods. They randomly select a fraction of the data as train data and use the remaining data as test data, and repeat this process multiple times.

Bootstrap validation models build the classification model on several training sets that are drawn from the original data with resampling and then apply this model to the original data. As each bootstrap sample has observations in common with the original training sample that is used as the test set, a factor 0.632 is applied to correct the optimistic resulting performance. For this reason, the method is also referred to as 0.632-bootstrap [13].

We note that when using CV methods, stratification of the data is an important aspect. One should make sure that the data is divided such that the class distribution of the whole data set is also reflected in the separate folds. Otherwise, a so-called data shift [39] can occur. This is especially important when working with imbalanced problems. The easiest way is to split each class in K parts and to assign a part to each fold. A more involved approach described in [7] attempts to direct similar instances to different folds. This so-called Unsupervised-CV (UCV) method is deterministic, that is, each run will return the same results.

A data shift cannot only occur in the class distribution, one should also be aware that the input features follow the same distribution to prevent a covariate data shift [39, 40, 49]. One solution is to use Distribution-Balanced-Stratified-CV (DB-SCV, [56]), which divides nearby instances to different folds. An improved solution is Distribution-Optimally-Balanced-Stratified-CV (DOB-CV, [49]).

We conclude this section with recommendations on the use of validation schemes. When choosing a validation scheme, three aspects should be kept in mind: variance, bias, and computational cost. In most situations, K-CV schemes can be used safely. The choice of K depends on the goal of the evaluation. For model selection, low variance is important, so a LOOCV scheme can be used. To assess the quality of a classifier, the bias is more important, so $K = 5$ or $K = 10$ is a good choice. For imbalanced datasets or small datasets, $K = 5$ is often recommended. K-CV schemes with higher K values are computationally more expensive, especially LOOCV, which is mostly too complex to use for large datasets. The advantage of MCCV over K-CV is that more combinations of instances are used, but of course this comes with a high computational cost. For small datasets, 0.632-bootstrap is highly recommended.

TABLE 24.1: Example of a Discrete Classifier c for a Multi-Class Classification Problem.

x	$r(x)$	$c(x)$
x_1	A	B
x_2	B	B
x_3	C	A
x_4	A	A
x_5	B	B
x_6	B	B
x_7	B	A
x_8	A	C
x_9	C	B
x_{10}	A	A

Note: In the second column, the real class $r(x)$ of the instance x in the first column is given, while $c(x)$ in the third column denotes the class returned by c for x.

24.3 Evaluation Measures

Once a validation scheme is chosen, the classifier can build its model on the training data and classify the test instances based on this model. Afterwards, evaluation measures are needed to describe how well this classification is done. We first discuss accuracy related measures that describe how similar the predicted classes of the test instances are to the real classes. Next, we consider additional measures that are needed for a good evaluation of the classifier.

24.3.1 Accuracy Related Measures

In this section we study how we can assess the quality of a given classifier based on a list of the real classes of instances and the predicted classes.

In the following we first consider discrete classifiers, for instance, K Nearest Neighbors, which return an actual class for each instance. The second type of classifiers, for instance Bayesian classifiers, are so-called probabilistic classifiers. Instead of returning a class they return the probability that they belong to a certain class. Both types of classifiers require different evaluation measures.

24.3.1.1 Discrete Classifiers

We consider a finite set $X = \{x_1, \ldots, x_n\}$ of n instances, and for each instance $x \in X$ we know its real class $r(x)$ and the classification $c(x)$ returned by the classifier c. As c is a discrete classifier, $c(x)$ takes values in the same set $C = c_1, \ldots, c_k$ of classes as the real classes $r(x)$. An example of a multi-class classification problem is given in Table 24.1 and an example of a binary classification problem is shown in Table 24.2. Note that we use the capitals P and N for the binary classification problems, referring to the positive and negative class, respectively, which is common use in binary classification problems.

A tool that is very useful when evaluating a discrete classifier is the so-called confusion matrix M. The dimension of the squared matrix M is the number of classes, and the entry M_{ij} denotes how many times an instance with real class c_i was classified as c_j. The confusion matrix corresponding to the example in Table 24.1 (respectively, Table 24.2) is given in Table 24.3 (respectively, Table 24.4). In the binary case, we denote by True Positives (TP) and True Negatives (TN) the number of correctly classified positive and negative instances, respectively. False Negatives (FN) stands for the number of instances that are predicted negative but that are actually positive, while the False Positives (FP) are the number of instances that are falsely classified to the positive class.

TABLE 24.2: Example of a Discrete Classifier c for a Binary Classification Problem.

x	$r(x)$	$c(x)$
x_1	P	P
x_2	P	P
x_3	N	P
x_4	P	P
x_5	N	N
x_6	P	P
x_7	P	N
x_8	N	N
x_9	P	P
x_{10}	N	P

Note: In the second column, the real class $r(x)$ of the instance x in the first column is given, while $c(x)$ in the third column denotes the class returned by c for x.

Based on the confusion matrix, many metrics can be defined. The most well-known one is classification accuracy, denoted by acc here. It is defined as the ratio of correctly classified instances, which can also be expressed as the sum of the diagonal elements in the confusion matrix:

$$acc(c) = \sum_{i=1}^{k} M_{ii}. \tag{24.1}$$

It is a general measure that gives an idea of the overall performance of the classifier, in the example in Table 24.1, $acc(c) = 0.5$, while for the classifier in the example in Table 24.2, $acc(c) = 0.7$.

Other well-known evaluation metrics are only defined for binary classifiers. For instance, the *recall* (also referred to as true positive rate or sensitivity) is the true positives compared to the number of truly positive instances:

$$recall(c) = \frac{TP}{TP + FN}. \tag{24.2}$$

The *precision* is the true positives compared to the number of instances predicted positive:

$$precision(c) = \frac{TP}{TP + FP}. \tag{24.3}$$

Another metric is the *specificity* (also called true negative rate), defined as the number of correctly classified negative instances divided by the number of truly negative instances:

$$specificity(c) = \frac{TN}{FP + TN}. \tag{24.4}$$

Finally, the false alarm (written as *falarm* here, also known as the false positive rate) is the false positives compared to the number of negative instances:

$$falarm(c) = \frac{FP}{TN + FP}. \tag{24.5}$$

In the example in Table 24.2, $recall(c) = 0.83$, $precision(c) = 0.71$, $specificity(c) = 0.5$, and $falarm(c) = 0.5$.

Based on precision and recall, the F-measure (F_1-score) can be used, which is the harmonic mean of *precision* and *recall*:

$$F_1(c) = 2 \frac{precison * recall}{precision + recall}. \tag{24.6}$$

TABLE 24.3: Confusion Matrix Corresponding to the Classifier Represented in Table 24.1

	A	B	C
A	2	1	1
B	1	3	1
C	1	0	0

TABLE 24.4: Confusion Matrix Corresponding to the Classifier Represented in Table 24.2

	P	N
P	5 (TP)	1 (FN)
N	2 (FP)	2 (TN)

and is some weighted average of *precision* and *recall*. It is a generalisation of the F_β measure, defined as:

$$F_\beta(c) = (1+\beta^2)\frac{precision * recall}{(\beta^2 precision) + recall} \tag{24.7}$$

The higher β, the more emphasis is put on recall, the lower β, the more influence precision has. For instance, in the example in Table 24.2, it holds that $F_1(c) = 0.77$, while $F_2(c) = 0.80$, which is closer to the recall, and $F_{0.5}(c) = 0.76$, which is closer to the precision.

All metrics above are defined for binary classification problems, but they can easily be used for multi-class problems. A common practice is to calculate the measure for each class seperately and then to average the metrics over all classes (one vs. all). In the case of the three-class classification problem in Table 24.1, it means that first the binary classification problem that has A as first class and B or C as second class is considered, and that the corresponding metric is calculated. This is repeated for class B vs. class A and C and for class C vs. class A and B. At the end, the three measures are averaged. For example, in the example in Table 24.1, the recall is $recall(c) = \frac{0.5+0.2+1}{3} = 0.57$. Another metric that can handle multi-class problems is Cohen's kappa [1], which is an agreement measure that compensates for classifications that may be due to chance, defined as follows:

$$\kappa(c) = \frac{n\sum\limits_{i=1}^{k} M_{ii} - \sum\limits_{i=1}^{k} M_{i.}M_{.i}}{n^2 - \sum\limits_{i=1}^{k} M_{i.}M_{.i}} \tag{24.8}$$

where $M_{.i}$ is the sum of the elements in the ith column of M and $M_{i.}$ the sum of the elements in the ith row of M.

There is no simple answer to the question of which evaluation metric to use, and in general there is no classifier that is optimal for each evaluation metric. When evaluating general classification problems, the accuracy is mostly sufficient, together with an analysis of Cohen's kappa. Of course, when problems are imbalanced, one should also take into account the F-measures to check if there is a good balance between recall and precision. When considering real-world problems, one should be careful when selecting appropriate evaluation metrics. For instance, when there is a high cost related to classifying instances to the negative class, a high false alarm is problematic. When it is more important not to oversee instances that are actually positive, a high recall is important. In some cases it is recommended to use multiple evaluation metrics, and a balance between them should be aimed for.

24.3.1.2 Probabilistic Classifiers

In this section we consider probabilistic classifiers, that is, classifiers for which the outcome is not a class, but for each class a probability that the instance belongs to it. An example is provided

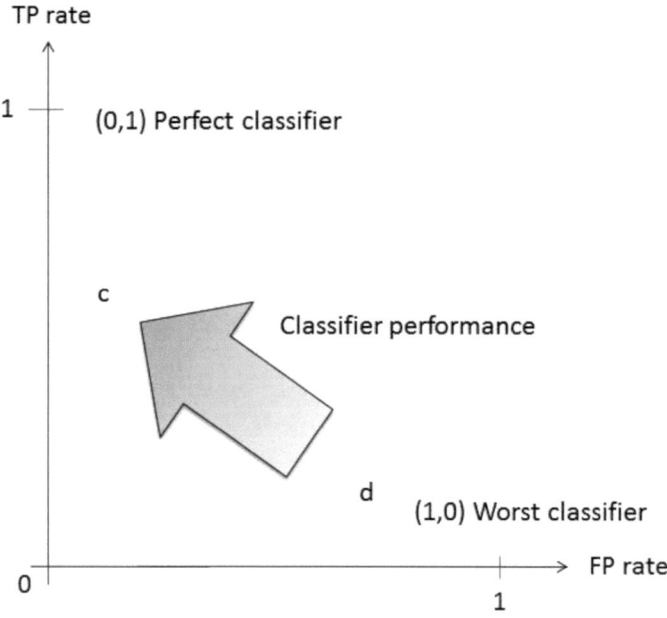

FIGURE 24.1: ROC-space with two classifiers c and d. The arrow indicates the direction of overall improvement of classifiers.

in Table 24.5. These probabilities are generated by the classifiers themselves, and contain more information than discrete classifiers. As such, there would be a loss of information if one would ignore the probabilities and discretize the probabilities to obtain a discrete classifier.

We first discuss binary classification problems. The most important evaluation metrics for probabilistic classifiers are related to Receiver Operating Characteristics (ROC, [38]) analysis. These techniques place classifiers in the ROC space, which is a two-dimensional space with the false positive rate on the horizontal axis and the true positive rate on the vertical axis (Figure 24.1). A point with coordinates (x, y) in the ROC space represents a classifier with false positive rate x and true positive rate y. Some special points are $(1, 0)$, which is the worst possible classifier, and $(0, 1)$, which is a perfect classifier. A classifier is better than another one if it is situated north-west of it in the ROC space. For instance, in Figure 24.1, classifier c is better than classifier d as it has a lower false positive rate and a higher true positive rate.

Probabilistic classifiers need a threshold to make a final decision for each class. For each possible threshold, a different discrete classifier is obtained, with a different TP rate and FP rate. By considering all possible thresholds, putting their corresponding classifiers in the ROC space, and drawing a line through them, a so-called ROC curve is obtained. In Figure 24.2, an example of such a stepwise function is given. In this simple example, the curve is constructed by considering all probabilities as thresholds, calculating the corresponding TP rate and FP rate and putting them in the plot. However, when a large dataset needs to be evaluated, more efficient algorithms are required. In that case, Algorithm 24.1 can be used, which only requires $O(n \log n)$ operations. It makes use of the observation that instances classified positive for a certain threshold will also be classified for all lower thresholds. The algorithm sorts the instances decreasing with reference to the outputs $p(x)$ of the classifier, and processes each instance at the time to calculate the corresponding TP rate and FP rate.

ROC curves are a great tool to visualize and analyze the performance of classifiers. The most important advantage is that they are independent of the class distribution, which makes them interesting to evaluate imbalanced problems. In order to compare two probabilistic classifiers with each

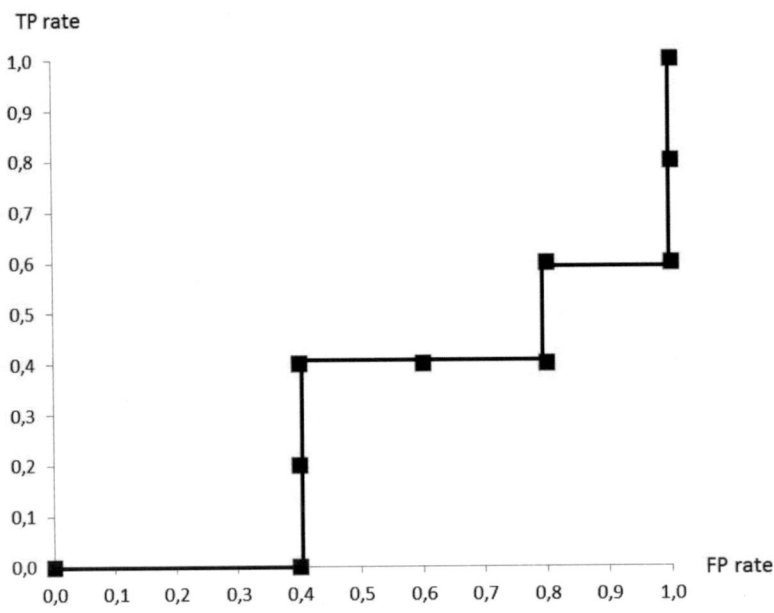

FIGURE 24.2: ROC-curve corresponding to the classifier in Table 24.5.

Algorithm 24.1 Efficient generation of ROC-curve points.

INPUT: n **instances** $X = \{x_1, \ldots, x_n\}$ **estimated probabilities returned by the classifier** $\{p(x_1), \ldots, p(x_n)\}$, **true classes** $\{r(x_1), \ldots, r(x_n)\}$, **number of positive instances** n_P **and number of negative instances** n_N.

$X_{sorted} \leftarrow X$ sorted decreasing by $p(x)$ scores
$FP \leftarrow 0, TP \leftarrow 0$
$R = \langle \rangle$
$p_{prev} \leftarrow -\infty$
$i \leftarrow 1$
while $i \leq |X_{sorted}|$ **do**
    if $p(x_i) \neq p_{prev}$ **then**
        push $(FP/n_N, TP/n_P)$ onto R
        $p_{prev} \leftarrow p_{x_i}$
    end if
    if $r(X_{sorted}[i]) = P$ **then**
        $TP \leftarrow TP + 1$
    else
        $FP \leftarrow FP + 1$
    end if
    $i \leftarrow i + 1$
end while
push $(1, 1)$ onto R
OUTPUT: R, **ROC points by increasing FP rate.**

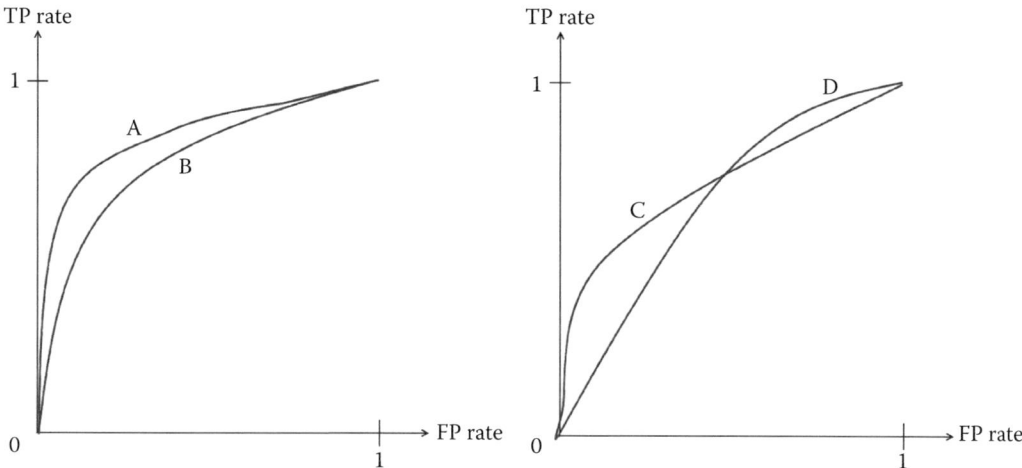

FIGURE 24.3: Comparing classifiers using ROC-curves.

other, the two ROC curves can be plotted on the same ROC space, as illustrated in Figure 24.3. On the left-hand-side, it is easy to see that classifier *A* outperforms *B*, as its ROC-curve is north-west of the ROC-curve of *B*. This means that for each threshold, the TP rate of *A* is higher than the TP rate of *B*, and that the FP rate of *B* is higher for each threshold. On the right-hand-side, the situation is less obvious and it is hard to say if *C* is better than *D* or the other way around. However, some interesting information can be extracted from the graph, that is, we can conclude that *C* is more precise than *D* for high thresholds, which means that it is the best choice if not many false positives are allowed. *D* on the other hand is more appropriate when a high sensitivity is required.

Although this visualization is a handy tool that is frequently used, many researchers prefer a single value to assess the performance of a classifier. The Area-Under the Curve (AUC) is a measure that is often used to that goal. It is the surface between the ROC curve and the horizontal axis, and is a good measure to reflect the quality, as ROC curves that are more to the north west are better and have a bigger surface. AUC values are often used as a single evaluation metric in imbalanced classification problems. In real-world problems, however, one should keep in mind that ROC curves reveal much more information than the single AUC value.

An important point when evaluating classifiers using ROC curves is that they do not measure the absolute performance of the classifier but the relative ranking of the probabilities. For instance, when the probabilities of a classifier are all higher than 0.5 and the threshold is also 0.5, no instance will be classified as negative. However, when all probabilities of the positive instances are higher than the probabilities of the negative instances, the classifier will have a perfect ROC curve and the AUC will be 1. This example shows that determining an appropriate threshold for the final classifier should be done appropriately.

When evaluating probabilistic classifiers for multi-class problems [38], a straightforward option is to decompose the multi-class problems to multiple two-class problems and to carry out the ROC analysis for each two-class problem. Understanding and interpreting these multiple visualizations is challenging, therefore specific techniques for multi-class ROC analysis were developed.

The first approach is discussed in [41], where ROC analysis is extended for three-class classification problems. Instead of an ROC curve, an ROC surface is plotted, and the Volume Under the Surface (VUS) is calculated analogously to the AUC. ROC surfaces can be used to compare classifiers on three-class classification problems by analyzing the maximum information gain on each of them.

ROC analysis for multi-class problems gains more and more attention. In [10], the multi-class problem is decomposed in partial two-class problems, where each problem retains labels of the in-

TABLE 24.5: Example of a Probabilistic Classifier.

	$p(x)$	$r(x)$
x_1	0,27	P
x_2	0,12	N
x_3	0,67	N
x_4	0,66	P
x_5	0,73	N
x_6	0,08	N
x_7	0,94	P
x_8	0,51	P
x_9	0,49	N
x_{10}	0,94	P

Note: In column $p(x)$ the probability that the instance belongs to the positive class is given, $r(x)$ denotes the real class.

stances in one specific class, and the instances in the remaining classes get another label. The AUC's are calculated for each partial problem and the final AUC is obtained by computing the weighted average of the AUC's, where the weight is determined based on the frequency of the corresponding class in the data. Another generalization is given in [19], where the AUC's of all combinations of two different classes are computed, and the sum of these intermediate AUCs is divided by the number of all possible misclassifications. The advantage of this approach over the previous one is that this approach is less sensitive to class distributions. In [14], the extension from ROC curve to ROC surface is further extended to ROC hyperpolyhedrons, which are multi-dimensional geometric objects. The volume of these geometric objects is the analogue of the VUS or AUC. A more recent approach can be found in [20], where a graphical visualization of the performance is developed.

To conclude, we note that computational cost is an important issue for multi-class ROC analysis; mostly all pairs of classes need to be considered. Many researchers [31–34] are working on faster approaches to circumvent this problem.

24.3.2 Additional Measures

When evaluating a classifier, one should not only take into account accuracy related measures but also look into other aspects of the classifier. One very important feature of classifiers is time complexity. There are two time measurements. The first one is the time needed to build the model on the training data, the second measures the time needed to classify the test instances based on this model. For some applications it is very important that the time to classify an instance is negligible and meanwhile building the model can be done offline and can take longer. Running time is extremely important when working with big data.

Time complexity can be measured theoretically, but it is good to have an additional practical evaluation of time complexity, where the time is measured during the training and testing phases.

Another important measure is storage requirements of the model. Of course, the initial model is built on the entire data, but after this training phase, the model can be stored elsewhere and the data points are no longer needed. For instance, when using a classification tree, only the tree itself needs to be stored. On the other hand, when using K Nearest Neighbors, the entire dataset needs to be stored and storage requirements are high.

When using active learning algorithms, an additional metric that measures the number of required labels in the train data is needed. This measure is correlated with the time needed to build the model.

24.4 Comparing Classifiers

When evaluating a new classifier, it is important to compare it to the state of the art. Deciding if your new algorithm is better than existing ones is not a trivial task. On average, your algorithm might be better than others. However, it may also happen that your new algorithm is only better for some examples. In general, there is no algorithm that is the best in all situations, as suggested by the *no free lunch* theorem [53].

For these and other reasons, it is crucial to use appropriate statistical tests to verify that your new classifier is indeed outperforming the state of the art. In this section, we give an overview of the most important statistical tests that can be used and summarize existing guidelines to help researchers to compare classifiers correctly.

A major distinction between different statistical tests is whether they are parametric [48] or non-parametric [5, 18, 37, 48]. All considered statistical tests, both parametric and non-parametric, assume independent instances. Parametric tests, in contrast to non-parametric tests, are based on an underlying parametric distribution of the (transformed) results of the considered evaluation measure. In order to use parametric statistical tests meaningfully, these distributional assumptions on the observations must be fulfilled. In most comparisons, not all these assumptions for parametric statistical tests are fulfilled or too few instances are available to check the validity of these assumptions. In those cases, it is recommended to use non-parametric statistical tests, which only require independent observations. Non-parametric statistical tests can also be used for data that do fulfill the conditions for parametric statistical tests, but it is important to note that in that case, the non-parametric tests have in general lower power than the parametric tests. Therefore, it is worthwhile to first check if the assumptions for parametric statistical tests are reasonable before carrying out any statistical test.

Depending on the setting, pairwise or multiple comparison tests should be performed. A pairwise test aims to detect a significant difference between two classifiers, while a multiple comparison test aims to detect significant differences between multiple classifiers. We consider two types of multiple comparison tests; in situations where one classifier needs to be compared against all other classifiers, we speak about a multiple comparison test with a control method. In the other case, we want to compare all classifiers with all others and hence we want to detect all pairwise differences between classifiers.

All statistical tests, both parametric and non-parametric, follow the same pattern. They assume a null hypothesis, stating that there is no difference between the classifiers' performance, e.g., expressed by the mean or median difference in performance. Before the test is carried out, a significance level is fixed, which is an upper bound for the probability of falsely rejecting the null hypothesis. The statistical test can reject the null-hypothesis at this significance level, which means that with high confidence at least one of the classifiers significantly outperforms the others, or not reject it at this significance level, which means that there is no strong or sufficient evidence to believe that one classifier is better than the others. In the latter case, it is not guarenteed that there are *no* differences between the classifiers; the only conclusion then is that the test cannot find significant differences at this significance level. This might be because there are indeed no differences in performance, or because the power (the probability of correctly rejecting the null hypothesis) of the test is too low, caused by an insufficient amount of datasets. The decision to reject the null hypothesis is based on whether the observed test statistic is bigger than some critical value, or equivalently, whether the corresponding p-value is smaller than the prespecified significance level. Recall that the p-value returned by a statistical test is the probability that a more extreme observation than the observed one, given the null hypothesis, is true.

In the remainder of this section, the number of classifiers is denoted by k, the number of cases (also referred to as instances, datasets, or examples) by n, and the significance level by α. The calculated evaluation measure of case i based on classifier j is denoted Y_{ij}.

24.4.1 Parametric Statistical Comparisons

In this section, we first consider a parametric test for a pairwise comparison and next consider a multiple comparison procedure with a discussion on different appropriate post-hoc procedures, taking into account the multiple testing issue.

24.4.1.1 Pairwise Comparisons

In parametric statistical testing, the paired t-test is used to compare the performance of two classifiers ($k = 2$). The null hypothesis is that the mean performance of both classifiers is the same and one aims to detect a difference in means. The null hypothesis will be rejected if the mean performances are sufficiently different from one another. To be more specific, the sample mean of the differences in performance $\bar{D} = \sum_{i=1}^{n} (Y_{i1} - Y_{i2})/n$ of the two classifiers is calculated and then a one-sample t-test is performed to test if $\bar{D}$ significantly differs from zero. The test statistic is given by the studentized sample mean $T = \bar{D}/(S/\sqrt{n})$ where S is the standard deviation of the differences. Assuming the differences are normally distributed, the test statistic T follows a t-distribution with $n - 1$ degrees of freedom. The null hypothesis is rejected if $|T| > t_{n-1,\alpha/2}$ where $t_{n-1,\alpha/2}$ is the $1 - \alpha/2$ percentile of a t_{n-1}-distribution.

To check whether the normality assumption is plausible, visual tools like histograms and preferably QQ-plots can be used. Alternatively, one can use a normality test. Three popular normality tests, which all aim to detect certain deviations from normality, are (i) the Kolmogorov-Smirnov test, (ii) the Shapiro-Wilk test, and (iii) the D'Agostino-Pearson test [4].

24.4.1.2 Multiple Comparisons

When comparing more than two classifiers ($k > 2$), the paired t-test can be generalized to the within-subjects ANalysis Of VAriance (ANOVA, [48]) statistical test. The null hypothesis is that the mean performance of all k classifiers is the same and one aims to detect a deviation from this. The null hypothesis will be rejected if the mean performances of at least two classifiers are sufficiently different from one another in the sense that the mean between-classifier variability (sytematic variability) sufficiently exceeds the mean residual variability (error variability). For this purpose, define $\bar{Y}_{..} = \sum_{j=1}^{k} \sum_{i=1}^{n} Y_{ij}/nk$, $\bar{Y}_{i.} = \sum_{j=1}^{k} Y_{ij}/k$ and $\bar{Y}_{.j} = \sum_{i=1}^{n} Y_{ij}/n$. The mean between-classifier variability $MS_{BC} = n \sum_{j=1}^{k} (\bar{Y}_{.j} - \bar{Y}_{..})^2/(k-1)$ measures the mean squared deviations of the mean performance of the classifiers to the overall mean performance. The mean residual variability $MS_{res} = \sum_{j=1}^{k} \sum_{i=1}^{n} \{(Y_{ij} - \bar{Y}_{..}) - (\bar{Y}_{i.} - \bar{Y}_{..}) - (\bar{Y}_{.j} - \bar{Y}_{..})\}^2/(n-1)(k-1)$ measures the variability that is truly random. The test statistic is given by the F-ratio

$$F = \frac{MS_{BC}}{MS_{res}}.$$

If this quantity is sufficiently large, it is likely that there are significant differences between the different classifiers. Assuming the results for the different instances are normally distributed and assuming sphericity (discussed below), the F-ratio follows an F-distribution with $k - 1$ and $(n - 1)(k - 1)$ degrees of freedom. The null hypothesis is rejected if $F > F_{k-1,(n-1)(k-1),\alpha}$ where $F_{k-1,(n-1)(k-1),\alpha}$ is the $1 - \alpha$ percentile of an $F_{k-1,(n-1)(k-1)}$-distribution. The normality assumption can be checked using the same techniques as for the paired t-test. The sphericity condition boils down to equality of variance of all $k(k-1)/2$ pairwise differences, which can be inspected using visual plots such as boxplots. A discussion of tests that are employed to evaluate sphericity is beyond the scope of this chapter and we refer to [29].

If the null hypothesis is rejected, the only information provided by the within-subjects ANOVA is that there are significant differences between the classifiers' performances, but not information is provided about which classifier outperforms another. It is tempting to use multiple pairwise comparisons to get more information. However, this will lead to an accumulation of the Type I error coming from the combination of pairwise comparisons, also referred to as the Family Wise Error Rate (FWER, [42]), which is the probability of making at least one false discovery among the different hypotheses.

In order to make more detailed conclusions after an ANOVA test rejects the null hypothesis, i.e., significant differences are found, post-hoc procedures are needed that correct for multiple testing to avoid inflation of Type I errors. Most of the post-hoc procedures we discuss here are explained for multiple comparisons with a control method and thus we perform $k-1$ post-hoc tests. However, the extension to an arbitrary number of post-hoc tests is straightforward.

The most simple procedure is the Bonferroni procedure [11], which uses paired t-tests for the pairwise comparisons but controls the FWER by dividing the significance level by the number of comparisons made, which here corresponds to an adjusted significane level of $\alpha/(k-1)$. Equivalently, one can multiply the obtained p-values by the number of comparisons and compare to the original significance level α. However, the Bonferonni correction is too conservative when many comparisons are performed. Another approach is the Dunn-Sidak [50] correction, which alters the p-values to $1-(1-p)^{1/(k-1)}$.

A more reliable test when interested in all paired comparisons is Tukey's Honest Significant Difference (HSD) test [2]. This test controls the FWER so it will not exceed α. To be more specific, suppose we are comparing classifier j with l. The Tukey's HSD test is based on the studentized range statistic

$$q_{jl} = \frac{|\bar{Y}_{\cdot j} - \bar{Y}_{\cdot l}|}{\sqrt{MS_{\text{res}}/n}}.$$

It corrects for multiple testing by changing the critical value to reject the null hypothesis. Under the assumptions of the within-subjects ANOVA, q_{jl} follows a studentized range distribution with k and $(n-1)(k-1)$ degrees of freedom and the null hypothesis that the mean performance of classifier j is the same of classifier l is rejected if $q_{jl} > q_{k,(n-1)(k-1),\alpha}$ where $q_{k,(n-1)(k-1),\alpha}$ is the $1-\alpha$ percentile of a $q_{k,(n-1)(k-1)}$-distribution.

A similar post-hoc procedure is the Newman-Keuls [28] procedure, which is more powerful than Tukey's HSD test but does not guarentee the FWER will not exceed the prespecified significance level α. The Newman-Keuls procedure uses a systematic stepwise approach to carry out many comparisons. It first orders the sample means $\bar{Y}_{\cdot j}$. For notational convenience, suppose the index j labels these ordered sample means in ascending order: $\bar{Y}_1 < \ldots < \bar{Y}_k$. In the first step, the test verifies if the largest and smallest sample means significantly differ from each other using the test statistic q_{n1} and uses the critical value $q_{k,(n-1)(k-1),\alpha}$, if these sample means are k steps away from each other. If the null hypothesis is retained, it means that the null hypotheses of all comparisons will be retained. If the null hypothesis is rejected, all comparisons of sample means that are $k-1$ steps from one another are performed, and q_{n2} and $q_{n-1,1}$ are compared to the critical value $q_{k-1,(n-1)(k-1),\alpha}$. In a stepwise manner, the range between the means is lowered, until no null hypotheses are rejected anymore.

A last important post-hoc test is Scheffé's test [46]. This test also controls the FWER, but now regardless, the number of post-hoc comparisons. It is considered to be very conservative and hence Tuckey's HSD test may be preferred. The procedure calculates for each comparison of interest the difference between the means of the corresponding classifiers and compares them to the critical value

$$\sqrt{(k-1)F_{k-1,(n-1)(k-1),\alpha}}\sqrt{\frac{2MS_{\text{res}}}{n}}$$

where $F_{k-1,(n-1)(k-1),\alpha}$ is the $1-\alpha$ percentile of an $F_{k-1,(n-1)(k-1)}$-distribution. If the difference in sample means exceeds this critical value, the means of the considered classifiers are significantly different.

24.4.2 Non-Parametric Statistical Comparisons

In the previous section, we discussed that the validity of parametric statistical tests heavily depends on the underlying distributional assumptions of the performance of the classifiers on the different datasets. In practice, not all these assumptions are necessarily fulfilled or too few instances are available to check the validity of these assumptions. In those cases, it is recommended to use non-parametric statistical tests, which only require independent observations. Strictly speaking, the null hypothesis of the non-parametric tests presented here state that the distribution of the performance of all classifiers is the same. Different tests then differ in what alternatives they aim to detect.

Analogous to the previous section about parametric statistical tests, we make a distinction between pairwise and multiple comparisons.

24.4.2.1 Pairwise Comparisons

When comparing two classifiers ($k = 2$), the most commonly used tests are the sign test and the Wilcoxon signed-ranks test.

Since under the null hypothesis the two classifiers' scores are equivalent, they each win in approximately half of the cases. The sign test [48] will reject the null hypothesis if for instance the proportion that classifier 1 wins is sufficiently different from 0.5. Let n_1 denote the number of times classifier 1 beats classifier 2. Under the null hypothesis, n_1 follows a binomial distribution with parameters n and 0.5. The null hypothesis will then be rejected if $n_1 < k_l$ or $n_1 > k_u$ where k_l is the biggest integer satisfying $\sum_{\ell=0}^{k_l-1} \binom{n}{\ell} 0.5^n \leq \alpha/2$ and k_u is the smallest integer satisfying $\sum_{\ell=k_u+1}^{n} \binom{n}{\ell} 0.5^n \leq \alpha/2$. A more simple strategy is based on the asymptotic approximation of the binomial distribution. That is, when the number of instances is sufficiently large, under the null hypothesis, n_1 approximately follows a normal distribution with mean $n/2$ and variance $n/4$. The null hypothesis is rejected if $|n_1 - n/2|/(\sqrt{n}/2) > z_{\alpha/2}$ where $z_{\alpha/2}$ is the $1-\alpha/2$ percentile of the standard normal distribution.

An alternative paired test is Wilcoxon signed-ranks test [52]. The Wilcoxon signed-ranks test uses the differences $Y_{i1} - Y_{i2}$. Under the null hypothesis, the distribution of these differences is symmetric around the median and hence we must have that the distribution of the positive differences is the same as the distribution of the negative differences. The Wilcoxon signed rank test then aims to detect a deviation from this to reject the null hypothesis. The procedure assigns a rank to each difference according to the absolute value of these differences, where the mean of ranks is assigned to cases with ties. For instance, when the differences are $0.03, 0.06, -0.03, 0.01, -0.04$, and 0.2, the respective ranks are $3.5, 6, 3.5, 1, 5$, and 2. Next, the ranks of the positive and negative differences are summed separately, in our case $R^+ = 12.5$ and $R^- = 8.5$. When few instances are available, to reject the null hypothesis, $\min(R^+, R^-)$ should be less than or equal to a critical value depending on the significance level and the number of instances; see [48] for tables. Example, when $\alpha = 0.1$, the critical value is 2, meaning that in our case the null hypothesis is not rejected at the 0.1 significance level. When a sufficient number of instances are available, one can rely on the asymptotic approximation of the distribution of R^+ or R^-. Let T denote either R^+ or R^-. Under the null hypothesis, they both have the same approximate normal distribution with mean $n(n+1)/4$ and variance $n(n+1)(2n+1)/24$. The Wilcoxon signed rank test then rejects the null hypothesis when

$$\frac{|T - n(n+1)/4|}{\sqrt{n(n+1)(2n+1)/24}} > z_{\alpha/2}$$

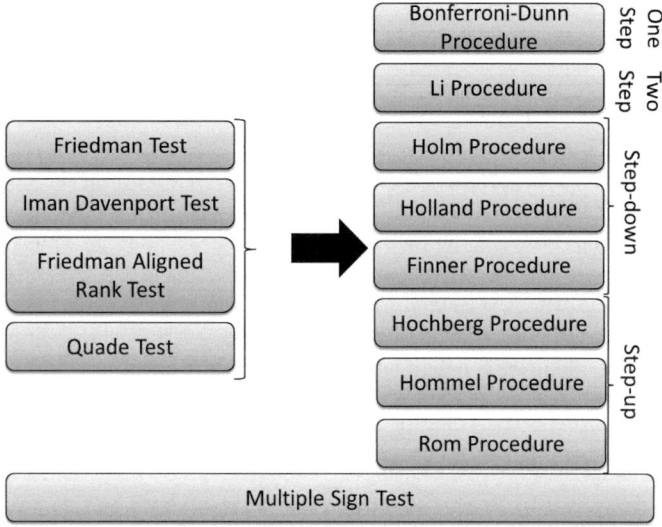

FIGURE 24.4: Non-parametric statistical tests for multiple comparisons.

where $z_{\alpha/2}$ is the $1 - \alpha/2$ percentile of the standard normal distribution. An important practical aspect that should be kept in mind is that the Wilcoxon test uses the continuous evaluation measures, so one should not round the values to one or two decimals, as this would decrease the power of the test in case of a high number of ties.

24.4.2.2 Multiple Comparisons

The non-parametric statistical tests that perform multiple comparisons we will discuss in this section are depicted in Figure 24.4. The lower test can be used as a stand-alone method, while the four upper-left tests need a post-hoc procedure to finish the comparison.

Friedman's test [16, 17] and the Iman Davenport's test [27] are similar. For each instance, the classifiers are ranked according to the evaluation measure (let r_i^j denote the corresponding rank) and next the average rank R_j for each classifier over the different instances is calculated, $R_j = \sum_{i=1}^{n} r_i^j / n$. The best method gets rank 1, the second best method gets rank 2, and so on. Under the null hypothesis, all classifiers are equivalent and hence the average ranks of the different classifiers should be similar. Both the Friedman test and Iman Davenport test aim to detect a deviation from this and can detect if there are significant differences between at least two of the methods. The test statistic of the Friedman test is given by

$$\chi^2 = \frac{12n}{k(k+1)} \left(\sum_{j=1}^{k} R_j^2 - \frac{k(k+1)^2}{4} \right).$$

For a sufficient number of instances and classifiers (as a rule of thumb $n > 10$ and $k > 5$), χ^2 approximately follows a chi-square distribution with $k - 1$ degrees of freedom. If χ^2 exceeds the critical value $\chi^2_{k-1,\alpha}$, where $\chi^2_{k-1,\alpha}$ is the $1 - \alpha$ percentile of the chi-square distribution with $k - 1$ degrees of freedom, the null hypothesis is rejected. For a small number of data sets and classifiers, exact critical values have been computed [48, 55]. The test statistic for the Iman and Davenport's test, which is less conservative than Friedman's test and should hence be preferred, is given by

$$F = \frac{(n-1)\chi^2}{n(k-1)-\chi^2}$$

and (assuming a sufficient number of instances and classifiers) follows approximately an F-distribution with $k-1$ and $(n-1)(k-1)$ degrees of freedom. If $F > F_{k-1,(n-1)(k-1),\alpha}$ with $F_{k-1,(n-1)(k-1),\alpha}$ the $1-\alpha$ percentile of an $F_{k-1,(n-1)(k-1)}$-distribution, the null hypothesis is rejected. In both cases, if the test statistic exceeds the corresponding critical value, it means that there are significant differences between the methods, but no other conclusion whatsoever can be made.

Next we discuss two more advanced non-parametric tests that in certain circumstances may improve upon the Friedman test, especially when the number of classifiers is low. The Friedman Aligned Rank [22] test calculates the ranks differently. For each data set, the average or median performance of all classifiers on this data set is calculated and this value is substracted from each performance value of the different classifiers to obtain the aligned observations. Next, all kn aligned observations are assigned a rank, the aligned rank R_{ij} with i referring to the data set and j referring to the classifier. The Friedman Aligned Ranks test statistic is given by

$$T = \frac{(k-1)\left[\sum_{j=1}^{k} \hat{R}_{\cdot j}^2 - (kn^2/4)(kn+1)^2\right]}{[kn(kn+1)(2kn+1)/6] - \sum_{i=1}^{n} \hat{R}_{i\cdot}^2/k}$$

where $\hat{R}_{\cdot j} = \sum_{i=1}^{n} R_{ij}$ equals the rank total of the jth classifier and $\hat{R}_{i\cdot} = \sum_{j=1}^{k} R_{ij}$ equals the rank total of the ith data set. For a sufficient number of data sets, T approximately follows χ^2-distribution with $k-1$ degrees of freedom. If $T > \chi^2_{k-1,\alpha}$, the null hypothesis is rejected with $\chi^2_{k-1,\alpha}$ the $1-\alpha$ percentile of a χ^2_{k-1}-distribution.

The Quade test [43] is an improvement upon the Friedman Aligned Ranks test by incorporating the fact that not all data sets are equally important. That is, some data sets are more difficult to classify than others and methods that are able to classify these difficult data sets correctly should be favored. The Quade test computes weighted ranks based on the range of the performances of different classifiers on each data set. To be specific, first calculate the ranks r_i^j as for the Friedman test. Next, calculate for each data set i the range of the performances of the different classifiers: $\max_j Y_{ij} - \min_j Y_{ij}$, and rank them with the smallest rank (1) given to the data set with the smallest range, and so on where average ranks are used in case of ties. Denote the obtained rank for data set i by Q_i. The weighted average adjusted rank for data set i with classifier j is then computed as $S_{ij} = Q_i\left[r_i^j - (k+1)/2\right]$ with $(k+1)/2$ the average rank within data sets. The Quade test statistic is then given by

$$T_3 = \frac{(n-1)\sum_{j=1}^{k} S_j^2/n}{n(n+1)(2n+1)k(k+1)(k-1)/72 - \sum_{j=1}^{k} S_j^2/n}$$

where $S_j = \sum_{i=1}^{n} S_{ij}$ is the sum of the weighted ranks for each classifier. T_3 approximately follows an F-distribution with $k-1$ and $(n-1)(k-1)$ degrees of freedom. If $T_3 > F_{k-1,(n-1)(k-1),\alpha}$ with $F_{k-1,(n-1)(k-1),\alpha}$ the $1-\alpha$ percentile of an $F_{k-1,(n-1)(k-1)}$-distribution, the null hypothesis is rejected.

When the null hypothesis (stating that all classifiers perform equivalently) is rejected, the average ranks calculated by these four methods themselves can be used to get a meaningful ranking of which methods perform best. However, as was the case for parametric multiple comparisons, post-hoc procedures are still needed to evaluate which pairwise differences are significant. The test statistic to compare algorithm j with algorithm l for the Friedman test and the Iman Davenport's test is given by

$$Z_{jl} = \frac{R_j - R_l}{\sqrt{k(k+1)/6n}}$$

with R_j and R_l the average ranks computed in the Friedman and Imand Davenport procedures. For the Friedman Aligned Ranks procedure, the test statistic to compare algorithm j with algorithm l is given by

$$Z_{jl} = \frac{\hat{R}_j - \hat{R}_l}{\sqrt{k(n+1)/6}}$$

with $\hat{R}_j = \hat{R}_{.j}/n$ and $\hat{R}_l = \hat{R}_{.l}/n$ the average aligned ranks. Finally, for the Quade procedure, the test statistic to compare algorithm j with algorithm l is given by

$$Z_{jl} = \frac{T_j - T_l}{\sqrt{[k(k+1)(2n+1)(k-1)]/[18n(n+1)]}}$$

with $T_j = \sum_{i=1}^{n} Q_i r_i^j / [n(n+1)/2]$ the average weighted ranks without average adjusting as described in the Quade procedure. All three test statistics Z_{jl} all approximately follow a standard normal from which an appropriate p-value can be calculated, being the probability that a standard normal distributed variable exceeds the absolute value of the observed test statistic.

A post-hoc procedure involves multiple pairwise comparisons based on the test statistics Z_{jl}, and as already mentioned for parametric multiple comparisons, these post-hoc tests cannot be used without caution as the FWER is not controlled, leading to inflated Type I errors. Therefore we consider post-hoc procedures based on adjusted p-values of the pairwise comparisons to control the FWER. Recall that the p-value returned by a statistical test is the probability that a more extreme observation than the current one is observed, assuming the null hypothesis holds. This simple p-value reflects this probability of one comparison, but does not take into account the remaining comparisons. Adjusted p-values (APV) deal with this problem and after the adjustment, these APVs can be compared with the nominal significance level α. The post-hoc procedures that we discuss first are all designed for multiple comparisons with a control method, that is, we compare one algorithm against the $k-1$ remaining ones. In the following, p_j denotes the p-value obtained for the jth null hypothesis, stating that the control method and the jth method are performing equally well. The p-values are ordered from smallest to largest: $p_1 \leq \ldots \leq p_{k-1}$, and the corresponding null hypotheses are rewritten accordingly as $H_1, \ldots, H_{k-1}$. Below we discuss several procedures to obtain adjusted p-values: one-step, two-step, step-down, and step-up.

1: One-step. The Bonferroni-Dunn [11] procedure is a simple one-step procedure that divides the nominal significance level α by the number of comparisons $(k-1)$ and the usual p-values can be compared with this level of significance. Equivalently, the adjusted value in this case is $\min\{(k-1)p_i, 1\}$. Although simple, this procedure may be too conservative for practical use when k is not small.

2: Two-step. The two-step Li [36] procedure rejects all null hypotheses if the biggest p-value $p_{k-1} \leq \alpha$. Otherwise, the null hypothesis related to p_{k-1} is accepted and the remaining null hypotheses H_i with $p_i \leq (1 - p_{k-1})/(1-\alpha)\alpha$ are rejected. The adjusted p-values are $p_i/(p_i + 1 - p_{k-1})$.

3: Step-down. We discuss three more advanced step-down methods. The Holm [24] procedure is the most popular one and starts with the lowest p-value. If $p_1 \leq \alpha/(k-1)$, the first null hypothesis is rejected and the next comparison is made. If $p_2 \leq \alpha/(k-2)$, the second null hypothesis H_2 is also rejected and the next null hypothesis is verified. This process continues until a null hypothesis cannot be rejected anymore. In that case, all remaining null hypotheses are retained as well. The adjusted p-values for the Holm procedure are $\min[\max\{(k-j)p_j : 1 \leq j \leq i\}, 1]$. Next, the Holland [23] procedure is similar to the Holm procedure. It rejects all null hypotheses H_1 tot H_{i-1} if i is the smallest integer such that $p_i > 1 - (1-\alpha)^{k-i}$, the adjusted p-values are $\min[\max\{1 - (1-p_j)^{k-j} : 1 \leq j \leq i\}, 1]$. Finally, the Finner [15] procedure, also similar, rejects all null hypotheses H_1 to H_{i-1} if i is the smallest integer such that $p_i > 1 - (1-\alpha)^{(k-1)/i}$. The adjusted p-values are $\min\left[\max\{1 - (1-p_j)^{(k-1)/i} : 1 \leq j \leq i\}, 1\right]$.

4: Step-up. Step-up procedures include the Hochberg [21] procedure, which starts off with comparing the largest p-value p_{k-1} with α, then the second largest p-value p_{k-2} is compared to $\alpha/2$ and so on, until a null hypothesis that can be rejected is found. The adjusted p-values are $\max\{(k-j)p_j : (k-1) \geq j \geq i\}$. Another more involved procedure is the Hommel [25] procedure. First, it finds the largest j for which $p_{n-j+\ell} > \ell\alpha/j$ for all $\ell = 1,\ldots,j$. If no such j exists, all null hypotheses are rejected. Otherwise, the null hypotheses for which $p_i \leq \alpha/j$ are rejected. In contrast to the other procedures, calculating the adjusted p-values cannot be done using a simple formula. Instead the procedure listed in Algorithm 24.2 should be used. Finally, Rom's [45] procedure was developed to increase Hochberg's power. It is completely equivalent, except that the α values are now calculated as

$$\alpha_{k-i} = \frac{\sum_{j=1}^{i-1} \alpha^j - \sum_{j=1}^{i-2} \binom{i}{k}\alpha_{k-1-j}^{i-j}}{i}$$

with $\alpha_{k-1} = \alpha$ and $\alpha_{k-2} = \alpha/2$. Adjusted p-values could also be obtained using the formula for α_{k-i} but no closed form formula is available [18].

Algorithm 24.2 Calculation of the adjusted p-values using Hommels post-hoc procedure.

INPUT: p-values $p_1 \leq p_2 \leq p_{k-1}$

for all $i = 1 \ldots k-1 : APV_i \leftarrow p_i$ **do**
    for all $j = k-1, k-2, \ldots, 2$ **do**
        $B \leftarrow \emptyset$
        for all $i = k-1, \ldots, k-j$ **do**
            $c_i \leftarrow jp_i/(j+i-k+1)$
            $B \leftarrow B \cup c_i$
        end for
        $c_{min} \leftarrow \min\{B\}$
        if $APV_i < c_{min}$ **then**
            $APV_i \leftarrow c_{min}$
        end if
        for all $i = 2, \ldots k-1-j$ **do**
            $c_i \leftarrow \min\{c_{min}, jp_i\}$
            if $APV_i < c_i$ **then**
                $APV_i \leftarrow c_i$
            end if
        end for
    end for
end for
OUTPUT: Adjusted p-values $APV_1, \ldots, APV_{k-1}$

In some cases, researchers are interested in all pairwise differences in a multiple comparison and do not simply want to find significant differences with one control method. Some of the procedures above can be easily extended to this case.

The Nemenyi procedure, which can be seen as a Bonferroni correction, adjusts α in one single step by dividing it by the number of comparisons performed $(k(k-1)/2)$.

A more involved method is Shaffer's procedure [47], which is based on the observation that the hypotheses are interrelated. For instance, if method 1 is significantly better than method 2 and method 2 is significantly better than method 3, method 1 will be significantly better than method 3. Shaffer's method follows a step-down method, and at step j, it rejects H_j if $p_j \leq \alpha/t_j$, where t_j is the maximum number of hypotheses that can be true given that hypotheses $H_1, \ldots H_{j-1}$ are false. This value t_i can be found in the table in [47].

Finally, the Bergmann-Hommel [26] procedure says that an index set $I \subseteq \{1, \dots M\}$, with M the number of hypotheses, is exhaustive if and only if it is possible that all null hypotheses H_j, $j \in I$ could be true and all hypotheses $H_j, j \notin I$ are not. Next, the set A is defined as $\cup\{I : I \text{ exhaustive}, \min(p_i : i \in I) > \alpha/|I|\}$, and the procedure rejects all the H_j with $j \in A$.

We now discuss one test that can be used for multiple comparisons with a control algorithm without a further post-hoc analysis needed, the multiple sign test [44], which is an extension of the standard sign test. The multiple sign test proceeds as follows: it counts for each classifier the number of times it outperforms the control classifier and the number of times it is outperformed by the control classifier. Let R_j be the minimum of those two values for the jth classifier. The null hypothesis is now different from the previous ones since it involves median performances. The null hypothesis states that the median performances are the same and is rejected if R_j exceeds a critical value depending on the the nominal significance level α, the number of instances n, and the number $k - 1$ of alternative classifiers. These critical values can be found in the tables in [44].

All previous methods are based on hypotheses and the only information that can be obtained is if algorithms are significantly better or worse than others. Contrast estimation based on medians [9] is a method that can quantify these differences and reflects the magnitudes of the differences between the classifiers for each data set. It can be used after significant differences are found. For each pair of classifiers, the differences between them are calculated for all problems. For each pair, the median of these differences over all data sets are taken. These medians can now be used to make comparisons between the different methods, but if one wants to know how a control method performs compared to all other methods, the mean of all medians can be taken.

24.4.2.3 Permutation Tests

Most of the previous tests assume that there are sufficient data sets to posit that the test statistics approximately follow a certain distribution. For example, when sufficient data sets are available, the Wilcoxon signed rank test statistic approximately follows a normal distribution. If only a few data sets are available for testing certain algorithms, permutation tests [12], also known as exact tests, are a good alternative to detect differences between classifiers. The idea is quite simple. Under the null hypothesis it is assumed that the distribution of the performance of all algorithms is the same. Hence, the results of a certain classifier j could equally likely be produced by classifier l. Thus, under the null hypothesis, it is equally likely that these results are interchanged. Therefore, one may construct the exact distribution of the considered test statistic by calculating the test statistic under each permutation of the classifier labels within each data set. One can then calculate a p-value as the number of permuted test statistic values that are more extreme than the observed test statistic value devided by the number of random permutations performed. The null hypothesis is rejected when the p-value is smaller than the significance level α. A major drawback of this method is that it is computationally expensive, so it can only be used in experimental settings with few datasets.

In the case where we want to perform multiple comparisons, however, permutation tests give a very elegant way to control the FWER at the prespecified α. Suppose for simplicity we want to perform all pairwise comparisons. For this purpose, we can use the test statistics Z_{jl} defined above in the following manner. Recall that the FWER equals the probabilty of falsely rejecting at least one null hypothesis. Thus, if we find a constant c such that

$$P\left(\max_{j,l} |Z_{jl}| > c | H_0\right) = \alpha$$

and use this constant c as the critical value to reject a certain null hypothesis, we automatically control for the FWER at the α level. To be more specific, we reject those null hypotheses stating that classifier j performs equally well as classifier l if the corresponding observed value for $|Z_{jl}|$ exceeds c. This constant can be easily found by taking the $1 - \alpha$ percentile of the permutation null distribution of $\max_{j,l} |Z_{jl}|$.

24.5 Concluding Remarks

To conclude, we summarize the considerations and recommendations as listed in [5] for non-parametric statistical tests. First of all, non-parametric statistical tests should only be used given that the conditions for parametric tests are not true. For pairwise comparisons, Wilcoxon's test is preferred over the sign test. For multiple comparisons, first it should be checked if there are significant differences between the methods, using either Friedman's (aligned rank) test, Quade's test or Iman Davenport's test. The difference in power among these methods is unknown, although it is recommended to use Friedman's aligned rank test or Quade's test when the number of algorithms to be compared is rather high. Holm's post hoc procedure is considered to be one of the best tests. Hochberg's test has more power and can be used in combination with Holm's procedure. The multiple sign test is recommended when the differences between the control method and the others are very clear. We add that if only a few datasets are available, permutation tests might be a good alternative.

Bibliography

[1] A. Ben-David. Comparison of classification accuracy using Cohen's weighted kappa. *Expert Systems with Applications*, 34(2):825–832, 2008.

[2] A. M. Brown. A new software for carrying out one-way ANOVA post hoc tests. *Computer Methods and Programs in Biomedicine*, 79(1):89–95, 2005.

[3] A. Celisse and S. Robin. Nonparametric density estimation by exact leave-out cross-validation. *Computational Statistics and Data Analysis*, 52(5):2350–2368, 2008.

[4] W. W. Daniel. *Applied nonparametric statistics, 2nd ed.* Boston: PWS-Kent Publishing Company, 1990.

[5] J. Derrac, S. García, D. Molina, and F. Herrera. A practical tutorial on the use of nonparametric statistical tests as a methodology for comparing evolutionary and swarm intelligence algorithms. *Swarm and Evolutionary Computation*, 1(1):3–18, 2011.

[6] L. Devroye and T. Wagner. Distribution-free performance bounds for potention function rules. *IEEE Transactions in Information Theory*, 25(5):601–604, 1979.

[7] N. A. Diamantidis, D. Karlis, and E. A. Giakoumakis. Unsupervised stratification of cross-validation for accuracy estimation. *Artificial Intelligence*, 116(1-2):1–16, 2000.

[8] T. Dietterich. Approximate statistical tests for comparing supervised classification learning algorithms. *Neural Computation*, 10:1895–1923, 1998.

[9] K. Doksum. Robust procedures for some linear models with one observation per cell. *Annals of Mathematical Statistics*, 38:878–883, 1967.

[10] P. Domingos and F. Provost. Well-trained PETs: Improving probability estimation trees, 2000. CDER Working Paper, Stern School of Business, New York University, NY, 2000.

[11] O. Dunn. Multiple comparisons among means. *Journal of the American Statistical Association*, 56:52–64, 1961.

[12] E. S. Edgington. *Randomization tests, 3rd ed.* New York: Marcel-Dekker, 1995.

[13] B. Efron and R. Tibshirani. Improvements on cross-validation: The .632+ bootstrap method. *Journal of the American Statistical Association*, 92(438):548–560, 1997.

[14] C. Ferri, J. Hernández-Orallo, and M.A. Salido. Volume under the ROC surface for multi-class problems. In *Proc. of 14th European Conference on Machine Learning*, pages 108–120, 2003.

[15] H. Finner. On a monotonicity problem in step-down multiple test procedures. *Journal of the American Statistical Association*, 88:920–923, 1993.

[16] M. Friedman. The use of ranks to avoid the assumption of normality implicit in the analysis of variance. *Journal of the American Statistical Association*, 32:674–701, 1937.

[17] M. Friedman. A comparison of alternative tests of significance for the problem of m rankings. *Annals of Mathematical Statistics*, 11:86–92, 1940.

[18] S. García, A. Fernández, J. Luengo, and F. Herrera. Advanced nonparametric tests for multiple comparisons in the design of experiments in computational intelligence and data mining: Experimental analysis of power. *Information Sciences*, 180:2044–2064, 2010.

[19] D. J. Hand and R. J. Till. A simple generalisation of the area under the ROC curve for multiple class classification problems. *Machine Learning*, 45(2):171–186, 2001.

[20] M.R. Hassan, K. Ramamohanarao, C. Karmakar, M. M. Hossain, and J. Bailey. A novel scalable multi-class roc for effective visualization and computation. In *Advances in Knowledge Discovery and Data Mining*, volume 6118, pages 107–120. Springer Berlin Heidelberg, 2010.

[21] Y. Hochberg. A sharper Bonferroni procedure for multiple tests of significance. *Biometrika*, 75:800–803, 1988.

[22] J. Hodges and E. Lehmann. Ranks methods for combination of independent experiments in analysis of variance. *Annals of Mathematical Statistics*, 3:482–497, 1962.

[23] M. C. B. S. Holland. An improved sequentially rejective Bonferroni test procedure. *Biometrics*, 43:417–423, 1987.

[24] S. Holm. A simple sequentially rejective multiple test procedure. *Scandinavian Journal of Statistics*, 6:65–70, 1979.

[25] G. Hommel. A stagewise rejective multiple test procedure based on a modified Bonferroni test. *Biometrika*, 75:383–386, 1988.

[26] G. Hommel and G. Bernhard. A rapid algorithm and a computer program for multiple test procedures using logical structures of hypotheses. *Computer Methods and Programs in Biomedicine*, 43(3-4):213–216, 1994.

[27] R. Iman and J. Davenport. Approximations of the critical region of the Friedman statistic. *Communications in Statistics*, 9:571–595, 1980.

[28] M. Keuls. The use of the studentized range in connection with an analysis of variance. *Euphytica*, 1:112–122, 1952.

[29] R. E. Kirk. *Experimental design: Procedures for the Behavioral Sciences, 3rd ed.* Pacific Grove, CA: Brooks/Cole Publishing Company, 1995.

[30] R. Kohavi. A study of cross-validation and bootstrap for accuracy estimation and model selection. In *Proceedings of the 14th International Joint Conference on Artificial Intelligence—Volume 2*, IJCAI'95, pages 1137–1143, 1995.

[31] T. C. W. Landgrebe and R. P. W. Duin. A simplified extension of the area under the ROC to the multiclass domain. In *17th Annual Symposium of the Pattern Recognition Association of South Africa*, 2006.

[32] T. C. W. Landgrebe and R. P. W. Duin. Approximating the multiclass ROC by pairwise analysis. *Pattern Recognition Letters*, 28(13):1747–1758, 2007.

[33] T. C. W. Landgrebe and R. P. W. Duin. Efficient multiclass ROC approximation by decomposition via confusion matrix perturbation analysis. *IEEE Transactions on Pattern Analysis and Machine Intelligence*, 30(5):810–822, 2008.

[34] T. C. W. Landgrebe and P. Paclik. The ROC skeleton for multiclass ROC estimation. *Pattern Recognition Letters*, 31(9):949–958, 2010.

[35] S. C. Larson. The shrinkage of the coefficient of multiple correlation. *Journal of Educational Psychology*, 22(1):45–55, 1931.

[36] J. Li. A two-step rejection procedure for testing multiple hypotheses. *Journal of Statistical Planning and Inference*, 138:1521–1527, 2008.

[37] J. Luengo, S. García, and F. Herrera. A study on the use of statistical tests for experimentation with neural networks: Analysis of parametric test conditions and non-parametric tests. *Expert Systems with Applications*, 36:7798–7808, 2009.

[38] M. Majnik and Z. Bosnić. ROC analysis of classifiers in machine learning: A survey. *Intelligent Data Analysis*, 17(3):531–558, 2011.

[39] J. G. Moreno-Torres, T. Raeder, R. A. Rodriguez, N. V. Chawla, and F. Herrera. A unifying view on dataset shift in classification. *Pattern Recognition*, 45(1):521–530, 2012.

[40] J. G. Moreno-Torres, J. A. Sáez, and F. Herrera. Study on the impact of partition-induced dataset shift on k -fold cross-validation. *IEEE Transactions on Neural Networks and Learning Systems*, 23(8):1304–1312, 2012.

[41] D. Mossman. Three-way ROCS. *Medical Decision Making*, 19:78–89, 1999.

[42] T. Nichols and S. Hayasaka. Controlling the familywise error rate in functional neuroimaging: A comparative review. *Statistical Methods in Medical Research*, 12:419–446, 2003.

[43] D. Quade. Using weighted rankings in the analysis of complete blocks with additive block effects. *Journal of the American Statistical Association*, 74:680–683, 1979.

[44] A. Rhyne and R. Steel. Tables for a treatments versus control multiple comparisons sign test. *Technometrics*, 7:293–306, 1965.

[45] D. Rom. A sequentially rejective test procedure based on a modified Bonferroni inequality. *Biometrika*, 77:663–665, 1990.

[46] H. Scheffé. A method for judging all contrasts in the analysis of variance. *Biometrika*, 40:87–104, 1953.

[47] J. Shaffer. Modified sequentially rejective multiple test procedures. *Journal of American Statistical Association*, 81:826–831, 1986.

[48] D. J. Sheskin. *Handbook of Parametric and Nonparametric Statistical Procedures, 4th ed.*, Boca Raton, FL: Chapman and Hall/CRC, 2006.

[49] H. Shimodaira. Improving predictive inference under covariate shift by weighting the log-likelihood function. *Journal of Statistical Planning and Inference*, 90(2):227–244, 2000.

[50] Z. Sidak. Rectangular confidence regions for the means of multivariate normal distributions. *Journal of the American Statistical Association*, 62(318):626–633, 1967.

[51] M. Stone. Cross-validatory choice and assessment of statistical predictions. *Journal of the Royal Statistics Society*, 36:111–147, 1974.

[52] F. Wilcoxon. Individual comparisons by ranking methods. *Biometrics Bulletin*, 6:80–83, 1945.

[53] D. H. Wolpert. The supervised learning no-free-lunch theorems. In *Proceedings of the Sixth Online World Conference on Soft Computing in Industrial Applications*, 2001.

[54] Q. S. Xu and Y. Z. Liang. Monte Carlo cross validation. *Chemometrics and Intelligent Laboratory Systems*, 56(1):1–11, 2001.

[55] J. H. Zar. *Biostatistical Analysis*. Prentice Hall, 1999.

[56] X. Zeng and T. R. Martinez. Distribution-balanced stratified cross-validation for accuracy estimation. *Journal of Experimental and Theoretical Artificial Intelligence*, 12(1):1–12, 2000.

Chapter 25

Educational and Software Resources for Data Classification

Charu C. Aggarwal

IBM T. J. Watson Research Center
Yorktown Heights, NY
charu@us.ibm.com

25.1 Introduction

This chapter will summarize the key resources available in the literature on data classification. While this book provides a basic understanding of the important aspects of data classification, these resources will provide the researcher more in-depth perspectives on each individual topic. Therefore, this chapter will summarize the key resources in this area. In general, the resources on data classification can be divided into the categories of (i) books, (ii) survey articles, and (iii) software.

Numerous books exist in data classification, both for general topics in classification and for more specific subject areas. In addition, numerous software packages exist in the area of data classification. In fact, this is perhaps the richest area of data classification in terms of availability of software. This chapter will list all the possible resources currently available both from commercial and non-commercial sources for data classification.

This chapter is organized as follows. Section 25.2 presents educational resources on data classification. This section is itself divided into two subsections. The key books are discussed in Section 25.2.1, whereas the survey papers on data classification will be discussed in Section 25.2.2. Finally, the software on data classification will be discussed in Section 25.3. Section 25.4 discusses the conclusions and summary.

25.2 Educational Resources

Educational resources are either generic in the form of books, or more focussed on the research communities in the form of surveys. Books are generally more useful from a practical perspective, whereas survey articles are more useful for an academic perspective. The survey articles are often focussed on special areas of data classification such as decision trees, rule-based methods, and neural networks.

25.2.1 Books on Data Classification

The most well known books on on data classification include those by Duda et al [19], Hastie et al [23], Mitchell [29], Murphy [31], and Bishop [11]. Many of these books discuss different aspects of data classification in a comprehensive way, and serve as useful resources on the subject. The book authored by Duda et. al [19] is written in a simple way, and serves as a particularly useful guide.

Numerous books have also been written in the context of specific areas of data classification. An excellent book on feature selection in classification may be found in [27]. Methods for probabilistic classification are discussed in [97]. For example, the topic of decision trees has been covered comprehensively in [34]. Decision trees are closely related to rule-based classifiers, and both topics are discussed in detail in [34]. The topic of Support Vector Machines is discussed comprehensively in [16, 21, 38–41]. A guide to instance-based classification may be found in [17]. Kernel methods are discussed in detail in [39]. Neural networks are discussed in detail in [10, 22].

In addition, a number of books exist, that address classification from the perspective of specific data types. Methods for text classification are discussed in detail in [5]. Methods for multimedia data classification are discussed in [28]. Network classification is discussed in detail in [3,4]. Methods for streaming classification are addressed in the book on data streaming [2]. The big data framework is explained in detail in [18, 42], though no book currently exists, that specifically addresses the problem of classification within this framework.

Numerous books also exist on different variations of classification algorithms. The problem of rare class learning is addressed in a general book on outlier analysis [1]. Active learning is addressed in the book by Burr Settles [36], whereas semi-supervised learning is addressed in [15, 46]. Visual data mining is addressed in [37], while the topic of evaluation of classification algorithms is addressed in [25].

25.2.2 Popular Survey Papers on Data Classification

Numerous surveys also exist in the area of data classification. Two popular surveys may be found in [24, 26]. Since data classification is a rather broad area, the books and surveys in this area generally cover a very specific area of classification. This is because the general area of data classification is too broad to be covered comprehensively by a survey paper. The most well known survey on decision trees is provided in [32]. Some other surveys may be found in [12, 30]. An excellent tutorial on support vector machines may be found in [13]. Methods for instance-based learning are discussed in [8,9,43]. Two surveys on text classification may be found in [6,35]. Time-series classification methods are addressed in [44].

An overview on rare class methods may be found in [14]. An excellent survey on transfer learning may be found in [33]. The topic of distance function learning is addressed in [45]. The topic of network classification is addressed in two survey chapters in [4] and [3], respectively. An overview on visual data mining methods is provided in [7].

25.3 Software for Data Classification

A significant amount of software is available for data classification. Interestingly, most of the sophisticated software on data classification is open-source software, which is freely available at different Web sites. On the other hand, most of the commercial software comprises implementations of simpler and more classical algorithms. This is possibly because open source software is often in the form of research prototypes, and created by researchers, which reflect more recent advances in the field. The KDD Nuggets site [51] provides a link to the important free and commercial tools for data classification.

The most well-known *general purpose* site offering open source software is the *WEKA* machine learning repository [50]. This is a general purpose repository that contains software not just for classification, but also for other data mining related tasks such as data pre-processing, clustering, and visualization. However, the classification problem seems to be the predominant one addressed by the software in this repository.

Numerous software packages for also available for different types of data classification techniques. An excellent overview on methods for Bayes classification or other graphical models may be found in [97]. The original C4.5 decision tree classifier is available from Ross Quinlan's site at [86]. The random forests implementation from Leo Brieman is available at [87]. The well-known rule-based system that uses CBA for classification is available at [88]. The family of rule-based systems such as *SLIPPER, RIPPER,* and *WHIRL* are available from William Cohen's site at [89]. For SVM classification, the *LIBSVM* library [90] is a well-known one, and it is also well-documented for easy use. A very popular implementation of SVM known as *SVM-light* [95] is provided by Thorsten Joachims for data classification, based on a simple and intuitive version of the SVM classifier. A general site that is devoted to implementations of kernel-machine classifiers such as SVM may be found in [96]. For neural networks, the neural network FAQ list [98] provides an excellent resource on both software and books in this area. A list of numerous other classification methods based on rough sets, genetic algorithms, and other methods may be found in [102].

For *text data classification*, the most well-known one is the BOW toolkit, which may be downloaded from [81]. This package implements many popular classification algorithms such as Bayes methods, nearest neighbor methods, etc. The UCR time-series [48] repository provides links to a significant amount of software for time-series classification. Some pointers are also provided for sequence classification at this site. The SNAP repository at Stanford [74] provides pointers to numerous software resources on network classification.

Many mathematical tools such as MATLAB [54] from *MathWorks* come with built-in methods for data classification. Many of these methods are fairly simple tools for data classification. This is because these are very general-purpose tools, and the classification capability is not the main purpose of such software. However, numerous other commercial tools have been constructed, with a specific goal of classification of different kinds of data. *MathWorks* also provides an excellent neural network toolbox [100] for data classification. Some third-party toolboxes such as LS-SVM [93] have been constructed on top of MATLAB, and can provide effective methods for SVM classification.

The KDD Nuggets site [51] provides a link to some of the more popular forms of software in this domain. This site provides pointers to software developed by other vendors such as *Oracle, IBM,* and *SAS* rather than its own dedicated software. One of the most well-known ones is the *IBM SPSS data mining workbench* [52]. An example of this is the *SPSS Answer Tree,* which implements decision tree methods. In addition, the neural network toolbox from the IBM SPSS Workbench [101] is a popular option. Oracle has an advanced analytics platform [105], which extends the Oracle database software to an analytics suite. Another well-known package is the *SAS Enterprise Modeler* [53]. Salford systems [85] provides an excellent implementation of CART for classification and regression. KXEN [91] provides SVM components, based on Vapnik's original work in this area.

Tiberius [92] provides tools based on SVM, neural networks, decision trees, and a variety of other data modeling methods. It also supports regression modeling, and three-dimensional data visualization. Treparel [94] provides high-performance tools based on the SVM classifiers for massive data sets. The software NeuroXL can be used from within the Excel software for data classification [99]. This is particularly helpful, since it means that the software can be used directly from spread-sheets, which are commonly used for data manipulation within corporate environments.

Software has also been designed for evaluation and validation of classification algorithms. The most popular among them is *Analyze-it*, which [104] provides a comprehensive platform for analyzing classification software. It is capable of constructing numerous metrics such as ROC curves for analytical purposes. The platform integrates well with Excel, which makes it particularly convenient to use in a wide variety of scenarios.

25.3.1 Data Benchmarks for Software and Research

Numerous data sets are available for testing classification software and research. Some of the data sets and Web sites are general-purpose, whereas other Web sites are tailored to specific kinds of data. The most well-known among the general-purpose Web sites is the UCI machine learning repository [20]. This resource is generally intended for classification, though many of the data sets can also be used for classification. A general purpose data market for different kinds of data is provided at [103].

The KDD Nuggets Web site [49] also provides access to many general-purpose data sets. This can be considered a meta-repository, in that it provides pointers to other sites containing data sets. Similar to this, the other popularly known meta-repositories that give a wide range of links to other data repositories are STATOO [67] and David Dowe's data links [68]. The latter also provides several online resources on data mining case studies, and competitions.

More large-scale datasets are available from the KDD Cup dataset repository [69] and other data mining competitions [83]. These datasets are not only large-scale but are also directly collected from complex real-world problems and hence provide several challenges. Most of these data sets are labeled, and can therefore be directly used for classification. For the statistics community, Statlib dataset archive [70] is a widely used data collection.

For specific data types, numerous resources are available. For *time-series classification*, the UCR time-series Web site [48] provides access to very long continuous series of data for classification. Another time-series data library is available at [80]. For the case of *network data*, the SNAP repository [74] hosted by Jure Leskovec at Stanford University provides access to a large number of network data sets.

For testing the performance of *text document classification* algorithms, there are several publicly available text document repositories available. Some of the most popular text document collections include Reuters [71], 20NewsGroups [72], TREC [82], and Cora [73].

To evaluate the *biological data classification*, there are a plethora of Web sites that host gene expression datasets, which can be used to evaluate the classification algorithms. Gene Expression Omnibus (GEO) repository [64] contains a comprehensive collection of gene expression datasets along with raw source files used to generate this data. Gene Expression Model Selector provides a simple repository of the most widely studied gene expression datasets [65] and several other cancer-specific gene expressions are available at [66]. Similarly, to test the performance of *biological network classification* algorithms, there are plenty of databases that contain protein-protein interactions. The most popular ones are DIP (Database of Interacting Proteins) [62], BioGRID [63], STRING (Search Tool for the Retrieval of Interacting Genes/Proteins) [61], and MIPS (Mammalian Protein-Protein Interaction Database) [60]. The last resource contains links to several other interaction network databases.

To evaluate the *sequence data classification* algorithms, several biological sequence database repositories are available at [59]. The most widely used repositories are GenBank [57] and EMBL

[58] for nucleic acid sequences and Protein Information Resources (PIR) [56] and UniProt [55] for protein sequences.

In the context of *image applications*, researchers in machine learning and computer vision communities have explored the problem extensively. ImageCLEF [77] and ImageNet [78] are two widely used image data repositories that are used to demonstrate the performance of image data retrieval and learning tasks. Vision and Autonomous Systems Center's Image Database [75] from Carnegie Mellon University and the Berkeley Segmentation dataset [76] can be used to test the performance of classification for image segmentation problems. An extensive list of Web sites that provide image databases is given in [79] and [84].

25.4 Summary

This chapter presents a summary of the key resources for data classification in terms of books, surveys, and commercial and non-commercial software packages. It is expected that many of these resources will evolve over time. Therefore, the reader is advised to use this chapter as a general guideline on which to base their search, rather than treating it as a comprehensive compendium.

Since data classification is a rather broad area, much of the recent software and practical implementations have not kept up with the large number of recent advances in this field. This has also been true of the more general books in the field, which discuss the basic methods, but not the recent advances such as big data, uncertain data, or network classification. This chapter is an attempt to bridge the gap in this rather vast field, by creating a book, that covers the different areas of data classification in detail.

Bibliography

[1] C. Aggarwal. *Outlier Analysis*, Springer, 2013.

[2] C. Aggarwal. *Data Streams: Models and Algorithms*, Springer, 2007.

[3] C. Aggarwal. *Social Network Data Analytics*, Springer, Chapter 5, 2011.

[4] C. Aggarwal, H. Wang. *Managing and Mining Graph Data*, Springer, 2010.

[5] C. Aggarwal, C. Zhai. Mining Text Data, *Springer*, 2012.

[6] C. Aggarwal, C. Zhai. A survey of text classification algorithms, In *Mining Text Data*, pages 163–222, Springer, 2012.

[7] C. Aggarwal. Towards effective and interpretable data mining by visual interaction, *ACM SIGKDD Explorations*, 3(2):11–22, 2002.

[8] D. Aha, D. Kibler, and M. Albert. Instance-based learning algorithms, *Machine Learning*, 6(1):37–66, 1991.

[9] D. Aha. Lazy learning: Special issue editorial. *Artificial Intelligence Review*, 11(1–5): 7–10, 1997.

[10] C. Bishop. *Neural Networks for Pattern Recognition*, Oxford University Press, 1996.

[11] C. Bishop. *Pattern Recognition and Machine Learning*, Springer, 2007.

[12] W. Buntine. *Learning Classification Trees. Artificial Intelligence Frontiers in Statistics*, Chapman and Hall, pages 182–201, 1993.

[13] C. Burges. A tutorial on support vector machines for pattern recognition. *Data Mining and Knowledge Discovery*, 2(2): 121–167, 1998.

[14] N. V. Chawla, N. Japkowicz, and A. Kotcz. Editorial: Special issue on learning from imbalanced data sets, *ACM SIGKDD Explorations Newsletter*, 6(1):1–6, 2004.

[15] O. Chapelle, B. Scholkopf, and A. Zien. *Semi-Supervised Learning. Vol. 2*, Cambridge: MIT Press, 2006.

[16] N. Cristianini and J. Shawe-Taylor. *An Introduction to Support Vector Machines and Other Kernel-based Learning Methods*, Cambridge University Press, 2000.

[17] B. V. Dasarathy. *Nearest Neighbor (NN) Norms: NN Pattern Classification Techniques*. IEEE Computer Society Press, 1990,

[18] J. Dean and S. Ghemawat. MapReduce: A flexible data processing tool, *Communication of the ACM*, 53(1):72–77, 2010.

[19] R. Duda, P. Hart, and D. Stork, Pattern Classification, *Wiley*, 2001.

[20] A. Frank, and A. Asuncion. UCI Machine Learning Repository, Irvine, CA: University of California, School of Information and Computer Science, 2010. http://archive.ics.uci. edu/ml

[21] L. Hamel. *Knowledge Discovery with Support Vector Machines*, Wiley, 2009.

[22] S. Haykin. *Neural Networks and Learning Machines*, Prentice Hall, 2008.

[23] T. Hastie, R. Tibshirani, and J. Friedman. *The Elements of Statistical Learning: Data Mining, Inference, and Prediction*, Springer, 2013.

[24] A. Jain, R. Duin, and J. Mao. Statistical pattern recognition: A review. *IEEE Transactions on Pattern Analysis and Machine Intelligence*, 22(1:4–37, 2000.

[25] N. Japkowicz, M. Shah. *Evaluating Learning Algorithms: A Classification Perspective*, Cambridge University Press, 2011.

[26] S. Kulkarni, G. Lugosi, and S. Venkatesh. Learning pattern classification: A Survey. *IEEE Transactions on Information Theory*, 44(6):2178–2206, 1998.

[27] H. Liu, H. Motoda. *Feature Selection for Knowledge Discovery and Data Mining*, Springer, 1998.

[28] R. Mayer. *Multimedia Learning*, Cambridge University Press, 2009.

[29] T. Mitchell. *Machine Learning*, McGraw Hill, 1997.

[30] B. Moret. Decision trees and diagrams, *ACM Computing Surveys (CSUR)*, 14(4):593–623, 1982.

[31] K. Murphy. *Machine Learning: A Probabilistic Perspective*, MIT Press, 2012.

[32] S. K. Murthy. Automatic construction of decision trees from data: A multi-disciplinary survey. *Data Mining and Knowledge Discovery*, 2(4):345–389, 1998.

[33] S. J. Pan, Q. Yang. A survey on transfer learning. *IEEE Transactons on Knowledge and Data Engineering*, 22(10):1345–1359, 2010.

[34] J. R. Quinlan, Induction of decision trees, *Machine Learning*, 1(1):81–106, 1986.

[35] F. Sebastiani. Machine learning in automated text categorization, *ACM Computing Surveys*, 34(1):1–47, 2002.

[36] B. Settles. *Active Learning*, Morgan and Claypool, 2012.

[37] T. Soukop, I. Davidson. *Visual Data Mining: Techniques and Tools for Data Visualization*, Wiley, 2002.

[38] B. Scholkopf, A. J. Smola. *Learning with Kernels: Support Vector Machines, Regularization, Optimization, and Beyond* Cambridge University Press, 2001.

[39] B. Scholkopf and A. J. Smola. *Learning with Kernels*. Cambridge, MA, MIT Press, 2002.

[40] I. Steinwart and A. Christmann. *Support Vector Machines*, Springer, 2008.

[41] V. Vapnik. *The Nature of Statistical Learning Theory*, Springer, 2000.

[42] T. White. *Hadoop: The Definitive Guide*. Yahoo! Press, 2011.

[43] D. Wettschereck, D. Aha, T. Mohri. A review and empirical evaluation of feature weighting methods for a class of lazy learning algorithms, *Artificial Intelligence Review*, 11(1–5):273–314, 1997.

[44] Z. Xing, J. Pei, and E. Keogh. A brief survey on sequence classification. *SIGKDD Explorations*, 12(1):40–48, 2010.

[45] L. Yang. Distance Metric Learning: A Comprehensive Survey, 2006. http://www.cs.cmu.edu/~liuy/frame_survey_v2.pdf

[46] X. Zhu, and A. Goldberg. *Introduction to Semi-Supervised Learning*, Morgan and Claypool, 2009

[47] http://mallet.cs.umass.edu/

[48] http://www.cs.ucr.edu/~eamonn/time_series_data/

[49] http://www.kdnuggets.com/datasets/

[50] http://www.cs.waikato.ac.nz/ml/weka/

[51] http://www.kdnuggets.com/software/classification.html

[52] http://www-01.ibm.com/software/analytics/spss/products/modeler/

[53] http://www.sas.com/technologies/analytics/datamining/miner/index.html

[54] http://www.mathworks.com/

[55] http://www.ebi.ac.uk/uniprot/

[56] http://www-nbrf.georgetown.edu/pirwww/

[57] http://www.ncbi.nlm.nih.gov/genbank/

[58] http://www.ebi.ac.uk/embl/

[59] http://www.ebi.ac.uk/Databases/

[60] http://mips.helmholtz-muenchen.de/proj/ppi/

[61] http://string.embl.de/

[62] http://dip.doe-mbi.ucla.edu/dip/Main.cgi

[63] http://thebiogrid.org/

[64] http://www.ncbi.nlm.nih.gov/geo/

[65] http://www.gems-system.org/

[66] http://www.broadinstitute.org/cgi-bin/cancer/datasets.cgi

[67] http://www.statoo.com/en/resources/anthill/Datamining/Data/

[68] http://www.csse.monash.edu.au/~dld/datalinks.html

[69] http://www.sigkdd.org/kddcup/

[70] http://lib.stat.cmu.edu/datasets/

[71] http://www.daviddlewis.com/resources/testcollections/reuters21578/

[72] http://qwone.com/~jason/20Newsgroups/

[73] http://people.cs.umass.edu/~mccallum/data.html

[74] http://snap.stanford.edu/data/

[75] http://vasc.ri.cmu.edu/idb/

[76] http://www.eecs.berkeley.edu/Research/Projects/CS/vision/bsds/

[77] http://www.imageclef.org/

[78] http://www.image-net.org/

[79] http://www.imageprocessingplace.com/root_files_V3/image_databases.htm

[80] http://datamarket.com/data/list/?q=provider:tsdl

[81] http://www.cs.cmu.edu/~mccallum/bow/

[82] http://trec.nist.gov/data.html

[83] http://www.kdnuggets.com/competitions/index.html

[84] http://www.cs.cmu.edu/~cil/v-images.html

[85] http://www.salford-systems.com/

[86] http://www.rulequest.com/Personal/

[87] http://www.stat.berkeley.edu/users/breiman/RandomForests/

[88] http://www.comp.nus.edu.sg/~dm2/

[89] http://www.cs.cmu.edu/~wcohen/#sw

[90] http://www.csie.ntu.edu.tw/~cjlin/libsvm/

[91] http://www.kxen.com/

[92] http://www.tiberius.biz/

[93] http://www.esat.kuleuven.be/sista/lssvmlab/

[94] http://treparel.com/

[95] http://svmlight.joachims.org/

[96] http://www.kernel-machines.org/

[97] http://www.cs.ubc.ca/~murphyk/Software/bnsoft.html

[98] ftp://ftp.sas.com/pub/neural/FAQ.html

[99] http://www.neuroxl.com/

[100] http://www.mathworks.com/products/neural-network/index.html

[101] http://www-01.ibm.com/software/analytics/spss/

[102] http://www.kdnuggets.com/software/classification-other.html

[103] http://datamarket.com

[104] http://analyse-it.com/products/method-evaluation/

[105] http://www.oracle.com/us/products/database/options/advanced-analytics/overview/index.html

Index